III Quality Management and Reliability of Technical Systems

AA007462

IOP Conference Series: Materials Science and Engineering
Volume 986

St. Petersburg, Russia
27 - 29 August 2020

ISBN: 978-1-7138-4797-7
ISSN: 1757-8981

Printed from e-media with permission by:

Curran Associates, Inc.
57 Morehouse Lane
Red Hook, NY 12571

CURRAN ASSOCIATES INC.
proceedings
.com

Some format issues inherent in the e-media version may also appear in this print version.

Institute of Physics
Dirac House, Temple Back
Bristol BS1 6BE UK

Phone: 44 1 17 929 7481
Fax: 44 1 17 920 0979

techtracking@iop.org

Additional copies of this publication are available from:

Curran Associates, Inc.
57 Morehouse Lane
Red Hook, NY 12571 USA
Phone: 845-758-0400
Fax: 845-758-2633
Email: curran@proceedings.com
Web: www.proceedings.com

TABLE OF CONTENTS

IOP Conf. Series: Materials Science and Engineering **986** (2020) 011001 doi:10.1088/1757-899X/986/1/011001

Quality Management and Reliability of Technical Systems

Ensuring the high quality and reliability of complex technical systems in the modern world remains the sharpest and most relevant topic. High dynamics of changes in technical products, such as automobiles, military equipment, marine vessels, electronic industry products, machine tools and equipment, dictates the conditions for revising methods and approaches to ensure quality, reliability and competitiveness. Serious changes in production technologies require a review of the fundamentals of statistical process control, since the effectiveness and efficiency of quality control methods aimed at directly influencing a technological operation remain in doubt.

In this regard, fundamental research is becoming particularly relevant. The international scientific-practical conference "Quality Management and Reliability of Technical Systems" is held with the aim of increasing the effectiveness of scientific research, developing regional and international scientific cooperation, creating conditions for scientists to exchange research results, systematizing current problems and identifying trends in scientific research in the field of quality management and reliability of technical systems. Fundamental scientific research aimed at addressing the issues of machines and energy complexes safety, reliability, survivability is of particular relevance; control systems optimization methods and intellectualization and processes; adaptive management; control systems modeling and identification; information interaction in complex systems; modeling of promising structures, materials and technologies; new methods and approaches to ensure the quality and reliability of modern industries study; modern standardization problems; complex technical systems and information management systems; control in deterministic, stochastic systems and in conditions of uncertainty; ergonomics and biomechanics of systems "man - machine - environment".

The issue was prepared with organizing, informational and technical support of the National Technology Initiative Center for Advanced Manufacturing Technologies based on the Institute of Advanced Manufacturing Technologies of Peter the Great St.Petersburg Polytechnic University.

The issue was prepared with financial support of the grant from the Program Competitiveness Enhancement of Leading Russian Universities among Global Research and Education Centers, supported by the Ministry of Science and Higher Education of the Russian Federation of Peter the Great St.Petersburg Polytechnic University, Project 5-100-2020.

All papers published in this volume of *IOP Conference Series: Materials Science and Engineering* have been peer reviewed through processes administered by the Editors. Reviews were conducted by expert referees to the professional and scientific standards expected of a proceedings journal published by IOP Publishing."

	Scientific Committee
Chair of the Organizing Committee	**Klochkov Yury** – Dr. (Prof.), Head of Academic Development Department, Peter the Great St. Petersburg Polytechnic University, Russia
Deputy Chairman of the Organizing Committee	**Gazizulina Albina** – Ph.D., Director of Center for Monitoring Science and Education, Peter the Great St. Petersburg Polytechnic University, Russia
Organizing Committee and Speakers	**Grechnikov, Fedor** - Dr., Acad. RAS, Prof., First Deputy. Chairman of the SSC RAS; Head of the Department of Metal Forming of Samara University, Russia.
	Papic Lubish - Dr., (Prof.), Technical University Kragujevac, Head of the Center for Reliability and Quality, Serbia.
	Ilya Frenkel - Ph.D., Head of the Reliability and Risk Management Center, Sami Chamoun Engineering College, Israel.
	García Andrés Carrión - Ph.D., Professor, Polytechnic University of Valencia, Head of the Center for Quality and Change Management, Spain.
	Cordoba Maria del Pilar Rodriguez - Ph.D., professor at the National University of Colombia, Department of Industrial Engineering (Manizales), Colombia.
	Polyakova Marina- Dr., Professor, Department of Materials Processing Technologies, Magnitogorsk State Technical University named after G.I. Nosova, Russia.
	Kumar Khatri Sunil - Ph.D., professor at Amity University, director of the Institute of Information Technology, India.
	Leventsov Valery - Ph.D., professor at National Technology Initiative Center for Advanced Manufacturing Technologies based on the Institute of Advanced Manufacturing Technologies of Peter the Great St.Petersburg Polytechnic University

IOP Conf. Series: Materials Science and Engineering **986** (2020) 011003 doi:10.1088/1757-899X/986/1/011003

Peer review declaration

All papers published in this volume of *IOP Conference Series: Materials Science and Engineering* have been peer reviewed through processes administered by the Editors. Reviews were conducted by expert referees to the professional and scientific standards expected of a proceedings journal published by IOP Publishing.

- **Type of peer review: Double-blind**
- **Conference submission management system:**
  Own http://qmrts.com/participate
- **Number of submissions received: 211**
- **Number of submissions sent for review: 209**
- **Number of submissions accepted: 63**
- **Acceptance Rate (Number of Submissions Accepted / Number of Submissions Received X 100): 29%**
- **Average number of reviews per paper: 2**
- **Total number of reviewers involved: 16**
- **Any additional info on review process:**
  To review the articles, 16 reviewers were invited: 4 for each conference's On average, each expert reviewed 13 articles, which made it possible to carefully and accurately examine each of them in accordance with the evaluation criteria.
  In the process of peer review by experts, the following form was filled out:

	Please evaluate the following	Poor	Below average	Average	Above Average	Outstanding
1	Problem formulation					
2	Organisation (logical flow, use of headings)					
3	English language					
4	Appropriateness of the research/study method					
5	Relevance and clarity of drawings, graphs and tables					
6	Contribution to discipline					
7	Use and number of keywords/key phrases					
8	Reference list, adequate and correctly cited					
9	Conclusions, Direction for future research and recommendations					

Please, provide your necessary comments for the Authors (mandatory)

IOP Conf. Series: Materials Science and Engineering **986** (2020) 011003 doi:10.1088/1757-899X/986/1/011003

Please, provide your confidential comments for the Scientific Committee (optional)

Please, provide your final decision

- ☐ Accept.
- ☐ Accept with minor revisions (reviewer does not intent to see manuscript again).
- ☐ Revise, major revisions (if properly revised, paper is publishable).
- ☐ Revise, major revisions – risky (it is not clear that adequate revisions can be made; however author should be allowed to try).
- ☐ Reject.

- **Contact person for queries: Dr. (Prof.) Yury Klochkov (klochkov_yus@spbstu.ru)**

Analysis of the requirements for foundry coke

V A Ivanova

Yaroslavl State Technical University, Yaroslavl, Russia, ivanova-waleriya@mail.ru

Abstract. Foundry coke is part of the mixture for melting of cast irons in coke oven and coke oven gas furnaces, which make up more than 37% of the domestic fleet of cast iron furnaces. It is obtained by sieving coal coke. Performing the function of fuel and providing the necessary heat, foundry coke has a significant impact on the stability of the melting process and the temperature of the cast iron. The share of energy and fuel in the cost of iron casting is 55-60%, so the reduction in coke consumption is of great importance for the energy efficiency of melting and the cost of castings. The requirements for foundry coke are currently set in the standard GOST 3340-88 "Foundry coke. Technical specifications ", in which five indicators are normalized: mass fraction of total moisture, W_t^r , %; ash content of analytical sample, Ad, %; mass fraction of total sulfur, S_t^d, %; strength according to MIKUM, M40, %; fraction of pieces, less than 40 mm.

1. Introduction

Foundry coke is part of the mixture for melting of cast irons in coke oven and coke oven gas furnaces, which make up more than 37% of the domestic fleet of cast iron furnaces. It is obtained by sieving coal coke. Performing the function of fuel and providing the necessary heat, foundry coke has a significant impact on the stability of the melting process and the temperature of the cast iron. The share of energy and fuel in the cost of iron casting is 55-60%, so the reduction in coke consumption is of great importance for the energy efficiency of melting and the cost of castings.

The requirements for foundry coke are currently set in the standard GOST 3340-88 "Foundry coke. Technical specifications ", in which five indicators are normalized: mass fraction of total moisture, W_t^r , %; ash content of analytical sample, Ad, %; mass fraction of total sulfur, S_t^d, %; strength according to MIKUM, M40, %; fraction of pieces, less than 40 mm.

2. Literature overview

In the literature, the requirements for coke include: reactivity [1-3]; ash content [1, 4-7]; humidity [8-10]; strength [1-6]; porosity, density [1-6]; combustibility [5]; sulfur content [1-6]; the size of the pieces [1]; uniformity in size of pieces [2]. For example, it is believed that high-quality coke should have a piece size 10-12 times smaller than the cupola diameter [2]. As a rule, foundry coke has a piece size of more than 40 mm, and for cupolas with a diameter of more than 1.2 m - more than 80 mm to 150 mm [1,8-10]. Abroad, the requirements for the size of foundry coke are almost the same.

Two methods are currently used to assess the reactivity of coke: according to GOST R 54250-2010 "Coke. Determination of coke reactivity index (CRI) and coke strength after reaction (CSR)" and according to GOST 10089-89 "Coal coke. The method of determining the reactivity" (Table 1). Coke reactivity index (%) (NSC method of Nippon Steel Corporation) is used in foreign practice and currently in Russia.

Table 1: Comparative characteristics of methods for assessing the reactivity of coke

Distinction	Methods for determining reactivity	
	GOST P 54250-2010	GOST 10089-89
Indicator designation, units	CRI, %	K, cm3/g·s
Formula for calculating an indicator	$CRI = 100 \cdot \frac{m_0 - m_1}{m_0}$, where m0 – sample weight before reaction, g; m1 – sample weight after reaction, g.	$K = \frac{V \cdot T}{m \cdot T_1} R$, where V – carbon dioxide feed rate, cm3/s; T – test temperature according to test thermocouple, K; T1 – carbon dioxide temperature, K; m – mass of carbon in a sample of coke, g; R – reagent gas conversion.
Indicator values	20-35 [11], 45,8; 38,0; 38,6; 36,8; 30,3 (ERCI)	0,51-0,76 [12]
Sample requirements for analysis	Weight 200±2 г g with a particle size of 19,0-22,4 mm	Weight 7-10 g with a particle size of 1-3 mm
Analysis conditions	Temperature 1100±3 0C; carbon dioxide feed rate 10±0,25 dm3/min; test time 120 min	Temperature 1000±5 0C; heating rate not less than 100 0C/min; carbon dioxide feed rate 3 cm3/s; test time 15 min

Reactivity determines the efficiency of coke use in the cupola [11-12]. The gasification reaction contributes to coke overspending and cupola cooling, besides, the formed carbon oxide cannot be used [13], which generally reduces the melting efficiency. Therefore, coke with the lowest possible reactivity (< 0,5 cm3/(g·s) is required for foundry [14]. The best varieties of domestic foundry coke have a reactivity of 0,4-0,5 cm3/(g·s) [2].
The ash content (Aa, %) is an indicator of the quality of foundry coke according to the requirements of GOST 3340-88 and depends on the brand of foundry coke. The ash content should not exceed 12.0% for coke of the KL-1 brand, 11.0% - for coke of the KL-2 brand, and 11.5% - for coke of the KL-3 brand. The composition of ash in coke is determined according to GOST 10538-87 "Solid fuel. Methods for determining the chemical composition of ash" and GOST R 54237-2010 "Solid mineral fuel. Determination of the chemical composition of ash by the method of atomic emission spectrometry with inductively coupled plasma", ash content according to GOST R 55661-2013 "Solid mineral fuel. Ash determination".

3. Main part

The coke ash in Russian plants is significantly higher than in European ones. This is explained by an increase in the ash content of coal concentrates from 9.6 to 10.5% according to GOST R 51588-2000 "Coals and anthracites of the Kuznetsk and Gorlovsky basins for technological purposes. Technical specifications", as well as the lack of price differentiation depending on the ash content of coal [15-18]. Hence, the ash content in coke for foundry is in the range of 11-12%, and in some cases even higher (13-14%) [2].
The moisture content according to the requirements of GOST 3340-88 is also an indicator of the quality of foundry coke. The share of total moisture in the working condition of the fuel (Wp, %) does not depend on the brand of coke and should not exceed 5.0%. The moisture content in coke is determined according to the requirements of GOST 27588-91 "Coal coke. Method for determination of total moisture", GOST 27589-91 "Coke. Method for determination of moisture in an analytical sample".
Humidity of coke in the plants of the European Union varies in a significant range (7.0-3.5%). Particularly high humidity is in some varieties of coke imported from China (9.5%). Minimum

humidity (0.1-0.5%) is obtained by dry quenching of coke [4, 16, 19-21]. In Russia, the moisture content in coke is from 0.14-1.26 to 4.3-5.0%.

The sulfur content according to the requirements of GOST 3340-88 is an indicator of the quality of foundry coke and depends on the brand of foundry coke. The mass fraction of total sulfur should not exceed 0.6% in the KL-1 brand foundry coke, not more than 1.0% - in the KL-2 brand coke, and not more than 1.4% - in the KL-3 brand coke. In the supplied domestic foundry coke, the total sulfur content (S_t^a, %) is on average 0.46-0.56%. Sulfur in coke passes from coal during coking. Methods for determining the forms of sulfur are presented in GOST 30404-2013 "Solid mineral fuel. Determination of sulfur forms", the total sulfur content in coke - GOST 2059-95 "Solid mineral fuel. Method for determination of total sulfur by burning at high temperature", GOST 8606-2015 "Solid mineral fuel. Determination of total sulfur. Ash's Method".

Phosphorus enters coke from a coal charge and is determined according to the requirements of GOST 1932-93 "Solid fuel. Methods for the determination of phosphorus". The phosphorus content in coke is very different and depends on the phosphorus content in the initial coals [4]. Mass fraction of phosphorus in coke ash at various metallurgical enterprises is presented in Table 2 [17, 21, 24-26].

Table 2: Mass fraction of phosphorus in coke ash at various metallurgical enterprises (P2O5), percent [17]

Nizhny Tagil	Chelyabinsk	Magnitogorsk	Orsk-Khalilovsky	Kuznetsk	West Siberian	Cherepovetsky	Novolipetsk number 1-4	Novolipetsk number 5-6	Kemerovo coking coal plant	Kaliningrad coke plant
0,35	0,65	1,00	0,70	1,10	0,65	0,25	0,60	0,60	0,31	0,55

Volatile substances are substances formed during the decomposition of coal under heating without access to air [18]. Volatile substances consist of adsorbed gases, especially air oxygen and carbon oxide [13, 27-28]. The content of volatile substances in coke is determined according to the requirements of GOST R 55660-2013 "Solid mineral fuel. Determination of the yield of volatile substances".

The amount of volatile substances of foundry coke (Vdaf, %) is from 0.5 to 1.5% [2, 16]. According to [20, 21, 29], the recommended amount of volatiles in foundry coke should not exceed 1.2%. The content of volatile substances in the supply of foundry coke according to GOST 3340-88 is not indicated. However, when technical specifications (TS) are developed, the developer usually adds the value of this indicator to the quality indicators for foundry coke. For example, according to the requirements of TS 0761-027-00187852-10 "Foundry coke. Specifications" the yield of volatile substances is normalized by a value of not more than 1.0%, and according to the requirements of TS 0761-028-00187852-10 "Foundry coal, coke produced by PAO "Mechel". Specifications" for coke with a grain size of 40 mm or more - not more than 1.2%.

Foundry coke strength is of great importance for the effectiveness of melting iron in a cupola furnace. For cupola melting, the strength of foundry coke against abrasion, discharge, shock loads, and also strength at high temperatures is of interest.

The following values of strength indicators are characteristic of domestic coke: M40 – 69-77 %; M25 – 83.7-92.3 %; M10 – 6.0-11.0 %; CSR – 32.6-58.2 %. For foundry coke, the strength value (M40, %) is established in GOST 3340-88 and ranges from 73 to 78% depending on the size class.

The size of the coke provides charge permeability to gases and liquid melting products. The size of foundry coke according to the requirements of GOST 3340-88 is regulated by an indicator that limits the proportion of pieces smaller than 40 mm.

Coke porosity is determined by various methods and it is characterized by the parameters of porosity (P_r, %), apparent porosity (Pa, %) and bulk density in a large container (Z, t/m3).
Bulk mass in a large container according to the requirements of GOST ISO 1013-95 "Coke. Method for determination of bulk density in a large container" (Z, t/m3) is defined as the mass of coke per container volume. The bulk density depends on both the actual density and the apparent density and porosity. For cokes from Donetsk coals, the bulk density is 0.43-0.45 [16, 20], and 0.46-0.48 t/m3 - for cokes from eastern coals [20].
For foundry coke, the value of apparent porosity (Pa, %) according to various sources, ranges from 35 to 45% [2, 4].
The degree of completeness of the molecular and supramolecular structure of coke determines its properties as a semiconductor, characterized by a certain amount of electrical resistance. The value of electrical resistance can serve as an indirect characteristic of a number of properties of coke [22, 31]. For example, in [23], the relationship was established between the number of volatile substances characterizing the degree of "readiness" of coke and the completeness of thermomechanical transformations during coking, and the coke electrical resistance (ρ, Ω·m). The reactivity of coke is known to be reflected in the value of electrical resistivity: the smaller the ρ value of coke samples, the lower the reactivity index K [12, 24]. The data presented in [25] indicate the presence of a connection between the specific electrical resistivity and the valid density of coke (the correlation coefficient was -0.82).
The analysis of data on the rationing of requirements for foundry coke (Table 4) indicates the following:
- foundry coke has a large number of properties that affect each other;

Table 3: Foundry coke standardization

Number	Requirement	Designation	Units	Optimum values
1	2	3	4	5
	Reactivity	K	cm3/g·s	< 0,4-0,5
1	Reactivity after reaction with carbon dioxide	CRI	%	Not established
2	Ash content	Aa	%	< 12,0
3	Moisture of analytical test	Wp	%	< 5,0
4	Total sulfur content	S_t^a	%	< 0,6
5	Phosphorus content	P	%	< 1,2
6	Devolatilization	Vdaf	%	< 1,2
	Abrasion of coke after reaction with carbon dioxide	AV	%	Not established
7	Coke strength after reaction with carbon dioxide	CSR	%	Not established
	The strength of coke according to MIKUM	M40	%	≥ 73 ≥ 78
	Strength dropping	Shl	mm	Not established
8	Mass fraction of pieces less than 40 mm	-	%	≤ 5
	Porosity	P_r,	%	Not established
9	Apparent porosity	Pa	%	< 45
	Bulk density in a large container	Z	t/m3	Not established
10	Electrical resistivity	ρ	Ω·m	Not established

4. Conclusion

Insufficient attention to foundry coke indicates the absence of standardized requirements for foundry coke indicators such as reactivity, strength and abrasion of coke after reaction with carbon dioxide, dropping strength, porosity, bulk density in a large container, electrical resistivity.

References

[1] Ivanov, E.B. Coke production technology / E.B. Ivanov, D.A. Muchnik. - M .: Publishing Association "Vishcha Shkola", 1976. - 232 p.

[2] Nefedov, P. Ya. Quality and efficiency of the use of foundry coke in cupola / P.Ya. Nefedov, V.M. Strakhov // Coke and Chemistry. - 2003. - No. 7. - S. 16-26.

[3] Dolotov, G.P. Furnaces and dryers of foundry production: Textbook for metallurgical technical schools / G.P. Dolotov, E.A. Kondakov. - 3rd ed., Rev. and add. - M .: Mashinostroenie, 1990 .- - 304 p.

[4] Hu, Y., Li, S., Li, D., & Vadim, S. (2020). Yield criteria and strength conditions considering comprehensive mechanical and acoustic emission characteristics of tension–torsion composite deformation of nylon materials. Composite Structures, 243 doi:10.1016/j.compstruct.2020.112278

[5] Ravich, M.B. Metallurgical fuel: a reference book / M.B. Ravich, S.S. Dvorin, A. Ya. Lenkov, S.I. Pevzner; ed. I.N. Sushkina, G.F. Knorre, T.A. Zikeeva. - M .: Metallurgy, 1965 .-- 653 p.

[6] Lazarenko, A. Ya. About zealous production of blast-furnace and non-blast coke / A.Ya. Lazarenko // Coke and Chemistry. - 2002. - No. 8. - S. 38-44.

[7] Lipnitsky, A.M. Plavka of cast iron and non-ferrous metal alloys / A.M. Lipnitsky. - L .: Mechanical engineering, 1973 .-- 192 p.

[8] Grachev, V.A. Modern methods of cast iron melting / V.A. Grachev, A.A. The black. - Saratov: Privolzhskoe book publishing house, 1973. - 342 p.

[9] Pluzhnikov, A.I. Criteria for assessing the quality of blast furnace coke. Prospects for its improvement at Ispat-Karmet OJSC / A. I. Pluzhnikov, O. G. Rifert, A. A. Trembach // Coke and Chemistry. - 2001. - No. 1. - S. 8-10.

[10] Mizin, V.G., Zinovieva L.A., Klyukin S.N., An integrated approach to the assessment of metallurgical coke produced in the conditions of OAO NLMK. Koks i Khimiya. - 2009. - No. 9. - P. 44-50.

[11] Zolotukhin, Yu. A. Requirements for the quality of coke for blast furnaces operating with different specific consumption of pulverized coal / Yu. A. Zolotukhin, NS Andreichikov, Ya. B. Kukolev // Koks chemistry. - 2009. - No. 3. - S. 25-31.

[12] Lazarenko, A. Ya. Theoretical aspects of coke gasification in blast furnace and non-blast technological processes / A.Ya. Lazarenko, V.E. Kononenko, E.L. Sorokin, A.P. Tolstoy // Coke and Chemistry. - 2003. - No. 6. - S. 14-18.

[13] J. Pufpaff, T. Buhles, J. Jahz. Coke properties and behaviors in blast furnace // The 5 European coke and iromaking congress. 2005 / Stockholm, Sweden. Proceedings. - V. 1. - P. Mo 2: 3-1-Mo 2: 3-14.

[14] Drobintsev, P., Voinov, N., Kotlyarova, L., Selin, I., & Aleksandrova, O. (2020). Optimization of technological processes at production sites based on digital modeling doi:10.1007/978-981-15-2341-0_75 Retrieved from www.scopus.com

[15] Panyukov, D., Kozlovsky, V., & Klochkov, Y. (2020). Development and research FMEA expert team model. International Journal of Reliability, Quality and Safety Engineering, doi:10.1142/S021853932040015X

[16] Redkov, A. V., & Kukushkin, S. A. (2020). Development of burton-cabrera-frank theory for the growth of a non-kossel crystal via chemical reaction. Crystal Growth and Design, 20(4), 2590-2601. doi:10.1021/acs.cgd.9b01721

[17] Svatovskaya, L., Shershneva, M., Baydarashvily, M., Sychova, A., Sychov, M., & Gravit, M. (2015). Geoecoprotective properties of cement and concrete against heavy metal ions. Paper presented at the Procedia Engineering, , 117(1) 345-349. doi:10.1016/j.proeng.2015.08.171

[18] Loison, R. Cox / R. Loison, P. Foch, A. Boyer. - M .: Metallurgy, 1975 .-- 520 p.

[19] Levy M.A., Marienbach L.M. - M .: Mechanical Engineering, 1970 .-- 496 p.

[20] Stepanov, Yu.V., Gilyazetdinov R.R., Popova N.K., Makhortova L.A. The influence of optimization of the composition of the charge and its ash content on the quality indicators of coke / Yu. V. Stepanov, R.R. - 2005. - No. 7. - S. 14-18.
[21] Wegman, E.F. Iron metallurgy / E.F. Wegman, B.N. Zherebin, A.N. Pokhvisnev, Yu.S. Yusfin. - M .: Metallurgy, 1978 .-- 480 p.
[22] Populyakh, L.A. Investigation of the behavior of phosphorus in a blast furnace in order to obtain cast irons with a reduced content of impurity elements: dis. ... Cand. tech. Sciences: 05.16.02 / Populyah Larisa Aleksevna. - M., 2009 .-- 145 p.
[23] Kazmin, V.V. Reducing the sulfur content of coke / V.V. Kazmina // Coke and Chemistry. - 1971. - No. 6. - S. 25-28.
[24] Loison, R. Cox / R. Loison, P. Foch, A. Boyer. - M .: Metallurgy, 1975 .-- 520 p.
[25] Ostroukhov, M. Ya. Blast furnace master / M.Ya. Ostroukhov, L. Ya. Schnarber. - M .: Metallurgy, 1977 .-- 304 p.
[26] Miroshnichenko, D.V. The influence of technological factors of preparation and coking of coal on the reactivity of coke / D.V. Miroshnichenko // Coke and Chemistry. - 2009. - No. 2. - S. 37-42.
[27] Pinchuk, S.I. On the quality of coke, problems of ecology and resource saving / S.I. Pinchuk // Coke and Chemistry. - 2006. - No. 4. - S. 6-13.
[28] Obukhovsky, Ya.M. Test methods and ways to improve the quality of blast furnace coke / Ya.M. Obukhovsky. - Donetsk: Regional book publishing house, 1947. - p. 35-48.
[29] Onusaytis, B.A. Formation and structure of coal coke / B.A. Onusaytis. - M .: Publishing house of the USSR Academy of Sciences, 1960 .-- 419 p.
[30] Kurapova, O. Y., Glukharev, A. G., Borisova, A. S., Golubev, S. N., & Konakov, V. G. (2020). Phase formation, stability and heat capacity of ternary TiO2–CeO2–ZrO2 solid solutions. Materials Chemistry and Physics, 242 doi:10.1016/j.matchemphys.2019.122547
[31] Soldatenko, E.M. Changes in the structure and properties of molded and layered coke during high-temperature processing / Problems of molded coke production. Thematic industry collection / E.M. Soldatenko, N.A. Walters. - M .: Metallurgy, 1983 .-- S. 41-45.

IOP Conf. Series: Materials Science and Engineering **986** (2020) 012002 doi:10.1088/1757-899X/986/1/012002

Influence of the chemical composition of cast iron on the temperature of loss of magnetic properties when heating by high-frequency currents

N.V. Kosareva, A.A. Chesnokova

Yaroslavl State Technical University, Yaroslavl, Russia, nataliacosareva@yandex.ru

Abstract. One of the important tasks of modern materials science concerning technical issues is the need to ensure the surface properties of hardened products.An increase in the resource of machines and an increase in reliability are limited by the wear resistance of their components and assemblies: the less wear of the main rubbing pairs, the longer the service life [1]. For example, according to [1], a piston internal combustion engine accounts for 36-52% of the total number of machine failures.Modern technologies have a large number of hardening methods that can increase almost any characteristic of the structural strength of the material. Surface hardening of parts can be carried out in two ways - by changing (modifying) the state of the surface layer or by coating.

Introduction

The cylinder liner is one of the main most loaded engine parts. Widespread in automotive and tractor internal combustion engines of domestic and foreign designs received sleeves of gray cast iron with surface hardening of the working mirror.

The cast-iron cylinder liner of the engines is a thin-walled cylinder with a variable thickness along the height. Working together with pistons and rings, it forms a volume in which the thermal energy of the fuel combustion process is converted into mechanical energy [1]. Under operating conditions, the liner experiences high heating temperatures, tremendous pressure from the combustion of a mixture of fuel and gases, the effects of wear and tear friction [2].

Abrasive wear resistance [3] and macrogeometry conservation of the sleeve mirror during operation [2, 4] are taken as the main indicator determining the durability of the liner, especially diesel engines. The issue of obtaining cast iron sleeves with high macrogeometry stability during operation and at the same time with high wear resistance of the working mirror is still relevant.

The practice has shown that the best results on hardening cast-iron cylinder liners of engines are obtained by surface hardening with heating by high-frequency currents (HFC). It was shown in [5, 6] that with such hardening, a sufficient depth of the layer allows reshaping of the sleeve to repair dimensions, which increases the resource of its operation. However, during hardening of the high-frequency part of the working surface of the liners, significant stresses, deformation of the mirror geometry and insufficient hardness can occur. These factors cause increased wear on the cylinder liners. Therefore, to prevent these defects, it is necessary to select the optimal hardening conditions for HFC of each chemical composition of cast iron.

Main part

Heating using high frequency currents is a complex thermal process. The result of induction heating depends on a large number of factors and is associated with the properties of the electromagnetic system, the basis of which is the heated object. The part is placed in an inductor connected to a high-frequency current generator [7]. An alternating magnetic field creates eddy currents in a thin surface layer of the metal, and heating is carried out due to the resistance of the metal to the flow of these currents. Immediately after heating, which lasts seconds, the part is placed in a sprayer for cooling.

The essence of the surface hardening of HFC is that the surface layers of cast iron parts quickly heat up to hardening temperatures, and then cool at a speed greater than critical. In the work [7] on the kinetics of induction heating, I.N. Khidin showed that a large power per unit of the heated volume at low temperatures decreased sharply when the surface of the product reached a temperature loss of magnetic properties, and the heating of the layer surface slowed down because of this. A diagram of the changes in the rates of high-frequency surface heating of steels and cast irons is presented in Figure 1.

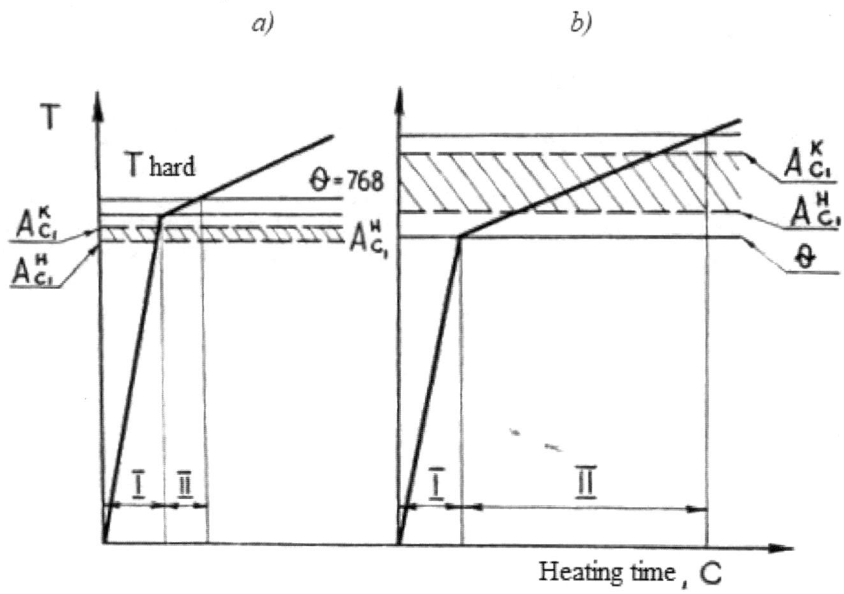

Figure 1. The scheme of the kinetics of induction heating:
a – eutectoid and hypereutectoid steels; b – pearlite cast irons

As can be seen from Figure 1, the surface is heated in 2 stages. At the first stage, heating is carried out to a temperature of loss of magnetic properties of 768 oC, while the heating rate is an order of magnitude higher than at the second stage. The second stage is characterized by heating from the temperature of loss of magnetic properties to the temperature of hardening. The total heating time for hardening is equal to the sum of the duration of the first and second stages. The temperature of the loss of magnetic properties in steels is located above the equilibrium temperature range of the eutectoid transformation Ac1н-Ac1к (Fig. 1, a). The heating time for hardening is short; therefore, stress and deformation after hardening are also insignificant.

Due to the high content of carbon and manganese, cast irons have a wider eutectoid transformation interval located at higher temperatures than steel (Fig. 1, b). In this case, the temperature of the loss of magnetic properties may be lower than the temperature at which the perlite to austenite begins to transform. Therefore, in cast irons, an increase in the duration of the second stage is observed, which practically determines the time of high-frequency heating of cast irons to reach the hardening temperature. Long-term heating of cast iron products gives greater deformations and lower surface layer quality, including lower hardness than expected.

As can be seen, the reduction in the time of heating cast irons during surface hardening with heating by HFC requires a temperature loss of magnetic properties above the temperature values of the critical interval of the eutectoid transformation. This will reduce the total heating time, which has a positive impact on the results of hardening.

Optimization of the hardening mode of cast iron cylinder liners of engines is possible due to changes in the chemical composition of cast irons. The optimal chemical composition should provide a state in which the temperature of the loss of magnetic properties would be higher than the temperatures of the critical interval of the eutectoid transformation. We studied 30 melts of the following chemical composition: 2,8-3,8 % C; 0,5-2,3 % Si; 0-1,2 % Al; 0,5-1,2% Mn; 1,5-4,5 % Co. Comparison was made with serial sleeves of gray cast iron type SCh22, the chemical composition of which is 3,3 % C; 2,2 % Si; 0,6% Mn; 0,2 % P.

Taking into account the results of [8], graphitizer - aluminum was introduced as a substitute for silicon in low-silicon cast iron to obtain lamellar graphite.

In all melts, a structure with uniformly distributed rectilinear plate graphite with a metal base consisting of plate perlite P-P96 was obtained [11].

To record temperature changes, an N041UN.2 light-beam oscilloscope with M001-IA type galvanic inserts was used, which records instantaneous values of current and voltage. The presence in the oscilloscope of six speeds of motion of the film made it possible to obtain a scan of the fast-moving process of high-frequency heating of the surface.

The calibration of thermocouple and the oscilloscope were calibrated using the reference platinum rhodium – platinum thermocouple and the secondary fixed point of the international practical temperature scale (antimony solidification temperature) according to the procedure [9] with five repetitions. The comparison of the results showed good reproducibility.

When repeating the experiments, a chromel-alumel thermocouple with a diameter of thermo electrodes of 0.17 · 10-3 m was used. The wires were fastened, the thermocouple was homogeneous, the circuit resistance was prepared to a predetermined value, the wires were shielded, and the filters were turned on according to the method proposed in [7]. The thermal inertia of the thermocouple was checked using acceleration curves [10], which reflect the reaction of the thermocouple to step disturbances.

The result of checking the selected thermocouples showed that with a step perturbation from 0 oC (water + snow) to 100 0 oC (boiling water), the delay of the thermocouple readings occurs in just 0.014 seconds. The duration of the entire disturbance at the selected heating rate of HFC at the first stage to the inflection point (Curie point) for cast iron of experimental melts is 1.2-1.3 sec (Fig. 2). Therefore, the inertia of the thermocouple can be neglected, since the delay time is small compared with the perturbation time. The temperature on the kinetic curves can be considered true at any time.

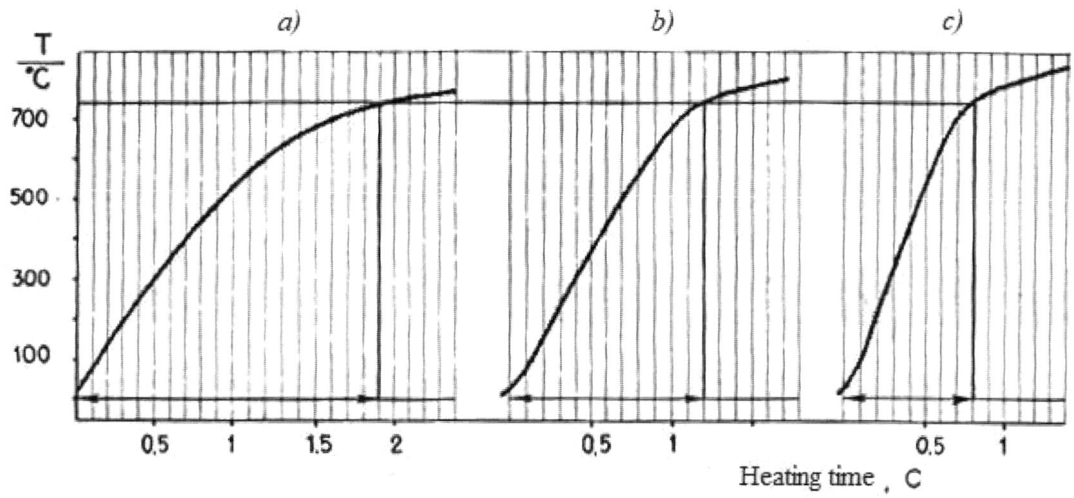

Figure 2. Influence of the heating rate of HFC on the character cast iron kinetic curves:
a – 500 deg/sec; b – 700 deg/sec; c – 900 deg/sec

The average temperature of the inflection point on the kinetic heating curve (Curie point) for serial cast iron of the SCh 22 type was determined with a 15-fold repetition of the experiment and corresponds to a temperature of 751±2 oC (Fig. 2) with a confidence probability of P=0,95.

Table 1 presents the chemical composition of several melts for a series of experiments presented in Figure 3.

Table 1: The chemical composition of cast iron melts

Melt number	Chemical composition, %				
	C	Si	Al	Mn	Co
Etalon	3,3	2,2	-	0,6	-
4	3,8	0,8	0,6	1,2	-
7	3,3	1,4	0,6	1,1	-
23	3,2	0,8	0,6	0,6	-
28	3,0	0,5	1,2	1,0	1,5
30	3,0	0,5	1,2	0,9	4,5

Figure 3 shows a series of kinetic heating curves for high-frequency samples for cast irons of various chemical compositions.

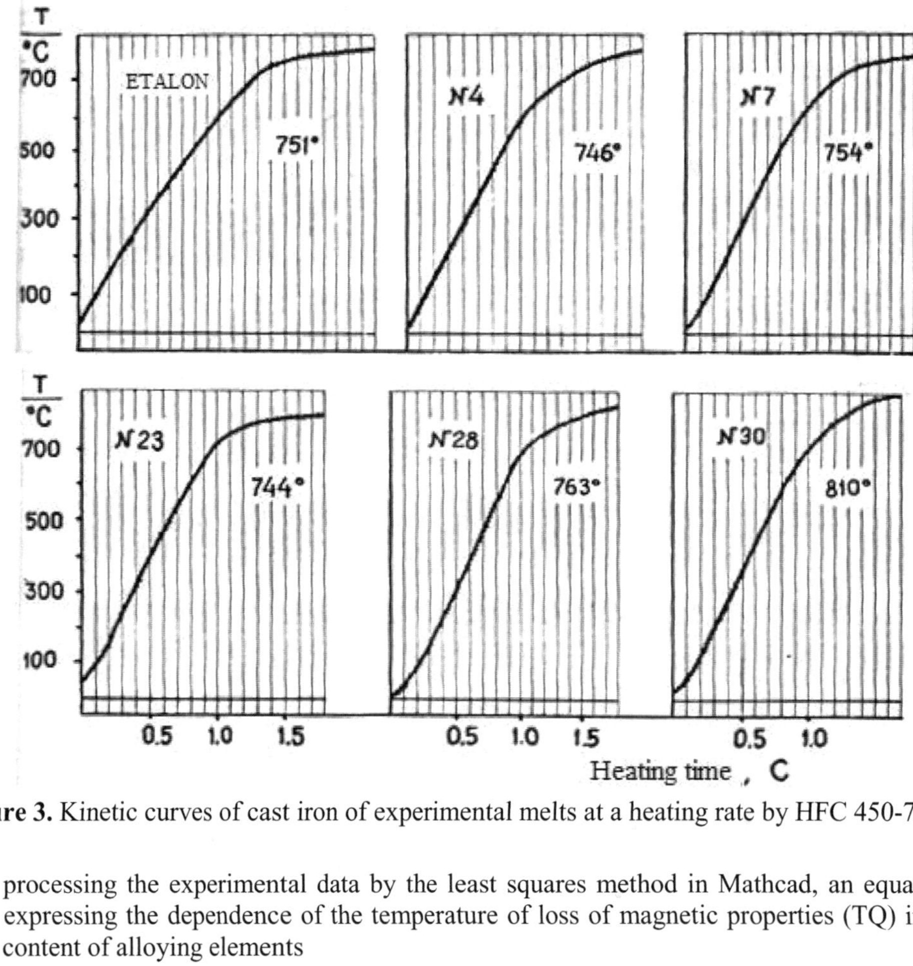

Figure 3. Kinetic curves of cast iron of experimental melts at a heating rate by HFC 450-700 deg/s

When processing the experimental data by the least squares method in Mathcad, an equation is obtained expressing the dependence of the temperature of loss of magnetic properties (TQ) in cast iron on the content of alloying elements

$$TQ= 747 + 1{,}6[C] - 1{,}0[Si]+ 1{,}0[Mn]- 6{,}6 [Al]+14{,}6[Co]. \quad (1)$$

As can be seen, the main elements of carbon, silicon, manganese do not have an impact on the temperature loss of magnetic properties in cast irons. The effect of aluminum at its low content, apparently, can be associated with an increase in the dispersion of perlite, which, as shown in [7], slightly decreases the Curie point. Cobalt has a positive effect on increasing the temperature of the loss of magnetic properties, but negatively affects the quality of hardening and the efficiency of the manufacturing process.

Conclusion

The experiments showed that when induction heating with high-frequency currents in the same mode, corresponding to the capabilities of the production facility for hardening by HFC, the value of the loss of the point of magnetic properties in cast iron of experimental melts and serial case cast iron is approximately the same and corresponds to a temperature of 750 oC. Therefore, it is not practical to change the relationship between the temperature of the point of loss of magnetic properties and the temperature interval of the eutectoid transformation in gray pearlite cast irons by increasing the Curie point.

To reduce the heating time by high frequency currents during surface hardening of cast iron products, it is necessary to select the chemical composition of cast iron in such a way as to lower the temperature of the interval of critical transformations relative to the Curie point. In turn, this leads to an increase in the hardness of the surface layer and the dimensional stability of the finished products.

References

[1]. Grigor'yev M.A., Ponomarev N.N. *Iznos i dolgovechnost' avtomobil'nykh dvigate-ley.* – M.: Mashinostroyeniye, 1976. – 247 s.

[2]. Bilik SH.M. Makrogeometriya detaley mashin. – M.: Mashinostroyeniye, 1973. – 346s.

[3]. Astashkevich B.M., Larin T.V. *Optimal'nyy sostav i struktura chugunov tsilindro-vykh vtulok teplovoznykh dvigateley.* – Liteynoye proizvodstvo, 1970, №6, 30-32s.

[4]. Trautwien R., Bechman telbildung. – MAHLE Technische Informationen. Folge B, September, 1973.

[5]. Pozdnyakov N.G., Saychuk A.V., Kokhan I.R. *Analiz metodov uprochneniya i vosstanov-leniya gil'z tsilindrov avtomobil'noy tekhniki.*//Vestnik KHNTUSG, Vypusk 96. – Khar'kov: KHNTUSG, 2010, 475-483s.

[6]. Golovin G.F. *Ostatochnyye napryazheniya, prochnost' i deformatsii pri poverkhnostnoy zakalke tokami vysokoy chastoty.* – L.: Mashinostroyeniye, 1973. – 144 s.

[7]. Kidin I.N.*Fizicheskiye osnovy elektromagnitnoy obrabotki metallov i splavov.* M.: Metallurgizdat. 1969. – 375 s.

[8]. Bobro YU.G. *Legirovannyye chuguny.* M.: Metallurgiya, 1976. - 286 s.

[9]. Preobrazhenskiy V.P. Teplotekhnicheskiy izmereniya i pribory. – M.: Energiya, 1978. – 703 s.

[10]. Lineveg F. *Izmereniye temperatury v tekhnike: Spravochnik.* – M.: Metallurgiya, 1980. – 543 s.

[11]. GOST 3443-87 *Otlivki iz chuguna s razlichnoy formoy grafita.* Metody opredele-niya struktury. - M.: Standartinform, 2005.

[12]. Povysheniye nadezhnosti dizeley YAMZ i avtomobiley KrAZ. – M.: Mashinostroye-niye, 1974. – 288 s.

IOP Conf. Series: Materials Science and Engineering **986** (2020) 012003 doi:10.1088/1757-899X/986/1/012003

Influence of transportation in railway cars of foundry coke on its quality

E. O. Pobegalova, A. J. Ivanova

Yaroslavl State Technical University, Yaroslavl, Russia, pobegalovaeo@gmail.com

Abstract. The granulometric composition of the bulk density of foundry coke is one of the most important characteristics that determine its consumer properties [1]. In the process of melting cast iron in a cupola, foundry coke has to provide sufficient temperature and drainage capacity. The latter is determined by the size of the foundry coke, the uniformity of its particle size distribution, as well as both "hot" (CSR, %) and "cold" (M40, I40, etc.) strength. The dependence of the strength of coke on the carbon structure, the parameters of the porous structure, size, amount and composition of ash is established [2]. There is data [3] indicating a decrease in the "hot" strength of coke with an increase in the content of basic oxides in the ash: an increase in the SO3 content by 1% reduces the CSR by 20-24%.

Introduction

The granulometric composition of the bulk density of foundry coke is one of the most important characteristics that determine its consumer properties [1]. In the process of melting cast iron in a cupola, foundry coke has to provide sufficient temperature and drainage capacity. The latter is determined by the size of the foundry coke, the uniformity of its particle size distribution, as well as both "hot" (CSR, %) and "cold" (M_{40}, I_{40}, etc.) strength. The dependence of the strength of coke on the carbon structure, the parameters of the porous structure, size, amount and composition of ash is established [2]. There is data [3] indicating a decrease in the "hot" strength of coke with an increase in the content of basic oxides in the ash: an increase in the SO_3 content by 1% reduces the CSR by 20-24%.

Quality requirements for foundry coke are standardized by GOST 3340-88 [4] and various specifications [5]. Quality indicators are determined by the supplier and during transportation and storage may vary [6-9, 13-16].

Coke is transported to the consumer in batches (wagons) according to GOST 22235-2010 [10] and should ensure the safety of the values of quality indicators obtained during production. For each batch of foundry coke, the manufacturer issues a certificate of compliance with the requirements of GOST 3340-88 or TS. However, the particle size distribution of foundry coke changes during transportation [1]. During transportation, coke is subjected to abrasion and is destroyed. Studies have shown [6] that the consumer in coke content of the class more than 80 mm in 2,2-11,4 times, and the class 60-80 mm in 1,3-2,4 times less than that of the manufacturer. The coke content of fineness classes of less than 40 mm also increases. Cokes are especially crushed, characterized by low strength. In them, the coke content with a piece size of less than 40 mm is 30-37%, and more than 60 mm required for cupola furnaces is only 25-30%.

Main part

According to GOST R 51908-2002 [11-12], the transportation of goods by rail belongs to the field of especially light conditions, and the main mechanical influence factor is vibration loads. In this regard,

studies were conducted to determine the degree of influence of vibration loads on the quality of foundry coke during transportation by rail
For testing, samples of foundry coke of three batches of various sizes were selected: 40-60 mm (15 samples), 60-80 mm (15 samples), 80 mm and more (15 samples). Samples of batches 1 and 2 met the requirements of GOST 3340-88, and batches 3 - TS 0761-032-00187852-2015. Certificate data are presented in Table 1

Table 1. Quality indicators and data of quality certificates of foundry coke samples.

Quality certificate data	Foundry coke batch		
	1	2	3
Manufacturer	JSC "Gubakhinsky coke"	JSC "Moskoks"	JSC "Moskoks"
Requirement document	GOST 3340-88	GOST 3340-88	TS 0761-032-00187852-2015
Size class	40 mm and more	60 mm and more	60 mm and more
Quality indicators	Quality indicators		
	1	2	3
Mass fraction of total moisture in working conditionW_t^r, %	1	2,5	1,10
Ash content A^d, %	10,6	11,5	11,1
Mass fraction of total sulfur S_t^d, %	0,58	0,6	0,6
Devolatilization V^{daf}, %	-	0,3	0,3
Strength indicator M_{40}, %	78,4	81,6	82,6
Mass fraction of pieces less than the lower limit, %	6	13,9	19,2
Including less than 40 mm, %	-	4,3	3,4

The tests were carried out on a UVE-100 / 5-300 vibrating electrodynamic bench, the maximum amplitude of vibration acceleration was 29.4 m/s^2 (3g), and the vibration frequency was 30 Hz. Samples of foundry coke of each batch were tested separately for 1 hour 10 minutes, which corresponds to transportation by rail over a distance exceeding 1000 km. Before and after the test, the mass of the sample was measured, according to the obtained data; the abrasion index was calculated by the formula:

$$A = \frac{m_1 - m_2}{m_1} \cdot 100\ \%, \quad (1)$$

where m_1 is the mass of foundry coke sample before testing, g; m_2 is the mass of foundry coke sample after testing, g.
Measurements m_1 and m_2 contain only the systematic component of the error associated with the measuring instrument, which, according to the instrument passport, is ±0,01 g. The results of indirect measurements of the abrasion of foundry coke samples were processed according to MI 2083-90 [12]. Given the non-linear nature of the dependence, the formula for finding the error of indirect measurements takes the form

$$\Delta A = \sqrt{\left(\frac{m_2}{m_1^2} \cdot \Delta m_1\right)^2 + \left(\frac{1}{m_1} \cdot \Delta m_2\right)^2} \quad (2)$$

where Δm_1 and Δm_2 are errors of direct measurements of the mass of foundry coke samples.
The average value of the abrasion of the samples ($\bar{A}$, %) for various size classes and their errors are presented in Table 2.

Table 2. The average abrasion values of foundry coke samples.

Size class, mm	Batch 1		Batch 2		Batch 3	
	$\bar{A}$, %	ΔA, %	$\bar{A}$, %	ΔA, %	$\bar{A}$, %	ΔA, %
40-60	3,59564	0,00031	5,59738	0,00028	3,97718	0,00020
60-80	4,38261	0,00008	4,35754	0,00007	3,79318	0,00011
80 and more	3,81169	0,00004	5,42267	0,00003	2,58090	0,00004

An analysis of the results indicates the presence of a correlation dependence of the average strength between the average values of abrasion and the ash content of foundry coke samples. The relationship is most pronounced for coke of particle size class 40-60 mm (correlation coefficient 0.9) and particle size class 80 mm or more (correlation coefficient 0.5).
The influence of strength (M_{40}, %) on the abrasion (A, %) of foundry coke depends on its size class. A strong inverse correlation is observed only for coke of size class 60-80 mm. However, the obtained data are consistent with the results of [12], where it was found that there is no dependence of the abrasion of foundry coke on the mass of samples and its relationship with the mechanical properties of the coke substance (Table 3).

Table 3. The average abrasion values of foundry coke samples.

Foundry coke batch	Strength indicator M_{40}, %	Ash content A^d, %	Abrasion index A, %	
			Size class 40-60 mm	Size class 60-80 mm
1*	73,2	12,0	12,775	13,975
2*	73,3	11,8	12,696	12,428
3*	73,3	11,7	13,807	11,525
4*	73,8	11,6	9,963	8,197

* batch number designation in the work [12]

Besides, during the tests, there was a change in the particle size distribution, an increase in the content of class 60-80 mm by 30% in batches 1 and 2 due to the destruction of samples of particle size class 80 mm or more, and the presence of particle size less than 40 mm, which is consistent with the data work [6].

Conclusion

1. When transporting in railway cars under vibration influence, the quality of foundry coke deteriorates, accompanied by a decrease in the mass of pieces due to abrasion.
2. Under loads equivalent to the transportation of foundry coke over a distance of more than 1000 km by rail, the mass loss (abrasion) of pieces of foundry coke ranged from 3.45 to 5.13% depending on the batch. At the same time, the granulometric composition of the batch of foundry coke changes in the direction of increasing the content of classes of smaller fineness that subsequently negatively affects the process of melting cast iron in a cupola.
3. To maintain the quality of foundry coke during transportation, the transportation distance should be minimized. If this is not possible, it is necessary to give preference to foundry coke with higher values of strength parameters.

References

[1] Muchnik D A and Babanin V I 2014 *Vozmozhnosti uluchsheniya kachestva koksa vne pechnoy kamery* (M.: Infra-Inzheneriya) p 368
[2] Ivanova V A 2014 Kachestvo liteynogo koksa: monografiya (Yaroslavl': Izd-vo YAGTU) p 144

[3] Ulanovskiy M L and Miroshnichenko D V 2005 *O metode firmy «Nippon Stil Korporeyshn» opredeleniya kachestva koksa* Koks i khimiya (№6) pp 18-21
[4] GOST 3340-88 1989 Koks liteynyy kamennougol'nyy. Tekhnicheskiye usloviya (Vved. 1990 01-01. M.: Izdatel'stvo standartov) p 4
[5] TU 761-027-00187852-10 2010 Koks liteynyy kamennougol'nyy. Tekhnicheskiye usloviya. p 13
[6] Nefedov P YA and Strakhov V M 2003 Kachestvo i effektivnost' ispol'zovaniya liteynogo koksa v vagrankakh Koks i khimiya (№ 7) pp 16-26
[7] Shchukin P A 1971 Issledovaniye svoystv metallurgicheskogo koksa (M.: Metallurgiya) p 184
[8] Muchnik D A and Ivanov Ye B 1968 Sortirovka koksa (M.: Metallurgiya) p 296
[9] Ivanova V A and Shamina Ye O 2018 Vliyaniye usloviy transportirovki na kachestvo liteynogo koksa Izvestiya Samarskogo nauchnogo tsentra RAN (t. 20. № 4.) pp 18-24
[10] GOST 22235-2010 2011 Vagony gruzovyye magistral'nykh zheleznykh dorogo kolei 1520 mm. Obshchiye trebovaniya po obespecheniyu sokhrannosti pri proizvodstve pogruzochno razgruzochnykh i manevrennykh rabot (Vved. 2011-05-01. M.: Standartinform) p 18
[11] GOST R 51908-2002 2002 Obshchiye trebovaniya k mashinam, priboram i drugim tekhnicheskim izdeliyam v chasti usloviy khraneniya i transportirovaniya (Vved. 2003-07-01. M.: IPK Izdatel'stvo standartov) p 20
[12] Ivanova V A and Vdovin K N 2016 Metod ispytaniya dlya opredeleniya istirayemosti i abrazivnoy sposobnosti liteynogo koksa Izvestiya vysshikh uchebnykh zavedeniy. Chernaya metallurgiya (T. 59. № 3.) pp 149-153
[13] Davydov, A. D., Erokhina, O. O., Ryaboshuk, S. V., & Kovalev, P. V. (2020). Analysis of the causes of cracks in the production of ingots and forgings from austenitic stainless steel 08X18H10T (aisi 321) doi:10.4028/www.scientific.net/KEM.854.16
[14] Konakov, V. G., Kurapova, O. Y., Novik, N. N., & Golubev, S. N. (2015). Microstructure of "copper-graphene" composites manufactured by electrochemical deposition using graphene suspensions stabilized by non-ionic surfactants. Materials Physics and Mechanics, 24(4), 382-390.
[15] Konakov, V. G., Kurapova, O. Y., Novik, N. N., Pivovarov, M. M., & Archakov, I. Y. (2015). Nanostructured ysz coatings on titanium. Reviews on Advanced Materials Science, 43(1-2), 94-101.
[16] Makovetskaya-Abramova, O., Lazarev, Y., Gravit, M., Silla, S., & Shakhova, M. (2020). Multiplicative method for creating the traffic monitoring base in a megapolis. Paper presented at the E3S Web of Conferences, , 164 doi:10.1051/e3sconf/202016403023

IOP Conf. Series: Materials Science and Engineering **986** (2020) 012004 doi:10.1088/1757-899X/986/1/012004

The specificity of forming requirements for the quality of information support of a high-tech enterprise

K. I. Porsev [1], M. F. Bulatov[2], P. I. Abramov [3]

[1] Yaroslavl State Technical University, Yaroslavl, Russia, ram_dva@mail.ru
[2] FSBSI Scientific and Technological Center for Unique Instrumentation of the Russian Academy of Sciences, Moscow, Russia
[3] JSC "Research Institute" POLYUS "them. M.F. Stelmakh "

Abstract. The level of the current state in the field of information support (IS) of the activities of high-tech enterprises is considered in this paper. The main problems are shown on the way to achieving the required level of IS quality in high-tech enterprises of the country. The need for the formation of a number of standardized requirements for the IS quality of a high-tech enterprise to improve its level is determined. A typical list of requirements for the IS quality is determined and a model of a system for accounting for requirements for the IS quality is developed using the example of Scientific and technological center of unique instrumentation of the Russian Academy of Sciences.

Introduction

The current stage in the development of high-tech industries of the country is characterized by a close dependence of the quality level of manufactured high-tech products on the effectiveness of information support (IS) of the activities of high-tech enterprise's personnel.

IS is a set of information resources and services necessary for solving managerial, creative and scientific - technical tasks in accordance with the stages of their implementation [1]. The quality of IS, according to GOST R ISO 9000-2015 [2], should be understood as the degree of compliance of the inherent characteristics of IS with the established requirements. The practical implementation of advanced achievements in the field of IS of high-tech enterprises should be aimed at providing continuous information support for the production processes of high-tech products, which becomes possible due to the introduction of advanced achievements of information technologies (IT) in the enterprise.

At present, a significant part of high-tech enterprises of the country is characterized by a number of problems directly related to the insufficient quality level of IS [3]. The most common problems include

- the duration and low certainty of information exchange processes between employees;
- low degree of implementation of electronic document management;
- high risk of irretrievable loss of information when it is repeatedly transmitted;
- low efficiency of information resource management processes, etc.

One of the main causes of these problems, in our opinion, is the universally observed phenomenon of uneven, local informatization of individual structural units and production processes of high-tech enterprises. The introduction of various automated control systems (ACS) and design (ADS) into the structure of a high-tech enterprise often does not bring the desired results due to the low consistency and limitedness of these measures within the framework of the exclusively technical side.

Methods

Despite the relative study of problems related to the general issues of informatization management in high-tech organizations [4-6], there are insufficient studies related to the development of methods for assessing and subsequently improving the current level of the IS quality in the context of creating conditions for highly effective information support of the activities of creating high-tech products. There is a general insufficiency of measures aimed at developing effective systems and methods for managing the computerization of science-intensive enterprises. All this allows to conclude that the problem under consideration is systematic within the high-tech industry and determines the relevance of research aimed at finding the implementation of highly effective methods for improving the quality level of IS, based on the application of advanced IT achievements in organizations.

The development of the most effective methods for improving the quality level of IS is possible only on the basis of a comprehensive assessment of the current level of quality. In this case, fundamental importance is given to the establishment of quality requirements based on a comprehensive analysis of the organizational and technical characteristics inherent in a particular high-tech enterprise.

In this regard, as the main object of study to determine the direct requirements to the quality level of IS, «Scientific and technological center of unique instrumentation of the Russian Academy of Sciences» (STC UI RAS) was selected which has a global priority in fundamental and applied research in the field of scientific instrumentation and optics [7].

To determine the basic requirements for the quality level of IS, STC UI RAS was considered as a complex organizational and technical system (OTS) with its inherent set of characteristics and active relationships between them.

Results

As a result of a comprehensive analysis of the external and internal characteristics of the development of OTS by the example of the enterprise under study, the main requirements for the level of quality of IS were identified and the final model of the accounting system for these requirements was built (Fig. 1).

It is established that the totality of all requirements for the quality of IS consists of 2 main groups, classified by degree of origin and level of influence, the quality of IS.

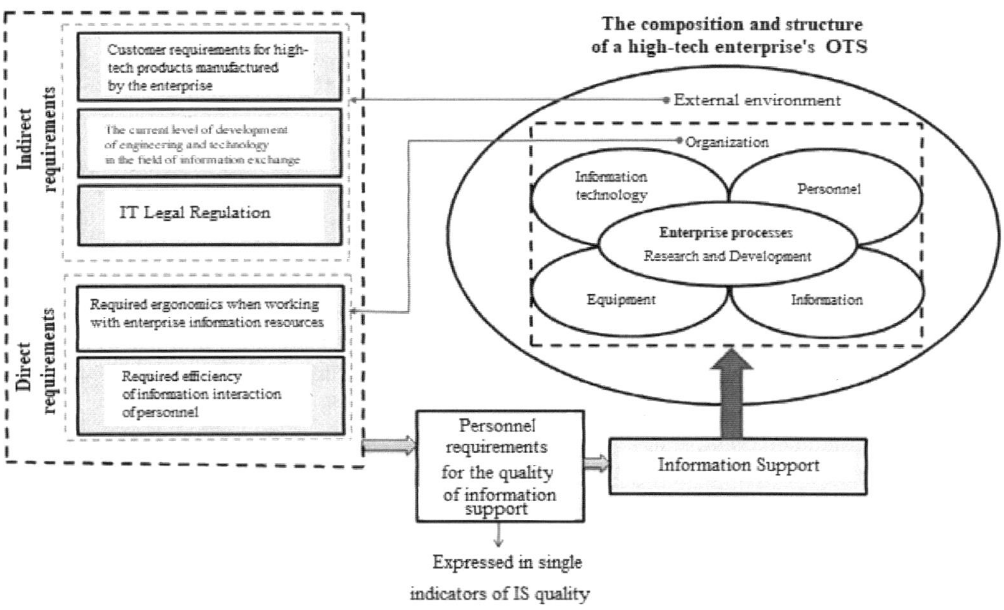

Figure 1. The model of the system for accounting the requirements for the IS quality of a modern high-tech enterprise.

The final integration of groups of indirect and direct requirements finds its final expression in the personnel requirements for the IS quality, since it is personnel that is the main consumer of IS of a high-tech enterprise.

In this regard, it becomes necessary to conduct an accompanying comprehensive study of personnel requirements for the IS quality at workplaces of specific stages, taking into account the specifics of creating high-tech products.

The final set of requirements for the required level of IS was formed as a result of the comprehensive analysis of the current specifics of the production of high-tech products within STC UI RAS, as well as the personnel requirements for the IS quality:

1. A high degree of compatibility between ACS and ADS in the integrated information environment (IIE) of a high-tech enterprise.

In modern high-tech enterprises, a large number of ACS and ADS of various functional and technical purposes are widely sold, the main purpose of which is to ensure the effective work of personnel with the information resources of the enterprise. In this regard, the implementation of this requirement becomes especially important, since the effective creation of high-tech products becomes impossible without a high degree of information compatibility of data on separately executed processes of the organization.

2. High efficiency of information monitoring of technological processes.

Information monitoring allows continuous monitoring of the emergence of new information about the processes of the product life cycle for given information flows in a fixed thematic field for analysis, management, and prediction of its development [8]. Fulfillment of this requirement becomes necessary in the context of the implementation of the ideas of continuous information support for the product life cycle, which is consistent with the main task of IS in modern high-tech organizations.

3. The maximum possible degree of implementation of simultaneous group work of personnel in ACS with large amounts of information.

Ensuring the joint work of personnel during the implementation of research and development works becomes the main factor in reducing time and costs in the development of high-tech products.

4. High speed of obtaining information about created high-tech products at any stage of its life cycle.

The speed, at which information is received by the organization's personnel, plays a crucial role in determining the competitiveness of an enterprise. The fulfillment of this requirement allows ensuring the effectiveness of the existing material and technical base and information flows in the enterprise involved in the processes of information transfer.

5. The prevailing position of electronic data interchanges in organizations.

The widespread use of electronic forms of documentation and the rejection of paper media is becoming one of the priority areas for the development of information technologies in the modern world [9-10]. The implementation of this indicator will help to form the basis of measures for material and technical equipping of the necessary ACS by the organization, as well as determine the need for organizational measures aimed at training personnel to work with the electronic documentation form.

Conclusion

Thus, accounting of the requirements given above, while informing high-tech industries, can serve as the development of widespread study and solution of existing problems of high-tech organizations' IS and ensuring the required level of quality of processes for creating high-tech products.

It should be noted the need to develop methods for quantitative assessment of requirements, as well as their possible supplement as further ways of research aimed at improving the quality of IS of high-tech organizations.

References

[1] Brezhnev VV and Minkina VA 2004 Information services: products and services provided by libraries and information services of enterprises (St. Petersburg: Profession) p 304

[2] ISO 9000: 2015 Quality management systems - Fundamentals and vocabulary

[3] Litvinenko OA 2011 Conceptual aspects of the enterprise informatization management system Scientific and technical bulletin of St. Petersburg State University ITMO (No. 1 (71) pp 120-123

[4] Carlock P G and Fenton R E 2001 System of Systems (SoS) enterprise systems engineering for information-intensive organizations Systems engineering pp 242—261
[5] Overby E, Bharadwaj A and Sambamurthy V 2006 Enterprise agility and the enabling role of information technology European Journal of Information Systems pp 120—131
[6] Durmusoglu S S and Barczakb G 2011 The use of information technology tools in new product development phases: Analysis of effects on new product innovativeness, quality, and market performance Industrial Marketing Management pp 321—330
[7] Balashov AA, Bulatov MF and Pozhar VE 2017 Spectral optical devices at the STC UP RAS Acousto-optic and radar methods of measurement and information processing: Proceedings of the 10th International Scientific and Technical Conference Russian NTORES them. A.S. Popova Suzdal Russia pp 18-21.
[8] Grum-Grzhimailo YV, Popov SV and Polyakov MN 2008 Information monitoring of the problem area structure to support managerial decision-making in the scientific and technical sphere Science. Innovation. Education (no. 6) pp 223-236
[9] Wolnowska A 2012 Implementation of qualitative rules in company's information management Management Systems in Production Engineering pp 21-25
[10] Kumar U D, Crocker J, Knezevic J and El-Haram M 2000 Reliability, Maintenance and Logistic Support – A Life Cycle Approach. Springer Science & Business Media 490 p

IOP Conf. Series: Materials Science and Engineering **986** (2020) 012005 doi:10.1088/1757-899X/986/1/012005

Properties of modifying additives for materials based on water polymer dispersions

A. E. Tereshko, E. A. Indeykin, I. V. Golikov

Yaroslavl State Technical University, Yaroslavl, Russia, tereshkoae@ystu.ru

Abstract. The properties of aqueous dispersions of petroleum paraffin and ceresin, which are developed as modifying additives for water-dispersion paints and varnishes, are investigated. It has been found that such dispersions have a narrow particle size distribution. The relative coagulation rate is calculated and the rheological properties of dispersions of paraffin and ceresin are studied.

Introduction

The use of aqueous dispersions of polymers in various fields of technology necessitates their modification to meet the requirements for adhesive films formed from them. Modification of polymers makes it possible to expand their scope and improve the operational properties of products using them. To impart hydrophobicity to adhesive polymer films, control the permeability of water vapor and other gases, reduce the stickiness of products during storage, increase the protective properties, etc., natural waxes of various types (beeswax, mountain, carnauba), synthetic (Fischer-Tropsch) can be used as modifiers, polymer (polyethylene and its derivatives), solid petroleum hydrocarbons (paraffins, ceresins, microcrystalline waxes), etc. Such modifications are considered in modern periodicals and discussed at international congresses devoted to polymer composite materials and coatings [1 - 7]. Micronized waxes or aqueous wax dispersions are used to modify water dispersible materials. [6 - 8].

One of the most affordable types of waxes is solid petroleum hydrocarbons. Their films have relatively low moisture, vapor and gas permeability.

The great advantage of solid hydrocarbons is that they are pure, often white or yellowish, odorless and tasteless substances that do not contain harmful pathogenic, carcinogenic and mutagenic compounds. Solid petroleum hydrocarbons and paraffin compositions are chemically stable inert substances □9□.

It is relevant to obtain their aqueous dispersions and study their properties in terms of manufacturability of using solid petroleum hydrocarbons. Aqueous dispersions of petroleum paraffin P-2 and ceresin 75 were developed in this paper [10]. The volumetric content of oil hydrocarbons in dispersions is ≈ 30%. Paraffin dispersion was stabilized using polyvinyl alcohol (PVA) with an average molecular weight (M_w) of 130,000 and an ether number of 130–150 mg KOH/g [11]. The dispersion process was carried out using a dissolver with a gear stirrer with a diameter of 40 mm and a rotational speed of 115 s^{-1}. Mixing at low speeds and dispersion was carried out at a temperature exceeding the melting point of hydrocarbons.

Experiment

It was experimentally established that the most stable dispersion of paraffin was obtained with a PVA content of 3.5% in the composition. For the dispersion under study, an analysis of variance was performed using ultramicroscopy. The particle size distribution curve is shown in

IOP Conf. Series: Materials Science and Engineering **986** (2020) 012005 doi:10.1088/1757-899X/986/1/012005

Figure 1, from which it follows that the most probable particle size of the paraffin dispersion is 3.3 μm.

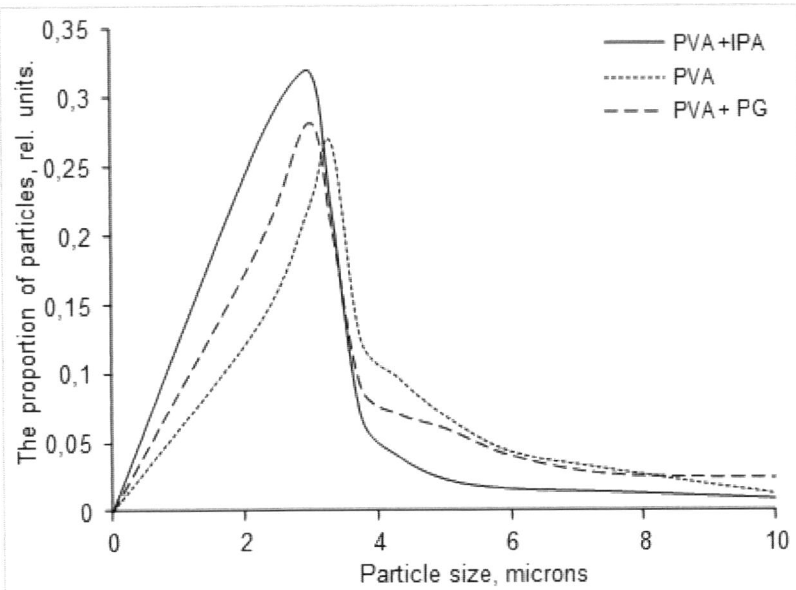

Figure 1. Particle size distribution of paraffin dispersions.

Thus, paraffin dispersions stabilized by PVA are quite stable, however, it was noted that a small amount of low-dispersed fractions floats during storage. It is known [12] that the behavior of surfactants that affects the size distribution of particles of the dispersed phase can be controlled by the introduction of hydrotropic additives into the composition of the dispersion medium. The effects of isopropyl alcohol (IPA) and propylene glycol (PG) were tested as such additives. They can effectively regulate the phase behavior of surfactants.

It was found that the most stable are dispersions stabilized by PVA in combination with hydrotropic additives: 10% IPA and 4.5% PG [10].

Particle size distribution curves (Fig. 1) showed that the most probable particle size for dispersions containing, in addition to the PVA solution, IPA and propylene glycol was 3 μm. The content of coarse fractions is the smallest in the case of using IPA as a hydrotropic additive. The content of finely divided fractions with the use of IPA increases. This is most likely due to its lower dielectric constant ($\varepsilon = 18.3$) compared to propylene glycol ($\varepsilon = 29.5 - 35.0$). [13]

Studies of the aggregative stability of these dispersions were evaluated by the change in the optical density of diluted dispersions over time (Fig. 2). Using the relative change in the optical density of the dispersions, the coagulation rates of paraffin particles were calculated (Table 1).

Table 1. Paraffin dispersion coagulation rate, $(1/h)\cdot 10^{-3}$.

The composition of the stabilizing system	PVA	PVA + IPA	PVA + PG
Coagulation rate, hour $^{-1}\cdot 10^{-3}$	14,9	1,8	5,8

Figure 2 and Table 1 data show that the coagulation rate of paraffin dispersions stabilized by PVA in combination with propylene glycol is greater than the coagulation rate of the dispersion obtained on the basis of a solution of PVA in combination with IPA. The appearance of a maximum in the dependences is associated with preliminary enlargement of particles whose sizes correspond to the Rayleigh region, with further coagulation and flotation.

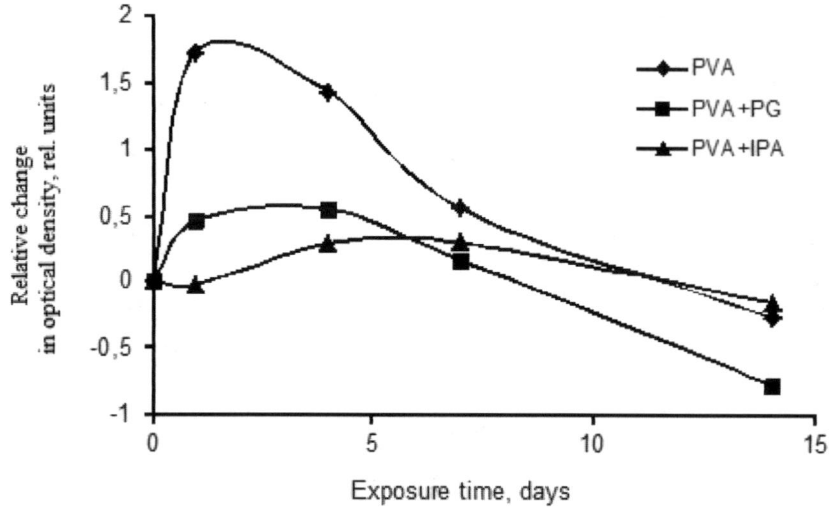

Figure 2. Relative change in optical density of paraffin dispersions over time.

Thus, we can conclude that the best dispersion conditions for paraffin correspond to the presence of PVA and IPA in the dispersion medium. Under these conditions, a narrower particle size distribution is also achieved.

Dispersion of ceresin related to microcrystalline petroleum waxes and characterized by a significantly higher melt viscosity than paraffin is possible at the same temperatures without the use of polyvinyl alcohol, but using an ionic surfactant with a high hydrophilic-lipophilic balance (HLB). For this purpose, we used monoethanolamine oleate, for which HLB is approximately 15 [10, 14]. It was found that with a minimum oleate content of 2.8%, stable aqueous dispersions of ceresin can already be obtained. The dispersions are stable with a higher content (more than 4%) but have a high viscosity, which is caused by the viscosity of the solution of this surfactant.

The ultramicroscopy method was used to perform the analysis of variance for ceresin dispersions with different surfactant contents (Fig. 3).

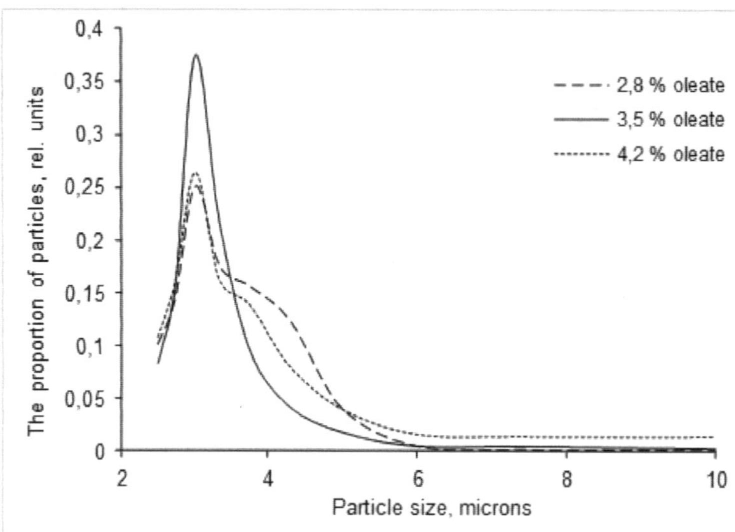

Figure 3. Particle size distribution of ceresin dispersions.

From the data presented in Figure 3, it can be seen that the average size of the dispersed particles is 3 μm (the same as for paraffin dispersions). Moreover, the largest proportion of particles with a size of 3 μm and the narrowest distribution curve corresponds to a dispersion containing 3.5% monoethanolamine oleate.

The results of studies of the aggregative stability of ceresin dispersions are presented in Figure 4 and Table 2

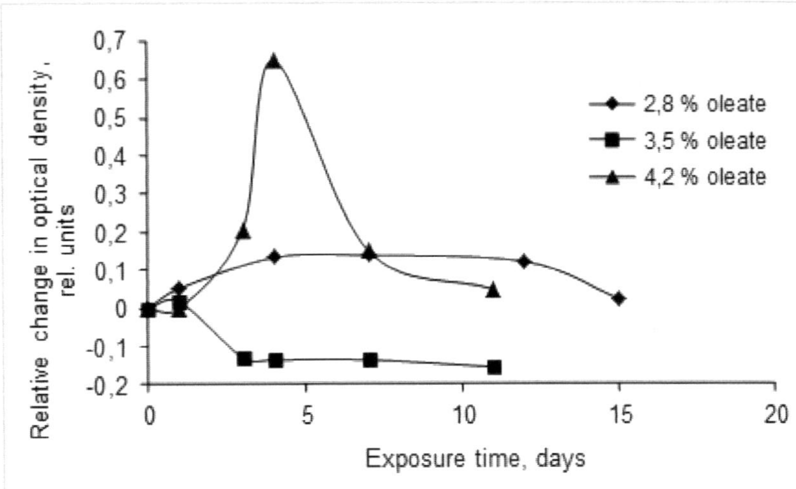

Figure 4. The change in optical density of ceresin dispersions.

Table 2. The coagulation rate of ceresin dispersions.

Monoethanolamine oleate content, %	2,8	3,5	4,2
Coagulation rate, hour$^{-1}\cdot10^{-3}$	1,8	1,4	9,0

For a ceresin dispersion (Fig. 4) stabilized with 3.5% oleate, a smooth achievement of the maximum change in optical density at an exposure time of 12 days is observed. The dispersion containing 2.8% oleate reaches the maximum change in optical density after 4 days of exposure, and the dispersion containing 4.2% after 3 days.

One of the most important properties of water-dispersion materials that determine the conditions for their use are rheological properties. The results of studies of these properties are presented in Figures 5 - 7.

Figure 5 shows that the use of PG as a hydrotropic additive has practically no effect on the viscosity of the dispersion of paraffin stabilized PVA, while the use of IPA for this purpose leads to a sharp increase in viscosity at all shear rates, which is most likely associated with differences in the dispersed composition of systems.

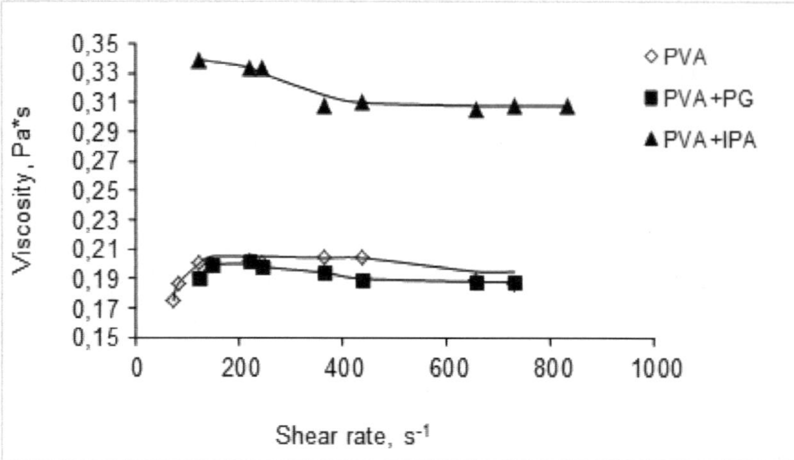

Figure 5. Dependence of viscosity of paraffin dispersions on shear rate.

It is also seen that paraffin dispersions stabilized by PVA and PG at low shear rates have a slightly dilatant flow pattern. For dispersion containing IPA, weak pseudoplastic nature of the flow is observed.

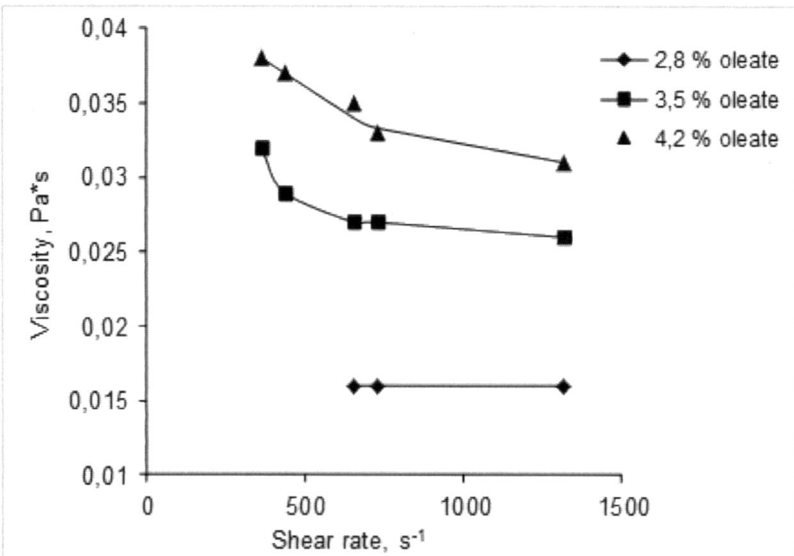

Figure 6. Dependence of the viscosity of ceresin dispersions on shear rate (exposure time - 6 days).

Conclusion

The results of the study of the rheological properties of ceresin dispersions are presented in Figure 7, from which it follows that with an increase in the oleate content, the nature of the flow in the region of low shear rates becomes plastic, and with an increase in the shear rate, it is again Newtonian. Moreover, over time, the viscosity of the dispersions increases, due to the formation of intermolecular bonds in the surfactant solution (Fig. 4).

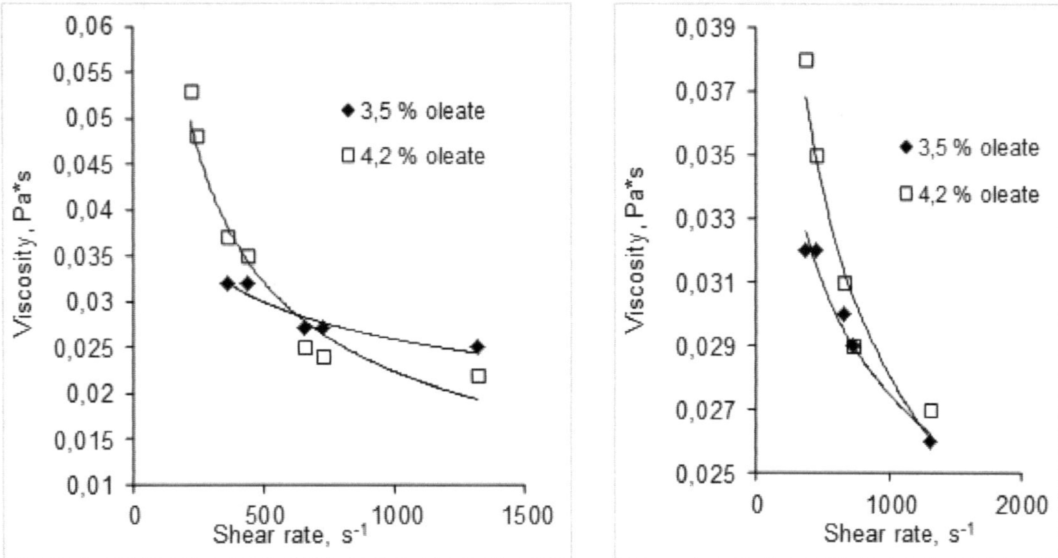

Figure 7. The change in viscosity of ceresin dispersions during storage:

a) the initial time, b) 2 days.

The low viscosity values of the dispersions under study at different shear rates, the pseudoplastic nature of their flow, and stability to shear deformations make them easy to use to modify polymer composite materials.

Thus, the results of studies of the fractional composition of dispersions of paraffin and ceresin, their aggregate stability, and rheological properties allow recommending them as modifying additives for materials based on aqueous dispersions of polymers for various purposes.

References

[1] Carroll J R, Bradley R M and Kalmikoff A 1994 Polymer Paint Colour J. pp 32 – 34
[2] Pushaw W, Giffleman D and Kensbock E 1994 Polymer Paint Colour J. (v. 189, N.4412) pp 18–21
[3] Bouvy A 2005 Polymer Paint Colour J (v. 195, N.4485) pp 22 – 26
[4] Volkmar H 2003 Welt Farben (№12) pp 17 – 20
[5] Holme J 2002 Surface Coat Int. (v. 85, №7) pp 278 – 279
[6] Tereshko A E, Golikov IV and Indeikin EA 2018 LKM and their application (No. 5) pp 15 – 19
[7] Golikov I V, Tereshko A E and Indeikin E A 2006 In book Recent Innovations and Future Challenges for the Coatings and Ink Industries. Curran Associates, Inc., Red Hook, NY, USA, pp 60 -608
[8] Tereshko A E, Golikov I V and Indeikin E A 2008 In book FATIPEC Conference, Full papers, (T2. Ghent, Belgium) pp 168-170
[9] Pereverzev AN 1990 Paraffin compositions. Overview information TsNIITE petrochem p 52
[10] Tereshko AE Golikov IV and Indeikin EA 2016 Journal of Applied Chemistry (T. 89. Issue 6) pp 808 - 811
[11] Finch C A 1992 Polyvinyl Alcohol — Developments Wiley, John and Sons Incorporated p 870
[12] Tolmachev IA and Petrenko NA 2010 Water-dispersive paints (Moscow: Paint-Media) p 106
[13] Rabinovich VA and Khavin ZYa 1991 Brief Chemical Handbook (L .Chemistry) p 432
[14] Lange KR 2007 Surfactants: synthesis, properties, analysis, application (SPb Profession) p 240
[15] Kumar U D, Crocker J, Knezevic J and El-Haram M 2000 Reliability, Maintenance and Logistic Support – A Life Cycle Approach Springer Science & Business Media p 490

Standardization of the new flaw detection material for magnetic powder method non-destructive testing

A. A. Chesnokova, V. A. Ivanova, S. Z. Kalaeva

Yaroslavl State Technical University, Yaroslavl, Russia, chesnokovaaa@ystu.ru

Abstract. The level of the current state in the field of information support (IS) of the activities of high-tech enterprises is considered in this paper. The main problems are shown on the way to achieving the required level of IS quality in high-tech enterprises of the country. The need for the formation of a number of standardized requirements for the IS quality of a high-tech enterprise to improve its level is determined. A typical list of requirements for the IS quality is determined and a model of a system for accounting for requirements for the IS quality is developed using the example of Scientific and technological center of unique instrumentation of the Russian Academy of Sciences.

Introduction

The magnetic powder method is a magnetic non-destructive testing method based on the use of magnetic powder as an indicator [1], used to control the surfac-es of products made of ferromagnetic materials during their production and opera-tion.

When the magnetic powder method of control is used, powders, suspension, and pastes magnetoluminescence are applied as radiographic material. Currently, magnetic powders are widely used in industrial enterprises, on the basis of which suspensions are prepared.

Requirements for the method of magnetic particle inspection, equipment, and applied flaw detection materials are established in national (GOST R), interstate (GOST) standards, industry standards, rules, recommendations and guidance doc-uments, as well as in standards of organizations (STO) and technical specifications (TS).

National standards give general requirements for the control method. To de-termine the specific values of the characteristics of the material, an analysis of in-dustry requirements is necessary.

The analysis of regulatory documents [2-7] governing the magnetic particle inspection method made it possible to formulate universal characteristics of the used flaw detection materials. Table 1 shows the characteristics of the flaw detec-tion material during monitoring at hazardous production facilities, trunk pipelines, gas pipelines, as well as in the railway and automotive industries [8]. It is allowed to fulfill certain requirements, depending on the type of flaw detection material.

Materials and methods

However, not all currently used flaw detection materials meet the requirements specified in Table 1, especially in terms of material stability, since the flaw material (suspension) usually has maximum stability of 120 hours. Therefore, it has to be manufactured and tested for suitability for each new control procedure. Besides, enterprises place high demands on the sensitivity level of the method, which de-pends on the average particle size of the ferromagnetic material, namely, the small-er the particle size, the higher the detecting ability of the flaw detection material [8]. In this regard, it is necessary to develop a material that meets the requirements of various industries, is universal, and also increases the efficiency of the method by improving the stability and average particle size.

Table 1. Requirements for magnetic flaw detection materials [8].

№	Indicator	Characteristic (value) of the indicator
1	Appearance	Black, colored, luminescent magnetic powders in dry form or as part of suspensions
2	Colour	Maximum contrast to the color of the surface to be monitored
3	Detecting ability, not less, %	Method applied magnetic field - 98, Method residual magnetization - 95
4	Mass fraction of the main substance, %,	For dry powders – 80-85; For use in suspensions – 20-30
5	Concentration of hydrogen ions (pH), not less	8
6	Average particle size, microns, no more than	Dry application - 150 microns; Use in suspensions - 50 microns
7	Viscosity, m^2/s, no more than	$36 \cdot 10^{-6}$ (36 cSt)
8	Dynamic viscosity at (20±2)°C, no more, MPa • s	5
9	The concentration of powder in suspension, g/l	(25 ± 5) - for black or colored (non-luminescent) powder; (4 ± 1) - for fluorescent
10	Suspension stability, not less than, h	120
11	Heat resistance	Materials must not degrade after 5 minutes of heating to maximum operating temperature
12	Corrosion ability	Non-corrosive
13	Foam generation	Resistance to foaming
14	Sulfur and halogen content	Sulfur content should be less than (200±10) parts per million; The halogen content should be less than (200 ± 10) ppm (halogens are understood as chlorine and fluorine)

The new flaw detection material for the magnetic particle inspection method is magnetic fluid. Due to the unusual combination of the properties of magnets, liq-uids, and colloidal solutions, magnetic fluids (MF) are promising materials and can be used in non-destructive testing [9].

Magnetic fluid is a stable ultrafine colloidal system of highly dispersed parti-cles of magnetic material (ferro- or ferrimagnetic substances) with an average size of 5 to 50 nm, dispersed in various liquids and performing intensive Brownian mo-tion, stabilized by surface-active substances, which is able to interact with a mag-netic field and in many respects behaves like a homogeneous liquid [10]. Magnetic fluids can be obtained from iron-containing wastes (ICW) using technology [11-12].

To obtain magnetic fluid, three components are necessary: magnetic particles of colloidal size (magnetite), a liquid base, and a stabilizer that prevents the adhe-sion of colloidal particles. Each component has to satisfy certain requirements; it is possible to obtain a magnetic fluid suitable for use in a particular direction only un-der this condition.

It is known that the interaction of aqueous solutions of ferrous and ferric salts and their joint precipitation results in the formation of magnetite in the form of highly dispersed particles

$$Fe^{2+} + 2Fe^{3+} + 8OH^{-} = Fe_3O_4\downarrow + 4H_2O \quad (1)$$

Iron-containing waste from industrial enterprises can be used as a source of ferrous and trivalent iron. The technology for producing magnetic fluids [11-12] consists of two main stages:
1) obtaining highly dispersed particles of magnetite (dispersed phase of MF);
2) stabilization of magnetite in a carrier fluid using a surfactant that prevents aggregation of magnetite particles in a carrier fluid and ensures the stability of MF.
The kerosene-based magnetic fluid obtained by the technology [11-12] is a black mobile fluid. By structure, it is a colloidal solution of magnetite in kerosene with particles stabilized by oleic acid (Table 2). Technical requirements for kero-sene-based magnetic fluids are given in Table 3.

Table 2. Composition and content of components of a magnetic fluid based on kerosene.

Component	Content of the component, mass. hours	Component Assignment
Kerosene brand MRTU-12N43-63	65 – 80	liquid base
Magnetite	15 – 20	colloidal particles
Oleic acid according to TS 6-09-5290-86	5 - 15	surface-active substance

Table 3. Technical requirements for magnetic kerosene fluids.

Indicator	Value
Density, not less	980 kg/m^3
Volume fraction of magnetite, not less	6 %
Saturation magnetization, not less	10 kA/m

Studies have shown that magnetic fluids from ICW meet the requirements for flaw detection materials and provide the necessary level of sensitivity.
To ensure organizational unity and create conditions for the use of a new flaw detection material – magnetic fluids in magnetic particle inspection, appropriate regulatory and technical documentation is necessary to develop.
According to Federal Law No. 162 "On Standardization in the Russian Feder-ation" (dated June 29, 2015), the technical specifications and standards of the or-ganization are documents in the field of standardization and have the same status. Technical specifications are a type of organization standard approved by the manu-facturer of the product or the contractor of the work, service [13].
It is most expedient to develop technical conditions for a new flaw detection material - magnetic fluid. Technical specifications (TS) are developed according to GOST 2.114-2016 [14]. Specifications for a new flaw detection material - magnetic fluid - should consist of an introductory part and the following sections:
- technical specifications;
- safety requirements;
- environmental protection requirements;
- acceptance rules;
- methods of control (testing);
- instructions for use;
- manufacturer's warranty.

The contents of the technical specifications can include terms and definitions, designations and abbreviations.

In the water part of the technical specifications indicate the name of the prod-uct, its name, scope and operating conditions, for example, "These specifications apply to flaw detection material (MF DF - 001) - a magnetic fluid based on kero-sene, used in the magnetic particle non-destructive testing method".
The requirements, norms, and characteristics that determine the quality indica-tors and operational characteristics should be given in the section "Technical re-quirements".
Flaw detection material (MF DF - 001) - kerosene-based magnetic fluid must meet the requirements given in Table 2, 3. Operational characteristics are presented in Table 1. The section, as a rule, should consist of the following subsections:
- the main parameters and characteristics (properties);
- requirements for raw materials;
- completeness;
- marking;
- packaging.

In the “Safety Requirements” section, requirements are established that contain all types of permissible hazards in such a way as to ensure the safety of the product during its service life. For flaw detection material, the section should indicate:
- fire safety requirements;
- explosion safety requirements;
- safety requirements from exposure to chemical and polluting substances, in-cluding maximum permissible concentrations of substances or its constituent com-ponents;
- safety requirements for servicing the product, including safety requirements in case of erroneous actions of service personnel and spontaneous malfunction;
- requirements for protective equipment and safety measures.

The section “Environmental Protection Requirements” sets out the require-ments for the prevention of harm to the environment and human health during test-ing, storage, transportation, operation, and disposal. For flaw detection material, the section should indicate:
- requirements for the permissible level of chemical impacts on the environ-ment;
- requirements for the stability of polluting, toxic substances in the environ-ment (water, atmospheric air, soil);
- requirements for utilization and disposal sites.

In the section “Acceptance rules”, it is necessary to indicate the procedure for product control, the procedure and conditions for presentation and acceptance by the organization’s technical control bodies and the consumer (customer), the size of the presented lots, the accompanying bearer documentation, as well as the proce-dure for processing acceptance results.
In the section "Control Methods", it is necessary to establish programs, meth-ods, and modes of control (testing, measurement) of parameters, norms, require-ments and characteristics, the need for control of which is provided in the section "Acceptance Rules". Programs and methods of control should be objective, clearly formulated, and accurate and should provide consistent and reproducible results. For each program or control method (tests, measurements), depending on the spe-cifics of the conduct, should be established:
- sampling methods;
- equipment, materials, and reagents;
- preparation for control (testing, measurement);
- carrying out control (tests, measurements);
- processing of results.

In the section "Instructions for use", it is necessary to establish the require-ments for operation and application at the place of their operation. This section may include subsections establishing requirements for transportation, storage, and dis-posal.
In the section “Manufacturer's warranties”, it is necessary to establish the rights and obligations of the manufacturer under warranties according to applicable law.
The development of technical specifications (TS) would allow for the use of a new flaw detection material of magnetic fluids for magnetic particle inspection at various enterprises.

IOP Conf. Series: Materials Science and Engineering **986** (2020) 012006 doi:10.1088/1757-899X/986/1/012006

Conclusion

The final integration of groups of indirect and direct requirements finds its final expression in the personnel requirements for the IS quality, since it is personnel that is the main consumer of IS of a high-tech enterprise.

In this regard, it becomes necessary to conduct an accompanying comprehensive study of personnel requirements for the IS quality at workplaces of specific stages, taking into account the specifics of creating high-tech products.

The final set of requirements for the required level of IS was formed as a result of the comprehensive analysis of the current specifics of the production of high-tech products within STC UI RAS, as well as the personnel requirements for the IS quality:

1. A high degree of compatibility between ACS and ADS in the integrated information environment (IIE) of a high-tech enterprise.

In modern high-tech enterprises, a large number of ACS and ADS of various functional and technical purposes are widely sold, the main purpose of which is to ensure the effective work of personnel with the information resources of the enterprise. In this regard, the implementation of this requirement becomes especially important, since the effective creation of high-tech products becomes impossible without a high degree of information compatibility of data on separately executed processes of the organization.

2. High efficiency of information monitoring of technological processes.

Information monitoring allows continuous monitoring of the emergence of new information about the processes of the product life cycle for given information flows in a fixed thematic field for analysis, management, and prediction of its development [8]. Fulfillment of this requirement becomes necessary in the context of the implementation of the ideas of continuous information support for the product life cycle, which is consistent with the main task of IS in modern high-tech organizations.

3. The maximum possible degree of implementation of simultaneous group work of personnel in ACS with large amounts of information.

Ensuring the joint work of personnel during the implementation of research and development works becomes the main factor in reducing time and costs in the development of high-tech products.

4. High speed of obtaining information about created high-tech products at any stage of its life cycle.

The speed, at which information is received by the organization's personnel, plays a crucial role in determining the competitiveness of an enterprise. The fulfillment of this requirement allows ensuring the effectiveness of the existing material and technical base and information flows in the enterprise involved in the processes of information transfer.

5. The prevailing position of electronic data interchanges in organizations.

The widespread use of electronic forms of documentation and the rejection of paper media is becoming one of the priority areas for the development of information technologies in the modern world [9-10]. The implementation of this indicator will help to form the basis of measures for material and technical equipping of the necessary ACS by the organization, as well as determine the need for organizational measures aimed at training personnel to work with the electronic documentation form.

Thus, accounting of the requirements given above, while informing high-tech industries, can serve as the development of widespread study and solution of existing problems of high-tech organizations' IS and ensuring the required level of quality of processes for creating high-tech products.

It should be noted the need to develop methods for quantitative assessment of requirements, as well as their possible supplement as further ways of research aimed at improving the quality of IS of high-tech organizations.

References

[1] GOST R 55612-2013 2014 Non-destructive magnetic testing. Terms and Definitions. (M .: Standartinform) p 12

[2] GOST 56512-2015 2016 Non-destructive testing. Magnetic particle method. Typical technological processes (Moscow: Standartinform) p 60

[3] GOST R ISO 9934-1-2011 2013 Non-destructive testing. Magnetic particle method Part 1. Basic requirements (Moscow: Standartinform) p 20
[4] GOST R ISO 9934-2-2011 2013 Non-destructive testing. Magnetic particle method. Part 2. Defectoscopic materials (Moscow: Standartinform) p 20
[5] RD 32.159-2000 2000 Magnetic particle non-destructive testing of car parts p 120
[6] RD-08.00-60.30.00-KTN-046-1-05 2005 Non-destructive testing of welded joints during the construction and repair of main oil pipelines p 131
[7] RD 13-05-2006 2006 Methodological recommendations on the procedure for conducting magnetic particle inspection of technical devices and structures used and operated at hazardous production facilities p 58
[8] Chesnokova AA, Kalayeva SZ and Ivanova VA 2017 Development of a flaw detection material for the magnetic particle method Journal of Physics: Conference Series (Volume 881. International Conference on Innovations in Non-Destructive Testing (SibTest) 27–30 June Novosibirsk , Russian Federation)
[9] Chesnokova AA, Vershinina NI and` Kalaeva SZ 2013 Application of magnetic fluids in flaw detection The Sixty-sixth All-Russian Scientific and Technical Conference of Students, Master's and Postgraduates of Higher Educational Institutions with International Participation (Yaroslavl: sb.tez.dokl. Pp 155 ...
[10] Taketomi S and Chikazumi S 1993 Magnetic fluids 272 p
[11] Kalaeva S Z 2009 Development of nanotechnology for the utilization of iron-containing waste in magnetic fluids Izvestiya TulGU. Natural Sciences. Ser. Earth Sciences Vol. 5 pp 73-76
[12] Chesnokova AA, Denisova EA, Terzi ME and Kalaeva SZ 2014 Optimization of parameters for obtaining magnetic fluids for use in flaw detection The 67th All-Russian Scientific and Technical Conference of Students, Master's and Postgraduates of Higher Educational Institutions with International Participation (Yaroslavl: Sat .tez.dokl Yaroslavl: YAGTU Publishing House) p 166
[13] Federal Law No. 162 of June 29, 2015 "On standardization in the Russian Federation" (as amended on July 3, 2016)
[14] GOST 2.114-2016 2016 Unified system for design documentation (ESKD). Technical conditions. (M .: Standartinform) p 15

IOP Conf. Series: Materials Science and Engineering **986** (2020) 012007 doi:10.1088/1757-899X/986/1/012007

Study of the influence of carbide inhomogeneity on operational and technological properties of high-speed tool steels

T. D. Stoyanova, V. A. Ivanova

Yaroslavl State Technical University, Yaroslavl, Russia, ivanova-waleriya@mail.ru

Abstract. The level of the current state in the field of information support (IS) of the activities of high-tech enterprises is considered in this paper. The main problems are shown on the way to achieving the required level of IS quality in high-tech enterprises of the country. The need for the formation of a number of standardized requirements for the IS quality of a high-tech enterprise to improve its level is determined. A typical list of requirements for the IS quality is determined and a model of a system for accounting for requirements for the IS quality is developed using the example of Scientific and technological center of unique instrumentation of the Russian Academy of Sciences.

Introduction

The destruction of the cutting tool made of high-speed steel during operation is a serious problem in modern production. Factors causing the destruction of any product during operation are:
- constructor errors;
- violations of technology and assembly;
- quality of the material;
- improper operation.

The state of the material structure as an element of quality that affects the operational and technological properties of tool steel is considered in the paper.

The structure of high-speed tool steel consists of martensite and carbides. A high level of operational and technological properties is ensured by very fine-needle martensite with finely dispersed, uniformly distributed carbides [1, 2].

It was found that carbides of two groups can form in steels [1]:

Group 1 – Fe3C; Mn3C; Cr23C6; Cr7C3; Fe3Mo3C; Fe3W3C;

Group 2 – Mo2C; W2C; WC; TiC; NbC; TaC; Ta2C; ZrC.

For some technological reasons, carbides in high-speed tool steel take both an uneven distribution over the volume of the product and take a different shape. This phenomenon is called carbide inhomogeneity [1, 3].

Level requirements and scales for determining the carbide inhomogeneity score are regulated by GOST 19265-73.

The following types of carbide inhomogeneity are distinguished [1, 3]:
- carbide lines;
- broad clusters of carbides;
- carbides deposited at the grain boundaries in the form of a grid;
- eutectic carbides arranged in a cellular structure

- large angular carbides.

Materials and methods

In this regard, studies of the microstructure of the cutting tool after destruction during the operation were carried out (Table 1, Fig. 1-6).

Analysis of the results of the study of fractures, properties, and microstructure of the cutting tool are presented in Table 2.

Thus, studies of the microstructure of the cutting tool indicate that:

1) One of the main reasons for the destruction of the cutting tool is carbide in-homogeneity;

2) Carbide inhomogeneity reduces both operational properties and technological properties;

3) The causes of destruction of the cutting tool associated with carbide inhomo-geneity are:

- inhomogeneity of the chemical composition due to the uneven distribution of carbides in the total volume and the micro volumes of the part, leading to inho-mogeneity of the microstructure and the appearance of additional internal stresses during quenching;
- stress concentrators on the cutting edge, formed by the output of carbide lines crushed during grinding [5];
- stress concentrators on the cutting edge, formed as a result of chipping large carbides, weakly associated with the matrix [6, 7];
- a steel strength decrease as a result of carbide precipitation along grain boundaries.

Table 1. Study of fracture, mechanical properties, and microstructure of the cutting tool.

Tool	Nature of destruction	Standard requirements	Research results			
			Hardness HRC	Microstructure by GOST 8233-56	Grain number by GOST 5639-82 scale	Carbide inhomogeneity by GOST 19265-73
Countersink 1	The first step of the countersink is destroyed in the longitudinal direction. The fracture is fine-grained, with an oxidized surface	- P6M5 steel; - hardness 62-65HRC	58-63 HRC	In areas with accumulations of eutectic carbides - medium-needle martensite and troostite. The main structure is crypto-needle martensite with carbides (Fig. 1, b). In the microsection plane, the fracture surface is oxidized	In areas with accumulations of eutectic carbides - 7-8 points; main structure - 10 points (Fig. 1, b)	6-7 points (scale No. 1) with eutectic carbides (Fig. 1, a)
Countersink 2	There collapsed in the transverse direction. The fracture is fine-grained without signs of oxidation. When checking on a magnetic flaw detector, deposits of magnetic powder in the form of thin longitudinal lines were found on the surface of the grooves for chip exit (Fig. 2, a, 2, b)	P6M5 steel; Hardness of the working part – 62-65HRC	64 HRC	Extremely fine-grained martensite 2 points of GOST 8233-56 scale with carbides. Non-uniform etchability (Fig. 2, c)	10 points	Carbide inhomogeneity - 4A point scale No. 1. Coarse lines of large angular carbides are observed with chipping of carbide particles merging into long sections (Fig. 2d)

Tap	Destruction in the transverse direction of the working part of the countersink	Steel - P18 - working part, Hardness 62-64HRC	64 HRC – working part	Fine-needle martensite 4 points of the GOST 8233-56 scale with carbides largely located along the grain boundaries (Fig. 3, a). The microstructure is not uniformly etched.	7-8 points	The real score of carbide inhomogeneity is not possible to determine because carbides are located at the grain boundaries. In the plane of the thin section, cavities from outgrown carbides are observed (Fig. 3, b, 3, c).
Broach	Destruction in the transverse direction along the working part. Some of the fractures adjacent to the surface have oxides (Fig. 4, a). When checking on a magnetic flaw detector, longitudinal deposits of magnetic powder were found on the surfaces of broaches (Fig. 4, c)	- P6M5 steel - the cutting part; - hardness 63-65HRC	63-64 HRC	Extremely fine martensite with carbides. In the plane of a longitudinal microsection, longitudinal deposits of magnetic powder are clusters of angular carbides arranged in strips painted during machining. (Fig. 4, b). A crack was detected on the surface (Fig. 4, d)	9-10 points	Carbide inhomogeneity corresponds to 4 points of scale No. 2 GOST 19256-73
Drill	Mass breakdown. When checking with a magnetic flaw detector, deposits of magnetic powder in the form of thin longitudinal strips are observed on the surfaces for chip exit (Fig. 5, a)	- P6M5 steel; - hardness of the working part - 63-65 HRC	62-65HRC	Martensite with carbides. Carbides are large, form wide clusters, elongated in the direction of rolling. Martensite in the area of carbide accumulations - medium needle 5 points, in other places - hidden needle 1 point of the scale (Fig. 5, c, 5, d)	In areas of carbide accumulation, austenitic grains have a value of 7-8 points. In other places - 10-11 points (Fig. 5, d)	Carbide inhomogeneity - 6A. In the thin section plane, magnetic powder deposits, observed on the surface of the chip exit grooves, are cracks passing along longitudinal accumulations of carbides (Fig. 5, b; 5, c)
Gear cutter	Chipping of the working edge is observed on the gear cutter (Fig. 6, a)	- P18 steel; - hardness of the cutting part - 62-64HRC; carbide inhomogeneity - not more than 3 points	62-64HRC	Very fine-needle martensite 2 points of the scale with eutectic carbides (Fig. 6, b; 6, d). Cracks passing through the accumulations of large carbides were found on the cutting edges (Fig. 6, c)	10 points	Carbide inhomogeneity has a slightly deformed cellular structure, consisting of clusters of unbroken eutectic (Fig. 6, d)

Table 2. Influence of types of carbide heterogeneity on the operational properties and production technology of cutting tools.

Types of carbide inhomogeneity	The effect of carbide inhomogeneity of the cutting tool	
	Operational properties	Production technology
Carbide lines 1. Carbides form lines that come to the surface during machining crumble 2. In the region of carbide lines, an uneven etching of the structure is observed, which indicates the inhomogeneity of the chemical composition	Coming to the surface, they are stress concentrators, which leads to the development of cracks	A noticeable anisotropy of mechanical (strength and toughness) and physical properties appears in the longitudinal and transverse directions. Heat treatment of such steels is accompanied by increased deformation.
Broad clusters of carbides Uneven structure. Around the accumulations of carbides, areas of medium-needle martensite with a large amount of residual austenite are observed, forming areas with reduced microhardness. Uneven grain sizes	1) the connection of large carbides with the matrix is reduced, which facilitates their chipping in the working edge; 2) going to the surface of areas with reduced microhardness reduces the fatigue strength of the cutting edge; 3) chipping carbides acting as abrasives	1) The concentration of stresses during hardening increases due to the inhomogeneity of the chemical composition of austenite. In this case, the steel has a grain of not the same size. In this regard, the sensitivity of steel to cracking during hardening of steels with an uneven distribution of carbides increases. 2) Chipping carbides hinder grinding and polishing tool
Carbides deposited at the grain boundaries in the form of a grid	Strength decreases	-
Eutectic carbides arranged in a cellular structure	Large carbides crumble, reaching the surface of the working edge, forming stress concentrators. Cracks develop in large accumulations of carbides.	1) The stress concentration increases during quenching due to heterogeneity of the austenite chemical composition 2) Chipping carbides hinder grinding and polishing tool

a)

b)

Figure 1. The microstructure of the countersink 1.
a) medium-needle martensite and troostite in areas with accumulations of eutectic carbides; b) carbide inhomogeneity 6-7 points (scale No. 1) with eutectic carbides

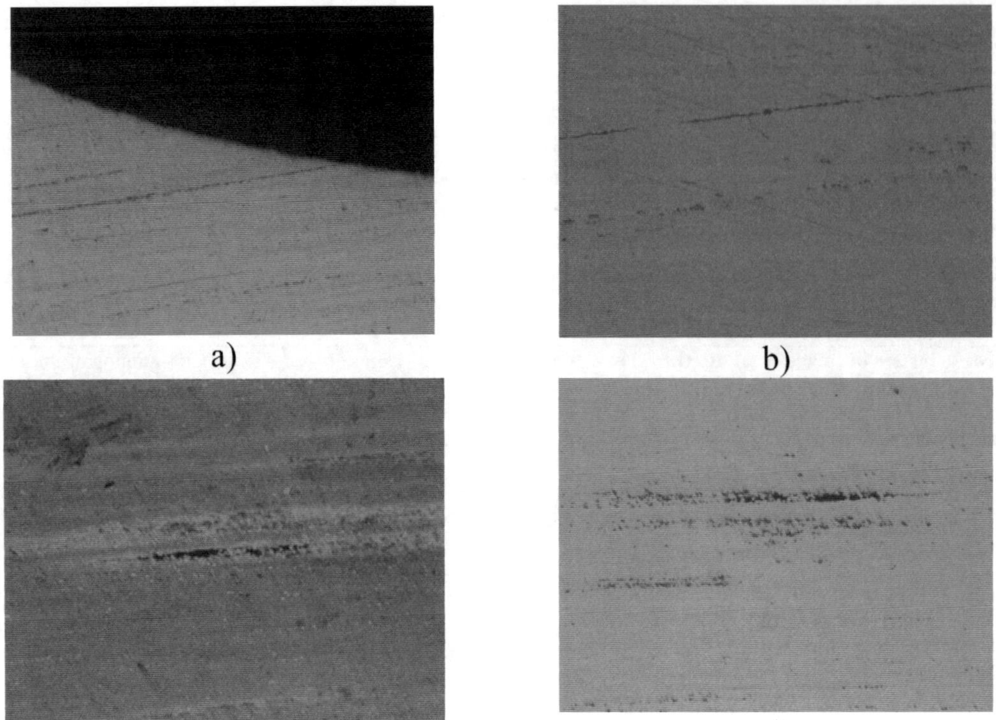

a) b)

c) d)

Figure 2. The microstructure of the countersink 2.
a) deposits of magnetic powder in the form of thin longitudinal lines on the surface of the grooves for chip exit; b) deposits of magnetic powder in the form of thin longitudinal lines; c) uneven etching of the microstructure; d) coarse lines of large angular carbides with chipping carbide particles merging into long sections of length.

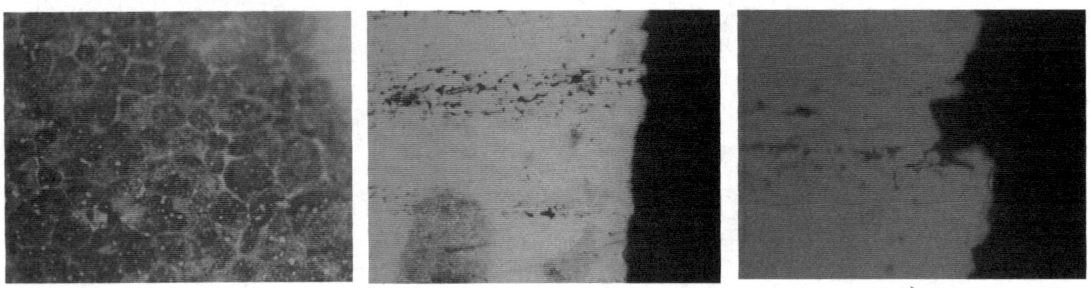

a) b) c)

Figure 3. The microstructure of the tap.
a) Fine-needle martensite 4 points with carbides in a significant part located at the grain boundaries; b) cavities from spun out carbides x100; c) cavities from spun out carbides x500.

IOP Conf. Series: Materials Science and Engineering **986** (2020) 012007 doi:10.1088/1757-899X/986/1/012007

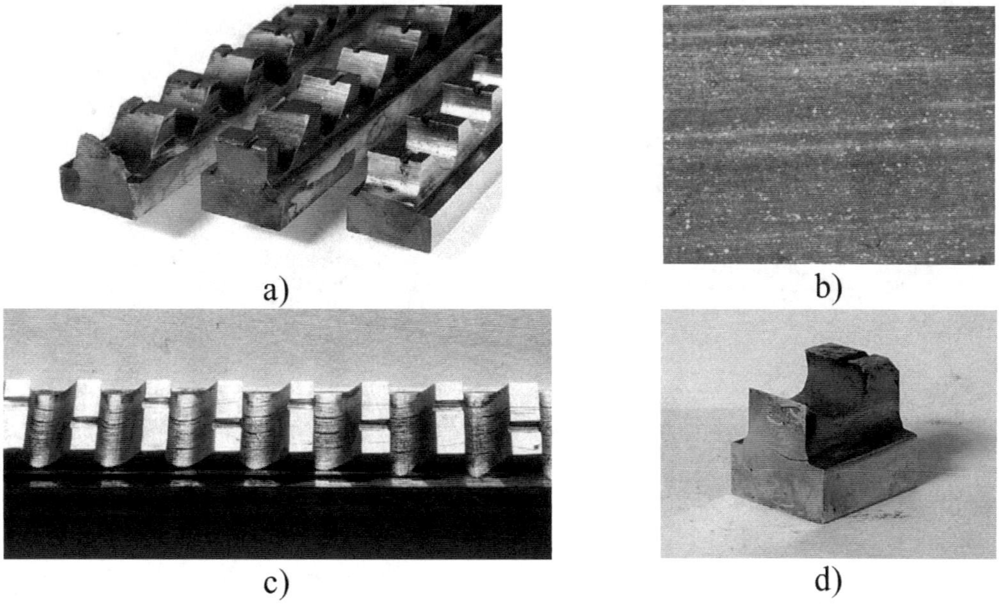

Figure 4. The microstructure of the broach.
a) oxides on the surface of fractures; b) accumulations of angular carbides lined up in strips; c) longitudinal deposits of magnetic powder on the surfaces of broaches; d) surface crack.

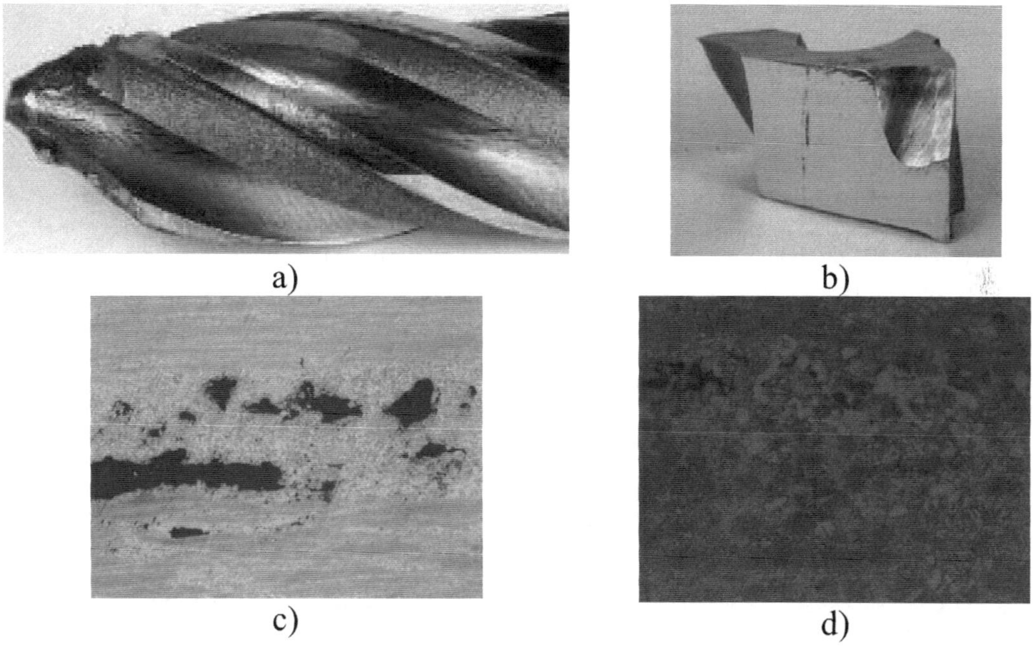

Figure 5. The microstructure of the drill.
a) deposits of magnetic powder in the form of thin longitudinal strips; b) cracks passing through longitudinal accumulations of carbides; c) a crack in longitudinal accumulations of carbides; d) austenitic grains 7-8 points in the area of accumulation of carbides.

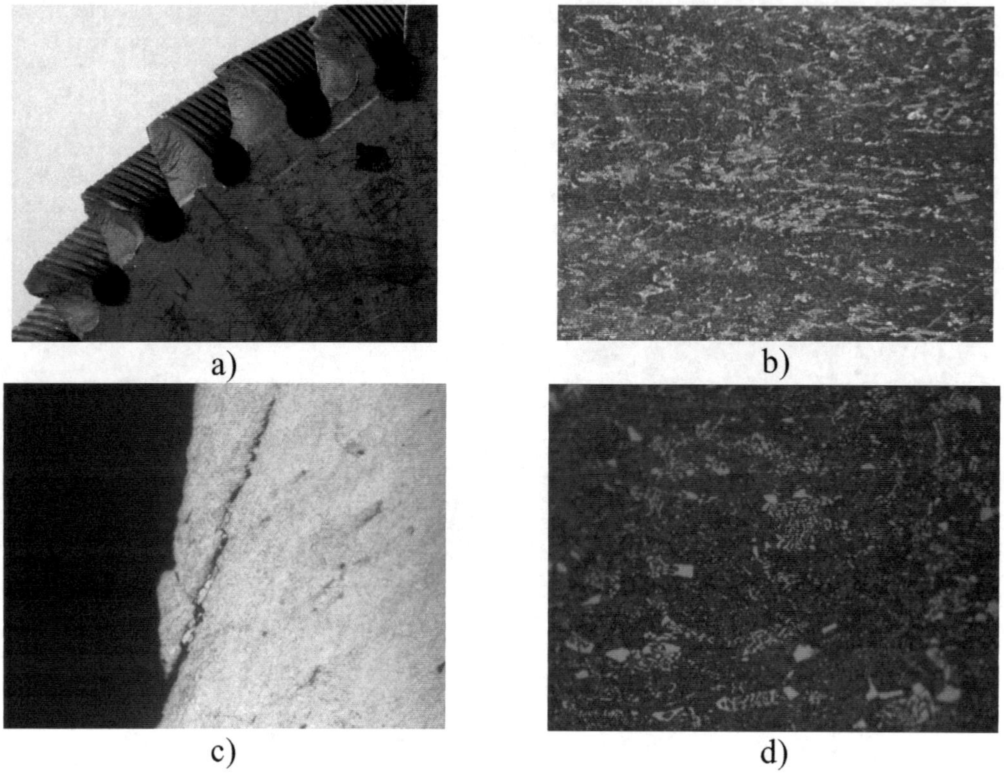

Figure 6. The microstructure of the gear cutter.
a) chipping of the working edge; b) fine-needle martensite with eutectic carbides; c) cracks on the cutting edges passing through the accumulations of large carbides; d) the carbide inhomogeneity has a labially deformed cellular structure consisting of accumulations of unbroken eutectics.

References

[1] Artinger I 1982 Tool steels and their heat treatment Handbook: trans. with Hungarian. M .: Metallurgy 312 p

[2] Skrynchenko Yu M and Pozdnyak LA 1977 Serviceability and properties of tool steels Kiev "Naukova Dumka" 168 p

[3] Geller Yu A 1983 Instrumental Steel 5th ed., Rev. and add. M .: Metallurgy 527 p

[4] GOST 19265-73 2003 Bars and strips from high speed steel. Technical requirements (with Amendments 1-6) M .: Publishing house of standards 87 p

[5] Bondar OV and Smirnov AN 2015 Solutions for reducing carbide inhomogeneity and improving the properties of cold deformation stamped steel Collection of materials of the VII All-Russian scientific-practical conference of young scientists with international participation Kemerovo, Kuzbass State Technical University named after T.F. Gorbachev 649p

[6] Antipov VI, Vinogradov LV, Lukina Yu A, Kolmakov AG, Doronin DI, Baranov E E and Mukhina Yu E 2017 Influence of carbide heterogeneity in the structure of high-speed steels R6M5 and 130R12M3F3K10-Sh on their physical and mechanical properties VII International Conference "Deformation and Fracture of Materials and Nanomaterials". Moscow. November 7-10, 2017 p 951

[7] Girshov VL Korotkikh MT, Krupnova IV and Korobeinikov NV 2011 High-performance tool made of powder high-speed steels with dispersed structure Metalworking, publishing house "Polytechnica" (St. Petersburg, ISSN: 1684-6702, №4 (64) pp 44 - 48

[8] GOST 8233-56 2004 Steel. Microstructure standards (Moscow: IPK Publishing house of standards) p 5

[9] GOST 5639-82 2003 Steels and alloys. Methods for detecting and determining grain size (Moscow: IPK Publishing house of standards) 16 p
[10] GOST R ISO 9934-1-201.1 2013 Non-destructive testing. Magnetic particle method. Part 1. Basic requirements (Moscow: Standartinform) 16 p
[11] GOST 9013-59 2001 Metals. Rockwell hardness measurement method (with Amendments No. 1, 2, 3 with Amendment M .: IPK Publishing house of standards) 7 p

Assessment of resources and development factors of the scientific and educational subsystem

I N Tomshinskaya, N V Ivanova and E B Vinogradova

Peter the Great St.Petersburg Polytechnic University, St.Petersburg, Russia, tomshinskaya_in@spbstu.ru

Abstract. This paper touches upon the scientific and educational subsystem, since the formation and development of this particular subsystem are demanded by the conditions of post-industrial transformations, the process of increased spatial competition. This need applied to the economy of a megalopolis is reinforced by a contradictory combination of relatively high investment attractiveness, the presence of several modern growth points, on the one hand, and the "obsolescence" of competencies that the human factor has, on the other hand. Significant problems in the development of the scientific and educational subsystem of St. Petersburg are considered in the paper. The individual components of the desired subsystem with the identification of strengths and weaknesses are analyzed. An approach to the analysis of the development factors of a promising subsystem of a scientific and educational profile is proposed: representing the integral result of this subsystem in the space-time continuum of the regional economy as a production function. The authors see the direction of further research in the allocation of groups of factors-arguments of the production function.

Keywords: Scientific and educational subsystem, innovative economy, advanced technologies, SWOT analysis, competencies

1 Introduction

There are unique organizations of higher education aimed at graduating specialists with modern key competencies for the leading sectors of the modern economy in St. Petersburg. The modern scientific complex of St. Petersburg, unlike other regions, has strong traditions and is focused on the needs of modernization and development of innovative reproduction.

There are about 40 research institutes and their branches in St. Petersburg. The share of the employed persons in research and development grows.

The volume of research and development costs concerning the regional gross product in the North-West Federal District is 1.4 times higher than the average for Russia. As of January 1, 2017, in St. Petersburg, there are 4164 organizations with the type of activity "Research and Development".

The productivity of using the existing scientific potential can be demonstrated by the issuance of patents, citation indices, and the presence of international publications (Tables 1 and 2).

Table 1. Creating and using advanced technologies in the economy of St. Petersburg

Indicator	2007	2009	2011	2013	2015
Number of created advanced manufacturing technologies	85	155	190	218	160
Number of used advanced	2985	3790	5122	7128	8099

manufacturing technologies					
Patent applications for inventions, utility models, industrial designs and the issuance of security documents					
applications for issuance	2920	2744	3192	2944	3014
granted patents	2513	2926	2323	2500	2518

The data in table 1 indicate that the scientific potential that is present and growing in the region is in demand by the development of territorial reproduction. About 3,000 patent applications are filed annually in St. Petersburg, and many applications have a positive response from the Federal Service for Intellectual Property, Patents, and Trademarks. The negative impact is exerted by a very large gap from the moment of applying for the grant of a patent: the minimum term is one year.
The results of the index evaluation of scientific publications of workers in the field of education and science of St. Petersburg are given in Table 2.

Table 2. Indicators of the Russian science citation index

Indicators	Russian Federation	St. Petersburg
Total authors on elibrary.ru	777756	77829
The largest number of publications	23841	1262
The largest number of citations	78121	22844
The largest H-index	87	67
The largest number of publications in foreign journals	6079	1039

Summarizing the results of the analysis presented in Table 2, we conclude that almost 10% of the entire Science Index information-analytical system is provided by the authors of scientific publications in the region. The scientific potential of the region is also characterized by a high H-index (67).
In St. Petersburg, there is a developed network of training scientific personnel; scientific councils have been created at research institutes, universities, which have provided relevant results in the field of certification of the scientific potential of the city (Fig. 1).

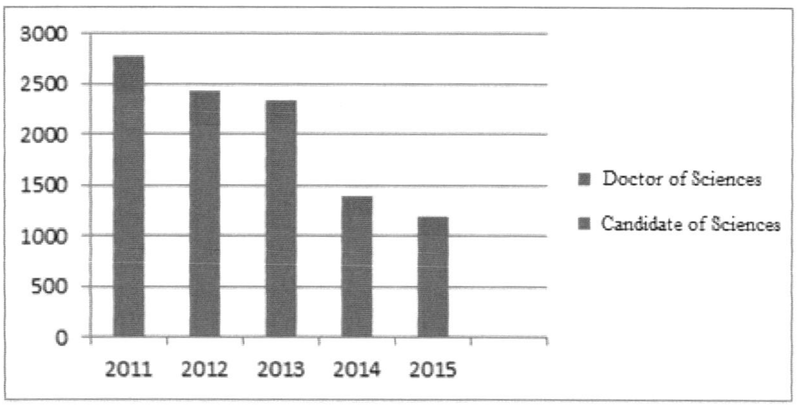

Figure 1. Defense candidate and doctoral dissertations in St. Petersburg, units

It is necessary to mention the low commercialization of scientific projects and defended dissertations in comparison with developed economies, at the same time, high - compared with the average Russian data. There is a significant gap between the available resources (mainly scientific and technical potential) and the effect of their use is production and export of high-tech products and technologies.
Commercialization of scientific projects is carried out through innovative activities related to the transformation of ideas into technologically new or improved products or services introduced on the market, also into new or improved technological processes or methods of production (transfer) of services used in practice. The highest rate of innovation activity (the proportion of organizations implementing innovations in the total number of surveyed) compared to other regions is 18.9% in St. Petersburg, twice the average (9.9%) for Russia. According to the analysis of the organization's

innovative activity, it can be concluded that only 10% of scientific developments find their application, besides, the most active users of scientific developments are the production of electric equipment, electronic and optical equipment, the production of coke and petroleum products, and chemical production.

Thus, the basis for the development of high-performance sectors of the new economy may be the unity of the processes of market consolidation, systemic organization and systemic development of the research and educational potentials available in the region with the support of the state.

2. Problem statement

It is legitimate to conclude that in the economy of a megalopolis there are complexes of industry and academic science, educational institutions, research centers that are just beginning to move towards synthesis into an integrated scientific and educational subsystem [1-3]. Let us single out the gaps that prevent this synthesis:

- between fundamental and applied science;
- demanded and the actual structure of personnel training;
- the theme of scientific development and the structure of production requirements.

To accelerate this synthesis, support is needed for competition between organizations in the relevant field, as well as support for the forces of specification and the process of forming clusters that combine scientific centers and manufacturing enterprises.

The development of the analyzed prospective subsystem is demanded by the functionality of informative tools:

- estimates of the prospective results of the formation in the internal environment of a promising subsystem of capital combinations, including forecast estimates of their synergistic effect; these predictive estimates would allow the scientific and educational subsystem to integrate into the process of strategic development of the innovative economy of the region;
- a comprehensive assessment of the adequacy of promising subsystems to the imperatives of post-industrial transformations and the relevant goals established in the process of strategic development of the innovative economy of the region; this assessment is necessary to ensure the competitive interaction of several promising subsystems and the distribution of territorial reproduction development resources between them.

3. Research methodology

We focus on the scientific and educational subsystem of St. Petersburg, since the formation and development of such a subsystem are demanded by the conditions of post-industrial transformations, the process of intensification of spatial competition [4]. As applied to the economy of St. Petersburg, such a need is reinforced by a contradictory combination of relatively high investment attractiveness, the presence of some modern growth points, on the one hand, and the “obsolescence” of competencies that the human factor has, on the other hand.

A conceptual idea of the scientific and educational subsystem is formed. In the first approximation, this is a scientific and educational localization that has formed in the space of the regional economy. [5] However, such a definition is not enough, because it does not take into account the effects of market forces in the economic space, the possibility of specification, as well as the priority of scientific and educational activities that create the key assets of the new economy: intellectual capital; intangible assets.

Based on the above provisions, it is legitimate to define the scientific and educational subsystem of the regional economy as the result of the specification of assets, contracts and behaviors of participants in the local market of scientific and educational services, integrated into the region’s innovative economic system and providing its needs for the formation of key competencies of the human factor and intangible assets adequate to accelerate post-industrial transformation.

Based on the fact that in the scientific and educational subsystem there are scientific and educational branches, four interrelated components in its structural organization are distinguished: general cultural, vocational training, research, and implementation [6-7]. The first two components are related to the educational branch and the next - to the scientific.

The necessary explanations are given. The general cultural component lays the necessary foundation for subsequent training. At the same time, taking into account the stated scientific problem, the focus of our research is the structural component of professional training, which includes educational institutions of the highest and middle levels. At the same time, this component presents not only educational activities, but also scientific activities (the provision of scientific services under the state order, the implementation of contracts with market participants, etc.).
Having determined the scientific and educational subsystem and having highlighted its main structural components, we proceed to the analysis of the scientific and educational subsystem of St. Petersburg, starting with the assessment of the dynamics of the main parameters of the component of vocational training (Table 3).

Table 3. Dynamics of the main parameters of the professional training component of the scientific and educational subsystem of St. Petersburg (higher education)

	2007	2009	2011	2013	2015
Number of institutions of higher professional education (HPE), units.	101	101	97	95	82
Non-state educational institutions HPE	45	45	41	41	41
The number that performed research and development, units	47	44	44	39	45
Number of graduates	84 333	90 132	91 441	90 345	75 391
The number of professorial teaching staff	30339	30078	28592	26 947	24 345
Doctor of Sciences	4923	4789	4628	4 607	4 479
Candidate of Sciences	15403	15473	14947	14 336	13 365

Analysis of table 3 allows to draw the following conclusions:
- the number of higher education institutions is declining. St. Petersburg corresponds to this trend, which is typical for all Russia. At the same time, the process of unification of educational institutions is expanding [8]. It is legitimate to assume that foreign experience is being copied here. Thus, in Germany, the initiative to unite 15 universities has shown itself positively. But there it came from the organizations themselves, with clearly defined and tangible positive consequences for each of the parties [9]. In our country, the process of merging universities does not come from the universities themselves, that is, it takes place outside of contracts, along the administrative vertical, which affect its results;
- The initial decrease in the number of universities occurred due to washing out of the market of a part of non-governmental higher education institutions. In 2012, the reduction began within the public sector of this sector, which is associated with increased competition and the emergence of budgetary restrictions;
- The number of universities actively conducting research and development remains at the same level. The activation of the sphere's organizations in this regard was not affected by the implementation of the federal target program "Scientific and scientific-pedagogical personnel of innovative Russia 2009-2013", adopted to improve the quality level and professionalism of domestic specialists, creating a clear system to stimulate the flow and consolidate young promising scientists [10]. The target indicators of this program were: the wage fund of young scientists, co-financing not less than 20% of the total funding, as well as the publication activity of participants in scientific projects and the creation of a scientific and technical product. Under budgetary constraints associated with the unfavorable situation in the economy, expenditures on education and science should be a priority, and their share in the total expenditures of the budget system should be increased.
This program is aimed at solving the tasks set by the Strategy for Innovative Development of the Russian Federation until 2020 to develop human resources in the field of science, education, technology, and innovation, as well as to create a balanced and sustainable developing research and development sector. At the same time, the Program aims to provide conditions for the modernization of the personnel policy of the Russian sector of research and development, including by creating mechanisms to attract young specialists in science and innovative activities.

As the historical experience of Germany (1930s) shows, strict centralization of management had a detrimental effect on the state of science and education. Due to the administrative pressure that has arisen, a significant part of scientists emigrated. A similar threat exists in Russia in the case of tightening administration in science and education [11-12].

One of the main reasons for the merger of universities is the inefficiency of many of them, as evidenced by international ratings of the best hundred universities in the world. But administrative mergers and market connections are two fundamentally different processes. Positive international experience has been accumulated with regard to the market connection, for example, merging two well-known universities in the UK - Victoria University of Manchester and UMIST - into a single University of Manchester, which has become the largest one in this country, the experience of the National University of Singapore, formed in 1980, as a result of the merger of the University of Singapore and Nanyang.

To enter leading universities in international ratings, the Ministry of Education and Science of the Russian Federation held a competition for the right to obtain special subsidies for the implementation of activities that will promote universities in international ratings. 15 Russian universities won the competitive selection of universities. There are three St. Petersburg's universities among them: St. Petersburg State Polytechnic University, St. Petersburg Electrotechnical University "LETI", St. Petersburg National Research University of Information Technologies, Mechanics and Optics. Let us make the criteria for competitive selection:

- level of development of scientific activity;
- attractiveness of educational programs for applicants;
- involvement in the international academic market;
- positions in international university ratings.

The decrease in the number of universities results in the reduction of scientific and educational personnel. There is a normative ratio of the number of faculty to the student population - 1:12 for state institutions of higher education. Based on this ratio, the planning of the volume of the study load at the university takes place, except universities having the status of a national research university (NRU) and the status of a federal university.

But in a real situation, these institutions move to a more enslaving relationship to "survive". This process is influenced by two factors:

- universities have reduced funding in connection with the introduction of regulatory funding for the provision of public services in the areas of training (specialties), on the one hand;
- an increase of the average salary of teachers of educational institutions of higher professional education and scientific staff to 200% of the average salary in the corresponding region in 2018, on the other hand.

The structural component of the professional training of the analyzed subsystem from the angle of spatial competition is evaluated. Table 4 shows the most powerful components in this respect.

Table 4. Comparative characteristics of the structural components of professional training in the scientific and educational subsystem of some Russian regions, 2015

Indicator	Moscow	Saint Petersburg	Sverdlovsk Region	Republic of Tatarstan	Samara Region	Rostov Region	Novosibirsk Region
Number of HPE institutions, units	256	82	32	29	28	25	24
Non-governmental educational institutions of HPE	150	41	14	10	11	11	9
The number of students in	1028,4	381,7	175,2	190,7	141,7	194,512	141,3

educational institutions of higher professional education, thousands of people.							
Accepted students in educational institutions of higher education, thousands of people	174,1	89,5	40,6	32,8	31,0	40,3	29,1
Specialists graduated from educational institutions of higher professional education, thousands of people	262,1	75,4	41,4	31,4	33,2	41,2	30,0
Students per 10,000 people of the population	805	759	409	499	440	473	525

The following conclusions can be drawn after analyzing Table 4:
- Moscow and St. Petersburg are natural leaders in the number of universities in Russia's regions; there is a tendency to develop large educational centers here;
- the separation of Moscow and St. Petersburg from the next region in the ranked series - the Sverdlovsk region - is very significant.

Let us try to evaluate the quality of higher education. Today, there are many university rankings compiled by the Expert RA Rating Agency, the Higher School of Economics National Research University in conjunction with RIA Novosti, the All-Russian public organization "Business Russia", the Federal Agency for Education (FAO), and the publishing house "Kommersant", an independent agency "Reitor". We note their weak comparability - some ratings are built on the basis of statistical indicators, others - on the basis of exam scores of received, others - on the basis of assessing the demand for graduates in the labor market, etc.

It seems that three criteria can be used as the basis for a rating assessment of the quality of educational activity [13]:

a) the quality of staff, which is determined by the degree of qualification, the competencies of teachers and researchers. The quality of personnel through the quality of educational programs in the combination of teaching and research, subject to the conditions of their compliance with public demand, determines the quality of the content of training;

в) the quality of preliminary training of students, which is determined by the gap between general secondary and vocational education, as well as the mechanisms of educational and vocational guidance and motivation of students;

c) the quality of the infrastructure and the "physical educational environment" of educational institutions, covering the entire set of conditions for their functioning, including computer networks and modern libraries, which can be ensured by adequate infrastructure investment.

Given the above criteria, universities in cities are in a privileged position under clustering and the creation of scientific and educational centers. Therefore, stably the first places according to different ratings are occupied by M.V. Lomonosov Moscow State University (Lomonosov MSU), Bauman Moscow State Technical University (BMSTU), St. Petersburg State University (SPbSU), St. Petersburg Polytechnic University (SPbPU), etc.

International ratings also confirm the leading position of universities in cities and metropolitan areas [14]. What is the reason for the loss of leadership positions as a professional education? Here are a few factors. The first of them is an ill-conceived, alienated from socio-cultural roots education reform, a blow to all its components, and above all, to high school [15-16].

The second factor is the nationalization of education and science. The share of the private sector of vocational education in the regions in 2012 was in the range of 35-45% (St. Petersburg was at the upper limit of this interval - 43%). Over the past three years, this indicator has changed downward, going into the interval of 30-40%, which indicates a weakening of the competitive conditions for the interaction of scientific and educational organizations. The ratio of the private and public sectors in the research and implementation structural components of the emerging scientific and educational subsystem is even worse. The private sector share is about 20% in the scientific field.

Taking into account the meso-level, we also single out the factor of regional representativeness of students per 10,000 people, which characterizes the promising possibility of replenishment of personnel of territorial organizations by qualified specialists. The average indicator in Russia is 454. Its highest values are in the Yamal-Nenets autonomous district. Analyzing this indicator, we note that most regions are close to the average value for Russia, except Moscow and St. Petersburg, where this indicator is 805 and 759, respectively.

Focusing on higher education, let us highlight the characteristic "problem factors" of the quality of education:

- gap between full-time and part-time forms affecting competencies;
- separation of training from the real demand of the labor market (inertial continuation of graduation of specialists in legal, managerial and economic profiles);
- imbalance of general and special training components;
- corruption destruction of the education process;
- bureaucratization of the educational and scientific processes in higher education, leading to a decrease in the quality and crowding out of the content components of the activity by purely formal ones [17-19].

Results

A SWOT-analysis is conducted to determine ways to increase the competitiveness and efficiency of higher education organizations and their strengths and weaknesses are identified (Table 5).

Table 5. SWOT- analysis of a typical organization of higher education in St. Petersburg

Evaluation Options	Internal factors	
	Strengths	Weaknesses
University management	Systematized organizational structure Clear distribution of authority Democratic management style	Low interest of employees in the development of the university
Finance	Government subsidies Income-generating activities	Reduction of financing according to the standards for performance of the state task Lack of free current assets University budget deficit
Personnel	Timely payment of wages Social Security Highly qualified personnel Performance evaluation motivation Ranked wage system	Weak career growth Staff turnover Very high age of teaching staff
Educational process	Development of a system of continuing professional education Expanding the range of training areas	Distance learning is not adjusted The difficulty of introducing innovations
International cooperation	Participation in international events Academic mobility of students	No educational tours for students
Research activities	The implementation of research work, the organization and conduct of international and national conferences, round tables, the development of research activities of students	A small share of financial income from applied and basic research

Summary of the analysis results presented in table 5 allows to identify the main opportunities and threats of the scientific and educational subsystem of St. Petersburg. Let us present them in accordance with the selected areas of assessment: organizational mechanisms, competition, obtaining impulses of market demand, government regulation, investment, technology (Table 6).

Table 6. The main opportunities and threats of the scientific and educational subsystem of St. Petersburg

Assessment directions	External factors	
	Opportunities	Threats
Organizational arrangements	Training of highly specialized personnel	Changing the preferences of applicants
Competition	Cooperation with the business environment	Strengthening the position of universities of a similar profile
Impulses of market demand	Cooperation with employers	An overabundance of humanitarian specialists, lack of experience in the specialty
State regulation	State interest in developing narrowly specialized universities and raising the global rating of national higher education	Administrative merger of universities
Investment	Financial autonomy of universities	Reduction in government funding The imaginary independence of universities
Technology	Application of modern technologies, equipment and programs	Applying new technologies by competitors

In recent years, some innovations have appeared in Russian higher education, which has led to structural and functional transformations. These innovations, in most cases, were initiated not by educational organizations, but by the relevant federal ministry. Moreover, regional features of the development of scientific and educational activities were not taken into account.

The change in the status of organizations in the field of higher education is evaluated. On January 1, 2011, Law 83-FZ "On Amending Certain Legislative Acts of the Russian Federation in Connection with the Improvement of the Legal Status of State (Municipal) Institutions" entered into force. With the entry into force of Law 83 –FZ, educational institutions are allowed to follow the path of commercialization. At the same time, motivating factors for ensuring the quality and accessibility of educational services should be:

- formation of tasks for each institution with indicators of the volume and quality of its implementation;
- control over task performance;
- the establishment of a direct relationship between the volume of financing of the assignment and the results of the institution;
- great economic independence of institutions and the responsibility of their leaders for the financial performance of the institution;
- unambiguous determination of the list of services financed from the corresponding budget;
- clear requirements to the procedure of providing paid services to institutions and determining their value for the consumer.

The implementation of this law has led to the emergence of three new organizational forms of educational institutions:

- government institutions;
- budgetary institutions;
- autonomous institutions.

The financing of the activities of public institutions at the expense of budgetary funds is carried out on the basis of budget estimates. Budgetary and autonomous institutions - in the form of subsidies for the fulfillment of state tasks provided on the basis of a plan of financial and economic activity. Budgetary and autonomous institutions received the direction of commercialization, independence in deciding on the spending of funds from income-generating activities and partially subsidies for the state (municipal) assignment, at the disposal of movable property, except especially valuable movable property. But this is a declared picture; in fact, the independence of budgetary organizations is very conditional, as the competition between sectors of the sphere. The necessary arguments are given.

The Ministry of Education and Science conducts open public competitions among educational institutions of higher education that have state accreditation, with the distribution according to the results of the competition, control figures for admitting citizens to areas (specialties) for training at the expense of the federal budget are set. But at the same time, the number of private organizations participating in such a distribution is limited; the share of budgetary reception falling on them, and the competition is conducted by an interested party - the state body, which is in charge of most of the participants.

Further, the standard costs for the implementation of the basic professional educational programs of higher education are determined, established by specialties (areas of training) and differentiated by levels - bachelor degree, specialist training, and magistracy. Besides, funding standards vary by specialty (area of preparation), depending on whether they require laboratory equipment or not. Standards for specialties with a special ratio of the number of teachers and students (1: 4, 1: 5, 1: 6) are separately established, as well as for priority specialties. However, it is unclear how these standards take into account significant differences in the property complexes of universities. Let us evaluate the ways of stimulating the choice of the analyzed organizational forms. The state is interested in choosing a form of autonomous organization and pushes state universities to switch to this particular form. For example, participants have to go into the status of autonomous organizations to participate in the program of inclusion in international rankings. The process of redistributing financial resources in favor of strong organizations is complicated, burdened with social consequences. Economic methods alone ("money follows the student") cannot be promoted. Weak universities, the funding of which is decreased, have to finish up the students who have already arrived, maintain state property, which creates a threat to their insolvency. Besides, the preferences of applicants may not coincide with the priorities of the state educational policy and the predicted structure of the labor market; therefore, some organizations in the sphere have to be maintained, despite low demand.

That targeted funds are required to allocate to weak organizations that need restructuring and a change of management to carry out the necessary transformations, which casts doubt on the financing mechanisms of the type of state registered financial obligations that would ensure the priority of strong universities in obtaining budget funds. The principle of "money follows the student" is effective where several universities are competing with each other and can increase supply and improve the quality of educational services, following increased funding [8].

The dominant vector in the direction of the development of higher education — the restructuring of the network of universities with the aim of liquidating inefficient institutions, the introduction of a financing mechanism based on the principle of "money follows the student" – predetermines the investigated subsystem to long-term administrative quasi-establishment, while the incapable business entities of the subsystem have to go through bankruptcy mechanisms and joining.

The next component of the scientific and educational subsystem is research. Let us evaluate the dynamics of the main parameters of the scientific component (Table 7).

Table 7. Dynamics of the main parameters of the research component of the scientific and educational subsystem of St. Petersburg

Indicator	2013	2014	2015
Number of organizations performing research and development, units	302	299	299
Including	85	84	87
public sector	174	173	164
business sector	39	40	45

professional higher education private sector	4	4	3
Number of employees performing research and development, people	78 704	78 783	79 076
Including researchers	43 828	43 390	42 956
technicians	5 064	4 861	5 104
support staff	17 444	17 793	18 149
other	12 368	12 739	12 867
Of the total number Doctors of Science	2 852	2 795	2 767
Candidates of Sciences	9 013	8 833	8 656

According to Table 7, the following conclusions can be drawn:
- the number of organizations performing research and development declines. This is due to the restructuring of the scientific and educational subsystem as a whole both in St. Petersburg and in Russia. Moreover, the decrease is insignificant if the whole subsystem is analyzed;
- infrastructural change is taking place in the public sector. This is due to one of the criteria for the effectiveness of both state organizations and higher education is the attraction of funds and it is precisely through a scientific line;
- it is characteristic that the number of workers performing research and development remains at the same level. But the total number of researchers - doctors and candidates of sciences decreases; the number of attendants grows. This means that among our scientists, the number and proportion of those with a scientific degree rapidly fall. Modern scientific trends follow the post-Soviet trend.

The implementation component is a dynamically developing system link in the scientific and educational subsystem. The introduction component of Peter the Great St. Petersburg Polytechnic University is characteristic in this work. Technopark Polytechnic created at the university. The main activities are the creation and implementation of innovative programs in the scientific and technological field; the development of innovative infrastructure at the university, the development of the implementation component of scientific research, the promotion of the spread of modern technologies for innovative products, etc. The technopark has been successfully operating for 20 years and is actively developing and helping residents of the technopark of commercialization and technology transfer. The Polytechnic business incubator promotes the development of business ideas for scientific youth. As part of the business incubator, Fablab Polytech works, which works not only with students, but also with schoolchildren.

Conclusions

The scientific and educational subsystem is generally recognized to function in a crisis economy. The implementation component of the three components actively develops, this is correct. Any scientific research should be implemented. However, the implementation component is not possible without the educational and research components. Since they are now relegated to the background, there will be a tendency to slow down in a few years.

Under accelerating post-industrial transformations, the needs for training specialists dynamically change. In the near future, specialists will be in demand, which no one is taught today, because the corresponding specialties have not yet appeared. In such a situation, the most effective way to develop a vocational education system is to work closely with innovative businesses. The mechanism of public-private partnership (PPP) provides increased efficiency in spending budget funds and improves the quality of public services.

The main mechanisms of public-private partnerships are implemented in continuing education. High-tech companies create laboratories of the future; provide full equipment for teaching children, not only in schools, but also in higher education institutions. In some universities of St. Petersburg, basic departments have been created, which is a mechanism of public-private partnership.

Further research areas

The scientific and educational subsystem is inscribed in increasingly complex socio-economic transformations, feeling economic instability, increasing uncertainty and social risks. At the same time, the requirements of society to the level of education, the competence of the individual in various areas of social practice are increasing. In these conditions, ensuring its effectiveness plays an important role, for which relevant indicators are in demand. [9].

Further research needs to determine the approach to such an assessment. We proceed from the urgent need to study competition in the scientific and educational subsystem and to develop clear competitive strategies adapted to the Russian conditions of the new system. Many regional economies are not sufficiently prepared for open competition. Therefore, the central moment is the formation by their scientific and educational subsystems of qualitatively new competitive advantages.

This problem can be solved, basing on the formation of key competencies. G. Hamela and K. Prahalada can be distinguished among the scientists who made a significant contribution to the development of this problem. They have developed a concept of key competencies and the corresponding strategy: "To take advantage of foresight and gain leadership in key competencies, the company has to take over important global markets before competitors. The goal is not to get into the market as quickly as possible, but to create a global lead in the market" [10].

The scientific and educational subsystem is designed to form key competencies. The more such competencies are created for it, the higher its contribution to the increase in the competitiveness of the regional economic system is.

References

[1] Leonidova G.V. Integratsiya nauki, obrazovaniya, biznesa kak resheniye strategicheskikh zadach v podgotovke kvalifitsirovannykh spetsialistov // Materialy mezhregional'noy nauchno-prakticheskoy konferentsii «Formirovaniye upravlencheskogo rezerva kadrov: regional'nyy opyt» [Elektronnyy resurs]. – Rezhim dostupa: http://www.ags-vologda.ru.
[2] Evseeva, S., Evseeva, O., & Kalinina, O. (2020). HR staff leasing: Opportunities in the digital economy environment (case of russia). Paper presented at the E3S Web of Conferences, , 175 doi:10.1051/e3sconf/202017513041 Retrieved from www.scopus.com
[3] Klochkov, Y. S., & Tveryakov, A. M. (2020). Approaches to the improvement of quality management methods. International Journal of Systems Assurance Engineering and Management, 11, 163-172. doi:10.1007/s13198-019-00939-x
[4] Murav'yova M. Minobrnauki restrukturiruyet desyatuyu chast' podvedomstvennykh vuzov [Elektronnyy resurs]. – Rezhim dostupa: http://www.strf.ru/material.aspx?CatalogId=221&d_no=47352.
[5] Avanesov V. Sliyaniye rossiyskikh vuzov – ocherednaya popytka reformirovaniya obrazovaniya [Elektronnyy resurs]. – Rezhim dostupa: http://testolog.narod.ru/Education73.html
[6] Reforma i razvitiye vysshego obrazovaniya. Programmnyy dokument. — Parizh: YUNESKO, 1995.
[7] Moskvin, D., Zegzhda, D., & Nikolskiy, A. (2015). Secure processor architecture modeling for large-scale systems. Automatic Control and Computer Sciences, 49(8), 684-689. doi:10.3103/S0146411615080301
[8] Sulyaev, Y. S., Alexandrov, E. V., Burdakov, A. V., Ivantsivskiy, M. V., Zemlyanskiy, Y. N., Pischinskiy, K. V., . . . Modestov, V. S. (2020). Engineering calculations and preparation for manufacturing of ITER equatorial port #11. IEEE Transactions on Plasma Science, 48(6), 1631-1636. doi:10.1109/TPS.2020.2985113
[9] Zegzhda, D. P., & Stepanova, T. V. (2015). Approach to APCS protection from cyber threats. Automatic Control and Computer Sciences, 49(8), 659-664. doi:10.3103/S0146411615080179

[10] MGU snova v reytinge // Moskovskiy komsomolets. – 2009. – 2 noyabrya.
[11] Svetlova, Ye. Komu on nuzhen, etot Pushkin // Moskovskiy komsomolets. – 2012. – 30 okt.
[12] Rossiyskiye reformy v tsifrakh i faktakh [Elektronnyy resurs]. – Rezhim dostupa: // http://refru.ru
[13] Klyachko T.L., Sinel'nikov-Murylev S.G. O reformirovanii sistemy finansirovaniya vuzov http://institutiones.com/general/2228-o-reformirovanii-sistemy-finansirovaniya-vuzov.html
[14] Finansovoye upravleniye razvitiyem ekonomicheskikh sistem: Monografiya /Dzyubenko I.B. i dr. Novosibirsk: OOO «Tsentr razvitiya nauchnogo sotrudnichestva», 2015. 242 s.
[15] Khamel G., Prakhalad K. Klyuchevaya kompetentsiya korporatsii. M.: «Harvard Business Review», 1990.
[16] Chernov, A. Y., & Konoplev, A. S. (2015). The use of virtualization technology in the dynamic analysis of software code. Automatic Control and Computer Sciences, 49(8), 834-837. doi:10.3103/S0146411615080234
[17] Davydov, A. D., Erokhina, O. O., Ryaboshuk, S. V., & Kovalev, P. V. (2020). Analysis of the causes of cracks in the production of ingots and forgings from austenitic stainless steel 08X18H10T (aisi 321) doi:10.4028/www.scientific.net/KEM.854.16
[18] Drobintsev, P., Voinov, N., Kotlyarova, L., Selin, I., & Aleksandrova, O. (2020). Optimization of technological processes at production sites based on digital modeling doi:10.1007/978-981-15-2341-0_75
[19] Makovetskaya-Abramova, O., Lazarev, Y., Gravit, M., Silla, S., & Shakhova, M. (2020). Multiplicative method for creating the traffic monitoring base in a megapolis. Paper presented at the E3S Web of Conferences, , 164 doi:10.1051/e3sconf/202016403023

IOP Conf. Series: Materials Science and Engineering **986** (2020) 012009 doi:10.1088/1757-899X/986/1/012009

Recommended initial pressure for the correct operation of vibrofuge based on the conducted research

V V Krokhmal

Institute mechanical engineering, materials and transport, Peter the Great St.Petersburg Polytechnic University, Saint-Petersburg, street Polytechnicheskaya, 29, Russia, w.krohmal@gmail.com

Abstract. The article presents the operation of vibrofuge test stand. The main research of the stand operation is discussed. The mode of piston motion in gas spring and gas behavior inside the working spring chamber is shown. Based on the research the recommended initial pressure is calculated, which must be set in spring chamber for the correct operation of the test stand.

1. Introduction

In modern industry test research cannot be excluded from the design life cycle of new products and technical devices. There are special test stands for test research. Test stand is a laboratory equipment that reproduces specialized control tests of any products [1].

Today test stands are divided into load categories. For example, shock tables test products on impact resistance. There are also stands that test technical objects on speed, vibration, compression, bending, temperature, etc.

Complex test stands are of the highest interest. Such stands are capable of reproducing various types of load on the tested objects. This helps to save time and money.

In this work the object of study is a test stand of a centrifuge, on rotor of which an air spring with an electromechanical vibrator - vibrofuge is fixed. Thanks to this design solution, on such a stand it is possible to reproduce loads not only on the centrifugal force, but also on a vibration component. Objects that are usually exposed to stresses are part of the structure of air vehicles (airplanes, helicopters, rockets, etc.).

The main object of this article is to show the main studies carried out in modeling the operation of a vibrofuge and to calculate the recommended level of the initial gas pressure inside the working air spring chamber, depending on the maximum constant speed of rotor rotation.

2. Literature review

On such complex test stands studies were mainly carried out on the parameters that affect the operation facility inaccuracy. For example, an assessment of the influence of Coriolis accelerations and forces on an object under the action of a vibration load in the field of linear accelerations was made [2]. A study was also carried out on the rotor length change during its rotation. The rotor length can be varied through minimal bending due to gravity, therefore it is important to predict the possible rotor length change, since such a prediction can significantly reduce the error [3, 4].

An important factor for the correct operation of the complex test stand is the stable balancing of the moving masses. There are several methods of stable balancing, which are included in design solutions both with the use of pneumatics and without it. [5, 6]. One of the solutions is the major theme of this article.

3. Research of the processes occurring during the operation of the vibrofuge

Within this article a model of a vibrofuge test stand is investigated. Its design features were mentioned in introduction. Using the SolidWorks software an approximate model of such a stand were designed (Figure 1).

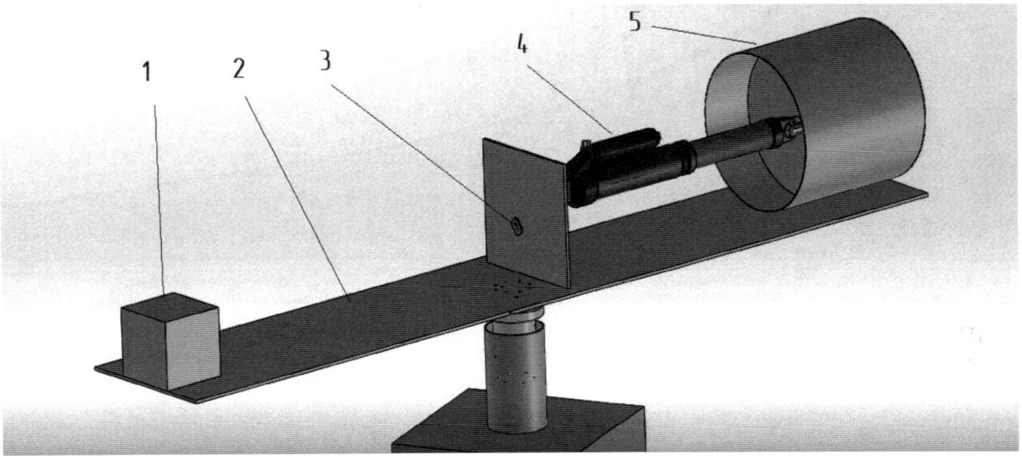

Figure 1. Vibrofuge test stand model in SolidWorks.

Figure 1 shows the following positions: 1 – gravity weight, 2 - test stand rotor, 3 - test object, for example, accelerometer, 4 - electromechanical vibrator, 5 - gas spring.

The main characteristics of the stand, which were specified in the following calculations: - an electromechanical vibrator with a maximum force F = 5000 N and a maximum speed of 2 m / s, the vibration frequency of vibrator f = 50 Hz in the form of a sinusoidal wave;

- the mass of vibrator, piston and table on which the accelerometer is mounted is M = 25 kg;
- gas spring chamber length is L = 0,5 m;
- maximum constant rotor speed is ω = 52,36 rad/s;
- vibrofuge rotor length from the axis of rotation is R = 2,5 m;
- gas spring piston area $S = \frac{\pi}{4}d^2$ (d = 0,320 m – piston diameter). As part of the main research of vibrofuge operation (Figure 1), an analysis of the operation of a pneumatic system with an electromechanical vibrator was carried out when the rotor rotates at a constant speed. To identify the nature of piston movement inside the gas spring chamber, we have the following formula:

$$dL = v_{i-1}dt + \frac{a(dt)^2}{2}, \quad (1)$$

where dL – piston movement, m, v_{i-1} – speed at the previous moment of computation time (at the initial moment of time the speed is equal to zero), m/s, dt – constant, constant computing step, a - $\frac{d^2L}{dt^2}$ – mechanical system acceleration (electromechanical vibrator, piston), m/s^2.

Using the given characteristics of the operation facility, formula and the MatLab software, a graph of the piston displacement in gas spring during rotor rotation at maximum and constant speed and during operation of the electromechanical vibrator is built (Figure 2) We also found the ranges of characteristics, at which the correct operation of the operation facility is observed (Figure 3).

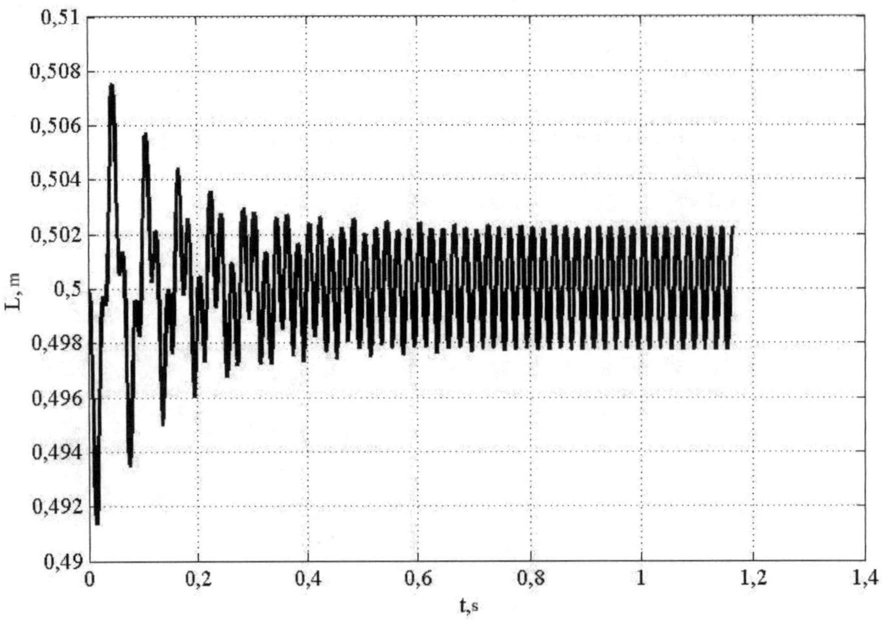

Figure 2. The nature of piston displacement in gas spring chamber.

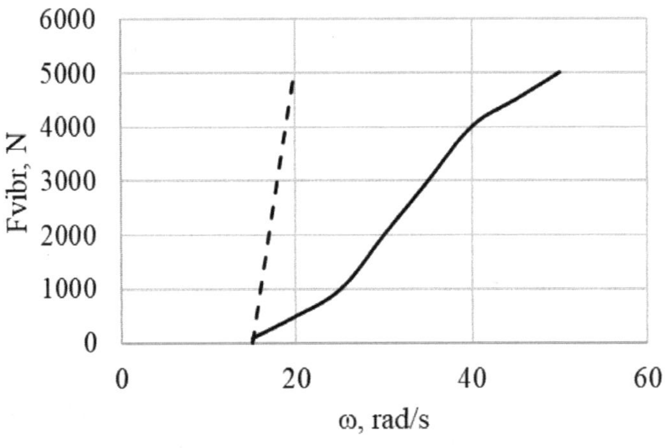

Figure 3. Areas of vibrofuge parameters.

In Figure 3, the correct operation of the operation facility can be achieved at any values of electromechanical vibrator forces and rotor speed in the right area of the graph to the solid thick line [7, 8]

In the framework of further studies of the operation facility functioning, it is necessary to find out the behavior of the gas inside the gas spring chamber during the rotation of the rotor at the maximum constant speed. Based on the showed mathematical model, a formula for calculating the gas pressure at each point along the entire length of the pneumatic spring chamber is derived:

$$p = p_1 \; e^{\frac{\omega^2 M (r^2 - r_1^2)}{2RT}}, \tag{2}$$

where p_1 – reference pressure, Pa, ω – rotor speed, m/s, M – gas molar mass (equal for air 29*10-3$\frac{\text{kg}}{\text{mole}}$), R – absolute gas constant (equal 8.314$\frac{\text{J}}{\text{kg*K}}$), T – absolute gas temperature, K, r_1 – rotor length from the axis of rotation to the first end of the gas spring chamber, m, r – parameter for setting a discrete section where the pressure level is calculated, m.

Using the MatLab software, graphs of pressure distributions along the entire length of gas spring chamber were built depending on the maximum constant rotor speed (Figure 4).

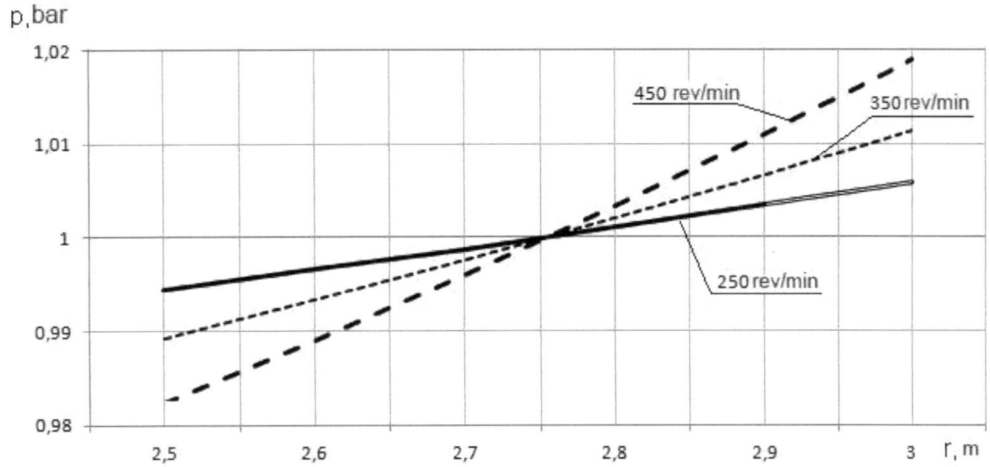

Figure 4. Pressure distribution inside the gas spring chamber at different rotor speeds.

It should be mentioned that under certain conditions, a vacuum may appear at the first end of the chamber. [9, 10].

The study carried out on calculating the gas pressure in the working chamber of a gas spring, when the rotor rotates at a constant speed, should be taken into account when working with the stand on Figure 1. So it is necessary to calculate the recommended level of the reference gas pressure before starting the operation facility.

The reference pressure level depends on rotor speed. For example, if a small gas pressure is pumped into the gas spring chamber and the rotor is strongly accelerated, then there will be no pressure left to resist the centrifugal force at the chamber inlet, which can lead to structural failure. The reference pressure formula is the following (2):

$$p_1 = \frac{p}{e^{\frac{\omega^2 M(r^2 - r_1^2)}{2RT}}}, \tag{3}$$

where p is calculated by dividing the centrifugal force value by piston cross-sectional area. The only variable that we set ourselves is the rotation speed ω.

So we have table №1, which shows the values of speeds and recommended reference pressure.

Table 1. Reference pressure at different speeds.

ω, rad/s	Recommended reference pressure, Pa
60	2812522.40
70	3835494.11
80	5020707.57
90	6370267.15
100	7886571.82

According to the Table №1, we can build a graph of rotation speed dependence on the reference pressure. The graph is shown in Figure 5.

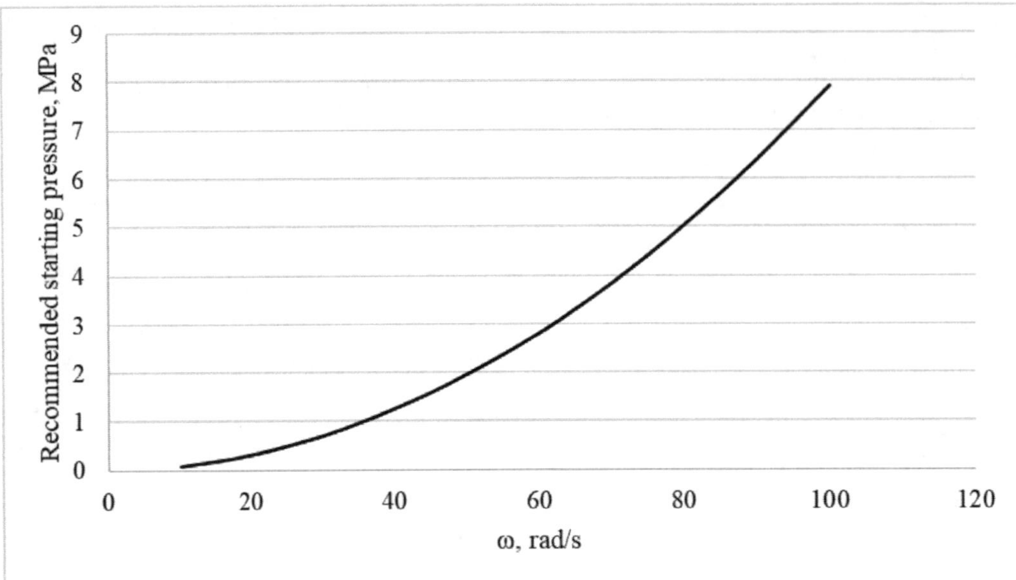

Figure 5. Dependence of reference gas pressure in air spring chamber on the vibrofuge rotor rotational speed.

The relationship shown in Figure 5 is exponential.

At the given values of the mass of the electromechanical vibrator with an air spring piston and the length of the vibrofuge rotor and at a rotor speed of 52.36 rad / s (500 rpm), it is recommended to set a reference pressure of 2109730.14 Pa, which is optimal for testing. The maximum value of the reference pressure that can be pumped into the air spring also depends on the material of the air spring and in general on the capabilities of the compressor.

4. Results

As a result of vibrofuge operation studies using software we have found the following:

1) Parameters of correct operation of the facility (Figure 3);
2) the pressure in working gas spring chamber is redistributed depending on the speed of rotor rotation (Figure 4);
3) the method for calculating the recommended level of reference pressure in the working gas spring chamber before starting the operation facility is shown (Figure 5).

5. Conclusion

In this article the operation of vibrofuge test stand was studied, on the rotor of which a gas spring and an electromechanical vibrator are installed. Such test stand can simulate complex vibration force and centrifugal acceleration tests.

The main studies of the operation facility were carried out to set recommendations for the correct operation of the test stand.

References

[1] Popov A N Polishchuk M N and Timofeev Al N 2016 Stands for mechanical testing: textbook. - *SPb.: Publishing house of Polytechnic. University* p 200

[2] Andrienko P A, Karazin V I, Kozlikin D P and Khebosolov I O 2019 On Vibrational Movements in a Testing Bench, Advances in Mechanical Engineering *Selected Contributions from the Conference «Modern Engineering: Science and Education»* (doi: 10.1007/978-3-030-39500-1_2) pp 11-17

[3] Karazin V and Khisamov A V 2014 Reducing the error of reproduction of a given acceleration on centrifugal benches *Week of Science SPbGPU: materials of the scientific and practical*

conference with the participation of RSM. Institute of Metallurgy, Mechanical Engineering and Transoprt. Part 1 (SPb. : Publishing house Polytechnical university) pp 167-170

[4] Andrienko P A, Kozlikin D P, Hisamov A V and Hlebosolov I O 2013 About the using of non-contact measurement systems in centrifugal srands. Modern mechanical engineering *Science and Education: a scientific edition of the MSTU. N.E. Bauman* (T. 3) p 623

[5] Karazin V I, Kozlikin V I and Khebosolov D P 2007 Balancing inertial forces in vibratory centrifuges *Theory of mechanisms and machines* (Volume 5 SPb Polytechnic University Publishing House) pp 63-71

[6] Karazin V I, Kozlikin D P, Sukhanov A A, Tereshin V A and Khlebosolov I O 2017 Some ways of stable counterbalancing in respect to moving masses on centrifuges *Lecture Notes in Mechanical Engineering* (doi:10.1007/978-3-319-53363-6_9) pp 73-85

[7] Krokhmal V V and Klyukin V Yu 2019 Area determination of correct operation of the vibrofuge. *IOP Conf. Series: Materials Science and Engineering* (doi:10.1088/1757-899X/666/1/012011)

[8] Klyukin V Yu, Kozlikin D P and Krokhmal V V 2017 Research of work of the pneumatic «spring» of vibrofug *Week of Science SPbPU: materials of a scientific conference with international participation. Institute of Metallurgy, Mechanical Engineering and Transport. Part 2* (SPb Publishing house Polytechnical university) pp 9-11

[9] Krokhmal V V, Klyukin V Yu and Sorokina E V Distribution of the pressure of gas under the action of centrifugal forces *AER-Advances in Engineering Research* (vol. 157 doi: 10.2991 / aime-18.2018.62) pp 319-324

[10] Sorokina E V, Klyukin,V Yu and Krokhmal V V 2018 Redistribution of gas under the action of centrifugal forces in pressure vessels, Mechanical Engineering and Technosphere of the XXI Century *Proceedings of the XXV International Scientific and Technical Conference in Sevastopol September 10-16*, (In 2 volumes. - Donetsk: DonNTU T. 2) pp 139 - 147

IOP Conf. Series: Materials Science and Engineering **986** (2020) 012010 doi:10.1088/1757-899X/986/1/012010

Analyzing the main casting defects and their mold parameters

A T Kozlova

Federal State Budgetary Educational Institution of Higher Education «Kazan National Research Technical University named after A.N. Tupolev-KAI», e-mail: dashenka.fedagina@gmail.com, alsu_kozlova@mail.ru

Abstract. At present, the manufacturer and the buyer find themselves in the market, and consumers pay great attention to the price and quality of the product. This means that the fate of the manufacturer depends on it. Casting gets closer and closer to the part, and a number of surfaces are the ones of the future part. Requirements for roughness, design, the competitiveness of the part increase, and the question of the quality and cost of the product is raised. All this determines the relevance of the chosen topic. In the foundry, the most popular technology for printing sand synthesis molds, due to inexpensive materials and quality of products. Casting quality directly depends on the quality of the synthesis molds; therefore, constant monitoring of the parameters of the sand molds is required. [1-3]

Introduction

Casting defects are formed when the metal is crystallized. They are formed outside and inside the casting. Everyone knows that defects in production are the source of defects. For a modern organization, this results in financial losses, additional costs for rework in casting.
There are many causes that provoke the defects to occur. Let us build a causal diagram for a visual representation of possible causes (Fig. 1).

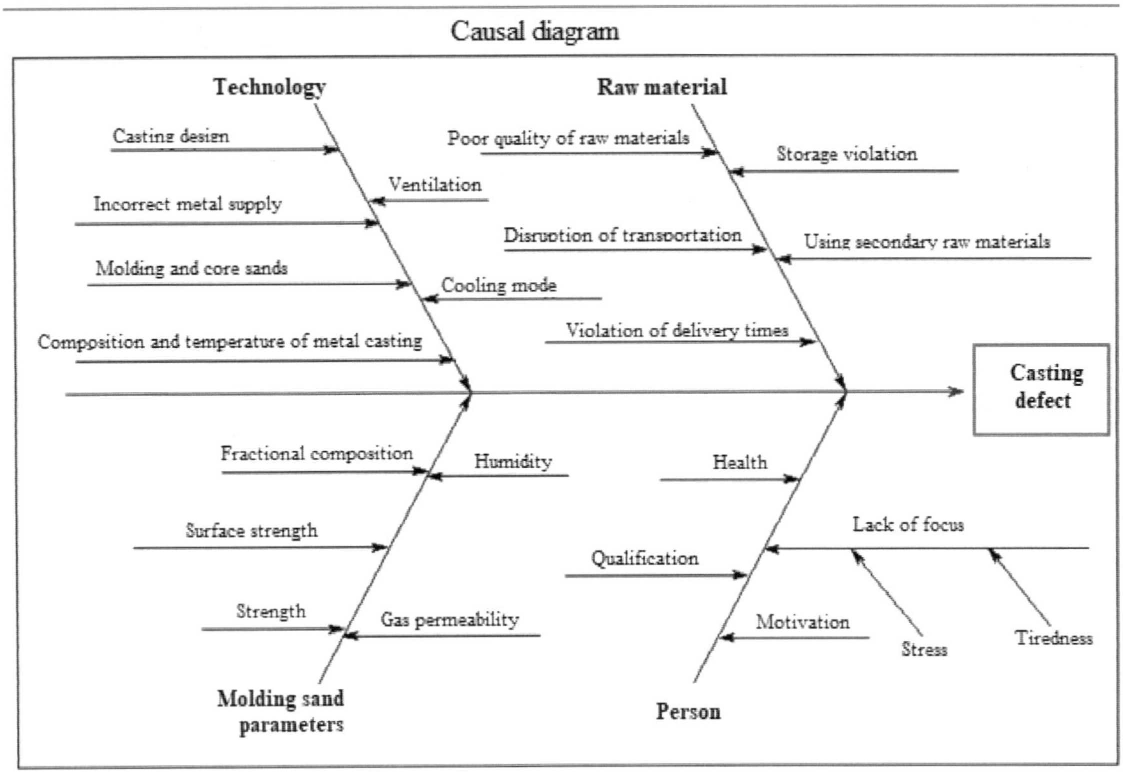

Figure 1. Causal diagram of casting defects.

Methods and results

There are many different defects in metallurgy. Here are the most common ones:
1) Warping is a change in the size and shape of the metal under the influence of shrinkage stresses [4]. Possible causes:
- the irrationality of the casting design (for example, the difference in wall thickness), which leads to the formation of internal stresses;
- incorrect supply of metal which adversely affects the uniformity of cooling;
- improper composition or temperature of the metal casting causing excessive shrinkage;
- incorrect cooling mode of the casting and insufficient flexibility of the mold and cores.

2) Blowhole defects are cavities formed in the metal due to poor gas permeability of the shape, non-compliance with the temperature regime, or poor quality of materials [5].
3) Sand spots are cavities in the metal filled with molding material due to the low strength of mold and core sand or the low-tech design of the casting.
4) Porosity is the presence of intercrystalline voids in the metal, in other words, pores. Reasons: uneven cooling, improper metal casting, and its temperature [6].
6) Hot and cold cracks are local failures in the metal or its rupture. The causes:
- design errors;
- the composition of the molding sand;
- metal casting;
- crystallization temperature;
- the high strength of molds.

7) Cold lap is a prematurely solidified metal in the form of streams [7-8].
8) Bulging is a shape distortion of a part [9].

The Pareto chart was used to analyze the causes of defects in this study. The set of defects was monitored for two months. Then, the table reflecting the list of defects and their number was built using the Statistica program (Fig. 2).

	Defects	Amount	Percentage,%	Accumulated %
1	Cracks	29	27	27
2	Blowholes	24	22	49
3	Sand holes	23	21	70
4	Warping	16	14	84
5	Bulging	5	5	89
6	Porosity	3	3	92
7	Cold lap	1	1	93
8	Others	9	7	100
9	Total	110		

Figure 2. The defects table.

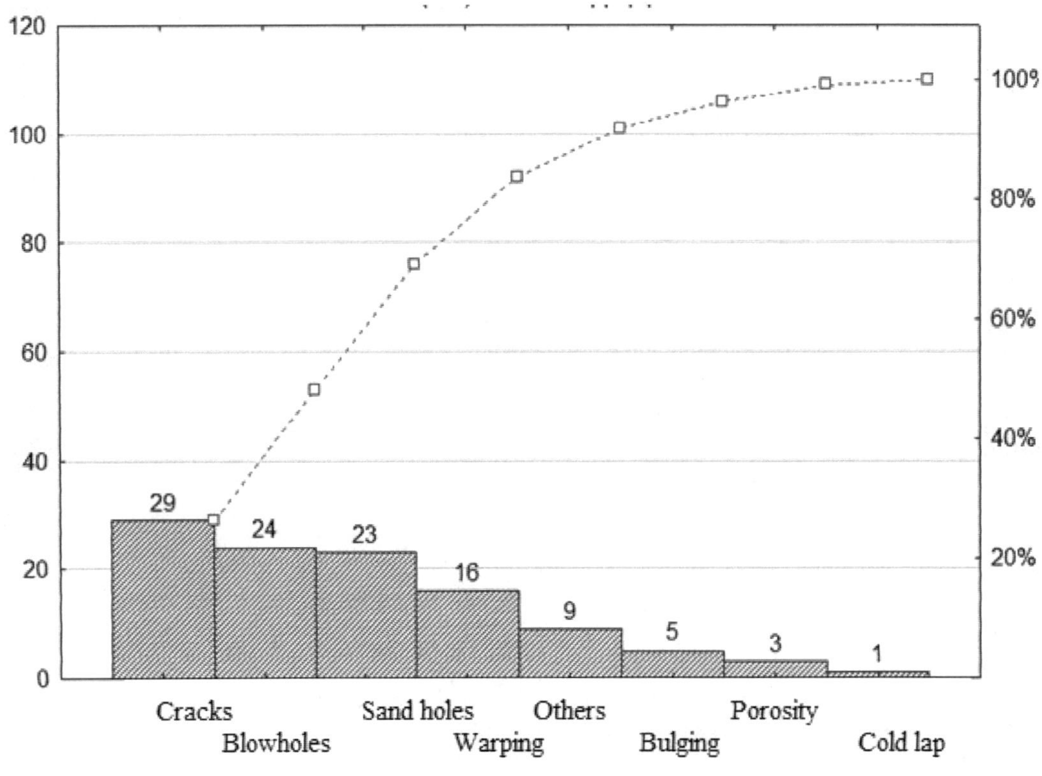

Figure 3. The Pareto chart by performance.

According to Figure 3, 70% of nonconformities are defects such as cracks, gas and sand holes. For this reason, first of all, special attention should be paid to the control of these defects. Next, the parameters of the molding sand and their relationship with the quality of the casting will be considered. [10-11] Thus, the analysis showed that the most important defects are cracks, gas and sand holes. These defects largely depend on the parameters of the molding sand. They will be discussed below.

Casting quality and its rejects largely depend on the parameters of the molding materials; therefore, they have to be carefully controlled.

Depending on the working conditions and the technological process of manufacturing the casting, molding and core mixtures must have the following basic properties:

1) strength - the ability of molds and cores not to collapse under stress;

2) surface strength - resistance of the surface layer of the mold or the core to abrasion forces;

3) gas permeability - the ability of a mixture to pass through itself the gases released by the mold and liquid metal; if the gas permeability of the mixture is insufficient, then the gases enter the metal causing casting defects. Gas permeability of the molding (core) sand depends on the moisture content, size and uniformity of sand grains, on the clay content and the degree of compaction of the mixture.

Gas permeability is the ability of the mold to pass gases [12]. When filling the mold with metal, as well as during the period of its crystallization, gases are released. Blowholes appear in the casting when the metal crust on the casting and poor gas permeability of the mold at the moment is formed and the gas pressure is maximum.

There is a standard technique to get rid of this defect. All castings are scanned with a tomograph for the presence of holes. If they are found, the cavities are drilled out and then welded. In this case, the casting becomes suitable - this is a normal practice. [13-14]

Besides, the problem of poor gas permeability can be solved by providing additional ventilation of the casting.

Strength is the ability of the mold (core) not to collapse from external forces. Mold (core) should not be destroyed by their weight and from shocks during assembly and transportation. Therefore, the ultimate strength of the form is of particular importance.

The surface of the mold and its inner part are not in the same conditions. It also provides control of such a parameter as surface strength.

Surface strength (crumbling) is the ability of a mold (core) to resist the abrasive forces of a moving metal jet. Sand holes are formed at low surface strength. The surface strength of the mold may decrease with a long wait for casting. In this case, the mold is covered with special refractory paint.

Conclusion

The control of the molding sand is carried out in the laboratory using special equipment. The main types of tests performed for controlling the mixture are incoming sand control, gas permeability and mixture strength tests.

The quality of the molding sand directly depends on the materials, in particular, on the quality of the sand. For example, the strength, ductility, and flexibility of the mixture directly depend on the quality of the sand, namely, on the average grain size, uniformity, mass fraction of moisture and mass fraction of the clay component.

Thus, it is necessary to perform the following tests during the control input: determination of the uniformity of sand and the average grain size; determination of the clay component; determination of moisture content in sand.

References

[1] GOST 2138-91 Molding sand. General technical conditions. - Used in incoming inspection to determine the quality of sand.

[2] Konakov, V. G., Kurapova, O. Y., Novik, N. N., & Golubev, S. N. (2015). Microstructure of "copper-graphene" composites manufactured by electrochemical deposition using graphene suspensions stabilized by non-ionic surfactants. Materials Physics and Mechanics, 24(4), 382-390.

[3] Ovid'Ko, I., Sheinerman, A., Skiba, N., Krasnitiskiy, S., & Smirnov, A. (2015). Twin boundary migration and nanocrack generation in ultrafine- grained materials with nanotwinned structure. WSEAS Transactions on Applied and Theoretical Mechanics, 10, 204-212.

[4] Redkov, A. V., & Kukushkin, S. A. (2015). Surface defects formation on strained thin films growing via chemical reaction: A model. Paper presented at the Journal of Physics: Conference Series, , 643(1) doi:10.1088/1742-6596/643/1/012005

[5] GOST 23409.0-78 Molding sand, molding and core sands. General requirements for test methods.

[6] GOST 29234.5-91 Molding sand. Moisture determination method.

[7] GOST 23409.6-78 Molding sand, molding and core sands. Gas permeability determination method.

[8] GOST 23409.7-78 Molding sand, molding mixtures and cores. Methods for determination of compressive, tensile, bending and shear strength.

[9] GOST 23409.18-78 Sands and molding mixtures. Method for determining the clay component.

[10] GOST 29234.3-91 Molding sand. Method for determination of average grain size and uniformity coefficient.

[11] Types of defects, measures to prevent and correct it. Classification of casting defects [Electronic resource]: http://delta-grup.ru/bibliot/21/121.htm

[12] Kalinin, M. O., & Pavlenko, E. Y. (2015). Increasing the fault tolerance and availability of software defined networks using network equipment control based on multiobjective optimization by service quality parameters. Automatic Control and Computer Sciences, 49(8), 673-678. doi:10.3103/S014641161508026X

[13] Klochkov, Y. S., & Tveryakov, A. M. (2020). Approaches to the improvement of quality management methods. International Journal of Systems Assurance Engineering and Management, 11, 163-172. doi:10.1007/s13198-019-00939-x

Using material science tools to monitor of the qualitative condition of hull steel of ships and vessels

G T Kazanov[1], M V Nagaeva[2] and A E Romanuk[1]

[1]Pacific Higher Naval School named after S.O. Makarova Kamsky per., 6, Vladivostok, e-mail: kazgentim1@mail.ru
[2]Far Eastern State Technical Fisheries university, e-mail: NagaevaMV@mail.ru

Abstract.
The article discusses two methods for monitoring the structural state of steel in ship hulls. The first method is based on the use of fragments extracted from the longitudinal stiffeners of the deck. The second method of monitoring of the technical condition of the ship's hull involves using overhead sheets as "surveillance specimens" of structural damage to the hull steel, contacting (jointly) working with the hull structures, in which the highest stresses from the transported goods and waves are developed. The total number of surveillance specimens should be so that it is possible to organize systematic monitoring of the structural damage that gradually accumulates in the hull steel throughout the entire operational period of the vessel. During the operation of a ship or a vessel, hull steel is exposed to time-varying wave loads and degrades. This is reflected in the growth of structural damage to the hull steel, the steel gradually looses, and its quality state is changed.

1. Introduction
According to Wikipedia [10], many experts suggest that most of the Arctic water space will be completely free of ice in the 21st century summer, and this opens up new perspectives for sea transportation of goods. Intensive operation of ships and vessels in this and other areas is a real prospect that causes increased attention to issues in the design, manufacture, operation of ships and vessels, including ice class soon. From this point of view, the future practical application, as well as the subsequent study of such issues, has reason to be considered as very important and highly demanded.
The state of our environment is of great importance for the Earth's climate system. Its pollution can contribute to the development of very serious negative consequences. Therefore, the ice of the Arctic reflects the sun's rays and thus prevents the planet from overheating. They play a large role in the systems of water circulation in the oceans, but the area of ice cover has been rapidly decreasing in recent years. The operation of ships and vessels, other structures with the danger of large scale pollution should not enhance the dynamics of these processes.

2. Methods and tools for monitoring of the technical condition of steel hulls
An unambiguous relationship between the structure of a material and its resistance to fracture has been proven [1, 2]. Therefore, when the total and the residual resources of the ship's hull and its structures are forecasted, it is necessary to have appropriate ideas about the structural damage and the structure of the hull steel.
In [1], two methods of monitoring of the structural state of steel in ship hulls are considered. The first method is based on using fragments extracted from the longitudinal ribs of the deck stiffness,

participating in the general buckling of the ship's hull and providing its longitudinal strength. The removed fragments are replaced with new ones, which are painted in bright light colors: white or yellow, indicating the date of the control fragment removal for the diagnosis of the structural state by metallography methods.

The second method of monitoring of the technical condition of the ship's hull involves using overhead sheets as "surveillance specimens" of structural damage to the hull steel, contacting (jointly) working with the hull structures, in which the highest stresses from the transported goods and waves are developed. In the appointment of the sizes of "surveillance specimens ", their peculiarities of working under load as part of the ship's hull and during fatigue tests, as well as the possibility of making specimens for metallographic studies and static tensile tests, are taken into account. Places of installation of surveillance specimens in the ship's hull are structures located in the middle part of the ship's hull, sheer strake, deck stringer, cargo hatch coamings (Fig. 1) [1, 8].

The total number of surveillance specimens should be so that it is possible to organize systematic monitoring of the structural damage that gradually accumulates in the hull steel throughout the entire operational period of the vessel. The frequency of checks of changes in the structural state of the hull steel every 5 years predetermines the number of surveillance specimens of 4 - 6, depending on the selected order of diagnostics, placed symmetrically on the left and right sides of the ship.

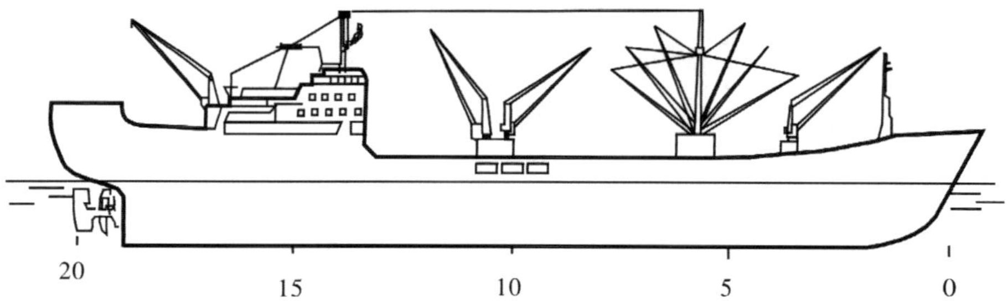

Figure 1. Installation places of surveillance specimens of structural damage to the steel of the ship's hull symmetrically for each side, depending on the chosen order of diagnostics.

Specially prepared schedules set the required dimensions of shell doubler, working as surveillance specimens of structural damage to the shell plate, with longitudinal and transverse set systems.

To service ships, the hulls of which include surveillance specimens, special research centers should be created in the east and west of our country, studying, through surveillance specimens, the quality condition or the degree of degradation of the hull steel of ships and vessels.

The proposed methods relate to practical methods for investigating the reliability and forecasting the technical state of a ship's hull for ships with a length of 80 meters or more using non-destructive diagnostics based on a systematic study of structural damage to hull steel caused by material fatigue. In the practice of operating ships and vessels, they are not used yet. The lack of the necessary information support for navigators, technical services for the fleet operation does not allow making informed decisions; conditions for catastrophic accidents are created [4].

During operation, the ship's hull experiences a cyclic action of wave loads, and changes in the structural damage of the hull steel are caused by the manifestation of material fatigue. These changes are difficult to track and quantify, despite new acoustic, electronic methods have appeared opening up wider possibilities in technical diagnostics. By tracking fatigue structural damages over time, a selective function of these damages is obtained, the consideration of which makes it possible to describe the process of accumulation and development of structural damages in hull steel at different stages of its operation. Ensuring the safe operation of ship hulls should be determined by the permissible level of structural damage to the hull steel, this requires appropriate technical control and diagnostics. At the same time, it can be said that now they are not only far from perfect, but practically absent.

3. Improving quality control tools for ship hulls and vessels

Fatigue failures of materials in structures, especially material-intensive ones, predetermined the creation of new diagnostic tools in the form of special devices [6] of portable devices [5]. These devices include the electronic small-sized portable TEMP-2, designed for Express measurement of the hardness of steels, alloys, and their welded joints on the Brinell (HB), Rockwell (HRC), shore (HSD), Vickers (HV) scales, as well as determining the tensile strength (kgf/mm2) according to GOST 22761-77 for carbon steels of the perlite class.

The control of the structural steel state of ships and vessels using materials science and, in particular, metallography is supplemented by classical methods of coercive force [3]. To carry out such studies, a portable coercimeter and a metallographic microscope are used [5].

The main advantage of such approaches in improving the tools for monitoring of the technical condition of the ships and vessels' hulls is the possibility of an operational assessment of the mechanical characteristics of steel without putting the ship out of action and without cutting out specimens from its hull. Thus, the unification of diagnostic measurements and their output to a new level of efficiency is realized.

The appearance of non-contact mine and torpedo weapons, and then magnetic detectors (magnetometers) of submarines in a submerged position, reacting to the magnetic field of the ship, led to the development and creation of methods and tools for protecting ships. The main method of passive protection is ship demagnetization.

Demagnetization is performed in two different ways - winding and without winding. These names should be understood as conventional, since the demagnetization of ships is performed by one or the other method using windings supplied with current (Fig. 2).

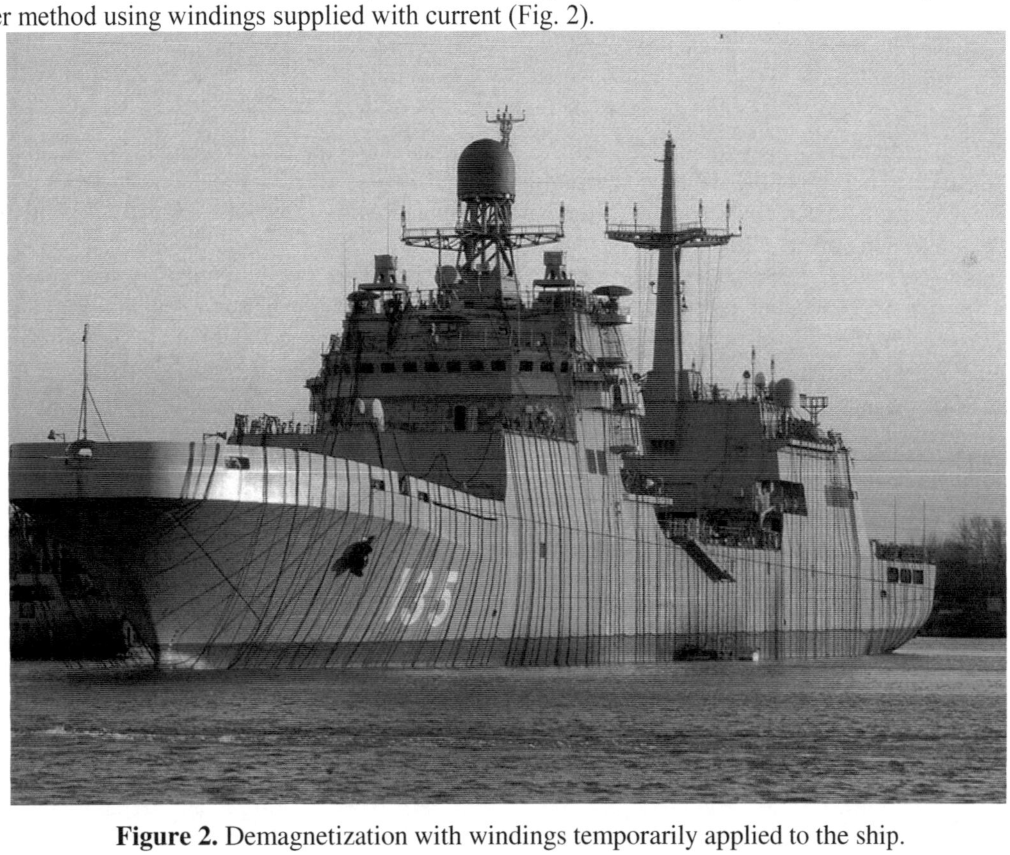

Figure 2. Demagnetization with windings temporarily applied to the ship.

In the first case, the windings are applied to the ship's hull temporarily, only for the period of demagnetization, or they are generally placed outside the ship, on the ground. When the second method is used, the windings are permanently mounted on the ship and turned on while traveling through hazardous areas.

The essence of demagnetization or electromagnetic processing (EMO) is to create in a certain way a magnetic field, opposite in sign to the ship's field. The process of such an artificial decrease in the magnetic field of the ship gives an idea of the tension or coercive force, which increases with the exhaustion of the metal resource or its degradation.

Coercimetric diagnostics [3], as well as diagnostics using overhead sheets [1, 2] performing the functions of “surveillance specimens”, are carried out on existing structures. The magnitude of the coercive force of the controlled metal depends on its current state. Therefore, the methods under consideration of all non-destructive testing ones closely reflect the mechanical state of the material of a real structure. Corrections in strength calculations based on these methods will make such calculations most adequate to the true current state of the hull steel.

Currently, calculated dependencies have been obtained that determine the endurance limit for some characteristics of the metal, including the characteristics of its structural damage and structural state [1, 2].

The well-known classical fatigue tests [9] (Fig. 3) allow simulating these processes and preparing them for use in practice.

In the test scheme for circular bending of a cantilevered specimen (Fig. 4), the head of a cylindrical specimen is fixed in a cartridge. At the other free end of the specimen, there is a ball bearing suspension that creates a vertical load. The bending moment will be distributed unevenly (in the form of a shaded triangle), and the maximum value of the moment is reached at the base of the console (Fig. 3, section a — a). The greatest alternating stress σmax in this section is obtained by the formula

$$\sigma_{\max} = \frac{Pl}{W} = \frac{32Pl}{\pi d^3} \tag{1}$$

where P — vertical load; l,d — shoulder and a cross-sectional diameter of the specimen; W — a moment of resistance to bending of the specimen section.

Hardness measurements in this case are considered as preferable ones, since they directly depend on the degree of loosening of the steel.

Alternating stresses in this test are achieved by rotating the specimen around its axis together with the chuck. Rotation provides a sinusoidal voltage change from to for each complete revolution of the specimen with a constant vertical load P. The level of alternating stresses varies along the length of the specimen from 0 to the right of the loading bearing to the maximum at the base of the console. The structural damage of the metal specimen will change accordingly. If the degradation of the hull steel of ships and vessels is studied, the specimen is made from it.

When test results are processed, the specimen is divided into cylinders of small height, they are numbered, preferably from the free end of the specimen, the number indicates the position of the cylinder in the specimen during testing. Each such a part of the specimen is subjected to metallographic studies, measurements with a coercimeter, and a hardness tester.

According to the metallographic parameter of structural damage of 3% the probability is $l_{3\%}$, the results of measurements with a hardness tester are set according to the formulas [1, 2], the corresponding fatigue limits. All results of measurements, calculations are displayed graphically: $l_{3\%}$ is plotted along the horizontal axis, together with the number of hardness and coercive force, they are grouped for each selected part of the specimen; vertical axis fatigue limits.

IOP Conf. Series: Materials Science and Engineering **986** (2020) 012011 doi:10.1088/1757-899X/986/1/012011

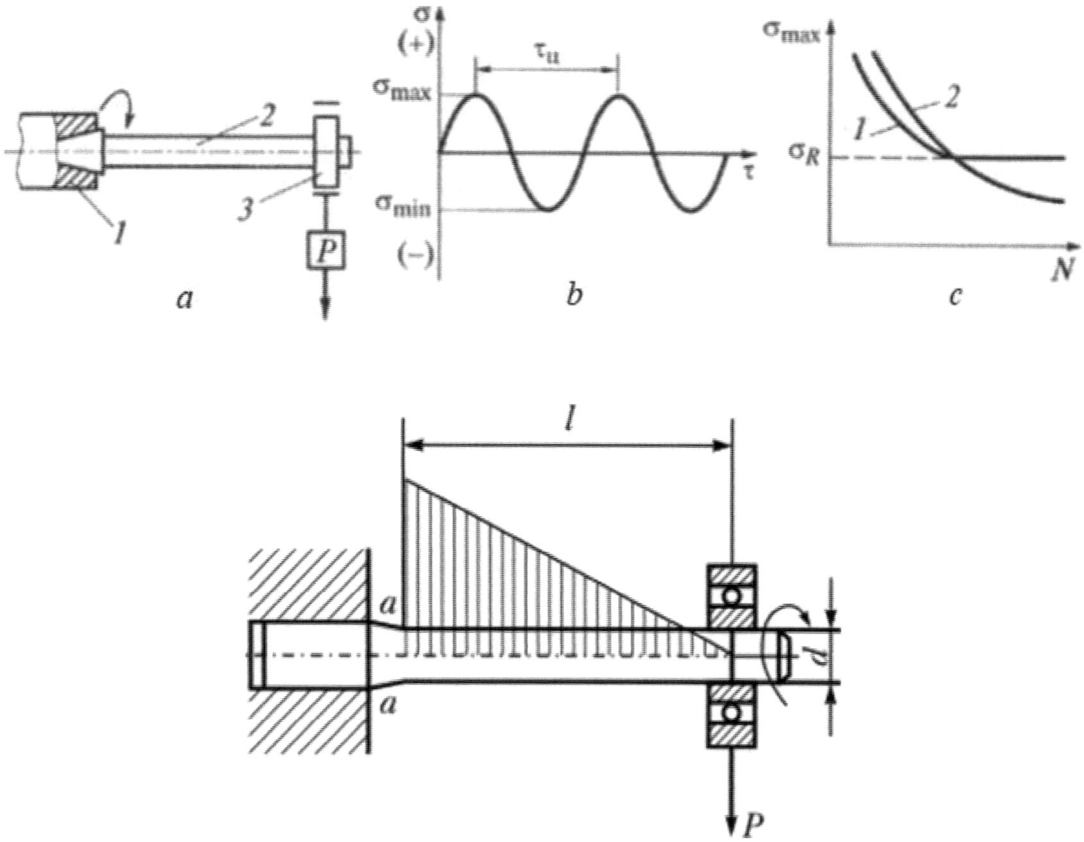

Figure 3. Fatigue test scheme.
a — specimen loading scheme: *1* — rotating spindle (arrow shows the direction of rotation); *2* — specimen; *3* — loading bearing; *P* — load gravity; b — cyclical change of voltage in the specimen: σ - voltage; τ - time; σ_{max} and σ_{min} - maximum and minimum cycle voltage respectively; τ_c — cycle time; *c* — fatigue diagram for materials with (1) and without (2) physical endurance limit σ_R

4. Conclusion

During the operation of a ship or a vessel, hull steel is exposed to time-varying wave loads and degrades. This is reflected in the growth of structural damage to the hull steel, the steel gradually looses, and its quality state is changed. Such qualitative changes in the hull steel determine the technical condition of the ship's hull and require system control to ensure its reliability and survivability.

The combination of metallography in the control of the quality state of the hull steel of ships and vessels match the values of the coercive force, hardness numbers with the endurance limit, metallographic characteristics of the metal, including the characteristics of its structural damage and structural state.

The effectiveness of the informative parameters of the devices allows supplementing and automating the classical diagnostics of the technical condition of the ships and vessels' hulls. A real opportunity is created to move to a strategy of reasonable and timely repair measures, corrections of operating conditions.

The operability rises and the cost of diagnostics decreases by reducing the volume of flaw detection, increasing the accuracy of conclusions about the current technical condition of the ships and vessels' hulls, and the forecast of the development of this state.

The economic effect is achieved by the prevention of fatigue accidents of hull structures, which are practically inevitable today because of the lack of monitoring of the technical condition of the steel hulls of ships and vessels by fatigue type.

References

[1] Kazanov G T, Novikov V V and Turmov G P 2014 Stress concentration and other features of the stress state of ship hull structures: monograph (scientific. ed. G.Yu. Illarionov Vladivostok: Far Eastern Federal University) p 178

[2 Turmov G P, Kazanov G T and Novikov V V 2015 Stress concentration in ship hull structures: monograph (scientific. ed. G.Yu. Illarionov Saarbrucken, Deutchland, Verlag, LAP LAMBERT Academic Publishing) p 212

[3] Bezlyudko G Ya, Popov B E, Solomakha R N and Karabin V V 2014 The main features of the method of coercive force as a new level of efficiency and culture of tracking fatigue and resource of metal structures and equipment *In the world of non-destructive testing* (№ 3 (65)) pp 67–73

[4] Casanov G T 2015 Justification of the terms and modes of operation in solving the problems of a trouble-free technical condition of ship hulls *Fundamental and applied problems of natural science: materials of the 58th All-Russian Scientific and Methodological Conference* (vol. III Vladivostok, TOVVMU them. S.O. Makarova) pp 117–121

[5] Romanyuk A E 2020 Control of technical condition of hulls of ships and vessels using materials science *International competition of military scientific works of students and cadets of higher educational institutions of the member states of the Commonwealth of Independent States dedicated to the 75th anniversary of Victory in the Great Patriotic War of 1941-1945* (Scientific adviser Kazanov G.T. Vladivostok, TOVVMU them. S.O. Makarova) p 24

[6] Romanyuk A E 2020 Control of the technical condition of the hulls of ships and vessels with instruments based on lux meters *Abstracts of the conference VNOK* (Scientific publication. Scientific adviser Kazanov G.T. Vladivostok, TOVVMU them. S.O. Makarova)

[7] Development of proposals for the development of solutions for the release of BMT "V. Kolechitskiy " 2005 *at sea according to the protocol of the decision of the technical meeting of the State Technical University of the Navy dated July 14, 2005 Report on the research work of FSUE TsNII them. A.N. Krylov. Agreement with the Federal State Unitary Enterprise "178 SRZ" of the Ministry of Defense of the Russian Federation* (No. 113-31 / 05. Stage number 1. -Saint Petersburg) p 117

[8] Patent No. 2475404 (RU); IPC B63B 9/00 (2006.01); priority 11/15/2011. Method for assessing the technical condition of the ship's hull / Kazanov G.T. (RU) // print. Inventions. Useful models: Official Bulletin of Rospatent (M .: FIPS, 20.02.2013. - No. 5)

[9] https://studme.org/243126/tehnika/ispytaniya_ustalost

[10] https://ru.wikipedia.org/wiki

[11] Kazanov G T and Nagaeva M V 2011 Methodological features of determining the resource of steel of ship hulls / *Fundamental and applied issues of natural science: Materials of the 54th All-Russian Scientific and Methodological Conference, December 8* (vol. III. - Vladivostok: TOVMI VUNC Navy "VMA") pp 99–102.

IT security in special purpose networks

N.V. Kireeva[1], T.V. Filippova[1], O.A. Karaulova[1] and M.V. Aleshin[2]

[1]Volga State University of Telecommunications and Informatics Samara, Samara, Russian Federation, zeppelinsn@yandex.ru
[2]National Technology Initiative Center for Advanced Manufacturing Technologies based on the Institute of Advanced Manufacturing Technologies of Peter the Great St. Petersburg Polytechnic University Polytechnicheskaya, 29, St.Petersburg, 195251, Russia

Abstract. The article highlights the problem of ensuring secure communications in special purpose networks, in particular, the insufficient status of domestic telecommunication services market development. To ensure secure communication the measures must be sufficient for the strict functionality and information security requirements and they must have the respective certificates. Certificating is a time taking process and the issue of using foreign made items remains the issue of the day. The article covers burning issues of cyberattack countermeasures, the ways of solving IT security issues on essential IT segments of a special purpose network.

Keywords: cyber security/ IT security, dedicated network, essential IT segments, cyberattack countermeasures

1. Introduction

The most important component of information security technology is the cyberattack countermeasures on essential IT segments. Essential IT segments are those subsystems and research institute objects, which breach of database causes significant damage to the economics, national security and national interests of the country.

1.1 Secure communications ensuring in special purpose networks

Dedicated networks are needed for government authorities, the national defense demands, national security and enforcement of law and order. The need of integration of various types of services, such as telephone, fax communications, video communication, data transfer, has led to the plain old telephone service and the multiservice communication networks development. Traditional public telecommunications are built on the basis of TDM technology (Time Division Multiplexing) [1]. At the present stage of the telecommunications development digital TDM-PBX (Private branch exchange) is losing its market dominance due to the possibility of providing a wide range of services from systems based on packet switching mode while data transferring [2].

IP packet switching is implemented in the so-called Next Generation Networks (NGN), which are designed to provide telecommunication services and to use several wideband transport technologies with QoS on. The basis of NGN networks, originally NGN IPCC (International Packet Communication Consortium), are flexible switches such as Soft switch and IP-ATS. NGN is an upgraded TDM network with the ability to transmit IP traffic, as well as with additional capabilities to provide services to the end user.

Currently TISPAN is replacing the IPCC in NGN, which key element is IMS (IP Multimedia Subsystem). The main difference between NGN TISPAN (IMS) and NGN IPCC is the flexibility and scalability of such networks. Here the session control and routing control functions are performed by the CSCF (Call Session Control Function), which replaced the Soft switch. On the one hand, NGN IPCC / TISPAN (IMS) networks provide a wide range of additional services (so-called VAS – value added services) that provide more efficient network management capabilities comparing to the TDM architecture. On the other hand, this range of functionality increases the number of insecurities and gives rise to sources of information security threats (Fig. 1).

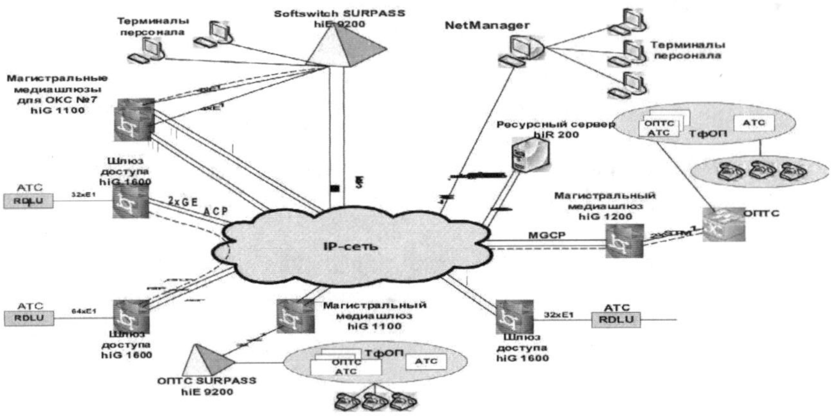

Figure 1. NGN project implementation

Taking into account the nature of the information transferred through the dedicated network, the issue of using secure patching facilities is especially relevant [4]. In order for the means of communication to be used in dedicated network it must meet the requirements of the Federal Security Service of Russia (FSB), FSTEC (Federal Service for Technology and Export Control) or the Ministry of Defense of Russia and must have the appropriate certificates. The security requirements for ATS and ATS control systems has a number of restrictions on advanced features of equipment, hacking protection requirements and signaling if it occurs, event registration (logging), access control and management.
Consequently, the certification of IP ATS, including those currently available on the market, is quite a challenge taking into account the above mentioned security requirements.
The issue of ensuring secure communication in special-purpose networks today is the issue of the day, it requires the development of telecommunication market (especially the domestic communications industry development), which must meet strict requirements of functionality and information security.

1.2 Cyberattack countermeasures in special purpose networks.

The main tasks of essential IT segments are: — master data transmit and receive for computational analysis; — logging, data storage and delivery of information at control point; — delivery, logging and management data delivery to control centers subscribers; — information exchange between subscribers of points and control centers; — organization of operational interaction and management between essential IT segments through communication channels. The topology, structure and functions of essential IT segments are formed on the basis of unified information and telecommunication facilities in accordance with the requirements for controlled objects [3,4]. Methods and protocols of information interaction of essential IT segments subscribers for each specific control object are implemented by organizing communication with subscribers, combining measuring, logging, data processing and transmit. Each essential IT segment can have an optional switching mode, which considers the specificity of hardware and software.
Currently under conditions of cyberattacks, the effectiveness of essential IT segments as the main important and controlling component of automated systems is largely determined by the stability of essential IT segments of logging, processing and transmission of information [5,6]. With the old-fashioned computer facilities and computer systems replacement in essential IT segments, cyberattacks became possible in principle because of critical areas in digital communication equipment and using

software in remote access mode through transparent networks (both for a system operator and attacker) based on standard data transfer protocols. The consequence of cyberattacks can be the control information blocking and the false data introduction, violation of logging fixed regulations, processing and transmission of information in control complexes, failures of essential IT segments, as well as discrediting of information received by consumers [7].

The applicability of the proposal development, which will neutralize the cyberattacks and at the same time maintain the stability of essential IT segments during the management production run, is due to the following factors:

— the intensive technologies development and cyberattacks implementation tools;
— danger of loss of essential IT segments stability during cyberattacks;
— the forced necessity to implement foreign-made hardware and software into essential IT segments;
— the impossibility to stop the cyberattacks increase by traditional methods and information protection tools;
— imperfection of cyberattacks countermeasures.

The development of cyberattacks countermeasures on essential IT segments is based on the multipurpose use of the information protection theory, the stability and reliability of automated systems, complex system modeling, automatic control, information quality and reliability, cyberattacks discovery.

Conclusion

The results of modern approaches analysis of information protection, ensuring the reliability and stability of essential IT segments functioning show that they have not fully solved the following issues:

— issues related to the methods and models development for countering cyberattacks on the tools of logging, processing and transmitting information from essential IT segments, intelligent tools for monitoring the state of the information computation process and the organization of an active cyberattacks countermeasures have not been considered;
— most of the known cyberattack detecting methods and models do not take into account the interconnected dealing with the cyberattacks countermeasures and the stability of essential IT segments during the management production run;
— there are practically no complex methods, models and algorithms where along with signature attacks analysis and essential IT segments anomalies detection the detection of attacks is carried out by functional analysis of controlled parameters and identification of stability state of essential IT segments;
— methods and models of decision making including statistical approach to ensure active cyberattack countermeasures are not well developed.

In addition, the currently used methods and models for countering cyberattacks on essential IT segments do not resolve a number of contradictions:

— between increasing the efficiency and stability of the control system, on the one hand, and decreasing of management processes under the influence of computer attacks on essential IT segments, on the other hand;
— between reducing the number of vulnerabilities factors of essential IT segments, requiring the elimination of extra functions and parameters, on the one hand, and complicating the information and computing monitoring process in essential IT segments and detecting attacks, on the other hand;
— between the requirements for the timeliness of the control technology cycle, on the one hand, and the time required to detect and neutralize the source of the attack, on the other hand.

One of the possible ways to create cyberattack protected essential IT segments is the essential IT segments information technologies integration with cyberattack countermeasures sensors set up in its structure, in compliance with the mandatory interaction of administrative tools with cyberattack countermeasures control programs.

At the same time, essential IT segments security will be achieved when the complex use of organizational and technical measures, the maximum use of administrative and network monitoring tools, antivirus tools, certified software, computer and switching equipment are used along with cyberattack countermeasures components.

Besides, it is necessary to generate a list of extreme monitored parameters of essential IT segments, to determine their sensitivity to cyberattacks, to experimentally verify the consistency of essential IT segments components, cyberattack countermeasures, means of protecting information from unauthorized access when performing real technological cycles of cyberattack control, and to develop (specify) technical solutions based on changes in the controlled parameters. This will help us to solve the issue of information security ensuring and information protection on essential segments of the special-purpose network.
We can distinguish two contradictory approaches to ensuring information security.
Russia like China sticks to the position of the need of total informational space demilitarization. According to the Russian Federation the arms race in the informational space can disrupt existing agreements on disarmament and international security.
The United States like the EU maintains the position that the main threats to cybersecurity are cyberterrorism and cybercrime, and the issues of interstate confrontation in cyber space should be regulated within the international humanitarian law.

References

[1]. Druk E.V. Fundamentals of organization and management of VPN-networks of modern information systems for special purposes // T-Comm: Telecommunications and transport. 2019.Vol. 13.No. 6.P. 19. (In Rissian)
[2]. Goldstein B.S. Communication networks: Textbook for universities. SPb.: BHV Saint Petersburg, 2018.S. 29. (In Rissian)
[3]. V.V. Karganov To prevent and ensure the required level of information security of a special-purpose information-computing network from unauthorized impact // Questions of defense technology. Series 16: Technical means of countering terrorism. 2018.No. 2.P. 78. (In Rissian)
[4]. Baklanov I.G. NGN: principles of construction and organization. M .: Eco-Trends, 2017.S. 28 (In Rissian)
[5]. Zegzhda, D. P., Lavrova, D. S., & Pavlenko, E. Y. (2020). Management of a dynamic infrastructure of complex systems under conditions of directed cyber attacks. Journal of Computer and Systems Sciences International, 59(3), 358-370. doi:10.1134/S1064230720020124
[6]. Zegzhda, D. P., & Stepanova, T. V. (2015). Approach to APCS protection from cyber threats. Automatic Control and Computer Sciences, 49(8), 659-664. doi:10.3103/S0146411615080179
[7]. Zegzhda, P. D., Zegzhda, D. P., & Stepanova, T. V. (2015). Approach to the construction of the generalized functional-semantic cyber security model. Automatic Control and Computer Sciences, 49(8), 627-633. doi:10.3103/S0146411615080192

IOP Conf. Series: Materials Science and Engineering **986** (2020) 012013 doi:10.1088/1757-899X/986/1/012013

Forecasting models of agricultural process based on fuzzy time series

V. E. Parfenova, G. G. Bulgakova, Yu. G. Amagaeva, K. V. Evdokimov, V. I. Samorukov

Saint Petersburg State Agrarian University, St. Petersburg, Russia. ekvmanager@mail.ru

Abstract. Today, the problem of increasing the validity and accuracy of forecasts based on the analysis of time series under conditions of uncertainty is very important. The models and methods used to predict the dynamics of agricultural processes are built on quantitative information and are implemented as part of a statistical approach. In this approach, time-based forecasting models are constructed on several requirements for the initial data, the main of which are the requirements of comparability, sufficient representativeness to reveal regularity, uniformity, and stability. Only keeping these requirements, uncertainty can be interpreted in terms of randomness and appropriate statistical forecasting methods can be applied. However, the real dynamic processes taking place in agriculture are represented by time series, for which these requirements are rarely feasible, due to the great uncertainty of the factors determining their dynamics. The problem of forecasting such series is particularly relevant for agricultural science and practice. The article touches upon the possibilities of using fuzzy modeling tools to predict the dynamics of processes in the agricultural sector.

1. Introduction

Agricultural production is mainly carried out under conditions of a high degree of uncertainty under market economy conditions. Uncertainty arises because of the influence of many factors, not only industrial, economic and social, characteristic of any production, but as a result of the influence of natural and biological factors to a large extent. The combined influence of all these factors leads to agricultural production indicators fluctuate both in space and in time. Therefore, the problem of making managerial decisions under uncertainty in the field of agricultural production is extremely important and relevant. One of these problems is increasing the validity and accuracy of forecasts.

The models and methods that are currently used in predicting agricultural processes taking place under uncertainty are built on quantitative information and are implemented as part of a statistical approach. At the same time, the statistics that serve as initial information for forecasting are either spatial (regression models) or temporal (time series models). Regression modeling of agrarian processes was discussed in previous work [1]. Here we will focus on time series and forecasting the dynamics of processes based on them.

In the statistical approach, the forecasting models are based on the number of source data requirements. The main requirements are comparability, sufficient representativeness to identify patterns, uniformity, and stability. Only observing these requirements, uncertainty can be interpreted in terms of randomness and the appropriate statistical forecasting methods can be applied. However, the real dynamic processes taking place in agriculture are represented by time series, for which these requirements are rarely feasible, which is associated with a large uncertainty of the factors determining their dynamics [2]. Non-statistical uncertainty of the time series of real dynamic processes is expressed in a combination of such properties as:

1) short row length (less than 30 - 50);
2) unsteady time-series behavior;
3) incompleteness and inaccuracy of the values of the series;
4) unknown probabilistic characteristics of the stochastic process describing the time series;
5) nonlinear nature of the desired relationship between the levels of the series.

Given these properties, time series is said to have a high degree of uncertainty, reflecting the dynamics of poorly structured processes. [3]. The problem of forecasting precisely such series is particularly relevant for agricultural science and practice, since traditional statistical approaches to solving this problem are not suitable.

2. Problem definition and decision methods

A new direction has been actively developing the last two decades on the basis of methods and models of artificial intelligence, in particular, using fuzzy models. The development of this trend in relation to time series with a high degree of uncertainty has led to the creation of fuzzy time series models, which are based on fuzzy time series [4].

In 1993, the Song and Chissom first defined fuzzy time series models in [5, 6] and discussed their practical use. From that moment, the formation of the theory and practice of modeling fuzzy time series (FTS) began, which was developed in the works of such foreign scientists as S. Chen, Q. Tanaka, M. Sah, V. Pedrich, V. Novak, A. Jilani, Q. Hwang et al. Among domestic scientists in the field of fuzzy time series models are the studies of N. Yarushkina, I. Batyrshin, S. Kovalev, Yu. Kudinov, K. Degtyarev and others.

Fuzzy time series (FTS) is defined as a sequence of observations ordered at equally spaced points in time if the values that a certain value takes at the observed point in time are expressed using a fuzzy label. A fuzzy label is a fuzzy set defined on a universal set.

Two possibilities occur in practice: historical data (available statistics) is set either in the form of linguistic quantities or as real numbers. In both cases, the available data can be represented by fuzzy time series based on the fuzzy approximation theorem [7].

Currently, various models of fuzzy forecasting are proposed. Fuzzy first-order models are considered in this article. The first-order FTS model is defined by the equation:

$$F(t)=F(t-1)\circ R(t,t-1), \tag{1}$$

where F(t) – the forecast value of the indicator during a time period t, F(t–1) – the real value during a time period t–1, R – fuzzy relationship on successive instants.

There are two basic models among the set of fuzzy models proposed by various authors for forecasting, which differ by computational methods for determining the fuzzy ratio R in the model (1). The first model involves calculating R using the operation of fuzzy sets defined as the Mamdani implication. The second model is based on using arithmetic operations [8,9].

The common characteristic of most fuzzy approaches is the sequential performance of the main three procedures:

1) definition of the universe and fuzzification of historical data;
2) identification of internal fuzzy relationships and the definition of their groups;
3) finding fuzzy forecasts, and their defuzzification.

In this article, the basic basis of fuzzy forecasting is the method proposed in [7], using the second model.

Let us consider the procedure for fuzzy time series forecasting on a specific example of a time series of potato yields. Potatoes are one of the leading crops in Russian agricultural production. Potato is not only a food product, but also an industrial raw material and feed for agricultural animals. Russia currently lags behind developed countries in terms of potato yields. However, the experience of these countries shows that Russia has good prospects for the development of this industry [10]. The correct forecast of the most important indicators of the potato industry, in particular, productivity, is important. So let us turn to the example.

The example. There are historical data on potato productivity in agriculture of the Russian Federation (Table 1) [11].

Table 1. Potato yields.

Year	c/ha	Year	c/ha
2005	124	2012	136
2006	133	2013	147
2007	132	2014	153
2008	138	2015	164
2009	144	2016	158
2010	100	2017	163
2011	150	2018	170

The traditional model for forecasting time series is first applied to conduct a comparative analysis. Let us consider the simplest model of 1st-order autoregression:

$$y=ay_{t-1}+\varepsilon, \qquad (2)$$

where t – the period number, y_t (y_{t-1}) – yields during a time period t (t–1), a – the model parameter, ε – the random component.

Parameter a was calculated using the least-squares method, which gave the following results:

$$a=1{,}0158 \rightarrow \hat{y}=1{,}0158\cdot t. \qquad (3)$$

The parameter turned out to be greater than unity (a>1) in the model (3), which indicates the unsteady behavior of the investigated time series. The average forecasting error rate (AFER) was used to assess the quality of the constructed equation (3):

$$AFER=\frac{\sum_{i=1}^{n}\left|(y_i-\hat{y}_i)/y\right|}{n} \qquad (4)$$

where y_i – the real value of the indicator at the *i-th* time, $\hat{y}_i$ – the calculated value at the same time, n – the number of observations.

The average forecasting error rate for the model (3) was 9.8%.

Now let us turn to the construction of a fuzzy time series model. In the fuzzy version of the model (3), there corresponds a fuzzy model of the first order (1):

$$F(t)=F(t-1)\circ R(t,t-1). \qquad (5)$$

To build the forecast, a method based on the analysis of the fuzzy time series, proposed in [8], was chosen. Here is a step-by-step description of the algorithm for constructing a fuzzy time series and obtaining predictive estimates based on it.

Step 1. Definition of a universal set U, on which all fuzzy sets will be determined. For this, the maximum (D_{max}) and minimum (D_{min}) values are determined on the set of historical data. Based on them, a universal set $U=[D_{min}-D_1, D_{max}+D_2]$ is constructed, where D_1, D_2 are two positive real numbers. In our example, $D_{min}=100$, $D_{max}=170$. For the convenience of calculations, we take $D_1=4$, $D_2=6$, then $U=[100–4, 170+6]=[96, 176]$.

Step 2. The choice of the number of intervals and the partition of the universal set into intervals of equal length u_i.

The following method was proposed to select the optimal value for the length of the interval. The first differences are calculated based on table 1. The average modulus of the difference between most of the

obtained difference values is approximately 8 units. We accepted this value as the length of the intervals. When choosing such an interval length, the number of intervals is (176 – 96)/8=10.
The following partition U was obtained: u_1=[96, 104], u_2=[104, 112], u_3=[112, 120], u_4=[120, 128], u_5=[128, 136]. u_6=[136, 144], u_7=[144, 152], u_8=[152, 160], u_9=[160, 168], u_{10}=[168, 176].
The average values of the intervals were calculated to perform the following steps: u^1_{av}=100, u^2_{av}=108, u^3_{av}=116, u^4_{av}=124, u^5_{av}=132, u^6_{av}=140, u^7_{av}=148, u^8_{av}=156, u^9_{av}=164, u^{10}_{av}=172.

Step 3. The definition of fuzzy sets A_i (i=1÷10) on the universal set U. A_i is determined using the formula: $A_i=(d_{i-1}, d_i, d_{i+1}, d_{i+2})$, i=1÷10, where d_{i-1} – the lower bound u_i reduced by the length of the interval, d_i is the lower bound u_i, d_{i+1} is the lower bound u_{i+1}, d_{i+2} is the upper bound u_{i+1}. Moreover, A_i (i=1÷10) has a linguistic interpretation (from a very small value to a very large value), i.e. are the linguistic terms of the set U. The following fuzzy sets are obtained:

A_1 = [88, 96, 104, 112],
A_2=[96, 104, 112, 120],
A_3=[104, 112, 120, 128],
A_4=[112, 120, 128, 136],
A_5=[120, 128, 136, 144],
A_6=[128, 136,144, 152],
A_7 = [136, 144, 152, 160],
A_8= [144, 152, 160, 168],
A_9= [152, 160, 168, 176],
A_{10}=[160, 168, 176, 184].

Step 4. Fuzzification of historical data. It is carried out according to the rule: if the historical value falls into the interval u_i, i=1÷10, then it belongs to the fuzzy set A_i. The results of historical data belonging to fuzzy sets are presented in Table 2.

Table 2. Identification of historical data with fuzzy sets.

Year	c/ha	A_i	Year	c/ha	A_i
2005	124	A_4	2012	136	A_6
2006	133	A_5	2013	147	A_8
2007	132	A_5	2014	153	A_8
2008	138	A_6	2015	164	A_{10}
2009	144	A_7	2016	158	A_9
2010	100	A1	2017	163	A9
2011	150	A8	2018	170	A10

Step 5. The formation of fuzzy logical dependencies. To build dependencies (relations), sequential fuzzified data of the table are considered in Table 2.
If the value of the time series $F(t – 1)$ is fuzzy as A_j and $F(t)$ as A_k, then A_j is associated with A_k. This relationship is denoted as $A_j \rightarrow A_k$, where A_j is the current state and A_k is the following one.

Table 3. Fuzzy logical dependencies.

$A_4 \rightarrow A_5$	$A_5 \rightarrow A_5$	$A_5 \rightarrow A_6$	$A_6 \rightarrow A_7$	$A_7 \rightarrow A_1$	$A_1 \rightarrow A_8$	$A_8 \rightarrow A_6$
$A_6 \rightarrow A_8$	$A_8 \rightarrow A_8$	$A_8 \rightarrow A_{10}$	$A_{10} \rightarrow A_9$	$A_9 \rightarrow A_9$	$A_9 \rightarrow A_{10}$	

Step 6. Formation of groups of fuzzy logical dependencies. Dependencies with the same left parts are grouped.

Table 4. Fuzzy link groups.

№	Fuzzy link	№	Fuzzy link	№	Fuzzy link	№	Fuzzy link
1.	$A_4 \to A_5$	3.	$A_6 \to A_7$, A_8	5.	$A_1 \to A_8$	7.	$A_9 \to A_9$, A_{10}
2.	$A_5 \to A_5$, A_6	4.	$A_7 \to A_1$	6.	$A_8 \to A_6$, A_8, A_{10}		

Step 7. Definition of fuzzy forecasts and their defuzzification. If $F(t-1)=A_j$, then forecasting $F(t)$ is performed using the following heuristic rules.
Rule 1. If the group of heuristic fuzzy logical dependencies (FLD) is empty, i.e. $A_j \to 0$, then the forecast estimate is defined as A_j, i.e. midpoint of the interval $[u_{j-1}, u_j, u_{j+1}, u_{j+2}]$.
Rule 2. If the group of heuristic fuzzy logical dependencies is displayed "one to one", i.e. $A_j \to A_p$, then the forecast estimate is found as A_p, i.e. defined by the midpoint of the interval $[u_{p-1}, u_p, u_{p+1}, u_{p+2}]$.
Rule 3. If a group of heuristic FLD displays "one to many", then the predictive estimate is found as $\frac{A_{p_1} + A_{p_2} + \ldots + A_{p_k}}{k}$, i.e. calculated as the midpoint of the interval

$$\left[\frac{u_{p_1-1} + u_{p_2-1} + \ldots + u_{p_k-1}}{k}, \frac{u_{p_1} + \ldots + u_{p_k}}{k}, \frac{u_{p_1+1} + \ldots + u_{p_k+1}}{k}, \frac{u_{p_1+2} + \ldots + u_{p_k+2}}{k}\right]$$

Observed and predicted values are shown in Table 5.

Table 5. Historical and projected potato yields.

Year	The historical value (c/ha)	Predictive estimate (c/ha)	Year	The historical value (c/ha)	Predictive estimate (c/ha)
2005	124		2013	147	148
2006	133	132	2014	153	148
2007	132	132	2015	164	156
2008	138	136	2016	158	164
2009	144	148	2017	163	164
2010	100	100	2018	170	168
2011	150	156			
2012	136	156		Forecast for 2019	172

To assess the quality of the forecast, the average forecasting error rate (AFER) was used (similar to formula (4)):

$$AFER = \frac{\sum_{i=1}^{m} |(F_i - T_i)| / T_i}{n} \cdot 100\%, \tag{6}$$

where F_i and T_i are the predicted and real values for the i-th period, n is the number of values of the time series. The lower this estimation, the better the model.

Table 6. Calculation of the average forecasting error.

Year	Historical data	Forecast value	Average error
2006	133	132	0,00752
2007	132	132	0
2008	138	136	0,01449
2009	144	148	0,02778
2010	100	100	0
2011	150	156	0,04
2012	136	156	0,14706
2013	147	148	0,00680
2014	153	148	0,03268
2015	164	156	0,04878
2016	158	164	0,03797
2017	163	164	0,00613
2018	170	168	0,01176
Total			0,38098
AFER			0,38098/13· ·100=2,9%

The average forecasting error rate is 2.9%.

Forecasting results. A time series of forecast values were obtained as a result of the study of the fuzzy model. The maximum forecast error was approximately 3%, which is three times lower than in the forecast according to the classical autoregression model. In addition to the above calculations, fuzzy time series constructed with a different number of intervals u_i were analyzed. The calculation results are given in Table. 7. The best predictive model was the fuzzy model with the number of intervals equal to 10.

Table 7. Quality rating of different models.

Rating	Autoregression	u=7	u=8	u=9	u=10
AFER	9,8%	4,8%	4,4%	3,7%	2,9%

Additional calculations of the same example using the first model were performed in order to make sure that the fuzzy model (the second model [8]) gives better results compared to the model based on the Mamdani implication.

Let us show the results obtained when dividing the universal set into seven intervals. The algorithm also consists of several steps, but the computational operations are different.

Step 1. Universal set U=[96, 180]

Step 2. Partitioning a universal set into intervals:

u_1=[96, 108], u_2=[108, 120], u_3=[120, 132], u_4=[132, 144], u_5=[144, 156], u_6=[156, 168], u_7=[168, 180] with average values:

$u^1_{av}=102$, $u^2_{av}=114$, $u^3_{av}=126$, $u^4_{av}=138$, $u^5_{av}=150$, $u^6_{av}=162$, $u^7_{av}=174$.

Step 3. Fuzzy sets A_i (i=1÷7) on the universal set U are determined by the rule: $A_r=0{,}5/u_{r-1}+1/u_r+0{,}5/u_{r+1}$, r=1÷m is the number of intervals (other terms have zero membership function values). For m=7:

$$A_1=1/u_1+0{,}5/u_2+0/u_3+\ldots+0/u_6+0/u_7,$$
$$A_2=0{,}5/u_1+1/u_2+0{,}5/u_3+0/u_4+\ldots+0/u_7,$$
$$\ldots\ldots\ldots\ldots\ldots\ldots\ldots\ldots\ldots\ldots\ldots$$
$$A_7=0/u_1+0/u_2+\ldots+0/u_5+0{,}5/u_6+1/u_7.$$

Step 4. Fuzzification of historical data. It is carried out in two stages. First, the belonging of historical data to the fuzzy sets A_i is determined (Table 8). Then we define real data as fuzzy sets with membership degrees determined from the formula (7). The results are shown in Table 9.

$$\mu_i(x_t)=\frac{1}{1+k(x_t-u_{av}^i)^2}; \quad (7)$$

where x_t — the actual value of the time series during a time period t;

u_{av}^i – the midpoint of interval i;

k (in the example k=0,01) is a positive constant reflecting the membership of x_t in the interval u_i.

Table 8. Identification of historical data with fuzzy sets.

Year	c/ha	A_i	Year	c/ha	A_i
2005	124	A_3	2012	136	A_4
2006	133	A_4	2013	147	A_5
2007	132	A_4	2014	153	A_5
2008	138	A_4	2015	164	A_6
2009	144	A_5	2016	158	A_6
2010	100	A_1	2017	163	A6
2011	150	A_5	2018	170	A7

Table 9. Degrees of belonging of historical data to fuzzy sets.

Year	A_1	A_2	A_3	A_4	A_5	A_6	A_7
2005	0,7	0,9	1	0,8	0,6	0,4	0,3
2006	0,5	0,7	0,9	1	0,8	0,5	0,4
2007	0,5	0,7	1	1	0,7	0,5	0,4
2008	0,4	0,6	0,9	1	0,9	0,6	0,4
2009	0,4	0,5	0,7	1	1	0,7	0,5
2010	1	0,8	0,6	0,4	0,3	0,2	0,1
2011	0,3	0,4	0,6	0,9	1	0,9	0,6
2012	0,5	0,7	0,9	1	0,8	0,6	0,4
2013	0,3	0,5	0,7	0,9	1	0,8	0,6
2014	0,3	0,4	0,6	0,8	1	0,9	0,7
2015	0,2	0,3	0,4	0,6	0,8	1	0,9
2016	0,2	0,3	0,5	0,7	0,9	1	0,8
2017	0,2	0,3	0,4	0,6	0,8	1	0,9
2018	0,2	0,2	0,3	0,5	0,7	0,9	1

Step 5. Based on the results of table 8, fuzzy logical relationships are formed: $A_3 \rightarrow A_4$; $A_4 \rightarrow A_4$; $A_4 \rightarrow A_1$; $A_1 \rightarrow A_5$; $A_5 \rightarrow A_4$; $A_4 \rightarrow A_5$; $A_5 \rightarrow A_5$; $A_5 \rightarrow A_6$; $A_6 \rightarrow A_6$; A6$\rightarrow A_7$.

Step 6. The formation of the resulting relationship R is the union of the logical relationships obtained in step 5.

Let us denote: $R_1=A_3^T$ x A_4; $R_2=A_4^T$ x A_4; $R_3=A_4^T$ x A_1; $R_4=A_1^T$ xA_5; $R_5=A_5^T$ xA_4; $R_6=A_4^T$ xA_5; $R_7=A_5^T$ xA_5; $R_8=A_5^T$ xA_6; $R_9=A_6^T$ xA_6; $R_7=A_6^T$ xA_7. When evaluating R_j, (j=1÷10), the following rule is used: the ratio D=$B\circ C$ for arbitrary vectors B and C is calculated using the formula: $d_{ij}=b_i^T$ x $c_j=min(b_i,c_j)$.

Having calculated all R_j, R is defined as $R=\bigcup_{j=1}^{10} R_j$. We obtain:

$$R=\begin{pmatrix} 0 & 0 & 0 & 0{,}5 & 1 & 0{,}5 & 0 \\ 0 & 0 & 0{,}5 & 0{,}5 & 0{,}5 & 0{,}5 & 0 \\ 0{,}5 & 0{,}5 & 0{,}5 & 1 & 0{,}5 & 0 & 0 \\ 1 & 0{,}5 & 0{,}5 & 1 & 0{,}5 & 0{,}5 & 0{,}5 \\ 0{,}5 & 0{,}5 & 0{,}5 & 0{,}5 & 1 & 1 & 0{,}5 \\ 0 & 0 & 0 & 0{,}5 & 0{,}5 & 1 & 1 \\ 0 & 0 & 0 & 0 & 0{,}5 & 0{,}5 & 0{,}5 \end{pmatrix}$$

Step 7. Definition of fuzzy forecasts and their defuzzification. Fuzzy forecast is carried out using the rule:

$$A_i=A_{i-1}\circ R,$$

where A_{i-1} – the fuzzy historical value during the period $t-1$,
A_i – the fuzzy forecast value during the period t in terms of degrees of belonging to these sets, ◦ - operation «max-min».
The results of the operation are given in Table 10.

Table 10. Fuzzy forecast values in terms of μ.

Year	A_1	A_2	A_3	A_4	A_5	A_6	A_7
2006	0,8	0,5	0,5	1	0,7	0,6	0,5
2007	1	0,5	0,5	1	0,8	0,8	0,5
2008	1	0,5	0,5	1	0,7	0,7	0,5
2009	1	0,5	0,5	1	0,9	0,	0,6
2010	1	0,	0,5	1	1	1	0,7
2011	0,5	0,5	0,5	0,6	0,5	0,5	0,3
2012	0,9	0,5	0,5	0,9	1	1	0,9
2013	1	0,5	0,5	1	0,8	0,8	0,6
2014	0,9	0,5	0,5	0,9	1	1	0,8
2015	0,8	0,5	0,5	0,8	1	1	0,9
2016	0,6	0,5	0,5	0,6	0,8	1	1
2017	0,7	0,5	0,5	0,7	0,9	1	1
2018	0,6	0,5	0,5	0,6	0,8	1	1

For defuzzification (translation of fuzzy forecast values to fuzzy), the following rules apply:
1) If there is only one maximum in the row of the table 10, then the midpoint of the corresponding interval u_i is taken as the forecast.
2) If there are several consecutive maxima in the line, then the average value of the midpoints of the corresponding intervals is taken as the forecast.
3) In all other cases, the data lines of the table 10 are normalized, and the forecast is carried out as a weighted average of the midpoints of all intervals.
The results of defuzzification are shown in table 11. The forecast (columns 4, 5) estimated using the formula (4) is also given there.

Table 11. Crisp values of forecast values and their assessment.

Year	Real value, c/ha	Forecast, c/ha	Calculation using the formula (4)	Rating *AFER*
2006	133	138	0,037594	
2007	132	130	0,015152	
2008	138	152	0,101449	
2009	144	147	0,020833	
2010	100	81	0,19	
2011	150	140	0,066667	
2012	136	137	0,007353	
2013	147	140	0,047619	
2014	153	156	0,019608	
2015	164	158	0,04878	
2016	158	168	0,063291	
2017	163	168	0,030675	
2018	170	168	0,011765	
Total			0,660786	0,05083=5%

It can be seen from the table that the forecast estimate is higher (5%) than the same estimate for the second model (4.8%) (Table 7), but lower than for the autoregressive model [12].

3. Conclusion

The fuzzy time series can serve as a new tool for modeling and describing the dynamics of agrarian processes operating in conditions of "non-stochastic" uncertainty.

The number of fuzzy sets used to describe FLD significantly influences the accuracy of the forecast. The modification of the initial method proposed in this paper can be recommended for practical use in the short-term forecasting of time series.

Forecasting methods based on fuzzy TS models allow us to give the most adequate estimate for predicting future changes under uncertainty.

References

[1] Parfenova V E, Bulgakova G G, Amagaeva Yu G and Evdokimov K V 2019 Fuzzy modelling for tasks of management of the agricultural-industrial complex *2019 IOP Conf. Ser.: Mater. Sci. Eng.* (666 011001)

[2] Aytasova A, Selezneva Z, Belinskaia I and Evdokimov K 2019 Development of the process map «research and development» for agricultural organizations *IOP Conference Series: Materials Science and Engineering* (Volume 666, Number 1) p 012072

[3] Yarushkina N G. et al. 2010 Intellectual analysis of time series study guide *UlSTU* p 324

[4] Bozhuk S G, Maslova T D, Pletneva N A and Evdokimov K V 2019 Improvement of the consumers' satisfaction research technology in the digital environment *IOP Conference Series: Materials Science and Engineering* (Volume 666, Number 1) p 012055

[5] Song Q and Chissom B S 1993 Fuzzy forecasting of deposits with fuzzy time series. Part 1. Fuzzy sets and systems (54) pp 1–9

[6] Song Q and Chissom B S 1993 Fuzzy time series and their models. Fuzzy sets and systems (54) pp 269–277

[7] Kosko Bart 1991 Neural Networks and Fuzzy Systems *Englewood Cliffs, NJ: Prentice-Hall* p 449

[8] Chen S M 1996 Forecasting enrollments based on fuzzy time series *Fuzzy Sets and Systems* (N 81) pp 311 – 319

[9] Chen S M 2002 Forecasting enrollments based on high-order fuzzy time series *Cybernetics and Systems: an International Journal* (N 33) pp 1 – 16
[10] Bozhuk S G, Evdokimov K V, Maslova T D and Pletneva N A 2019 Problems and perspectives of sustainable marketing development on the Russian food market *IOP Conference Series: Materials Science and Engineering* (Volume 666, Number 1) p 012040
[11] Statistical collections "Russia and the countries of the world (M.: 2014, 2016 and 2018)
[12] Klochkova E, Evdokimov K, Klochkov Y and Samorukov V 2018 Methodology for reducing risk of underperformance of personnel functions *Engineering for Rural Development 17th International Scientific Conference* pp 1213-1222

Research of social media channels as a digital analytical and planning technology of advertising campaigns

E Isaenko[1], E Makrinova[1], I Rozdolskaya[1], E Matuzenko[1] and S Bozhuk [2]

[1]Belgorod University of Cooperation, Economics and Law, Sadovaya st. 116-A, Belgorod, 308023, Russia E-mail: pror-nr@bukep.ru
[2]Peter the Great St. Petersburg Polytechnic University, Russia, E-mail: bojuk.svetlana@yandex.ru

Abstract. In Russia social media, in particular instant messengers, have advanced rapidly in recent years. The most popular ones are WhatsApp, Viber, Skype, Telegram. The purpose of this article is to systematize factors of market positioning of social media channels as a digital analytical and planning technology of advertising campaigns. A systematic approach and comparative analysis were used to study this problem.The authors attempted to develop a methodological approach towards analyzing and planning of advertising campaigns in social media based on the analysis of the factors characterizing quantitative parameters of the Internet platform development and its target audience distribution.The proposed approach was tested by the authors for determining the development features of Telegram messenger, analyzing its target audience (users with higher education engaged in science, technique, information technology, etc.) and its promotional product (posts in the form of a native format, etc.). The analysis concluded that Telegram platform is suitable for analyzing and planning of promotion campaigns in any business field.

1. Introduction

Telegram, as an advertising campaign planning tool, is a relatively young online platform. In July 2017, it started as a channel catalog with the ability to track reposts; currently it is positioned as a tool for analyzing and planning of advertising campaigns with a monthly attendance of more than 600 thousand people. It should be noted that in 2018 the number of active service users increased sixfold (compared with 2017), and in December 2018 alone, they viewed more than 2.5 million pages.

A high popularity of the application, in our opinion, is due to the applied encryption technology, which guarantees users' anonymity and restricts access to personal data. According to the results of the Project Group study, focused on the analysis of using Telegram by the central government, the messenger is convenient for information dissemination: access to any Telegram channel can be purchased for a certain amount of money and the customer remains anonymous.

A potential Telegram's advantage is its user-friendly interface and the targeted nature of advertising, while a high conversion percentage is ensured by placing information on relevant channels and focusing on a potentially interested consumer. Therefore, manufacturers are actively using the application for

IOP Conf. Series: Materials Science and Engineering **986** (2020) 012014 doi:10.1088/1757-899X/986/1/012014

business purposes, and channel owners make a profit of advertising placement, which contributes to the steady growth of the messenger's popularity.

Finally, popular channels owners profit successfully by selling them. A promoted channel with a large number of subscribers can cost several hundred thousand rubles.

Currently, the number of Telegram users is growing, and along with the growth of the target audience, political and business structures are starting to view it as an effective PR mechanism, an advertising and propaganda tool.

2. Research methodology

Cabinet research methods were used in the analysis: analysis of scientific and methodological literature on the topic under study, generalization and classification of the investigation, overview of the existing foreign and Russian experience in the field of marketing [1, 2] and Internet advertising. A systematic approach and comparative analysis were used for examining this problem.

This study is a continuation of the authors' previous researches [3, 4, 5, 6] on the development issues and features of social media in the digital marketing technology system. The purpose of this article is to investigate the results of using social media for both analyzing and planning of advertising campaigns [7, 8].

In contrast to previously published works (Arshad, 2018 [9]; Cavaliere, 2018 [10]; Jain, Subhash, 1999 [11]; Moskaleva et al, 2018 [12]; Neumann, 2018 [13]), devoted to digital technologies development, the authors attempted to work out a methodological approach to analyzing and planning of advertising campaigns in social media based on the analysis of factors, characterizing quantitative parameters of the Internet platform development and its target audience distribution.

The proposed approach was tested by the authors when determining developmental characteristics of Telegram messenger, analyzing its target audience (users with higher education engaged in science, technique, information technology, etc.) and an advertising product (posts in the form of a native format, etc.) [14].

3. Results obtained

A study of various sources on the use of Telegram messenger for planning advertising campaigns allowed the authors to conclude that currently there are no methodological approaches satisfactory for such studies. The available statistics, as well as the polls conducted on Telegram messenger development, do not provide a holistic picture for making managerial decisions regarding planning an advertising campaign by using this online platform [15].

In that context, the authors attempted to systematize the factors of Telegram messenger market positioning in order to use it as an advertising platform for business promotion. Moreover, the authors systematized market-positioning factors in two groups (Fig. 1).

The first group gives an idea of Telegram's development trends. It includes the following criteria: distribution geography, distribution of channels on subject matters, number of channel subscribers, channel citation index, number of channel links to media sites and social networks. The indicators of the second factor group of Telegram messenger market positioning give an idea about its target audience according to gender, age, occupation, interests, objectives of using, number of contacts, number of channels read by users, share of other Internet services usage, frequency of commercials on various platform channels.

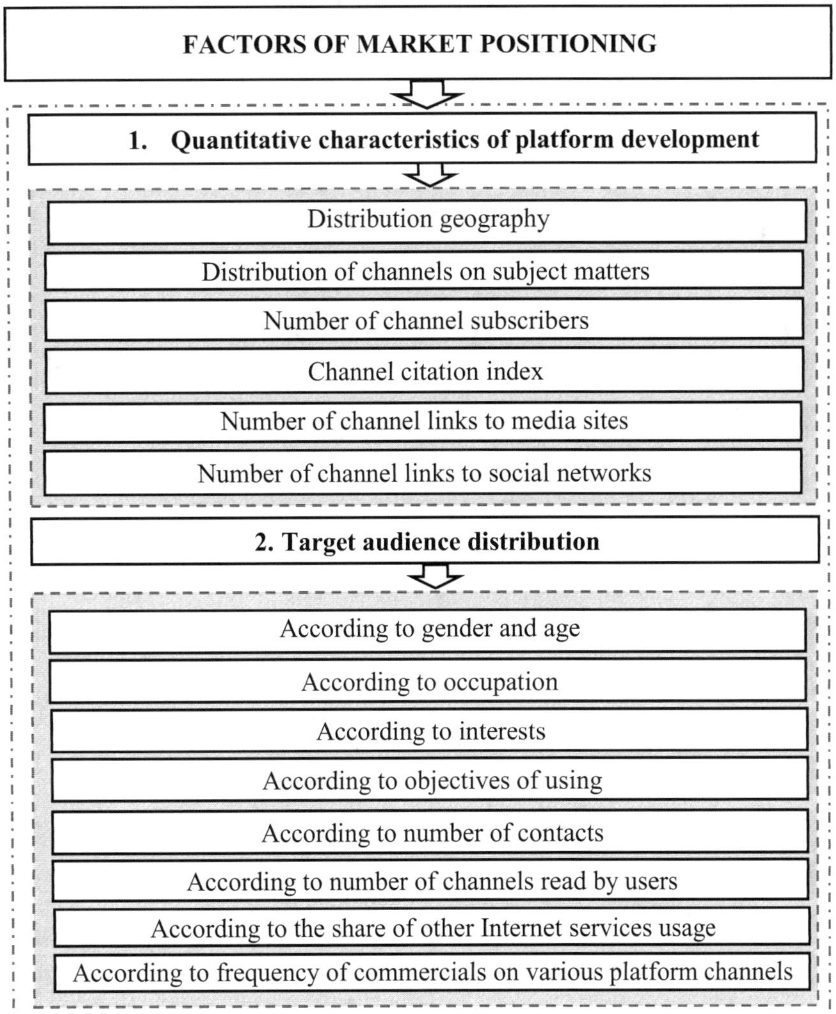

Figure 1. Systematization of social media factors of market positioning.

Let us consider these factors in more detail.

Initially, Telegram focused on the Russian market (63 thousand channels at the end of 2018), and now the Russian-language segment remains the main priority, but geography of the channel penetration in other countries is expanding (Fig. 2).

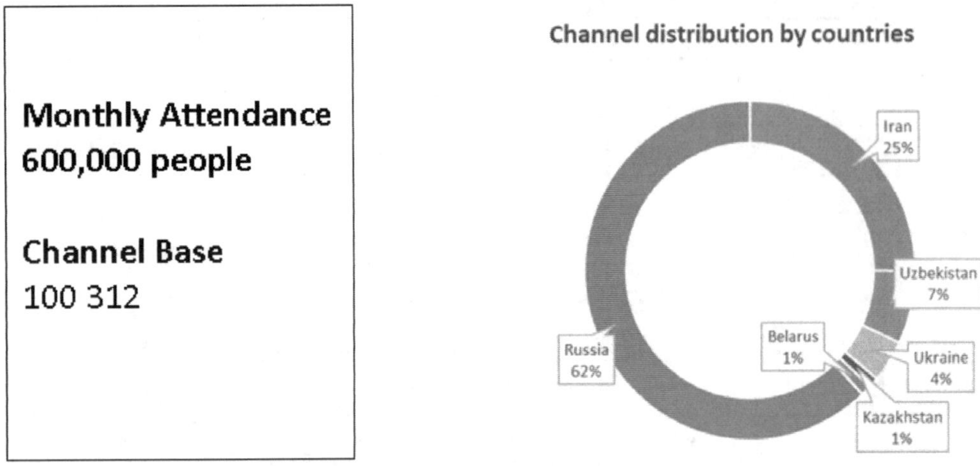

Figure 2. Distribution geography of Telegram in 2018, %.

Currently, users from 6 countries communicate in Telegram messenger, whereof 62% are Russians, 25% are citizens of Iran, 7% are from Uzbekistan, 4% are Ukrainians, and citizens of Kazakhstan and Belarus –1% each. On average, 600 thousand people visited the channel per month, using 100312 channels; in Russia 856 channels link to Telegram.

Analyzing Russian Telegram users by region, we can conclude that a large share is constituted by residents of Moscow (36%) and St. Petersburg (15%); large cities range from 1.5% to 2.7%. The share of Telegram users in the central part of Russia accounts for almost 50% of the total.

During 2018, the number of Telegram channels increased and by the end of the year, almost 40 thousand new Russian-language channels were indexed.

In 2018, interests of Telegram users changed (Fig. 3). Despite the fact that almost every major entertainment public in the social network Vkontakte has a Telegram channel, according to criteria number of channels the author's blogs regained the first place (at the end of 2017 humorous channels were on top for a short time), and the second place belongs to business issues; music channels have risen in ranking by several times. Cryptocurrencies are currently not as popular as they were at the end of 2017, so this subject matter is in the sixth place on Telegram. Moreover, Technology category channels, which took the leading position a year and a half ago, are now not included in the top 10, giving way to Health and Sports channels as well as to Art and Photo channels.

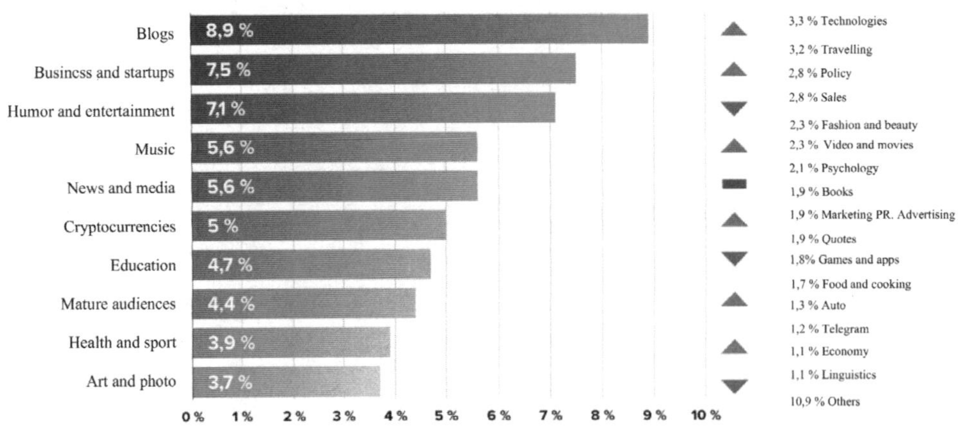

Figure 3. Telegram channels share by subject matters, %.

Let us consider the most popular channels of Telegram platform by the number of subscribers (Fig. 4). Mash channel is at the top with 470.1 thousand subscribers, in second place is KB with 376.5 thousand users, and third place is taken by MDK.

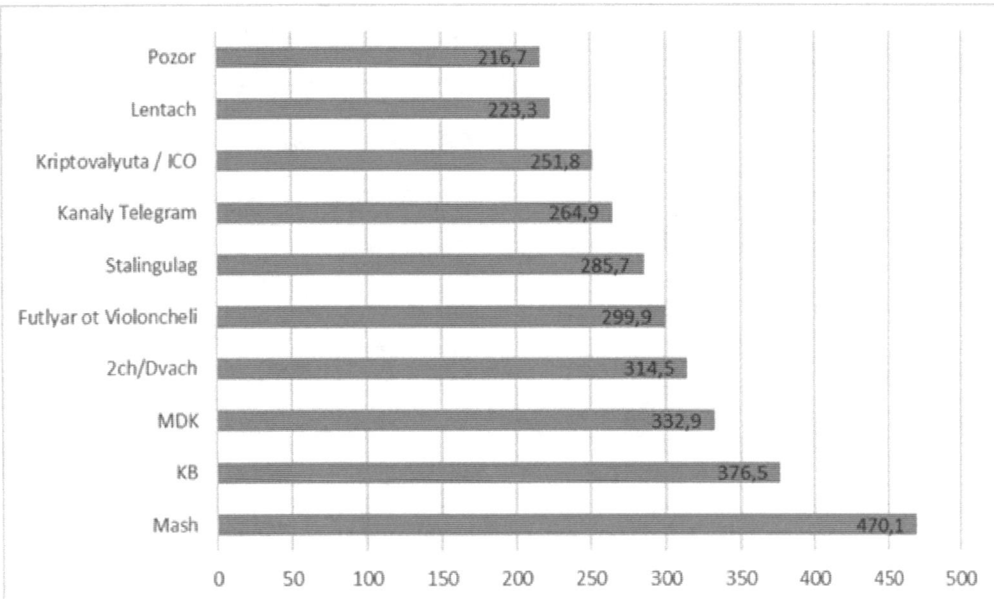

Figure 4. The number of Telegram subscribers in 2018 by channels, thousand people.

Compared to 2017, the number of subscribers in 2018 increased in the following way: Mash channel – more than 3.5 times, Futlyar ot Violoncheli channel increased the number of subscribers fivefold, Dvach and MDK channels doubled the number.

In 2018, specialists of Telegram platform changed the formula for citation index calculating because this figure did not reflect reality due to a large-scale purchase of commercials at low-quality channels. At present, it is possible to realistically assess the most influential and popular Telegram channels inside the messenger at year's end (Fig. 5).

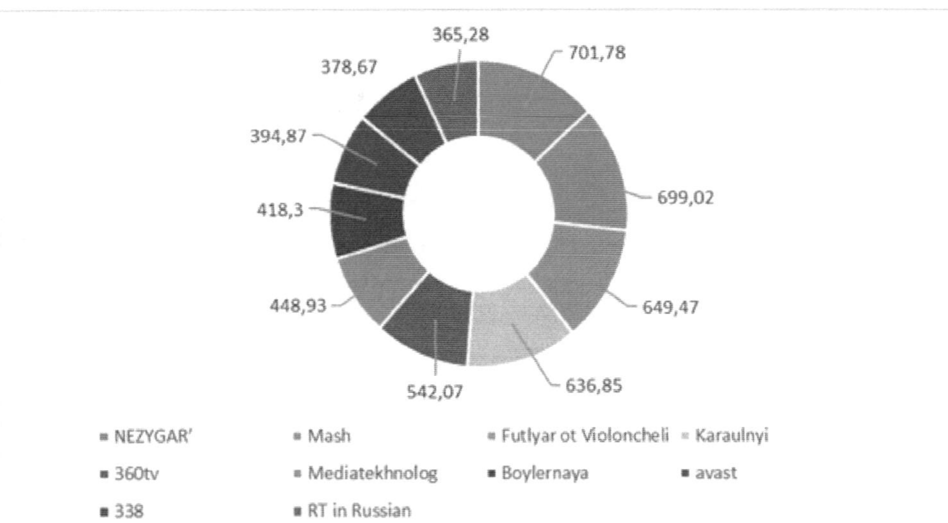

Figure 5. Telegram channels by citation index.

IOP Conf. Series: Materials Science and Engineering **986** (2020) 012014 doi:10.1088/1757-899X/986/1/012014

The figure illustrates that the citation number of Nezygar channel ranks first, Mash channel is in 2nd place, and Futlyar ot Violoncheli is in 3rd place. The last two channels have significantly increased the number of subscribers by the end of 2018.

Let us estimate the rating on number of channel links to media sites in Telegram in 2018 (Table 1). Medusa is at the top with its own six popular channels (they account for 80% of the entire link mass).

The five most mentioned media also includ TJournal, vc.ru, Kommersant and RBC. If to sort the top channels descending, it is as follows: 1st place is taken by Medusa, RBC is in 2nd place and Vc.ru is the 3rd one.

Table 1. Number of channel links to media sites in Telegram in 2018.

№ п/п	Mass media	Site	Number of views of publications with a link to a media site	Number of publications with a link to a media site	Number of channels with at least one link to a media site
1	Medusa	medusa.io	286677786	29175	3506
2	TJournal	tjournal.ru	103433963	16333	2101
3	VC	vc.ru	82231848	22676	2719
4	Kommersant	kommersant.ru	71412407	29437	2663
5	RBC	rbc.ru	68948175	36355	2915
6	Dozhd	tvrain.ru	67983505	10253	963
7	RIA News	ria.ru	55539747	18240	2276
8	TASS	tass.ru	35555541	14982	2082
9	Znak	znak.com	34561384	46591	1168
10	Life	life.ru	33587035	12170	858
11	Lenta.ru	lenta.ru	33151456	44711	2121
12	Fontanka	fontanka.ru	32880892	73898	718
13	Nozh	knife.media	30463505	5892	960
14	Snob	snob.ru	29208256	4245	845
15	Mediazona	zona.media	28620544	9234	929
16	Vedomosti	vedomosti.ru	28558058	9773	1605
17	BBC	bbc.com	26920512	9453	1763
18	RTVI	rtvi.com	22438291	5711	482
19	Interfax	interfax.ru	22281463	10652	1315
20	Rossiyskaya gazeta	rg.ru	21475200	7911	1090

Figure 6 shows the services most often mentioned by Telegram. YouTube is at the top with more than half a million posts in almost 17 thousand channels (30% of the entire Russian-language Telegram). Social network VKontakte ranks second; 3rd place is taken by social network Twitter; Instagram is in 4th place and Facebook is the fifth one.

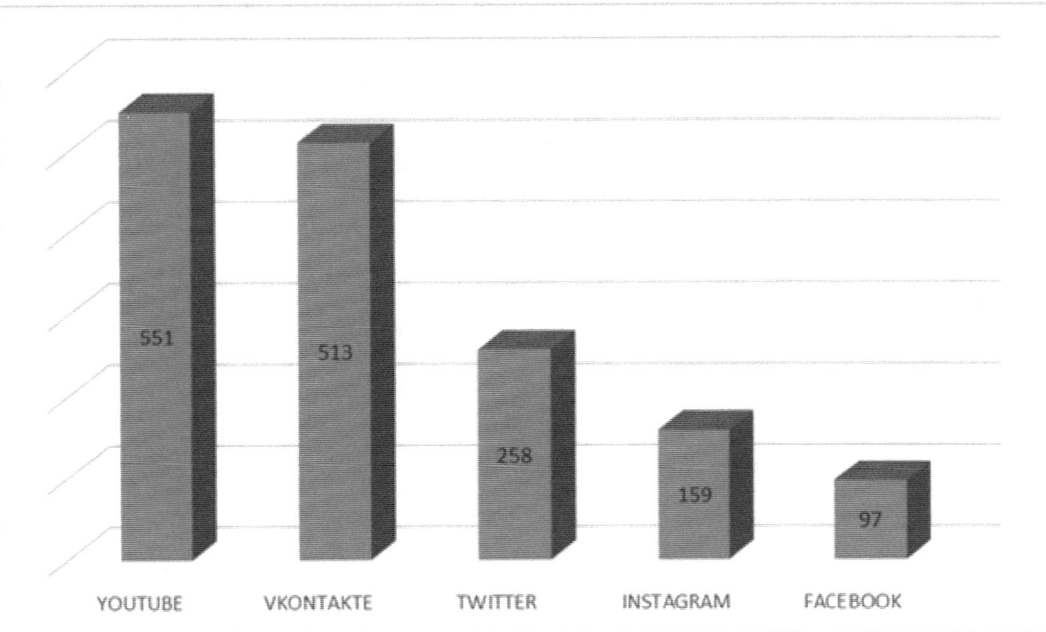

Figure 6. Services mentioned in Telegram in 2018, thousand units.

In order to correctly promote business in Telegram, it is necessary to study the target audience and their needs. Entrepreneurs, executives, freelancers, and Internet marketing specialist are popular users of Telegram.

Let us consider the Telegram's audience based on a survey conducted by @aboutSMM channel. Currently, men constitute the majority of the messenger users (57%).

Telegram's target audience, depending on gender and age, is shown in Fig. 7. The figure depicts that 48% of women aged from 18 to 24 are users of the platform, and 40% constitute men aged from 25 to 34. Men and women aged from 45 to 64 and schoolchildren aged from 12 to 17 use Telegram the least. Therefore, we can conclude that the active audience of Telegram messenger is in the age range from 18 to 34 years.

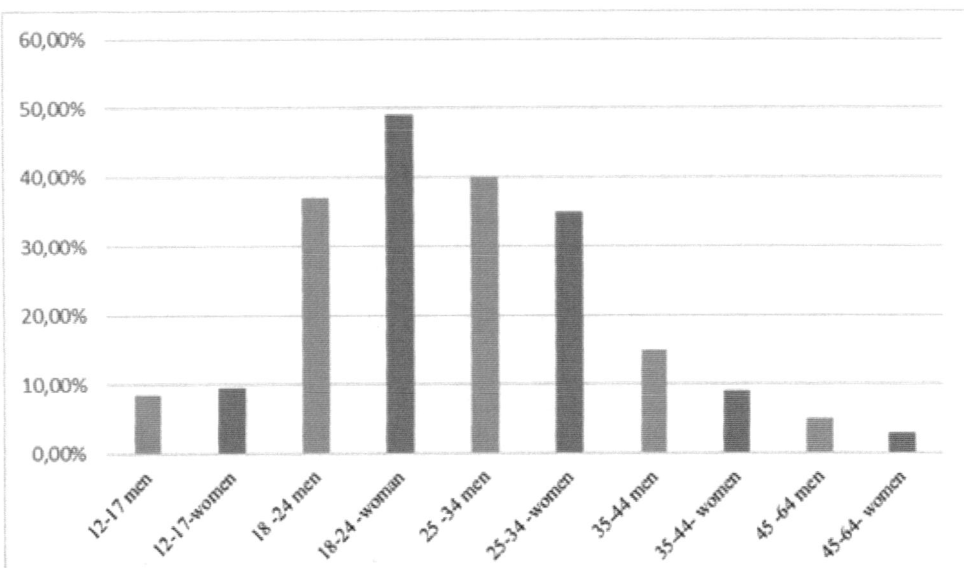

Figure 7. Distribution of Telegram target audience according to gender and age, %.

The most active users of Telegram are "experienced specialists" (35%), "students" (25%) and "young specialists" (17%), a small proportion of Telegram users constitute schoolchildren and unemployed population (5% each). The rest of the audience is made up of mid- and top-level executives; 57% of users have higher education, which proves a high quality of the target audience.

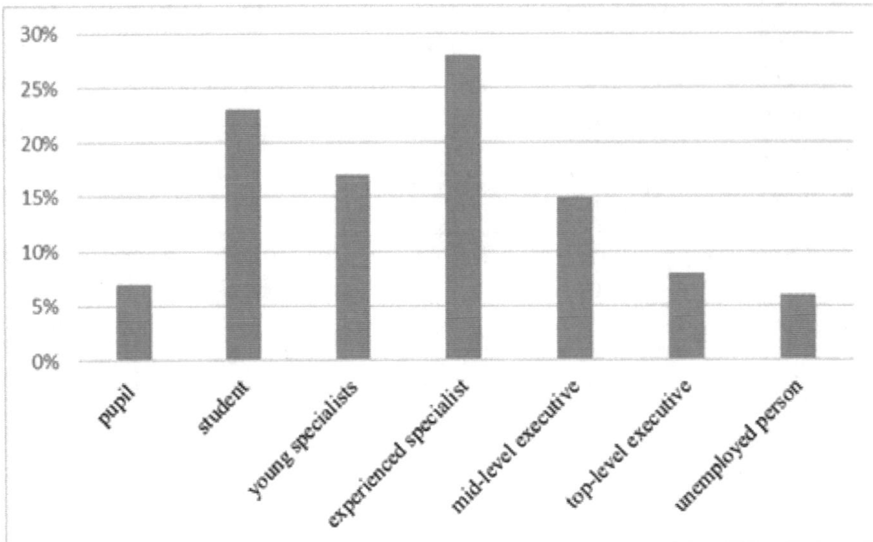

Figure 8. Distribution of Telegram target audience according to occupation, %.

Figure 9 shows interests of Telegram users. Almost half of the messenger users (47%) are interested in information technology, 42% are interested in science and technology, 41% – watch movies and TV shows (Fig. 9). 39% of Telegram users enjoy reading news on large media channels and 35% are interested in reading books. The proportion of Telegram users subscribed to entertainment channels amounts to 29%.

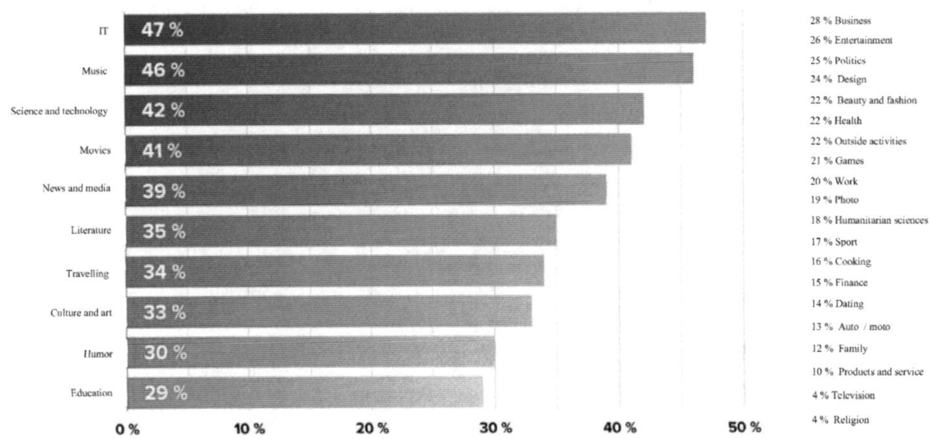

Figure 9. Distribution of Telegram target audience according to interests, %.

The largest share of Telegram target audience use the messenger via Androids (60%), about 40% of users get access to the platform via iPhones, and 2% of the audience use Windows platform.
In the course of the survey, it turned out that more than 90% of Telegram users enjoy the messenger to see something they need, 88% – use the platform for correspondence purposes, 45% of users prefer chat rooms (Fig. 10). The number of Telegram messenger's authors in 2018 reached 16%, and the number of subscribers using bots was 45%.

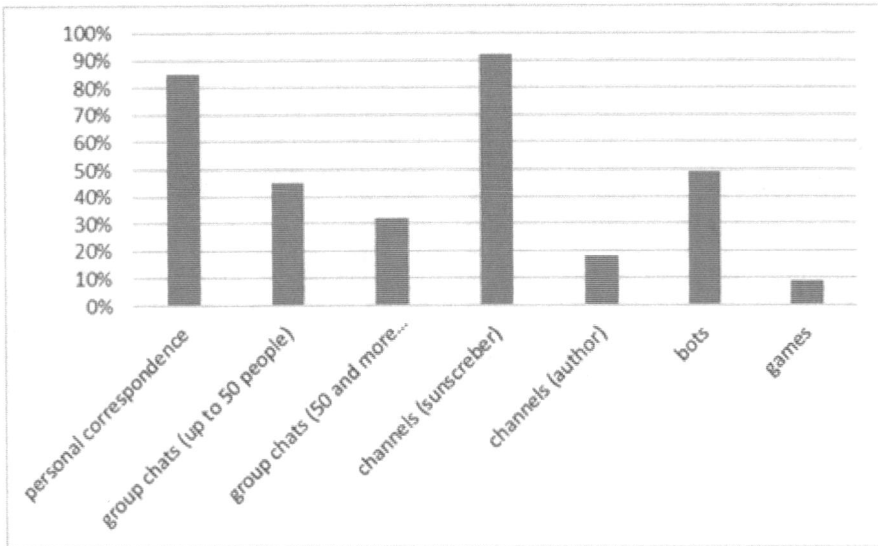

Figure 10. Distribution of Telegram target audience according to the objectives of using, %.

Any social network or messenger user has a list of contacts (friends). Unlike social networks, Telegram messenger opens chat on adding a contact. Therefore, most users have only 5-10 contacts of their closest friends and partners with whom they often communicate (Fig. 11).

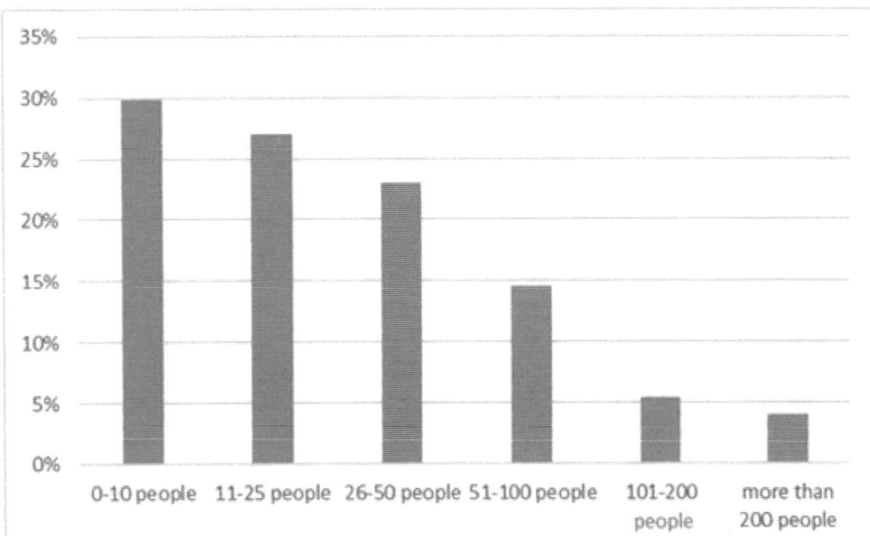

Figure 11. Telegram target audience distribution according to the number of contacts,%.

Approximately one third of Telegram messenger users (30%) are subscribed to 10-25 channels, 27% of users – to 5-10 channels, and 24% – to 1-5 channels. Those who are not subscribed to any channel and those who are subscribed to more than 50 channels amount to less than 5% (Fig. 12).

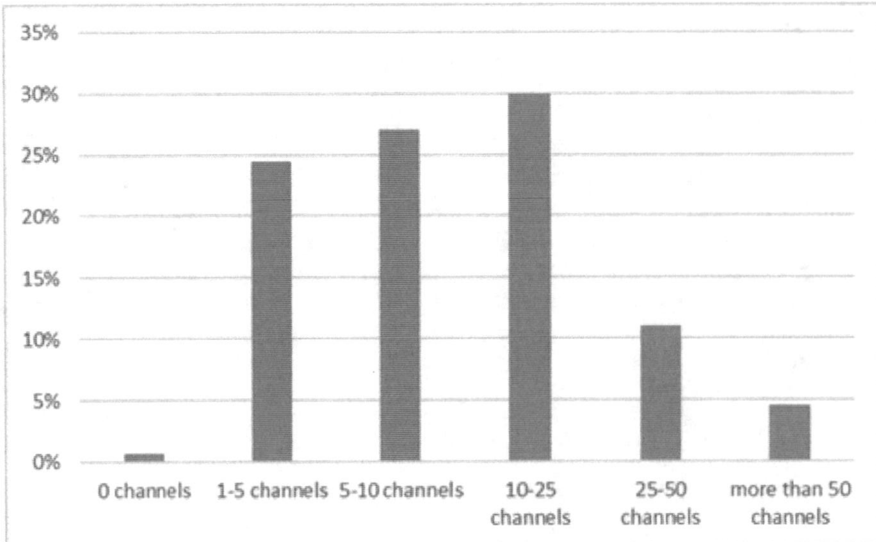

Figure 12. Telegram target audience distribution according to number of channels read by users, %.

Almost 70% of Telegram audience are either registered in the social network VKontakte or often use it, 56% of them chat in WhatsApp, 30% use Viber, 30% use the social network Facebook, 19% use iMessage via iPhone (Fig. 13).

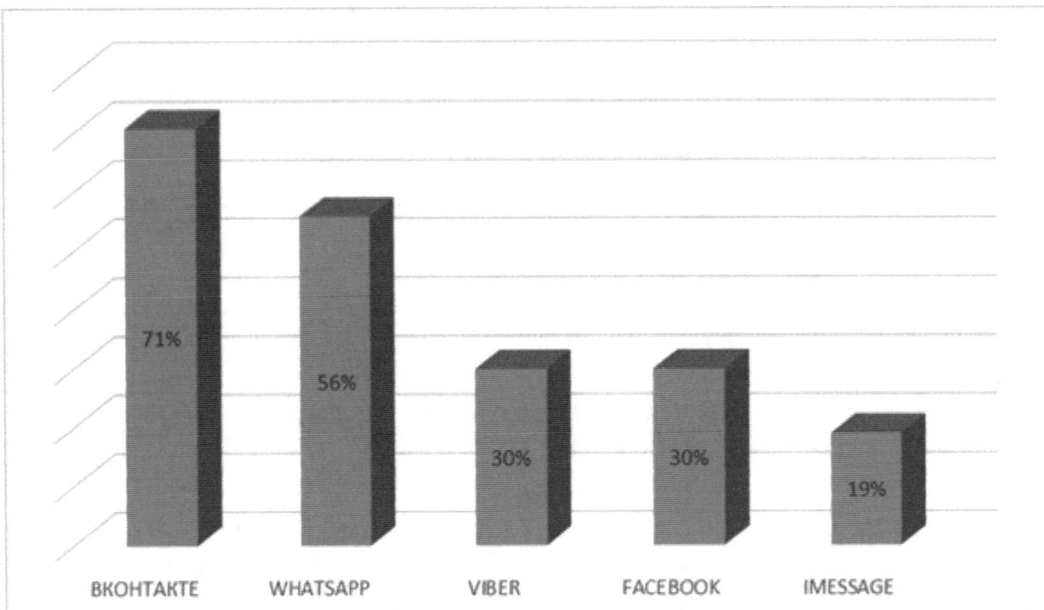

Figure 13. Distribution of Telegram target audience according to the share of other Internet services usage,%.

In Telegram messenger an advertisement is placed as a post or a native format. Therefore, 25% of Telegram users do not see advertising on the messenger channels, 30% see it very rarely and 25% see it only occasionally.

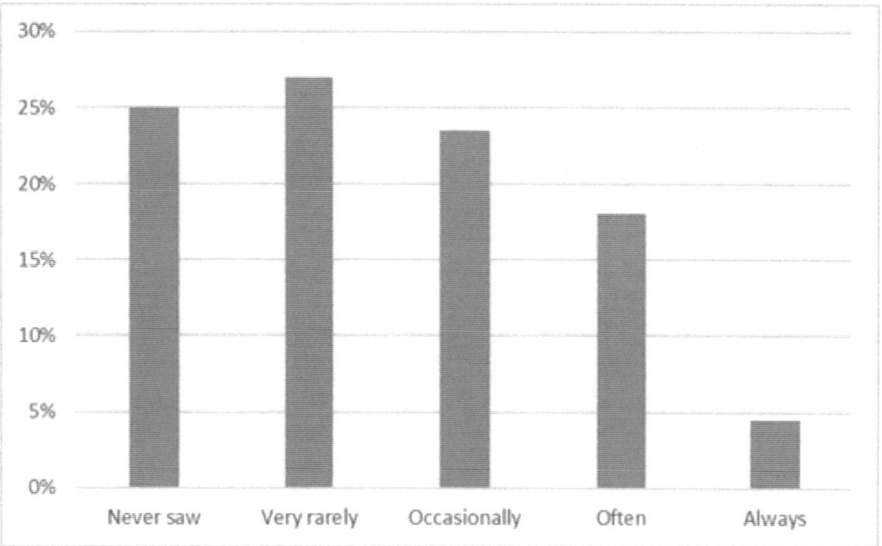

Figure 14. Distribution of Telegram target audience according to frequency of commercials on various platform channels, %.

Approximately 17% of users see advertisements very often and 5% see them constantly. Apparently, they are subscribed only to entertainment channels.

Telegram has become a popular messenger and its audience is very diverse. Now the messenger is used not only by entrepreneurs and IT specialists, but also by different categories of users.

Currently, there is no fierce competition on this platform; there are no clear leaders in every economic realm. Therefore, nearly any author with minimal skills can develop their channel and get first profit in 2-3 months.

To evaluate Telegram target followers' feedback we use the ER indicator (engagement rate of followers'), which is calculated as follows:

ER = the sum of all engagements (likes, reposts, comments) / number of subscribers * 100%.

However, in the judgment of the authors, the drawback of this indicator is that it does not take into account coverage, involvement on specific days and interaction with individual posts.

Therefore, in our opinion, to evaluate Telegram target followers' feedback, it is necessary to use several subtypes of ER.

1. Reach engagement rate (ERR) = number of engagements / coverage * 100%.

This indicator shows the approximate share of those who have seen community publications and reacted to them in one way or other.

2. Engagement rate per day (ER Day) = number of engagement per day / number of subscribers * 100%.

Daily engagement shows how many times a day the average subscriber showed his/her activity.

3. Engagement rate per post (ER Post) = the amount of involvement per 1 publication / number of subscribers at the date of publication * 100%.

The indicator allows evaluating interest in specific publications, comparing them in terms of effectiveness.

4. View engagement rate (ER View) = number of engagements per 1 publication / number of views * 100%.

This indicator shows the specific share of the target audience who made comments on a specific advertisement viewing.

Social media analytics tool "Popsters" is used for running community analysis in social networks and messengers. It automatically calculates all of the listed above indicators.

IOP Conf. Series: Materials Science and Engineering **986** (2020) 012014 doi:10.1088/1757-899X/986/1/012014

Thus, in order to enable effective advertising communications in messengers, many advertising specialists calculate rates of both return on investment and target followers' feedback, while advertising on several platforms to achieve appropriate goals and decide on the format for each target audience.

4. Conclusion

The study on Telegram's market positioning factors for analyzing and planning of advertising campaigns for business operations leads the authors to the following conclusions.

The active Telegram messenger audience is in the age range from 18 to 34 years. 57% of users have higher education, which proves the high quality of the audience. 47% of the messenger users are interested in information technology, 42% are interested in science and technology, 41% watch movies and TV shows, 39% read news channels of major media companies, and 35% read books. Almost 70% of the Telegram audience are either registered in the social network VKontakte or often use it, 56% of them chat in WhatsApp, 30% use Viber, 30% use the social network Facebook, 19% use iMessage via iPhone. In Telegram messenger an advertisement is placed as a post or a native format. Therefore, 25% of Telegram users do not see advertising on the messenger channels, 30% see it very rarely and 25% see it only occasionally.

Based on the above stated, we can conclude that Telegram messenger is currently an effective platform for analyzing and planning advertising campaigns in any business field.

References

[1] Doyle P 2002 Marketing management and strategy *Financial Times Prentice Hall* p 446

[2] Dunnett A 1987 Understanding the market (Longman) p 197

[3] Makrinova E I, Svyataya E O, Strukov G N, Grigoryeva V V and Pereverzeva E S 2017 Positioning of hotel services manufacturers in regional market according to ranking of marketing approaches: methodology and empirical studies *Contributions to Economics* (9783319606958) pp 529-540

[4] Teplov V I, Tarasova E E, Alyabieva M V, Isaenko E V and Pidoimo L P 2016 International Review of Management and Marketing (T. 6. № 5) pp 8-15

[5] Makrinova E I, Matuzenko E V and Kolesnikova E V 2019 Modern trends and technologies for building advertising communications in social media *Herald of the Belgorod University of Cooperation, Economics and Law* (No. 2 (75)) pp 313-321

[6] Rozdolskaya I V, Mozgovaya Yu A and Chicherin Yu A 2018 Interaction of offline and online marketing strategies as an innovative tool to strengthen competitive positions in the interactive environment of enterprise functioning *Herald of the Belgorod University of Cooperation, Economics and Law* (No. 1 (68)) pp 31-240

[7] Parfenova V E, Bulgakova G G, Amagaeva Y V and Evdokimov KV 2019 Fuzzy modelling for tasks of management of the agricultural-industrial complex *IOP Conference Series: Materials Science and Engineering* (Volume 666, Number 1) p 012067

[8] Aytasova A, Selezneva Z, Belinskaia I and Evdokimov K 2019 Development of the process map «research and development» for agricultural organizations *IOP Conference Series: Materials Science and Engineering* (Volume 666, Number 1) p 012072

[9] Arshad M, Ahmad Sh and Siddiqui F A 2018 Social media technologies and their benefits in higher education institutions: a survey of Jazan university *Information technology and learning tools* (65, 3) pp 91-102

[10] Cavaliere F 2018 Discursive mechanisms of news media - investigating attribution and attitudinal positioning *Bulletin of People's Friendship University of Russia. Series: Linguistics* (22, 2) pp 338-356

[11] Jain Subhash C 1999 Marketing planning and strategy *South-Western College Publishing* p 925

[12] Moskaleva N B, Zaitseva N A, Vinogradova M V, Larionova A A, Kuljamina O S, Dashkova E V and Popovich A E 2018 The use of digital technologies to improve the quality of the "Social taxi" services *Modern Journal of Language Teaching Methods* (8, 5) pp 38-48

[13] Neumann K 2018 Digital economy: for the benefit of users, business and the state *Bussiness Excellence* (2) pp 22 -29

IOP Conf. Series: Materials Science and Engineering **986** (2020) 012014 doi:10.1088/1757-899X/986/1/012014

[14] Larionova A A and Zaitseva N A 2019 Using the information and communication technologies in the hospitality industry enterprises' strategies' implementation *Advances in Intelligent Systems and Computing* (726) pp 229-23
[15] Bozhuk S G, Maslova T D, Pletneva N A and Evdokimov K V 2019 Improvement of the consumers' satisfaction research technology in the digital environment *IOP Conference Series: Materials Science and Engineering* (Volume 666, Number 1) p 012055

IOP Conf. Series: Materials Science and Engineering **986** (2020) 012015 doi:10.1088/1757-899X/986/1/012015

The role of automation in improving the quality of enterprise business processes

A V Bataev[1] and I S Davydov[2]

[1]Peter the Great St. Petersburg Polytechnic University, Polytechnicheskaya, 29, St.Petersburg, 195251, Russia, bat_a68@mail.ru
[2]National Technology Initiative Center for Advanced Manufacturing Technologies based on the Institute of Advanced Manufacturing Technologies of Peter the Great St. Petersburg Polytechnic University Polytechnicheskaya, 29, St.Petersburg, 195251, Russia

Abstract. The enterprise has to determine and manage numerous and interrelated activities for successful work and to maximize profits. According to the international standard ISO 9001-2011, the activity using resources and managed in order to transform inputs to outputs can be considered as a business process. Using a process system in an organization along with their identification and interaction, as well as process management aimed at obtaining the desired result, can be defined as a "process approach". The object of this study is a business process that is one of the key concepts of the international standard ISO 9001-2011. This standard establishes the requirements for quality management systems and describes the principles of applying the process approach, the essence of which is continuous process management. The article deals with the individual business processes in the enterprise and their subsequent task automation with using information technology. An automated application was proposed within the study that provides automatic search for part sizes that do not correspond to the tolerance field, automatic construction of control charts of average values and ranges, as well as calculation of process reproducibility indices.

1. Introduction

Enterprise performance requires special attention to business process management. To do this, models and process charts are made up, audits are conducted and their performance is analyzed. As practice shows, there are problems associated with the coordination of functions between divisions, document management, and the high cost of time for performing the same operations. The solution to such problems is often included in the list of short-term goals of the enterprise. [1-2]

Employees of the enterprise, who have to constantly work with large amounts of data, take direct participation in the creation, maintenance, control of all processes. It is possible to reduce the workload of employees aimed at supporting business processes through automation. One can automate both one process and several processes, while creating a single information space at the enterprise.

Employees of the enterprise, without which it is impossible to manufacture any products, take direct participation in the creation, maintenance, control of all processes but the degree of human participation in business processes can be reduced by automation.

Automation is the application in the production of technical facilities, methods, and control systems that exempt a person from direct participation in production processes.

The purpose of automation is to increase productivity and labor efficiency, improve product quality, and human working conditions. [3]
The tasks of automating business processes become increasingly relevant. This is due to analyzing the effectiveness of business processes and determining the causes of identified inconsistencies in most cases, it turns out that the main problem of modern enterprises is the lack of automation and databases to systematize the accumulated information [4]. A large percentage of routine operations, which are manual input, search, data analysis, become a source of errors, loss of time, and a decrease in employee motivation in general.
The main means to solve the problems of automating business processes is the implementation of information technology and the development of special programs and modules.
The implementation of such programs in various functional units allows making the information accumulated at the enterprise clear, accessible, and structured, as well as reducing the share of the human factor and unnecessary paperwork. [5]

2. Key research findings

2.1. Tasks for automating business processes

The main criterion for the functioning of the process approach is the continuous management of processes and their optimization.
Business processes are divided into three groups:

1) Main processes. They serve as the basis of the enterprise. Processes add value to the consumer. For example, manufacturing, logistics, marketing and sales, procurement, etc.
2) Supporting processes. They provide the activity of the main processes. For example, technical support, accounting, staffing, equipment maintenance, etc.
3) Control processes. The main function is the management of the enterprise. For example, economic planning, strategic development, paperwork, etc.

Automation covers all business processes operating at the enterprise. It can be used to solve both local tasks at the level of one small business function, and larger tasks covering a whole chain of interconnected business processes. [6-8]
Automation of accounting, client accounting, document management was most widespread, since they are precisely connected with the processing of large volumes of information.
Procedures for automating business processes can be divided into three types, which are presented in Figure 1.

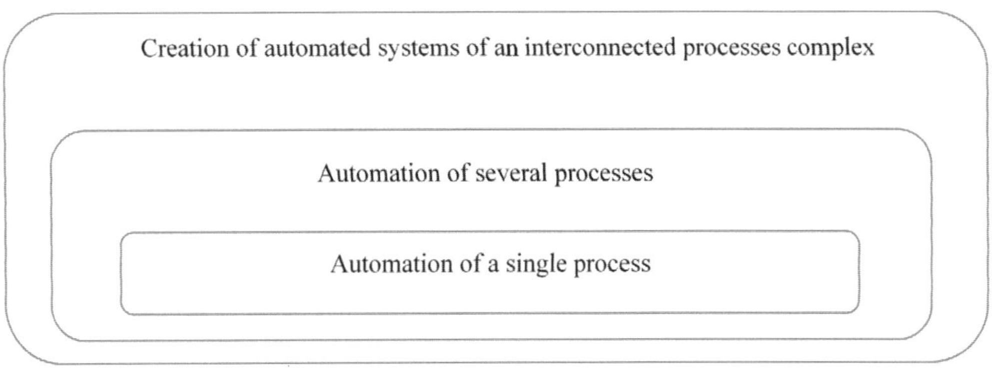

Figure 1. Types of business process automation.

1) Automation of the process.
In this case, one or two specific tasks are solved within one department, related to the simplification of work with individual documents. For example, the development of software modules for graphing,

tooltips, formatting spreadsheet cells depending on the specified parameters, generating reports, moving folders to the archive, sending letters at the click of a button, etc.

2) Automation of several processes.

This type of automation covers two or more processes and is achieved through specially designed programs, which usually include a database for accumulating and storing information, input and data exchange forms, and reports for analyzing these data.

Enterprises independently develop client-server applications for order management, procurement, internal audit documents, statistical data processing, production planning, etc.

3) Automation based on the creation and implementation of special systems covering a complex of interconnected business processes.

Such automation systems allow us to make all processes transparent and to monitor the progress of production in real-time and competently manage the enterprise, while receiving maximum profit. [9]

2.2. Creation of automated applications within a single business process

A stable production process ensures the quality of engineering products. The stability control of the technological process is implemented on the basis of the standard GOST R 50779.42-9.

To improve the quality of manufacturing parts at a machine-building enterprise, a procedure for selective dimensional control was created, by which eight samples of castings obtained in each of the nests of the mold were periodically selected. Measurement data was entered in the checklists - MS Word format files. However, the post-processing of the results was not automated. [10]

To solve this problem, an application was created that automatically searched for dimensions of parts that did not correspond to the tolerance field, automatically built control charts of average values and ranges, as well as calculated the reproducibility indices of the processes Cp and Cpk. [1, 11]

The application for building control charts is an MS Excel file. The first sheet of the file - a table with the results of measurements of the controlled parts obtained in each nest of the foundry mold (control sheets), is manually filled by the operator. Figure 2 shows a view of the control sheet for slot 1.

11–14	Pos.	Size / Specification			Measurement Results — Completed by the supplier — Number of mold and sample socket: 1/1	1/2	1/3	1/4	1/5	1/6	1/7	1/8	Fin	Unfin
15		Main projection	LTL	UTL										
16	1	Ø11,8		0,3	11,933	11,945	11,904	11,912	11,933	11,921	11,912	11,93	+	
17	2	Ø11,8		0,3	11,895	11,942	11,947	11,884	11,921	11,825	11,892	11,883	+	
18	3	R34,5			34,064	34,832	34,268	34,691	34,499	34,724	34,322	34,254		-
19	4	50,0	-0,2	0,2	50,179	50,283	50,149	50,189	50,305	50,152	50,161	50,18		-
20	5	53,6	-0,2	0,2	53,71	53,763	53,718	53,765	53,797	53,629	53,754	53,735	+	
21	6	30,50	-0,2	0,2	30,472	30,502	30,492	30,484	30,491	30,478	30,482	30,481	+	
22	7	13,10	-0,2	0,2	13,24	13,335	13,274	13,492	13,367	13,155	13,234	13,336		-
23	8	7,50			7,874	6,918	7,695	7,12	7,368	7,125	7,365	7,556		-
24	9	M10x1,25			ok	ok	ok	ok	ok	ok	ok	ok	+	
25	10	M10x1,25			ok	ok	ok	ok	ok	ok	ok	ok	+	
26	11	(∟90°)	⊥-0,3	⊥0,3	90,069	90,115	90,106	90,128	90,183	90,228	90,123	90,109	+	
27	12	(∟90°)	⊥-0,3	⊥0,3	90,16	90,16	90,256	90,194	90,142	90,192	90,154	90,15	+	
28	13	⏥0,2			0,006	0,006	0,004	0,008	0,002	0,009	0,005	0,01		-
29	14	⌖0,4			0,372	0,125	0,285	0,301	0,142	0,327	0,268	0,159		-
30	15	Rz 30min												
31	16	37		0,3										

Figure 2. Type of checklist.

The process of detecting defects after entering measurement results and constructing control charts is automated using the VBA programming language. When entering dimension values in the control sheet, this value is compared with the controlled size recorded in three cells. The nominal dimension value is indicated in the cell of the "Main Projection" column, the lower tolerance field - in the "LTL" column cell, and the upper tolerance field - in the "UTL" column cell. If the value of the monitored parameter is outside the tolerance field, the corresponding MS Excel table cell is highlighted in red. [12]

Control charts are built for each controlled size. The values of average sizes and ranges for building charts are calculated according to measurements of controlled sizes.
Figure 3 shows the algorithm for constructing charts of average values and ranges.

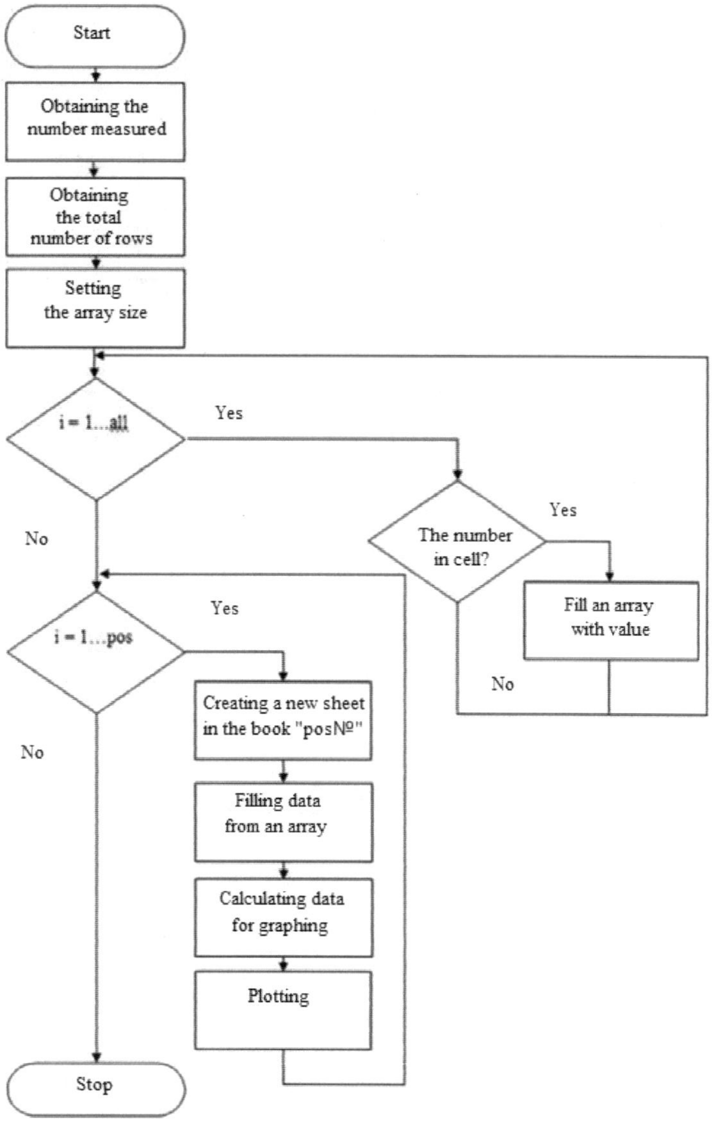

Figure 3. An algorithm for constructing control charts.

When the "Create charts" button is clicked, the number of controlled sizes is automatically determined by the data of the checklists [13]. Next, an array of the form array (1 To pos, 1 To all) is formed for each size, where pos is the number of positions and all is the number of sizes obtained during control. The first three numbers of each line of the array correspond to the nominal size, upper and lower tolerance fields, respectively.
To build control charts for each controlled size, a new sheet is automatically created in the MS Excel workbook with a name that matches the position number indicated in the list of controlled part sizes. After the array is formed, sheets with the name "pos." are automatically created, into which an empty template table is copied for entering the control results. The data for calculating the average values and ranges for the groups corresponding to the nests of the foundry mold are entered in the template columns, 8 values - in each one. To this, the data array of the first sheet "Sheet1" is transposed. Values of nominal

sizes, upper and lower boundaries of the tolerance field are also automatically entered in the corresponding cells. The resulting table of input data for calculating the average values and range for groups is presented in Figure 4.

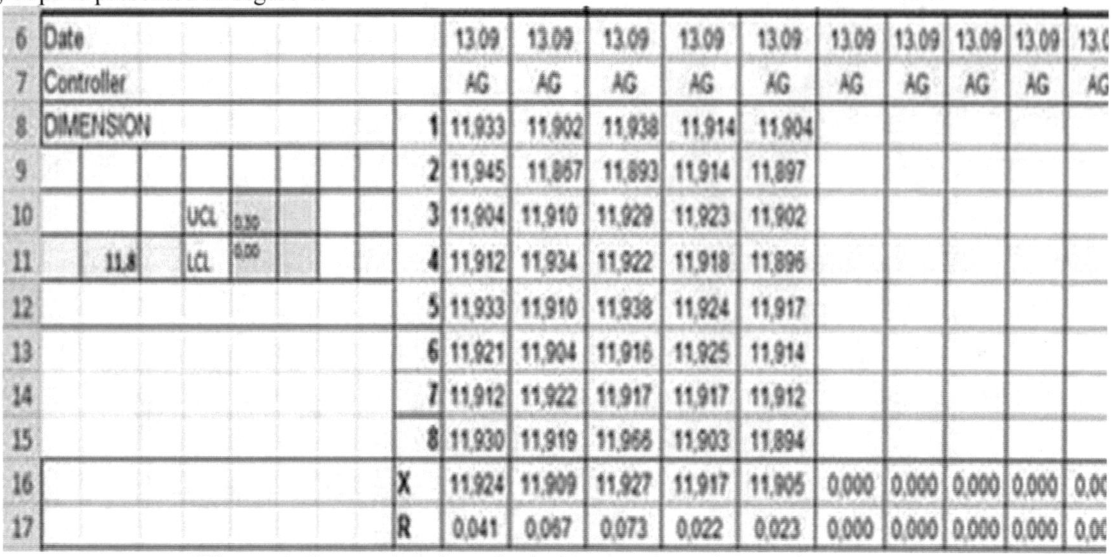

6	Date					13.09	13.09	13.09	13.09	13.09	13.09	13.09	13.09	13.09	13.0
7	Controller					AG	AG	AG	AG	AG	AG	AG	AG	AG	AG
8	DIMENSION				1	11,933	11,902	11,938	11,914	11,904					
9					2	11,945	11,867	11,893	11,914	11,897					
10			UCL	0,30	3	11,904	11,910	11,929	11,923	11,902					
11		11,8	LCL	0,00	4	11,912	11,934	11,922	11,918	11,896					
12					5	11,933	11,910	11,938	11,924	11,917					
13					6	11,921	11,904	11,916	11,925	11,914					
14					7	11,912	11,922	11,917	11,917	11,912					
15					8	11,930	11,919	11,966	11,903	11,894					
16					X	11,924	11,909	11,927	11,917	11,905	0,000	0,000	0,000	0,000	0,00
17					R	0,041	0,067	0,073	0,022	0,023	0,000	0,000	0,000	0,000	0,00

Figure 4. Source data for calculating average values and ranges.

The program fills in the templates of sheets "pos№" and the construction of control cards in sequence. After filling in the template, the number of groups is automatically calculated and the average value, range, and standard deviation for the entire data sample are calculated for each group. These values are automatically recorded in a table cell.
Based on the obtained data, the values of C_p and C_{pk} are calculated by the formulas:

$$C_p = \frac{UTL-LCL}{6\sigma} \tag{1}$$

where *UTL* is the upper tolerance limit; LTL is the lower tolerance limit; σ is the standard deviation.

$$C_{pk}(P_{pk}) = \min[PPU; PPL] \tag{2}$$

PPU and PPL are upper and lower process suitability indicators.
PPU and PPL are calculated using the formulas:

$$PPU = \frac{UTL-\bar{x}}{3\sigma} \tag{3}$$

$$PPL = \frac{\bar{x}-LTL}{3\sigma} \tag{4}$$

where UTL is the upper tolerance limit; LTL is the lower tolerance limit; $\bar{x}$ is the average value; σ is the standard deviation.
The results of the program are presented in Figure 5.

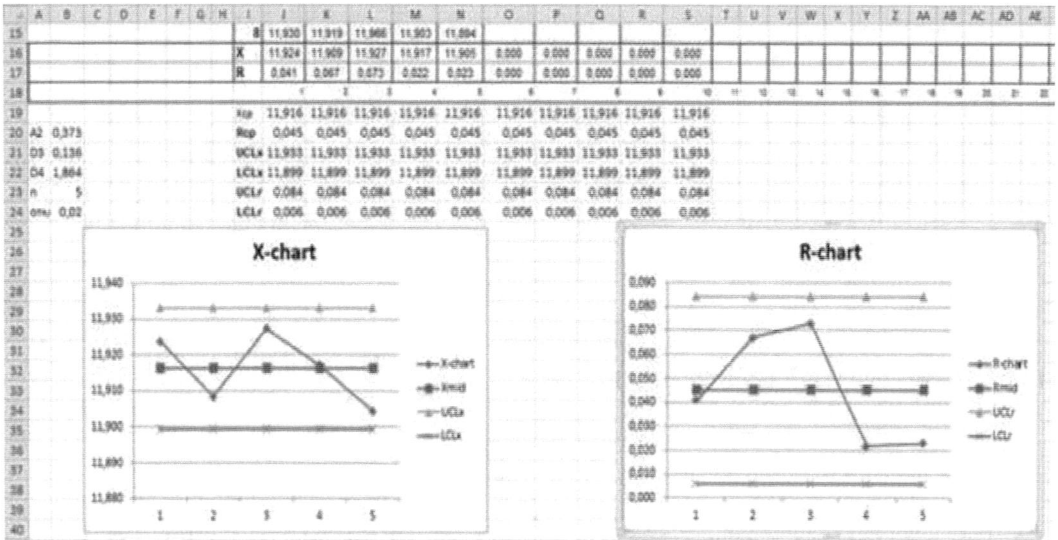

Figure 5. Automated application results.

3. Conclusion

Today, information technology is implemented to raise all the processes of managing an industrial enterprise to a new qualitative level and ensure the achievement of new horizons in all areas of activity;

When the work is done in accordance with the requirements of the standard GOST R 50779.42-99 (ISO 8258-91), automated statistical control of the production of automotive components was organized at the machine-building enterprise;

The results of the calculations and the constructed control cards can be used to interpret the casting process;

The introduction of an automated application has improved the quality of manufacturing parts of automotive components by preventing defects;

The cost-effectiveness analysis allows us to conclude the feasibility of introducing an automated application that would pay off within twelve months.

References

[1] Bataev, A. V. (2018). Analysis of the application of big data technologies in the financial sphere. Paper presented at the *Proceedings of the 2018 International Conference "Quality Management, Transport and Information Security, Information Technologies", IT and QM and IS 2018,* 568-572. doi:10.1109/ITMQIS.2018.8525121

[2] Klochkov, Y., Gazizulina, A., & Ostapenko, M. (2019). Development of QFD methodology. Paper presented at the *IOP Conference Series: Materials Science and Engineering, , 618*(1) doi:10.1088/1757-899X/618/1/012083

[3] Aytasova, A., Selezneva, Z., Belinskaia, I., & Evdokimov, K. (2019). Development of the process map "research and development" for agricultural organizations. Paper presented at the *IOP Conference Series: Materials Science and Engineering, , 666*(1) doi:10.1088/1757-899X/666/1/012072

[4] Klochkov, Y. S., & Tveryakov, A. M. (2020). Approaches to the improvement of quality management methods. *International Journal of Systems Assurance Engineering and Management, 11*, 163-172. doi:10.1007/s13198-019-00939-x

[5] Alekseev, A. P., Efremov, V. V., Potekhin, V. V., Zhao, Y., & Du, H. (2020). *Digital twin analytic predictive applications in cyber-physical systems* doi:10.1007/978-3-030-34983-7_35

[6] Babkin, A. V., Kuzmina, S. N., Oplesnina, A. V., & Kozlov, A. V. (2019). Selection of tools of automation of business processes of a manufacturing enterprise. Paper presented at the *Proceedings of the 2019 IEEE International Conference Quality Management, Transport*

and Information Security, Information Technologies IT and QM and IS 2019, 226-229. doi:10.1109/ITQMIS.2019.8928302

[7] Kobicheva, A., & Safonova, A. (2020). Specialists' training by integration of the high school and business structures. Paper presented at the *E3S Web of Conferences, , 164* doi:10.1051/e3sconf/202016412016

[8] Pupentsova, S., Leventsov, V., Livintsova, M., Alexeeva, N., & Vodianova, S. (2019). Assessment of the internet of things projects on the real estate market. Paper presented at the *IOP Conference Series: Materials Science and Engineering, , 618*(1) doi:10.1088/1757-899X/618/1/012041

[9] Zegzhda, D. P., Kalinin, M. O., & Levykin, M. V. (2019). Actual vulnerabilities of industrial automation protocols of an open platform communications series. *Automatic Control and Computer Sciences, 53*(8), 972-979. doi:10.3103/S0146411619080339

[10] Bolsunovskaya, M., Shirokova, S., & Loginova, A. (2019). State of current data storage market and development of tools for increasing data storage systems reliability. Paper presented at the *E3S Web of Conferences, , 135* doi:10.1051/e3sconf/201913504076

[11] Drobintsev, P., Voinov, N., Kotlyarova, L., Selin, I., & Aleksandrova, O. (2020). *Optimization of technological processes at production sites based on digital modeling* doi:10.1007/978-981-15-2341-0_75

[12] Gutman, S., & Rytova, E. (2020). *Indicators for assessing the development of smart sustainable cities* doi:10.1007/978-3-030-60080-8_4

[13] Ivanov, D., Kalinin, M., Krundyshev, V., & Orel, E. (2020). Automatic security management of smart infrastructures using attack graph and risk analysis. Paper presented at the *Proceedings of the World Conference on Smart Trends in Systems, Security and Sustainability, WS4 2020,* 295-300. doi:10.1109/WorldS450073.2020.9210410

Business incubator assessment model

N S Tsygankov[1], A E Petrunina[1], A K Moskalev[1] and Y E Valkova[2]

[1]Institute of Engineering Physics and Radioelectronics, Siberian Federal University, Krasnoyarsk, Russia, ak_moskalev@mail.ru
[2]Department of the English language and Professional Communication, Financial University under the Government of the Russian Federation, Moscow, Russia

Abstract. At the initial stage of economy innovative potential formation, the leading role belongs to small venture firms that provide market development of the intellectual activity results. Support for small businesses of this type is usually provided through the development of a business incubators system initiated by regional authorities and implemented through targeted budget funding. In this regard, it is important for incubators to develop an indicators system for evaluating the effectiveness of budget allocations and private investment. The paper proposes a two-level system of business incubator functioning and develops a management structure that allows analyzing the current state of the organization and the results of its work using a mathematical model. The incubator activity is determined by the state of the system at the time of starting the next cycle of work, the state of the incubated companies at each stage of incubation (pre-incubation, incubation, post-incubation) and planned tasks, taking into account resource and technological limitations. The proposed model can be used in the development of methods for continuous deviations monitoring from the planned performance indicators of the organization with the ability to pinpoint the problem and respond to it. The advantage of the proposed model is that it can be scaled up to the level of the innovation structure of the region.

1. Introduction

In today's Russia, various measures are being developed and policies are adopted to promote economic development and increase the innovation potential of the economy. At the same time, the leaders in this context are small venture capital firms that facilitate market development of intellectual activity results. However, the majority of small firms are known to have failed and passed away, once got into the "valley of death" for various reasons of managerial, organizational and financial nature.

According to scientific articles [1, 2], business incubation programs for innovative companies enhance the survival rate of venture firms and reduce the number of unsuccessful innovation projects. Therefore, they have a significant role to play in the innovation infrastructure.

Practical indicators of incubator performance in Russia do not allow us to draw any conclusions about how successfully the correspondent tasks are solved. This is primarily due to the lack of unified approaches to understanding the efficiency of infrastructure facilities, as well as the absence of a system of quantitative indicators that would provide such an assessment. In Russia, this becomes particularly important because the creation of business incubators is usually initiated by state authorities or universities. Accordingly, their business model often implies targeted financing from the founders [3], i.e. budget financing. That is why the issues of effective use of these funds seem to be exceptionally relevant.

2. Theoretical framework

According to the definition of the state fund of funds and a development institute "Russian Venture Company" (hereinafter referred to as RVC) [4-6], a business incubator is an organization established to support entrepreneurs at an early stage of their business (under 3 years of activity). This organization shall offer premises for rent and provide services essential for business activities, including consulting, accounting and legal services. Statistics show that only 5% of all incubators in Russia are owned by individuals or companies, while the rest have been created and financed from the federal or regional budget.

The majority of business incubators are considered, on the one hand, as organizations for "nursing" fast-growing innovative companies and, on the other hand, as an instrument of economic development that promotes the spread of innovations, job creation, as well as the business realization of intellectual activity in the scientific sector. The main services of the business incubator include generating employment and infrastructure, creating educational programs and expert consulting, as well as assistance in team building, PR, and preparation for market entry (including foreign markets).

There are different business models and sources of income for incubators. As mentioned above, most of them are formed on the initiative of state authorities and budgetary educational institutions, so the main source of income for them is funding from stakeholders. In addition, the following sources of income can be singled out: rent for the use of premises and equipment, consulting services to residents, payment for educational and incubation programs, payment for services to attract financing.

As a rule, the specifics of a business incubator imply that only business units that have achieved certain success may become residents. As noted by Russian and foreign researchers [1, 5-9], the ultimate goal of the incubation process is boosting the survival rate of the incubated company and transforming the startup into a successful stable business.

3. Statement of the problem

In this regard, the main purpose of this work is to form a system of performance indicators for business incubators, which can be used by management bodies of the business incubators in assessing their activities and the efficiency of the use of budgetary allocations and private investments.

4. Methodology

In this regard, the main purpose of this work is to form a system of performance indicators for business incubators, which can be used by management bodies of the business incubators in assessing their activities and the efficiency of the use of budgetary allocations and private investments.

5. Discussion

5.1. Approaches to evaluate the efficiency of the incubators.

A considerable amount of scientific works is devoted to the issues of evaluation of business incubators' performance. Given the budget nature of financing Russian incubators, the most remarkable are the requirements for their efficiency on the part of the authorities. For example, the Ministry of Finance has developed methodological recommendations to the executive authorities of the constituent entities of the Russian Federation and local governments, which help to increase the revenue base of the budgets of the constituent entities of Russia and municipal units in the country (8). Among other measures, this document states the need for constant evaluation of the effectiveness of the innovative economic infrastructure components. The recommended methodology is based on periodic surveys aimed at collecting quantitative and qualitative information about both the infrastructure facilities themselves and their residents. The following areas are suggested for analysis: human capital, organizational structure and economic contribution. Meanwhile, specific performance indicators are supposed to be set at regional levels of management independently (Figure 1).

These indicators have a recommendatory character and can be used by executive authorities of the constituent entities of the Russian Federation and local authorities both for comparative assessment of various business incubators and for studying the dynamics of a particular object development.

Thus, within the state task of the Krasnoyarsk Regional Innovation and Technology Business Incubator (hereinafter referred to as KRITBI) for 2019 and for the planning period of 2020-2021, the Ministry of Economy and Regional Development of Krasnoyarsk Krai sets the following targets, which determine the quality and efficiency of KRITBI activities by main types of services and works performed (Table 1)[10-11].

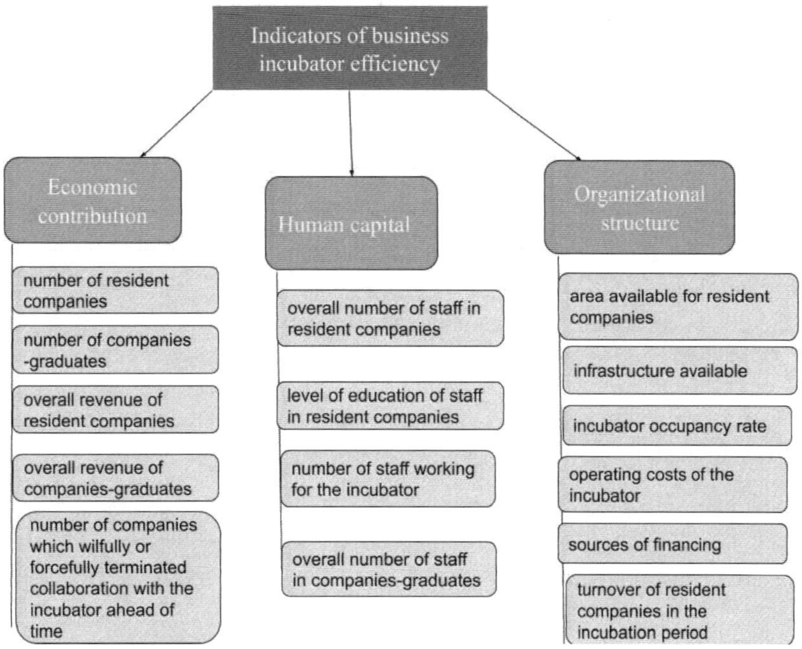

Figure 1. Indicators of business incubator efficiency.

Table 1. Information on works performed and services rendered by KRITBI according to the state task.

Data on rendered state services		
Indicator characterizing the content of the state service	Quality indicator of the state service	Amount indicator of the state service
Provision of information and consulting support to small and medium-sized businesses		
Providing services for the development of business plans, concepts, technical and economic specifications and investment projects to be implemented in a particular constituent entity of the Russian Federation	Satisfaction indicator of the service recipients, %	Number of small business enterprises which got the service
Preparation of presentation and information materials on the activities of small businesses for mass media and for publications	Satisfaction indicator of the service recipients, %	Number of small business enterprises which got the service
Data on works being performed		
Indicator characterizing the content of the work	Indicator of the work quality	Indicator of the work amount

Organization of events in the field of youth policy aimed at involving young people in innovative, entrepreneurial and voluntary activities, as well as events for the development of civic engagement and healthy lifestyle in youth	Indicator of satisfaction with the quality of the works performed, %	Number of events
	Number of people involved in the events	
	Number of technological startups created by young people who used state financing	
Maintenance (operation) of property in state (municipal) ownership		
Ensuring operational and technical maintenance of facilities and premises; upkeep of these facilities and premises, equipment and adjacent territory in due condition	Uninterrupted supply of heat, water and power, %	Area in operation, m^2
	Maintenance of intangible property in proper sanitary condition, %	Number of facilities serviced (operated)

As can be seen from Table 1, for evaluating the quality of work and services provided they use metrics to assess the incubator's efficiency in its key area of activity. However, these metrics do not take into account output criteria of incubator residents, and, accordingly, cannot assert that the main purpose of incubation has been achieved. Besides, the main criteria include the results of the incubator's economic activity on a par with the indicators of incubation of innovative projects, which seems incorrect.

Also of interest are the results of research and quality analysis of business incubators and accelerators in Russia, conducted in 2017 by UBI Global on behalf of RVC and the Business Incubator of the National Research University "Higher School of Economics" (HSE {Inc}) [12]. According to this study, the key performance indicators of business incubators can be presented in three enlarged groups of criteria: value for ecosystem, value for clients, and attractiveness of programmes (Figure 2).

The main advantage of this methodology is that it takes into consideration a wide range of indicators recognized and used by many incubators in Russia and abroad. Therewithal, the methodology provides for the assignment of a weighting factor to each key performance indicator that reflects the relative importance of the respective indicator, which was not born in view in KRITBI.

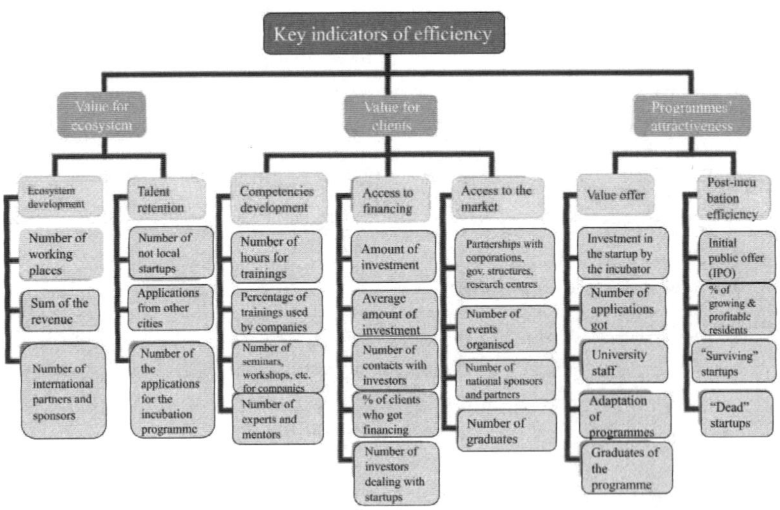

Figure 2. Key indicators of efficiency measured by UBI GLOBAL.

Although the methodology covers a wide range of incubator performance indicators and is intuitive and easy to implement, we can but highlight its significant limitations:

- due to differences in the sectoral specialization of business incubators, it is not possible to create a universal system of weighting criteria under this methodology;
- due to the relatively young age of some business incubators and incubation programs, as well as their current and former residents, it may be difficult to collect information on all the above criteria.

Therefore, despite all the advantages of this methodology, it cannot be considered the only true and universal one.

Another approach for assessing the performance of the business incubator is presented in [13]. From the author's point of view, the system of evaluation indicators concerning business incubator performance consists of three enlarged groups: indicators of the business incubator development, indicators of the residents' development, and indicators of the business incubator's impact on residents' development. The main indicators in the above three areas are shown in Figure 3.

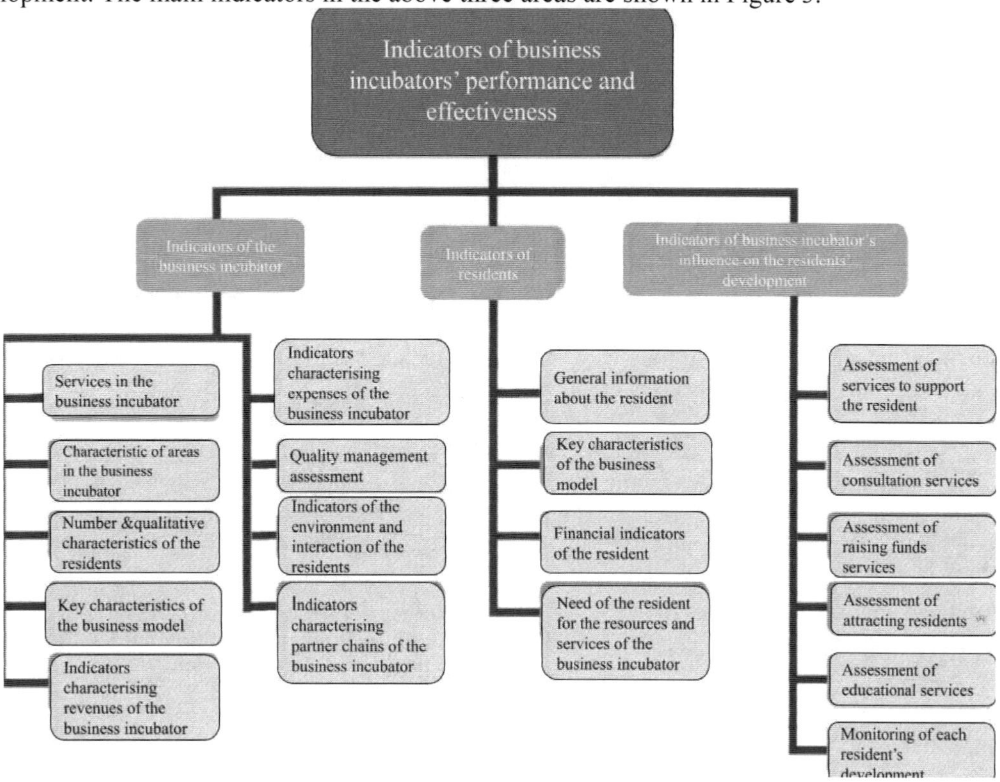

Figure 3. Indicators of business incubators' performance and effectiveness / efficiency.

As the author remarks, each of the indicators presented in Figure 3 can be characterized by one or more numerical values, but these parameters are specific only for incubators established at universities, so this methodology seems very narrow and specialized.

In the foreign literature, the main performance indicators of business incubators are usually divided into following levels of activity: pre-incubation, incubation as it is, and post-incubation (after graduation from incubation). For example, paper [14] presents the author's methodology, which allows for the assessment of incubator performance at all three incubation stages by so-called "hard" (quantifiable) and "soft" (non-quantifiable) criteria (Figure 4). According to this approach, the business incubator's performance is evaluated by the progress of incubated companies, so the main performance indicators are reduced to mere assessment of residents at these three stages.

"Hard" criteria		
✓ Amount of sales ✓ Profitability ratio ✓ Rate of enterprise's growth ✓ Independent sales' realization Competitive	✓ Amount of sales ✓ Profitability ratio ✓ Rate of enterprise's growth ✓ Independent sales' realization ✓ Competitive advantages ✓ State financing ✓ Client retention ✓ Incubation area	✓ Amount of sales ✓ Number of employees ✓ Net profit ✓ Competitive advantages ✓ State financing ✓ Attraction and retention of clients ✓ Incubation area
Pre-incubation	**Incubation**	**Post-incubation**
✓ Business acumen and reputation ✓ Team's professionalism ✓ Innovative character of the business offered ✓ Strategic goals ✓ Knowledge of target sphere	✓ Improving business acumen and reputation ✓ Improving team's professionalism ✓ Improving customer loyalty ✓ Enlarging partner network ✓ Improving knowledge of target sphere ✓ Cost saving due to automatization and BI tools	✓ Professional attitude to clients ✓ Wide business partner network ✓ Mass advertisement ✓ Introduction of innovative products ✓ Contribution to economic development of the region
"Soft" criteria		

Figure 4. Conceptual framework of "hard" and "soft" assessment criteria before, during and after the incubation process.

Another approach described in [15] implies a complex system of performance indicators (Figure 5), consisting of two enlarged groups: internal indicators of the business incubator itself (14 categories of indicators in total), and indicators of incubated companies, which in turn are divided into two groups: indicators of the company at the "input" level of the incubation program, and at the "output" level. Such elaborate system of indicators expediates evaluating the efficiency of the incubator as an infrastructure facility from the point of view of its main function, i.e. increasing the survival of companies. Nevertheless, this system is very complicated in terms of monitoring and collecting values. Besides, the work [16] does not offer a methodology for obtaining the final efficiency value for the listed indicators.

IOP Conf. Series: Materials Science and Engineering **986** (2020) 012016 doi:10.1088/1757-899X/986/1/012016

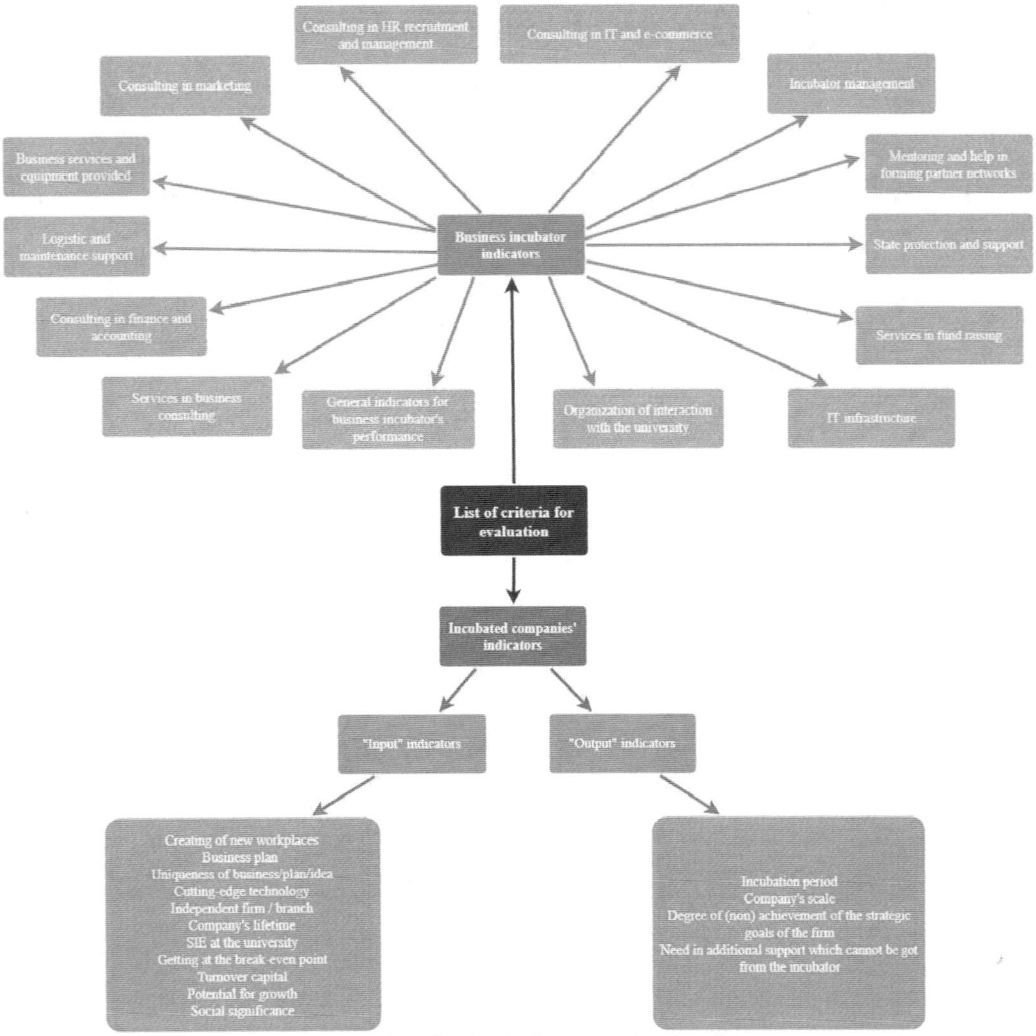

Figure 5. List of criteria for assessing incubators.

In general, it is possible to deduce that due to significant differences between Russian and foreign innovation infrastructure, as well as differences in legislation and regulation of innovation activities, orientation towards foreign experience may be but limited due to differences in the environment in which business incubators operate.

5.2. Two-tier business incubator model.

The process of incubating as regards innovation projects can be presented as a two-tier model, consisting of a management body and a set of business incubators at the lower tier. The interaction between the tiers is characterized by the degree of centralization of management, the degree of separation concerning the financial flow and the extent of isolation.

Figures 6 and 7 are structural diagrams of the two-tier model with distinct variants of relationships between the levels. These models can be used for business incubators with diverse types of management.

The autonomy of business incubators lies in their ability to independently determine their specialization, prices for their services, order and methods of work with projects.

In a model with a high level of centralization (Figure 2), business incubators serve only as executors of orders coming from the managing body, which distributes them among business incubators according to the preset specializations.

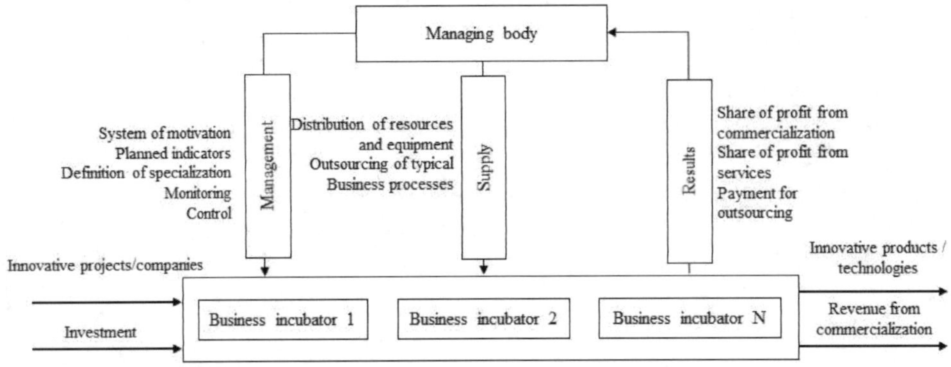

Figure 6. The model of interaction for business incubators with high autonomy.

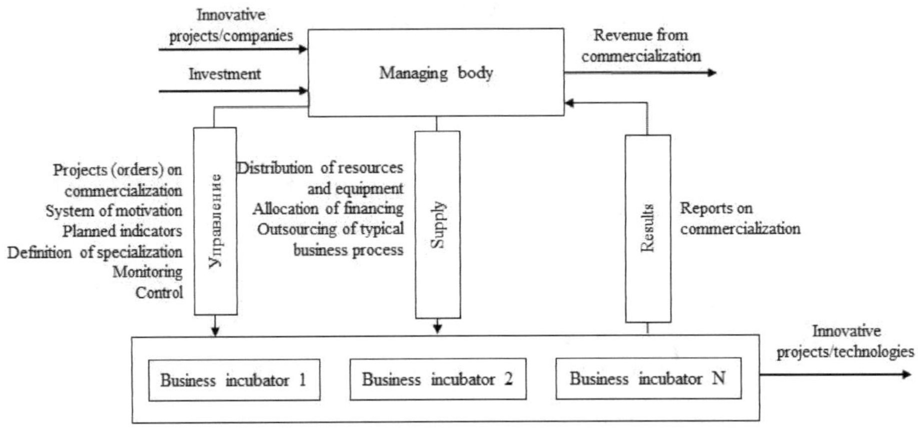

Figure 7. Model of interaction with a high level of centralization at the level of managing body.

The result of the two-tier model is innovative products and technologies that have already been brought to market and which are beginning to generate revenue. Simultaneously, this model also fits into the chain of commercialization with the possibility of subsequent transfer of the innovation project to the corresponding object of innovative infrastructure.

The managing body defines the main criteria for efficiency and direction of business incubators, moreover, within its competence it monitors and adjusts the works of the lower tier. In general, the managing body performs the following functions:

- determination of targets and goals of business incubators for the planned period, including explicitation of their specialization;
- allocation of financing, or making advance payments for planned tasks;
- allocation of available resources and provision of access to equipment;
- situational management;
- development and management of a system aimed at motivating and stimulating business incubators;
- distribution of incoming innovative projects and companies by directions according to the specialization of business incubators.

Business incubators, in turn, work with innovative projects and companies that select and commercialize innovations to generate new sources of income by getting a share in the incubated companies. Further we are going to examine innovative projects and companies as applications for the business incubator.

The process of functioning of each business incubator is usually broken down into 3 stages [16-18]:

- pre-incubation. This stage includes the initial selection of incoming applications from innovative companies or teams and the determination of criteria for this selection;
- incubation. At this stage, the main work is done with innovative companies to commercialize innovations and generate income from them;
- post-incubation. This is a final stage of work with innovative companies, which involves steps to transfer the company to another object of innovative infrastructure or to establish the company's foothold in the market.

However, this approach does not take into consideration the preparatory stage of the next cycle for a business incubator and the stage of efficiency assessment of the current cycle at its completion, which, accordingly, does not form an integral part of the system of the incubators' performance indicators. This reduces the possibility of assessing the efficiency of the next cycle of the business incubator shorn of information at the beginning and end of the cycle.

5.3. Mathematical model of the business incubator state.

Functioning of business incubators can be effective only in case of a systemic approach to their management. For this purpose, the authors have designed a system of indicators for evaluating business incubators, which takes into account the two-tier model of their functioning. At any time business incubator's activity is calculated by 1) the state of the system at the moment when the next work cycle starts, 2) the state of incubated companies at each stage (pre-incubation, incubation, post-incubation) and 3) compliance with the planned tasks (given resource-related and technological limitations). The overall state of each business incubator can be presented as following:

$$A = F(f_0(t_0), f_1(X_1(t), X_2(t), X_3(t)), f_2(r, a, t)) \quad (1)$$

where $f_0(t_0)$ – a vector of indicators characterizing the state of the business incubator at the beginning of t_0 – the period under review,
f_1 – a vector of indicators characterizing the incubation of companies at the time t during the pre-incubation stage X_1, incubation stage X_2, post-incubation stage X_3,
f_2 – a vector of indicators characterizing the implementation of the planned tasks r aimed at commercialization of innovations with available resources a at the moment t.
As for the set of indicators f0, let us take the following in general cases (Table 2).

Table 2. Indicators of the initial state of the business incubator.

Indicator	Procedure of calculation	Designation
1 Autonomy (Ability to independently determine specialization, prices for the services, order and methods of work with projects)	$P_1 * 30\% + P_2 * 40\% + P_3 * 30\%$	P_1 – Separate character of business incubator P_2 – Self-determination in decision-making for functioning of the business incubator P_3 – Self-organization of the business incubator (ability to develop owing to internal potential)
1.1 Separate character (Level of internal supply as regards equipment, technology, etc.)	$(1 - \frac{\sum A_{com}}{\sum A}) * 60\% + (1 - \frac{\sum L_{com}}{\sum L}) * 40\%$	Acom – An asset of the business incubator which is shared with another company A – An asset of the business incubator itself L_{com} – Liabilities of the business incubator which are shared with another company L – Liabilities of the business incubator

Indicator	Procedure of calculation	Designation
1.2 Self-determination (Level of independent decision-making capability)	$\left(1-\frac{F_{st}+F_{in}}{F}\right)*30\%+$ $+\left(\frac{\sum_s I_{of}}{\sum_s I-\sum_s I_p}-\frac{\sum_s I_{st}}{R}\right)*$ $*70\%$	F – Total number of founders F_{st} – Founders from state enterprises F_{in} – Founders from large industrial companies s – Number of startups which got investment I – Total amount of investment in startups I_{of} – Sum of investment in startups at the expense of the own funds I_p – Sum of private investment in startups I_{st} – Sum of state investment in startups R – revenue from the business incubator
1.3 Self-organization (Internal growth level of the business incubator)	$\frac{CO}{CO+CB}$	CO – Number of projects formed from within CB – Number of projects introduced from outside
2 Level of development of organizational communications contributing to the informative value of interaction	$\frac{(CM-CM_r)*2}{n*(n-1)}*30\%+$ $+\frac{\sum_c TC}{\sum_c TW}*25\%+\frac{CI_r}{CI}*15\%+$ $+ST*10\%+LL*20\%$	CM – Number of existing communication channels between the business incubator and innovative infrastructure facilities CM_r – Number of repetitive channels of communication n – Number of participants in the process of communication TC – Time spent on communication process in providing services to 1 client c – Client TW – Time spent on the work with one client CI – Amount of works on collection and analysis of information CI_r – Amount of works on collection and analysis of information on identical (repetitive) topics ST – Speed of transfer and spread of information LL – Level of information loss

Indicator	Procedure of calculation	Designation
3 Level of demand for services contributing to development	$\frac{O_{R\&D} + O_D}{O_{total}} * 60\% + \frac{P_{sold}}{P_{got}} * 40\%$	O_{total} – Total number of orders $O_{R\&D}$ – Total number of orders for research and development from private sector O_D – Total number of orders for development works from private sector P_{sold}– Number of patents sold through the business incubator P_{got} – Number of patents sold through the business incubator
4 Level of engagement with academia and the private sector	$\frac{P_j}{N} * 65\% + \frac{E_{sc}}{RS} * 35\%$	P_j – Number of joint projects N – Total number of new small innovative enterprises E_{sc} – Share of research scientists employed in small innovative enterprises RS – Optimum share of research scientists employed in small innovative enterprises (it is determined by the business incubator according to the directions of projects)
5 Share of private financing R&D	$\frac{Ir\&d_{depart} + Ir\&d_{sc}}{Ir\&d}$	$Ir\&d_{depart}$ – Sum of private investment in R&D departments $Ir\&d_{sc}$ – Sum of private investment in R&D departments of scientific institutes Ir&d – Total sum of investment in R&D, including grants
6 Efficiency of the network uniting business incubator's enterprises	$\frac{NP}{N} * 25\% + \frac{PA}{NP} * 50\% + \frac{JE}{NP} * 15\% + \frac{S_{free}}{S} * 10\%$	PA – Number of partner agreements NP – Number of network participants JE – Количество совместных предприятий S_{free} – Number of free services within the network S – Number of services provided within the network N – Number of newly registered small innovative enterprises
7 Availability of access to private investors	$(NET * 40\% + SI * 60\%) * 50\% + 50\% * \left(\frac{D}{N} * 70\% + \frac{D}{B} * 30\%\right)$	NET – Net present value of startups SI – Satisfaction with the investments' security B – Number of investors in the database D – Number of agreements on investment N – Number of startups which need investment

Indicator	Procedure of calculation	Designation
7.1 Net present value of startups	$\sum_{i} \mathrm{NPV}$	NPV – Net present value of startups of *i*-startup
7.2 Satisfaction with investment security	$\frac{\sum_i \mathrm{I}}{\mathrm{N}}$	I – Opinion of the i-investor asked N – Number of investors asked
8 Efficiency of IP commercialization	$\frac{\mathrm{IP_c}}{\mathrm{IP_b} - \mathrm{IP_s}} * 65\% + \frac{\mathrm{IP_s}}{\mathrm{IP_b} - \mathrm{IP_c}} * 35\%$	IP_c – Number of commercialized objects of intellectual property (IP) IP_s – Number of commercialized objects of intellectual property sold IP_b – Number of commercialized objects of intellectual property which were created with the help of the business incubator
9 Efficiency of the allocation of the business incubator's own funds	$\frac{\mathrm{I_1} + \mathrm{I_2} + \mathrm{I_3}}{\mathrm{N}}$	I_1 – Sum of preferable terms for incubation I_2 – Sum of direct investments I_3 – Sum of indirect investments (grants, competition, etc.) N – Revenue of enterprises which got financing

As for the indicators characterizing the work on incubation of companies, it may be preferrable to use the previously considered indicators (Figures 6 and 7). They include general economic indicators and the results of technological audit.

Planned tasks for the business incubator stand for a plan for the number of commercialized products and technologies at a given time period, as well as the revenue of the business incubator, the number of activities carried out, etc. At the same time, the orientation of the projects the business incubator works with is not constant and can change during its work.

Planned targets may be achievable with sufficient resources and may be unachievable with their absence. In this case, this group of indicators can be presented as below:

$$f_2(r,a) = \begin{cases} f_2^+, \ r|r(a) \neq \emptyset \\ f_2^-, \ r|r(a) = \emptyset \end{cases} \quad (2)$$

In the first case, the business incubator is able to fulfill all the planned tasks with resources available, so the task becomes optimizational one entailing search for the most effective way of implementation against the background of abundant and overlapping planned tasks and possible results (the available resources are undoubtedly counted). In case of lack of resources for full implementation of planned tasks, the solution is to identify the minimum distance between *r* and *r(a)*.

Thus, to manage the business incubator it is feasible to determine the set of planned values *A* in accordance with the number of planned periods. At the end of each period, the deviations for each of the components are analyzed and corrective measures are developed. In order to achieve a single numerical value of deviation from the plan for each of the components and individual indicators, we quantify weight coefficients pursuant to the strategy and development priorities of the specific business incubator.

6. Conclusion

The following results were obtained following the work done:

1. There was built a two-tier model of business incubator functioning. A simple structure of this model enables analysis of business incubator's work and it can be easily integrated into a more complex model, for example, the model for work of the innovation infrastructure as a whole. The proposed model can be of avail in studying both private and public business incubators and those run by universities.
2. A mathematical model has been made to describe the current state of the business incubator. This model can be applied to develop methods for continuous monitoring of business incubator activity and increasing the speed of response to deviations from the plan with the possibility of point solution of a particular problem.

Acknowledgements

The reported study was funded by RFBR, project number 19-37-90023.

References

[1] Posza A 2019 Business incubation analysis with the help of real option theory *International J. of Multidiscipl. in Business and Science* (5(8)) pp 64-72.

[2] Zegzhda, D. P., & Stepanova, T. V. (2015). Thermodynamic approach to modeling multi-agent systems in cyberspace. Automatic Control and Computer Sciences, 49(8), 622-626. doi:10.3103/S0146411615080398

[3] Zegzhda, P. D., Zegzhda, D. P., & Stepanova, T. V. (2015). Approach to the construction of the generalized functional-semantic cyber security model. Automatic Control and Computer Sciences, 49(8), 627-633. doi:10.3103/S0146411615080192

[4] Vasin S M and Gamidullaeva L A 2013 Povyshenie effektivnosti upravleniia biznes-inkubatorom kak sotsial'no-ekonomicheskoi sistemoi [Improving the efficiency of business incubator management as a socio-economic system] (Penza: Penza State University) p 233

[5] Guidelines for improving the efficiency of business incubators and accelerators 2017 *Available at: https://www.rvc.ru/upload/iblock/c5e/guidelines.pdf* (accessed 8 April 2020).

[6] Problems and solutions: business incubators and technology parks in Russia 2014 *Available at: https://www.rvc.ru/upload/iblock/f7d/201403_Business_incubators.pdf* (accessed 7 April 2020).

[7] Rogova E M 2013 Effektivnost' funktsionirovaniia biznes-inkubatorov kak elementa spin-off strategii universitetov [The effectiveness of business incubators as the element of the universities' spin-off strategy] *Innovations* (10(180)) pp 41-46

[8] Stephens S and Onofrei G 2012 Measuring business incubation outcomes *Entrepreneurship and innovation* (13(4)) pp 277-285

[9] Trivedi H and Asrani R 2019 Best Incubation Practices: An In-depth Framework for Indian Business Incubators *J. of The Gujarat Research Society* (21(8)) pp 176-185.

[10] Drobintsev, P., Voinov, N., Kotlyarova, L., Selin, I., & Aleksandrova, O. (2020). Optimization of technological processes at production sites based on digital modeling doi:10.1007/978-981-15-2341-0_75

[11] Moskvin, D., Zegzhda, D., & Nikolskiy, A. (2015). Secure processor architecture modeling for large-scale systems. Automatic Control and Computer Sciences, 49(8), 684-689. doi:10.3103/S0146411615080301

[12] Panyukov, D., Kozlovsky, V., & Klochkov, Y. (2020). Development and research FMEA expert team model. International Journal of Reliability, Quality and Safety Engineering, doi:10.1142/S021853932040015X

[13] Methodological recommendations to Executive authorities of the subjects of the Russian Federation and local self-government bodies that contribute to increasing the revenue base of the budgets of the subjects of the Russian Federation and municipalities 2018 *Available at: https://www.minfin.ru/* (accessed 6 April 2020).

[14] State task of the regional state autonomous institution "Krasnoyarsk regional innovation and technology business incubator" for 2019 and for the planning period 2020-2021 (2019). *Available at: http://kritbi.ru/* (accessed 4 April 2020).

[15] Assessment of University-Linked Business Incubators and Accelerators 2017 *Available at: https://www.rvc.ru/upload/iblock/4b1/UBI_Global-Russia-Impact_Assessment_University-Linked_Business_Incubators_Accelerators_RU.pdf* (accessed 7 April 2020)

[16] Gozali L, Masrom M, Zagloel T Y M, Haron H N, GarzaReyes J A, Tjahjono B, Irawan A P, Daywin F J, Syamas A F, Susanto S, Aliwarga H K K and Marie I A 2020 Performance Factors for Successful Business Incubators in Indonesian Public Universities *Int. J. of Technology* (11(1)) pp 155-166.

[17] Poltavtseva, M. A. (2015). Support of the integrity of administrative information in intensive data exchange with RDBMS clients. Automatic Control and Computer Sciences, 49(8), 690-695. doi:10.3103/S0146411615080143

[18] Stepanova, T. V., & Zegzhda, D. P. (2015). Large-scale systems security evolution: Control theory approach. Paper presented at the ACM International Conference Proceeding Series, , 08-10-Sep-2015 doi:10.1145/2799979.2799993

IOP Conf. Series: Materials Science and Engineering **986** (2020) 012017 doi:10.1088/1757-899X/986/1/012017

Neural network modelling for determining the priority areas of regional development

A K Moskalev, A E Petrunina, N S Tsygankov

Siberian Federal University, 28, Kirensky St., Krasnoyarsk 660074, Russia

Email: ak_moskalev@mail.ru

Abstract. Neural network modeling based on self-organizing Kohonen maps was carried out in order to cluster the economy of one of the administrative districts of the Krasnoyarsk Territory. The research is based on statistical data on the economy of about one hundred enterprises of various spheres operating in the region. It is shown that in the process of clustering four groups of enterprises can be clearly distinguished according to the types of economic activity. Having applied the neural network modelling method, we identified three enterprises within the framework of this cluster, which can be considered as "growth areas" of the district economy. Self-organizing maps that is trained using unsupervised learning make it possible to get an idea not only of promising areas of development, but also to identify key parameters that ensure the leadership and competitiveness of the region.

1. Introduction

The evolution of the cluster theory in economics has happened for several decades already, but this topic still does not lose its relevance [1]. Today in Russia, cluster approach is the most effective method of development of the regions, since as a result of the functional integration, the businesses have not only increased their profitability, but also, due to the active cluster interactions, improved and multiplied competitive advantages for both cluster members and the region as a whole [2].

Cluster formations, being independent and self-organized economic units, are able to ensure not only quantitative growth of the economy, but also to act as stimulators of innovative activity in the region and certain interrelated industries [2-4]. The cluster approach as a technology of regional development management and, accordingly, the cluster itself as an element of innovation infrastructure, create opportunities for the region and businesses not only to survive, but also to develop during the transition to an innovative model of development [5-6].

There are different approaches to the definition of a cluster. Analyzing them taken together, we can say that they have one common feature, namely the integration of any objects that are allocated from a large population on the grounds of a certain identical group characteristic. Emphasizing the distinctive features of clusters in each of the sciences, one shall note that the universal features of clusters can be the following: homogeneity of components for the mutual efficient exchange of available resources; the combination of centripetal and centrifugal forces; the fuzziness and mobility of boundaries and configuration; the shared resource base.

The following methods are used to determine cluster specialization [7-9]:

- analytical method: the study of data on the territory, its natural, labour and financial resources of the productive forces.
- method of indicative planning of competitiveness: quantitative and qualitative analyses of the present competitiveness and determination of the prospective one.
- modelling method: the study of cluster objects by building and studying models of real organizations, processes or phenomena.

IOP Conf. Series: Materials Science and Engineering **986** (2020) 012017 doi:10.1088/1757-899X/986/1/012017

- the “input-output” method: the study of the circulation of goods and services between industries and other sectors of the economy.

2. Methods

In this study, we used a neural network method to model the specialization of the cluster, and Kohonen self-organizing maps of as a clustering tool. The simulation was carried out using special software – analytical platform Deductor Studio to build a neural network trained by unsupervised learning method based on the structure of the input data.

The classification method based on self-organizing Kohonen maps is as follows:

- determination of the input data vector based on the results of the training experiment;
- determination of parameters for building a network;
- analysis of the results with finding an estimate of the classification accuracy.

To organize a network, you must define the following construction parameters:

- the size of the training and test set;
- network size (number of cells);
- an example recognition criterion (the value of the recognition error);
- learning speed and radius;
- the number of clusters.

Since the sample size is small, the training set should be as large as possible and sufficient for the correct organization of the network. In this case, the test set should be sufficient for testing the network with 100% traffic. Based on the stated conditions, the size of the training set was 78 copies (91% of the sample), the size of the test set was 8 copies (9% of the sample).

The size of the network was chosen in such a way as to exclude the appearance of "dead" neurons; such neurons, the vector of synaptic weights of which is significantly removed from the vector of input data. Such neurons are not used when training the network, because cannot win competition from nearby neurons. This situation leads to the fact that the input data is interpreted by a smaller number of neurons than was originally specified, which introduces distortions in the final result. To study the presented samples, networks with a dimension of 16x12 cells were selected.

The type of cells is selected as hexagonal: it more correctly displays the Cartesian distance between the centers of the cells. The rest of the parameters for constructing the Kohonen network are presented in table 1.

Table 1 – Network parameter

Network parameter	**Specified Criterion**
Recognition error rate	An example is considered recognized if the recognition error is less than 0,05%
Number of eras	250
Learning speed	At the beginning of training: 0,4 At the end of the training: 0,005
Training radius	At the beginning of training: 2 At the end of the training:: 0,2
Clustering	Fixed number of clusters: 4 [6]

Kansky district of the Krasnoyarsk Krai was chosen as an object of clustering for several reasons:

- need for the development of territories, including districts of the Krasnoyarsk Krai, is determined by the strategy of social and economic development of Siberia until 2020;
- agriculture and forestry sectors prevailing in the region are designated as priority areas in this programme;
- low indicators of efficiency of economy of the area in comparison with nearby territories of Krasnoyarsk Krai (Nazarovsky, Achinsky, Sharypovsky Districts).

The analysis of the economy made on the basis of the study of the economic activities of the leading enterprises of the Kansky District showed that there are four main areas of activity:

- meat and dairy production;
- plant growing;
- forestry;

- other companies (commodity and resource-providing producers).

The database contains economic indicators of 86 enterprises of the Kansky District with different spheres of activity, including:

- 12 enterprises producing meat and dairy products;
- 15 enterprises producing crop products;
- 46 enterprises in the sphere of forestry;
- 13 resource-providing enterprises.

Kansk, the administrative centre of Kansky District, is considered to be the core of the future cluster.

During the simulation, the following parameters were selected as input:

- cost of sales (RUB);
- net profit (RUB);
- distance from the cluster core (km);
- number of employees (people);
- type of activity.

3. Results

The result of modelling self-organizing maps is presented in Figure 1. A fairly homogeneous structure is visible, which corresponds to the sectoral affiliation of the District's economy.

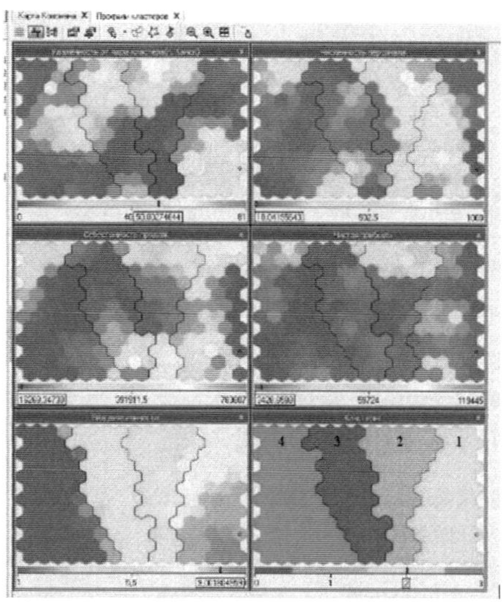

Figure 1. Visualization of clustering results. Add legend and describe the process.

The X axis shows the correspondence of the attribute value and the color gradient depending on its value. Cells in the figure characterize the number of neurons used in the neural network (192). Boundaries, in black, characterize the selection of values that fall into one cluster. The top left card is the Distance from the cluster core indicator, the top right card is the Number of employees indicator, the middle left card is the Cost of sales indicator, the middle right card is the Net profit indicator, the lower left card is the indicator "Type of activity", bottom right map - general map of the distribution of the resulting neurons into clusters.

Neural network modelling showed the presence of four distinct clusters, the sizes of which differ slightly. Taking into account the topological similarity of companies, the indicator of the type of activity was singled out as the main one.

Figure 2 shows the significance of the indicators for creating modelling maps. The first cluster combines companies producing meat and dairy products, and processing companies with the highest competitiveness. Also, the results of modelling show that the companies of this sector have a similar number of staff and the cost of production, so these indicators are also set as having a high importance.

The second cluster includes companies producing crop products and companies of the processing industry. The indicators that determine the specifics of the companies of this cluster include the type of activity, the number of staff and the annual net profit of the company.

Table 2 - Cluster profile

Field	Indicator	Clusters				
		1	2	3	4	TOTAL
Number of enterprises in cluster	Number	46 (53,5%)	17 (19,8%)	15 (17,4%)	8 (9,3%)	86 (100%)
Type of activity	Significance	100,0%	100,0%	50,3%	93,8%	100,0%
Number of employees	Significance	97,1%	94,2%	82,1%	99,8%	100,0%
Net profit	Significance	93,6%	72,8%	63,4%	98,7%	99,72%
Cost of sales	Significance	93,8%	93,2%	15,5%	64,8%	97,42%
Distance from the cluster core	Significance	86,8%	43,4%	73,7%	78,7%	91,62%

The third cluster includes resource-providing companies. These companies provide:
- water supply, power supply and heat supply;
- equipment supply.

In this cluster, the type of activity is no longer decisive, as the key ones are the distance from the core of the cluster and the number of employees.

In the fourth cluster, companies related to forestry were combined. This cluster is formed by the largest number of companies, so this direction is the main candidate for leadership.

However, analyzing the requirements and recommendations for the structure of cluster formations [2], it can be concluded that in addition to industry companies, the clusters should include:

- suppliers of raw materials and equipment;
- financial institution;
- research institutes;
- transport and trade intermediaries;
- waste processing and recycling facilities;
- social infrastructure companies.

On this occasion, it can be concluded that the agricultural and industrial cluster is the most promising tendency of economic development of the Kansky District.

Based on the received information and the requirements for companies that should join the cluster, we may offer to introduce into the agricultural cluster of Kansky District the companies from the first group, which form the core of a cluster, companies from the second group, which supplement a system of agricultural products, as well as some companies of the third group, supporting the activities of major companies.

It is imperative to note that the indicator of territorial remoteness from the centre of the cluster was not identified as determining what is adequate in modern conditions of digital transformation and widespread development of communication technologies.

Based on the requirements for the companies that must be included to the cluster we find it necessary to add the following companies to the above mentioned ones:

- veterinary institutions (3 organizations in the District);
- scientific and educational institutions (more than 10 institutions in the District);
- banking institutions (branches of more than 15 different banks).

In the cluster formed during the modelling, it is possible to identify key "growth areas", i.e. companies that are mostly important for this cluster. For this purpose, using statistical data only of the first group companies, a modelling procedure was carried out, the results are presented in figure 2.

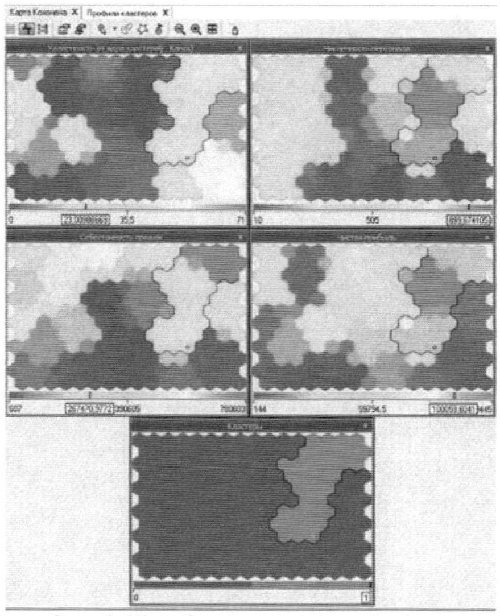

Figure 2. Visualization of modelling for determining the key enterprise for the District.

The X axis shows the correspondence of the attribute value and the color gradient depending on its value. Cells - characterize the number of neurons used in the neural network (192). Boundaries, in black, characterize the selection of values that fall into one cluster. Top left map - Distance from the cluster core, upper right map - Number of employees, left middle map - Cost of sales, right middle map - Net profit, bottom map - general distribution map final neurons per clusters.

4. Discussion of the results

As a result of modelling, three companies of the Kansky District with the criterion of maximum net profit were selected, and they can be the key/flagship enterprises of the cluster.

This research allows us to conclude that the use of the neural network approach in the formation of the list of priority areas of economic development of regions has great prospects. Self-organizing maps that support the algorithm of teaching without a teacher, enable one to get an idea not only about the promising directions of development, but also to identify the crucial parameters that ensure the leadership and competitiveness of the region.

The application of the considered approach of using a neural network for clustering enterprises within a region can be used in the development of comprehensive programs for regional innovative development, as well as in the development of a model for the functioning of a regional innovation infrastructure in terms of methods for selecting enterprises and projects to support individual objects of innovation infrastructure

Acknowledgments

The reported study was funded by RFBR, project number 19-37-90023.

References

[1]. Markov L S 2015 *Theoretical and methodological foundations of the cluster approach* (Novosibirsk, SB RAS Publ.) 300

[2]. Abramov R A, Podchufarov S K 2017 Regional innovation potential of industrial clusters. *Reg. Econ.: Theory and Practice* **15 5** 881-895

[3]. Kotliarov N N 2014 Foreign experience in the formation of cluster systems *World Econ. and Int. Econ. Relations* **10** 105-110

[4]. Tonkova S, Burmatova O P 2016 Bulgaria's experience in the formation of cluster policy *Region: Econ. and Sociology* **2** 237-252

[5]. Burmatova O P 2016 Clusters as a form of local production systems *Econ. and Management World* **16 4** 93-108

[6]. Zemtsov S P, Bukov D V 2016 Methods for identifying clusters of small and medium-sized businesses *Regional Econ.: Theory and Practice* **3** 104-117

[7]. Glukhov, V. V., & Balashova, E. (2015). Operations strategies in info-communication companies doi:10.1007/978-3-319-23126-6_48

[8]. Ivanov, D., Sokolov, B., Serova, E., & Yusupov, R. (2020). Combined approach to the complex objects control and stability analysis of management decisions. International Journal of Risk Assessment and Management, 23(1), 106-118. doi:10.1504/IJRAM.2020.106175

[9]. Shestak, O., Shcheka, O. L., & Klochkov, Y. (2020). Methodological aspects of use of countries experience in determining the directions of the strategic development of the russian federation arctic regions. International Journal of Systems Assurance Engineering and Management, 11, 44-62. doi:10.1007/s13198-019-00805-w

Imitation Simulator for PID Controller Adjustment

M E Kremenetskaya[1], A A Shkromado[2] and E A Bulkaeva[3]

[1] Control Engineering and System Analysis Department, Samara State Technical University, 244 Molodogvardeyskaya Street, Samara, 443100, Russia, mme82@mail.ru
[2] Electromechanics and Industrial Automation Department, Syzran Branch of Samara State Technical University, 45 Sovetskaya Street, Syzran, 446001, Russia
[3] Refining Engineering Department, All-Russian Research Institute for Oil Refining, 6/2 Aviamotornaya Street, Moscow, 111116, Russia

Abstract. The article contains the description of the structure and functionality of the developed imitation simulator of PID controllers adjustment, which makes it possible to setup their typical modifications on the example of a hypothetical technological process taking into account the peculiarities of real contollers and restrictions, introduced by actuators and control signals.

1. Introduction

The quality of technological processes of the continuous and hybrid productions is largely determined by the quality indicators of the transition processes of technological parameters. Despite the obvious progress and achievements in the field of automatic control theory, for the implementation of the required technological modes and achievement of quality parameters are still used primarily proportional-integral-differential (PID) controllers as relatively easy to configure and providing the necessary result [1-2].

The principles of construction, modification of PID controllers and methods of their adjustment, including experimental ones, are well enough researched and described in a wide range of sources, for example, in [2-6]. Programs of automated calculation of parameters, such as, for example, P.I.D.-Expert or PIDTune, and also the whole environments of dynamic modelling, such as MatLab Simulink, containing tools of autotuning taking into account restrictions, render invaluable help in adjustment of controllers.

However the first of them implement the "strict" model of adjustment which is limited by the underlying methods of calculation and often does not consider features of real controllers and restrictions [6-8], made by actuators and control signals; the second provide rich functionality, but require the development of specialized models, accounting for the features of the object and the actuator, the complexity and cost of which can exceed the effect obtained.

Therefore, to improve the quality of process control, it is important to develop a compromise application. On the one hand, it is built on the modular principle and allows to account for some or other (by connecting the appropriate modules) typical limitations in the setting and features of real PID controllers, depending on the process. And on the other, it is free from the need to develop full-featured models.

From the point of view of the authors, it is reasonable to combine the calculation module with the interface module, which will not simply provide the results of the setting in the form of numerical

values of parameters and transition diagrams, and additionally visualize the technological component of the process, in order to finally get a simulator for setting up PID controllers.
In the turn, simulators play a major role in the development of professional skills and competencies, as well as allow to fully master the nature of the processes that are necessary to acquire the profession.

2. Structure and functionality of the application

Two basic modules can be distinguished in the application: the calculation and the interface ones. As the calculation module assumes layout of conditions, based on the problem solved, it is a set of simulation subprograms, designed in the form of ready-made functional blocks that form the final program simulator [9-10].
In contrast to the dynamic simulation programs, these blocks are specified for specific application tasks (taking into account actuator limitations and a control signal type) and require a limited set of typical setting design parameters, the source of which are equipment data sheets and process guidelines.
Thus, the calculation module makes it possible to differentiate the PID controller setting process depending on the control signal and actuator type, together with significant actuator limitations. The differentiation involves the appropriate simulation subprogram [11].
For example, for a simple case where the control is carried out by an analogue set-point signal and the speed of the actuator is high enough, i.e. slightly impairs the dynamic performance of the system, it is possible to use the elementary simulation subprogram shown in Figure 1 (hereafter, for clarity, the object model is included in the simulation subprogram).

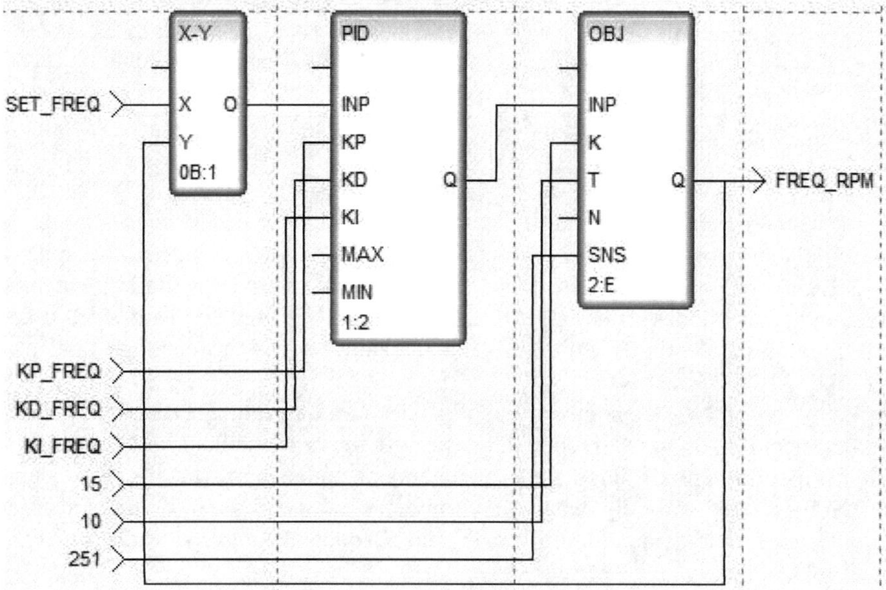

Figure 1 - Simulation program with no actuator limitations
and the analogue signal control

This program in the simulator is used to calculate the reference setpoint value, expressed as a percentage, used to control the pump motor by a frequency inverter.
The PID controller generates the output value according to the law (1)

$$Q_i = KP \cdot INP_i + \frac{KD \cdot (INP_i - INP_{i-1})}{\Delta t} + KI \cdot \Delta t \cdot \sum_{k=1}^{i} INP_k \ , \qquad (1)$$

where INP - the value received at the input of the controller; i - the current clock; KP, KD and KI - coefficients at proportional, differential and integral components, respectively; Δt - recalculation period of the block in seconds (duration of the clock).
To describe the technological control object the first order inertial transfer function with delay (2) is used

$$W(s) = \frac{K}{Ts+1} \cdot e^{-\tau s}, \qquad (2)$$

where K is the transfer coefficient, T is the time constant, τ is the delay of the control object [12].

In case of use of the PWM (pulse-width modulated) control signal and necessity to account for the resource limitation of the non-inertial actuator operation, it is possible to use a simulation program in Figure 2.

The program in the simulator is used to simulate PID temperature control in a reagent tank by controlling an electric heater with electric pulses of constant frequency and variable duration.

To convert the control signal of the PID regulator "PID" into a sequence of latitudinal-modulated pulses of single amplitude, the "IREG" unit is used. The pulse duration is determined by the formula (3)

$$T_{imp} = \frac{T_{\max} \cdot INP}{100}, \qquad (3)$$

where INP is the input value (INP) of the "IREG" block in the range [-100 ... 100]; Tmax is the maximum pulse duration in seconds. At Timp < Tmin, where Tmin - minimum pulse duration in seconds, the pulse is not generated [7].

The duration of the pause between pulses is defined as (4)

$$T_{pause} = T_{\max} - T_{imp}, \qquad (4)$$

Certainly, the choice of Tmin and Tmax values should be made primarily based on the characteristics of the actuator, which is also worked out in this simulation model.

As long as the value at the input (INP) is non-negative, pulses are generated at the output (Up) and at negative values - at the output (Dn) [7]. Since the heater has no reverse mode (i.e. cooling mode), all control signals are removed from the output (Up).

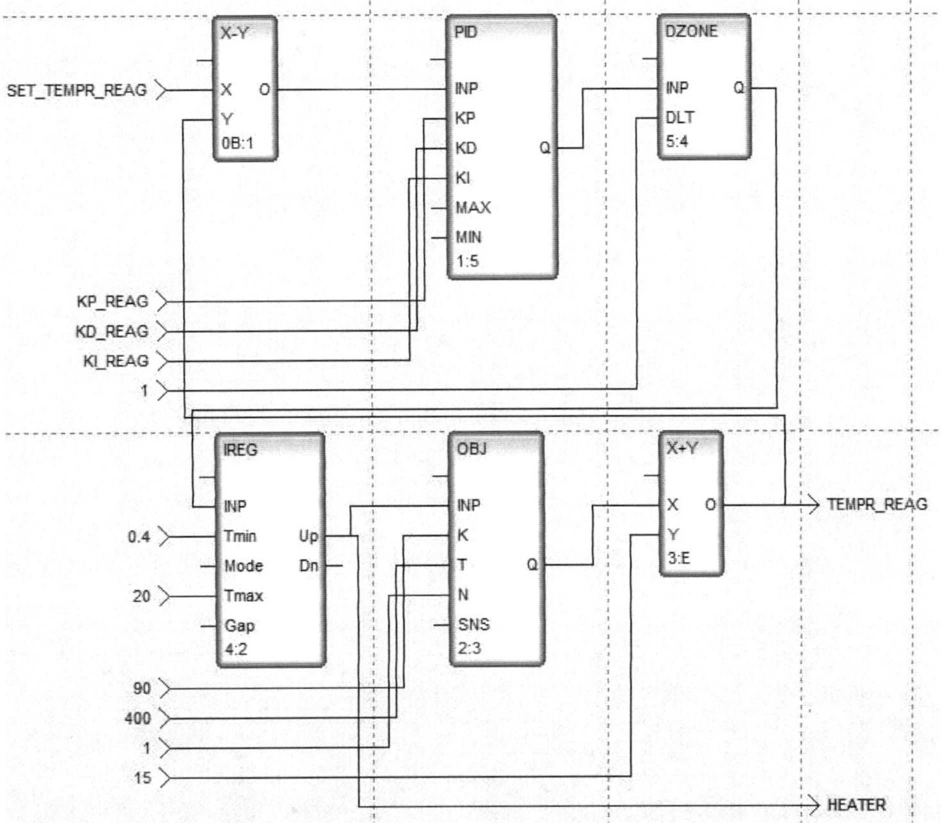

Figure 2 – Simulation program for accounting of the actuator triggering resource and PWM control

IOP Conf. Series: Materials Science and Engineering **986** (2020) 012018 doi:10.1088/1757-899X/986/1/012018

To register the influence of the actuator frequency and reduce the depreciation of the switching equipment (actuator) of the heater let us use the block "Insensitivity Zone" DZONE. This block can be installed both in front of the "PID" block (in this case the insensitivity zone by reagent temperature would be entered) and in front of the "IREG" pulse controller (ignoring control pulses of short duration). We use the second solution (Fig. 2) due to the presence of thermal asymmetry of the object. This implementation provides a reduction in the number of starts in the area of the established value due to the fact that the small values of the control signal of the PID controller (block "PID"), arising from an insignificant mismatch and which requires a brief switching on of the heater, are ignored, and thus are considered equal to zero. A side effect of this solution is a decrease in control accuracy.
The "Insensitivity Zone" unit works in the following way - as long as the value at its input (INP) is greater than the value of the insensitivity zone submitted to the input (DLT), the value itself from the input (INP) is transmitted to the output (Q). Otherwise, the output (Q) is zero [7].

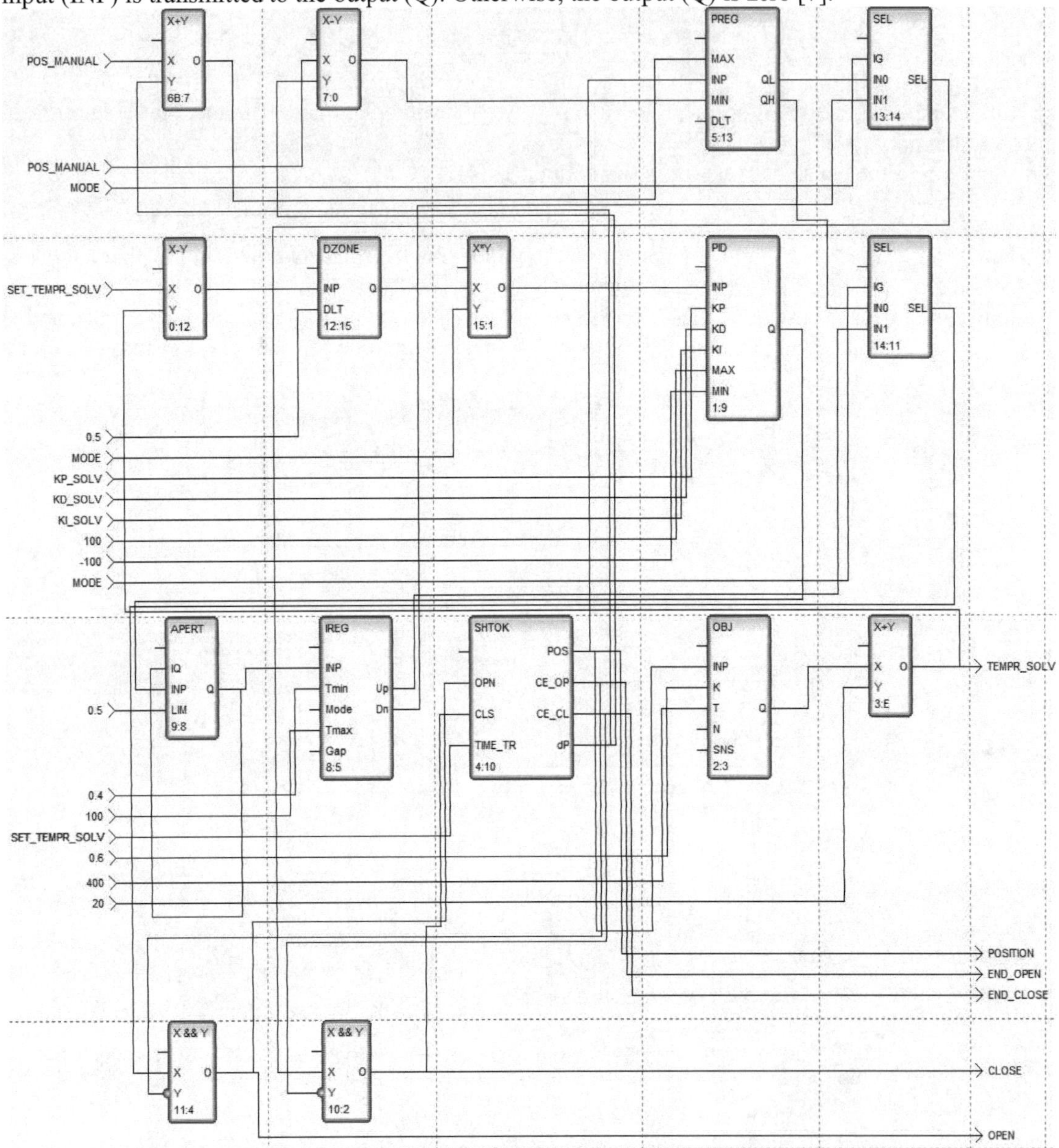

Figure 3 - Simulation program for cascade control mode

As one more example let's review the simulation program of adjustment of the PID controller, taking into account as restrictions the speed and stroke (the "saturation" restriction type) of the integrating actuator type, and also influence of aperture smoothing and a zone of insensitivity on regulated technological parameter and a zone of insensitivity of the actuator. In addition, this program allows to work out the setting of the PID controller in a cascade control, in which the PID controller performs regulation of the technological parameter of the external circuit, and the three-position controller provides regulation of the technological parameter of the internal circuit on a signal of the PID controller task. In this case it is possible to directly control the three-step regulator in manual mode with ignoring the task signal from the external loop controller, which allows to work out such a feature of "behavior" of real PID controllers as integral saturation.

The program simulates PID temperature control at the heat exchanger output by controlling the "More" and "Less" pulse width modulation of the reversing valve actuator that changes the fluid flow.

The SHTOK block is responsible for the simulation of the actuator operation. This block allows you to set the full stroke time of the control valve stem, the amount of stroke, to monitor the limit switches on the position of the stem, and calculates the "real" position of the stem.

The DZONE "Insensitive Zone" block allows for the implementation of a temperature insensitive zone. Thus, the mismatch values, smaller insensitivity zone values, will not arrive at the "PID" link input - the mismatch calculated by block "X-Y" №0 is zero and, accordingly, the controller will not generate an output signal. This will reduce the switching frequency of the executive device and thus increase its resource.

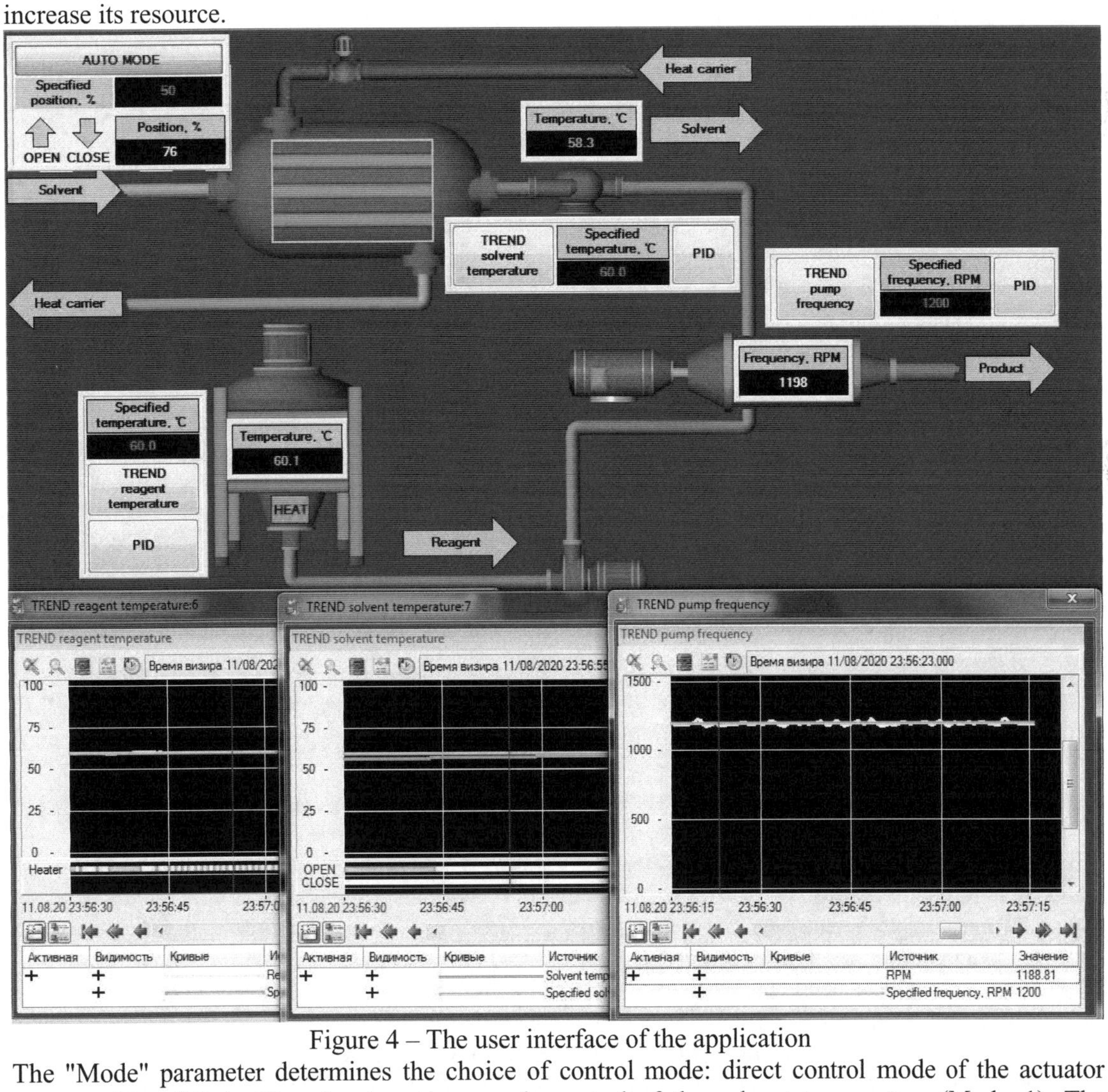

Figure 4 – The user interface of the application

The "Mode" parameter determines the choice of control mode: direct control mode of the actuator stem position (Mode=0) and automatic cascade control of the solvent temperature (Mode=1). The

IOP Conf. Series: Materials Science and Engineering **986** (2020) 012018 doi:10.1088/1757-899X/986/1/012018

value of the parameter is also used to stop the accumulation of integral component error and to prevent integral saturation in stem position control mode (automatic temperature control mode is disabled), in which Mode=0.

To filter small changes in the control signal of the "PID" unit, an APERT aperture block is used to reduce the switching frequency of the valve actuator and to prevent frequent changes in the direction of the stem movement (reversal). The block implements the IF algorithm |INP-IQ|<LIM THEN Q:=IQ ELSE Q:=INP [7].

Direct operator control of the stem position is implemented by the "Three-position Controller" PREG block based on threshold values compared to the current stem position (INP), calculated from the insensitivity zone on the valve stem position and the stem position setpoint (Handle position). Thus, control commands of the actuator are issued: at INP>MAX, the outputs QL=1, QH=0 (closing); at INP<MIN, the outputs QL=0, QH=1 (opening); at MIN<INP<MAX, the outputs QL=0, QH=0 (actuator stopped - value within the insensitive zone).

Since the program provides two control modes, the mode selection is organized by means of the "Choice of two" SEL blocks. This unit sends an output (SEL) value from the input ((IN0) or (IN1)) whose number is set at the input (IG) [7].

The interface part of the application makes it possible to evaluate the quality of setting of PID controllers for the considered typical applications (types of actuators, configurable limitations, applied control signals) and the overall process flow based on the analysis of numerical values of process parameters as well as tracking the dynamics of their change according to real-time graphs (trends).

In addition, the interface part increases the visibility of the process, thus simplifying the perception of information and increasing involvement in the training process, which has a positive impact on the results (quality) of training the trainee.

3. Conclusions

Therefore, the article considers the use of a simulator to improve the quality of process control, enabling to consider the necessary (by connecting the appropriate modules) typical limitations of actuators and features of real PID controllers and does not require the development of full-featured process models. The simulation program is formed, proceeding from a solved problem and conditions, in the form of the typical simulation subprograms designed as functional blocks.

The results of setting are visualized in the form of numerical values of parameters on the technological mnemonic scheme and graphs of their transition processes, which allows you to comprehensively assess the quality of PID controllers.

4. References

[1] Ang K H, Chong G and Li Y 2005 *PID control system analysis, design, and technology* vol 13, No 4 (IEEE Transactions on Control Systems Technology) pp 559-576

[2] Denisenko V V 2006 *PID controllers: principles of construction and modification* no 4 (Modern automation technology) pp 66–74

[3] Denisenko V V 2007 *PID controllers: principles of construction and modification* no 1 (Modern automation technology) pp 78–88

[4] Denisenko V V 2006 *PID controllers: questions of realization the modification* no 4 (Modern automation technology) pp 86–97

[5] Denisenko V V 2008 *PID controllers: questions of realization the modification* no 1 (Modern automation technology) pp 86–99

[6] Varlamov I and Zubov P 2017 *PID-expert — automation by automation* no 2(68) (Control engineering Russia) pp 96–99

[7] IEC 61131-3: *Programming Industrial Automation Systems. Concepts and Programming*

[8] Glukhov, V. V., & Balashova, E. (2015). Operations strategies in info-communication companies doi:10.1007/978-3-319-23126-6_48

[9] Kalinin, M. O., & Pavlenko, E. Y. (2015). Increasing the fault tolerance and availability of software defined networks using network equipment control based on multiobjective optimization by service quality parameters. Automatic Control and Computer Sciences, 49(8), 673-678. doi:10.3103/S014641161508026X
[10] Klochkov, Y. S., & Tveryakov, A. M. (2020). Approaches to the improvement of quality management methods. International Journal of Systems Assurance Engineering and Management, 11, 163-172. doi:10.1007/s13198-019-00939-x
[11] Pyayt, A. L., Shevchenko, D. V., Kozionov, A. P., Mokhov, I. I., Lang, B., Krzhizhanovskaya, V. V., & Sloot, P. M. A. (2015). Combining data-driven methods with finite element analysis for flood early warning systems. Paper presented at the Procedia Computer Science, , 51(1) 2347-2356. doi:10.1016/j.procs.2015.05.404
[12] Zharkovskii, A. A., & Pospelov, A. Y. (2015). Use of 3D methods for flow analysis, prediction of characteristics, and optimization of the shape of settings of hydraulic turbines. Power Technology and Engineering, 49(1), 27-32. doi:10.1007/s10749-015-0568-4

IOP Conf. Series: Materials Science and Engineering **986** (2020) 012019 doi:10.1088/1757-899X/986/1/012019

Assessment of digital maturity of enterprises

D.A. Levaniuk[1], M V Bolsunovskaya[2], S V Shirokova[2] and A M Gintciak[2]

[1]StudyFree Inc., Delaware, the USA, marina.bolsunovskaia@spbpu.com
[2]National Technology Initiative Center for Advanced Manufacturing Technologies based on the Institute of Advanced Manufacturing Technologies of Peter the Great St. Petersburg Polytechnic University Polytechnicheskaya, 29, St.Petersburg, 195251, Russia

Abstract. The project management methodology is an approach to the formation of a set of methods that structures the project management system and is reflected in the manuals. The aim of the article is to increase the efficiency of IT project management by adapting existing management methodologies considering the characteristics of projects implemented in the company. Performing the following tasks allows to achieve the goal:
- review of existing methodologies;
- description of the object of study;
- identification of key features of IT project management in the company OOO «ITSK»;
- IT project management guidelines;
- application of a corporate standard as part of an IT project in the oil and gas industry.

The relevance of the article is explained by the wide distribution of IT projects of various fields of application, including projects to automate the activities of an oil and gas industry company that require management throughout the entire life cycle. The application of the project management methodology allows you to record the goals and results of the project, give them quantitative characteristics, determine the time, cost and quality parameters of the project, create a realistic plan for the project, identify, assess risks and prevent possible negative consequences during the implementation of the project. In the course of the work, a brief analysis of the existing project management methodologies was carried out, the specifics of IT project management described, the adaptation of the methodology in the oil and gas industry and the description of the standard used within the company are reflected. Using an example of a project to automate monitoring of KPIs, the application of a corporate methodology based on best practices was demonstrated.

1. Introduction

In the 21st century, project management is one of the most popular ways to achieve the strategic goals of companies. Creating a unique result for a certain time interval with allocated resources allows you to better focus on the implementation task. Firstly, the effectiveness of the project participants is increasing, everyone is interested in achieving its success. Secondly, the predictability of the result is the advantage of project management. In the course of work, a schedule is compiled, divided into stages with specific start and end dates, possible risks. Thirdly, the company's activities become transparent, controlled and easily managed. When manag-ing, specific results are indicated, the resources that are already spent or needed at further stages are fixed.

Proper project management, following the recommendations of existing methodologies, using the regulatory framework of standards allows us to achieve successful results. To get the most efficient work in the midst of change, it is important not only to use best practices, but also to be able to adapt to the particularities of the organization of activities of enterprises in which projects are implemented.

2. Material and methods

The article is based on the analysis of existing approaches to the assessment of maturity and digital maturity and summarizes the data obtained from various sources. This section will briefly describe the methods that companies use in project management and strategy development.

2.1. Project management methodologies

The project management methodology is an approach to the formation of a set of methods that structures the project management system and is reflected in the manuals.

Standards in the field of project management are a combination of best practices and they are developed by examining successfully implemented projects, systematizing the information received and accumulating it in specialized reference books. This knowledge is formed sequentially over a long time [1].

Consider the main project management standards used in the world.

PMBoK, by far, is the most popular project management standard and is applied in almost all countries of the world. Historically, it is most popular in the countries of North and Central America (more than 70% of PMI members live in North America), as well as Southeast Asia, but it is no less popular and successfully used in Europe [2].

The PRINCE2 methodology is slightly less popular and is mostly used in European countries. Historically, it enjoys the greatest popularity in the UK, but is often used by American companies, especially in the field of oil production and refining[3].

The use of PMBoK in the PRINCE2 environment enriches PRINCE2 with valuable information, especially in the following areas that are missing or insufficiently strong in PRINCE2: Procurement Management, Human Resource Management, Earned Volume Method, Communication Management, Cost Management, Timing Management, Common Project Management Knowledge, including well-known terminology.

When using PRINCE2 in the PMBoK environment, PMBoK acts as a foundation on which the well-developed processes of the PRINCE2 methodology work, the latter can be considered as a structured checklist for project management.

Thus, an organization whose project managers use PRINCE 2, one way or another, needs additionally a more extensive methodology, such as, for example, the PMBOK Guide.

At the same time, the application of the PMBOK Guidelines requires a localized method for national and industry specifics, the same PRINCE 2 or another specialized standard.

ISO 21500: 2012 or its equivalent GOST R ISO 21500-2014 establishes more concise requirements that make it easier to develop an adapted corporate standard. At the same time, neither PMBOK nor PRINCE 2 ISO contradict.

Consider the distinguishing features of the most popular PMBoK and PRINCE2 standards used by the company for project management.

To apply PRINCE 2 and the PMBoK Guide at the corporate level, these standards require a process of adaptation to real conditions and an established management culture.

Thus, almost all standards applied to project management are based on two main PMBoK and PRINCE2. Companies are based on best practices and create their own corporate standards that suit their occupation and individual characteristics.

Despite the existence of a large number of different project management standards, all of them are aimed at fixing the purpose and results of the project, determining quantitative characteristics, determining the time, cost and quality parameters of the project, creating a realistic plan for the project, designating, assessing risks and preventing possible negative consequences during project implementation.

2.2. IT project management

The term "IT project" is usually used to refer to activities related to the use or creation of some information technology. This leads to the fact that IT projects cover a wide variety of areas: development of software applications, creation of information systems, deployment of IT infrastructure, etc.

A project is a set of efforts undertaken to obtain specific unique results within the allotted time and within the approved budget, which is allocated to pay for resources used or consumed during the project.

An IT project is a separate set of interrelated activities aimed at creating or substantially changing the totality of hardware and software tools aimed at automating business processes for managing an enterprise / company under temporary, resource, financial, and technical constraints [8].

On the other hand, they have well-known distinctive features:

- Separation at the ideology level of the customer and the contractor: the customer, as a rule, is a business, and the contractor is IT specialists, and there are difficulties in identifying requirements, expectations from the project, in the formation of technical specifications. There is also the problem of effective communications.
- Responsibility for the result of the project has a "joint" character. That is, one cannot assign responsibility for the success of the project only to the contractor, just as one cannot say that the customer is solely guilty of the failure of the project. The IT project must create certain conditions for the interaction of the parties, and the parties involved in it bear equal responsibility for the results of the project; · often the implementation of the IT project involves changing the existing organizational structures in the enterprise;
- Typically, many organizational units are involved in an IT project.
- There is a high likelihood of conflicts between the project manager, senior management, department heads, and organization personnel.
- Many IT projects have huge budgets. In large companies, the scale of project activities in the field of information technology (IT) is measured in millions of dollars, and the implementation of new projects is ongoing. If, for example, it is enough to build an industrial enterprise once — and it will work without requiring regular investments, then the development of IT infrastructure in growing companies requires large and regular investments. Large budgets, in turn, imply a greater level of responsibility and, accordingly, a greater level of competence of those people who manage these projects.

The implementation of IT projects, requires following such features as:

- Often, at the customer's company, several IT projects are carried out simultaneously.
- Priorities for the implementation of projects are constantly being adjusted.
- As the projects are implemented, the requirements and content of the projects are refined and adjusted.
- The influence of the human factor is great: the timing and quality of the project are mainly dependent on the direct performers and the communication between them.
- Each performer can participate in several projects; there are difficulties in planning creative activity; there are no uniform norms and standards.
- An increased level of risk remains, up to unpredictability of results.
- There is a continuous improvement of the technology of work [9].

Analysis of statistics shows that approximately 90 percent of IT projects are similar to those already completed. The project manager has experience in implementing such tasks and understanding of possible problems. In these cases, the hierarchical structure of the project and work (ISP / ISR) is formed using the Top-down approach (top to bottom), the typical structure of the project team is used, project plans (risk management plan, communication plan, etc.) are similar to the plans of previous projects [10]. However, 10 percent of projects are innovative, implemented from scratch and require creativity, innovative solutions and managerial courage. Decision-making in such projects is characterized by high risks, which requires the head of deep knowledge of the project management methodology and understanding of the features of its application in the field of information technology.

GOST 34 standard. "Development of an automated control system (ACS)"

One of the widely used is GOST 34. "Development of an automated control system (ACS).

This standard was created to record the results of the work performed.

According to GOST 34.601-90, the life cycle of the process of creating an ACS consists of certain stages [11]:

- Formation of requirements for speakers.
- Development of a speaker concept.
- Terms of Reference.
- Draft design.
- Technical project.
- Working documentation.
- Commissioning.
- Speaker support.

2.3. Description of the current standard for managing IT projects using the example of an oil and gas company

In the course of the work, referring to the enterprises of the oil and gas area, one of the companies developed its own corporate standard.

In this paper, we will consider an aspect that relates more to project planning. In cases of IT projects, the probability of technological changes is high, so the project manager should clearly understand what actions and at what stages will be undertaken. In such situations, planning and preparation become crucial tasks, along with managing the project participants, choosing a technological solution, forming and using the budget.

Awareness of possible changes during the project in advance helps to avoid the risk of a shift in the timing of the start or implementation of the project, which is perceived extremely negatively by stakeholders and users.

One of the reasons for the failure of the project is the poor communication between the participants. As a result, the project manager should be able to manage all aspects of the project, quickly and clearly interact with participants to quickly resolve any problems that arise [12].

IT project management aims to successfully complete and get a business effect. Each group of project processes begins with initiation and should be as clear and transparent as possible for the performers. In cases where adjustments to the technical solution, selection of another contractor, change of concept, or failure to meet deadlines by suppliers, it is also necessary to maintain transparency of the process. Each stage and milestone of the project must have an actual start or end date. Thus, management, in particular IT project, requires constant monitoring of compliance with the plan.

The methodology of the company's IT projects corresponds to:

- the approach adopted by the GPN to the implementation of projects;
- best practices of international oil companies;
- standard GOST 34 601-90.
- The best IT project management methodologies contain the following key elements:
- stages and gates;
- tasks and results of the stages;
- role model.

The focus on obtaining business benefits as part of a single and transparent project management process is key in the implementation of projects.

This standard describes the basic rules of IT project management for:

- focus on obtaining business effects, as part of a single and transparent project management process;
- increase the degree of satisfaction of business requirements;
- rational use of internal and external resources, as well as effective planning and implementation of project tasks;
- project risk and deviation management;
- increasing the guarantee of obtaining the declared results of projects;
- ensuring compliance with the terms, budget and scope of the project.
- To achieve these goals, the following guidelines have been introduced:
- Improving the manageability of the project by breaking it into successive stages;

- Phased implementation of the project and reducing the level of uncertainty of project parameters (timelines, budget, volume of tasks);
- Phased detailing of requirements and control of their satisfaction;
- Use of external expertise to ensure the quality of project results;
- Use of previous experience;
- The materiality of deviations of project parameters.

Guidelines

Consider implementing the guidelines in more detail:

1. Improving the manageability of the project by breaking it into successive stages.

To increase the controllability of achieving results, all project activities are divided into stages. The stages of the project are carried out sequentially. This provides an opportunity to stop the implementation of the project at any stage, if its further implementation is deemed inappropriate.

Each stage is implemented as part of the Deming cycle (PDCA approach): stage work planning ("Plan"), stage work ("Do"), verification ("Check"), improvement, decision making ("Act").

Each stage of the project ends with the adoption of one of the decisions:

- Finalize the project results;
- Pause the project;
- Cancel the project;
- Go to the next stage of the project.

The work of the stage begins after a decision is made on the transition of the project to this stage.

2. Phased implementation of the project and reducing the level of uncertainty of project parameters.

Staging is an uncertainty management tool. The work of the stages is aimed at increasing the value of the project and reducing the level of uncertainty of project parameters (time, budget, amount of work. Achieving the required level of uncertainty of the project parameters (timelines, budget, volume of tasks) and the project value is confirmed by the results of the stage.

3. Phased detailing of requirements and control of their satisfaction.

Particular attention in the project should be given to the conceptual development of an IT solution. choosing the right project implementation concept is critical to project success.

At the stages preceding the direct implementation of the IT solution, a consistent study of the requirements for the IT solution delivered by the Project is carried out.

When forming requirements, it is important to determine the circle of stakeholders of the project, reflecting them in the organizational structure of the project, and take into account the requirements of all project participants.

At the stage of "1. Evaluation "defines the objectives of the IT project according to SMART criteria (specific, measurable, achievable, relevant, time-limited), business requirements, the process and organizational scope of the IT project, at stage" 2. Choice "solution architecture with approval by the Technical Council, requirements for the subsystems and components of IT solutions, at the stage of" 3. Definition »detailed requirements for an IT solution.

Throughout the project life cycle, the compliance of requirements of different levels to each other is monitored.

Compliance with the requirements of all levels should be monitored (the requirements for examinations are described in clause 6. "Project implementation process").

4. Using external expertise to ensure the quality of project results.

In most cases, external experts (in relation to the project team) are involved to ensure the quality of the project results and individual stages of the project. Based on the results of the examination of the project results, either quality is confirmed, or recommendations are given for their refinement

5. Use of previous experience.

In the course of the project activity, the experience and results of already implemented projects are actively used, including the replication of best practices, the use of ready-made results as a basis for developing and preventing the repetition of mistakes.

6. The materiality of deviations of project parameters.

If there are significant deviations from the project, the Project Manager submits the project for reapproval to the level of the body that made the decision to launch the project - IC.

Stage works that do not lead to significant deviations are not suspended unless otherwise approved by the decision of the UKP.

Significant deviations from the project include any of the following:

- by time: increase the duration of the project / project phase by a period exceeding the possible assumptions (the ratio of the planned duration of the project to the projected duration (in days)).
- by budget: increasing the ratio of the planned budget of the project to the forecast budget of the project.
- in terms of the volume of tasks to be solved: changing / expanding the goals of the project, changing the functional or organizational volume of the project (the volume of tasks to be solved).

Choosing the right project implementation concept is critical to project success.

The ability to control project costs varies significantly from stage to stage. A significant influence is exerted in the early stages, when major design decisions are made and actual costs are small.

At the "implementation" stage, the bulk of the actual costs occur, and the ability to influence them practically disappears. Project life cycle.

When carrying out projects, four consecutive stages are distinguished:

- Evaluation;
- Choice;
- Definition;
- Implementation;
- Development and implementation;
- Experimental-industrial exploitation;
- Completion.

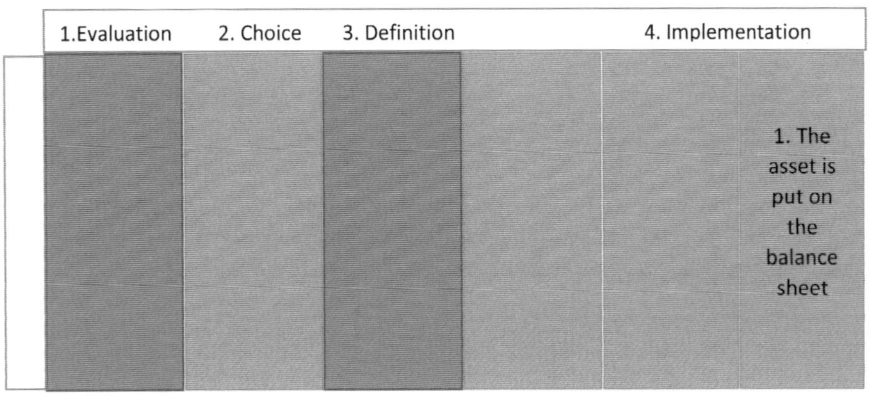

Figure 1. Each stage is aimed at solving a specific problem and corresponding solutions.

The key decisions are at stages "2. Choice "and" 4.1 Development and implementation ":

"2. Choice "- Making the final decision on investing;

"4.1 Development and implementation" - Decision-making on the readiness to use an IT solution.

Tasks and results of the stages.

Each stage has its own task and results, the receipt of which provides a solution to the task of the stage. The achievement of the results of the stage is documented in the form of a package of documents for decision-making.

Role model

In the applicable standard for project management, key roles are identified:

Single responsible person: the employee responsible for the implementation of the project and for achieving the goals and key performance indicators of the project.

Stakeholder of the project: the employee, whose functions and business processes can positively or negatively affect the results of the project and which can influence the project and its results.

Customer: a unit (manager) interested in the implementation of the project, initiating its formation for subsequent approval and carrying out strategic project management, responsible to the IC for achieving the project goals.
Investment body (IC): a permanent collegial body that makes decisions on the launch of the project (the feasibility of implementation, taking into account the volume, cost and timing).
Project Steering Committee: the collegial body of the project, monitoring the progress of the project and its strategic leadership.
Project manager: an employee of the organization responsible for the implementation of the work and activities provided for by the investment project, taking into account the planned time and the assigned budget and resources and endowed with the necessary authority.
Project team: a temporary team consisting of workers from various fields of activity, organized for the implementation of the project.
Group of expert assessment: an external (in relation to the project group) group of persons having sufficient competencies and authority to conduct an examination of the results of the project stage and to develop recommendations for improving their quality.

Table 1. Role model

After-sales service: an organization that provides technical support for software under a service agreement with the Company.Criterion	PMBoK	Company standard	PRINCE2	GOST.34
Stages (phases) of the life cycle	Initiation: Planning; Organization of execution; Control of execution; Completion.	Evaluation; Choice; Definition; Implementation; Development and implementation; Experimental-industrial exploitation. Completion;	Start of the project Initiation of the project Project management Stage boundary management Stage control Product Supply Management Closure of the project	Formation of requirements Concept development Terms of Reference Draft design Technical project Working documentation Commissioning Support
Key roles	Customer; Project manager; Project team; Contractor; Project management team; Sponsor; Sources of influence; Project Management Office.	Customer; Project Manager; Project team; Common responsible person; Investment committee (IC); Project Steering Committee; Stakeholder of the project; Peer Review Team; Service.group	Customer; Project manager; Design team; Consumer Supplier; Administrator; Project team..	-

After-sales service: an organization that provides technical support for software under a service agreement with the Company.Criterion	PMBoK	Company standard	PRINCE2	GOST.34
Application features	Doesn't give clear practical ones. Explains which parts of the process may be included depending on the situation, describes inputs and outputs, results and tools. Suitable for most projects as a universal guide. Redundancy of documentation; Difficulty for small projects; The need for adaptation to the field of application	Used to implement IT projects in a group of companies in the oil and gas industry.	Applies to all managed elements of the company. Three-level project management, a clear distribution of tasks between participants. A clear sequence of steps (steps) for a number of repetitive processes. It is used where constant monitoring of processes is required, due to large losses due to failure. in the implementation of large projects that are closely monitored by management in government agencies. Non-universality - the complexity of application in an environment with high uncertainty and poor communication between the management and the executors (for example, due to the independence of the executing team). Excessive bureaucratization is inherent.	It is used in the fields of automated systems (AS) and applies to AS used in various fields of activity (management, research, design, etc., including their combination), the content of which is information processing.

The materials of table 1, we can draw the following conclusions:

The project life cycle in the standard of the company considered during the work does not contradict those described in the PMBoK, PRINCE2 and GOST 34 methodologies. However, IT projects involve the creation / refinement / use of an information system, therefore it is advisable, to a greater extent, to use the stages from GOST. 34.

The definition of the project in the methodology used in the company is similar to the others and includes both the general features of the project (the concept of uniqueness) described in the American and British standards, as well as the features of the IT terminology used in GOST. 34.

The standard within the company is collective. The peculiarities of conducting IT projects are borrowed from GOST. 34, but the corporate management system considered to a greater extent is based on the existing PRINCE2 and PMBoK methodologies.

The description of the key roles in the standard applied by ITSK LLC represents the common between PRINCE2 and PMBoK, but also highlights the distinctive inherent roles of the enterprise, such as the Investment Committee, the Single Responsible Person (EOL).

IOP Conf. Series: Materials Science and Engineering **986** (2020) 012019 doi:10.1088/1757-899X/986/1/012019

PRINCE2 focuses on key roles and their responsibilities. The British standard provides step-by-step instructions on project management, concretizes the processes. This methodology is prescriptive.
Unlike PRINCE2, PMBoK is a recommendation only. A common body of knowledge that emphasizes project management tools and techniques. Provides a common understanding of process control in accordance with the entire spectrum of emerging situations.
The applied management processes in ITSK LLC are defined in the Charter of the project and comply with the recommendations on the application of the PMBoK management processes discussed in APPENDIX D.
Organization of IT project management in the company clearly demonstrates the application of the following areas of knowledge: Management of integration, content, project timing, cost, quality, human resources, communications, risks and procurement.
The distribution of processes occurs not only in areas of knowledge, but also in the phases of the life cycle: Assessment, Selection, Definition, Implementation. In addition, each region has a low, medium or high degree of importance at a certain stage of the LC project.
Despite the existence of minor differences, the corporate project management system is based on the base of best international practices.
Thus, the symbiosis of the methodologies considered provides new opportunities in managing IT projects in the Information Technology Service Company.

2.4. The key role of checkpoints in project management

The weakness of the project management can be considered a low level of control over compliance with the schedule, as well as insufficient control over the actions of the contractor, which also lead to delayed deadlines.
The company under consideration, using well-known methodologies, has key features that allow us to achieve our goals and easily control the processes in stages. One of them is the use of control points throughout the project.
Checkpoint (ChP) - a project event related to the receipt of any project result (isolated / product element of the project).
Milestones reflect project progress based on key deliverables.
Project management based on control points allows to:

- Improve the reliability of the evaluation of the intermediate results of the project.
- Focus on key project results.
- Assess the dynamics of the project to achieve intermediate results.
- Carry out an assessment of the risks of passing ChP and forecasting the timing of the implementation of ChP.
- Analyze the passage of ChP based on statistics.

Type ChP - allows to separate different types of ChP ("stage", "financial", "grocery", "atypical").

Table 2. Examples of types of ChP and their description are presented in.

Type	Description
Stage ChP	ChP stages of the life cycle of the project; required in every project; are formed at the time of the start of the project.
Financial ChP	comply with acts under the project contracts; contain information on planned and actual costs of the project and the deadlines for acceptance of work; are planned and updated from the "Assessment" stage through the project budget.
Product ChP	ChP of the project results; reflects the creation of a project product; the results of product ChP are material in nature and are confirmed by documents; initially planned at the Assessment stage.
Not typical ChP	ChP of the operational control of the project; can be formed at each stage without restrictions.

Management based on ChP allows to combine the traditional management approach based on the control of the stages of the project roadmap and budget with details regarding the creation of the project product. To shift the focus of management from fulfilling tasks to achieving project results.
The role model of ChP allows to provide differentiation of powers when planning, evaluating the forecast of terms for achieving ChP and accepting the results of ChP.
ChP life cycle - allows you to monitor the state of the ChP (“draft”, “in work”, “completed”, “completed”).
ChP weight - a sign of "importance" of the result associated with the control point ("high", "medium", "low", the greater the weight of ChP, the more important result is associated with this ChP).
The deadline for the implementation of ChP is the date of the planned / forecasted / actual implementation of the ChP.
Indicators - reflect the status of the implementation of the ChP (“in operation”, “forecast of failure to meet the deadlines”, “overdue”, “completed”, “confirmed”).
The distribution of key points in stages with deadlines, results, as well as additional information, for example, the number of contracts, applications, indicating the name of the counterparty forms a road map.
The ChP role model is based on several mandatory roles with its own set of ChP tasks:
Owner - launches ChP to work, manages the planned date of ChP, agrees the changes for ChP. Basically, the project administrator.
Responsible - a project participant who is responsible for obtaining a result related to ChP (forecast and fact of ChP implementation). In general, the project manager.
Receiver - an employee confirming receipt of the result by ChP (IT architect, Business Analyst of the project).
The life cycle ensures the sequential passage of the following conditions of “draft”, “in work”, “completed”, “completed” ChP.
Changing the planned dates, launched by ChP, is possible only through a change request with confirmation of the change by the ChP Owner.
To confirm ChP, one Receiver can be involved, as well as several Receivers within the framework of the Checklist for the approval of the results of ChP (see Figure 2.10).
Determination of achievement of the project result by control points:
Each “product” and “step-by-step” KTP creates a contribution to the achievement of the stage results taking into account its weight.
Contribution of one ChP to% of project completion = ChP weight * Stage Weight / Total Stage ChP Weight
% completion of the project = Σ Contributions of each ChP in% of completion of the project.
In the general case, the weight distribution of ChP at each stage of the LC is shown in Figure 1. It should be noted that weights may vary at the “Assessment” stage, depending on the project concept.

Table 3. Composite formulation.

Stage	Weight, %	ChP	Weight, 0-4	Status ChP	Result,%
1.Evaluation;	5	ChP№1	1	1	0,625
		ChP№2	3	1	1,875
		ChP№3	2	1	1,25
		First Stage ChP	2	1	1,25
2. Choice	10	Second Stage ChP	2	1	10,00
3. Definition	15	ChP№4	2	0	-
		ChP№5	3	0	-
		Third Stage ChP	2	0	-
4.1. Development and implementation	40	ChP№6	0	0	-
		ChP№7	3	0	-
		Fourth (1) Stage ChP	3	0	-

Stage	Weight, %	ChP	Weight, 0-4	Status ChP	Result,%
4.2. Experimental-industrial exploitation	20	ChP№8	2	0	-
		Fourth (2) Stage ChP	3	0	-
5.Completion	10	ChP№9	2	0	-
		Fifth Stage ChP	3	0	
Final result,%					15%

Based on the selected criteria for comparison, it was concluded that despite the existence of minor differences, the corporate project management system is based on the best international practices, but has key features. For example, the control points considered in more detail necessary to track the status of the project and create a picture of the implementation of each of the stages.
An analysis was carried out to identify the key features of the two most widely used: Code of Knowledge PMBoK and PRINCE2. Despite the existence of a large number of different project management standards, all of them are aimed at fixing the purpose and results of the project, determining quantitative characteristics, determining the time, cost and quality parameters of the project, creating a realistic plan for the project, designating, assessing risks and preventing possible negative consequences during project implementation. In addition, the standard GOST. 34, which is used in projects to create automated systems of various fields of activity, was described.

3. Conclusion

Comparison of methodologies in the main part showed that the standard operating within the company standard is a symbiosis of the best world practices, which takes into account the scope of the company and has individual characteristics. So, in addition to the guidelines, role model, life cycle stages, control points were described. Reference points allow to track the status of the project at any time and see a transparent picture of the project. In addition, ChP-based management allows to combine the traditional management approach based on the control of the stages of the project roadmap and budget with details regarding the creation of the project product. In the research work, examples of types of control points, responsible roles, life cycle were described. Based on the results of the analysis, a list of recommendations for managing IT projects implemented within the company methodology was formulated.

References

[1] Denisenko V I 2015 Project Management: Textbook. allowance *Vladimir: Publishing House of VlSU* p 108 .
[2] Guide to the Project Management Body of Knowledge (PMBOK Guide) (Sixth Edition - Project Management Institute) p 978.
[3] Andy Murray, Nigel Bennett, John Edmonds, Bob Patterson, Sue Taylor and Graham Williams 2017 Managing Successful Projects with PRINCE2 *TSO Publishing House* p 400.
[4] Joseph Higney 2018 Fundamentals of project management: Classical management *From Mann, Ivanov and Ferber* p 240
[5] Dzhuraev E Sh and Persod N L 2016 Mechanics of project management. Management Methodology *Research Institute of KPU* p 290
[6] Matveeva L G and Nikitaeva A Yu 2016 IT project management, textbook *allowance Publishing house of the Southern Federal University* p 288
[7] Afonin A M 2016 Project Management (M .: Forum) p 184
[8] Zharova M, Shirokova S and Rostova O 2019 Management of pilot IT projects in the preparation of energy resources *E3S Web of Conferences*

[9] Bolsunovskaya M, Shirokova S, Loginova A and Uspenskij M 2019 The development and application of non-standard approach to the management of a pilot project. 2019 *IOP Conference Series: Materials Science and Engineering*
[10] Kulikova L L 2017 Features of evaluating the effectiveness of IT projects *Bulletin of Irkutsk State Technical University* p 7
[11] GOST 34. Development of an automated control system (ACS).
[12] Knowledge Base: IT Project Management (5 challenges and their overcoming URL: https://it-guild.com/info/blog/upravlenie-it-proektami-5-vyzovov-i-ikh-preodolenie/ appeal date 05/11/2020)
[13] GOST 34.003-90 Automated systems. Terms and Definitions.
[14] Kazak A Yu 2018 Modern methods for assessing project risks: traditions and innovations *Vestnik Ural Federal University. Series Economics and Management* p 13

IOP Conf. Series: Materials Science and Engineering **986** (2020) 012020 doi:10.1088/1757-899X/986/1/012020

Virtual proving ground for aircraft structures

Yu A Gorskii, P A Gavrilov and A I Borovkov

Peter the Great St. Petersburg Polytechnic University, St. Petersburg, Russia, E-mail address: gorskij_yua@spbstu.ru

Abstract. This paper presents an approach to development of complexes and programs of virtual experimental investigation in order to reduce full-scale testing in the design process of new products via state-of-the-art computer simulation techniques. Generally Virtual Proving Ground (VPG) is to be understood as integrating system for modeling strategies, solver settings, modules of automatic post-processing and load cases that match typical airplane tests and operating conditions. In the VPG environment has been developed a set Virtual Test Beds corresponding to tasks from various technical areas:

- VTB «Static structural strength» for virtual static strength tests to determine the stress-strain state of the structure and identify potentially critical places;
- VTB «Fatigue» for estimating fatigue life under the influence of a multi-cycle load;
- VTB «Durability» for simulating crack growth in the most critical places based on VTB «Fatigue» and «Static structural strength» data;
- VTB «Birdstrike» for simulating tests of bird collisions with aircraft panels;
- VTB «Hailstrike» for simulating tests of hailstones collisions with aircraft panels;
- VTB «Aerodynamics» for determining the aerodynamic characteristics of an aircraft.

Preliminary estimation of structure is possible due to automatic post-processing module: visualization of virtual testing, documentation of results (graphs, pictures, key values) and union of all experimental data into a presentation. The application of presented approach is described on example of stringer panel of a real airplane.

1. Introduction

Over the past few years, more attention has been paid to the development of virtual test methods [1-4]. This fact emphasizes the relevance of the topic presented in the article.

A presented approach is a set of complex (using several interacting types of models) approaches for virtual testing, which allows computer simulation of testing of aircraft products, providing a low error in terms of target characteristics.

The research process includes both the tests themselves and the preparation of the model (preprocessing), as well as the processing of the results (postprocessing). Automation of pre- and postprocessing of mathematical models in addition to automatic interconnection of multidisciplinary tasks, allows to increase the speed and accuracy of engineering calculations and presentation of results.

Preprocessing includes preparation of finite element (or finite volume) models, boundary conditions, material models selection etc.

Effective use of virtual testing is possible only with proper planning of design cases. Boundary conditions (fixing and loading) should as fully as possible reflect the actual working conditions of the structure or the actual conditions of the experiments (including the rig influence in the experimental bench). Virtual tests based on experimental stands should include equipment models as well.

IOP Conf. Series: Materials Science and Engineering **986** (2020) 012020 doi:10.1088/1757-899X/986/1/012020

Another no less important step is the development of a matrix of targets and limitations. In relation to the design of aircraft structures it expressed in translating the technical task into digital form: formalizing requirements, determining the group of monitored parameters and results. A mandatory element is the correction of requirements, taking into account the characteristics and limitations of technological processes of production.

A presented approach to virtual testing of aircraft structures is well suited to study the following design scenarios, compiled on the basis of the possible design loads acting on the fuselage of the aircraft [5]:

- Inertial forces from the masses of bodies placed inside the fuselage and its own mass;
- Internal and external pressure on the airtight compartments of the fuselage, the loads from the aerodynamic effects of the incoming flow;
- Local impacts caused by collisions with birds and hail.

2. Virtual proving ground

As said before, Virtual Proving Ground is an integrating system for modeling strategies, solver settings, modules of automatic post-processing etc. It requires well-organized structure for data storage, transfer and management. VPG is based on Digital twin development and computer engineering management system CML-Bench™.

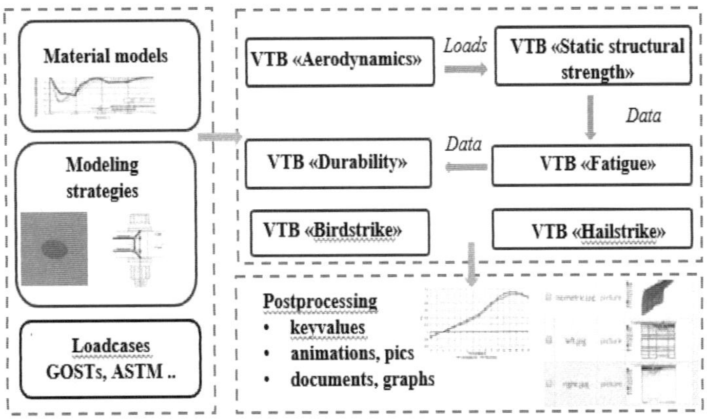

Figure 1. Scheme of data transfer in VPG.

Models and calculations should be carried out in a measurement system in which the derived units are obtained by multiplying or dividing the base units without using numerical factors. SI units form a coherent system; for example, the unit of force is a newton; kilogram, meter and second are the basic units of the system.

For user convenience FE-models may have modular or include structure:

- Linker: defines the reading order of submodels;
- Submodel of the material base;
- Submodels of FE mesh;
- Submodel of solver settings: BCs, solver cards, output cards;
- Additional submodels.

Usually, due to the size of aircraft structures, the construction of design models is carried out simultaneously by several employees. For the correct assembly of models into one, it is necessary to avoid the same numbers and indices for nodes, elements, sets, etc.

Compliance with the numbering rules allows you to change the FE-model of parts independently from each other without harm for subsequent assembly into a global model.

Often of interest is not only the general behavior of the structure, but also the specified stress-strain state of an individual part (for example determining the risk of cracks in loaded joints and fillets).

For Abaqus and Nastran solvers, scripts have been developed for reading items (nodes and elements sets) of interest for postprocessing. In this case numbering rules are good pointers.

The accuracy and adequacy of the calculation results directly depends on the finite element representation of the structure. Special attention should be paid for modelling strategies including shell-modeling, FE representation of bolted and riveted joints, different metrics for finite element quality etc. The problem of correct modeling of bolt and rivet joints was considered in articles [6-10].
The main tasks performed in postprocessing module are:

- Visualization of virtual tests of the object;
- documentation of the results (creation of graphs, figures, tables);
- combining the data obtained about the object of study into a presentation.
- Processing and recording of so-called keyresults allows engineer to make an express assessment of the design: strength, weight optimization, etc. For analysis of structural strength, keyresults may include:
- Movements in a given set of points;
- Maximum stresses in parts;
- Safety margin in details;
- Special flags of the boolean type (the fact of non-destruction or destruction).

A function is provided for comparing a key parameter with predetermined values (yield strength, safety factor, etc.). For the convenience and simplicity of the search in a detailed analysis, each result is recorded with an indication of an element or node.
Automated processing of the results is carried out in a post-processor program using session files — a sequence of commands that can be parameterized and executed without manual intervention. The functionality of postprocessors is accessed through the API (Application Programming Interface) using scripting languages such as Python.

3. Virtual test beds

Virtual Test Bed is a set of configuration files for the solver and post-processor, communications between modules are configured for one or a group of specific testings. In general, any software may be used as solvers and this approach allows you to use most suitable software solutions. In this work, the VPG is configured to work with the packages Nastran, Abaqus, nCode and CFX.

3.1. VTB «Static structural strength»

Carrying out virtual tests for static strength allows you to get the stress-strain state of the structure and identify potentially critical places.
For virtual tests were selected the stringer-frame section of the aircraft fuselage (Figure 2) loaded by pressure and the wing mechanization elements: aileron and spoiler loaded by testing rig. These parts are made of high-strength aircraft aluminum. VTB includes base of materials with basic mechanical properties and the material stress-strain curve.

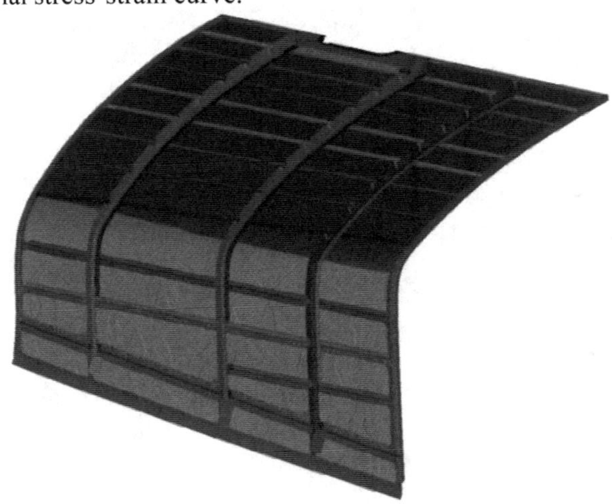

Figure 2. Airplane stringer-frame section.

Results of virtual testing are automatically post-processed and user can make a rapid analysis of structure. Basic available results are the section deflection forms (Figure 3) and stress fields (Figure 4); additionally, 3D-PDFs with displacement and stress fields are created for pre-selected parts, a video of the loading process, types of the part in several projections are recorded.

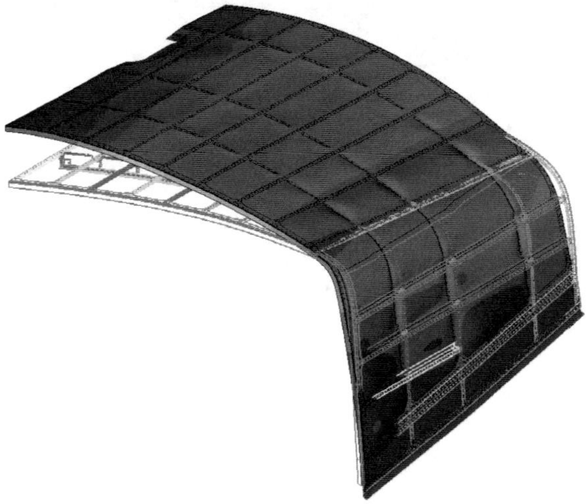

Figure 3. Deformed shape of stringer-frame section.

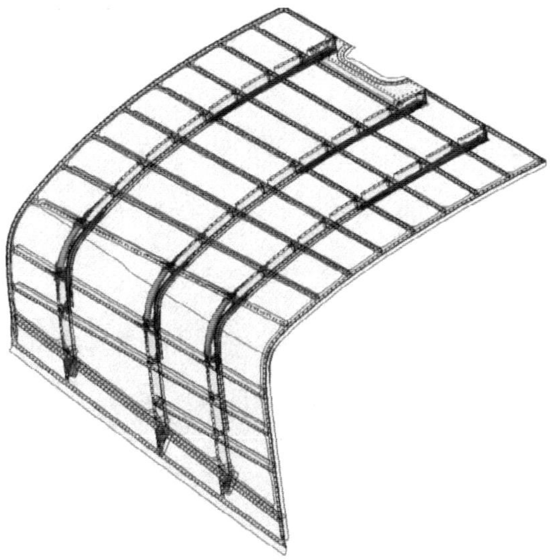

Figure 4. Stress state of stringer-frame section.

A static strength virtual testing of an aircraft wing mechanization elements allows you to evaluate the stress-strain state of the structure and obtain the value of the shear force in the earrings of the elements. The stand for full-scale tests is a lever design, therefore, in the VTB, the special FE-models of loading pads were developed (Figure 5).

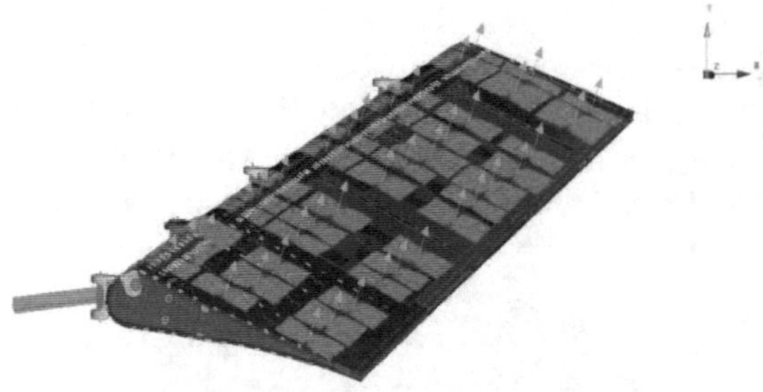

Figure 5. An aileron loading scheme.

As a result, stress-strain states of the aileron and interceptor are obtained. Deflection forms of the front and rear panel edges are defined. Potentially critical structural locations have been identified. The automatic postprocessing is configured, the main key results are displayed:

- Strength on the booster;
- Moving at the points of the leading edge;
- Moving at the points of the trailing edge;
- Strains in specific places (compared with voltages received from strain gauges).

All these values are automatically compared with the "targets" - the values obtained as a result of field tests.

3.2. VTB «Fatigue»

Fatigue virtual testing is performed by applying multi-cyclic loading in order to identify critical areas for crack growth. Testing is carried out on the basis of the stress-strain state obtained during the static structural analysis of VTB «Static structural strength».

VTB includes material fatigue characteristics: S-N type fatigue curves (Stresses - loading cycles).

The picture of damage accumulation in the central frame of the stringer-frame panel is shown in Figures 6 - 7.

In the areas highlighted in Figure 7, damages begin to appear after 3600 cycles.

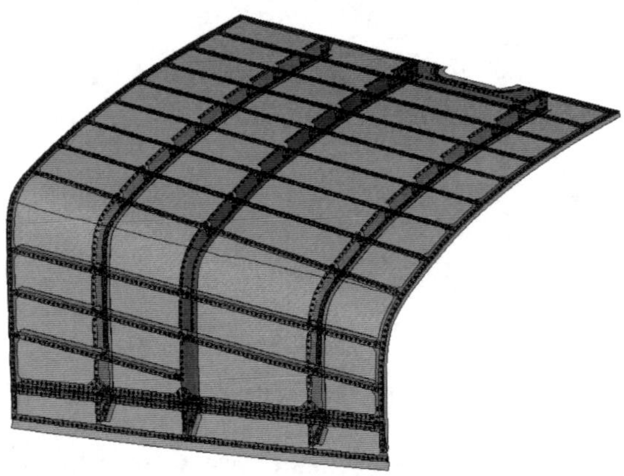

Figure 6. The most damaged frame.

IOP Conf. Series: Materials Science and Engineering **986** (2020) 012020 doi:10.1088/1757-899X/986/1/012020

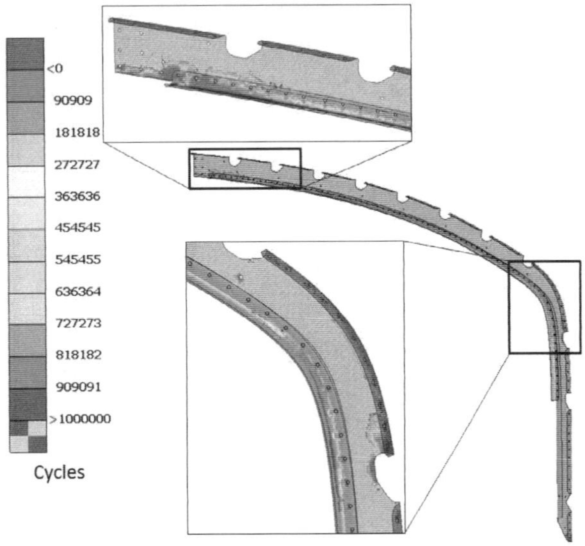

Figure 7. Damage accumulation.

3.3. VTB «Durability»

Durability calculations are based on the results of VTB «Static structural strength» and VTB «Fatigue». The most critical places of the structure are determined and graphs of crack growth in these places are obtained.

VTB uses the same loading cycles as were used for fatigue loading and includes material fracture toughness characteristics (generally obtained by field tests):

- fracture toughness;
- crack growth rate curves (Crack growth rate - stress intensity factor).

Based on the geometry of the section in which the crack will propagate, and the stress-strain state in this section, obtained as a result of static analysis, a mathematical model is selected to obtain the dependence of the stress intensity factor on the length of the crack.

Then, a cyclic loading is to be applied to the calculation model. The main measurable criterion is the growth of the crack length depending on the number of loading cycles applied.

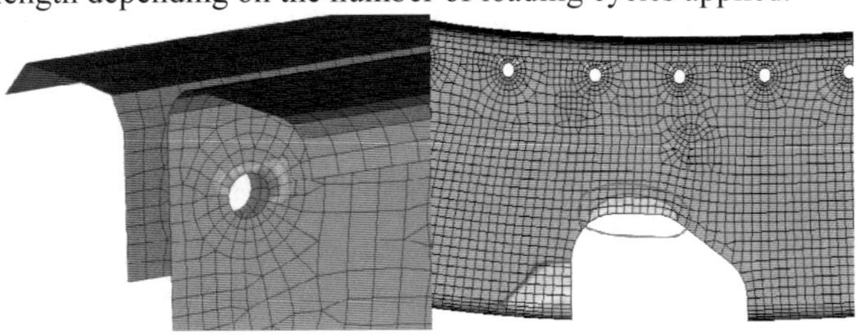

Figure 8. Potentially critical areas of central frame (marked with colour).

The most critical crack growth graph is shown on a figure 9. In the presence of an initial crack, structural failure will occur between 350-400 loading cycles.

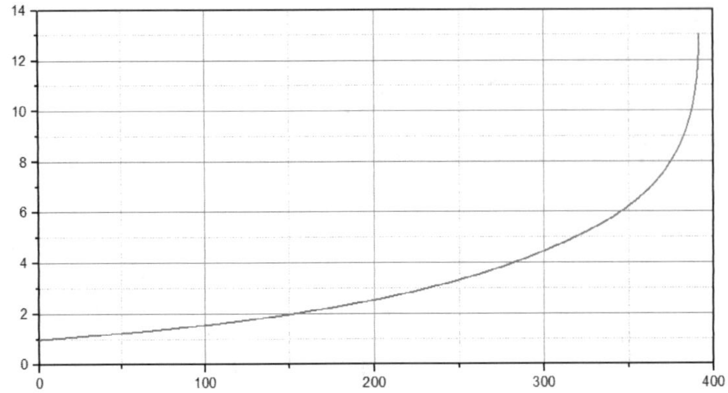

Figure 9. Most dangerous crack growth graph, mm - cycles.

3.4. VTB «Birdstrike»

Bird resistance testing is an essential part of aircraft development. Such testing refers to high-speed dynamic effects involving the "soft" body. This indicates that the stresses arising in the contact area significantly exceed the strength of the impactor (bird), but are lower in relation to the strength of the test structure (parts of the wings, tail, etc.). Due to the high deformability of the impactor, the shock propagates rapidly in a certain area at the moment of impact. In this regard, it is necessary to use special material models taking into account the influence of the strain rate.

It was shown in [11] that during high-speed processes, the bird's body behaves like a liquid. Therefore, it is proposed to use a hydrodynamic model of the material to describe the rather complex structure of the bodies of real birds under high-speed impacts.

A mathematical model of a bird impactor was developed and validated at SPbPU [12], which corresponds to 1.8 kg of bird impactors actually used in field tests. This mathematical model is an integral part of the VTB. As a solver, ABAQUS software version 6.14 of the developer Simulia was chosen, which for many industries is certified for solving problems of high-speed dynamic interaction, which includes the task of assessing bird resistance.

This VTB uses the Coupled Euler-Lagrange (CEL) approach. In this case, the target with which the contact interaction occurs is modeled by the standard Lagrange method. The impactor is modeled using the Euler method. In this case, the finite element mesh can move after the main part of the material and change its dimensions in the process of solving the problem.

Verification of the bird impactor model and analysis of modeling techniques were carried out in [12] using the parameters: maximum values of interaction forces and contact stresses, value of stationary flow pressure and Hugoniot pressure.

As part of the development of the VTB, a program was developed for solving the system of transcendental equations (48) - (61) from [11] in the Mathcad package. As input data, material properties, porosity (an indicator of the volume fraction of air in the material) and the initial impactor speed are set. The calculations for the following materials are implemented: water, gelatin, vulcanized rubber, "bird". The result of the solution (the equation of state coefficeints) from equation (1), are used to construct a model of the impactor material in a finite element formulation.

$$P = C_0 + C_1\mu + C_2\mu^2 + C_3\mu^3 \quad (1)$$

$$\mu = \rho/\rho_0 - 1$$

where P – pressure; ρ_0, ρ – material density at atmospheric pressure and in the shock wave formation phase; C_i (i = 0-3) – experimental polynomial coefficients.

Model is calibrated to simulate collisions with velocities up to 500 km/h. Figures 10-11 demonstrate the aftermath of bird strike in different areas of stringer-frame section – critical damaging with breaking of skin surface.

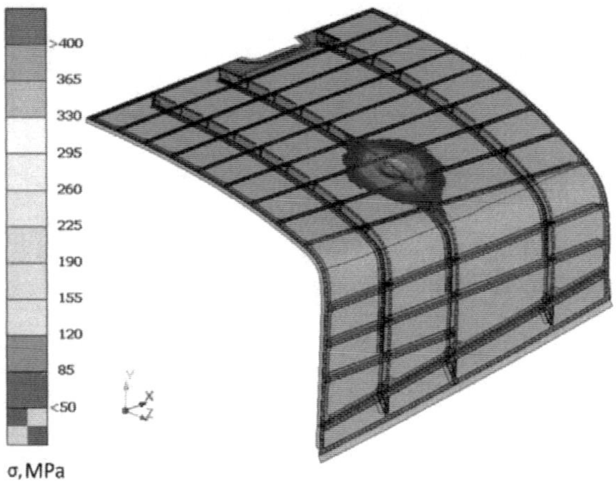

Figure 10. Stress state of stringer-frame section, collision in frame area.

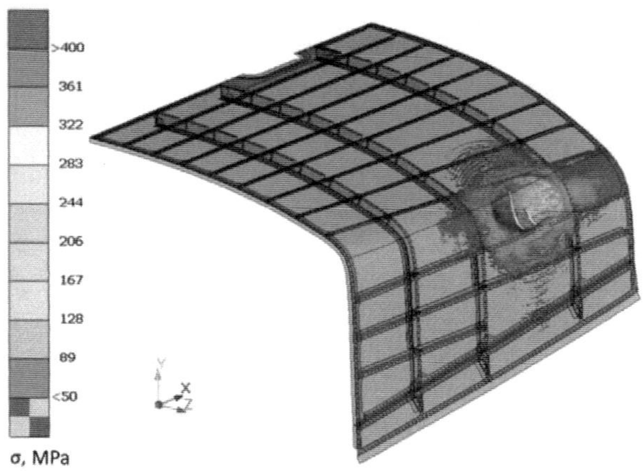

Figure 11. Stress state of stringer-frame section, collision in skin area.

3.5. VTB «Hailstrike»

In concert with Birdstrike, Hailstrike resistance testing is an essential part of aircraft development.
A mathematical model of a hail developed and validated at SPbPU is based on a series of experiments [12] on the collision of ice spheres with a titanium plate at high speeds (73.5-126 m/s). This mathematical model is an integral part of the VTB.
Currently, commercial FE packages do not have a suitable ice material template for modeling high-speed collisions, despite interest from the aerospace industry. It is also worth noting that ice is not one type of material with well-known properties; in the literature there is a description of about 13 different crystalline structures and two amorphous ones.
In this paper for ice is used a homogeneous elastoplastic material with linear hardening and yield strength, which depends on the strain rate.
The hail model is a sphere with a diameter of 50 mm. In contrast to birdstrike analysis the Lagrangian formulation is used, since hail behaves like a solid at high speeds. The hail material takes into account its destruction at stresses of 5.2 MPa. During the simulation, the hail completely collapses during a collision with a stringer-frame panel. The hail material also takes into account the plastic behavior of ice, which depends on the rates of plastic deformation [13, 14].
Model is calibrated to simulate collisions with velocities up to 500 km/h. Figures 12-13 demonstrate the aftermath of hail strike in different areas of stringer-frame section – plastic strains occurred.

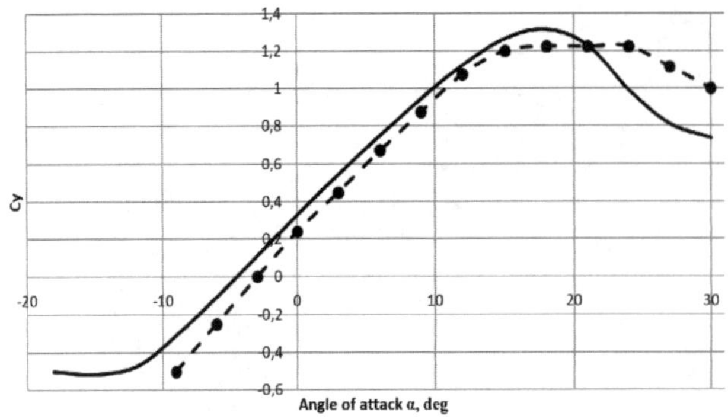

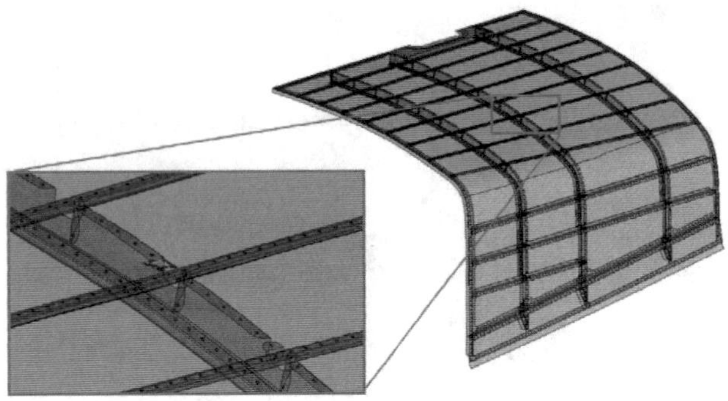

Figure 12. Highlighted elements with plastic strains.

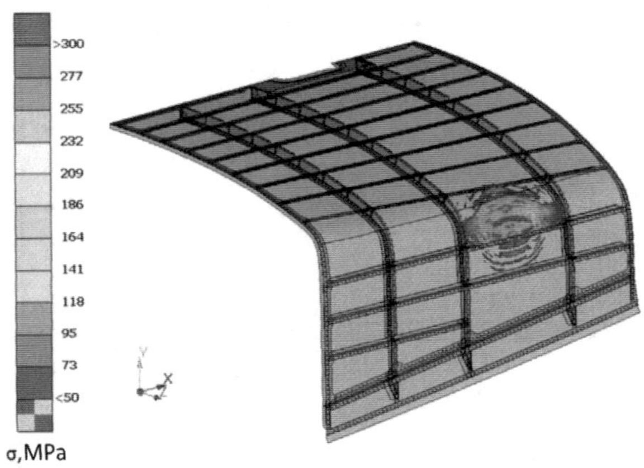

Figure 13. Stress state of stringer-frame section, collision in skin area.

3.6. VTB «Aerodynamics»

High efficiency and aerodynamic qualities of the aircraft are laid at the design stage of the appearance. The main geometric parameters, such as the length, width and height of the aircraft, as well as the wing span, profile selection, determination of the area of the control planes, on which its aerodynamic characteristics directly depend: lift coefficient, drag coefficient and others, must be determined at the development of any project. VTB «Aerodynamics» provides wide opportunities of virtual testing: simplifies the process of obtaining aerodynamic characteristics, allows ещ increase a number of tests without raising costs, consider all possible design options.

VTB was tested and models were calibrated on a set of airfoils: A-12, P-III, NASA-0018 [15] and ultralight aircraft A-8 at different angles of attack. Values of the lift and drag coefficients obtained during the TsAGI experiment practically coincide with the values obtained in the numerical solution (Figure 14).

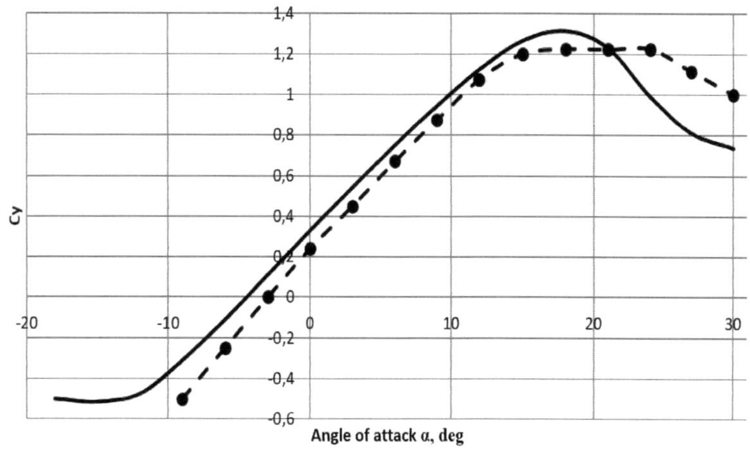

Figure 14. Lift coefficient, SLA A-8.

Special scripts have been developed for data transfer between multidisciplinary tasks of VTB «Static structural strength» and VTB «Aerodynamics». For example, one of them read pressure distribution (Figure 15) on aerodynamic faces and interpolate them to finite element mesh.

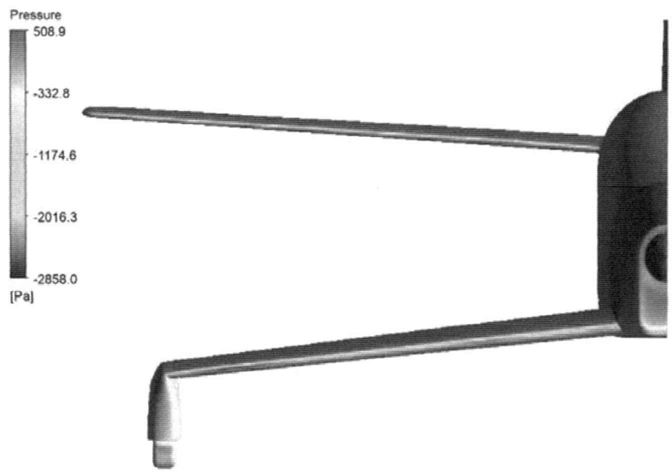

Figure 15. Pressure distribution, SLA A-8.

VTB «Aerodynamics» can be used for optimization purposes: searching for an optimal wing profile shape with multiple geometry changes. Automatic systems of pre- and postprocessing provide multiple speed-up in engineering work. Optimal shape searching history is shown on figure 16. The best modes with high lift and low drag coefficient are marked with numbers 1 and 2.

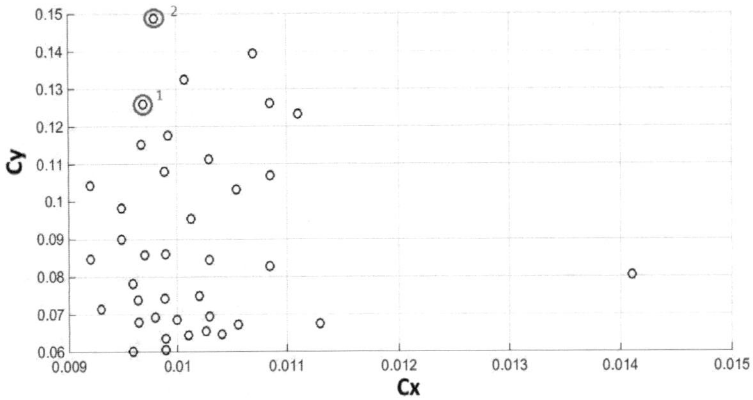

Figure 16. Optimal shape searching history.

4. Conclusion

The developed modeling methods of aircraft structures (FE-mesh criteria, model structure, numbering, features of connection models), material models, data transfer paths allow virtual testing of structures made of metal and composite materials.

The main advantage of this work is the use of a highly adequate and multidisciplinary approach for virtual testing and optimization of an aircraft elements, which gives a significant reduction in costs during its development.

Developed Virtual Proving Ground covers wide variety of technical areas and easily can be expanded with additional Virtual Test Beds due to flexibility of data management system.

References

[1] Ostergaard M G, Ibbotson A R, Roux O L and Prior A M 2011 Virtual testing of aircraft structures *CEAS Aeronautical Journal* (Vol.1) pp 81-103

[2] Chirme A., Parle D, Awati N, Udali R and Kulshrestha S 2012 Integrated Tool for Strain Extraction in Virtual Testing *SIMULIA Community Conference*

[3] Wang Ch 2016 A multidisciplinary design and analysis environment and its application to aircraft flight dynamics analysis *Journal of Industrial Information Integration* (Vol.1) pp14-19

[4] Nu X 2005 Digital Simulation of Full Scale Static Test of Aircraft *Chinese Journal of Aeronautics* (Vol. 18 №2) pp 138-141

[5] Chepurnyh I V 2013. Prochnost konstruktsiy letatelnykh apparatov [The strength of aircraft structures. Training manual] *FSFEI HPL "KnASTU"*

[6] Rutman A and Boshers C 2007 Fastener Modeling for Joining Parts Modeled by Shell and Solid *Americas Virtual Product Development Conference*

[7] Martins R, Ernani S P and Lorentz A 2017 Influence of Types of Discrete Modelling of Fasteners in FEM Models *NAFEMS World Congress*

[8] Rutman A, Viisoreanu A and Parady J 2000 Fasteners Modeling for MSC.Nastran Finite Element Analysis *SAE Transactions* (Vol. 109 Section 1: JOURNAL OF AEROSPACE) pp 1220-1237

[9] MSC Software. Case Study: Gulfstream 2015 (Retrieved from https://www.mscsoftware.com/sites/default/files/cs_gulfstream_ltr_w.pdf)

[10] Martins R, Ernani S P and Santos M 2018 Fastening Analysis Using Low Fidelity Finite Element Models *31st Congress of the International Council of the Aeronautical Sciences*

[11] Wilbeck J S 1974 Impact Behavior of Low Strength Projectiles *Final Report. Air Force Materials Lab*

[12] Carney K S, Benson D J, DuBois P and Lee R 2006 A phenomenological high strain rate model with failure for ice *International Journal of Solids and Structures* (Vol 43) pp 7820-7839

[13] Kim H and Kuene J N 2007 Compressive Strength of Ice and Impact Strain Rates *Journal of Material Science* (Vol.42) pp 2802-2806

IOP Conf. Series: Materials Science and Engineering **986** (2020) 012020 doi:10.1088/1757-899X/986/1/012020

[14] Sain T and Narasimhan R 2011 Constitutive modeling of ice in the high strain rate regime *International Journal of Solids and Structures* (vol. 48)
[15] Spravochnik aviatsionnykh profiley [Aviation airfoils handbook] (Retrieved from https://kipla.kai.ru/liter/Spravochnic_avia_profiley.pdf)

Methodological approaches to assessing the level of digitalization of countries and regions

A Kurilova[1] and D Antipov[2]

[1]Tolyatti State University, Togliatti, Russia, aakurilova@yandex.ru
[2]Samara University, Samara, Russia

Abstract. New digital technologies are dynamically penetrating the economies of various countries and regions, having a significant impact on the development of its infrastructure and economic efficiency. The effectiveness of the latest digital technologies is becoming a key element in international economic competition. The expanding use of modern information and communication technologies is becoming a prerequisite for the emergence and development of new effective management technologies, entrepreneurial practices, and successful businesses. Therefore, most countries and regions stimulate these processes at the state and regional levels. In order to ensure the systematic implementation and development of digital technologies, the development of methodological approaches to assessing the digitalization level of countries and regions is relevant.

1. Introduction

New digital technologies are dynamically penetrating the economies of various countries and regions, having a significant impact on the development of its infrastructure and economic efficiency. The effectiveness of the latest digital technologies is becoming a key element in international economic competition. The expanding use of modern information and communication technologies is becoming a prerequisite for the emergence and development of new effective management technologies, entrepreneurial practices, and successful businesses. Therefore, most countries and regions stimulate these processes at the state and regional levels.

There are several approaches to defining digitalization in the literature. According to Reddy and Reinartz (2012), digitalization has two definitions and one of them is the use of the Internet and the computer to create effective economic value relative to the traditional sense [12].

The other definition fully asserts the new technology's changes in operation, interaction, configuration, and wealth creation in relation to the broader meaning. Schallmo et al (2017) investigate the definition of digitalization and the proposed definition of digital transformation for business model research [13]. According to Schallmo et al. (2017) digitalization involves networking business, customers, segments and requires data analysis to achieve the information used to evaluate the options that performance in order to increase value [13].

Berman (2012) proposed the value of digitalization based on consumer value and transformational operations based on digital technology to achieve customer collaboration and interaction for successful digital transformation [2]. The digitalization of countries can be measured with six key attributes: ubiquity, accessibility, reliability, speed, convenience and skills [7].

Westerman et al. conducted an Executive study on digitalization in terms of three aspects: buyers, business models, and operational process [17].

In the literature, Schwarzmuller investigated how digital transformation affects organizations [14]. Majchrzak et al (2016) conducted a study on the implementation of digitalization[10]. Balyer and Oz (2018) investigated the impact of digitalization on education according to the point of view of academics [1]. Bouwman investigated the effects of digitalization on small and medium-sized businesses [4].
Motta also investigated the impact of digitalization on socio-economic development [11]. Lee investigated the ability of small and medium-sized businesses to digitally transform [8].
Loonam investigated some case studies on digital transformation [9]. Digitalization affects the world economy, contributes to the emergence of such concepts as the digital economy (UNCTAD, 2017) [16].
Russia, like other countries that have a significant impact on the global and political arena, has entered the digital technology race. This is also facilitated by the fact that, according to most indicators characterizing socio-economic development, Russia is among the countries that are world leaders.
So according to the CIA Fact book 2018 Russia:
by occupied area - takes first place in the world;
by population - ranks ninth in the world;
in terms of GDP (purchasing power parity) the sixth place in the world, after Germany and Japan;
in terms of exports of products in 15th place in the world Singapore, Mexico and Canada;
by the number of Internet users in 6th place in the world after Japan and Brazil;
by the ratio of the number of Internet users to the total population in twenty-eighth place in the world, ahead of the United States and lagging behind countries such as Israel and Spain.
Despite the leading places in the world in the level and dynamics of indicators, Russia lags behind most countries in the dynamics of socio-economic development.
For example, in terms of happiness, Russia is in 68th place in the world after countries such as Peru, Portugal and Pakistan. Also, the residents of Russia have a rather low level of GDP per capita in terms of which, amounting to $ 27,900, Russia ranks 74th in the world [19].
This necessitates the speedy development of digital technologies in Russia, as their speedy implementation will significantly increase the efficiency of economic processes. The result of increased efficiency will be an increase in the level of happiness and income of the population.
The effective development of digital technologies and the planning of their development cannot be carried out without constant monitoring of the digitalization level of Russian society, both as a whole and by regions. Therefore, the development and improvement of methodological approaches to the evaluation of digitalization processes is relevant.

2. Conceptual Framework

Currently, there are many ratings and indices that characterize the digitalization of the economy of a country or business entity. Among these studies, the following research and development should be highlighted.
a. Digital Evolution Index rating developed by MasterCard and the School of Law and Diplomacy named after Fletcher at Tufts University.
The results of the study show the competitiveness and development potential of the digital economy in 60 countries. The Digital Evolution Index 2017 rating evaluates each state according to 170 unique parameters.
They describe 4 main factors that determine the pace of digitalization:
- level of supply (availability of Internet access and degree of development infrastructure);
- consumer demand for digital technology;
- environment (state policy, legislation, resources);
- An innovative climate (investments in R&D and in digital startups).

The MasterCard study proposed a way to study and measure the level of "digital trust", as well as assess the state and speed of digital development.
In addition, the Digital Evolution Index considers various examples of the introduction of digital services around the world, which allows each state to study the practices of other countries.
The disadvantage of this rating is that this rating reflects the progress in the development of the digital economy throughout the country, in terms of integrating the global network into people's lives.

b. ICT Development Index, an IDI that is measured annually by the International Telecommunication Union, a specialized UN entity.
The index consists of 11 statistical indicators reflecting the availability and use of ICT, as well as practical skills in the use of ICT by the population of 190 countries
The IDI is comprehensive and is based on the following three basic factors (sub-indexes).
The basic factor (sub-index) "Access" (Access): the number of subscribers of fixed telephone lines per 100 inhabitants, people .; the number of mobile phone owners per 100 inhabitants, people; width of the cross-border channel in terms of one Internet user, Bit / s; share of households with a computer,%; share of households with Internet access,%.
The basic factor (sub-index) "Use": the percentage of Internet use,%; fixed wired access per 100 inhabitants, people; the number of active subscribers to mobile broadband access per 100 inhabitants, people
Basic factor (sub-index) "Skills": average number of years of study, year; primary education enrollment rate,%; indicators of secondary and higher education coverage,%.
The main drawback of this rating is that it provides the basic parameters for the development of ICT in the country (regions), but does not answer many questions related to the introduction of new technologies, such as cloud computing, the Internet of things, big data, which significantly determine the degree advancement of the country (regions) on the path to building a digital economy.
c. PwC's The Future is Coming Rating
The study provides a rating of large cities in terms of their readiness to introduce new digital technologies, as well as their current capabilities, readiness and elaboration of the implementation of technological solutions and products in various social spheres: health, education, security, tourism and culture, transport, economy , Housing and communal services, urban planning, interaction with citizens.
The readiness of the city was assessed by a number of parameters: infrastructural readiness (availability of basic infrastructure), regulatory and technological readiness (availability of strategies, regulations and examples of innovative solutions), as well as the social readiness of citizens to use future technologies.
Index "Virtual services": "E-democracy and budgeting with the support of citizens' initiatives", "City as a service".
Digital Economy Index: A Shared Economy (a shared economy means the physical or virtual association of people to use resources and skills effectively, the sharing of cars (Carsharing) or group trips (Ridesharing) are typical examples of a shared economy); "Innovative financial technologies" (as examples of innovative financial technologies, the authors call decision-making technologies based on big data analysis in the financial industry, as well as block chain-based decisions); "Industry 4.0" (production based on robotics and process automation, including innovations such as the use of industrial IoT and 3D printing).
Infrastructure Readiness Index (readiness to implement 5G technology that provides data transmission with high bandwidth and minimal latency (including to support unmanned vehicles), and IoT networks with low power consumption and a large number of connected devices (including for personalized medicine), the development of data analysis centers using intelligent analytics and artificial intelligence, as well as providing online information about the city and its development based on data (received from city services, residents and companies): "Data Infrastructure"; "Data usage"; "Data analysis".
The Smart Home and Utilities Index (smart lighting systems, smart electricity networks, the use of sensors in housing infrastructure and utilities to monitor and optimize resource consumption): Smart buildings and systems for efficient use of resources; "Intelligent Electric Networks".
"Unmanned transport" index (the index determines the willingness to switch to unmanned electric vehicles, which will optimize travel time, reduce the number of traffic jams and incidents, improve the environmental situation in the city, as well as readiness for aerial drones, which will take on many functions of transport and monitoring services): "Unmanned ground transportation"; "Unmanned aerial vehicles".
Index "Digitalization of Culture and Tourism"
Smart Health Index
Open Adaptive Education Index

Proactive Security Index
Index "Virtual city".
One of the drawbacks of PwC's The Future is Coming rating is that the PwC rating assesses a greater degree of readiness to introduce technologies than the effect of their introduction; The rating is aimed at researching megacities.
d. Global Connectivity Index (GCI), Huawei
When calculating this index, the country's status in the field of digital economy development is determined, which is determined by investments, implementation, experience and potential in building the appropriate ICT infrastructure.
The GCI index takes into account the degree of development of such technologies in the country as big data, cloud technologies, data centers, the Internet of things and broadband data transfer. The degree of implementation and commercialization of these technologies is evaluated in various categories, including supply, demand, quality and potential.
The main drawback of the GCI index is the lack of official statistics for most countries and the world and their subjects. The development of the entire spectrum of digital technologies is not disclosed.
e. The Digital Russia Index, which allows one to evaluate each subject of the Russian Federation by a wide number of criteria using weighting factors, including based on:
assessing the level of public coverage in open sources of development in the constituent entity of the Russian Federation of normative regulation and administrative indicators of digitalization, ensuring the digitalization of the economy at a particular point in time;
assessment of the level of public coverage in open sources of development in the subject of the Russian Federation, specialized personnel and training programs that ensure the processes of digitalization of the economy at a particular time;
assessment of the level of public coverage in open sources of development in the constituent entity of the Russian Federation of the availability and formation of research competencies and technological groundwork, ensuring the processes of digitalization of the economy at a particular point in time;
assessment of the level of public coverage in open sources of development in the constituent entity of the Russian Federation, the direction of the information infrastructure that ensures the processes of digitalization of the economy at a particular point in time;
assessment of the level of public coverage in open sources of development in the subject of the Russian Federation, the direction of information security, ensuring the processes of digitalization of the economy at a particular point in time;
assessment of the level of public coverage in open sources of financial and economic efficiency of the development of digitalization in the subject of the Russian Federation at a particular point in time;
assessment of the level of public coverage in open sources of the social effectiveness of the development of digitalization in the subject of the Russian Federation at a particular point in time.
As conceived by the creators, the Digital Russia Index allows you to get an expert assessment of this parameter in the context of each subject of the Russian Federation, the accuracy and reliability of which corresponds to the accuracy and reliability of information obtained from open sources.
In our opinion, this approach to the level of digitalization has its drawbacks, as it is associated with the level of public coverage of one or another aspect of digitalization in the region of the Russian Federation. Also, this rating can be used to a limited extent to assess countries and regions of the world as a whole.

3. Conclusion
The study showed that at the present time the introduction of the latest digital technologies can improve the efficiency of the economic activity of the country in its regions and the world economy as a whole. Digital technologies, including indirectly affect the overall level of socio-economic development of the country, which makes them a relevant tool for the development of modern society.
Formation of plans for the development of digital technologies and their development is impossible without an objective assessment of the current state in a particular country and its regions, as well as in the world as a whole. Therefore, most of the participants in digitalization form their methodological approaches to assessing the level of digitalization.

The analysis above showed that each of the assessment methods has its own characteristics and disadvantages.
In our opinion, the main drawback of existing methods is the lack of assessment of the negative impact of digitalization on society. The most striking examples of the negative impact of digitalization include the emergence and explosive increase in the number of cybercrimes, cases of dependence on the Internet and computer games in children and adults [18].
Therefore, in our opinion, the complex of assessing the digitalization level of socio-economic objects (the subject of a particular country, country, region of the world and the world as a whole) should consist of several main blocks:
a. Assessment of the technological level of digitalization in a particular region.
b. Digitalization assessment in relation to socio-economic processes.
c. Assessing the negative impact of the digitalization process on society.
The following indicators can be used as indicators for assessing the negative impact: the number of cyber-attacks (theft of cash from credit and debit cards, suicide using social networks, theft of vehicles and real estate using electronic means of registration, theft of personal data), cases of dependence from the Internet and games, including mental illness, the average time spent by children and adults in social networks and games, in comparison with the standard, I provide general normal physical and mental condition of a person, without consequences for health and psyche.
The calculation of such an indicator will allow us to identify not only the overall development of digital technologies in a particular region or country, but also to assess the negative consequences of their implementation.

References

[1] Balyer A and Öz Ö 2018 Academicians' views on digital transformation in education. International *Online Journal of Education and Teaching* (5(4)) pp 809-830.
[2] Banker R D, Charnes A and Cooper W W 1984. Some models for estimating technical and scale inefficien-cies in data envelopment analysis *Management Science* (30(9)) pp 1078-1092.
[3] Berman S J 2012 Digital transformation: Opportunities to create new business models *Strategy and Leader-ship* (40(2)) pp 16-24.
[4] Bouwman H, Nikou S, Molina-Castillo F J and de Reuver M 2018 The impact of digitalization on business models *Digital Policy, Regulation and Governance* (20(2)) pp 105-124.
[5] Ceccobelli M, Gitto S and Mancuso P 2012 ICT Capital and labour productivity growth: A non-parametric analysis of 14 OECD countries *Telecommunications Policy* (36) pp 282-292.
[6] Digital Transformation Scoreboard 2018 EU businesses go digital: Opportunities, Outcomes and Uptake, Publications Office of the European Union
[7] El-Darwiche B, Singh M and Ganediwalla S 2012 Digitalization and prosperity *Strategy Business* (68) pp 2-10.
[8] Li L, Su F, Zhang W and Mao J Y 2018 Digital transformation by SME entrepreneurs: A capability perspective *Information Systems Journal* (28(6)) pp 1129-1157.
[9] Loonam J, Eaves S, Kumar V and Parry G 2018 Towards digital transformation: Lessons learned from traditional organizations *Strategic Change* (27(2)) pp 101-109.
[10] Majchrzak A, Markus M L and Wareham J 2016 Designing for digital transformation: Lessons for information systems research from the study of ICT and societal challenges *MIS Quarterly* (40(2)) pp 267-277.
[11] Motta G 2016 The Impact of Digitalization on Socio-economic Development: An Analysis at the European and Italian Levels (Thesis) Politecnico Di Milano School of Industrial and Information Engineering
[12] Reddy S and Reinartz W 2017 Digital transformation and value creation: Sea change ahead *Value in the Digital Era* (9(1)) pp 11-17.
[13] Schallmo D, Williams C A and Boardman L 2017 Digital transformation of business models — Best prac-tice, enablers and roadmap *International Journal of Innovation Management* (21(8)) pp 1-17.

[14] Schwarzmüller T, Brosi P, Duman D and Welpe I M 2018 How does the digital transformation affect organizations? Key themes of change in work design and leadership *Management Revue* (29(2)) pp 114-138.
[15] Shim W and Kantor P B 1999 Evaluation of digital libraries: A DEA approach *Proceedings of the ASIS Annual Meeting* (36) pp 5-15
[16] UNCTAD 2017 Information Economy Report 2017 Digitalization *Trade And Development, United Nations Conference on Trade and Development.*
[17] Westerman G, Calméjane C, Bonnet D, Ferraris P and McAfee A 2011 Digital Transformation: A roadmap for billion-dollar organizations *MIT Center for Digital Business and Capgemini Consulting* pp 1-68.
[18] Anderson C 2014 Makers: The new industrial revolution *New York, NY: Crown Business.*
[19] Kurilova A A 2018 Estimation of time spent by schoolchildren and students for work on the Internet, social networks and online games *Karelian Scientific Journal* (1 (22)) pp 137-140
[20] The world factbook 2019 Internet, *Official site CIA* ([Online], [Retrieved: 2.12.2019], https://www.cia.gov/library/publications/the-world-factbook/geos/us.html)

Impact of digital innovation on company performance

A Kurilova[1] and D Antipov[2]

[1]Tolyatti State University, Togliatti, Russia, aakurilova@yandex.ru
[2]Samara University, Samara, Russia

Abstract. Companies are concerned about the need to increase competitiveness and strengthen their positions in the market, based on innovation and the introduction of digital production. Digital innovations affect the performance and results of companies, changing business models, expanding markets and attracting new customers. At the same time, an active role is played by the impact on the modernization of operations at all stages of the life cycle of goods and the emergence of new business lines. At the same time, internal effects are associated with the improvement of various aspects of the company's life and its economic performance. External effects affect the relationship with customers, competitiveness and position of the company in the market.

1. Introduction

Worldwide, large companies embrace digital transformation through digital innovation. Digital innovation, that is, new combinations of digital and physical components for the production of new products [23], allow companies to strengthen their competitive advantages by differentiating from competitors. The introduction of digital manufacturing technologies involves the introduction of digital technologies in order to create value and improve the efficiency of the organization. Thus, the study of the impact of digital innovation on the economic performance of companies will be relevant.

Digital transformation leads to transformation of production, as well as a positive impact on the reputation of the company[1].

Organizations use digital technologies for establishing new products and services, to implement new business processes, or to operate new business models [15].

Digital innovation can act as the core and represent the central element of digital transformation, and is a domain where the use of digital technology is manifested in organizations [13,17]. In addition, we know that the implementation and implementation of digital innovation will not happen automatically, but requires a thoughtful approach that allows them to translate digital technology into digital innovation [13].

According to a number of authors, effective adoption of digital innovations can have three advantages: obtaining new customers and experience; modernization of operations; and the emergence of new business lines [6] (Fitzgerald, Kruschwitz, Bonnet, & Welch).

Through automation, the companies with the most success in implementing digital technologies in customer service and operations according to Fitzgerald [6] are distinguished by improved customer service.

On the consumer side, digital innovation has revolutionized the way consumers seek information, buy, consume, and share their experiences about goods and services. From a corporate perspective, digital innovation opens up new opportunities «to reach, inform, engage, sell, learn, and provide services to customers» [14].

Digital innovation is changing business models in the industry [24], leading to lower costs, cost of production and expansion of markets. Contactless payments, for example, have revolutionized banking. In addition, thanks to digital innovations, a huge amount of customer data is collected.

Since the adoption of digital innovation is associated with a high level of uncertainty for potential customers, it is especially important to understand how consumers may not accept digital innovation. Kuester, S., Konya-Baumbach, E., & Schuhmacher, M. C. [11] suggest that digital innovation should primarily signal reliability and usability.

Nambisan, S., Lyytinen, K., Majchrzak, A., & Song, M. [17] indicate that digital innovation refers to the use of digital technology in the process of innovation in production. Many innovations are based on digital technologies that enhance the human ability to acquire, produce, distribute and consume information at an unprecedented level and value [22]. In addition, value creation for companies is increasingly carried out through the production of digital information. This is due to digital production [20]. Schultz, D. E., & Peltier, J. [21] argue that for most companies, the challenge is not to develop and launch digital innovations, but to make them more attractive and useful to consumers. We understand that this can be achieved by finding new markets and reducing costs.

2. Conceptual Framework

The aim to introduce digital innovation in enterprises is to increase economic efficiency of enterprises and product quality, reduction of terms of creation and the start of production, optimizing the use of available resources of enterprises, the mobilization of the reserves of production capacity, competitiveness of products on world markets, formation of the modern business model of the enterprise.

The introduction of certain digital innovations can have a different impact on the performance of the enterprise. Table 1 shows the impact of digital innovation on the economic performance of the enterprise.

Table 1. Impact of digital innovation on enterprise performance.

Digital innovation	Influence on economic indicators of the enterprise
Additive technologies	Improving the design and use of materials The quality of the products Risk minimization Reduce development and prototyping time Reduction of production costs (labor costs, operating time of technological equipment, electricity, materials) Reducing the cost of the product by reducing the costs of the company
Rapid prototyping	Quality control Reduce product development and testing time Reducing the cost of goods
Cloudy computing	Optimizing the use of computing resources Access to the skills and abilities of professionals Improve customer experience Improving compliance with safety requirements Accelerate the introduction and deployment of applications Increase productivity, reduce costs for deployment and maintenance of it infrastructure Reduction of capital costs The decrease in expenses for purchase of software
Big data	The basis for the development of the Internet of things Operational monitoring of all systems and processes Analysis of unstructured data sets

Digital innovation	Influence on economic indicators of the enterprise
	Expansion of markets Reduction of costs for repair and maintenance of facilities Improving the quality of processed data for decision making The reduction of production costs The increase in profits due to emerging opportunities
Internet of things	Automation of tracking and forecasting of complex systems and objects Full implementation of object and process control automation Increased productivity Cost optimization and reduction of maintenance and repair costs Savings of the wage Fund
Digital double	Reduce design and prototyping errors Application of new production technologies Reduce development and prototyping time Reducing the cost of designing, testing and upgrading of products Reduction and elimination of losses from marriage Reducing the cost of finished products

3. Conclusion

Internal effects are associated with the improvement of various aspects of the life of the company and its economic performance.

These include the following effects:

1. Reduction of internal costs.
2. Increased productivity.
3. Acceleration of business processes (production, sales, service, development, etc.).
4. Increase efficiency in decision-making.
5. Improve the manageability of the company.
6. Improving awareness about the operation of equipment, transport, products.
7. Increase employee satisfaction with the work performed.
8. Attracting new qualified employees.

External effects include those related to customer relationships, competitiveness, market position:

1. Increase in sales.
2. The growing number of customers.
3. Increasing customer satisfaction
4. Improve customer awareness of the company's products/ services.
5. Better understanding of customer needs.
6. The emergence of fundamentally new products and services.
7. Reducing the backlog of competitors.

Thus, the introduction of digital innovations allows you to develop much faster, cheaper and with a much smaller number of employees, significantly reduces the risks of increasing time and reduces development costs, allows to significantly reduce the amount of field testing, finally allows to dynamically and efficiently manage changes of target characteristics and to consider new constraints that arise during design, production or operation.

References

[1] Anderso C 2014 Makers: The new industrial revolution *New York, NY: Crown Business*

[2] Chan Y, Huff S, Copeland D 1997 Assessing realized information systems strategy *J. Strateg. Inf. Syst.* (6 (4)) pp 273–298.

[3] Dörner K and Edelman D 2015 What 'digital' really means *McKinsey and Company*. (www.mckinsey.com/industries/technol-ogy- media- and- telecommunications/our-insights/what- digital- really- means)

[4] Elia G, Margherita A, and Petti C 2016 An operational model to develop technology Entrepreneurship "EGO- System." *International Journal of Innovation and Technology Management* (13(5)) pp 21–46. https://doi.org/10.1142/S0219877016400083

[5] Fichman R G, Dos Santos B L and Zheng Z E 2014 Digital innovation as a fundamental and powerful concept in the infor-mation systems curriculum *MIS Quarterly* (38(2)) pp 329–353.

[6] Fitzgerald M, Kruschwitz N, Bonnet D and Welch M 2013 Embracing digital technology: A new strategic imperative *MIT Sloan Management Review* (55(2)) pp 1–2.

[7] Henfridsson O, Nandhakumar J, Scarbrough H and Panourgias N 2018 Recombination in the open- ended value landscape of digital innovation *Information and Organization* (28(2)) pp 89–100.

[8] Ibarra D, Ganzarain J and Igartua J I 2018 Business model innovation through industry 4.0: A review *Procedia Manufacturing* (22) pp 4–10.

[9] Jonsson K, Mathiassen L and Holmström J 2018 Representation and Mediation in Digitalized Work: Evidence from Maintenance of Mining Machinery *Journal of Information Technology* (33(3)) pp 216–232.

[10] Kannan P K 2017 Digital marketing: A framework, review and research agenda International *Journal of Research in Marketing* (34(1)) pp 22–45.

[11] Kuester S, Konya- Baumbach E and Schuhmacher M C 2018 Get the show on the road: Go- to-market strategies for e- innovations of start- ups *Journal of Business Research* (83(February)) pp 65–81.

[12] Koch S and Bierbamerm M 2016 Opening your product: impact of user innovations and their distribution platform on video game success *Electronic Markets* 26(4)) 357–368.

[13] Kohli R and Melville N P 2018 Digital innovation: a review and synthesis *Information Systems Journal* (29(1)) pp 200–223.

[14] Lamberton C and Stephen A T 2016 A thematic exploration of digital, social media, and mobile marketing: Research evolution from 2000 to 2015 and an agenda for future inquiry *Journal of Marketing* (80(6)) pp 146–172.

[15] Legner C, Eymann T, Hess T, Matt C, Böhmann T, Drews P, et al. 2017 Digitalization: opportunity and challenge for the business and information systems engineering community *Business and Information Systems Engineering* (59(4)) pp 301–308.

[16] Margiono A, Zolin R and Chang A 2018 A typology of social venture business model configurations *International Journal of Entrepreneurial Behavior and Research* (24(3)) pp 626–650.

[17] Nambisan S, Lyytinen K, Majchrzak A and Song M 2017 Dig-ital innovation management: Reinventing innovation management research in a digital world *MIS Quarterly* (41(1)) pp 223–238.

[18] Nylén D and Holmström J 2015 Digital innovation strategy: A framework for diagnosing and improving digital product and service innovation *Business Horizons* (58(1)) pp 57–67.

[19] Sahut J M, Hikkerova L and Moez K 2013 Business model and performance of firms *International Business Research*, (6(2)), pp 64–76. https://doi.org/10.5539/ibr.v6n2p64

[20] Schallmo D, Williams, C A and Boardman L 2017 Digital transformation of business models: Best practice, enablers, and roadmap *International Journal of Innovation Management* 2(1(08)) pp 1–17, 1740014.

[21] Schultz D E and Peltier J 2013 Social media's slippery slope: Challenges, opportunities and future research directions *Jour-nal of Research in Interactive Marketing* (7(2)) pp 86–99.

[22] Teece D J 2018 Profiting from innovation in the digital economy: Enabling technologies, standards, and licensing models in the wireless world *Research Policy* (47(8)) pp 1367–1387.

[23] Yoo Y, Boland R J Jr, Lyytinen K and Majchrzak A 2012 Organizing for innovation in the digitized world *Organization Science* (23(5)) 1398–1408.

[24] Zott C and Amit R 2017 Business model innovation: How to create value in a digital world *GfK Marketing Intelligence Review* (9(1)) pp 18–23.

The role of quality management systems for educational institutions in the digital economy

A A Sherstobitova and M O Iskoskov

Togliatti State University, Togliatti, 445667, Russia, E-mail address: ya_anya@mail.ru

Abstract. One of the priority tasks of education in the modern conditions of economic management is to provide quality technological and personnel reserves for the implementation of the tasks set by the Government of the Russian Federation in the field of development of the digital economy of the country. Considering the approaches to the transformation of all types of activities into a digital platform, it is possible to note the stage-by-stage transformation of the information infrastructure over the last period of development of market relations in the direction of digital transfer of information, knowledge and technologies. One of the barriers preventing a faster transition to new challenges to reality is the lack of compliance of the level of personnel training with the requirements of QMS standards, in particular the requirements of ISO 21001-2018 and ISO 9001-2015. The article considers the peculiarities of the requirements of the regulatory framework for the organization of QMS of higher education institutions, presented in the form of tasks described in the roadmap. The novelty of the proposed approaches consists in the integration of QMS requirements in the construction of a competency model. The proposed solutions to improve the processes of personnel training will ensure that society can implement the Passport "National Program "Digital Economy of the Russian Federation".

1. Introduction. Basic concepts of digitalization and digital economy

The digital economy is understood as a system of socio-economic relations based on the use of digital means of communication and new information technologies [1].

The conceptual approach to the use of digital resources to improve the efficiency of organizations is interpreted in many documents as "digitalization", and this concept is nowadays used in a particularly wide way. For example, the Strategy for the Development of the Information Society in the country focuses on the mass introduction of domestic production tools in the field of digital interaction of complex systems, in particular, integrated production structures [2-4]. Higher education institutions are expected to train personnel that meet the requirements of the digital economy [5-7].

Many authors [10-14] consider digitalization as a process of transformation and presentation of various data in electronic form. Over the past three years, the regulatory framework has been significantly supplemented by new guiding documents, the essence of which reveals the opportunities for the development of individual areas of the digital economy and the country's economy as a whole, as well as its informatization.

Analysis of the essence of the content and requirements of the roadmap discussed in the Passport " National Program "Digital Economy of the Russian Federation " (hereinafter referred to as the CERF Passport) [8], and the Programs of Socio-economic Development of the country for the period up to 2030 [9] showed that they focus on the emergence of new concepts and definitions, new standards of digital data processing and tools for their practical implementation.

IOP Conf. Series: Materials Science and Engineering **986** (2020) 012023 doi:10.1088/1757-899X/986/1/012023

Standardization is also understood as the process of application of regulatory requirements and rules ensuring the quality of collection, processing, transmission and storage of information by the personnel, which allows to regulate the existing information flows and manage their application in all spheres of activity, for the purpose of making quality management decisions, and taking into account modern approaches of informatization and digitalization.

Therefore, one of the key tasks in higher education is the formation of a set of new competencies that meet the requirements of the FSES, which should ensure the implementation of the requirements of the CERF Passport and the existing professional standards of the application to a particular subject area, for example, in the field of management. The analysis of the requirements reflected in the CERF Passport for technological and human resources management has shown that there are pressing problems that hinder the development of digitalization in educational institutions (Fig. 1)

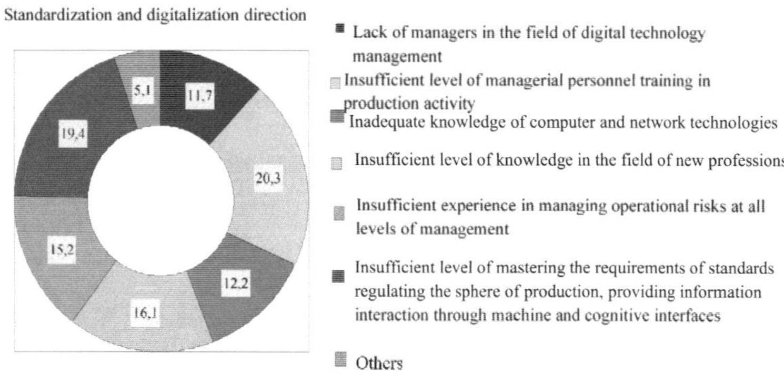

Figure 1. Analysis of the problems hindering the formation of the technological backlog of digitalization processes

From figure it is visible that one of the main problems for realization of directions of routing is a lack of experts who can operate processes of acceptance of administrative decisions on workplaces of users as do not have the special skills and the skills corresponding to requirements of time. It is necessary to be ahead of the mass demand, to bring to the market specialists who already have the required set of competencies.

Of course, the objective reality is a significant advance in the development of information technologies, computerization and communication tools before learning tools, and this is one of the problems of higher education. Another problem is the lack of time for retraining and professional development of teaching staff, who carry out the transfer of knowledge to students. Without going into revealing the peculiarities of the removal of these conflict situations, we will focus on the process of personnel training itself.

The article investigated the problems of organizing the process of training managers and their correlation with the identified needs of the country, identified within the framework of the problem of paragraph 2.1 of the «Road Map».

A fragment of the Roadmap [1, pp. 38-39] with the planned terms of its implementation and the key indicators being formed is shown in Table 1. Thus, for example, the concept of "Internet of Things" (IoT) has been used recently, the mastering and implementation of which in the practical management of real production contributes to the transition to a new stage of development of the Internet, as it allows to significantly expand the opportunities for the collection, processing and distribution of data, including the processing of knowledge [15].

Table 1. Roadmap fragment of the CERF Passport.

1. Ensuring the training of highly qualified personnel for the digital economy		
1.1 The concept of the basic model of digital economy competences, the list of key competences and the mechanism of their updating were developed	01.10.2019	Ilya Torosov, Deputy Minister of Economic Development of the Russian Federation
1.2 A venture fund was created to support promising educational technologies of the digital economy	30.12.2019	Ilya Torosov, Deputy Minister of Economic Development of the Russian Federation
1.3 Approves the open format of competence profiles of citizens, their development trajectories and procedures for their creation	30.06.2020	Alexey Sokolov, Deputy Minister of Digital Development, Communications and Mass Media of the Russian Federation

Table 2 shows a fragment of the Roadmap [1, p.72-73], which allows the educational institution to restructure its educational activities taking into account the qualitative components of project activities.

Table 2. Roadmap fragment of the CERF Passport.

1. Introduction of digital technologies and platform solutions in public administration and public service delivery, including for the benefit of the population and small and medium enterprises, including individual entrepreneurs		
1.1 Legislation has been drafted to ensure that government agencies primarily use a single e-government infrastructure	31.12.2019	Oleg Pak, State Secretary - Deputy Minister of Digital Development, Communications and Mass Media of the Russian Federation
1.2 The federal law and other normative legal acts fixing the target state of provision of state and municipal services have been developed and adopted	31.12.2020	Savva Shipov, Deputy Minister of Economic Development of the Russian Federation

It will allow to form competences regulating sphere of production of a full cycle of production and use of the technical decisions providing information interaction by means of machine and cognitive interfaces, including the Internet of things. In these economic conditions, one of the priority tasks of higher education institutions is the quality training of personnel, which is ensured through the introduction of quality management system in the educational institution. This stimulates the staff of higher education institutions to constantly improve all internal processes and thus allows achieving positive economic effects.

2. The role of quality training in the digitalization environment

Considering the quality of education, we mean the quality of the result of the educational process, that is, the set of characteristics that determine the ability to sufficiently meet the needs and expectations of external consumers of educational services. In other words, the quality of this or that characteristic of education is such a level that is in demand and meets the requirements of the consumer.

In our opinion, the diversity of definitions of the concept of "quality" is caused by the impossibility to express in a single definition absolutely such equal characteristics of this concept as absolute and relative (comparative) quality.

Absolute quality is a philosophical category characterized by a set of individual properties of an object. The most capacious is the definition of quality as compliance with the norm, standard.

Unity of needs in the properties of the object (compliance with the norm, standard) makes it possible to compare the real properties of different objects with each other. The use of the concept and features of relative quality makes it possible to compare the objects from the economic point of view.

Specificity of educational service is the presence of dualism in the concept of quality. On the one hand, as a provider of educational institution offers the applicant an educational program in conjunction with other (personnel, scientific and technical, etc.) components of its potential, and on the other hand, a specialist who has received a professional education, entering the labor market, faces the assessment of the quality of his professionalism by the employer.

Accordingly, there are significant differences in the quality of educational services, including the methods of their qualification and benchmarking assessments.

An important component of the marketing strategy of the university is adequate positioning of the services provided by it according to the price characteristic, namely: accessible education, elite, balanced combination of quality and price. It is necessary to bear in mind that the consumer of educational services will pay a higher price for them if the university can visually present to him additional competitive advantages of its educational programs, provided services (for example, quality of service), qualitatively distinguishing them from the programs or services of competitors.

In view of the above, we can propose the following definition of the content of educational services.

Educational service is a set of ordered information transmitted to the learner in the form of knowledge of educational, special and specialized nature, as well as some of the most important practical skills of their use in the chosen learning industry or specialty, the formation of positive personal characteristics necessary in professional activities [16].

The concept of "quality of education", which includes not only cognitive, social and cultural characteristics, is an integral characteristic of educational activity as a process and its outcomes.

The modern approach is characterized by a sharp increase in the requirements for quality assurance in higher education. In order to ensure the solution of this problem it is necessary to solve the issues with the modernization of the quality assurance system, for this purpose it is necessary to develop indicators, evaluation criteria, methodologies, as well as to solve the issues on the development of effective mechanisms to ensure rapid assessment, the possibility of eliminating the detected negative phenomena, the formation of measures to ensure the growth of quality indicators of the educational institution.

Quality management systems (QMS) are one such mechanism.

3. Quality management systems of educational institutions

In accordance with the terminology of ISO International Standards (IS) "Quality management system is a management system for the management and administration of an organization in relation to quality" [16].

The first international standards ISO 9000 series were developed and approved in 1987. In 1994 and 2000 the work on revision of ISO 9000 series was carried out taking into account the accumulated experience on their application and forecast of the world tendencies of quality problems development. In 2008, the international standard ISO 9001:2008 (ISO 9001-2008) was published. The latest version is the standard of quality management system ISO 9001:2015 (ISO 9001-2015) [17].

Modern trends in the field of quality assurance of education and the basic models of quality management in the university today continue to be considered on the basis of the model of quality management system that meets the requirements of the international standard ISO series 21001: 2018, the requirements of ISO series 9001:2015, the model of the European Foundation for Quality Management (EFQM), as well as various models of national and regional quality awards and their modifications for higher education.

Below is the structure of ISO 21001:2018 standard "Educational organizations - management systems for educational organizations - requirements with guidance for use" ("Educational organizations - Management systems for educational organizations - Requirements and guidelines for use"). Let's consider some of its features.

3.1. ISO 21001-2018 requirements for the activities of educational institutions

The standard applies to higher, secondary, secondary, specialized and primary educational institutions, as well as institutions providing additional education services. ISO 21001 provides users with a model that is designed to be as convenient and effective as possible for a variety of educational institutions. Confirmation of the implementation of ISO 21001 standard and obtaining the certificate will allow the educational organization to demonstrate to all stakeholders a high level of quality of educational services, as well as to obtain additional points related to the existence of a quality system.

It is important to note that the standard can be used by educational organizations not only to ensure the quality of educational processes, but also to organize research activities.

Figure 2 provides a brief description of the organizational structure of the standard.

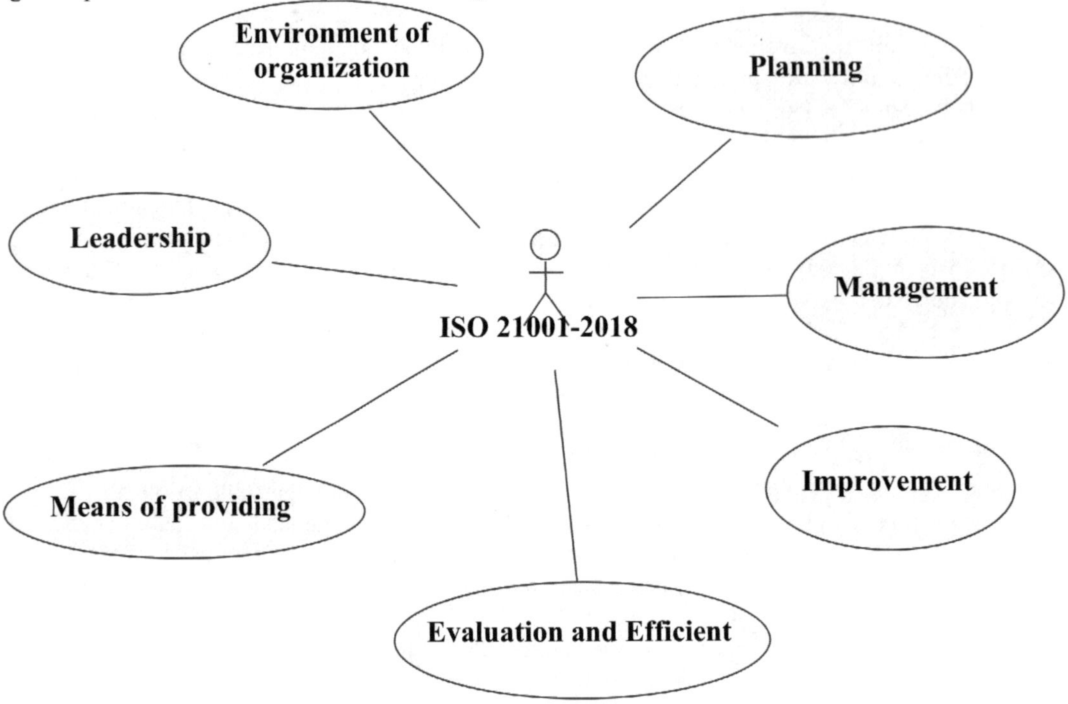

Figure 2. Organizational structure of the standard.

This standard is based on the following principles of education management:
a) The orientation of students and other beneficiaries;
b) Visionary leadership;
c) Involving people;
d) Process approach;
e) Improvement;
f) Informed decisions;
g) Relationship management;
h) Social responsibility;
i) Accessibility and equity;
j) Ethical behaviour in education;
k) Data security and protection

As an example, we will further look at the process approach (d) and evaluate the quality of the educational process. The emphasis will be on the fact that the actual completeness of the process can be assessed on the basis of a qualimetric approach.

3.2. ISO 9001-2015 requirements for the activity of educational processes

The version of the ISO 9001-2015 standard was developed in accordance with the annex to the ISO Annex SL (ISO/IEC Directives, Part 1 Consolidated ISO Supplement - Procedures specific to ISO) directive. The directive defines the requirements for regulatory documents for management systems. It establishes a new, uniform standard for the structure of management systems (not only ISO 9001, but also for other management systems).

According to this directive, all management system standards will be brought to a single structure and will have common section names.

The management model, based on the requirements of the international quality standards ISO 9001:2015, assumes compliance with the following quality management principles:

a) Consumer orientation;
b) Leadership by management;
c) People-to-people interaction;
d) Process approach;
e) system approach to management;
f) Continuous improvement of activities;
g) Evidence-based decision-making;
h) Relationship management.

One of the basic differences of GOST R ISO 9001-2015, in our opinion, is reflected in the requirement to measure the control points of the beginning and end of the processes of quality management systems [18].

As an example, we will further consider the process approach (d) and evaluate the quality of the educational process. The emphasis will be on the fact that the actual completion of the process can be estimated on the basis of the qualimetric approach.

4. Our achievements in quality management for the digital economy

4.1. Algorithm of educational process management on the basis of integration of standards GOST R ISO 9001-2015 and GOST R ISO 2001-2018

When managing business processes, based on the principles of TQM, we will focus on one of them "making decisions based on facts". It has a direct link with the requirements of GOST R ISO 9001-2015, as the measurement of key process indicators, namely the points of its beginning and end, in our view, ensures the implementation of this principle.

This is the first advantage acquired by the organization during the implementation of the standard.

The second basic advantage is that risks of process activity essentially decrease if the control and measurement of key points of process will be realized at system level.

Therefore, introducing in the activity requirements of GOST R ISO 9001-2015, top-management of the organization should put before the employees following problems. Their decision is presented by stage-by-stage algorithm of realization:

Stage 1: Identify and justify the main educational processes that are necessary for the quality management system of the organization and its development;

Stage 2: Identify the key start and end indicators in each of the processes ("required inputs and expected outputs") and describe their quantitative parameters, accepting these values as normative;

Stage 3: Determine the interrelated hierarchy of processes, justifying their sequence and conditions of interlevel interaction (transitions);

Stage 4: Determine the allowable risk zone for making a wrong decision about the possibility of transition from one process to another.

Stage 5: Modeling of educational processes based on the requirements of GOST R ISO 2001-2018.

Stage 6: Critical assessment of the “Entry” and “Exit” points of the process based on the requirements of GOST R ISO 9001-2015.

Stage 7: Risk assessment of non-achievement of goals on the basis of the risk management tool defined in the standard GOST R ISO 9001-2015 and the formation of corrective effects on the process.

It is impossible to solve the set tasks if the employees of the organization do not have the tools of standardization and skills of their practical implementation in the management of the quality of business processes, their monitoring and evaluation of the effectiveness of management decisions.

4.2. Example of implementation of author's approaches in the educational process

The Digital Economy of the Russian Federation Passport pays special attention to the formation of the personnel component and the intellectualization of knowledge.

"Knowledge" is a resource based on the practical experience of specialists and on information that allows to form new knowledge. The ability to acquire, integrate, accumulate and apply knowledge increases its competitiveness. Therefore, intellectual activity is considered as the main one in all regulatory documents of the Government of the Russian Federation related to the socio-economic development of the country [19]. As we can see from the roadmap (Table 1,2), the first requirement is to develop a model of competencies that could provide the labor market with demanded specialists. To solve this problem, the Institute of Finance, Economics and Management studied the requirements of labor standards and labor functions, which could be correlated with the challenges of the CEFF Passport. The most difficult, in our opinion, is the implementation of paragraph 2.1, which reflects the requirements for the development of tools for the certification of emerging competencies in the education system and the digitalization program. Let's assume that we need to carry out qualitative training of a manager to work in a full production cycle. According to the challenges of the Roadmap, knowledge of the Industrial Internet of Things (IIoT) is required.

As an example, let us consider the requirements of FSES 3+ for the preparation of bachelors in the direction of "manager". Table 3 shows the purpose of educational activities, which should be subsequently certified by evaluation funds.

Table 3. Baseline data for assessing the probability of achieving the objective.

Goal	Subobjectives	Tasks
Introduction of IIoT in the educational process of preparation	1. Formation of knowledge in the field of production automation	1.1 Study of peculiarities of production automation 1.2 Development of automated production management skills
	2. Formation of knowledge in the field of "big data" processing	2.1 Study of modern computer technologies for "big data" processing 2.2 Building "big data" management skills using modern computer tools
	3. Formation of knowledge about the processes of intermachine communication	3.1 Formation of knowledge base about the processes of intermachine communication 3.2 Building skills to manage intermachine communication processes

Let's introduce signs: let's say it's the probability of achieving the main goal. Let's set it at 95%.

It has been determined by experts that in order to achieve the main objective (1) it is necessary to ensure the achievement of three subobjectives (1.1, 1.2, 1.3) in terms of the required performance (EGP) of functioning.

Then, Table 4 will define quantitative requirements for the indicators of the completion of controlled processes to ensure the required level of quality of solution of the target task.

Table 4. Formation of normative indicators to ensure target parameters.

Target	Probability of collateral, %	Subtargets	Probability of collateral, %	Tasks	Probability of collateral, %
Implementation of IIoT in the training process	95	1. Formation of knowledge in the field of production automation	98,3	1.1 Study of peculiarities of production automation.	99,15
				1.2 Development of automated production management skills	99,15
		2. Formation of knowledge in the field of "big data" processing.	98,3	2.1 Study of modern computer technologies for "big data" processing	99,15
				2.2 Development of "big data" management skills based on modern computer tools	99,15
		3. Formation of knowledge about the processes of intermachine communication	98,3	3.1 Formation of knowledge base about the processes of intermachine communication	99,15
				3.2 Development of interpersonal communication management skills	99,15

Formed competences can be recommended to be studied in the process of building their own model of knowledge in the discipline under study and use of production rules for their extraction on the simplest examples.

5. Conclusion

The digital economy is represented by the following 3 levels, which in their close interaction affect the lives of citizens and society as a whole, this:
- markets and sectors of the economy (spheres of activity), where interaction of specific subjects (suppliers and consumers of goods, works and services) is carried out;
- Platforms and technologies where competencies for the development of markets and sectors of the economy (spheres of activity) are formed;
- Environment, which creates conditions for the development of platforms and technologies and effective interaction of subjects of markets and sectors of economy (spheres of activity) and covers normative regulation, information infrastructure, personnel and information security.

Due to the fact that the effective development of markets and industries (spheres of activity) in the digital economy is possible only in the presence of developed platforms, technologies, institutional and infrastructural environments, it is advisable to consider two lower levels. **Acknowledgement**
This work was supported by the Russian President's grant for state support of young Russian scientists having a doctor's degree (MK – 2246.2020.8).

References

[1] Passport of the national project "National Program "Digital Economy of the Russian Federation" (approved by the Presidium of the Presidential Council for Strategic Development and National Projects, Minutes of 04.06.2019 N 7)
[2] Decree of the President of the Russian Federation of 09.05.2017 N 203 "On the Strategy for the Development of the Information Society in the Russian Federation for 2017 - 2030"
[3] Order of the Government of the Russian Federation No. 1632-r [Electronic resource] (Source: http://static.government.ru/media/files//)
[4] Order of the Government of the Russian Federation of November 17, 2008, No. 1662-r On the Concept of Long-term Socio-Economic Development of the Russian Federation until 2020 (as amended and supplemented) [Electronic resource].
[5] Order of the Government of the Russian Federation of October 20, 2010 N 1815-r "On the State Program of the Russian Federation "Information Society (2011 - 2020)". (as amended and supplemented).
[6] Strategy (Memorandum) of IT industry development as the basis of digital economy [Electronic resource] (Access mode: http://www.apkit.ru/files/PWC_strategy_miiting2017.pdf/)
[7] Decree of the President of the Russian Federation of 09.05.2017 No. 203 "On the Strategy for the Development of the Information Society in the Russian Federation for 2017 - 2030" [Electronic resource] (Source: http://publication.pravo.gov.ru/Document/View/0001201705100002)
[8] Decree of the President of the Russian Federation of May 7, 2012, No. 596 "On Long-term State Economic Policy" [Electronic resource] (Source: http://www.kremlin.ru/news/15232)
[9] Alexeev I R and Alekseev I V 2017 Digital economy: features and trends of electronic interaction development [Text] *Actual directions of scientific research: from theory to practice*. - — № 4 (10). - — pp 42–45. - P.: Interactive Plus,
[10] Alekseev I R 2015 Information support of the management system of franchise enterprises [Text] *Economic and humanities* (№1 (276)) pp 105–110.
[11] Glukhova L V 2017 Mitrofanova Y S Digitalization of economy and peculiarities of its application in the activity of integrated production structures (in Russian) *Vestnik Povolzhsky State University of Service . Section "Economics"* p 4
[12] Glukhova L V and Nemtsev A D 2013 Management of changes and innovations: conceptual approach to the use of quality management tools [Text] *Vestnik Volzhskogo Universiteta n.a. V N. Tatishchev. Section "Economics"* (№2 [26]) pp 21-29

[13] Gugelev A V 2018 Digitalization in the quality management system (in Russian) *Prospects of development of the domestic enterprises in the conditions of formation of the digital economy: mat. Internar. (full-time and part-time) scientific practice. conf. Saratov* pp 158-161
[14] Petrosyan A 2017 What do we need to know about the digital economy and its development prospects [Electronic resource] (Access mode: http://www.kommersant.ru/doc/3063024 (access date: 14.06.2017)).
[15] Black Yu B 2016 Standardization - the main instrument of innovative development [Text] *Economics of quality* (№ 1(13)) pp C. 34-40
[16] Gorbashko E A, Bonushko N A and Semchenko A A 2017 Development of the quality management system in the cluster economy *monograph* (St. Petersburg: St. Petersburg State University of Economics) pp 159
[17] Glukhova L V, Syrotyuk S D, Sherstobitova A A and Pavlova S V 2019 Smart university development evaluation models *Smart Innovation, Systems and Technologies* (T. 144) pp 539-549
[18] Glukhova L V, Syrotyuk S D, Sherstobitova A A and Gudkova S A 2019 Identification of key factors for a development of smart organization *Smart Innovation, Systems and Technologies*. (T. 144) pp 595-607.
[19] Nemtsev A D and Glukhova L V 2018 Concept of the standardization of the managers' activity in the conditions of the digital economy *Bulletin of the Volga University named after V.N. Tatishchev* (T. 2. № 1) pp 165-175
[20] Antipov D V, Akhmetzhanova G V, Antipova O I, Gazizulina A U and Sharov R 2017 Organizational models of teal organizations *6th International Conference on Reliability, Infocom Technologies and Optimization: Trends and Future Directions, ICRITO 2017*
[21] Kozlovsky V N, Lysov V E, Ermakov V V, Antipov D V and Skripnuk D F 2019 Calculation and statistical experiment on the Monte Carlo method when assessing the stability of the technical characteristics of the automobile generator set in mass production *Proceedings of the 2019 IEEE Conference of Russian Young Researchers in Electrical and Electronic Engineering, ElConRus*

Digital Technologies and QMS of Industrial Enterprise: Focus on Efficiency

D V Antipov[1],E Yu Kuznetsova[2] and A Aytasova[3]

[1]Samara National Research University academician S.P. Koroleva, Samara, Russia. E-mail: con-expert@mail.ru
[2]Togliatti State University, Togliatti, Russia. E-mail: elenatgus@yandex.ru
[3]Peter the Great St. Petersburg Polytechnic University, St. Petersburg, Russian Federation

Abstract. The relevance of the problem is due to the fact that currently no aspect of the functioning and development of society can be imagined without the use of digital technologies. And if at first they were identified and associated with the use of communications, then as they developed, the scope of their application expanded due to significant competitive advantages of both the private and global order: leveling of interstate and territorial borders in the transmission, dissemination and receipt of information, including large volumes, adaptation of best practices, the possibility of faster adoption and implementation of management and technical and technological decisions. Digital technologies have become an integral part of the modern economic formation: in the household sphere, the education and health system, transport and other spheres of human life. This ultimately contributed to the natural process of diffusion of digital technologies, production systems and business in general, i.e. the so-called "digitalization of industry". It is important to note that the successful operation of the enterprise, regardless of the level of progressiveness of the industrial technologies used by it, should be ensured by the production of products or services that meet certain market needs, approved standards and specifications and, finally, the requirements of the current legislation. At the same time, the quality of products and services that meet the needs of consumers cannot be ensured through the one-off events' introduction. It must be a systematic approach to quality management that covers all aspects and stages of the full life cycle of a product or service. This involves the development and implementation of quality management systems at the enterprises. Therefore, digital technologies in the QMS should be considered as an important tool for analysis, control and quality assurance of the final product and, accordingly, the competitiveness of the economic entity. The purpose of the article is to develop an economic and mathematical model for obtaining a regression dependence between the amount of revenue from sales and the factors of its formation. The leading method of research was the method of correlation and regression analysis. The main results of the article: the main factors of QMS and digitalization that affect the income of the enterprise through the quality of its products are identified; the dependence of the influence of various factors on revenue is determined, as well as the factors themselves characteristic of the enterprises of the automotive components industry on the example of JSC «Togliatti plant of automotive units». The materials of the article can be useful for any economic entities engaged into the production of goods, as well as focused on achieving high indicators of its quality through the introduction of digital technologies in the manufacturing process and QMS. The obtained results allow us to assess the profitability of the digitalization of production processes by taking into account the influence of factors of direct (on the amount of sales volume) and indirect (on the amount of costs for quality assurance) impact.

IOP Conf. Series: Materials Science and Engineering **986** (2020) 012024 doi:10.1088/1757-899X/986/1/012024

1. Introduction

A fait accompli can be considered a new technological revolution, a characteristic feature of which is the digital transformation. In essence, this means that analog and control technologies, where the decision-maker was a person, are gradually being replaced by artificial intelligence and machine control technologies. Digital transformation expands the spheres of influence and covers not only an increasing number of economic entities, but also the functional spheres of its application. In such circumstances, it is not an exaggeration to say that a high level of digitalization becomes synonymous with the competitiveness and prospects of companies, industries and national economies [6; 14; 15], and its neglect is a sign of insolvency in competition and satisfaction of the consumers' needs [3; 13].

Quality management problems are no exception. Rather, on the contrary: the introduction of digital technologies is focused on improving the quality of a product or service, compliance with regulatory requirements, improving the decision-making process, increasing efficiency while reducing risk [4; 10]. Therefore, speaking about the effectiveness of digital technologies for enterprises, it is firstly necessary to note its exogenous nature, i.e. external competitiveness due to the increase in production volumes induced by the growth of labor productivity. However, at present, the endogenous, i.e. internal competitiveness of the economic entity becomes the goal of digitalization of production. Its formation is carried out on the basis of tracking the parameters of all production processes in real time and their corresponding adjustments online. This means a preventive problems' solution, including in terms of product quality management, when instead of eliminating defects in already manufactured products, i.e., «fight against defects», enterprises immediately set up technological processes, ensuring their convergence and reproducibility, preventing or reducing almost to zero the amount of losses from defects. In other words, ensuring the external competitiveness of the company is directly related to the internal competitiveness. It is here that the connection and feasibility of using digital technologies in quality management and embedding them in the QMS of business entities is formed.

2. Research structure

Research structure included the following stages: as the object of the study was selected close corporation «Togliatti plant of auto units» (plant «TZA»), which is the official supplier of the main conveyor of JSC «AVTOVAZ» - one of the largest domestic car manufacturers in Russia; developed a system of indicators to assess the efficiency of the industrial enterprise, taking into account the functioning of its QMS and the introduction of digital technologies; identified and quantified the factors that affect the amount of revenue and costs to ensure product quality; an econometric model based on the assumption that it is possible to establish a regression relationship between sales volume and factors affecting it on the part of QMS and digitalization of production.

3. Representativeness of the object of study

Since its foundation in 1995 and up to the present time the plant «TZA» provides the needs of both the main conveyor of JSC «AVTOVAZ» (AVTOVAZ-RENAULT-NISSAN) and the spare parts market with its products.

The quality management system of the company complies with the requirements of international standards ISO 9001:2015, IATF 16949: 2016, ISO 14001, as well as the requirements of the supplier evaluation standard of the Renault-Avtovaz Alliance (ASES-Alliance Supplier Evaluation Standart). All production of the enterprise corresponds to requirements of technical regulations of the Customs Union [19].

All products supplied to the spare parts market have branded packaging with the use of anti-counterfeit products. The range of products of «TZA» is constantly updated, their nomenclature extends, the let-out production is modernized according to the clients' requirements. Currently the plant produces the following product groups:

• cooling system;
• lubrication system;
• engine elements;

• transmission;
• steering;
• technical liquids.

In 2012, within the framework of the Russian Federation government's program aimed at the introduction of energy-saving technologies, as well as to meet the growing demand for energy-efficient lighting devices, serial production of led lamps was launched at its own production facilities.

In 2017, the production of new goods — technical liquids: antifreeze and antifreeze cooling liquids, as well as summer glass-breaking liquid and glass-breaking non-freezing liquid for the winter season for cars of domestic and imported production was started.

Close corporation «TZA» has resources for the development and production of modern automotive components [19]:

• design management and pilot production;
• machining of aluminum and cast iron alloy parts;
• assembly production of main and auxiliary units;
• modern semi-automatic Assembly centers of European production;
• automatic line of production and filling of technical liquids;
• testing laboratory;
• quality control department;
• logistics management.

The high quality of automotive components produced by close corporation «TZA» is confirmed by the presence of certificates of the international standard of the quality management system of the automotive industry [19]:

International standard ISO 9001: 2015 No. 84503 / AA / 0001 / SM / RUS dated 16.11.2017
International standard IETF 16949: 2016 No. 84503/A / 0001/SM/RUS dated 16.11.2017
Renault-Avtovaz Alliance supplier Evaluation standard (ASUS-Alliance Supplier Evaluation Standart):
- No. R-RU T 0007-0-F1921 from 27.05.2019 g, the final result of the assessment-rank B
- SHC-Supplier Health Check score no. R-RU T-0020-0-H 1820 of 23.10.2018 (assigned category H1).
Technical regulations of the Customs Union.

Close corporation «TZA» is the patentee of more than 20 utility models and inventions. In addition, all technological processes used at the enterprise are approved and meet all the requirements of JSC «AVTOVAZ» and the agreed specifications for the products.

Therefore, the object of research can be considered involved in the process of digitalization. However, like most of Russian enterprises, close corporation «TZA» implements separate digital technologies into the production process and its management (automatic lines that allow online transmission and processing of information about the parameters of production, the level of its stability; the consumers' feedback collection by using a questionnaire on the website and the formation of a database on the problems; electronic document management, etc.). Partial digitalization is due to the high cost of complex solutions, including Russian developers, and the amount of costs for servicing borrowed funds in the case of their involvement due to the insufficient volume of its own resources, the lack of personnel of the necessary qualifications. In addition, «automation of production sites and business processes in Russian companies, as a rule, do not have common links, which means that it is not possible to organize information links at all levels of the production process, from design and preparation of production, ending with logistics support of the manufactured product and sales management» [1; 8; 9; 18].

All this allows us to consider close corporation «Togliatti plant of auto units» as a typical representative of Russian enterprises engaged in the production of automotive components with similar problems, and the probability of the impact of digital technologies and QMS on financial results and costs to ensure product quality is high.

The authors propose to build an economic and mathematical model to assess the effectiveness of the QMS and the introduction of digital technologies in the activities of close corporation «TZA». Econometric correlation-regression model system of interrelated characteristics of the studied aggregate is a regression equation that includes the factors influencing the variation of resultant variable in the aggregate, possesses a high value of determination coefficient (not less than 0,5), reliable and correctly interpreted in accordance (in sign and order of magnitude) with the theory of the system under study with the regression coefficients, and these properties are suitable for the evaluation of population units and for forecasting.

In general, the linear regression equation can be represented as follows [5]:

$$y = \beta_1 + \beta_2 \cdot x_2 + \beta_3 \cdot x_3 + \cdots + \beta_k \cdot x_k + \varepsilon \quad (1)$$

where y – dependent variable (the resulting feature); $\beta_{1, 2...k}$ – regression coefficients showing the average value of the change in the resulting feature when changing the variables x_k per unit of measurement; $x_{1, 2...k}$ – independent variables (factors of influence); ε – random errors of the multiple regression equation.

The construction of a model for assessing the effectiveness of the introduction of digital technologies in the activities of close corporation «TZA» is based on the establishment of a relationship between the amount of revenue from sales of products (the resulting indicator) and specific performance indicators of the quality management system and digitalization of production (independent variables), which requires the implementation of appropriate calculations.

Stages of construction of the specified economic and mathematical model include:

1. Development of the indicators' system to assess the effectiveness of the QMS and digitalization of the enterprise;
2. Analysis of private indicators of the level of product quality;
3. The analysis of private indicators of the QMS' effectiveness;
4. Analysis of product quality assurance costs;
5. Analysis of private indicators of the enterprise's digitalization level;
6. Construction of a multiple regression model of the dependence between close corporation «TZA»'s sales volume (the resulting indicator) and private efficiency indicators of the quality management system and digitalization of production (independent variables).

Stage 1. To assess the effectiveness of the QMS and the introduction of digital technologies in the activities of the enterprise, the authors developed a system of indicators, the structure of which is presented in figure 1.

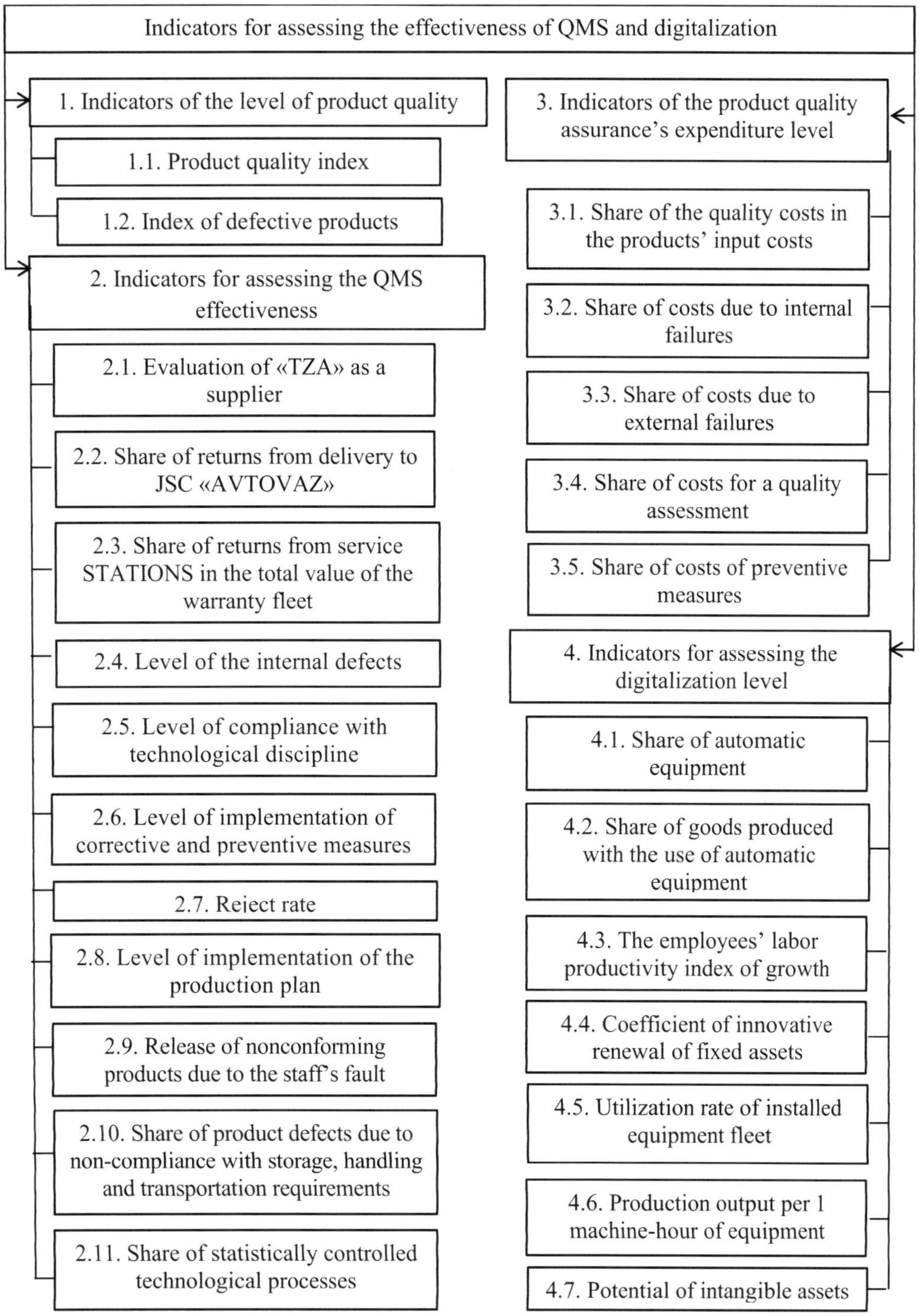

Figure 1. The indicators to measure the effectiveness of the QMS and implementation of digital technologies in the company's activities.
Source: developed by the authors

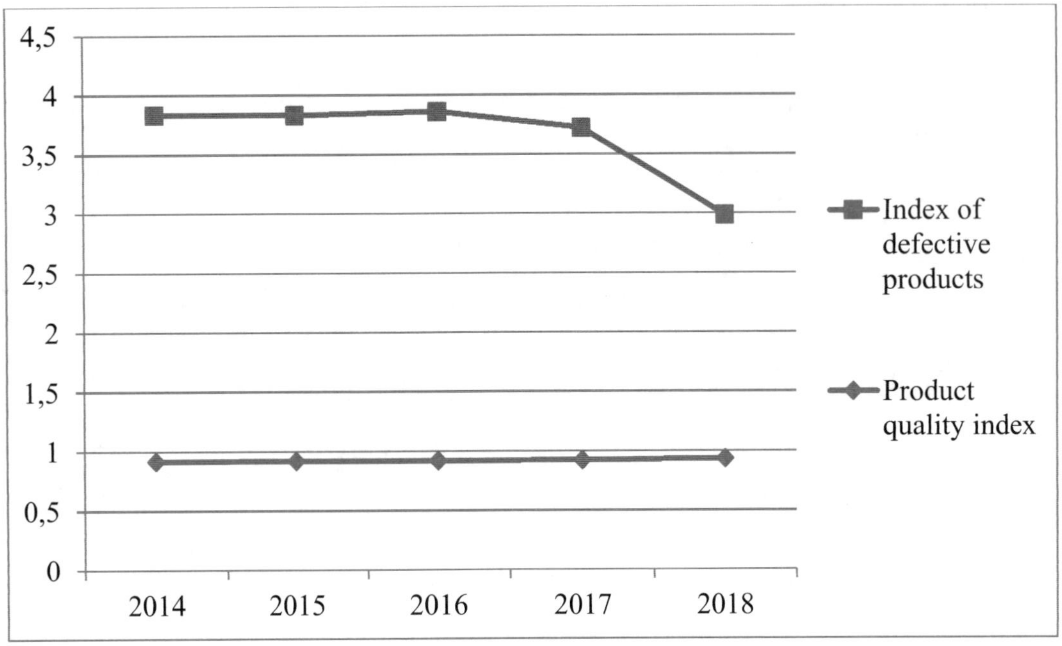

Figure 2. Dynamics of quality and defect indices of close corporation «TZA» products for 2014-2018
Source: developed by the authors

Stage 2. The analysis of the level of product quality is carried out using two indicators:
1) product quality index, defined as the weighted average of relative values of quality indicators of heterogeneous products of the enterprise for a certain period;
2) index of defective products, defined as the weighted average of the defect coefficients of heterogeneous products of the enterprise for a certain period. The defect rate in this case is calculated taking into account the dynamics of its level, i.e. takes into account the trend of change.
Dynamics of quality and defect indices of close corporation «TZA» products for 2014-2018 is presented in figure 2. Based on their dynamics, it follows that the level of quality of the company's products increased by 0,018 points, and the level of defects, on the contrary, decreased by 0,866 points. This trend is positive, as it indicates an improvement in the quality of the plant's products.
Stage 3. The analysis of private indicators of the QMS' effectiveness.
Most of the products of close corporation «TZA» are made for the needs of JSC «AVTOVAZ». In this regard, JSC «AVTOVAZ» included «TZA» in the list of its main suppliers and regularly conducts a point quality assessment of its deliveries [17; 19].
As part of this research, based on internal audit data, the analysis of close corporation «TZA» as a supplier of the main conveyor of JSC «AVTOVAZ» was carried out in accordance with the scale of table 1. The rating calculations' results of the studied enterprise as a supplier are presented in figure 3.

Table 1. Evaluation criteria of close corporation «TZA» as a supplier

Supplier evaluation	Evaluation criterion
Excellent supplier	85 or more points
Reliable supplier	80-84 points
Unreliable supplier	50-79 points
Unsatisfactory supplier	49 or less points

Source: developed by the authors

As can be seen from figure 3, the company maintains and controls the supply's efficiency and therefore the rating is growing.

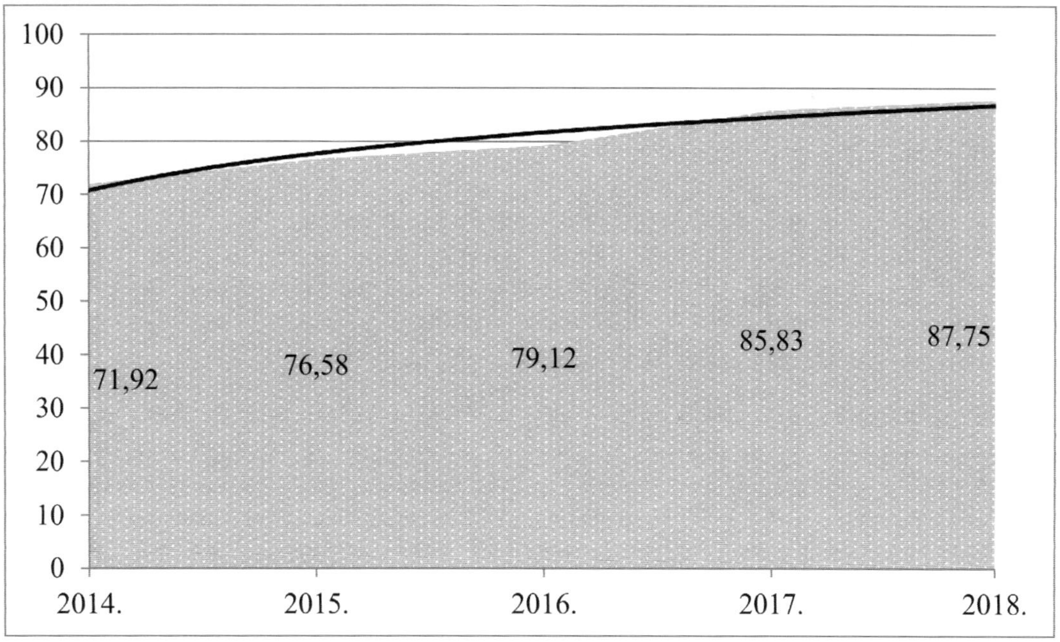

Figure 3. Dynamics of rating of «TZA» as a supplier
Source: developed by the authors

In 2001, the standardization Bureau of close corporation «TZA» and the technical control Department together with consultants of JSC «AVTOVAZ» developed a quality management system. At the same time, the first certification of the quality management system for compliance with ISO 9001 version 2000 was carried out and the certificate of the German certification body TÜV CERT was obtained.

In order to obtain timely objective evidence about the need to reduce, eliminate and prevent QMS inconsistencies, the company conducts internal (auditors of the technical control Department) and external (representatives of the German certification body TIC) audits.

According to the authors, the enterprise's effectiveness of the quality management system can be assessed by using indicators of defects, carrying out the planned activities to improve product quality, and coverage of technological processes of the enterprise advanced (statistical) methods of quality control. The results of calculations of the QMS efficiency indicators of close corporation «Togliatti plant of auto units» are presented in Table 2.

Table 2. Indicators of efficiency of functioning of the QMS of close corporation «TZA» for 2014-2018.

Indicators	Value per year				
	2014	2015	2016	2017	2018
1. Evaluation of TZA as a supplier, score	71,92	76,58	79,12	85,83	87,75
2. Share of returns from delivery to JSC «AVTOVAZ», %	0,19	0,17	0,18	0,16	0,16
3. Share of returns from service STATIONS in the total value of the warranty fleet, %	0,09	0,104	0,093	0,080	0,087

IOP Conf. Series: Materials Science and Engineering **986** (2020) 012024 doi:10.1088/1757-899X/986/1/012024

4. Level of the internal defects (in relation to the number of machined parts), %	1,6	1,6	1,78	1,53	0,93
5. Level of compliance with technological discipline, %	98,65	98,74	99,16	99,24	99,33
6. Level of implementation of corrective and preventive measures, %	97,82	98,61	100	93,75	100
7. Reject rate, PPM	920	960	840	749	580
8. Level of implementation of the production plan, %	82,57	89,96	91,36	94,49	98,26
9. Release of nonconforming products due to the staff's fault, %	0,63	0,62	0,84	0,79	0,63
10. Share of product defects due to non-compliance with storage, handling and transportation requirements, %	0,55	0,56	0,56	0,85	0,95
11. Share of statistically controlled technological processes, %	92,12	92,47	93,15	93,69	94,59

Source: developed by the authors

Consequently, there are minor problems in the quality management system of the enterprise, as during the analyzed period there is an increase of the product defects' share due to non-compliance with the requirements for storage, loading and unloading and transportation. It also must be mentioned that the share of returns from service STATIONS in the total value of the warranty fleet of cars in 2018 increased by 0,007% in comparing to 2017.

Stage 4. This stage is associated with the research of the enterprise's costs to ensure the products' quality. In accordance with the reasons for the formation the quality assurance costs are classified into the following groups: 1) costs due to internal failures associated with the identification and possible correction of defects found before the products' transfer to the consumer; 2) costs due to external failures due to the identification and possible correction of defects found after the products' transfer to the consumer; 3) the costs of quality assessment associated with the determination of the degree of conformity of products with quality requirements; 4) the costs of preventive measures designed to carry them out in order to minimize losses from failures and the costs of quality assessment.

Dynamics of the quality expenses' shares by categories in their total amount of close corporation for 2014-2018 is presented in figure 4.

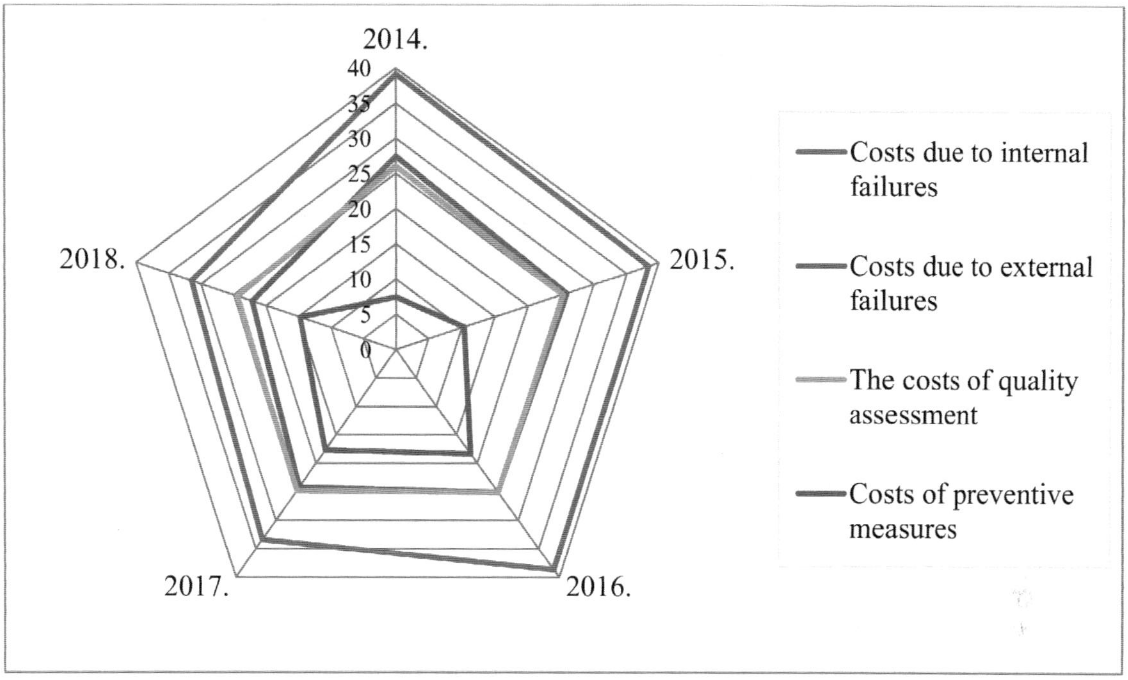

Figure 4. Dynamics of the quality expenses' shares by categories in their total amount, %
Source: developed by the authors

Positive dynamics of the quality expenses' shares by categories in total, presented in figure 4, we should recognize as the perfection of their structure, namely the decline in the share of expenditures for internal and external cracks, contributing to the formation of the reputation of the manufacturer, and the growth of precautionary costs, indicating the strengthening of control of the company for the level of product quality, the development and implementation of preventive measures. Thus, in general, during the period under review, the total share of non-productive expenses (expenses due to internal and external failures) decreased by 12,95%, while preventive expenses increased by 7,38%. This indicates a shift in emphasis in the field of quality management of close corporation «TZA» from solving problems due to the release of reject rate to its prevention.

In addition to the structure, an important aspect of the analysis of costs for product quality assurance is their share in the total cost of production and its dynamics (figure 5).

From the data of figure 5 it follows that the share of quality costs in the production costs of close corporation «Togliatti plant of auto units» decreased by the end of the period by 0,71%. Other things being equal, this fact should be considered as a positive trend.

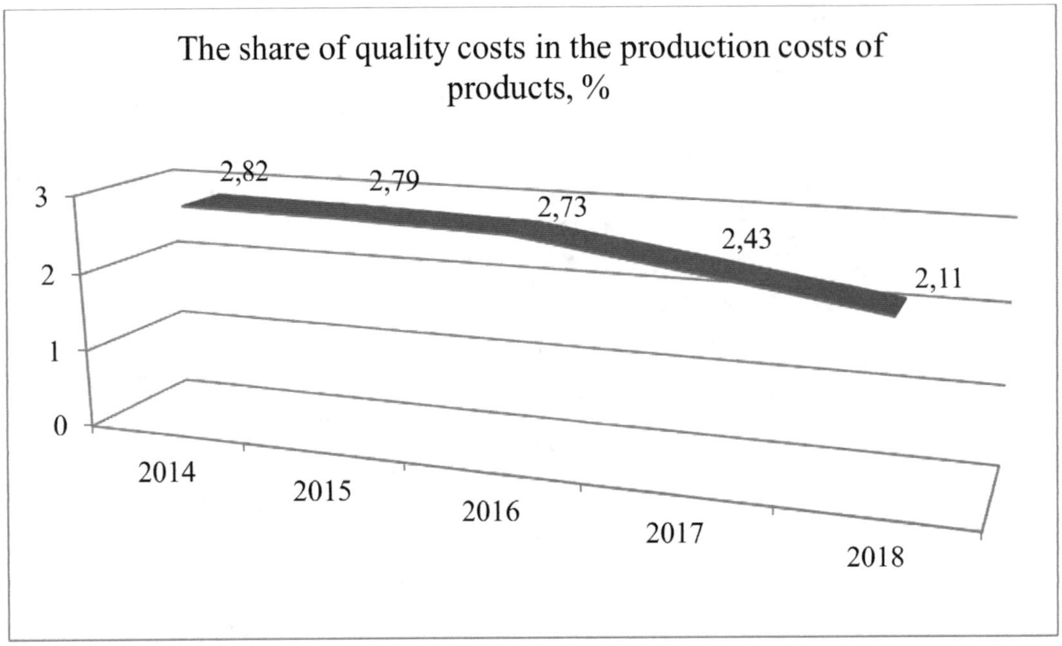

Figure 5. Dynamics of the share of quality costs in the production costs of close corporation «Togliatti plant of auto units» for 2014-2018, %
Source: developed by the authors

Stage 5. Construction of a model of multiple regression of the dependence of the efficiency of the enterprise on the particular performance indicators of QMS and the use of digital technologies.

The ultimate goal of this study is to build an econometric model for assessing the effectiveness of the QMS of an economic entity through the introduction of digital technologies. Typical performance criteria are time, money, compliance with requirements (regulatory, legislative, technological) and the level of performance of planned tasks. Given that in a market economy, any activity is evaluated in terms of the ability to generate income, the effectiveness of the QMS, including through digitalization, it is advisable to analyze its value. In other words, the authors propose to consider revenue as an independent variable that serves as a criterion for the performance of the enterprise to ensure the quality of its products.

Data on the sales amount of close corporation «Togliatti plant of auto units», as well as the calculated private indicators of the level of product quality, QMS efficiency, costs for product quality and the level of digitalization of business for 2014-2018 can be presented in the form of initial data of this model (table 3) [19].

Table 3. Table of source data for the calculation of multiple regression equation

Indicators	Letter of the designation	Value per year				
		2014	2015	2016	2017	2018
1	2	3	4	5	6	7
Sales amount, million rubles	y	539,224	593,412	627,132	604,705	684,045
1. Indicators of the level of product quality						
1.1. Product quality index	x_1	0,916	0,917	0,919	0,920	0,934
1.2. Index of defective products	x_2	2,918	2,914	2,943	2,803	2,052
2. Indicators for assessing the QMS effectiveness						

Indicators	Letter of the designation	Value per year				
		2014	2015	2016	2017	2018
1	2	3	4	5	6	7
2.1. Evaluation of «TZA» as a supplier, score	x_3	71,92	76,58	79,12	85,83	87,75
2.2. Share of returns from delivery to JSC «AVTOVAZ», %	x_4	0,19	0,17	0,18	0,16	0,16
2.3. Share of returns from service STATIONS in the total value of the warranty fleet, %	x_5	0,09	0,104	0,093	0,080	0,087
2.4. Level of the internal defects (in relation to the number of machined parts), %	x_6	1,6	1,6	1,78	1,53	0,93
2.5. Level of compliance with technological discipline, %	x_7	98,65	98,74	99,16	99,24	99,33
2.6. Level of implementation of corrective and preventive measures, %	x_8	97,82	98,61	100	93,75	100
2.7. Reject rate, PPM	x_9	920	960	840	749	580
2.8. Level of implementation of the production plan, %	x_{10}	82,57	89,96	91,36	94,49	98,26
2.9. Release of nonconforming products due to the staff's fault, %	x_{11}	0,63	0,62	0,84	0,79	0,63
2.10. Share of product defects due to non-compliance with storage, handling and transportation requirements, %	x_{12}	0,55	0,56	0,56	0,85	0,95
2.11. Share of statistically controlled technological processes, %	x_{13}	92,12	92,47	93,15	93,69	94,59
3. Indicators of the product quality assurance's expenditure level						
3.1. Share of the quality costs in the products' input costs, %	x_{14}	2,82	2,79	2,73	2,43	2,11
4. Indicators for assessing the digitalization level						
4.1. Share of automatic equipment (index)	x_{15}	0,07	0,12	0,12	0,19	0,19
4.2. Share of goods produced with the use of automatic equipment, %	x_{16}	8,56	9,12	10,87	15,23	16,92
4.3. The employees' labor productivity index of growth	X_{17}	1,0568	1,0677	1,0712	1,1255	1,1638
4.4. Coefficient of innovative renewal of fixed assets	X_{18}	1,95	3,57	0,94	0,39	0,90
4.5. Utilization rate of installed equipment fleet	X_{19}	0,92	0,93	0,93	0,94	0,94
4.6. Production output for 1 machine-hour of equipment operation, thousand rubles	x_{20}	1,37	1,40	1,41	1,39	1,40
4.7. Potential of intangible assets, thousand rubles	x_{21}	154	307	279	409	361

Source: developed by the authors

To form an econometric model, first of all, it is necessary to establish a close relationship between the dependent variable and the expected impact factors. For this purpose was held the analysis of multiple correlation. According to this analysis the criterion of influencing factors' selection is the closeness of communication which is expressed by coefficient more than 0,7. The results of the calculations are presented in the table 4.

The sign of the correlation coefficient indicates the direction of the factor's influence on the resulting indicator. Thus, its values for the index of defective products, the share of returns from delivery to JSC «AVTOVAZ», the reject rate and the share of quality costs in the products' input costs are negative, therefore, the growth of these indicators leads to a decrease of the enterprise's amount of sales. Based on the content of these factors, the value of the correlation coefficient is calculated adequately.

Table 4. Correlation coefficients of indicators of efficiency of quality management system and digitization of the enterprise

The factor	Letter of the designation	Value of the correlation coefficient
Product quality index	x_1	0,8774
Index of defective products	x_2	-0,79
Share of returns from delivery to JSC «AVTOVAZ», %	x_4	-0,702
Level of compliance with technological discipline, %	x_7	0,8485
Reject rate, PPM	x_9	-0,828
The level of implementation of the production plan, %	x_{10}	0,9259
Share of product defects due to non-compliance with storage, handling and transportation requirements, %	x_{12}	0,7008
Share of statistically controlled technological processes, %	x_{13}	0,9072
Share of the quality costs in the products' input costs, %	x_{14}	-0,813
The employees' labor productivity index of growth	x_{17}	0,7956
Utilization rate of installed equipment fleet	x_{19}	0,7935
Production output for 1 machine-hour of equipment operation, thousand rubles	x_{20}	0,7322

Source: developed by the authors

Analysis of all the identified factors having the strongest relationship with the resulting indicator from the point of view of determining the enterprise's QMS effectiveness, including at the expense of digitization. The use of the software product Excel allowed the authors to obtain the model value of revenue:

$$y = 294{,}32 - 52{,}95 \cdot x_2 - 0{,}192 \cdot x_9 + 8{,}6931 \cdot x_{10} - 257{,}1 \cdot x_{12} \quad (2)$$

where y – is the amount of revenue from sales, mln rubles; x_2 – an index of defective products; X_9 – the reject rate, PPM, X10 – the level of implementation of the plan of production, %; X_{12} – share of defective products by reason of non-compliance to storage, handling and transportation, %.

The value of R-square, which characterizes the quality of the model, in our case is 1 or 100%. So it can be made a conclusion that the model has an excellent level of quality.

The «Y-intersection» and the «Coefficients» column is «294,32». Therefore, the value of Y, i.e. the value of the company's sales volume with all other factors equal to zero, will assume its formation at the level of 294,32 million rubles.

Interpreting the model based on revenue from sales, it is worth noting that an increase in the index of defective products by 1 point, the reject rate at 1 PPM and the proportion of defective products by reason of non-compliance to storage, handling and transportation by 1%, revenue will decline accordingly on 52,95; 0,192; 257,1 million rubles. At the same time, the increased level of implementation of the production plan by 1% will lead to revenue growth for 8,6931 million rubles.

IOP Conf. Series: Materials Science and Engineering **986** (2020) 012024 doi:10.1088/1757-899X/986/1/012024

The resulting model (formula 2) reflects the natural regression coefficients. However, to ensure the possibility of their comparison with each other and ranking of factors by the power of influence, it is necessary to obtain a multiple regression equation in a standardized form (formula 3) [11].

$$t_y = -0{,}38272 \cdot t_{x2} - 0{,}5534 \cdot t_{x9} + 0{,}96496 \cdot t_{x10} - 0{,}9345888 \cdot t_{x12} \quad (3)$$

From formula 3 it follows that the greatest influence on the resulting characteristic (revenue) is rendered by factors of the production plan's implementation level and the reject rate due to non-compliance to the execution of logistics operations, since their module values are close and much higher than levels of defects and number of claims. However, these factors are quite strong (regression coefficients greater than 0,35) [16]. Therefore, the increase in profitability of the enterprise directly depends on reducing the level of defects and negative consumer feedback, as well as compliance with the requirements of the main and auxiliary production processes.

In this case, the established directions of the factors' influence on revenue can be interpreted as logical and justified from the point of view of the considered indicators' content.

It is also important to note that, despite the close connection of a number of factors of digitalization with the revenue of the studied enterprise (in accordance with the calculated correlation coefficients), they were not included in the parameters of the regression model. This allows us to conclude that the influence of these factors is indirect. At the same time, it is scientifically proved that the deployment of digital technologies in manufacturing and QMS as a system-regulator of production and technological processes leads to an increase in the productivity of personnel and equipment, and therefore, reduce the costs of the business entity and increase its income on this basis [14].

The category that unites the sphere of production, controlling its system in terms of quality assurance of products and services (QMS) and the enterprises' financial results are the costs of quality assurance and their share in the production costs. In order to establish the impact of the company's level of digitalization on this indicator, it is advisable to build an economic and mathematical model.

First, it was established that the share of quality assurance costs in the production costs (dependent variable) was closely related to the expected impact factors (independent variables). The results of the analysis of multiple correlation of interdependent factors are presented in table 5.

Table 5. Correlation coefficients for the share of quality assurance costs in the production costs and digitalization of the enterprise

The factor	Letter of the designation	Value of the correlation coefficient
Share of automatic equipment (index)	x_{15}	-0,87852
Share of products produced with the use of automatic equipment, %	x_{16}	-0,96902
The employees' labor productivity index of growth	x_{17}	-0,99517
Utilization rate of installed equipment fleet	x_{19}	-0,84383

Source: developed by the authors

From the data of table 4 it follows that only 4 from 7 factors of digitalization of the production activity of close corporation «Togliatti plant of auto units» have a close feedback with the share of quality assurance costs in the production costs, since the values of the correlation coefficients are higher than 0,8 points. The opposite effect of the digital technology's level implementation is concluded in increasing the level of automation of information processing and execution of manufacturing operations leads to lower costs of providing quality and share in cost of production. This fact should be assessed as a positive economic effect.

Having carried out calculations with the help of Excel software, the authors obtained the following model of dependence of the share of expenses for quality assurance in the production costs of close corporation «TZA» in the natural form:

$$y = 34{,}36357 + 6{,}447341 \cdot x_{15} - 0{,}01928 \cdot x_{16} - 7{,}5045 \cdot x_{17} - 25{,}9773 \cdot x_{19} \quad (4)$$

where y – the share of costs for quality assurance in the production costs,%; x_{15} – the share of automatic equipment (index); x_{16} – the share of products manufactured with the use of automatic equipment,%; x_{17} – The employees' labor productivity index of growth; x_{19} – the utilization rate of the installed equipment. The R-squared value for this model is also 1 or 100% and characterizes its quality as excellent.

The «y-intersection» and the «Coefficients» column is «34,36357», i.e. the share of quality assurance costs in the production costs with all other factors equal to zero will assume its formation at the level of 34,36357 %. The regression coefficients for the factors are negative, since, as it was noted, the growth of automation and digitalization of production leads to a decrease in the costs of the economic entity to conduct business.

The obtained model of the dependence of the share of quality assurance costs in the production costs on the factors of digitalization (formula 4) in order to ensure the possibility of their comparison with each other and ranking of factors by the power of influence should be transformed into a multiple regression equation in a standardized form (formula 5).

$$t_y = 1{,}099075 \cdot t_{x15} - 2{,}37877 \cdot t_{x16} - 1{,}13555 \cdot t_{x17} - 0{,}71699 \cdot t_{x19} \quad (5)$$

Formula 5 shows that the greatest impact on the resulting feature (the share of costs for ensuring product quality in the production costs) has a share of products produced with the use of automatic equipment. Automatic equipment and the employees' labor productivity index of growth also have a strong impact on the value of the dependent variable, because their module values are close and higher than the value of the utilization rate of the installed equipment fleet.

Thus, the level of profitability of the enterprise directly depends on factors of the effectiveness of the company QMS and indicators of digitalization, namely the reduction of defects and scrap, of compliance of main and auxiliary production processes, ensure the specified level of implementation of the plan of production, automation of production processes and increase production by using automatic equipment.

4. Conclusion

The intensive development of the industry and the formation of a new economic situation on the principles of digitalization, orientation to the best satisfaction of the consumers' needs caused the public interest to the problems of introduction of QMS and digital technologies in the activities of enterprises. This interest has a very specific practical value: increasing the competitiveness of both private companies and national economies due to GDP indicators [2; 7; 12].

However, this raises the problem of assessing the effectiveness of these areas and their impact on the financial results of economic entities. One possible solution to this problem can be developed by the authors of the methods of evaluating the effectiveness of the QMS and digitalization based on the use of objective statistical data of industrial enterprises and the mathematical apparatus, allowing to evaluate and objectively interpret actual indicators and achievements in the field of quality and level of digital technologies' implementation in the activities of economic entities.

The models developed by the authors to estimate the sales volume and the share of costs for quality assurance can be used in the practice of companies to analyze the effectiveness of QMS and digitalization processes. The identification of priority impacts and strength of their influence should be considered as a basis for improving the efficiency of the company by reducing the number of complaints and quality defects of products, and interrelated growth of productivity and income.

References

[1] Digital production in Russia: opinion of professionals ([Electronic resource]. – URL: http://integral-russia.ru/2017/05/20/tsifrovoe-proizvodstvo-v-rossii-mnenie-professionalov/)

[2] Digital Russia: a new reality *The McKinsey Report* ([Electronic resource]. – URL: www.tadviser.ru/images/c/c2/Digital-Russia-report.pdf)

[3] Digital technologies in production – what is the advantage ([Electronic resource]. – URL: https://tsa.su/news/cifrovye-tehnologii-v-proizvodstve/)
[4] Digital transformation in quality management ([Electronic resource]. – URL: https://www.comindware.com/ru/blog-%D1%86%D0%B8%D1%84%D1%80%D0%BE%D0%B2%D0%B0%D1%8F-%D1%82%D1%80%D0%B0%D0%BD%D1%81%D1%84%D0%BE%D1%80%D0%BC%D0%B0%D1%86%D0%B8%D1%8F-%D0%B2-%D0%BC%D0%B5%D0%BD%D0%B5%D0%B4%D0%B6%D0%BC%D0%B5%D0%BD%D1%82/)
[5] Furmanov K K Measurement and visual representation of the practical significance of regression relations ([Electronic resource]. – URL: https://www.hse.ru/data/2014/05/29/1325341336/%D0%98%D0%B7%D0%BC%D0%B5%D1%80%D0%B5%D0%BD%D0%B8%D0%B5%20%D0%B8%20%D0%BD%D0%B0%D0%B3%D0%BB%D1%8F%D0%B4%D0%BD%D0%BE%D0%B5%20%D0%BF%D1%80%D0%B5%D0%B4%D1%81%D1%82%D0%B0%D0%B2%D0%BB%D0%B5%D0%BD%D0%B8%D0%B5%20%D0%B7%D0%BD%D0%B0%D1%87%D0%B8%D0%BC%D0%BE%D1%81%D1%82%D0%B8.pdf/)
[6] Global Economic Prospects: Managing the Next Wave of Globalization ([Electronic resource]. – URL: http://bookfi .net/book/792994)
[7] Index of global competitiveness ([Electronic resource]. – URL: https://www.weforum.org/reports/the-global-competitiveness-report-2017-2018)
[8] Khoo H H and Tan K C 2002 Critical success factors for quality management implementation in Russia *Industrial and Commercial Training* (Vol. 34. No 7) pp 263–268.
[9] Korobenkov A Digital production management system – an important step to «Industry 4.0» [Electronic resource] (URL: https://controlengrussia.com/internet-veshhej/tsifrovaya-sistema-upravleniya-proizvodstvom-vazhny-j-shag-k-industrii-4-0/)
[10] Measuring the Information Society 2017 [Electronic resource] (URL: https://www.itu.int/en/ITU)D/Statistics/Documents/publications/misr2017/MISR2017_Volume1.pdf)
[11] Nathans L L, Oswald F L and Nimon K 2012 Interpreting Multiple Linear Regression: A Guidebook of Variable Importance [Electronic resource] *Practical Assessment, Research & Evaluation* (Vol. 17. No.9 URL: http://pareonline.net/getvn.asp?v=17&n=9).
[12] Rating of doing business [Electronic resource] (URL: www.doingbusiness.org/content/dam/doingBusiness/media/.../DB2018-Full-Report.pdf.
[13] Scantlebury S. Designing digital organizations [Electronic resource] (URL: https://www.bcg.com/ru-ru/about/bcg-review/design-of-digital-organizations.aspx)
[14] Shirinkina E V 2018 Features of functioning of industrial enterprises in the digital economy *Economy in the industry* (Vol. 11. No. 2) pp 143–150.
[15] Sirkin H, Zinser M and Rose J 2015 The Robotics Revolution *The Next Great Leap in Manufacturing* (Boston: BCG) p 28
[16] Soofi E S, Retzer J J and Yasai-Ardekani M A Framework for Measuring the Importance of Variables with Applications to Management Research and Decision Models *Decision Sciences* (Vol. 23. No. 3) pp 595-625.
[17] AVTOVAZ. Official site [Electronic resource] (URL: http://info.avtovaz.ru/pages/section_5641/5727.html/)
[18] Information portal «Production Management» [Electronic resource] (URL: http://www.up-pro.ru/)
[19] TZA. Official site [Electronic resource] (URL: https://tza11.ru)

IOP Conf. Series: Materials Science and Engineering **986** (2020) 012025 doi:10.1088/1757-899X/986/1/012025

Information support of a quality management system in the context of digitalization of business processes

D Antipov[1], A Smagina[2] and N Klassen[3]

[1]Department of Aircraft Production and Quality Management in Mechanical Engineering, Samara National Research University Academician S.P. Koroleva, con-expert@mail.ru
[2]Undergraduate department, Togliatti State University, anastasiasmgn@rambler.ru
[3]Peter the Great St. Petersburg Polytechnic University, St. Petersburg, Russian Federation, nellieklassen@gmail.com

Abstract. In conditions of improving production processes, the organization and implementation of a quality management system is one of the most important issues in the management system of individual business processes and the organization as a whole. The formation and implementation of control over the organization of production, product quality, labor protection and other business processes and management objects is carried out in accordance with the accumulated theoretical, methodological and practical experience. However, at present, in the conditions of intensive development of technologies that allow us to move to a qualitatively new level of production, traditionally established approaches to managing business processes are losing their relevance and require additions, including in terms of enhancing the automation of information support for the quality management system. Therefore, as part of this study, it is proposed to pay attention to the development of tools of intelligent quality management systems in terms of performing control and analytical tasks.

1. Introduction

Currently, the digital economy and the areas of activity that are associated with it are rapidly developing in the world, namely: the public sector and business, citizens' livelihoods, ensuring information security, telecommunication network capabilities, etc. This process involves activities in which one of the factors (resources) of production are digital data that acts both as a means and as a result. Such activities contribute to the development of the information society by:

- creating an information space that takes into account the needs of citizens and society as a whole in high-quality and reliable information,
- development of information infrastructure in Russia;
- the creation and use of Russian information and telecommunication technologies.

Information is one of the most important components of reality surrounding us. Now the understanding of information should be interpreted much more widely than previously done. Information is not only information necessary for the user about an object, but also a new asset, which, unlike other groups of assets, has a unique value, the essence of which is that the information can be used for new purposes and for implementation of new activities.

The use of information currently characterizes it as a resource, the result of intellectual work, along with labor, material, financial and other resources.

In this regard, the question arises of the relationship between the concepts of "knowledge" and "information". Discussions of scientists regarding whether it is possible at the moment to argue that information is an ordered knowledge of any object comes down to a positive statement. In accordance with this, knowledge is a category that is close to information, focused on a specific user to make a specific decision.

One of the main goals of the quality management system is to confirm the buyer's confidence that the product he purchases meets all quality standards and also fully satisfies all his needs. Currently, in the scientific literature the topic of digitalization of individual business processes is widely discussed, however, specific approaches to applying digitalization tools to existing management systems are often absent or are fragmented, even contradictory. Therefore, the transformation of the information support of the quality management system in the context of digitalization of business processes is one of the important problems requiring immediate solution in terms of not only supplementing the theoretical experience, but also improving the practical aspects of introducing the formulated recommendations into the enterprise management system as a whole and individual business processes in particular.

2. Methods and materials

To solve the problem of developing an approach to the transformation of information support of the quality management system in the context of digitalization of business processes, the following methods were used:

- systematization - the collection and grouping of information from scientific literature and other sources that led to the conclusion that there is no comprehensive approach to the transformation of information support for the quality management system in the context of digitalization of business processes;
- modeling - building a model of information support for a quality management system in the context of digitalization of business processes;
- the method of complexity - research and proposal for the transformation of information support of the quality management system on the basis of an integrated approach by integrating information about all objects of management in a single intellectual environment;
- the dialectic method - the study of the content of the concepts of "information" and "knowledge" as the main tools of communication and interaction in modern society.

3. Discussion

The study of information issues, its role in society, as well as issues of promising information management systems was carried out by such scientists as A. Esatiani, E.J. Pentinen, Strouhal J., Orak J., Boksova J. and several other scientists.

The development of approaches to the transformation of control systems in the context of digitalization of data and business processes was studied by scientists Limonova E., Stankin A., Kopnov V., Correa P., Fernandes A., Uregian C. and others.

Based on a study of the results of research by economists regarding digitalization, it was concluded that digitalization means the transition from an analog form of transmitting information to digital. At the same time, digital transformation involves the use of modern technologies to dramatically increase the productivity and value of enterprises. However, currently in the economic literature there is no comprehensive approach to the transformation of information support of the quality management system in the context of digitalization of business processes. Therefore, this direction was chosen as the goal of this study.

4. Results

Production activities are more focused on achieving effective indicators: maximizing profits, improving product quality, reducing costs, using more modern materials for production, switching to automated production and management, etc. As a result of the study, it was found that digital technology is changing the operating model of companies, especially in the banking and telecommunications sectors, increasing cost efficiency and identifying new opportunities in the market.

In the service sector, the benefits of digital transformation are cost savings of up to 60 percent and significant acceleration of the launch of new products on the market. Another result is an increase in customer loyalty and the frequency of interaction with them, leading to an increase in customer base profitability. Partnerships or partnerships with companies in related fields also provide access to new customer segments. For industry, the use of robotics, artificial intelligence, big data systems allows us to produce high-tech products that are in demand on the market.

According to research, the best companies, the "digital elite" (Digirati) of our time, combine digital technology and strong leadership, making the transition from just using IT to transforming a business.

Automation allows the company to reorient its team to solving strategic problems, gain more free time to improve managerial skills, concentrate on research and creativity, and not waste time on repetitive processes. This significantly contributes to the rapid formation of data streams that can be useful in mining.

At the individual level, work is essentially virtualized - the work process is no longer associated with a physical workplace. Corporate collaboration tools and networking tools allow employees to communicate with colleagues wherever they are. A number of tools that allow you to work remotely become powerful means of exchanging information. Sales managers and front-office employees are beginning to actively use tools for collaboration, thanks to which they can quickly establish communication with experts and receive answers to questions in real time. They also gain access to a single, global point of interaction between the company and their customers.

Transactional systems give managers the opportunity to better understand products, regional characteristics and customer needs, which allows them to make informed decisions based on real data, not assumptions. This applies to both internal processes and client processes. The level of detail is also growing, which allows managers to compare current statuses and redistribute production capacities in ways that were previously unavailable. In addition to better awareness, digital transformation provides a change in the strategic decision-making process.

The indicated advantages of using digitalization in a quality management system are reduced to the need to build a high-quality new system of its information support, the model of which is presented in a simplified form in Figure 1.

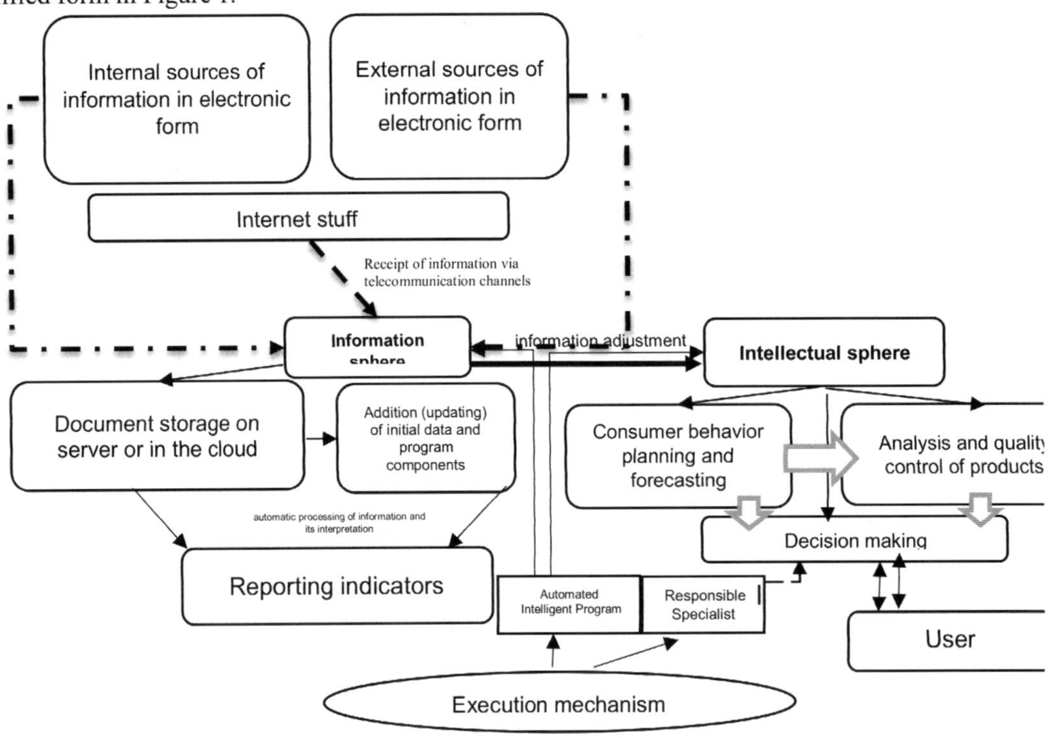

Figure 1. A model of information support for a quality management system in the context of digitalization of business processes.

IOP Conf. Series: Materials Science and Engineering **986** (2020) 012025 doi:10.1088/1757-899X/986/1/012025

The presented model contains two blocks in the basis: the information sphere and the intellectual sphere. The information sphere is understood as a unified system of external and internal data necessary for product quality management, which inevitably transforms into digital form or is created in such a form, stored on a dedicated server or on the cloud, processed in a single automated intellectual program and can be interpreted and detailed in the most user-friendly form. The intellectual sphere is a set of rules, laws, parameters, algorithms, processing and interpretation of information about the objects of the quality management system, which allows without significant user intervention to develop the most effective solution or system of solutions to a given problem.

The proposed model of information support for the quality management system in the context of digitalization of business processes using an intelligent information processing system allows avoiding the direct participation of a specialist in quality management and increasing the efficiency of processing large volumes of unsystematized and fragmented data to make effective management decisions.

Computerization of enterprise management assumes that this should be a coordinated support of the functional and process component of management activity using information technology and automation.

What is common in information systems is that they are designed to support management processes. Possible problems with the digitalization of the quality management system, in particular: Недостаточность финансирования для полной автоматизации и информатизации предприятия.

1. The lack of qualified specialists who can professionally build a real process model of an enterprise management system.
2. Lack of a sufficient number of properly qualified specialists who are able to work with BigDate.
3. Given the fact that the “digitized” enterprise assumes complete transparency of the personnel’s activities and minimization of the negative impact of the human factor, the implementation of digitalization measures will meet with opposition from the employees, most of whom seek to conceal “failures” in quality (concealment of marriage, excessive time, financial and material costs in the production process).

Due to the fact that the digitalization of business processes is a fairly new and not fully understood phenomenon, the quality management system is characterized by risks associated with the transformation of its information support. The systematization of the risks of digitalization of the quality management system is presented in Figure 2.

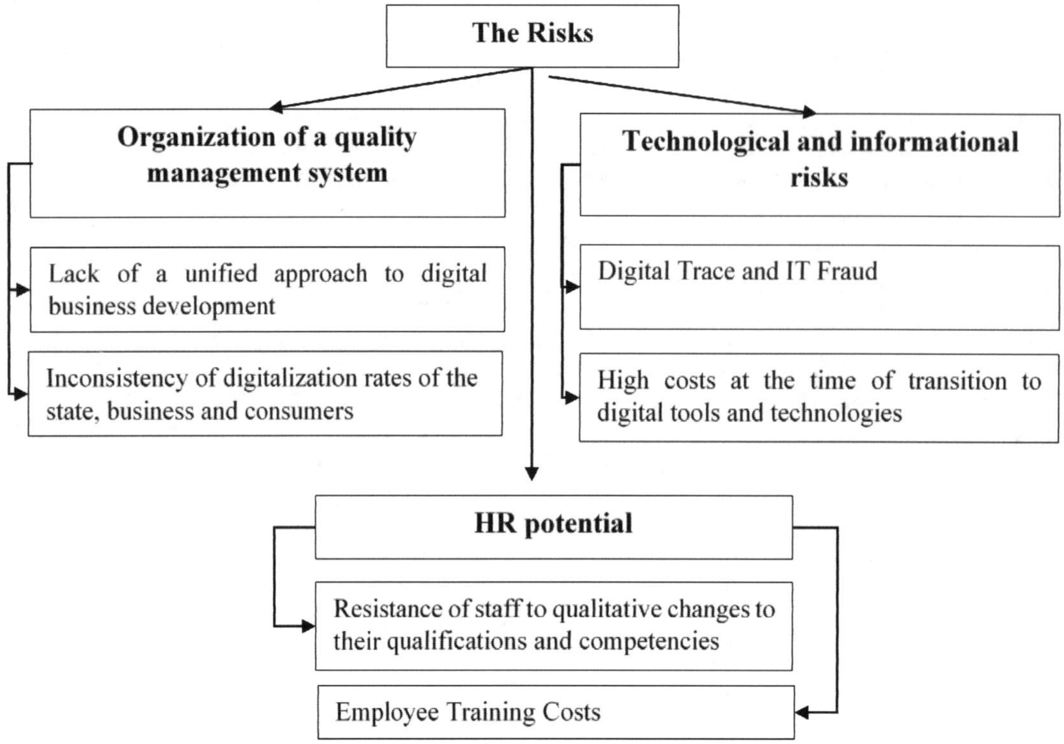

Figure 2. Risks of digitalization of the quality management system.

With regard to information support, one of the important risks in the context of digitalization of business processes is the development of IT fraud, an increase in the risk of unauthorized access to confidential information. Therefore, to protect the company from online fraud, it is necessary to use the latest means of protecting information.

Based on the risks presented, as well as the proposed model of information support for the quality management system in the context of digitalization of business processes, we determine how the principles presented to the quality management system in terms of digitalization should look like:

1. Consumer orientation - organizations depend on their consumers and therefore must understand their current and future needs, fulfill their requirements, taking into account the increase in sales in Internet commerce, the use of Internet things, which will most quickly and efficiently satisfy consumer demand.
2. Leadership of the leader - leaders ensure unity of purpose and direction of the organization. Digital transformation requires strong leadership - only it can be a driver of major changes. A clear understanding of which parts of the company should be transformed is also required.
3. Personnel involvement - employees of all levels form the basis of the organization, therefore their full involvement in solving problems in the context of digitalization, as well as continuous training, enables the organization to use their abilities to the benefit.
4. Process approach - the desired result is achieved more efficiently when activities and related resources are managed as a process.
5. A systematic approach to management - the identification, understanding and management of interconnected processes as a digital system contribute to increasing the effectiveness and efficiency of the organization in achieving its goals.
6. Continuous improvement - improving the organization's activities, taking into account its innovative development.
7. Making decisions based on facts - effective decisions should be based on the analysis of data and information, increasing the efficiency of its processing and presentation of a user-friendly form, which

is ensured by the proposed system of information support for the quality management system in the context of digitalization of business processes.

8. The principle of flexibility: the quality management system should be easily changed if there is a potential to improve the quality of services, labor productivity, profitability, and reduce costs.

9. The principle of manufacturability: the quality management system should use digitalization and robotization tools to improve the quality of service.

The presented traditional principles presented to the quality management system, supplemented taking into account modern digitalization requirements, allow you to build your own path of development of the enterprise in a turbulent economy and the lack of transparency of the future business environment.

5. Conclusions

Thus, in the course of the study, a comprehensive approach to the transformation of information support of the quality management system in the context of digitalization of business processes was applied to solve the problem. Modern technological, economic and social conditions dictate the requirements of the business as it should be. One of the essential qualities today is fast, informational, open, digital. If an enterprise does not make a qualitatively new leap to the digital future, or at least does not supplement traditional approaches and tools for managing modern attributes, it is no longer able to compete in the modern market.

References

[1] Asatiani A and Penttinen E J 2015 Managing the move to the cloud – analyzing the risks and opportunities of cloud-based accounting information systems *Journal of Information Technology Teaching Cases* (5 (1)) pp 27–34

[2] Correa P, Fernandes A and Uregian C 2010 Technology adoption and the investment climate: firm-level evidence for Eastern Europe and Central Asia *The World Bank Economic Review* (24 (1)) pp 121-147

[3] Gschwantner S and Hiebl M R W 2016 Management Control systems and organizational ambidexterity *Journal of Management Control* (27 (4) doi:10.1007/s00187-016-0236-3) pp 371–404

[4] Limonova E 2018 Digital companies: definitions and attributes *Economic security and quality* (1 (30)) pp 37-40 [in Rus]

[5] Mahadeen B, Hani Al-Dmour R, Yousef Obeidat B and Tarhini A 2016 Examining the Effect of the Organization's Internal Control System on Organizational Effectiveness: A Jordanian Empirical Study International *Journal of Business Administration* (7 (6) doi:10.5430/ijba.v7n6p22) pp 22-41

[6] Ragulina J, Suglobov A and Melnik M 2018 Transformation of the role of a man in the system of entrepreneurship in the process of digitalization of the Russian economy Quality - *Access to Success* (2) pp 171-175

[7] Rybina G 2017 Intellectual technology for constructing training integrated expert systems: new opportunities. *Open education* (4, 43-57. https://dx.doi.org/10.21686/1818-4243-2017-4-43-57) [in Rus].

[8] Strouhal J, Horák J and Bokšová J 2017 Harmonization in V4-Countries and its Impact on Financial Data *International Advances in Economic Research* (23 (4)) pp 431–432

[9] Stankin A and Kopnov V 2011 Proposals to address the causes of low efficiency of quality management systems of enterprises in the Russian Federation *Quality of Life Technologies* (11(1)) pp 51-62 [in Rus]

A study of the ice-resistant protective coatings effect on the deterioration of the outdoor covering of ice vessels and icebreakers

M V Kitaev[1], P O Pastukhov[2] and V A Surov[1]

[1]Department of Ship building and Ocean engineering, School of Engineering, Far Eastern Federal University, Vladivostok, Russia, kitaev.mv@dvfu.ru
[2]Far Eastern State Technical Fisheries University, Vladivostok, Russia

Abstract. The study examines the regularities of the outer coatings deterioration of ice class vessels and icebreakers, as well as the effect of ice-resistant protective coatings on deterioration rates. As the initial data, the results of measurements of residual thicknesses from the DEFHULL computer database were used. A comparison of the normative values of the outer coatings deterioration rates for ice class vessels presented in the current version of the RMRS Rules and the values obtained as a result of processing the DEFHULL database data was carried out. The results and conclusions were used in developing the proposals for amending and supplementing the RMRS Rules regarding the purpose of corrosion and abrasion surcharges when determining the thickness of the outer coatings of ice class vessels.

1. Introduction

According to the current Rules of the Russian Maritime Register of Shipping (RMRS) [1], when determining the thickness of the outer coatings of ice class ships and icebreakers, it is possible to reduce the corrosion and abrasion surcharges by special consideration. However, the outer coatings deterioration rate is clearly independent of the intensity of the ice load, which differs for different areas of the vessel hull [1] that leads to heavier structures. If there is a lot of statistical information on the measurements of the residual thickness of the vessel coatings (defection), this statement is not obvious.
The goal of this study is to investigate the effect of ice-resistant protective coatings on the outer coatings deterioration of ice class vessels and icebreakers.
To achieve this goal, the following tasks were solved:
1. Analysis of actual deterioration rates of the vessel and icebreakers outer coatings in areas of ice reinforcement.
2. Analysis of the relations of the deterioration rates of the vessel and icebreakers coatings and the ice load intensity.
3. Assessment of the effect of protective ice-resistant coatings on the deterioration rate of the vessels and icebreakers outer coatings in areas of ice reinforcement.
To solve the tasks, we used the data of measurements of the residual thickness of the vessels and icebreakers outer coatings from the DEFHULL computer database [2, 3].

2. Analysis of actual deterioration rates based on the results of residual thicknesses

To analyze the actual deterioration rates, the results of measurements of the residual thickness of the outer coatings of vessels and icebreakers from the DEFHULL database were processed (Tables 1, 2).

Table 1. Defection data on ice class vessels.

Ice strength category	Number of vessels					The number of sheets of outer coating	Number of measure-ment	Average number of measure-ments on one sheet
	Dry cargo	Liquid	Fishing	Tugboat	Total			
Ice1	9	3	12	0	24	1275	4205	3.3
Ice2	38	13	109	9	169	15283	69302	4.5
Ice3	19	8	47	1	75	7212	32896	4.6
Arc4	30	6	6	1	43	7920	35477	4.5
Arc5	7	5	0	24	36	4167	18190	4.4
Arc6-Arc9	5	0	0	0	5	613	2430	4.0
TOTAL	108	35	174	35	352	36470	162500	4.5

Table 2. Defection data on icebreakers.

Ice strength category	Defection year	The number of sheets of outer coating	Number of measure-ments	Average number of measurements on one sheet
Icebreaker6	2005	284	829	2.9
	2010	161	483	3.0
	2016	376	1695	4.5
Icebreaker7	2008	353	1441	4.1
	2013	386	1497	3.9
Icebreaker8	2007	267	1129	4.2
	2012	400	1300	3.3
Icebreaker8	2011	427	1511	3.5
TOTAL		2654	9885	3.7

To automate the statistical data processing, Excel and Visual Basic for Applications were used. The algorithm for processing the statistical data (actual measurements of the residual thickness of the outer coatings) in the areas of ice reinforcement is as follows:

1. The formation of primary data array, taking into account the location of the sheets of the outer coating relative to the side, belt and location along the length of the vessel from the databank DEFHULL;
2. Processing the data array in accordance with the methodology for assessing the reliability of the results of measurements of residual thicknesses [4];
3. The determination of the values of indicators characterizing the deterioration rate by the formulas:

- the average value (AR) of the deterioration rate is determined by the formula

$$U_{CP} = \frac{1}{n}\sum_{i=1}^{n} U_i \tag{1}$$

- standard deviations (SD) are calculated by the formula

$$\sigma = \sqrt{\frac{\sum_{i=1}^{n}(U_i - U_{CP})^2}{n}} \tag{2}$$

- estimated deterioration rate

IOP Conf. Series: Materials Science and Engineering **986** (2020) 012026 doi:10.1088/1757-899X/986/1/012026

$$u = U_{CP}+k\sigma, \quad (3)$$

where k - coefficient characterizing the probability of residual thickness appearance into the analyzed range of values. It is further assumed for vessels of ice classes k = 2, and for icebreakers k = 3; n is the number of measurements in the selection.

4. Graphical imagination and analysis of the results.

3. Analysis of the dependence of the deterioration rate of the outer coverings on the vessel exploitation period

In order to analyze the dependence of the outer coatings deterioration rates on the exploitation period of the vessels, the data characterizing the results of measurements of residual thicknesses are divided into time intervals with an interval of 5 years, which corresponds to the frequency of inspection of vessels in operation.

Figures 1 to 4 show the results of processing the outer coatings deterioration rates for vesselss and icebreakers. As a result of the analysis, it was found that the outer coatings deterioration rates in areas of ice reinforcement increase up to 16–20 years of operation, and decrease after 20 years of operation. This pattern is clear for the frontal (Figure 1) and middle (Figure 2) areas of the outer deterioration of high ice class vessels. The decrease in deterioration rates in the period after 20 years is probably due to the implementation of repair work related to the replacement of sheets of the outer coatings. According to the results obtained, the outer coatings deterioration in the stern of the vessel is uniform (Figure 3).

A similar result, associated with a decrease in the deterioration rate of the outer coating sheets depending on the exploitation period, is also noted for icebreakers (Figure 4). At the same time, we note that when analyzing the results of measurements of residual thicknesses for vessels presented in Table 3, the facts of repair work on the replacement of individual sheets of the outer coating, which affected the decrease in deterioration rates, were defined.

As a result of processing the measurements of the residual thickness of the vessels outer coatings (Tables 1 and 2), the maximum calculated deterioration rates for vessels of different ice classes and regions along the length of the hull were determined by formula (3) (Table 3).

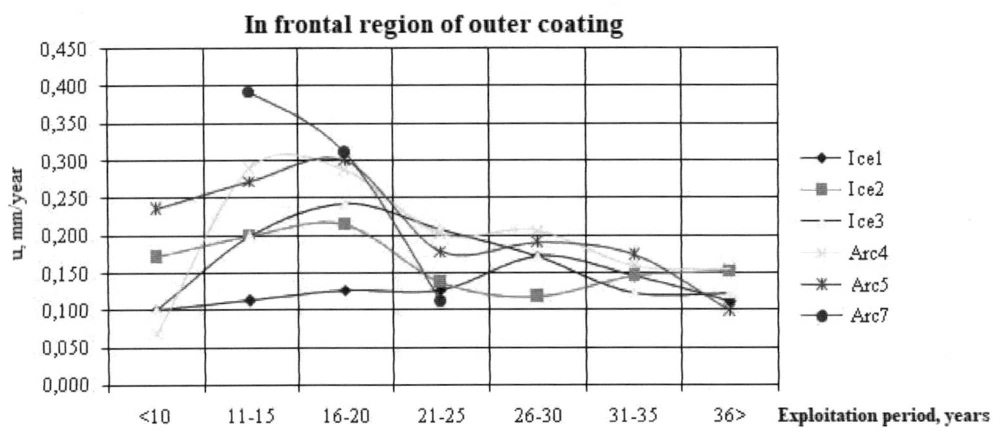

Figure 1. Deterioration rates of the outer coatings in the frontal region, mm/year.

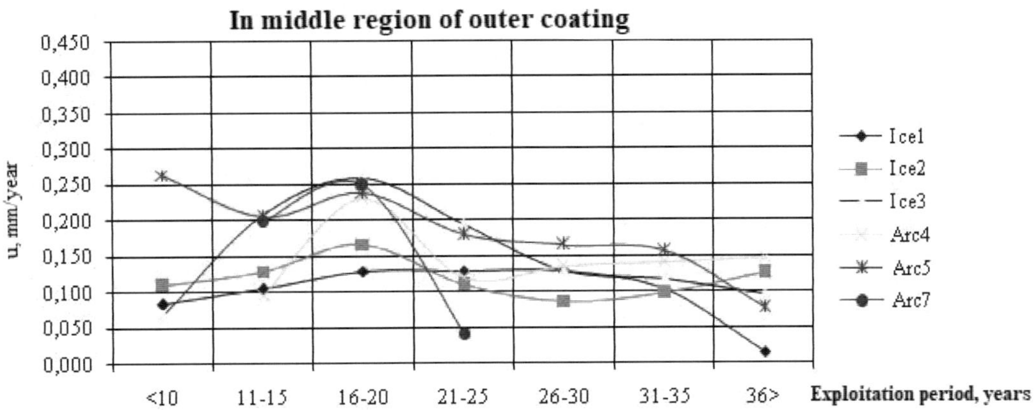

Figure 2. Deterioration rates of the outer coatings in the middle region, mm/year.

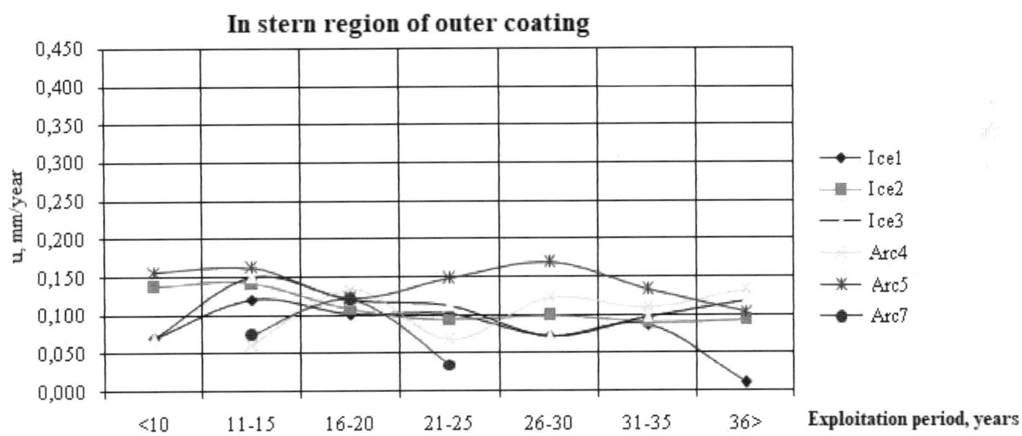

Figure 3. Deterioration rates of the outer coatings in the stern region, mm/year.

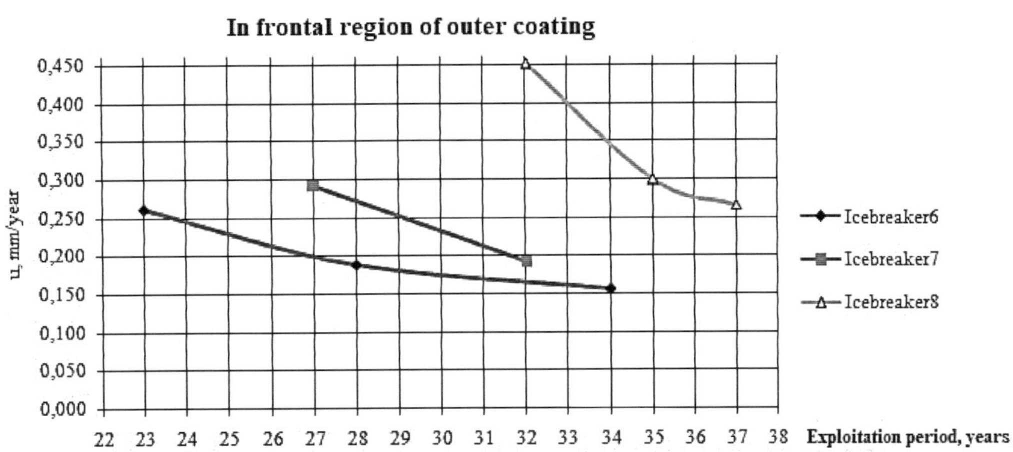

Figure 4. Deterioration rates of the icebreakers outer coatings in the frontal region, mm/year.

Table 3. Defection data on icebreakers.

Ice class	Deterioration rate u, mm/year for different areas (on vessel length)		
	Frontal	Middle	Stern
Ice1	0.172	0.130	0.122
Ice2	0.215	0.167	0.142
Ice3	0.243	0.209	0.150
Arc4	0.290	0.231	0.134
Arc5	0.300	0.264	0.168
Arc6-Arc9	0.394	0.250	0.121
Icebreaker6	0.260	0.126	0.058
Icebreaker7	0.291	0.171	0.111
Icebreaker8	0.450	0.359	0.279
Icebreaker9	No data		

4. Analysis of the relations between the deterioration rates of the outer coverings and the intensity of the ice load

The outer coatings of vessels are the subjects to significant deterioration when operating in ice conditions. It is due to both corrosion and abrasion of the coatings when the vessel moves in ice. With increasing the intensity of the ice load, the deterioration increases accordingly

The RMRS rules [1] declare the average annual thickness reductions only for three regions in length, and in fact only for two - frontal and intermediate (A and A1) and middle and stern (B and C). In the case of icebreakers, the number of regions reaches 16 (Figure 5). Formally, two values for the ice belt of the vessel, not related to its size, do not quite justifiably apply to all areas of the underwater part of the hull, overstating their weight and the cost of protective coatings

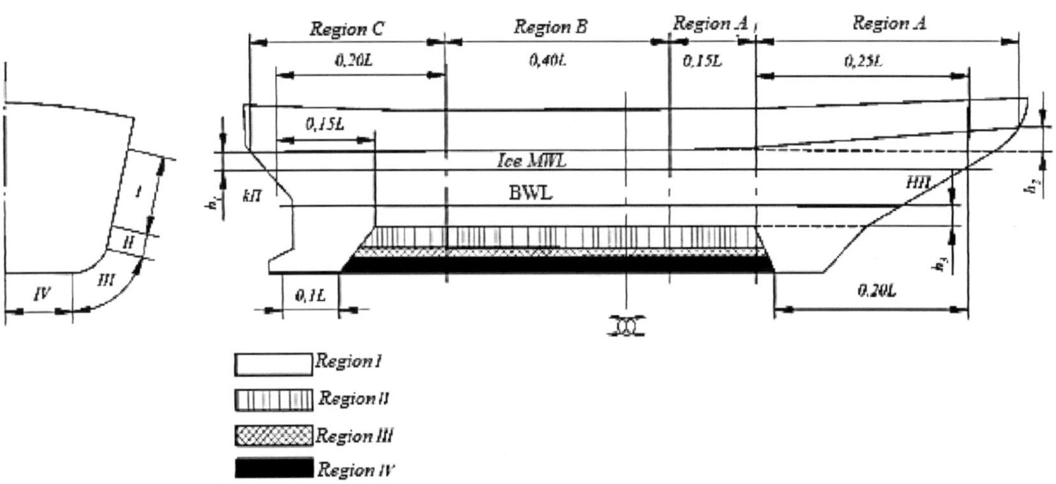

Figure 5. Areas of icebreaker ice strengthening.

Studies [5] allowed us to suggest a relations between the intensity of ice loads (MPa) and the average annual decrease in thickness (mm/year) of the outer coatings:

$$u = 0{,}1 \cdot (1 + p^{2/3}), \quad (4)$$

where p –ice load intensity according to the 1990 Rules.

Formula (4) requires adjusting the ice load values due to changes in the existing Rules compared to the 1990 Rules and allows differentially determining the average annual decrease in the thickness of the outer coatings, taking into account the characteristics and different areas of the vessel.

To correct formula (4), we processed the average annual decrease in the thickness of the outer coatings as a function of the intensity of the ice load. All vessels presented in Tables 1 and 2 were grouped depending on the intensity of the current ice load (Figure 6).

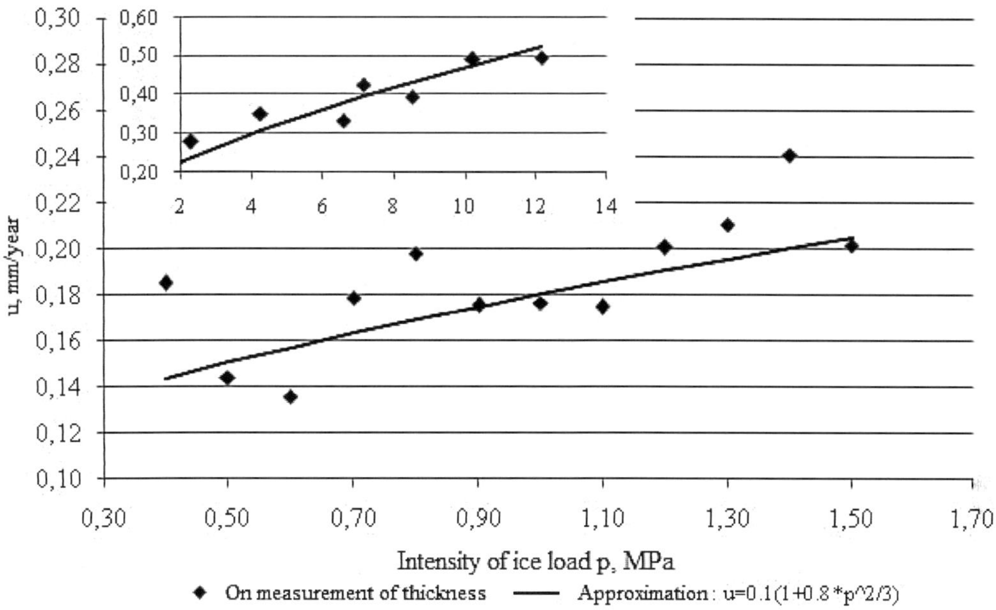

Figure 6. Average annual decrease in the thickness of the outer coatings as a function of the intensity of the ice load.

As a result of processing the statistical data, dependence (5) was obtained to determine the average annual decrease in the thickness of the outer coatings of vessels and icebreakers depending on the intensity of the ice load:

$$u = 0{,}1 \cdot (1 + 0.8 p^{2/3}) . \quad (5)$$

5. The effect of protective coatings on the deterioration rate of the outer coatings of vessels and icebreakers

The addition for corrosion deterioration of the coatings in the area of ice reinforcements in accordance with the current Rules is determined by the formula:

$$\Delta s_{u0} = 0{,}75 T u , \quad (6)$$

where T -. the planned exploitation period of the vessel, years.

The estimation of effect of protective coatings on the abrasion and corrosion deterioration of the outer coatings of vessels and icebreakers was fulfilled based on an analysis of the dependences of deterioration rates in the area of ice reinforcements on the durability of coatings. By this way we plotted the graphs (Figure 7), approximating the actual values of the deterioration of the sheets of the outer coatings of ice class vessels in the frontal region, and calculated by the formula (6) depending on the exploitation period. The initial deterioration rates are in Figure 1

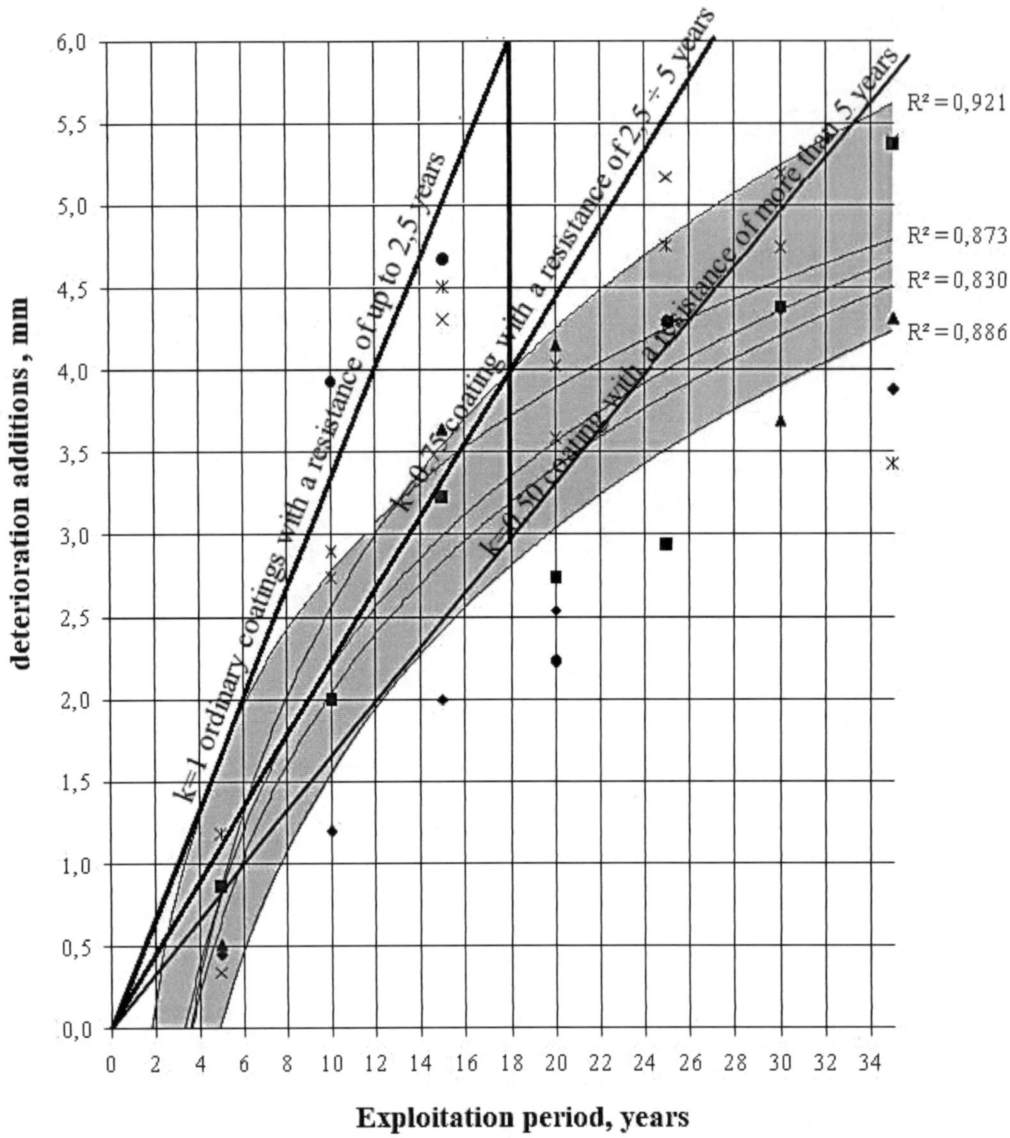

Figure 7. Assessment of the effect of coatings on corrosion additions.

Figure 7 shows the straight lines emerging from zero and being tangent to the curves approximating the actual values of deterioration of sheets of the outer coatings of ice vessels in the frontal region, and which are analogues of the average annual decrease in the thickness of the outer coatings according to the RMRS Rules. As a result, the formula for determining the deterioration of the outer coatings in the area of ice reinforcements, taking into account the use of protective coatings, can be represented as:

$$\Delta s_{u0} = 0{,}75Tu \cdot k\,, \tag{7}$$

where k – the coefficient of accounting for the presence and durability of protective coatings.

In Figure 7, the durability of the coatings is determined by the intersection of the approximating curves characterizing the actual values of the deterioration of the sheets of the outer coatings of vessels and icebreakers, with the abscissa axis. As a result, it was found that for icebreakers the coating resistance is up to 2 years, for vessels of the Arctic categories of 2,5 to 5 years, and for other vessels from 5 or more years. The analysis results shown in Figure 8 are confirmed by the practice of exploitation of ice class vessels and icebreakers [6, 7].

Based on the found dependences characterizing the actual values of deterioration, the coefficients are determined that take into account the change (decrease) in the deterioration rate of the outer coatings depending on the resistance of coatings or measures aimed at protecting the hull from corrosion deterioration and ice abrasion:

$k = 1.0$ - for ordinary coatings with a resistance of up to 2.5 years;

$k = 0.75$ - for coatings with a resistance of $2.5 \div 5$ years;

$k = 0.50$ - for coatings with a resistance of more than 5 years,

as well as for other effective measures to protect the outer coatings from corrosion deterioration and abrasion (ice-resistant coatings in conjunction with an electrochemical protection system; ice-resistant coatings in conjunction with an electrochemical protection system and clad steel, and other options).

The value of the coefficient $k = 0.50$ is consistent with the recommendations of the Rules DNV, IACS, LR, determining the structural or operational features of the vessel.

In the proposed version, in agreement with the Register of Ice Vessels, the range of decisions on the designation of additions for deterioration of the outer coatings in areas of ice reinforcement has been expanded taking into account the use of protective coatings, measures and remains with the designer. The designer must take into account the experience and data of the manufacturer of protective coatings (guarantees of certain properties of coatings and their durability in specified operating conditions). However, during the long-term operation of the vessel there is no guarantee that the owner/operator will always comply with the design conditions and protection recommendations. Therefore, the Register is invited to introduce a requirement in addition to the construction thicknesses of the outer coatings of ice class vessels on the design drawings to indicate the permissible residual thicknesses at which the vessel maintains compliance with the design ice category. Going beyond these values means the need to repair or lower the ice category of the vessel.

6. Conclusion

As a result of the study, the following conclusions can be made:

- measurements of the residual thickness of the outer coatings of vessels and icebreakers were processed, the dependences of the rate of deterioration of the coatings on the exploitation period were built;
- a formula is proposed for determining the deterioration rates of sheets of the outer coatings depending on the intensity of the ice load;
- an assessment of the effect of protective coatings on the deterioration rate in areas of ice reinforcement was made;
- the research results were taken into account when changing and supplementing the requirements of the RMRS in 2019 to the designation of corrosion and abrasion surcharges when determining the thickness of the outer coatings of vessels and icebreakers taking into account the protective coatings [6].

References

[1] Rules for the classification and construction of marine vessels 2020, II - Hull. Russian Maritime Register of Shipping (St. Petersburg. 2017)

[2] Kulesh V A, Zhitnikov A V and Surov O E 2010 The DEFHULL system is an innovative technology in managing the technical operation of the fleet *Electronic periodical "Bulletin of the Far Eastern State Technical University"* (Scientific journal, No. 3 (5), FESTU; Vladivostok) pp 23-34

[3] Kulesh V A, Pichugin O G and Surov O E 1997 Computer database and examination of the technical condition of the fleet (DEFHULL) *State Committee on fisheries of the Russian Federation. Safety of navigation and fishing* (St. Petersburg Vol. 1 (105)) pp 57-61

[4] Zinoviev P V, Kompanets V A and Surov O E 2015 Methodology for assessing the reliability of measurements of residual thicknesses during the detection of vessel constructions *Scientific and Technical Bulletin of the Russian Maritime Register of Shipping* (No. 40/41, St. Petersburg) pp 41-48

[5] Kulesh V A 1998 Estimated design and examination of the technical condition of vessel constructions in areas of extreme local loads *Dis. Doctor. Tech. Sciences* (FESTU, Vladivostok)

[6] Kitaev M V Kompanets V A Kulesh V A and Surov O E 2018 Development of proposals for the appointment of additions for the deterioration of the outer coatings of ice class vessels and icebreakers *Scientific and Technical Bulletin of the Russian Maritime Register of Shipping* (No. 52/53, St. Petersburg) pp 35-46

IOP Conf. Series: Materials Science and Engineering **986** (2020) 012027 doi:10.1088/1757-899X/986/1/012027

Tests of paint and varnish ice-resistant protective coatings of ice class vessels

V A Veselov [1], M V Kitaev [2], P O Pastukhov[2] and O E Surov [1]

[1]Department of Ship building and Ocean engineering, School of Engineering, Far Eastern Federal University, Vladivostok, Russia
[2]Far Eastern State Technical Fisheries University, Vladivostok, Russia,
kitaev.mv@dvfu.ru

Abstract. The study examines the regularities of the outer coatings deterioration of ice class vessels and icebreakers, as well as the effect of ice-resistant protective coatings on deterioration rates. As the initial data, the results of measurements of residual thicknesses from the DEFHULL computer database were used. A comparison of the normative values of the outer coatings deterioration rates for ice class vessels presented in the current version of the RMRS Rules and the values obtained as a result of processing the DEFHULL database data was carried out. The results and conclusions were used in developing the proposals for amending and supplementing the RMRS Rules regarding the purpose of corrosion and abrasion surcharges when determining the thickness of the outer coatings of ice class vessels.

1. Introduction

Protective materials and coatings are a promising area for improving the reliability, safety and functional efficiency of vessels and marine equipment. A large number of paint and varnish coatings have been developed and are being used in practice, which make it possible to protect the vessel hulls for various purposes from corrosion. In this case, the choice of protection system largely depends on the type and purpose of the vessel.

A use of traditional systems and paint and varnish coatings to protect against corrosion and abrasion does not fully solve this problem. On ice class vessels and icebreakers, the degree of destruction of the underwater hull, and especially in the area of the ice belt, is extremely high, which under the impact of ice leads to corrosion and erosion destruction of the sheets of the outer skin of the hull, lower ice permeability, increased fuel consumption and increased dock repair work.

The rapid deterioration of the paint and varnish coatings is due to the fact that the group of vessels under consideration shows a significant unevenness in the deterioration of the outer skin sheets both along the length and on the side height.

Only some types of coatings can provide effective protection. To protect the sheets of the outer skin of the icebreakers and ice class vessels against corrosion and abrasion, special coatings and protection systems have been developed and are used. Typically, such coatings have special properties, are resistant to prolonged abrasion and impact of ice and, as a rule, are certified by the classification societies as ice-resistant.

An ice-resistant coating is considered to be a coating that is capable of protecting the outer skin of the vessel hull from external influences in ice navigation conditions with characteristics that meet the requirements specified in the Rules for the Classification and Construction of Ships, Part XIII [1] of the Russian Maritime Register of Shipping (RMRS).

Table 1. Requirements of the RMRS Rules for ice-resistant coatings.

No	Indicators	Values			
		For icebreakers of all ice classes		For ice class vessels of Arc 4 and higher	
		Class I	Class II	Class I	Class II
1	Durability according to ISO 12944-6 for the corrosivity category Im2 according to ISO 12944-2	large		large	
2	Adhesion determined by the method of trellised notches according to ISO 2409 or the method of crosswise notches according to ISO 16276-2 after tests for resistance to low temperatures depending on the thickness and type of ice- Adhesive strength according to ISO 4624resistant coating	no more than 3 points		no more than 3 points	
3	Adhesive strength according to ISO 4624	more than 16 MPa	more than 10 MPa	more than 10 MPa	more than 8 MPa
4	Abrasion resistance after 1000 test cycles on Taber abrasimeter (CS-17 wheel)	no more than 80 mg	no more than 120 mg	no more than 120 mg	no more than 160 mg
5	Impact strength according to ISO 6272	not less than 5 J		not less than 5 J	
6	Resistance to cathodic peeling according to ISO 15711 (method A) for coatings compatible with cathodic protection	less than 5 mm after 3 months of tests, less than 8 mm after 6 months of tests		less than 5 mm after 3 months of tests, less than 10 mm after 6 months of tests	
7	Coefficient of friction on ice	no more than 0,03	no more than 0,08	no more than 0,03	no more than 0,08

This status also means that the use of such coatings will reduce the required thickness of vessel structures, which decreases the weight of the hull construction, reduces construction costs and increases the efficiency of the vessel.

The goal of this study is to test the strength and friction properties of ice-resistant paint and varnish coatings (IPVC).

The main tasks:

- Determination of friction indicators of samples with IPVP on ice for compliance with the requirements of part 2.5.7 of the RMRS Rules [1];

Determination of the strength indicators of IPVC during tensile and bending of the sample in order to determine the stresses at which cracking and/or peeling of the ice-resistant protective coating from a metal surface is observed.

2. Requirements for samples, conditions and means of testing

Tests of samples with IPVC were performed in a specialized laboratory equipped with a freezer (climatic) chamber and providing standard test conditions at a temperature of minus 20°C. For testing, coated samples were preliminarily held at a temperature of minus 20°C for at least 15 minutes, relative humidity of the surrounding air was not more than 85%, and atmospheric pressure was 1 bar.

For friction tests, metal samples with IPVC (plates) 250×130×3 mm in size were used. The shape and size of the samples comply with the requirements of 2.5.7 of XIII part ("Materials") of the RMRS Rules [1].

The shape and dimensions of samples with IPVC designed for tensile testing meet the requirements of 2.2.2 of XIII part ("Materials") of the RMRS Rules [1] and GOST (State Standard) 1497-84 [2].

The type and geometrical dimensions of samples for tensile testing with IPVC are in Figure 1.

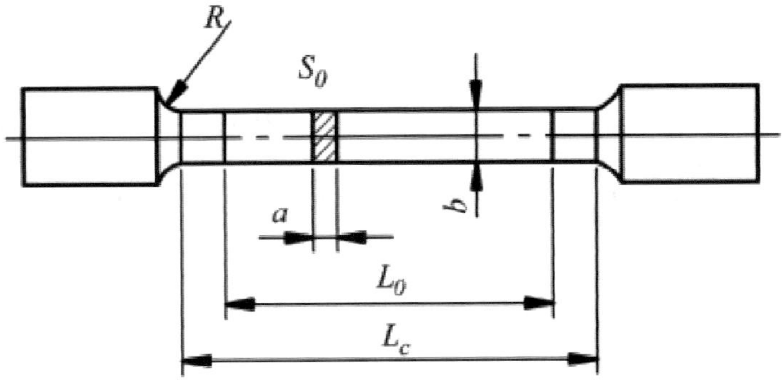

Figure 1. Sample for tensile testing.

Dimensions of the samples:
a = 3 mm - thickness of the working part of the sample;
b = 20 mm - width of the working part of the sample;
L_0 = 45 mm - length of calculated part of the sample;
L_c = 60 mm - length of working part of the sample;
R = 25 mm - radius of transition from the working part to the sample head.
For each of the three tests, it is necessary to have three coated samples.
The shape and dimensions of the samples for bending tests must comply with the requirements of 2.2.5 of XIII part ("Materials") of the RMRS Rules [1] and GOST 14019-80 [3].
Test samples were rectangular coated metal plates.
Dimensions of samples:
a = 3 mm - thickness of the working part of the sample;
b = 20 mm - width of the working part of the sample;
L = 70 mm - length of sample.
For each of the three tests, three coated samples were used.
The appearance of the samples intended for testing for friction, tension and bending is in Figure 2.

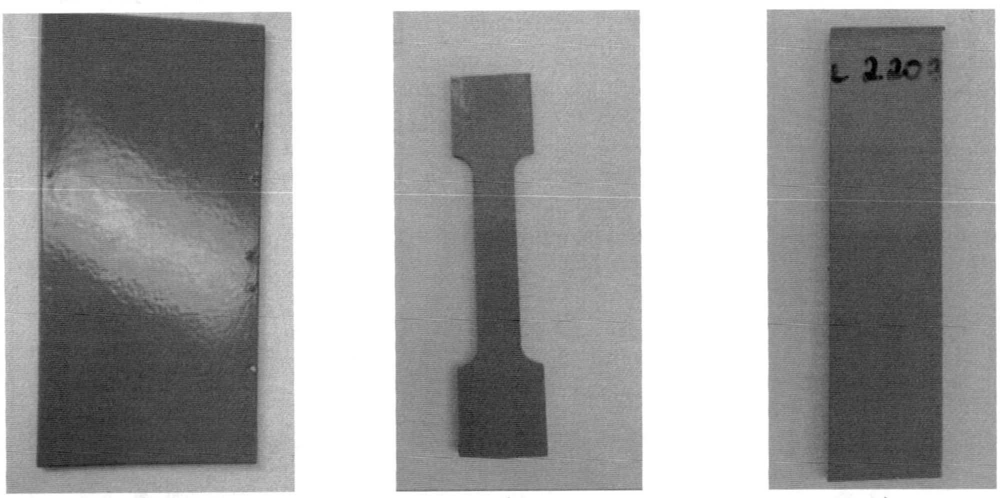

a) b) c)

Figure 2. Test samples:
a - to determine the coefficient of friction; b - for tensile tests; c - for bending tests.

For experiments to determine the friction indicators of samples coated on ice were used:
- climate chamber;

- a laboratory unit equipped with a compartment (reservoir) for placing the fresh ice, including a mechanism that provides a change in the vertical, pressing force on the sample coated on ice and providing a longitudinal displacement of the sample relative to ice at a given speed, a device that fixes the friction force during longitudinal movement of the sample relative to ice;
- a personal computer with installed software that records and fixes the vertical forces pressing the coated sample on ice and the friction force during longitudinal movement of the sample relative to ice.

To ensure experiments on determining the strength indicators of IPVC under tensile and bending of samples, the following were used:

- climate chamber;
- universal testing machine - AG-XPlus 100 kN;
- personal computer with specialized software;
- video recorder.

The appearance of the test equipment used to perform the experiments is in Figures 3, 4.

Figure 3. Equipment for determining the coefficient of friction of samples on ice.

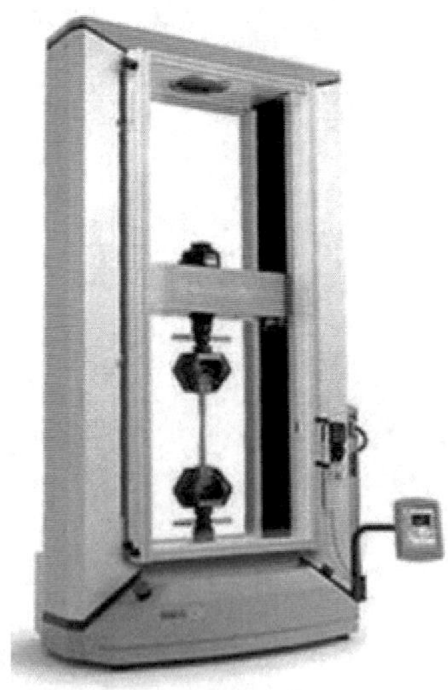

Figure 4. Universal testing machine - AG-XPlus 100 kN.

IOP Conf. Series: Materials Science and Engineering **986** (2020) 012027 doi:10.1088/1757-899X/986/1/012027

3. Indicators determined under the tests

When determining the parameters during the friction tests, the coated sample was fixed in a laboratory unit; a vertical force was applied to it, pressing it to the ice. Next, the mechanism of longitudinal movement of the sample with a pre-adjusted speed took place. Due to the frictional forces between the contacting surfaces of the sample and ice, they can remain stationary relative to each other until the force that moves the sample becomes equal to or exceeds the static friction force between the surfaces. The maximum initial value of the force was recorded, which is a component necessary for calculating the coefficient of rest friction.

The value of force was also recorded with constant movement of surfaces relative to each other at a distance of not less than 130 mm. This force is equal to the sliding friction force, and is needed to maintain the movement of surfaces relative to each other. Tests were carried out at least 3 times in accordance with 2.5.7.3 of XIII part ("Materials") of the RMRS Rules [1].

The registration and fixing of the vertical forces pressing the coated sample to ice and the friction forces during the longitudinal displacement of the coated sample relative to ice was carried out using specialized software.

The coefficient of rest friction is calculated as follows:

$$S=TS/N, \quad (1)$$

where TS - value on the scale of the device, corresponding to the beginning of movement, H; N - vertical force acting on the sample, N.

The sliding friction coefficient is calculated as follows:

$$K=TK/N, \quad (2)$$

where TK - average value on a scale corresponding to constant sliding, N.

The methodology for determining the error of experimental results when using the device is adopted in accordance with ISO 5725 [4].

After the tests, the sample was inspected.

The test results were issued in the form of a test report in the prescribed form.

During tensile and bending tests of the samples, the forces and deformations during coating failure were recorded in accordance with ISO 4628-4 [5] and ISO 4628-5 [6].

For each indicator for the characterization of cracking and/or peeling of coatings were fixed (Tables 2, 3):

- under the sample stretching - stress and relative elongation;
- Under the bending - stress and deformation of the sample.

Table 2. Diagram of indicators for the number of cracks and the area of exfoliation.

Indicator	Number of cracks according to ISO 4628-4	Peeling area,% according to ISO 4628-5
2	Few, i.e. small but significant number of cracks	0,3
3	Moderate number of cracks	1
4	Significant number of cracks	3
5	Dense crack structure	15

Table 3. Diagram of indicators for designating the size of cracks and designating the size of the areas of exfoliation.

Indicator	Crack size according to ISO 4628-4	Peel area size (largest size) according to ISO 4628-5
2	Barely visible with normal correction	Up to 3 mm
3	Clearly visible with normal correction	Up to 10 mm
4	Large cracks, usually with a width of up to 1 mm	Up to 30 mm
5	Very large cracks, usually with a width greater than 1 mm	More than 30 mm

IOP Conf. Series: Materials Science and Engineering **986** (2020) 012027 doi:10.1088/1757-899X/986/1/012027

After the test, the sample was inspected.

The test results were recorded in the form of protocols in the prescribed form.

4. Sample Modes and Tests

Friction tests were carried out for 3 values of vertical (clamping) force and at 3 speeds of longitudinal displacement of the coated sample relative to ice at 120, 150 and 180 mm/min.

In tensile and bending tests, the loading speed of the coated samples was constant, the possible deviations of the speed during the test were taken to increase without sharp changes, according to ISO 6892-84.

Examples illustrating the progress of testing the coated samples are in Figures 5-7.

a)

b)

Figure 5. Test Examples:
a - for friction; b - to bend.

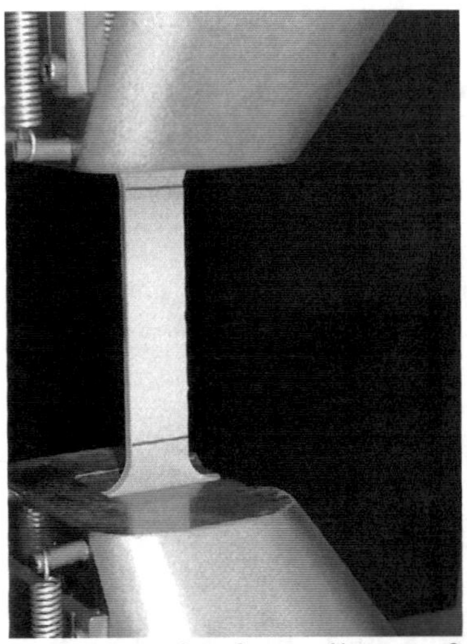

Figure 6. An example of a tensile test of sample.

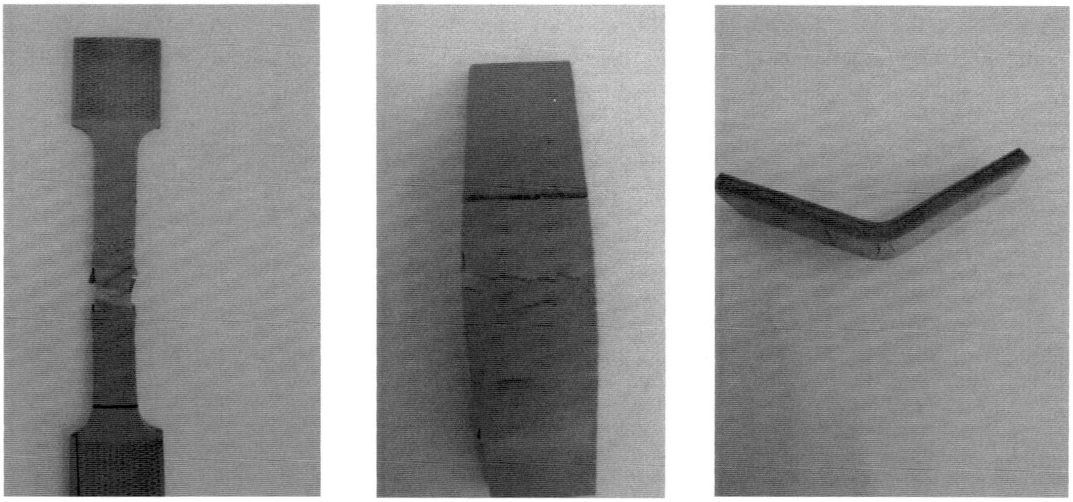

a) b) c)

Figure 7. Samples after testing
a - on tension; b, c - on bend.

5. The effect of protective coatings on the deterioration rate of the outer coatings of vessels and icebreakers

5.1 Determination of friction

The test results of the ice-resistant paint and varnish coating when driving on ice in full are in Tables 4 - 6 (summary data on the measured coefficients of initial friction of rest, as well as data on the coefficient of sliding friction).

Table 4. Definitions of friction coefficients for Sample No.1.

Indicators	The longitudinal displacement rate of the sample relative to ice, mm/min	Vertical (clamping) force N, N			Measurement error
		63,424	83,804	104,184	
		The actual value of the friction force, N			
Sample No.1					
Friction force corresponding to the beginning of the movement T_S, N	120	7,255	7,518	10,384	± 0,5%
	150	8,320	10,085	10,042	
	180	5,792	10,740	12,150	
Friction force corresponding to slip T_K, N	120	4,179	4,830	6,910	± 0,5%
	150	3,816	5,830	6,620	
	180	3,983	5,573	6,927	
The coefficient of friction corresponding to the beginning of movement $\mu_S=T_S/N$	120	0,114	0,090	0,100	± 0,5%
	150	0,131	0,120	0,096	
	180	0,091	0,128	0,117	
The coefficient of friction corresponding to the slip $\mu_K=T_K/N$	120	0,066	0,058	0,066	± 0,5%
	150	0,060	0,070	0,064	
	180	0,063	0,067	0,066	
t - Student's distribution, $t_{95\%}$					2,262
The average coefficient of friction corresponding to the beginning of movement μ_S					0,110
Standard deviation on general population μ_S					0,016
Confidence interval limits, $\pm\Delta\mu_S$					0,036

Average coefficient of friction under constant sliding μ_K	0,064
Standard deviation on general population μ_K	0,004
Confidence interval limits, $\pm\Delta\mu_K$	0,008

Table 5. Definitions of friction coefficients for Sample No. 2.

Indicators	The longitudinal displacement rate of the sample relative to ice, mm/min	Vertical (clamping) force N, N			Measurement error
		63,424	83,804	104,184	
		The actual value of the friction force, N			
Sample No.2					
Friction force corresponding to the beginning of the movement T_S, N	120	6,656	8,521	12,450	± 0,5%
	150	6,042	9,308	10,781	
	180	4,848	8,723	13,200	
Friction force corresponding to slip T_K, N	120	4,079	4,961	6,546	± 0,5%
	150	4,264	5,325	5,523	
	180	3,958	5,169	5,479	
The coefficient of friction corresponding to the beginning of movement $\mu_S=T_S/N$	120	0,105	0,102	0,120	± 0,5%
	150	0,095	0,111	0,103	
	180	0,076	0,104	0,127	
The coefficient of friction corresponding to the slip $\mu_K=T_K/N$	120	0,064	0,059	0,063	± 0,5%
	150	0,067	0,064	0,053	
	180	0,062	0,062	0,053	
t - Student's distribution, $t_{95\%}$					2,262
The average coefficient of friction corresponding to the beginning of movement □S					0,105
Standard deviation on general population μ_S					0,014
Confidence interval limits, $\pm\Delta\mu_S$					0,032
Average coefficient of friction under constant sliding μ_K					0,061
Standard deviation on general population μ_K					0,005
Confidence interval limits, $\pm\Delta\mu_K$					0,011

Table 6. Definitions of friction coefficients for Sample No. 3.

Indicators	The longitudinal displacement rate of the sample relative to ice, mm/min	Vertical (clamping) force N, N			Measurement error
		63,424	83,804	104,184	
		The actual value of the friction force, N			
Sample No.3					
Friction force corresponding to the beginning of the movement T_S, N	120	6,062	7,716	11,394	± 0,5%
	150	6,269	7,838	11,201	
	180	6,971	7,609	10,499	
Friction force corresponding to slip T_K, N	120	3,937	5,342	7,182	± 0,5%
	150	3,957	5,065	6,163	
	180	4,429	5,231	6,335	
The coefficient of friction corresponding to the beginning of movement $\mu_S=T_S/N$	120	0,096	0,092	0,109	± 0,5%
	150	0,099	0,094	0,108	
	180	0,110	0,091	0,101	
The coefficient of friction	120	0,062	0,064	0,069	± 0,5%

corresponding to the slip $\mu_K = T_K/N$	150	0,062	0,060	0,059	
	180	0,070	0,062	0,061	
t - Student's distribution, $t_{95\%}$					2,262
The average coefficient of friction corresponding to the beginning of movement μ_S					0,100
Standard deviation on general population μ_S					0,008
Confidence interval limits, $\pm\Delta\mu_S$					0,017
Average coefficient of friction under constant sliding μ_K					0,063
Standard deviation on general population μ_K					0,004
Confidence interval limits, $\pm\Delta\mu_K$					0,008

Summarized data on the results of tests of ice-resistant paint and varnish coating when the sample moves on ice (determination of the friction coefficient) are in Table 7.

Table 7. The results of the tests of IPVC when determining the coefficient of friction on ice.

Number of sample	Coefficient of friction	
	Friction of rest□□S	Sliding friction□□K
1	0,110±0,036	0,064±0,008
2	0,105±0,032	0,061±0,011
3	0,100±0,017	0,063±0,008
Average	**0,105±0,028**	**0,063±0,009**

As a result of the experiments, it was found that the tested IPVC in terms of the coefficient of sliding friction on ice corresponds to Class II in accordance with 6.5.3 of the Rules for the Classification and Construction of Ships, part XIII ("Materials").

5.2 Determination of strength indicators

Figures 8, 9 show examples of control of test results using the specialized software integrated into the working environment of a universal testing machine - AG-XPlus 100 kN.

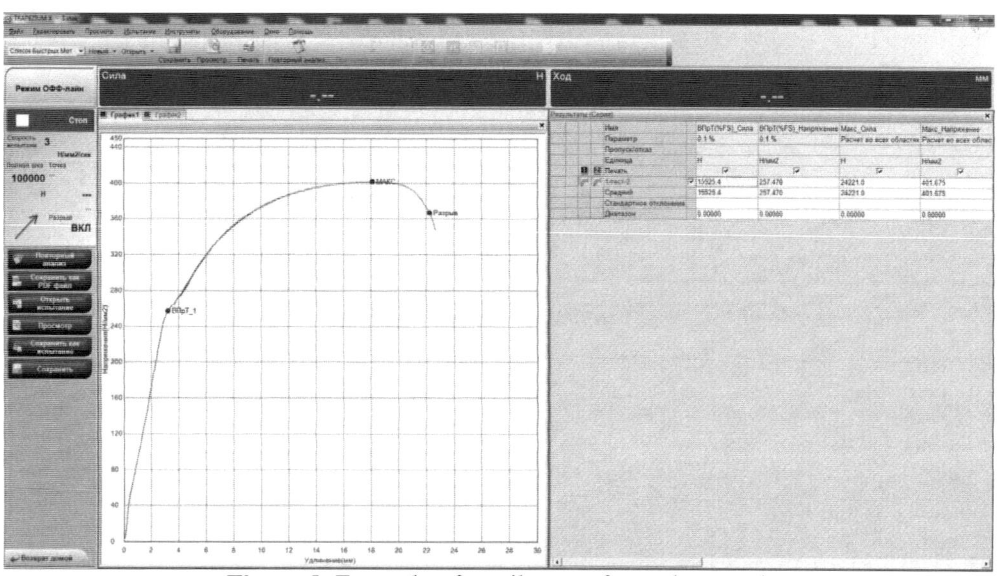

Figure 8. Example of tensile test of sample coated.

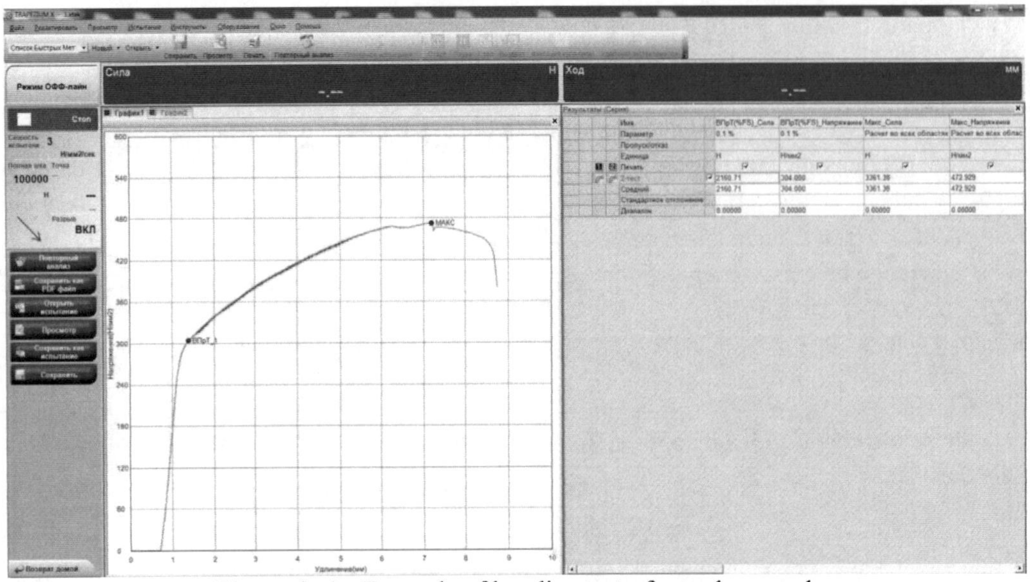

Figure 9. An Example of bending test of sample coated.

The results of the IPVC test under tension and bending of the samples are in Tables 8, 9.

Table 8. Results of stretching the coated samples.

Indicators	Crack and Peel Index	Actual values, MPa			
		Sample No.4	**Sample No.5**	**Sample No.6**	**Average**
Tension, MPa	2	279,7	278,5	276,4	278,2
	3	353,2	326,6	326,5	335,4
	4	381,1	381,2	377,7	380,0
	5	401,1	401,2	401,6	401,3
Relative elongation, %	2	6	6	6	6
	3	10	10	10	10
	4	15	15	15	15
	5	25	25	25	25

Table 9. Results of bending of coated samples.

Indicators	Crack and Peel Index	Actual values, MPa			
		Sample No.7	**Sample No.8**	**Sample No.9**	**Average**
Tension, MPa	2	313,5	311,2	318,8	314,6
	3	347,7	346,6	350,5	348,3
	4	452,7	454,7	456,2	454,6
	5	472,9	475,6	479,5	476,1
Bending angle of the sample, degrees	2	5	4	6	5
	3	17	16	19	17
	4	31	31	31	31
	5	45	45	42	44

We note that the strength parameters of the paint and varnish coating during tensile and bending of samples (determination of stresses at which cracking and/or peeling of the ice-resistant protective coating from a metal surface is observed) are currently not standardized and are not regulated by the requirements

of the RMRS Rules. The tests were performed in accordance with the Terms of Reference for tests for a complete study of the characteristics and quality of the developed coating.
The test results showed that the coating breaks at stresses close to the yield strength of the sample material.

6. Conclusion

The results of the study:

- a test program for the strength and friction properties of an ice-resistant paint and varnish coating has been developed;
- a complex of strength and friction tests was carried out;
- the coefficients of friction of samples with IPVC on ice were determined for compliance with the requirements of 2.5.7 of the Rules for the classification and construction of marine vessels, part XIII "Materials" of RMRS [1];
- the strength parameters of the paint and varnish coating during tensile and bending of the samples have been determined in order to identify the stresses at which cracking and/or peeling of the ice-resistant protective coating from a metal surface was observed.

As a result of the performed experiments, it was found that the tested IPVC on the coefficient of sliding friction on ice (0.063 ± 0.009) corresponds to Class II according to 6.5.3 of the Rules for the classification and construction of ships, part XIII "Materials" and can be used as an ice-resistant coating for the outer skin of vessels belonging to Group 1 (icebreakers of all ice classes) and Group 2 (vessels of ice classes Arc4 and above).
The results of the tests showed that the coating begins to collapse at stresses close to the yield strength of the material of the samples, which indicates the possibility of using the coating under stress, characteristic for the region of elastic deformation of the outer skin and arising from ice impact.

References

[1] Rules for the Classification and Construction of Ships 2020 (part XIII "Materials", Russian Maritime Register of Shipping)
[2] GOST 1497-84 (ISO 6892-84) 1987 Metals. Tensile test methods
[3] GOST 14019-80 1980 Metals. Bending test methods
[4] ISO 5725 Accuracy (correctness and precision) of measurement methods and results
[5] ISO 4628-4 Paints and varnishes. Assessment of the degree of destruction of coatings. Designation of the number and size of defects and the intensity of uniform changes in appearance. Part 4. Assessment of the degree of cracking
[6] ISO 4628-5 Paints and varnishes. Assessment of the degree of destruction of coatings. Designation of the number and size of defects and the intensity of uniform changes in appearance. Part 5. Assessment of the degree of exfoliation

IOP Conf. Series: Materials Science and Engineering **986** (2020) 012028 doi:10.1088/1757-899X/986/1/012028

The effect of ozonation on the wettability of wood

A Kh Safiullina[1], Sh R Mukhametzyanov[1] A R Shaikhutdinova[1] and M A Zhmaylo[2]

[1]Kazan National Research Technological University, Kazan, Russia, alphia50@gmail.com
[2]National Technology Initiative Center for Advanced Manufacturing Technologies based on the Institute of Advanced Manufacturing Technologies of Peter the Great St. Petersburg Polytechnic University Polytechnicheskaya, 29, St.Petersburg, 195251, Russia

Abstract. Ozone can be used to break down lignin and hemicellulose in many lignocellulosic materials. Ozone pre-treatment has the following advantages: it effectively removes lignin; as a result of such pre-treatment, toxic products are not formed for the subsequent process; the reaction is carried out at room temperature and atmospheric pressure. The purpose of this work is to study the effects of ozone on the surface of natural and thermally modified wood. Thermal treatment is an environmentally friendly method that allows you to change the color and performance properties of wood. This article analyses the effect of ozonation of the surface of natural and thermally modified wood in order to increase the adhesive properties during bonding. Thus, a more economical consumption of glue and a reduction in the emission of harmful volatile substances into the environment are possible.

1. Introduction

In the development of chemical technology, the main trend today is to increase production processes and reduce the number of by-products. Various oxidative processes occur in chemistry and technology [1]. One of the promising directions for obtaining industrially important products is the use of ozone. This substance is considered as a promising reagent from an environmental point of view, since it does not form toxic conversion products, and ozone reactions are carried out at room temperature and atmospheric pressure. In addition, ozone is used as an oxidizing agent, due to the fact that its reactivity allows interacting effectively with organic substances, especially unsaturated aromatic compounds. It is worth noting that the amount of ozone generated in atmospheric oxygen is sufficient for the destruction of various polymer products. Ozone is especially destructive to elastomers containing carbon bonds in the main chain of a macromolecule, which include various resin elastics and rubbers. In this case, the surface of these products instantly cracks and fails [1]. Under the influence of ozone, molecular weight decreases in polymer solutions and oxygen-containing functional groups become saturated: acids, ketones, peroxides, etc. This process proceeds at moderate temperatures, which is caused by the high reactivity of ozone.

Polymers that contain phenyl cycles in the main chain react slowly with ozone, and polymers with heteroatoms react much faster. The first group includes polystyrene - a thermoplastic polymer of linear structure, which is a product of styrene polymerization. The appearance and properties of polystyrene powder are changed significantly under the influence of ozone: a yellow color appears and the dielectric properties become worse. Analysis of the IR spectrum shows that an increase in the absorption intensity occurs in the region of 1000-1200 cm-1. At this point, the individual chains are structured and stitched together; most of the polymer becomes resistant to solubility after ozonation, forming gels. In addition, it is noted [2] that ozonation of low-molecular-weight analogues of polystyrene, such as diphenyl methane and isopropyl benzene, leads to the appearance of insoluble aromatic ozonizes, which then fall out.

The destruction of the structure of unsaturated polymers (in particular resin elastics and rubbers) under the action of ozone occurs with the formation of cracks on the surface of the products, loss of mechanical strength and destruction. The main reaction products are ozonizes, which retain a bond between fragments of the hydrocarbon chain on both sides of the double bond. When a stream of ozone is passed through a polymer solution in the initial stage of the reaction, ozone is absorbed almost completely. A noticeable leakage of ozone begins only at the moment of depletion of the double bonds. At the time of the early stages of the reaction, a decrease in viscosity is observed. According to the data obtained [2], the relative drop in molecular weight and the number of chain breakdowns per one reaction event are calculated. The nature of the polymer significantly affects the number of chain breakdowns, thus chloroprene has the lowest value, and gutta-percha and natural resin elastics has the largest one.

The development of ozone technologies and the emerging trend towards reducing the cost of ozone open up new opportunities for its use in the conversion of cellulose-containing plant materials, which are divided into woody and herbaceous plants. With direct ozone treatment of the plant substrate, lignin destruction is observed. Lignin is an aromatic part of plant materials. Studies are known [2], in which ozonation of various substrates was performed (straw, corn cobs, maize, coniferous wood, deciduous wood, etc.). Using IR and UV spectroscopy, it is shown that lignin is destroyed under the influence of ozone, while water-soluble lignin oxidation products (aromatic carbonyl and carboxyl compounds) are formed. The study shows that the main physicochemical characteristics of the ozone absorption process are determined by the conditions for the preparation of samples of plant substrates and the parameters of ozonation. The effectiveness of the ozonation process depends on the water content in the samples. The effect of ozone on wood has shown that mainly ozone dissolved in water (contained in the wood cells) takes part in reactions with wood.

Lignin is a companion of fiber in wood; its content is 20-30 %. It is known that ligninis a non-carbohydrate polymer in which phenolic compounds are monomer units. A large number of simple phenols are formed during chemical processing as a result of wood destruction. In industry, lignin oxidation is used for pulp bleaching.Industrial methods of oxidative wood delignification and new bleaching methods are being developed to create environmentally friendly and resource-saving technologies. The peculiarity of bleaching is the need to affect lignin that has already been subjected to chemical treatment during cooking (residual lignin), i.e., the most difficult to remove lignin the most difficult to remove lignin with a structure altered by cooking reagents[3]. Ozone is used for this purpose, due to the fact that it is a strong oxidizing electrophilic reagent that interacts with lignin in double bonds. In this regard, the study of the effect of ozone on wood for cellulose production is especially interesting, and an urgent task is to conduct studies of structural changes in wood under the influence of ozone.

This work is devoted to the study of the effect of ozone on birch wood. Due to the fact that under the influence of ozone the functional groups of wood are oxidized, and some low molecular weight products are adsorbed on the surface of ozonized wood, the aim of this work is to conduct the ozonation process to improve the adhesive properties of the surface layer of birch wood veneer, in order to reduce the consumption of glue when gluing, thereby reducing the emission of harmful substances.

2. Methods and materials

Experiments on ozonation of wood samples were conducted in the OZ-0500AC ozone chamber of GOTECH Testing Machines Inc. and UGNLab Co., Ltd. Testing Equipment (figure 1), designed to assess the resistance of materials to ozone. This equipment is equipped with a UV absorption analyzer that provides accurate measurement of ozone concentration. The ozone chamber makes it possible to evaluate the ozone aging of polymeric materials under static and dynamic conditions in the presence of a given ozone concentration. It is able to regulate ozone concentration, air flow rate, and a temperature in accordance with international standards.

The experimental studies were carried out as follows: samples of natural and thermally modified birch veneer with a moisture content of 7 % and a size of 120*60 mm were fixed to the ozonation chamber on a rotating platform [4-6]. Then the process parameters were set on a touch screen in the form of an ozone concentration of 1000 pphm and a medium temperature of 100 °C. The duration of the experiment on ozonation of the surface layer of wood samples was 6 hours.

Preliminary thermal modification of birch veneer samples was performed using a known method [7-15].

In order to analyze the effect of ozone treatment of wood samples, the control samples were simultaneously exposed to the laboratory drying oven LF-25/350-VS1 at a temperature of 100 ° C with a similar duration [16-18].

Figure 1. Ozone chamber

After a predetermined time, the test samples were removed from the ozonation chamber, and control samples from the drying chamber. Then the surface of the samples was examined for wettability. Wetting the surface determines the adhesive interaction between the wood and the binder. In this work, the sessile drop method was chosen for the experimental determination of the wetting angles of the birch veneer surface, as the most simple and convenient.

The test samples were placed on the table in a strictly horizontal position. Wettability tests were carried out for a control sample of natural and thermally modified wood, as well as for a sample of natural and thermally modified wood treated with ozone. During the study, three drops of round-shaped water of a strictly defined volume were applied to the veneer surface [19] by squeezing the liquid out of the syringe, and the drop was photographed every 2 seconds.

3. Results and discussion

Table 1 illustrates the changes in the size and shape of the drop on control samples of natural and thermally modified wood, as well as on samples subjected to ozone treatment.

Comparing the obtained images [20-21], it can be established that ozonation increases the wettability of the surface of cellulose-containing material.

Table 1. Wettability of wood samples.

Measurement time	Control sample aged in an oven		Sample treated with ozone	
	Natural wood T_{tr}= 100 °C τ_{tr} = 6 hours	Thermally modified wood T_{tr}= 220 °C τ_{tr}= 30 min	Natural wood T_{tr}= 100 °C τ_{tr} = 6 hours C= 1000pphm	Thermally modified wood T_{tr}= 220 °C τ_{tr}= 30 min C = 1000pphm
τ = 0 s.				

Measurement time	Control sample aged in an oven		Sample treated with ozone	
	Natural wood T_{tr}= 100 °C τ_{tr} = 6 hours	Thermally modified wood T_{tr}= 220 °C τ_{tr}= 30 min	Natural wood T_{tr}= 100 °C τ_{tr} = 6 hours C= 1000pphm	Thermally modified wood T_{tr}= 220 °C τ_{tr}= 30 min C = 1000pphm
τ = 10 s.		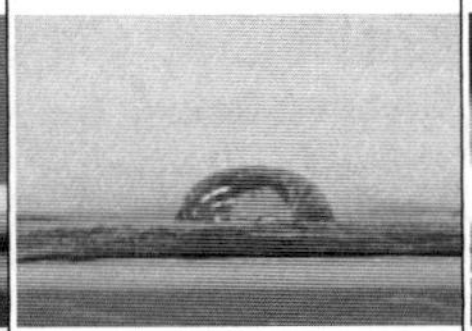	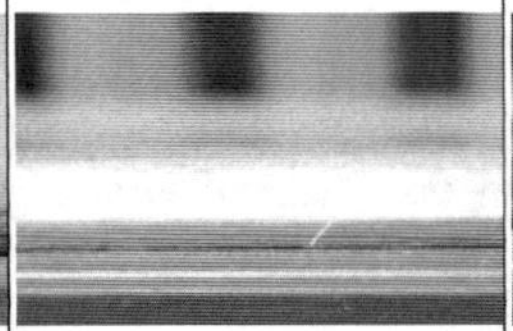	

According to the photograph of a control sample of natural wood, the drop does not spread on the surface of the veneer and remains at quiescent condition for another 180 seconds. While a sample of natural wood treated with ozone [22] shows rapid adhesion of the drop to the veneer surface. A drop of water applied to the surface spreads within 10 seconds [23] and covers the substrate with a thin layer. On the contrary, the photograph of a control sample of thermally modified wood shows that the drop does not spread on the surface of the veneer, but remains in this state for only 120 seconds. A change in the shape and size of the droplet is observed much faster on an ozonized sample that has previously undergone heat treatment [24]. In this case, we can say that the preliminary thermal modification of veneer partly prevents the spreading of the droplet on the surface of the sample.
Graphical dependences of the effect of the duration of ozone treatment on the limiting wetting angle of various samples are presented in Figure 2.

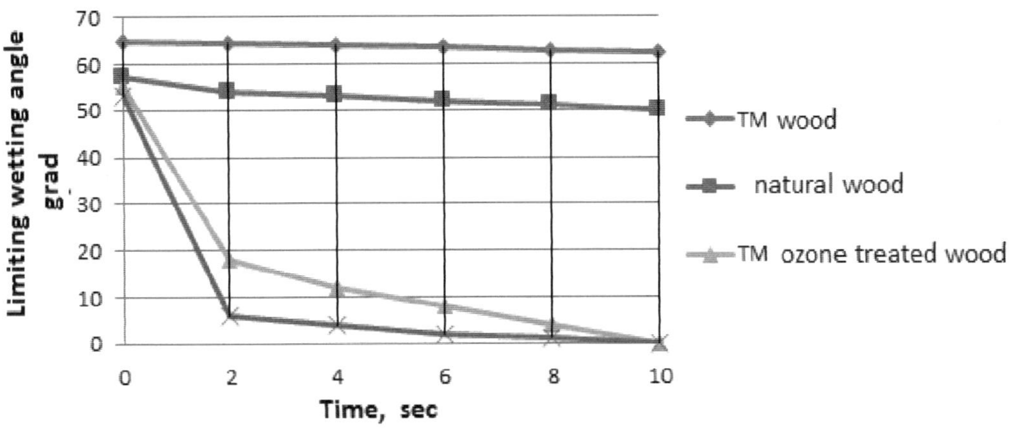

Figure 2. A graph of the duration of treatment with ozone at the limiting wetting angle.

The limiting wetting angle for samples of wood veneer at the initial stage is: for a control sample of thermally modified wood - 65 °, for a control sample of natural wood- 57 °, for thermally modified wood treated with ozone - 55 °, and for natural wood treated with ozone - 53 °. For 10 seconds, it is observed that the limiting wetting angle of the drop on the control sample of thermally modified wood [25] remains virtually unchanged, while the limiting wetting angle of the sample of thermally modified veneer treated with ozone rapidly decreases and reaches zero. Similar results are observed for natural wood veneer samples. However, it can be noted that the values of the limiting wetting angle of the veneer samples of thermally modified wood are higher than the values of the samples of natural wood veneer. This is due to the fact that extractive substances are removed and the number of hemicelluloses in the wood structure is reduced during high-temperature processing [26-27], thereby increasing the moisture resistance of the sample.

4. Conclusion

Based on the studies and analysis of visual measurements of the limiting wetting angle of wood samples, it can be concluded that the ozonation process improves the adhesive properties of the surface layer of veneer due to the reactive capacity of ozone to oxidize and degrade lignin-containing wood products. This allows further reducing the consumption of glue when gluing samples, thereby reducing the emission of harmful volatile substances.

Studies of ozonation of heat-treated veneer showed higher moisture resistance compared to natural wood veneer, which allows the production of layered composite materials with improved performance with lower adhesive consumption.

Acknowledgements

This work was performed with support of grant of the President of the Russian Federation for state support of young Russian scientists – PhD (MK-2246.2020.8)

References

[1] Kang G, Zhang Y, Ni Y and Van Heiningen A R P 1995 Influence of lignins on the degradation of cellulose during ozone treatment *J. Wood Chem. Technol.* (vol. 15(4)) pp 413-430

[2] Ventorim G, Colodette J, de Fatima Gomes A, and da Silva L 2008 Reaction rates of lignin and hexenuronic acids with chlorine dioxine, ozone, and sulfuric acid *Wood and Fiber Science* (vol. 40(2)) pp 190-201

[3] Masahiko Kobayashi, Toshiyuki Asano, Mikio Kajiyama and Bunichiro Tomita Effect of ozone treatment of wood on its liquefaction *Journal of Wood Science* (vol. 51) pp 348–356

[4] Galyavetdinov N R, Khasanshin R R, Safin R R, Safin R G and Razumov E Y 2015 The usage of wood wastes in the manufacture of composite materials *International Multidisciplinary Scientific GeoConference Surveying Geology and Mining Ecology Management, SGEM 15th* pp 779-786

[5] Safin R R, Khasanshin R R, Khakimzyanov I F, Mukhametzyanov S R and Kainov P A 2017 Increasing the Energy Efficiency of the Process of Oscillating Vacuum-Conductive Drying of Wood by Means of a Heat Pump *Journal of Engineering Physics and Thermophysics* (vol. 90(2)) pp 310-317

[6] Guojun Kang,Yijing Zhang,Yonghao Ni, Adriaan R P and van Heiningen 2006 Influence of Lignins on the Degradation of Cellulose During Ozone Treatment pp 413-430

[7] Pouyet F, Chirat C and Lachenal D 2013 On the origin of cellulose depolymerization during ozone treatment of hardwood kraft pulp *BioRes*. (vol. 8(4)) pp 5289-5298

[8] Safin R R, Mazurkin P M, Mukhametzyanov S R, Voronin A E and Salimgaraeva R V 2017 Dynamics of mass of a sample in the course of oscillating vacuum and conductive drying of timber *International Multidisciplinary Scientific GeoConference Surveying Geology and Mining Ecology Management* (vol. 17(62)) pp 97-104

[9] Safin R R, Safiullina A Kh and Nazipova F V 2017 Increase in energy efficiency of use of vegetable waste *IOP Conference Series: Earth and Environmental Science* (vol. 87(3) 032025)

[10] Safin R R, Galyavetdinov N R, Mukhametzyanov S R, Shaikhutdinova A R and Khasanshin R R 2019 Technology for producing high-quality natural biodegradable plastic for 3D printer *International Multidisciplinary Scientific GeoConference Surveying Geology and Mining Ecology Management, SGEM* (vol. 19/issue4.1) pp 503-509

Studies of the composition of the cryogenic ground chaga

E Yu Razumov[1], R R Safin[1], Sh R Mukhametzyanov[1], E I Baigildeeva[1], A V Safina[1] and D O Lebedev[2]
[1]Kazan National Research Technological University, Kazan, Russia, evgeny.razumov2011@yandex.ru
[2] National Technology Initiative Center for Advanced Manufacturing Technologies based on the Institute of Advanced Manufacturing Technologies of Peter the Great St. Petersburg Polytechnic University Polytechnicheskaya, 29, St.Petersburg, 195251, Russia

Abstract. The paper studies the properties and chemical composition of birch fungus - Chaga, which contains a unique complex of biologically active substances and is used as a raw material for the production of drugs and biologically active additives. The components listed prove their antioxidant, immunomodulating, antiviral, antidiabetic, and cytotoxic properties. The paper specifies the application of chaga in such areas as medicine, agriculture and food industry. It analyses the way the method of grinding of plant raw materials might influence the quality of the resulting drugs, including the sustaining the biological activity. The results determined that the cryogenic grinding of chaga, which allows to preserve valuable substances of the fungus due to its grinding at low temperature in liquid nitrogen to be the most reasonable. The paper gives the analysis of the obtained dispersed phase by high performance liquid chromatography and atomic absorption spectroscopy. The results of the studies indicate an increase in the content of particular valuable components of the fungus in extracts of raw materials that have gone through the cryogenic grinding, compared with the extracts of raw materials obtained by the traditional grinding method. The data acquired can be of practical importance when chaga is used in medicine.

1. Introduction

Chaga is a product of the sterile stage of activity of a tree-destroying fungus that parasitizes, most often, on the trunks of live birches. It grows mainly in humid climates such as forests of Russia, Ukraine, Belarus, and Poland [1]. The fruit body consists of large brown solid growths inside, while closer to the wood the fruit body is softer and lighter. Fungus hypha destroys wood and causes tree decay. Chaga contains a comprehensive set of substances that are essential for humans. In this context, the Chaga mushroom is widely used in the pharmaceutical industry.

Chaga containing drugs are generally used as anti-inflammatory and restorative. Chaga can also be used as part of the complex treatment of tumors of various localization, diseases of the gastrointestinal tract, it helps to mobilize the body's defenses by affecting metabolic processes [1-6]. Chaga-based drugs are biologically active. They are characterized by immunomodulatory, antiviral, antioxidant, antitoxic, radio protective, gene protective and adaptogene properties; these have the potential to regulate the activity of blood enzymes, as well as the activity of the cardiac, respiratory and nervous systems of the body [1-17].

Modern medicine uses chaga independently or as an ingredient of multicomponent drugs for internal and external use in the form of elixirs, decoctions, and ointments. Moreover, Chaga is used in animal husbandry; adding chaga to the diet of pigs stimulates the growth of piglets and weight gain. Extracted meal of Chaga can be good for crop production as a fertilizer to protect plants from late blight and stimulate growth. Food industry has applied Chaga extracts for preserving berry and vegetable juices [1].

Russian and foreign scientists are studying the biology of the fungus and its biosynthetic activity, performing chemical analysis of the fungus, researching the mechanisms of the formation of biologically active substances, methods to extract and purify them. In the middle of the XX century, Soviet scientists E.G. Platonova, G. A. Kuznetsova, A.N. Shivrina, P. A. Yakimov, N.L. Mattison, M. P. Berezina, P. K. Bulatov et al. laid the scientific foundation for the study of the properties of Inonotus obliquus, concentrates based on this fungus and the resulting drugs.

S. Vasser [18] summarized information about the pharmacological use of the fungi, their biologically active substances that can stimulate the immune responses of the human body. The results of studies of the chemical composition of Inonotus obliquus and its unique complexes are described by G. Britton [19]. Japanese scientists, in particular, Y. Nakajima [20] and I. Kuriyama [21] conduct research on the extraction of various components of Inonotus obliquus, determination of their structure, activity and mechanisms underlying their therapeutic action, and develop methods for cultivating the fungus. Chinese researchers, in particular, X. Lu [22], identify and study individual active fractions of Inonotus obliquus, defining potential applications in treatment of various diseases. Korean scientists, in particular, J.H. Lee and C. K. Hyun [23] analyze the chemical composition of Inonotus obliquus and determine the biological activity of the fungus components.

X-ray fluorescence method, along with gravimetric analysis and atomic absorption spectroscopy identified the following elements that Chaga consists of: carbon (39%), potassium (9 - 10%), hydrogen (3.6%), nitrogen (0.4%), magnesium (0.64%), calcium (0.37%), chlorine (0.33%), phosphorus (0.23%), sodium (0.05%), rubidium (0.04%), sulfur (0.02%), manganese (0.02%), iron, copper, zinc, vanadium, chromium, traces of nickel, selenium, iodine, etc. Hydrolysis products contain 15 amino acids, among which are glycine, aspartic and glutamic acids (40% of all the acids), as well as tyrosine, serine, threonine, alanine, leucine, methionine, lysine, histidine, arginine, tryptophan, cystine and proline [1]. High biological activity of amino acids helps human body to adopt more effectively drugs from medicinal raw materials.

Chaga contains flavonoids, including flavones, flavonones, anthocyanins, catechins (presented by apigenin, mariningin, corin, and quercetin), triterin and sterol compounds (6 - 8%), acid-resistant lignin (25 - 30%), dietary fiber (2%), hemicellulose (12.5%), and other derivatives of pteroylglutamic acid. A study of the acid composition of chaga also showed the presence of such organic acids as acetic, butyric, oxalic, formic [1].

A study of the chemical composition showed that the chaga contains a unique polyphenolic chromogenic complex, including humic pigments such as melanins, called chagic acid, which are protectors from the effects of carcinogenic and mutagenic factors [17]. The chromogenic complex has a rich macro- and microelement composition (K, Mg, Na, Ca, Fe, Si, Zn, Al, etc.), includes derivatives of pyrocatechol and pyrogallol, pteric, steroid and terpene derivatives, lignin, polysaccharides, organic acids [3], and also has a wide range of biological and physiological effects: a high antitoxic effect, an effect on enzymatic processes.

The analysis of physical and chemical properties proved that the pigments obtained from chaga belong to melanins, i.e. group of high molecular weight black and brown pigments forming at the oxidative polymerization of phenols. Chaga melanins show high antioxidant and gene-protective activity. Specific structural features define a wide spectrum of the biological activity of melanins. T. A. Kukulyanskaya et al. [24] found differences in the chemical structure and biological activity of melanin in Inonotus obliquus growing in-situ and cultivated.

Shashkina M.Ja. et al. [1] indicated that fraction of sterols and triterpenes extracted from chaga consist of lanosterol (a derivative of tetracyclic triterpen), inotodiol (triterpene alcohol), ergosterol, etc. These are insoluble in water; however, they are partially emulsified and extracted with hot water, and up to 80% remain in extracted meal while extracting. Inotodiol is known for an anti-blastoma activity.

Studies also note the content of lectins in Inonotus obliquus [1]. Lectins or agglutinins belong to the class of complex glycoproteins, which in some cases contain calcium, magnesium and other ions. They are capable of binding carbohydrates and participating in the processes of carbohydrates transporting and depositing, i.e. show the hypoglycemic effect, thus lowing the blood sugar level that people with diabetes suffer from. Furthermore, lectins can stimulate the growth and division of lymphocytes, participate in the regulation of immunological reactions, and block receptors of tumor cells, thus inhibiting their migration.

Kahlos K. [25] researched the composition of chaga terpenes, their biological activity, and antitumor effect in Finland. Taylor A. [26] from the United Stated studied the effects of aqueous chaga extracts in the treatment of adenocarcinoma. Chinese scientists, specifically Jiang J. [27] examine the effect of substances isolated from chaga on cancer cells. Polish scientists, specifically Mazurkiewicz W. [15, 16] study melanins secreted from chaga, as well as the effect mechanism of water extracts of birch fungus on the body in cancer treatment.

Chaga contains a complex of biologically active substances that are participating in the regulation of metabolism, as well as the correction and prevention of pathological disorders. In this context, it is crucially important to preserve these substances when receiving therapeutic agents. Technological problems related to the production of molecular ingredients that preserve the native structure and properties of the active complexes most important for the human body may arise while creating drugs based on biological raw materials [28].

Grinding is an essential technology in the pharmaceutical production. Conventional grinding has two significant drawbacks. The process involves the use of oxygen; hence, it causes intense oxidation and decomposition of substances of the processed raw materials. Moreover, heating of raw materials during grinding also results in the loss of valuable properties. Heating can be critical because of contact of the raw material with the surface of the working body of the grinder [29].

The technology for producing microdispersed materials will be considered effective only in case it ensures the reservation of the initial properties of the processed raw materials, in addition to fine grinding [30]. The solubility of the plant material in biological fluids improves with the increase in the degree of grinding, thus boosting the therapeutic activity of the preparations obtained, and consequently resulting in the decrease of the amount of the plant material use.

Grinding materials to obtain pharmaceuticals is carried out in air at elevated temperatures associated with prolonged mechanical stress, which accelerates the destruction and oxidation of raw materials. In this context, it is reasonable to grind dry plant materials in a crisp, chilled state in the environment of chemically inert gases, for example, nitrogen, in order to avoid losses of thermolabile chemical compounds [31]. Cryogenic processing of plant raw materials allows you to fully preserve the native structure of not only the vitamins contained in it, but also molecular complexes containing the widest spectrum of trace elements essential for humans.

Such scientists as Kapitsa P.L., Landau L.D., Arkharov A.M., Gersh S.Ya., Belyakov V.P., Mikulin E.I., Marfenina E.V. etc made the greatest contribution to the development of cryology in the 20th century. Among other scientists, there are such names as Linde K., Claude J., Geylandt P., Simon F., Nernst V., and Carrier V. Cryogenic grinding of plant materials is used in molecular biology, microbiology and biochemistry to isolate substances and certain organelles of a plant cell. Raw materials cooled to low temperatures are subjected to cryogenic grinding, which helps avoid oxidation, caramelization and aggregation processes. At a low temperature, the bonds of biologically active substances with protein molecules become crisp, and further they are destroyed. This has a positive effect as protein molecules, having a large size and weight, interfere with the absorption of beneficial substances by the body. Cryogenic grinding leads to a decrease in microbial contamination of medicinal raw materials. The resulting cryopowder has a long shelf life and is a natural concentrate of biologically active substances, including vitamins and trace elements.

Cryogenic grinding has been widespread abroad. This method was initially used to extract and recover valuable materials from the wastes of the electrical and automotive industries in the United States, Japan, Germany, and France. Later the method became popular in the production of powders from vegetables, spices, heat-sensitive raw materials. Menry Balfaur LTD (USA) developed the technology of cryogenic grinding of chocolate; Cryopowider Servise (Great Britain) began to process polymer materials and pharmaceuticals in the similar way [29]. The French concern Arkopharma, which specializes in the production of parapharmaceuticals, uses the technology of cryogenic grinding of raw materials using liquid nitrogen to obtain ingredients for biologically active additives. In a liquid state, liquid nitrogen is colorless, non-toxic, therefore, when freezing food, it can be used in direct contact with a food product.

IOP Conf. Series: Materials Science and Engineering **986** (2020) 012029 doi:10.1088/1757-899X/986/1/012029

Liquid nitrogen has a low boiling point (-1960C), it is chemically and biologically inert and is safe for working personnel. It makes a crushed mass fragility, protects it from overheating and deterioration. The presence of a dry inert atmosphere contributes to the preservation of the native properties of the source raw material.

Pavlyuk R.Yu. [29] indicated that as a result of cryogenic grinding of dried unconventional medicinal raw materials at a temperature of -100C and lower there is an increase in the concentration of biologically active substances, including beta-carotene, vitamin C, free amino acids, phenolic compounds with P-vitamin activity, compared to the source raw material. Such an effect can be explained by the degradation of bonds between biopolymers and low molecular weight compounds with the cleavage of the latter, and is also associated with a significant destruction of plant materials.

Pevneva O.P. and Schegolev A.A. point to the results of biochemical studies of cryogenically ground raw materials [31], during which it was found that 25–40% more essential oils are extracted from it than with traditional grinding and extraction methods [32]. Studies prove that low-temperature grinding of medicinal plant materials leads to an increase in the yield of extractive substances by 5–45% depending on the source raw material, including carbohydrates by 2–25%, organic acids by 15–35%, and vitamin C by 15–45%.

The following advantages of cryogenic grinding of herbaceous medicinal raw materials in comparison with the traditional method of grinding were revealed as a result of research by Pevneva O.P. and Shchegoleva A.A. [31]. Cryogenic grinding preserves biologically active substances of ground products; it helps obtain a homogeneous particle size distribution of the crushed product; it obtains powders with particle sizes not achievable with traditional methods; finally, it prevents aggregation of particles resulting from the accumulation of static electricity in the case of traditional grinding.

The authors consider the use of cryogenic grinding when receiving drugs from the Chaga fungus reasonable taking into account the advantages of the product obtained from cryogenic crushed raw materials, including the preservation of its biological value, the increase in bioavailability of biologically active substances and the improvement of their adoption by human body. This will allow the preservation of its biologically active substances, that are widely used both in phyto- and fungotherapy.

2. Methods and materials

The authors chose as an object of study a birch mushroom – a chaga harvested in the winter in the forest zone of the Mari El Republic, far from industrial and housing infrastructure, and highways. Furthermore, the authors grounded two samples of room-dried chaga to compare the impact of the method of chaga grinding on the content of biologically active substances in it. Preliminary grinding of raw materials was carried out on a rotary knife mill RM-120.

A drum-type ball cryogenic mill (Fig. 1) with grinding bodies in the form of metal balls with a diameter of 20 mm was used for grinding. The experimental unit consists of a metal drum mounted on the frame and having a cylindrical and conical parts. The drum is tilted during operation and during unloading of ground raw materials. The conical part is supplemented by a cover for loading and unloading material and a grid to prevent the loss of grinding bodies during unloading. The rotation of the drum is carried out with the help of a drive mechanism and a system of ball bearings, which are mounted on a frame using bearings.

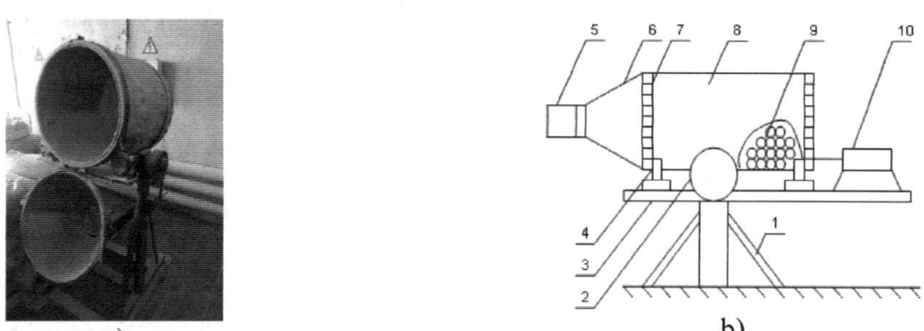

a) b)

Figure 1. Experimental ball cryogenic mill: a - appearance; b - scheme: 1 - frame; 2 - worm reducer; 3 - plate; 4 - plain bearing; 5 – uploading container for the ground product; 6 - conical cover; 7 - belt fixing the axis of the cryogenic mill and electric motor; 8 - cryogenic mill body; 9 - grinding bodies; 10 - electric motor.

The grinding of the first sample was carried out for 40 minutes in a dry form with a drum rotation speed of 70 rpm. The grinding of the second sample was carried out after feeding liquid nitrogen to the mill, in the medium of which the raw material was aged for 5 minutes in order to completely freeze the chaga particles. The ratio of raw materials and liquid nitrogen was 1:2. The grinding process proceeded at atmospheric pressure and intensive evaporation of nitrogen. Particles with a size of not more than 60 microns were obtained as a result of grinding.

The extraction of crushed raw materials was carried out with water (for organic acids), as well as ethanol and methanol (for other components) according to traditional methods. To determine ceresins insoluble in water and alcohols, 5 ml per 50 ml of benzene was added to the extract gross volume.

Analysis of aqueous and alcoholic extracts of samples obtained by conventional grinding and cryogenic grinding was carried out by high performance liquid chromatography (HPLC) with a ratio of raw materials to extractant 1:50 and atomic absorption spectrometry (AAS) with a ratio of raw materials to extractant 1:25. HPLC analysis was carried out on a Shimadzu LC-20 chromatograph, Prominence series (Japan) with a diode array detector using a S-18 reverse-phase column from Diasfer (Russia) (250x4 mm), injection volume 20 mql. The solvent systems used are water-acetonitrile/methanol and water-acetonitrile/ethanol. Registration of the separated substances was carried out at two wavelengths: 270 nm and 325 nm. The components were identified spectrally in the wavelength range of 220 - 630 nm. AAS was carried out on a Shimadzu model AA-7000 spectrometer.

3. Results

The authors carried out a comparative analysis of the results of extraction of samples obtained by conventional grinding and cryogenic grinding in order to determine the effect of cryogenic grinding of chaga on the content of biologically active substances in it. The results are given in Table 1.

Table 1. Results of High Performance Liquid Chromatography (HPLC) and Atomic Absorption Spectroscopy (ACC) of Chaga Extracts.

Item number	Structural class	Concentration, mcg%	
		Ground sample	*Cryogenic sample, ground*
1	Amino acids:		
	Glycine	not detected	12.77
	Glutamine acid	1.44	14.71
	Asparagine acid	2.06	9.35
	Methionine	1.83	10.60
	Leucine	not detected	7.94
2	Indole-3acetic acid	17.27	24.54
3	Flavonoids	3.91	7.68
4	Hemicellulose	5.11	11.64
5	Lignin	27.33	9.90
6	Polysaccharose	8.57	2.24
7	Acetic acid	not detected	4.66
8	Oxalic acid	4.28	15.06
9	Butyric acid	2.02	8.75
10	Thiols	3.13	11.68
11	Melanin	13.42	24.22
12	Magnesium	0.61	0.64
13	Calcium	0.21	0.36
14	Pentole-1	not detected	7.55
15	Glycerin	not detected	3.99
16	Ash	29.48	7.58

Item number	Structural class	Concentration, mcg%	
		Ground sample	*Cryogenic sample, ground*
17	Ceresin	not detected	1425

The analysis of amino acids in the samples indicates that biologically important compounds such as glycine and leucine were destructed in the process of conventional grinding. Glycine, which is one of the essential amino acids, is a metabolic regulator, normalizing and activating the processes of protective inhibition in the central nervous system, and increasing mental performance. Leucine is one of the most important amino acids, serves as a source of mental energy, stimulates growth hormone, and helps restore the body after injuries and operations. In the case of other amino acids, there is a significant increase in their content in the cryogenic ground sample, in particular, glutamine acid – by 10 times, asparagine acid – by 4.5 times. The amount of methionine, which is an essential amino acid, containing sulfur, which can produce other sulfur-containing molecules in the body and that is participating in the production of protein in the cells of the human body, increases by 6 times.

Indole-3acetic acid that is a substance of high physiological activity, and affects the processes of plant growth is found in both samples. However, the cryogenic ground samples contain 1.4 times more of that acid.

The amount of flavonoids and hemicelluloses increases in the cryogenic ground samples by two times. Flavonoids help to reduce pathologically increased permeability of capillaries, eliminate their fragility and ensure the preservation of ascorbic acid in the body, have a normalizing effect on the lymphatic flow and liver function, and have anti-inflammatory effects. Hemicellulose serves as a source of energy, affects lipid metabolism, plays the role of intestinal sorbent; it absorbs salts of heavy metals, and reduces cholesterol level.

The amount of lignin and polysaccharides in the cryogenic ground samples decreases almost three times. This effect can be explained by the destruction of the chaga lignin-carbohydrate complex in the process of freezing the raw materials.

The chaga components also include such organic acids as acetic, oxalic and butyric, which are crucial in normalizing the acid-base balance, and contribute to the elimination of toxins from the human body. No acetic acid was detected in the ground sample. All three organic acids are present in the cryogenic sample, while their content increased by 3.5 - 4.5 times.

The thiol content in the cryogenic sample increases almost 4 times. This component of chaga is an antioxidant. Thiol based drugs are used to increase the effectiveness of chemical and radiotherapy, and can stimulate the functioning of the human immune system.

The yield of the chromogenic complex is almost doubled, which can be explained by the stage of freezing the milled chaga. Moisture in raw materials under the influence of low temperatures turns into ice crystals, which destroy the cell walls of the fungus. Due to this, extractive substances, including melanins, become more accessible for extraction.

Among the selected components, magnesium and calcium are determined in both samples. However, the cryogenic ground samples show better results (by 5% and 71%, respectively). These elements are bound at the cellular level and, working together, contribute to optimal cellular metabolism.

Pentole is present only in the cryogenic ground sample. It serves as a carrier of a medicinal substance in therapeutic ointments, determines the speed and degree of absorption, affects the transport process through the skin, and contributes to the optimal therapeutic effect of the ointment.

Glycerin is determined only in the sample obtained by cryogenic grinding. It is an important component of the metabolic process in the body, that relates to nutritional supplements.

The ash content in the ground sample studied is more than 29%. As a result of cryogenic grinding, the ash content of raw materials is reduced by 4 times, which is a positive result. The high content of ash substances in chaga-based drugs can have a negative effect on the body, and therefore, reducing their content is a crucial objective.

Ceresin was found only in a cryogenic ground sample. With this component, metabolism and peripheral blood circulation are significantly increasing, which contributes to the absorbable, anti-inflammatory and antispasmodic effects of drugs containing it.

It was found that the sample obtained by cryogenic grinding of raw materials has a high content of valuable biologically active components as a result of a comparative analysis with high-performance liquid chromatography and atomic absorption spectroscopy of water and alcohol extracts of the fungus.

4. Conclusion

The complex of unique properties of chaga mushroom allows it to be widely adopted to obtain therapeutic and prophylactic preparations and food additives used to treat cancer and gastrointestinal diseases, diabetes, excretion of radionuclides, toxins, heavy metals, normalization of metabolism, etc. The following biologically active substances presented in chaga give it the therapeutic effect: a chromogenic polyphenol carbon complex (melanins), microelements, etc. The results of the study proved that the amount of therapeutic components of chaga-based products can be determined by the grinding method of the raw plant material. Thus, the problem of preserving the valuable components of the chaga can be solved by the use of cryogenic grinding technology.

The results of experimental studies showed that cryogenic grinding allows saving a larger number of biologically active substances of chaga compared to raw materials, ground by a conversional technology. Freezing raw materials in liquid nitrogen prevents the processes of oxidation, aggregation and caramelization. The bonds of biologically active substances with protein molecules are destroyed at low temperature, thus effecting a better absorption of the drug. Hence, chaga cryopowder is a product with a high content of valuable components. The data obtained can be of practical importance when using chaga as part of drugs and biologically active additives with targeted functional properties.

Acknowledgements

This work was performed with support of grant of the President of the Russian Federation for state support of young Russian scientists – PhD (MK-2246.2020.8)

References

[1] Shashkina M Ja, Shashkin P N and Sergeev A V 2006 Chemical and medico biological properties of chaga (Inonotus obliquus) *Pharmaceutical Chemistry Journal* (№10(40)) pp 37-44 (In Russ.)

[2] Korsun V F, Krasnopolskaya L M, Korsun E V and Avkhukova M A 2012 Antitumor properties of fungi (Moscow: Practical Medicine) p 210 (In Russ.)

[3] Shashkina M Ja, Shashkin P N and Sergeev A V 2005 Chaga in oncology *Russian Journal of Biotherapy* (№4 (4) pp 59-72 (In Russ.)

[4] Kuznetsova O Yu, Shaehov M F, Ziyatdinova G K and Budnikov H C 2019 Extracts and melanins of Inonotus obliquus obtained after plasma processing of raw materials *Proceedings of Kazan University. Ser. Nat. Sciences* (Vol. 161, Book 2) pp 211-221 (In Russ.)

[5] Safin R R, Razumov E Y, Khasanshin R R, Gubernatorov V V and Saerova K V 2018 Effect of growth conditions on qualitative characteristics of chaga mushroom International *Multidisciplinary Scientific GeoConference Surveying Geology and Mining Ecology Management, SGEM* (№3(18)) pp 781-788

[6] Safin R R, Mukhametzyanov S R and Gubernatorov V V 2019 Water vacuum-oscillating extraction of chaga *IOP Conference Series: Materials Science and Engineering* (№1 (666) 012086)

[7] Lee S H, Hwang H S and Yun J W 2009 Antitumor activity of water extract of a mushroom, Inonotus obliquus, against HT-29 human colon cancer cells *Phytother Res.* (V.23) pp 1784-1789

[8] Kim E J, Lee Y J, Shim H K and Yoon J H 2006 A study on the mechanism by which the aqueous extract of Inonotus oblicuus induces apoptosis and inhibits proliferation in HT-29 human colon cancer cells *J. Korean Soc. Food Sci. Nutr.* (№5(35)) pp 516-523

[9] Ham S S, Kim S H, Moon S Y, Chung M J, Cui C B, Han E K, Chung C K and Choe M 2009 Antimutagenic effects of subfractions of Chaga mushroom (Inonotus obliquus) extract *Mutation Research* (V. 672) pp 55-59

[10] Chung M J, Chung C K, Jeong Y and Ham S S 2010 Anticancer activity of subfractions containing purecompounds of Chaga mushroom (Inonotus obliquus) extract in human cancer cellsand in Balbcc mice bearing sarcoma-180 cells *Nutr. Res. Pract.* (V. 4) pp 177-182.

[11] Shin S H, Kim Y J and Park J 2013 Effect of Cell Proliferation and Apoptosis by Inonotus obliquus in Human Glioblastoma U-87 MG Cells *J. Korean Soc. Food Sci. Nutr.* (№7(42)) pp 1022-1028.

[12] Babitskaya V G, Shcherba V V and Ikonnikova N V 2000 Melanin complex of the fungus Inonotus obliquus Appl. Biochem. Microbiol (№4(36)) pp 377-381

[13] Lee I K, Kim Y S, Jung Y Wand Yun B S 2007 New antioxidant polyphenols from the medicinal mushroom Inonotus obliquus *Bioorg. Med. Chem. Letters* (V. 17) pp 6678-6681

[14] Zhong X,Wang L and Sun D 2011 Effects of Inotodiol Extracts from Inonotus Obliquus on Proliferation Cycle and Apoptotic Gene of Human Lung Adenocarcinoma Cell Line A549 *Chin. J. Integr. Med.* (V. 3) pp 218-223.

[15] Mazurkiewicz W, Rydel K, Podocki D, Lemieszek M K, Langner E and Rzeski W 2010 Separation of an aqueous extract Inonotus obliquus (Chaga). A novel look at the efficiency of its influence on proliferation of A549 human lung carcinoma cells *Acta Poloniae Pharmaceuyica* (№4 (67)) pp 397-406

[16] Mazurkiewicz W 2006 Analysis of aqueous extract of Inonotus obliquus *Acta Poloniae Pharmaceuyica* (№6 (63)) pp 497-501

[17] Kuznetsova O Yu 2016 Review of modern advanced medicinal products containing the biologically active components of chaga mushroom birch *Development and registration of medicines* (№1(14)) pp 128-141 (In Russ.)

[18] Vasser S P 2015 Science of Medicinal Pileate Fungi: Current Perspectives, Advances, Evidence, and Challenges *Biosphere* (№ 2(7)) pp 238-248 (In Russ.)

[19] Britton G 1983 The Biochemistry of natural pigments *Cambridge University Press* p 422

[20] Nakajima Y 2007 Antioxidant small phenolic ingredients in Inonotus obliquus (persoon) Pilat (Chaga) *Chem. Pharm. Bull (Tokyo)* (№2 (55)) pp 1222-1226

[21] Kuriyama I, Nakajima Y, Nishida H, Konishi T, Takeuchi T, Sugawara F, Yoshida H and Mizushina Y 2013 Inhibitory effects of low molecular weight polyphenolics from Inonotus obliquus on human DNA topoisomerase activity and cancer cell proliferation *Asian Pac J Cancer Prev* №3((14)) pp1571-1578

[22] Lu X, Chen H, Dong P, Fu L and Zhang X 2010 Phytochemical characteristics and hypoglycaemic activity of fraction from mushroom Inonotus obliquus *J Sci Food Agric* (№2(90)) pp 276-280

[23] Lee J H and Hyun C K 2014 Insulin-sensitizing and beneficial lipid-metabolic effects of the water-soluble melanin complex extracted from Inonotus obliquus *J Nat Prod* (№1(77)) pp 35-41

[24] Kukulyanskaya T A, Kurchenko N V, Kurchenko V P and Babitskaya V G 2002 Physicochemical properties of melanins produced by the sterile form of Inonotus obliquus («Chagi») in natural and cultivated fungus *Applied Biochemistry and Microbiology* (№1 (38)) pp 58-61 (In Russ.)

[25] Kahlos K, Kangas L and Hiltunen R 1986 Antitumor activity of triterpenes in Inonotus obliquus Planta Medica (№6(52)) pp 554-556

[26] Taylor A, McKenna G and Burlage H 1956 Anticancer activity of plant extracts *Texas Reports on Biology and Medicine* (№4(14)) pp 538-556

[27] Jiang J H, Dou Y, Feng Y J and Bondartseva M A 2007 The anti-tumor activity and MDR reversal properties of constituents from inonotus obliquus *Mycology and phytopathology* (№5 (41)) pp 455-460

[28] Osetsky A I, Grischenko V I, Goltsev A N, Kravchenko M A and Stryuchkova E V 2009 Cryogenic technologies in production of pharmaceutical, cosmetic, agrotechnical formulations and biologically active food additives *Problems of Cryobiology* (№4(19)) pp 488-499 (In Russ.)

[29] Pavlyuk R Yu 2000 Advanced technologies of dietary supplements from pollen and plant materials *Monograph* (Kiev) p133 (In Russ.)

[30] Prosvirnikov D B, Baigildeeva E I, Sadrtdinov A R and Fomin A A 2017 Modelling heat and mass transfer processes in capillary-porous materials at their grinding by pressure release

Industrial Engineering, Applications and Manufacturing (ICIEAM) 2017 International Conference on (IEEE) pp 1-7

[31] Pevneva O P and Schegolev A A 2015 Features of technology of microdispersion vegetable materials at negative temperatures and used in cosmeceuticals and industrial biotechnology *Forests of Russia and their economy* (№1(52)) pp 52-54 (In Russ.)

[32] Galyavetdinov N R, Safin R R, Voronin A E and Shaikhutdinova A R 2015 The mathematical modeling of the wood greenery extraction processes *Proceedings of 2015 International Conference on Mechanical Engineering, Automation and Control Systems, MEACS* (7414885)

IOP Conf. Series: Materials Science and Engineering **986** (2020) 012030 doi:10.1088/1757-899X/986/1/012030

Characterization of a 17-4 PH stainless steel obtained through metal powder Hot Isostatic Pressing process for automotive application

M Dabala[1] **and M Polyakova**[2]

[1]University of Padova, Department of Industrial Engineering, Via Marzolo 9, Padova 35131, Italy
[2]Nosov Magnitogorsk State Technical University, Department of Materials Processing, Lenin Avenue 38, Magnitogorsk 455000, Russia, m.polyakova-64@mail.ru

Abstract. Steels of the family 17-4 Precipitation Hardening stainless steels are often considered as potential candidates for applications requiring high levels of strength (typically above 600 MPa in tensile strength) and a moderate ductility. The aim of this work is the characterization of a 17-4 precipitation hardening stainless steel obtained through metal powder Hot Isostatic Pressing process. The main purposes of using Hot Isostatic Pressing process are the better chemical composition and microstructure homogeneity, the finer grain size (allowing to higher mechanical properties), and the possibility to obtain net shape products, avoiding subsequent finishing operations. The Hot Isostatic Pressing process temperature does not seem to significantly influence the tensile properties, while, the aging temperature is a key factor: lower aging temperature brings to higher strength and lower ductility. However, a higher influence of the Hot Isostatic Pressing process temperature was detected on the toughness. Improved toughness both at room and at low temperature was obtained by optimizing the aging treatments. An improvement of 32 % in impact energy was obtained by the optimization of the aging process thanks to a higher austenite content, which has shown a nanosized shape between the martensite platelets, in the steel structure.

1. Introduction

17-4 PH steel is a martensitic precipitation hardening (PH) stainless steel which offers a combination of high mechanical strength up to a temperature of 300°C, good corrosion resistance and good toughness, thanks to the possibility of a wide selection of heat treatments. It also exhibits excellent weldability and good workability. The great advantage of these steels is that they can be supplied in the "solution treated" condition. In this condition the steel is just machineable. Following machining, forming etc. the steel can be hardened by a single, fairly low temperature "ageing" heat treatment which causes no distortion of the component All these qualities make the 17-4 PH a valid solution in specific sectors: from aerospace, to chemical and petrochemical, to food, to the mechanical industry in general. [1]

The hot isostatic pressing process (HIP) consists in the simultaneous application of high pressure and high temperature to the component in special vessels. Because the pressure is applied through an inert gaseous medium, the process is denoted as isostatic pressing. This process is a valid alternative to more traditional processes, such as casting, and forging.

Among the main advantages over the latter are:

• better chemical composition homogeneity;
• better microstructural uniformity, which leads to isotropic properties;
• a finer grain microstructure, which improve mechanical properties.

Although process costs are higher, mainly due to the preliminary stages, part of these are recovered considering the minimum waste of associated material. In fact, with this technology, near net shape or even net shape products are obtained, thus avoiding to carry out all the finishing operations downstream of the traditional methods. Thanks to these advantages extended research was performed on the HIP process applied on many metals and alloys [2-5]. Moreover HIP is very effective for compacting different kinds of metal wastes in the form of granules and pellets [6-9]. Furthermore it was shown that high pressure and temperature during HIP enables to increase the level of mechanical properties of the forgings [10, 11].

Aim of this work is to characterize the mechanical properties and the microstructure of a 17-4 precipitation hardening stainless steel obtained through metal powder HIP and to investigate the effect of HIP process temperature and aging temperature on the properties of the steel.

2. Experimental part

In this study two kinds of 17-4 steel powders obtained by gas atomization were used. The powder A was obtained using argon as the gas atomizer while the powder B was atomized by nitrogen. The chemical composition of the two powders are summarized in Table 1.

Table 1. Chemical composition of 17-4 steel powders and of the massive steel.

Powder	C (%)	Cr (%)	Ni (%)	Cu (%)	Nb + Ta (%)	N (%)	Si (%)
A	0.05	16.1	3.6	3.5	0.31	0.01	0.48
B	0.06	17.1	4.5	3.2	0.35	0.03	0.58
Steel	0.05	16.7	4.6	3.4	0.33	0.01	0.55

Four cylindrical capsules of mild steel with 150 mm high and 180 mm in diameter were filled in a protected atmosphere, two with powder A and two with powder B. The filled capsules were degassed at temperatures of about 573 K and pressures of about 1 Pa. At the end of the degassing, the capsules were sealed using a crimping tool.

HIP process was performed at temperature of 1273 and 1313 K. After the corresponding pressing treatment, each billet was divided into 2 portions and each portion was submitted to aging at 2 different temperatures (813 and 873 K) for 4 hours.

The specimens necessary for the tests were extracted from each heat-treated portion: 3 specimens for the tensile test and 3 specimens for notched bar impact testing. Optical microscopy analysis was used to study the surface of the samples. The tensile tests were performed in a Schenck Hydropuls PSA 100 machine, at room temperature. The Charpy impact tests were performed with V-notched samples prepared as reported in accordance with standard ASTM E23. The grain size measurements were performed in accordance with standard ASTM E112.

3. Results and Discussion

The results of tensile tests for each material are summarized in Table 2 and compared with the same massive steel grade treated under the same conditions.

Table 2. Average results from tensile tests of hipped samples and massive steel.

Powder	HIP Temperature* (K)	Aging Temperature (K)	Yield Strength (0.2%) (MPa)	Tensile Strength (MPa)	Elongation (%)
A	1273	813	913 ± 4	983 ± 3	11.9 ± 0.1
A	1273	873	738 ± 1	898 ± 2	12.3 ± 0.2
A	1313	813	836 ± 3	903 ± 4	12.4 ± 0.1
A	1313	873	709 ± 3	862 ± 4	13.0 ± 0.1
B	1273	813	962 ± 3	1028 ± 5	11.8 ± 0.2
B	1273	873	849 ± 4	927 ± 7	12.9 ± 0.1
B	1313	813	955 ± 4	1032 ± 3	12.2 ± 0.1
B	1313	873	802 ± 1	931 ± 3	12.5 ± 0.1
Massive steel	1313	813	1000 ± 15	1070 ± 17	12.3 ± 0.4
Massive steel	1313	873	795 ± 17	965 ± 11	14.6 ± 0.3

* for massive steel the temperature reported in the first column is the solution temperature.

The dispersion of the tensile tests results is very low, especially the results concerning the yield and tensile strengths. This behaviour is frequent in materials obtained in powder metallurgy thanks to their high compositions homogeneity and to an almost absence of segregations and to the high microstructural isotropy intrinsic to the HIP process [11].

The mechanical properties obtained by HIP processing of the two powders are close to the mechanical properties of the massive steel. It can be noted that samples obtained from powder B shows higher resistance than the corresponding samples obtained from powder A. The reason can be due to the higher amount of alloying elements (especially nitrogen) in powder B, which acts as a reinforcement of the matrix [12]. Instead, the difference in elongation percentage is negligible.

The temperature of HIP processing does not affect significantly the mechanical properties. In fact, an increasing in HIP temperature produces a very slight drop in the range between 4-1.5% in yield strength and a small increase in the range between 4 and 0.7% in elongation percentage. These ranges can be considered negligible.

Aging temperature has a significant effect on the mechanical properties. In fact, an increase of 60 K in aging temperature produces an increase in yield strength of about 20% and a reduction in elongation percentage of about 9%.

The results of impact tests are summarized in Table 3 and, as for the results of tensile tests, they are compared with the same massive steel grade treated under the same condition.

Table 3. Average results from Chapry tests of samples after HIP and massive steel.

Powder	HIP Temperature* (K)	Aging Temperature (K)	Fracture Energy (Joule)
A	1273	813	40 ± 4
A	1273	873	33 ± 3
A	1313	813	51 ± 3
A	1313	873	44 ± 2
B	1273	813	39 ± 3
B	1273	873	31 ± 2
B	1313	813	52 ± 3
B	1313	873	45 ± 2
Massive steel	1313	813	56 ± 5
Massive steel	1313	873	68 ± 5

* for massive steel the temperature reported in the first column is the solution temperature.

In this case the impact resistance of the samples after HIP is lower than the resistance of the massive steel. Although by HIP it is possible to eliminate all the porosities, the interfacial bonds between the particles which are created during the process may not be sufficient to guarantee performance similar to massive material due to the impurities that can be present on the particles surface [13]. It can be noted that the difference between the samples produced by different powders is negligible. This is because the difference in alloying elements and impurities is not sufficient to generate different paths for the cracks growth.

The temperature of HIP processing affects the impact resistance of the materials. In fact, the treatment at higher temperature leads to the increase of the fracture energy, especially at room temperature. However, the aging treatment has a larger effect on the impact energy. In fact an increase of 60 K in aging temperature produces a reduction of fracture energy in the range between 20% and 30%.

The microstructure of the samples obtained from the powder A are reported in Fig. 1 and the microstructure of the samples obtained from powder B are reported in Fig. 2.

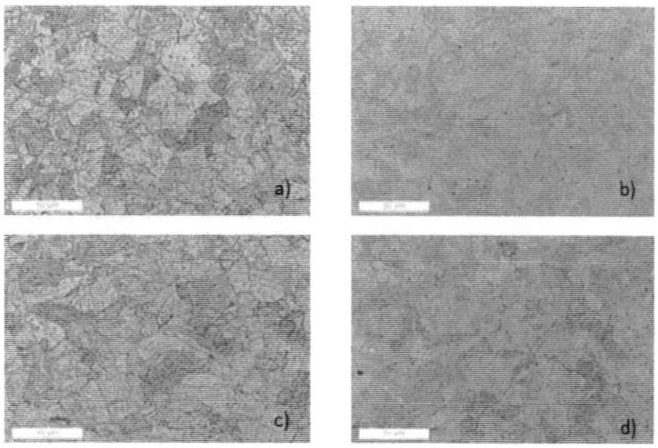

Figure 1. Microstructure of samples obtained from powder A, x1000:
(a) HIP temperature 1273 K, aging temperature 813 K;
(b) HIP temperature 1273 K, aging temperature 873 K;
(c) HIP temperature 1313 K, aging temperature 813 K;
(d) HIP temperature 1313 K, aging temperature 873 K.

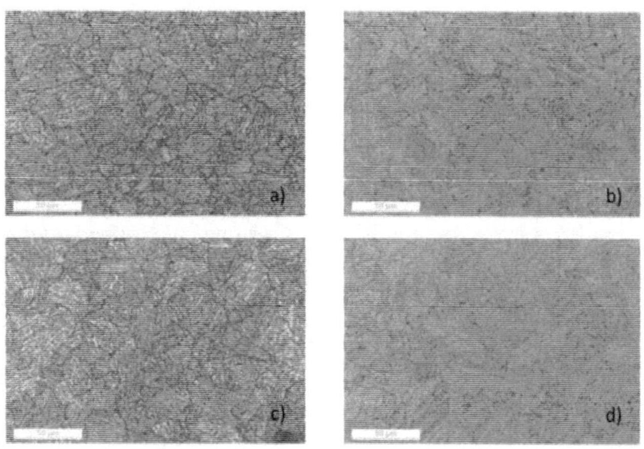

Figure 2. Microstructure of samples obtained from powder B, x1000:
(a) HIP temperature 1273 K, aging temperature 813 K;
(b) HIP temperature 1273 K, aging temperature 873 K;
(c) HIP temperature 1313 K, aging temperature 813 K;
(d) HIP temperature 1313 K, aging temperature 873 K (a) first picture; (b) second picture.

It can be noted that there is a total absence of porosity. As expected, the isostatic pressing treatment leads to a relative density of about 1. The microstructure is, in all the samples fully martensitic, characterized by small plate martensite, similar to the massive steel microstructure. There are no traces of delta-ferrite. The structure is equiassic and homogeneous. Samples obtained from powder B show a higher amount of precipitates at the martensite plates boundaries, probably due to a higher percentage of carbon. This justifies the increase of strength of the samples after HIP processing obtained from the powder B.

Average grain size analysis shows that the level of temperature in HIP processing do not influence on the microstructure, while the aging temperature is more important: in fact, the average grain size is increased from 70-80 μm to a range 130-150 μm with the increase of the aging temperature to 40 K. This correlates with the reduction in fracture energy observed in the samples aged at higher temperature.

4. Conclusions

The 17-4 PH stainless steel obtained through metal powder Hot Isostatic Pressing process, especially for the material obtained from powder B, have microstructure with higher amount of precipitates at the martensite plates boundaries.

Material treated at higher aging temperature is more deformable, while one treated at lower aging temperature has higher level of strength but reduction in ductility. Conversely, the temperature of HIP processing does not significantly influence the tensile properties.

Due to high chemical homogenization, low segregation, and high microstructural isotropy obtained after HIP process the results of tensile tests show minimal dispersions in mechanical properties, especially of tensile strength. The impact properties of 17-4 PH stainless steel obtained from powders are lower as compared with the massive material.

The influence of the HIP processing temperature is more relevant than the results of tensile tests. However, aging treatment at different temperatures has higher effect on impact properties. Fracture energy is higher to 20-30% for the material treated at lower temperature than for the same material treated at higher aging temperature. It can be explained by the increase of grain size at higher aging temperature.

References

[1] Jui-Hung Wu, Chih-Kuang Lin 2002 *Metall. Mater. Trans.* (33a(6)) pp 1715-1724.
[2] Atkinson H V and Davies S (2000) *Metall. Mater. Trans. A.* (31(12)) pp 2981-3000.
[3] Broeckmann C (2012) *Powder Metall.* (55) pp 176-179.
[4] . GentzbittelI J M and Chu H. Burlet J 2002 *Nucl. Mater.* (307(1)) pp 540-543.
[5] Swinkels F B, Wilkinson D., Arzt E., Ashby F.M. (1983) *Acta Metall* (31(11) pp 1829-1840.
[6] Zagirov N.N., Kovaleva A A, Olkhovik A Ya and Koval O E 2015 *Vestnik of Nosov Magnitogorsk State Technical University* (2) pp 59-64.
[7] Zagirov N N, Loginov Yu N and Ivanov E V 2016 *Vestnik of Nosov Magnitogorsk State Technical University* (4) pp 50-56.
[8] Belokopytov V I 2015 *Vestnik of Nosov Magnitogorsk State Technical University* (3) pp64-69.
[9] Zagirov N N, Ivanov E V,. Kovaleva A A and Anikina V 2013 *Vestnik of Nosov Magnitogorsk State Technical University.* (2(42) pp 64-68.
[10] Usynina G P, Zagirov N N and Anikina V I 2012 *Vestnik of Nosov Magnitogorsk State Technical University* (3(39) pp 9-13.
[11] Kazior J, Szewczyk-Nykiel A, Pieczonka T, Hebda M and Nykiel M 2013 *Adv. Mater. Res.* (811) pp 87-92.
[12] Laurent C, de Geuser F, Descoins M and Deschamps A 2016 *Mater. Des.* (107(5)) pp 416-425.
[13] Misra R D K, Prasad C Y, Balasubramanian T V and Rama Rao P 1987 *Scr. Metall.* (21(8)) pp 1067-1070.

IOP Conf. Series: Materials Science and Engineering **986** (2020) 012031 doi:10.1088/1757-899X/986/1/012031

A Simplified Yield Criterion of the Orthotropic Material Considering the Crystallographic Texture

Y Erisov [1,2], F Grechnikov [2] S Surudin[1] and V Razhivin [1]

[1]Metal Forming Department, Samara National Research University, Moskovskoye shosse 34, Samara, 443086, Russia, yaroslav.erisov@mail.ru (corresponding author)
[2]Samara Federal Research Center of Russian Academy of Sciences, Studencheskii pereulok 3A, Samara, 443001, Russia

Abstract. The proposed linearized yield function considers the crystal lattice constants and the parameters of crystallographic orientation of material and has simplified form, which makes it suitable for technological calculations of sheet metal forming processes. It is shown that the Lode coefficient is determined not only by the stress state, as in the case of an isotropic material, but also depends on anisotropy parameters, which reflects the elastic constants of the crystal lattice and the crystallographic orientation of the structure.

1. Introduction

The current development stage of the technologies for aircraft production, automobile construction, shipbuilding and other industries is characterized by the continuous improvement of the existing constructional materials and technological processes of their forming, as well as by the research and development of new ones. The development of a rational, science-based technology in the metal forming processes is primarily concerned with the need of a detailed study of the structure (texture) formation, the material properties and the most extensive application of these properties in engineering analysis.

One of the specific characteristics inherent in the majority of materials is the anisotropy of their properties, which is caused by the crystallographic structure and texture formation under high plastic strains [1]. However, the assumption of the material isotropy is still being used as one of the main hypotheses in the analysis and calculations of metal forming processes. Although this hypothesis facilitates the solution of numerous metal forming problems, it does not actually meet the real deformation conditions.

The abandonment of the assumption of the medium isotropy allows to generalize the metal forming theory and to use the anisotropy effectively in technological processes [2-3]. Despite the fact that over the recent years greater attention is paid to the theoretical and experimental research in the field of anisotropic plastic deformations, there is still a number of unsolved problems. These issues are related primarily to the further development of the anisotropic plasticity theory in a form suitable for the engineering and technological analysis.

Till nowadays the most widely used yield criterion was proposed by R. Hill [4]. Verification of these yield function proved its applicability only for steels [5], so for other materials, R. Hill proposed a non-quadratic yield criterion [6]. However, it does not take into account the shear stresses and, therefore, it is only applicable in the case when the direction of the principal stresses coincides with the anisotropy axes. This disadvantage was eliminated by F. Barlat [7].

For the general stress state, F. Barlat developed the yield function, in which the stress tensor is expressed in terms of the coordinates parallel to the anisotropy axes [8]. Using two linear transformations of stress tensor F. Barlat obtained the anisotropic criterion for plane and general stress state for the materials with the equal yield strengths of tension and compression [9-10].

Another approach is the replacement of initial isotropic invariants of the stress deviator by the similar generalized anisotropic invariants, derived on the basis of the theory of representations of tensor functions [11-12].

Thus, the modern yield functions in contrast to the earlier ones, which were derived to describe the behavior of any metal, allow taking into account the peculiarities of the specific materials. It should be noted that the high accuracy of the recently proposed criterions is achieved by a large amount of the anisotropy coefficients (up to 18), determination of which involves numerous mechanical tests at different stress states [13].

Though the applied anisotropy coefficients characterize the anisotropy of plastic deformations, they do not take into account the reason of anisotropy, i.e. the crystallographic texture. Thus, the mentioned yield functions, on the one hand, allow describing the plastic flow of anisotropic materials. On the other hand, they do not allow carrying out technological analysis of metal forming processes considering the crystallographic texture. As a result, it is impossible to determine the composition of crystallographic texture in terms of the requirements of certain metal forming operations.

As shown in [14], to embed the parameters of crystallographic texture into the yield function it can be used the hypothesis of proportionality of the elastic and plastic material deviators. Alternative approach is to use the specific distortion strain energy to derive the yield criterion and the constitutive equations, which consider the crystal lattice constants and parameters of the crystallographic texture [15-19]. The main practical significance of this yield criterion is the possibility to predict the effect of crystallographic texture on the metal flow during plastic deformation. Consequently, it allows designing the composition of crystallographic texture depending on the requirements of the metal forming processes.

However, in most cases, for the analysis of metal forming processes it is convenient to use a simplified form of the yield criterion, since, on the one hand, technological calculations do not require high accuracy, and, on the other hand, it is necessary to simplify the derivation of analytical dependencies. In this regard, in this paper it is proposed the linearized yield function for the orthotropic material with consideration of the crystal lattice constants and parameters of the crystallographic texture.

2. General anisotropic yield criterion

Let us use the yield criterion, in the basic equations of which the parameters of the material's structure are introduced [15]:

$$\sigma_i = \frac{1}{\sqrt{2}}\left\{ \eta_{12}(\sigma_{11}-\sigma_{22})^2 + \eta_{23}(\sigma_{22}-\sigma_{33})^2 + \eta_{31}(\sigma_{33}-\sigma_{11})^2 + \right.$$
$$\left. +4\left[\left(\frac{5}{2}-\eta_{12}\right)\sigma_{12}^2 + \left(\frac{5}{2}-\eta_{23}\right)\sigma_{23}^2 + \left(\frac{5}{2}-\eta_{31}\right)\sigma_{31}^2\right]\right\}^{1/2}, \quad (1)$$

where σ_i – equivalent stress; σ_{ij} – the components of the stress tensor; (i, j = 1, 2, 3; 1 – the rolling direction, 2 – transverse direction; 3 – the normal direction); η_{ij} – generalized anisotropy parameters:

$$\eta_{ij} = 1 - \frac{15(A'-1)}{3+2A'}\left(\Delta_i + \Delta_j - \Delta_k - \frac{1}{5}\right); \quad (2)$$

A' – the parameter of anisotropy of the crystal lattice:

$$A' = \frac{S'_{1111} - S'_{1122}}{2S'_{2323}}; \tag{3}$$

S'_{ijkl} – elastic constants of the crystal lattice; Δ_i – parameters of crystallographic orientation of the structure:

$$\Delta_i = \sum_{\{hkl\}\langle uvw\rangle} p^{\{hkl\}\langle uvw\rangle} \Delta_i^{\{hkl\}\langle uvw\rangle}; \tag{4}$$

$p^{\{hkl\}\langle uvw\rangle}$ – weight fraction of *i*-th component $\{hkl\}\langle uvw\rangle$; $\Delta_i^{\{hkl\}\langle uvw\rangle}$ – orientation factor of ideal crystallographic orientation $\{hkl\}\langle uvw\rangle$:

$$\Delta_i^{\{hkl\}\langle uvw\rangle} = \frac{h_i^2 k_i^2 + k_i^2 l_i^2 + l_i^2 h_i^2}{\left(h_i^2 + k_i^2 + l_i^2\right)^2}; \tag{5}$$

h_i, k_i, l_i – Miller indices determining the *i*-th direction in the crystal relative to the coordinate system associated with the sample.

Linearized anisotropic yield criterion

In solving specific problems of metal forming, a simplified linearized form of the yield criterion is widely used:

$$\sigma_{max} - \sigma_{min} = \beta \sigma_i, \tag{6}$$

where σ_{max} and σ_{min} are the major and minor principal stresses ($\sigma_{max} > \sigma_{av} > \sigma_{min}$). β is Lode coefficient, which characterizes the influence of the average stress σ_{av} and is determined by the magnitude of the stress state factor $\beta = f\left(\nu_\sigma\right)$:

$$\nu_\sigma = \frac{\sigma_{av} - \dfrac{\sigma_{max} + \sigma_{min}}{2}}{\dfrac{\sigma_{max} - \sigma_{min}}{2}}. \tag{7}$$

If an isotropic material is considered, then β does not depend on the choice of the directions of the principal axes and is determined only by the stress state.

In case of an orthotropic medium Lode coefficient is dependent on the direction of the stresses relative to the principal anisotropy axes. To derive a simplified expression of the yield criterion of an orthotropic medium in the form (6), it is necessary to consider possible options for the mutual arrangement of the principal stresses relative to the anisotropy axes. For the case $\sigma_1 > \sigma_2 > \sigma_3$ Eq. (7) can be rewritten in the following form:

$$\sigma_2 = \nu_\sigma \frac{\sigma_1 - \sigma_3}{2} + \frac{\sigma_1 + \sigma_3}{2}. \tag{8}$$

Substituting Eq. (8) into the yield criterion (1), we obtain:

$$\sigma_1 - \sigma_3 = \frac{2\sqrt{2}\sigma_i}{\sqrt{\eta_{12}\left(\nu_\sigma - 1\right)^2 + \eta_{23}\left(\nu_\sigma + 1\right)^2 + 4\eta_{31}}}. \tag{9}$$

Then the Lode coefficient is determined by the dependence:

$$\beta = \frac{2\sqrt{2}}{\sqrt{\eta_{12}(\nu_\sigma - 1)^2 + \eta_{23}(\nu_\sigma + 1)^2 + 4\eta_{31}}}. \quad (10)$$

Thus, the value of the Lode coefficient is determined not only by the magnitude of the stress state factor ν_σ, as in the case of an isotropic material, but also depends on parameters η_{ij}, which reflects the elastic constants of the crystal lattice and the crystallographic orientation of the structure.

Similarly, considering other combinations of arrangement of stresses σ_{max}, σ_{min} and σ_{av} with respect to the principal anisotropy axes 1, 2, and 3, we obtain the remaining expressions for calculating the Lode coefficient β (Table 1). The six cases of expressions for the Lode coefficient, which are indicated in Table 1, can be represented as a general expression:

$$\beta = \frac{2\sqrt{2}}{\sqrt{\eta_{max.av}(\nu_\sigma - 1)^2 + \eta_{av.min}(\nu_\sigma + 1)^2 + 4\eta_{min.max}}}, \quad (11)$$

here η_{ij} are the generalized anisotropy parameters, where the indices *i*, *j* are determined by the location of the stresses relative to the principal anisotropy axes. For example, when $\sigma_2 > \sigma_3 > \sigma_1$ the indices are following: $max = 2$, $min = 1$ and $av = 3$. The order of the indices does not matter: $\eta_{ij} = \eta_{ji}$. In the case of isotropic material ($\eta_{ij} = 1$), Eq. (11) takes the traditional form:

$$\beta = \frac{2}{\sqrt{3 + \nu_\sigma^2}}. \quad (12)$$

Table 1. Expressions for calculating the Lode coefficient

Stress state	$\beta = f(\eta_{ij}, \nu_\sigma)$
$\sigma_2 > \sigma_1 > \sigma_3$	$\beta = \frac{2\sqrt{2}}{\sqrt{\eta_{12}(\nu_\sigma - 1)^2 + \eta_{31}(\nu_\sigma + 1)^2 + 4\eta_{23}}}$
$\sigma_3 > \sigma_1 > \sigma_2$	$\beta = \frac{2\sqrt{2}}{\sqrt{\eta_{31}(\nu_\sigma - 1)^2 + \eta_{12}(\nu_\sigma + 1)^2 + 4\eta_{23}}}$
$\sigma_1 > \sigma_2 > \sigma_3$	$\beta = \frac{2\sqrt{2}}{\sqrt{\eta_{12}(\nu_\sigma - 1)^2 + \eta_{23}(\nu_\sigma + 1)^2 + 4\eta_{31}}}$
$\sigma_3 > \sigma_2 > \sigma_1$	$\beta = \frac{2\sqrt{2}}{\sqrt{\eta_{23}(\nu_\sigma - 1)^2 + \eta_{12}(\nu_\sigma + 1)^2 + 4\eta_{31}}}$
$\sigma_1 > \sigma_3 > \sigma_2$	$\beta = \frac{2\sqrt{2}}{\sqrt{\eta_{31}(\nu_\sigma - 1)^2 + \eta_{23}(\nu_\sigma + 1)^2 + 4\eta_{12}}}$
$\sigma_2 > \sigma_3 > \sigma_1$	$\beta = \frac{2\sqrt{2}}{\sqrt{\eta_{12}(\nu_\sigma - 1)^2 + \eta_{31}(\nu_\sigma + 1)^2 + 4\eta_{12}}}$

3. Results and discussion

Let us determine the value of the Lode coefficient for the most common cases of stress state (uniaxial compression – $\nu_\sigma = 1$; pure shear – $\nu_\sigma = 0$; uniaxial tension – $\nu_\sigma = -1$):

$$\begin{aligned}
&\nu_\sigma = 1,\ \beta = \frac{\sqrt{2}}{\sqrt{\eta_{av.\min} + \eta_{\min.\max}}};\\
&\nu_\sigma = 0,\ \beta = \frac{2\sqrt{2}}{\sqrt{\eta_{\max.av} + \eta_{av.\min} + 4\eta_{\min.\max}}};\\
&\nu_\sigma = -1,\ \beta = \frac{\sqrt{2}}{\sqrt{\eta_{\max.av} + \eta_{\min.\max}}}.
\end{aligned} \tag{13}$$

It follows from Eq. (13) that in the case of an orthotropic material, the value of the Lode coefficient β on a interval $\nu_\sigma \in [-1;1]$ can vary within other limits than for an isotropic medium, when $\beta \in \left[1; \frac{2}{\sqrt{3}}\right]$, and depends on the elastic constants of the crystal lattice and texture parameters.

When analyzing the sheet metal forming processes, in case of relatively small average stress, the β value is taken equal to the mean value of the Lode coefficient $\beta = f(\eta_{ij}, \nu_\sigma)$ on the interval $\nu_\sigma \in [-1;1]$:

$$\bar{\beta} = \frac{1}{2}\int_{-1}^{1} f(\eta_{ij}, \nu_\sigma)\, d\nu_\sigma. \tag{14}$$

Substituting Eq. (11) in (14), we obtain:

$$\bar{\beta} = \frac{\sqrt{2}}{\sqrt{\eta_{\max.cp} + \eta_{cp.\min}}} \ln \frac{\sqrt{\eta_{12}\eta_{23}\eta_{31}N + \eta_{cp.\min}^2} + \eta_{cp.\min}}{\sqrt{\eta_{12}\eta_{23}\eta_{31}N + \eta_{\max.cp}^2} - \eta_{\max.cp}}, \tag{15}$$

where $N = \frac{1}{\eta_{12}} + \frac{1}{\eta_{23}} + \frac{1}{\eta_{31}}$.

It is easy to show that the $\bar{\beta}$ value depends only on the location of the average stress σ_{av} relative to the principal anisotropy axes, the location of $\sigma_{\max}$ and $\sigma_{\min}$ can be arbitrary. Consider two cases:

1) $\sigma_1 > \sigma_2 > \sigma_3$, then Eq. (15) can be written as follows

$$\bar{\beta}_1 = \frac{\sqrt{2}}{\sqrt{\eta_{12} + \eta_{23}}} \ln \frac{\sqrt{\eta_{12}\eta_{23}\eta_{31}N + \eta_{23}^2} + \eta_{23}}{\sqrt{\eta_{12}\eta_{23}\eta_{31}N + \eta_{12}^2} - \eta_{12}}; \tag{17}$$

2) $\sigma_3 > \sigma_2 > \sigma_1$, then Eq. (15) can be written as follows

$$\bar{\beta}_2 = \frac{\sqrt{2}}{\sqrt{\eta_{12} + \eta_{23}}} \ln \frac{\sqrt{\eta_{12}\eta_{23}\eta_{31}N + \eta_{12}^2} + \eta_{12}}{\sqrt{\eta_{12}\eta_{23}\eta_{31}N + \eta_{23}^2} - \eta_{23}}. \tag{18}$$

Equating the right-hand sides of Eq. (17) and (18), we obtain that $\bar{\beta}_1 = \bar{\beta}_2$.

In sheet metals, crystallographic orientations are not random but show preferred orientations around one or several components. After rolling, for FCC alloy sheets, these components are mainly {112}<111> (copper), {110}<112> (brass), {123}<634> (S) and {100}<011> (rotated cube) orientations, with the relative amounts dependent on rolling conditions. After annealing, depending on the process conditions, the resulting texture mainly composed of the {100}<001> (cube) and {110}<001> (Goss) recrystallization components.

Consider the sheet from copper for which components of compliance tensor S'_{ijkl} are $S'_{1111}=15.0$ TPa^{-1}; $S'_{1122}=-6.30$ TPa^{-1} and $S'_{2323}=3.33$ TPa^{-1} [16], i.e. $A'=1.207$ (Eq. (3)). The parameters of crystallographic texture and the generalized anisotropy parameters calculated using Eq. (5) and (2) for stated components are listed in Table 2.

Table 2 The parameters of crystallographic texture and the generalized anisotropy parameters of single ideal components

Component		The parameters of crystallographic texture			The generalized anisotropy factors		
Name	Orientation	Δ_1	Δ_2	Δ_3	η_{12}	η_{23}	η_{31}
Copper	{112}<111>	0.333	0.250	0.250	0.924	1.019	0.924
Brass	{110}<112>	0.250	0.333	0.250	0.924	0.924	1.019
S	{123}<634>	0.281	0.278	0.250	0.938	0.973	0.970
Rotated cube	{100}<011>	0.250	0.250	0.0	0.828	1.115	1.115
Cube	{100}<001>	0.0	0.0	0.0	1.115	1.115	1.115
Goss	{110}<001>	0.0	0.250	0.250	1.115	0.828	1.115
Isotropy	-	0.20	0.20	0.20	1.0	1.0	1.0

Table 3 shows the results of calculating the mean Lode coefficient $\bar{\beta}$ for typical crystallographic orientations.

Table 3. The value of the mean Lode coefficient $\bar{\beta}$ for typical crystallographic orientations

Component		Average stress σ_{av}		
Name	Orientation	σ_1	σ_2	σ_3
Copper	{112}<111>	1,167	1,376	1,376
Brass	{110}<112>	1,376	1,167	1,376
S	{123}<634>	1,238	1,245	1,315
Rotated cube	{100}<011>	0,960	0,960	1,387
Cube	{100}<001>	0,842	0,842	0,842
Goss	{110}<001>	1,387	0,960	0,960
Isotropy	-	1,099	1,099	1,099

From the point of view of the analysis of sheet metal forming processes, the case where the average stress is directed along the normal to the sheet surface is of most interest. As can be seen from Table 3, in this case, the crystallographic orientations of the deformation type provide higher values of the mean Lode coefficient, and hence – higher flow stresses, in comparison with the isotropic case, and the orientations of recrystallization type – lower.

4. Conclusions

The phenomenological yield criterions applied in technological analysis of the metal forming processes do not take into account the crystallographic texture of materials. The proposed linearized yield function considers the crystal lattice constants and the parameters of crystallographic orientation of material and has simplified form, which makes it suitable for technological calculations of sheet metal forming processes. The linearized yield criterion is applicable to orthotropic materials with cubic crystal lattice in case when texture does not change, for example, under small elastic-plastic strains. The proposed yield criterion allows determining the characteristics of the metal forming processes considering the texture of sheet metal. Consequently, it allows designing the composition of crystallographic texture depending on the requirements of the metal forming processes or products performance.

The study was carried out as part of a state assignment (FSSS-2020-0016).

References

[1] Truszkowski, W. (2001), The Plastic Anisotropy in Single Crystals and Polycrystalline Metals, Springer, Netherlands.
[2] Engler, O. and Hirsch, J., (2002) Texture control by thermomechanical processing of AA6xxx Al-Mg-Si sheet alloys for automotive applications - a review, Materials Science and Engineering A, 336, 249-262.
[3] Hirsch, J. and Al-Samman, T. (2013) Superior light metals by texture engineering: Optimized aluminum and magnesium alloys for automotive applications, Acta Materialia, 61, 818–843.
[4] Hill, R. (1948), A theory of the yield and plastic flow of anisotropic metals, Proc. Roy. Soc. London. Ser A., 193, 281-297.
[5] Woodthorpe, J. and Pearce, R. (1970), The anomalous Behavior of Aluminum Sheet under Balance Biaxial Tension, Int. J. Mech. Sci., 12, 341-347.
[6] Hill, R. (1979), Theoretical plasticity of textured aggregates, Math. Proc. Cambridge Philos. Soc., 85, 179-191.
[7] Barlat, F. and Lian, J.A. (1989), Plastic Behavior and Stretchability of Sheet Metals. Part 1: Yield Function for Orthotropic Sheets under Plane Stress Conditions, Int. J. Plasticity, 5, 51-66.
[8] Barlat, F., Lege, D.J. and Brem, J.C. (1991), A six-component yield function for anisotropic materials, Int. J. Plasticity, 7, 693-712.
[9] Barlat, F., Brem, J.C., Yoon, J.W., Chung, K., Dick, R.E., Lege, D.J., Pourboghrat, F., Choi, S.-H. and Chu, E. (2003), Plane stress yield function for aluminum alloy sheet. Part 1: Theory, Int. J. Plasticity, 19, 1297-1319.
[10] Barlat, F., Aretz, H., Yoon, J.W., Karabrin, M.E., Brem, J.C. and Dick, R.E. (2005), Linear transformation-based anisotropic yield functions, Int. J. Plasticity, 21, 1009-1039.
[11] Cazacu, O. and Barlat, F. (2001), Generalization of Drucker's yield criterion to orthotropy, Mathematics and Mechanics of Solids, 6, 613-630.
[12] Cazacu, O. and Barlat, F. (2003), Application of representation theory to describe yielding of anisotropic aluminum alloys, Int. J. of Engng. Sci., 41, 1367-1385.
[13] Soare, S. and Banabic, D. (2008), About mechanical data required to describe the anisotropy of thin sheets to correctly predict the earing of deep-drawn cups, Int. J. Plasticity, 4, 34-37.
[14] Aryshenskii, V.Yu., Grechnikov, F.V. and Zaitsev, V.M. (1990), The determination of the material plastic deviator of the anisotropic medium by its texture parameters, Izvestia Akademii nauk SSSR. Metally, 4, 158-162. (in Russ.)
[15] Erisov, Y.A., Grechnikov, F.V. and Surudin, S.V. (2016) Yield function of the orthotropic material considering the crystallographic texture, Structural Engineering and Mechanics, 58(4), 677-687.
[16] Landolt-Bornstein (1966), Numerical data and functional relationships in science and technology. New Series. Group III: Crystal and solid state physics. Volume 1: Elastic, piezoelectric, piezooptic and electrooptic constants of crystals, Springer, Berlin, Germany.
[17] Anisimov, A. N., Egorova, Y. G., & Khromov, A. I. (2016). The odel of cutting problem subject to a coulombmohr yield condition. Materials Physics and Mechanics, 28(1-2), 67-71.

[18] Karpov, V. V., & Semenov, A. A. (2017). Numerical methods for calculating the strength and stability of stiffened orthotropic shells. Materials Physics and Mechanics, 31(1-2), 16-19.
[19] Li, C., Tie, Y., & Li, K. (2016). Analysis of the opening problem of the composite materials structures with boundary integral equation method. Materials Physics and Mechanics, 27(1), 108-112.

Intumescent coatings with improved properties for fireproofing of wooden building constructions

A. Ustinov[1], O. Zybina[1] and E. Kruglov[2]

[1]Fire Safety Department, Peter the Great St. Petersburg Polytechnic University, 195521 Polytechnicheskaya, 29, Saint-Petersburg, Russia, ozakata@mail.ru
[2]Academy of the state fire service of EMERCOM of Russia, 129366, 4 Boris Galushkin str., Moscow, Russia

Abstract. The paper overviews the mechanism of fire-retardant action of intumescent coatings in relation to wooden structures. A comparative assessment of fire-retardant performance of various intumescent compositions, their effectiveness in relation to wood and materials based on it, has been carried out. For this, tests were carried out using the «ceramic tube» method and tests in the cellulosic fire mode. As a result of tests, it was shown that glass microspheres, kaolin, and water-soluble sulfonated graphene are significantly effective functional additives in the intumescent composition. The mechanism of their influence on fire retardant efficiency is considered. It has been shown that compositions containing sulfonated graphene provide a delay in heating in the surface layer of wooden sample, whereas classical intumescent compositions more effectively provide a delay in heating in the depth of the sample. Almost all of the considered compositions demonstrated satisfactory performance when tested by the «ceramic tube» method, thus these compositions allow to make the wood hardly flammable and hardly combustible.

1. Introduction

The fireproofing of wooden constructions is an effective preventive procedure due to the aim to reduce the damage from possible fires. Fire protection of wood is considered to be a mandatory stage when putting a building into operation.

First of all, fire protection of wood is needed to solve the problem of minimizing the percentage of fires as a result of the occurrence of accidental fire sources, which are usually called low-calorie. They include a cigarette that has not been extinguished, short circuit, molten metal dripping during welding, etc. These are the cause of fire in 80% of cases based on the total number of fires. Wood is one of the most widely used materials in construction; at the same time, according to statistics, 75% of fires in the field of residential construction in the territory of the Russian Federation occur in the private sector – these are both wooden buildings themselves and a large number of structures inside buildings (beams, roofs, floors). The increased flammability of wood is the main reason why its fireproofing is one of the most pressing problems in the construction industry.

There are studies dedicated to the problem of wood fireproofing. For example, [1] describes the studies on intumescent compositions based on vinyl acetate copolymer (VAC) (29.15 wt.%) as a binder; ammonium polyphosphate (APP) (18.50 wt.%) as an acid source; pentaerythritol (PER) (9.25 wt.%) as a carbon source; melamine (9.25 wt.%) as a blowing agent; and water (23.85 wt.%). The varied amounts of industrial fillers, i.e., titanium dioxide (TiO_2) and aluminium trihydroxide ($Al(OH)_3$), and biological fillers, i.e., Rice Husk Ash (RHA) and Eggshell (ES) were incorporated into these formulations. Cone calorimetry was used to establish combustivity of wooden samples covered by intumescent compounds, and Raman spectroscopy was used to observe the char residues after the fire tests. The authors achieved the increase in back-surface temperature for the wood sample coated with the formulation F (combining two industrial and two bio-fillers) to be delayed to 990 s with the highest temperature reaching around 250 °C, compared to 530 °C maximum for the uncoated sample. The Raman spectroscopy showed that char residue obtained from the wood substrate protected with the coating A, containing titanium dioxide and RHA fillers, was characterized by the most ordered char structure, very similar to the one of unprotected wood, whilst the char from the sample coated with the formulation B, based on the industrial fillers only, was the most disordered.

In [2] the alder and spruce samples were covered by commercial intumescent paint containing titanium dioxide (TiO_2) and antimony trioxide (Sb_2O_3) as fireproofing additives. The fire performance of these chemicals combined with the fire retardant paint was evaluated in an oxygen index instrument (Dynisco Limiting Oxygen Index Chamber), which is generally applied to evaluate and classify the fire performance of a material, under nitrogen flowing upward in a test column according to ASTM-D 2863. Limiting oxygen index (LOI) values give information about the flammability of the material. For fire performance test by the in-house method, the blowtorch was kept at a distance of 5 cm from the coated and uncoated wood surfaces. The flaming angle was set to approximately 90°. After the flame was applied onto wood surface for 60 s, the observations were performed. The flame extinguishing time was recorded. The best LOI result was obtained from KA2 (42.5; alder, 2 wt.% of Sb_2O_3) for the alder solid wood surface whilst the maximum LOI value was achieved from LA2 (spruce, 2 wt.% of Sb_2O_3) and LT2 (spruce, 2 wt.% of TiO_2) (41) for the spruce wood surface. These findings showed the antimony trioxide and titanium dioxide enhanced physical properties as well as flame retardancy.

Most other researches [3-10] concentrate on fireproofing of metal (in particular steel) constructions; there is not much information particularly on the intumescent coatings for wood surfaces. Though the materials and approaches in development of fireproofing compositions for metal and wood are quite similar; in both cases, the researchers aim to apply different carbonaceous additives in intumescent compositions in order to improve their fireproofing and mechanical properties.

Our research is specifically devoted to fire retardant intumescent coatings for wood based on the melamine-pentaerythritol-ammonium polyphosphate composition, and the purpose of the work is to select the optimal composition in terms of fire retardant properties; a number of experiments were carried out to reach the aim of the research.

2. Materials and methods

2.1. Preparation of intumescent compositions

The original intumescent composition was a mixture of melamine (MA), pentaerythritol (PE) and ammonium polyphosphate (APP) in an aqueous dispersion of vinyl acetate copolymer [11,12]; titanium dioxide was used as a pigment [13,14]. Five compositions were prepared on the basis of composition 1; their formulations are presented in Table 1

Table 1. Intumescent compositions prepared for further investigation

Components, mas.%	Comp. No.1	Comp. No.2	Comp. No.3	Comp. No.4	Comp. No.5
Aqueous dispersion of vinyl acetate copolymer	20.00	20.00	20.00	20.00	20.00
APP	25.00	25.00	25.00	25.00	25.00
Melamine	9.00	9.00	9.00	9.00	9.00
Pentaerythritol	9.00	9.00	9.00	9.00	9.00
Titanium dioxide	6.00	6.00	6.00	6.00	6.00
Talc	2.00	2.00	2.00	–	2.00
Kaolin	–	–	4.00	–	–
Sulfonated graphene	–	–	–	–	0.05
Glass microspheres	–	1	–	–	–
Other functional additives	0.20	0.20	0.20	0.20	0.20
Water	25.80	24.80	21.80	27.80	25.75

IOP Conf. Series: Materials Science and Engineering **986** (2020) 012032 doi:10.1088/1757-899X/986/1/012032

The compositions were prepared by mixing the components in a metal beaker using a mechanical stirrer. At the stage of dispersion, the stirring speed was 1000-1200 rpm, then, with the addition of PFA, it decreased to 600 rpm. Composition 2 contains hollow glass microspheres (Fig. 1), composition 5 contains water-soluble sulfonated graphene [15] (manufactured by Laboratory of composite materials in Building, RUDN University, Russia).

a) *b)*

Figure 1. SEM images of microspheres used as an additive in composition No.2; *a)* x200; *b)* x500

2.2. Swelling coefficient

The swelling coefficient is a parameter which means the ability of intumescent coating to expand in volume when exposed to heating. To determine the swelling coefficient, the test the compositions were applied to clean and degreased metal plates with a size of 150 mm x 75 mm x 2 mm. The coating was dried at room temperature for 72 hours. The thickness of the dry layer of 1 mm was achieved on all samples.

The thickness of the initial coating layer was measured on the plate with the ETARI ET-11P magnetic thickness gauge in five points (results were averaged).

The samples with applied flame retardant compositions were placed in a furnace heated to 600°C and exposed for 5 minutes so that the charred layer could form.

Using a Zipower PM 4265 caliper, the thickness of charred layer was measured at five points. The final value was averaged. The swelling coefficient was defined as the ratio of the charred layer thickness to the thickness of the initial coating.

2.3. Thermal analysis

To assess the behavior of intumescent compositions in the process of thermolysis, the thermal analysis was used. Thermal analysis of the compositions was performed using SETSYS Evolution TG-DSC/DTA 1750 by SETARAM (France).

The following data was recorded during thermal analysis:

– values of weight loss at determined temperatures (in the range of 300-550 °C);

– ash or char residue, %, at the temperature of destruction completion;

– temperatures of maximum DTA peaks in the range of 150-550 °C.

The average of the three measurements for each flame retardant composition was analyzed.

2.4. Tests by «ceramic tube» method

After evaluating the fire-retardant properties of the compositions themselves, the further step is fire tests, in which the compositions are applied to wooden samples and their fire-retardant effectiveness in relation to wood is being assessed. One of the methods for assessing the effectiveness of fire retardants for wood is the «ceramic tube» method described in [16] (Fig. 2). The essence of the method for testing wood samples with a fire retardant coating in the «ceramic tube» installation is the following: a standard wood sample treated with an intumescent composition is weighed, then placed in a ceramic box, the walls of which are lined with aluminum foil; the sample is exposed to the flame of a gas burner from below (200 ± 5 °C, 2 min); the change in mass of the sample after the test is registered. The result of test is a difference in sample mass before and after testing, i.e. weight loss (in g and %): the lower the weight loss, the more effective fire protection is provided by the intumescent composition. The fire retardant composition belongs to the I group of fire retardant efficiency (hardly combustible) if the value of the weight loss is not more than 9%. If the weight loss of the sample is not more than 25%, then the composition is assigned the II group of fire retardant efficiency (hardly flammable) [16]. Preliminary comparative tests were carried out on three samples, expanded tests – on 10 samples. After the tests, the samples were re-weighed. The final test result was the arithmetic mean of three tests.

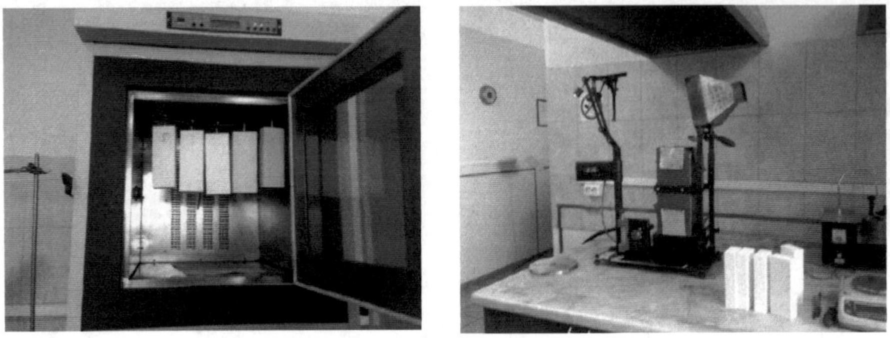

a) *b)*

Figure 2. *a)* Drying of wood samples with applied fire retardants; *b)* General view of the «ceramic tube» installation

2.5. Tests in cellulosic fire mode

However, the most complete picture emerges after conducting the real fire tests, which make it possible to evaluate the behavior of the sample under conditions close to those of a real fire. The tests were carried out in the Academy of State Fire Service of the Ministry of Emergencies, Moscow. The authors of the methodology are E. Kruglov, R. Aseeva.

The testing device was a furnace (Fig. 3) in which the conditions of a cellulosic fire were created in accordance with [17].

Samples of wood were placed in the furnace in such a way that heating was carried out only on one side. The unheated side was protected with intumescent coatings. The spots for thermocouples on a sample are shown in Figure 4.

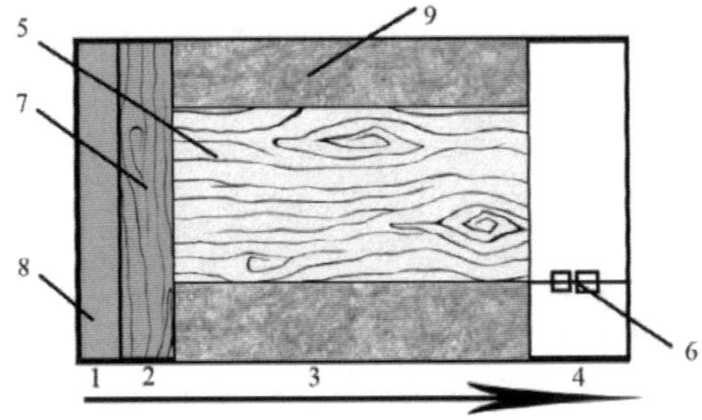

Figure 3. Furnace design 1, 2, 3, 4 - sequence of layers of the structure; 5 - frame beam; 6 - tongue; 7 - external isolating structure made of wood; 8 - gypsum; 9 - mineral wool (heat-insulating layer); → direction of heat transfer

During the tests, the main recorded indicator was considered the time of heating the sample to a temperature of 270 °C on the surface, and at a depth of 5, 10, 15 mm, respectively.

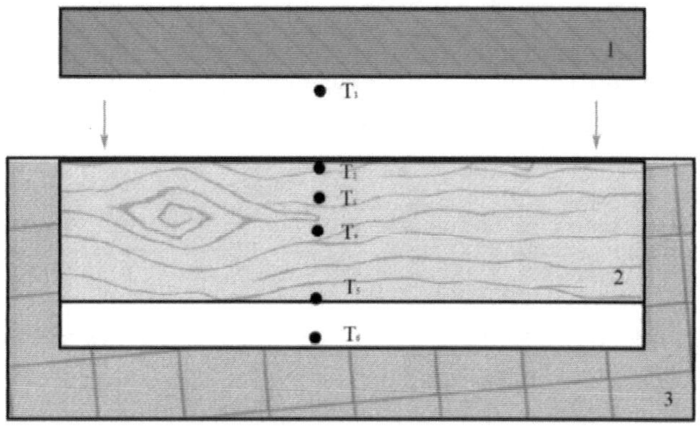

Figure 4. Placement of thermocouples on a wood sample: T_1, T_2, T_n - temperature measurement points: 1 – radiation panel, 2 – wood sample, 3 – heat-insulating material

3. Results and discussion

3.1. Results of determining the swelling coefficient of intumescent coatings

The test results showed that the swelling coefficient increases for all compositions, except for No. 2. For composition No. 5 with the addition of sulfonated graphene, an increase in the expansion of intumescent char was detected, which is corresponding with the data of previous studies [18,19]. A decrease in the value of the swelling coefficient for composition No. 2 may be due to the fact that the hollow microspheres act as a heat shield on the coating surface, delaying the heating of the underlying layers [20,21]. Values of swelling coefficient are presented in Table 2 (*Results of thermal analysis* subsection).

3.2. Results of thermal analysis

Results of thermal analysis are presented in Figures 5-9. Based on the thermal analysis data, the temperature values of the main stages of thermolysis and the characteristics of thermal stability of intumescent coatings (Table 2) were determined.

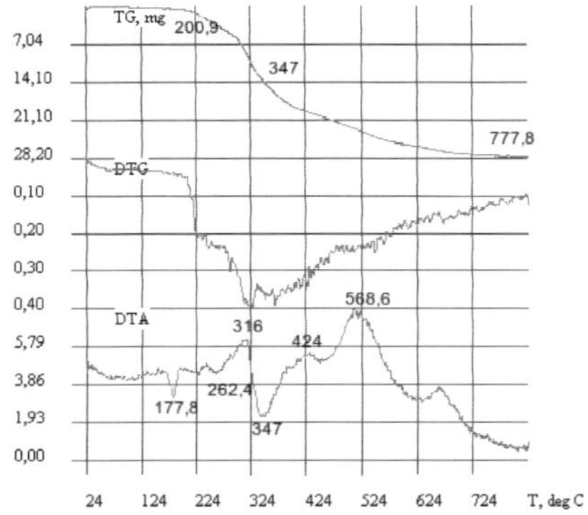

Figure 5. TA curves of composition No. 1

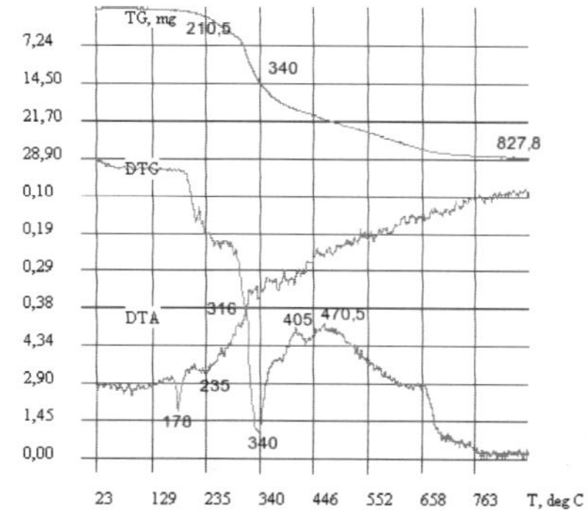

Figure 6. TA curves of composition No. 2

IOP Conf. Series: Materials Science and Engineering **986** (2020) 012032 doi:10.1088/1757-899X/986/1/012032

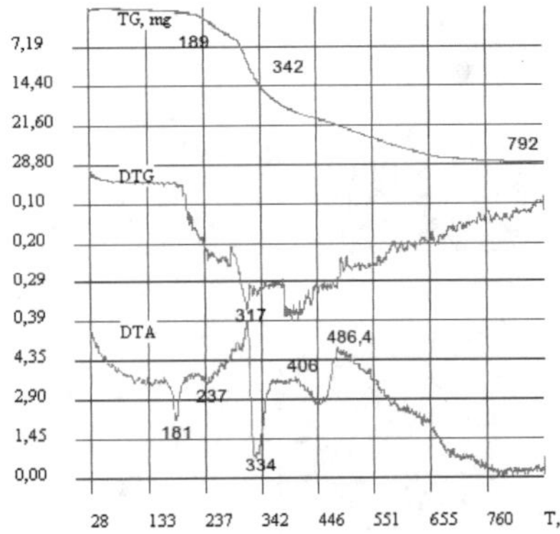

Figure 7. TA curves of composition No. 3

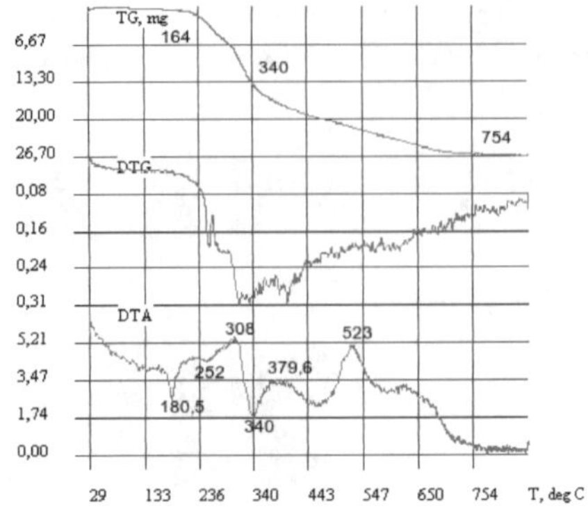

Figure 8. TA curves of composition No. 4

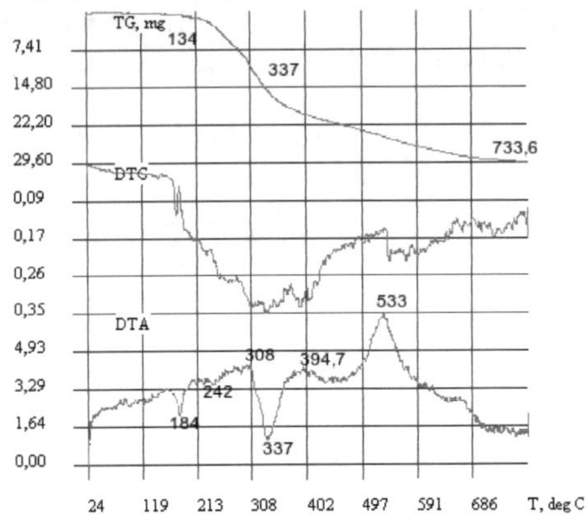

Figure 9. TA curves of composition No. 5

Table 2. Thermal analysis summary data

Stages:	Comp No. 1	Comp No. 2	Comp No. 3	Comp No. 4	Comp No. 5
T of start of pentaerythritols decay, °C	177.8	178	181	180.5	184
T of pentaerythritols full decay, °C	262.5	235	237	252	242
Synthesis of resin	316	316	317	308	308
T of ammonium polyphosphates decay, °C	347	340	334	340	337
T of start of weight loss, °C	200.9	210.5	189	164.4	134
T of 50% of weight loss, °C	347	340	342	340	337
T of end of weight loss, °C	777.8	827.8	792	754	733.6
Ash residue, %	21.8	21.1	21.2	23.3	20.4
Swelling coefficient	54	51	75	70	76

Based on the data presented in Table 2, the endothermic peaks corresponding to the thermal decay of the main components of intumescent composition (pentaerythritol and ammonium polyphosphate) are shifted along the temperature axis when various additives are included into the composition. So, for example, the synthesis of char-forming resin in compositions No. 4 and 5 begins earlier in comparison with others. Composition No.2 has the highest thermal stability, which is confirmed by the values of temperature of total weight loss and ash residue. The temperature of total weight loss for compositions No.4 and 5 is the lowest, which indicates an earlier formation of intumescent char.

Presented data confirms the fact that functional additives directly affect the process of thermolytic synthesis of the charred layer: in modified compositions No. 4 and 5 it occurs faster, which ultimately has a positive effect on their fire retardant efficiency.

3.3. Results of tests by «ceramic tube» method

First of all, comparative tests were carried out for each composition on three wooden samples; the results are shown in table. 3.

Table 3. Comparative parameters of fire tests of 5 compositions by «ceramic tube» method

№ of sample	Weight of a wooden sample before treatment, g	Weight of a treated wooden sample before the test, g	Approximate composition consumption, g/m^2	Weight of sample after the test, g	Mass loss, g	Mass loss, %
1	142.34	147.42		137	10.42	7.1
2	143.6	149.56		140.5	9.06	6.1
3	139.16	144.23	280-290	132.4	11.83	8.2
4	139.71	145.23		116	29.23	20.1
5	133.39	139.81		132.2	7.61	5.4

According to experimental data, all wooden bars protected by test compositions, except for composition No.4, demonstrated a weight loss which corresponds to group I of fire retardant efficiency (less than 9% weight loss).

Bars protected by compositions No. 2 and No. 5 showed the most significant results in all tests, therefore extensive tests were carried out for them using the «ceramic tube» method (on 10 samples). Results of these tests are presented in Tables 4 and 5.

Table 4. Results of extensive tests according to the «ceramic tube» method for composition No. 2

№	W, %	Weight of a wooden sample before treatment, g	Length of sample, mm	Δm	Weight of a treated wooden sample before the test, g	Approximate composition consumption, g/m^2	Weight of sample after the test, g	Mass loss, g	Mass loss, %
1	7	143.36	151.60	8.24	147.03	271.0	130.08	16.95	11.5
2	7.5	143.50	151.80	8.30	145.50	268.1	129.70	15.80	10.8
3	7	145.50	153.85	8.35	147.30	271.2	133.52	13.78	9.4
4	7.5	145.01	153.41	8.40	147.86	275.0	132.78	15.08	10.2
5	6	146.15	154.80	8.65	148.45	274.0	137.30	11.15	7.5
6	7.5	149.16	157.78	8.70	153.31	282.0	141.20	12.07	7.9
7	7	147.30	155.92	8.62	149.65	276.4	136.05	13.60	9.0
8	6	144.36	153.05	8.69	148.42	284.0	137.10	11.32	7.6
9	7.5	144.70	153.38	8.68	145.45	268.0	128.40	17.05	11.7
10	6	147.64	156.24	8.60	151.10	281.0	137.48	13.62	9.0
		Average value:				275.1		14.04	9.46

Table 5. Results of extensive tests according to the «ceramic tube» method for composition No. 5

№	W, %	Weight of a wooden sample before treatment, g	Length of sample, mm	Δm	Weight of a treated wooden sample before the test, g	Approximate composition consumption, g/m²	Weight of sample after the test, g	Mass loss, g	Mass loss, %
1	7.5	150.96	159.70	8.70	155.45	284	146.25	9.20	5.9
2	6	148.58	157.23	8.65	152.30	283	143.35	8.95	5.8
3	7.5	147.58	156.18	8.60	152.52	281	143.40	9.12	6.0
4	7	143.30	151.90	8.60	148.40	283	139.25	9.15	6.1
5	6	147.23	155.87	8.64	151.43	282	139.40	12.03	7.9
6	6.5	145.30	154.00	8.70	149.00	284	140.02	9.08	6.1
7	7	144.59	153.19	8.60	149.04	281	139.92	9.12	6.1
8	6.5	146.20	154.85	8.65	151.20	283	141.45	9.75	6.5
9	7	144.08	152.76	8.68	149.88	284	139.93	9.95	6.6
10	7.5	145.20	153.50	8.30	149.00	284	140.02	9.08	6.1
		Average value:				282.9		9.54	6.31

According to the data in Tables 4 and 5, the weight loss did not exceed 9% only for wooden samples treated with composition No. 5. It was decided to conduct further tests in a cellulosic fire mode with a composition No.5. Composition No. 1 was chosen as a control sample.

3.4. Test results under cellulosic fire conditions [17]

Three samples were tested: W1 – an untreated wooden block, W2 – a wooden block treated with composition No.1 and W3 – a wooden block treated with composition No.5. The results of tests under cellulosic fire conditions are presented on Figures 10-12 and in Table 6. Thermocouples were measuring the following temperatures (see Figures 10-12): T_1 – standard fire mode in the furnace (°C); T_2 is the temperature on the heated side of the sample; T_3 is the temperature in the wooden samples section at a depth of 5 mm; T_4 – temperature in the wooden samples section at a depth of 10 mm; T_5 – temperature in the section at a depth of 15 mm.

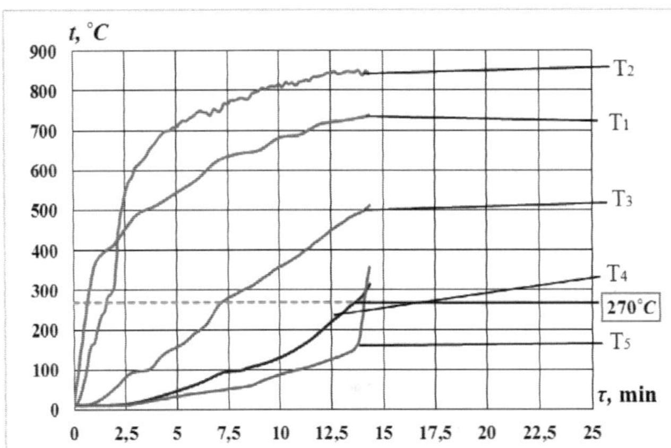

Figure 10. Heating of an untreated wooden sample W1 under the influence of cellulosic fire mode

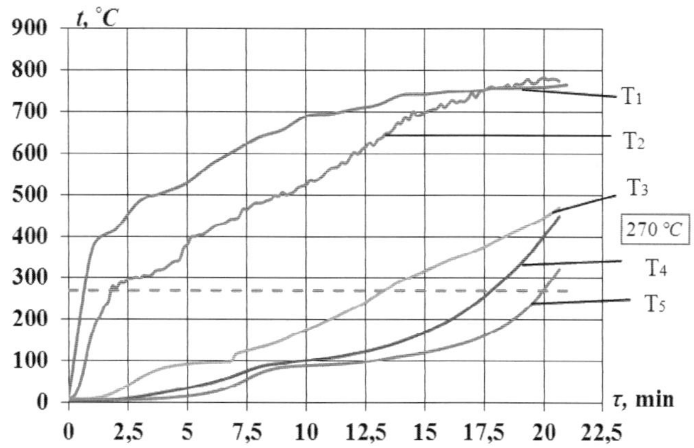

Figure 11. Heating of a wooden sample W2 covered by composition No.1 under the influence of cellulosic fire mode

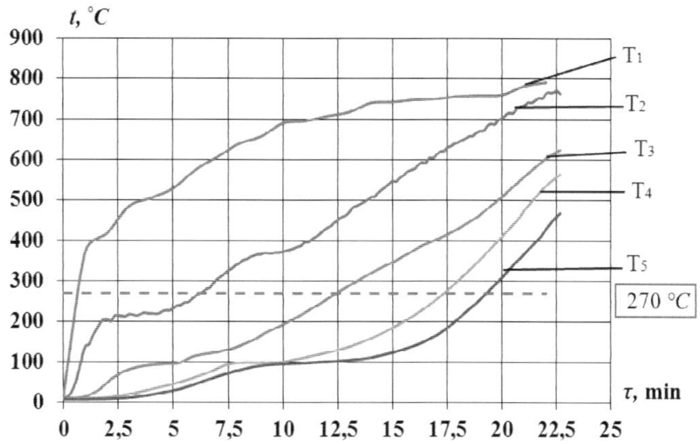

Figure 12. Heating of a wooden sample W3 covered by composition No.5 under the influence of cellulosic fire mode

Table 6 shows the generalized characteristics of the time of reaching a temperature of 270 °C on all samples surface, and at a depth of 5 mm, 10 mm and 15 mm, respectively.

Table 6. Comparison of data on the time to reach a temperature of 270 °C for all samples

	Depth of a section			
	δ = 0 mm	δ = 5 mm	δ = 10 mm	δ = 15 mm
Sample W1	1.66	7.33	13.66	14.17
Sample W2	1.83	13.33	17.83	**20**
Sample W3	**6.30**	**12.5**	17.33	**19.33**

According to the data in Table 6, the maximum time to reach a temperature of 270 ° C on the wood surface is for sample W3.

At a depth of 5 mm, the results for samples W2 and W3 are mostly similar, but significantly higher than those for untreated sample W1. This means that the samples treated with intumescent compositions are heating up more slowly.

At a depth of 10 mm, samples W2 and W3 are heated to a temperature of 270 ° C in 17.83 and 17.33 minutes respectively. Untreated wooden sample W1 – in 13.66 minutes. The time difference is stable around 4 minutes.

At a final depth of 15 mm, samples W2 and W3 have heating times of 20 and 19.33 minutes, respectively. For sample W1, this time was 14.17 minutes. The final difference in time between samples W2 and W1 is 5.83 minutes.

The results obtained show that the compositions belonging to the I group of fire retardant efficiency with a thickness of the applied layer of 1 mm provide a heating delay on average of 4-6 minutes. At the same time, the composition containing sulfonated graphene (sample W3) as an additive slows down heating on the wooden samples surface for 5 minutes longer in comparison with untreated wood (sample W1). Composition No.1 (Sample W2), which is less effective in the surface layer, turns out to be slightly more effective when a sample is heated at depths of 5, 10 and 15 mm. These results allow a more conscious approach to the choice of additives when solving specific problems of fire protection of wood and materials based on it.

4. Conclusions

In this work the behavior and fire-retardant characteristics of intumescent compositions for wood were investigated by various methods (including thermal analysis and tests in the «ceramic tube» installation). Comparative tests were carried out on five intumescent compositions based on the classical intumescent triad, some of them were modified with functional additives such as glass microspheres and water-soluble sulfonated graphene. Composition 1, 2, 3 and 5 demonstrated a satisfactory result in establishing the fire retardant efficiency according to weight loss parameter. Composition 5, modified with sulfonated graphene, demonstrated the best performance, providing the ability to transfer wood to the category of hardly combustible. The effectiveness of this additive from the point of view of improving the properties of the final charred layer, in our opinion, lies in its active catalytic action in the process of thermolytic synthesis of intumescent char. It was also shown that the process of thermolytic synthesis begins earlier, and this leads to the formation charred layer with bigger thickness.

This composition was tested under cellulosic fire conditions when applied to a wood sample. As a result of the tests, it was found that the fire-retardant efficiency of intumescent coatings under the conditions of a cellulosic fire for wood samples consists in a delay in the time of the beginning of wood heating, its charring and decomposition under the action of high temperatures and in charred layers ability to be an obstacle to the spread of flame along the wood surface. In comparison with untreated wood, the delay in heating is 5.8 minutes. In case of a real fire, this time can play a significant role. However, under the same conditions, composition No.1, which showed less efficiency in preliminary tests on determining the group of fire retardant efficiency, provides bigger values of the heating time under conditions of a cellulosic fire. The results of this work will be useful in solving problems of fireproofing of wood and materials based on it.

References

[1] Aqlibous A., Tretsiakova-McNally S. and Fateh T., Waterborne Intumescent Coatings Containing Industrial and Bio-Fillers for Fire Protection of Timber Materials, *Polymers* **12** (4) (2020).

[2] Torun Sevda B., Cavdar Ayfer D. and Turgay O., The synergistic effect of intumescent coating containing titanium dioxide and antimony trioxide onto spruce and alder wood species, *J. of Building Engineering* **31** (2020).

[3] Ullah S., Ahmad F., Shariff A. M., Rafi Raza M. et. al., The role of multi-wall carbon nanotubes in char strength of epoxy based intumescent fire retardant coating, *J. of Analytical and Applied Pyrolysis* **124** (2017).

[4] Ullah S., Ahmad F., Shariff A. M. and Bustam M.A., Synergistic effects of kaolin clay on intumescent fire retardant coating composition for fire protection of structural steel substrate, *Polymer Degradation and Stability* **110** (2014) 91-103.

[5] Dzulkaflia H. H., Ahmad F., Ullah S., Hussain P., Mamat O. and Megat-Yusoff P. S.M., Effects of talc on fire retarding, thermal degradation and water resistance of intumescent coating, *Applied Clay Science* **146** (2017) 350–361.

[6] Ullah S. and Ahmad F., Enhancing the Char Resistant of Expandable Graphite Based Intumescent Fire Retardant Coatings by Using Multi-Wall Carbon Nano Tubes for Structural Steel, *Solid State Phenomena* **185** (2012) 90-93.

[7] Baldissera A. F., Silveira M. R., Beraldo C. H., Tocchetto N. S. and Ferreira C. A., Evaluation of the expandable graphite/polyaniline combination in intumescent coatings, *Synthetic Metals* **256** (2019).

[8] Wang Y. C., Tang G., PingZhao J. and Han Y., Effect of flaky graphite with different particle sizes on flame resistance of intumescent flame retardant coating, *Results in materials* **5** (2020).

[9] Wang Y. C., PingZhao J. and Xiaojing M., Effect of expandable graphite on polyester resin-based intumescent flame retardant coating, *Progress in Organic Coatings* **132** (2019) 178-183.

[10] Zhang P., Hu Y., Song L., Ni J., Xing W. and Wang J., Effect of expanded graphite on properties of high-density polyethylene/ paraffin composite with intumescent flame retardant as a shape-stabilized phase change material, *Solar Energy Materials and Solar Cells* **94** (2) (2010) 360-365.

[11] Morys M., Illerhaus B., Sturm H. and Schartel B., Variation of Intumescent Coatings Revealing Different Modes of Action for Good Protection Performance, *Fire Technology* **53** (4) (2017).

[12] Morys M., Illerhaus B., Sturm H. and Schartel B., Size is not all that matters: Residue thickness and protection performance of intumescent coatings made from different binders, *Journal of Fire Sciences* **35** (2) (2017).

[13] Ustinov A., Zybina O. and Andreev A., On the Impact Caused by Titanium Dioxide of Different Trademarks on the Properties of Intumescent Fire-Protective Coatings, *Materials Science Forum* **945** (2019) 212-217.

[14] Li H., Hu Z., Zhang S., Gu X. et. al., Effects of titanium dioxide on the flammability and char formation of water-based coatings containing intumescent flame retardants, *Progress in Organic Coatings* **78** (2015) 318-324.

[15] Ponomarev A.N., Yudovitch M.E. and Kozeev A.A. Patent WO 2011/102754 A2 (2011) (in Russian)

[16] Sobur S.V. 2019 *Fire protection of materials and structures* (Moscow: PozhKniga) 208 p.

[17] European Committee for Standardization (CEN), "Eurocode 1: Actions on structures – Part 1-2: General actions – Actions on structures exposed to fire ", EN 1991-1-2, June 2007.

[18] M. Nikitina, A. Ustinov, V. Kiseleva, I. Babikov, New fire retardant compositions for fire-resistant automatic curtains, MATEC Web Conf. 245, 11004 (2018).

[19] A. Ustinov, O. Zybina, L. Tanklevsky, V. Lebedev, A. Andreev, Intumescent coatings with improved properties for high-rise construction, E3S Web Conf. 33, 02039 (2018).

[20] A. Ustinov, O. Zybina, A. Tomakhova, S. Pavlov, The enhancement of operating properties of intumescent fire-protective compositions, MATEC Web Conf. 245, 11008 (2018).

[21] O. Zybina, M. Gravit, Y. Stein, Influence of carbon additives on operational properties of the intumescent coatings for the fire protection of building constructions, IOP Conf. Ser. Earth and Environmental Science 90 (1) 012227 (2017).

IOP Conf. Series: Materials Science and Engineering **986** (2020) 012033 doi:10.1088/1757-899X/986/1/012033

Improving the printing process stability and the geometrical accuracy of the parts manufactured by the additive techniques

A I Borovkov[1], L B Maslov[1,2], K S Ivanov[1], E N Kovaleva[1], F D Tarasenko[1] And M A Zhmaylo[1]

[1]Center of National Technological Initiative, Institute for Advanced Manufacturing Technologies, Peter the Great St. Petersburg Polytechnic University, 29 Politekhnicheskaya, St. Petersburg, 195251, Russia, E-mail address: zhmaylo@compmechlab.com
[2]Department of Theoretical and Applied Mechanics, Ivanovo State Power Engineering University,34 Rabfakovskaya, Ivanovo, 153003, Russia

Abstract. Current research is devoted to the computer-aided engineering application for direct modeling of the additive manufacturing process and detailed analysis of the resulting shape of the formed part. The utilized approach is based on the idea that it is possible and efficient to perform a numerical simulation of the process of a layer-by-layer building of the analyzed part, considering all the required mechanical and thermal effects. The results of such simulation should allow obtaining the detailed information about the deformed shape of the printed part before running the printing process itself. Such approach allows avoiding the unsuccessful printing attempts, improve the integral stability of the production process and reduce its cost. The paper presents the methodology of preparing a "digital twin" of the additive manufacturing process performed with AlSi10Mg alloy using one of metal machines and running a simulation of the manufacturing process for a representative structural component. The study has shown that implementation of modeling of the manufacturing process before the printing process leads to significant shape improvements of the parts printed, especially in case of performing additional optimization of the support structures. The research outcomes can be well applied for improving the production processes in the companies working with additive manufacturing in order to increase the stability of the processes and accuracy of the printed parts.

1. Introduction

Additive methods as advanced manufacturing technologies allow producing the complex metal parts and assemblies, which have been previously unfeasible from the technological point of view [1]. However, regardless of all outstanding capabilities of metal additive manufacturing, this group of technologies has an important weak point, i.e. all metal additive manufacturing methods are characterized by the high possibility of deviations between the nominal shape of the part and its printed shape [2]. This specific feature causes additional limitations and introduces difficulties to the manufacturing process, which finally increase the cost of production and its duration.

Due to this fact, each of the companies, which produce or operate additive machines, generate their own knowledge base or "know-how", which allows them to improve the stability of the manufacturing process and increase the printing accuracy. Such bases, as well as the knowledge of the technological staff, are highly valued by the companies.

Current research is devoted to the application of computer-aided engineering for direct modeling of the additive manufacturing process and detailed analysis of the resulting shape of the formed part to increase the stability of the processes and accuracy of the printed parts.

2. Review of current researches

Most of the works devoted to modeling of the additive processes date back to 2012 and later years. There are some studies, which were done and published earlier, but the amount of such publications is much lower. It may be assumed that this is caused by the increase of the availability of the additive machines, but there is another important reason – the increase of the performance of the computers, which are required for solving the complex thermomechanical tasks related to the printing process.

One of the most detailed and recent works on the subject of modeling the additive processes is [3], which was published by the group of researchers led by Pan Michaleris, the founder of the pioneering company in the field of direct modeling of additive manufacturing, acquired by Autodesk. This book describes and discusses most of the issues related to the modeling of thermomechanical effects that arise, when modeling the process of selective laser melting with metal powders.

In their earliest studies [4-7] the group assessed the efficiency of the approach of modeling the additive manufacturing using the technique of "birth" and "death" of finite elements and studied how the surface thermal interaction affects the results of the simulations. It should be mentioned that at the moment the technique of "birth" and "death" is an industry standard of modeling the additive process of adding the new layers for the finite-element systems designed for performing the additive simulations.

But there is a significant difference between the physical layers, which are used during the printing, and the virtual layers that are used to simulate the additive manufacturing. The reason for the difference is that the layers in the simulations are the macro-layers with the thicknesses, which are much higher than the thickness of the real layers. The implementation of the macro-layers is discussed in [8, 9]. The researchers conclude that this approach is fully applicable and correct, but it is mentioned in the study that there are many difficulties with the simulation process related to the high computational cost of the thermomechanical problems.

However, despite the fact that the metal additive manufacturing is a relatively new technology, the basic principle of the technology, which is based on the application of layers of melted material, is well known for a long time. It is widely used in welding, and the technology of finite-element modeling of the welding process was discussed in [10] based on the results of a study carried out in 1975.

In addition to the numerical approaches, some groups of researchers make attempts of performing the simulations of the build process using hybrid approached, which are partially based on some analytical and pseudo-analytical methods [11, 12]. According to the study, the use of hybrid method can significantly accelerate the modeling process. Such acceleration could be extremely useful, since the existing approaches, which are based on numerical methods only, require a lot of time to perform the build simulations with high accuracy.

Along with the studies of the problems related to the modeling of the additive process itself, several groups of scientists focus their work on the studies of the procedure of generation of the optimal support structures [13, 14]. Such structures are required both for geometrical supporting the printed part and making it possible to print them and for the thermomechanical support that does not allow the parts to separate from the build plate and provides the conditions for the proper heat transfer. The developed approaches are based on such technologies as direct modeling of additive manufacturing and topology optimization.

The brief review of the current studies presented above makes possible to conclude that the area of application of computer-aided engineering to the modeling of the additive layer processes is still in the phase of intensive development, and the existing results and tools are very promising and allow improving the quality of the metal prints.

3. Simulation software

This article describes the process and the results of the simulation of additive manufacturing based on the finite-element modeling software called Amphyon, which is developed by Additive Works GmbH.

It is a relatively new tool, as well as all other systems working with this type of simulations. The user interface of the software is shown in Fig. 1.

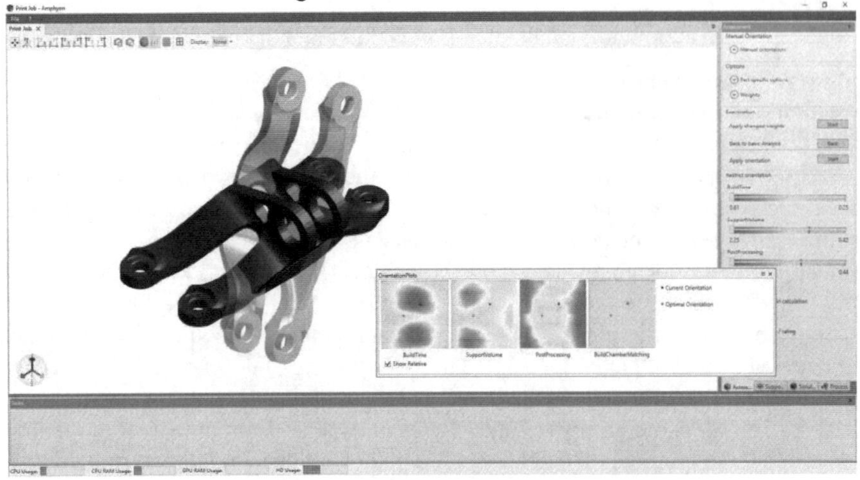

Figure 1. Graphical user interface of Amphyon by Additive Works.

The study was performed with Amphyon 2019, which has the following modeling capabilities:

- optimization of the orientation of the part on the printer platform;
- generation of support structures;
- optimization of support structures in terms of structural integrity and thermal effects;
- simulation of the 3D printing process using rational finite-element mesh models;
- generation of deformed and predeformed shapes of the parts allowing to print the analyzed parts with the nominal geometry;
- modeling of thermal fields for the printed parts.

The first and the last points from the list above were not used in current study.

It is important to mention that Amphyon is based on a specialized finite-element solver, unlike many other similar software tools. This feature allows using specific algorithms, which are optimized for solving a narrow class of numerical problems with a high efficiency both on CPU's and on GPU's with good parallelization, which is very important due to the high computational cost of the considered type of tasks.

4. Analyzed object

The modeling in the study is performed for a representative complex part, which is a service bracket that is installed on a jet engine [15]. The shape and the size of the bracket are explained in Fig. 2.

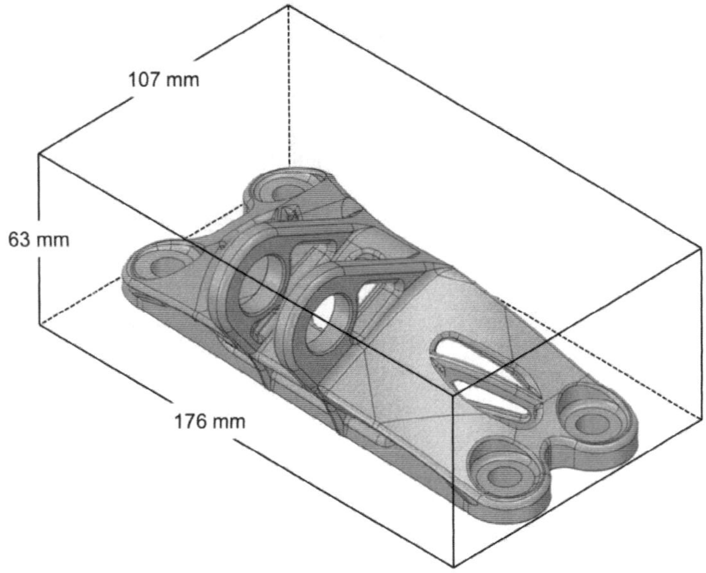

Figure 2. The analyzed bracket.

The bracket has four points that are used to mount the part onto the engine and two padeyes, which are loaded with some forces and torque.

The study considers virtual printing of the bracket in two problem statements and describes the comparison of the real prints. In order to make the results comparable and more representative, the orientation of the bracket on the printing platform is not optimized.

5. Modeling of the 3d printing process

5.1. Approach of modeling

Due to the high complexity of the physical processes, which are to be described to simulate the additive layer manufacturing, the numerical models in most of the software tools are based on the following parameters of the material and printing process:

- elastic modulus;
- Poisson ratio;
- yield strength;
- layer thickness;
- results of measurements performed for a series of calibration specimens.

There is a significant point in the list above, which requires performing a number of measurements for 3D printed tests specimens. It may seem strange that some additional measurements shall be done, but there is an important reason for that. The reason is that the results of additive manufacturing strongly depend on the material properties, machine type and printing parameters. It means that in case of printing two parts made of the same material, with the same machine, but with different parameters of the printing process, the resulting parts will have identical chemical composition, but they may have different microstructure and behave differently from the mechanical point of view.

This specific feature shows that proper modeling of the 3D printing process requires the typical material data, but it also requires the printing parameters, which can be obtained in the simplified and integrated form as the results of measurement of the printed calibration specimens.

Of course, it may seem reasonable that the simulation could be run without the calibration of the numerical algorithm, but with some additional data about the material and the printing process instead. This concept is correct, but there is an important fact that taking into account the additional parameters of the material and process requires carrying out additional measurements and natural experiments, which require printing more specimens.

This allows making a conclusion that the approach based on using several calibration specimens should be quite efficient to provide accurate modeling results.

5.2. Calibration specimens

Each of the developers of the software that is designed for the additive simulations suggest a specific shape of the calibration specimen and a specific approach of the calibration, but the general idea is the same.

Additive Works recommend running the calibration based on the results of printing and measuring of three calibration specimens.

The main concept is that the specimens are printed and partially separated from the platform in such a way that they can deform as it is shown in Fig. 3. The deformation is affected by a number of parameters of the process of additive manufacturing, so it can provide a kind of integral indicator. The analyzed value of deflection is shown in Fig. 4.

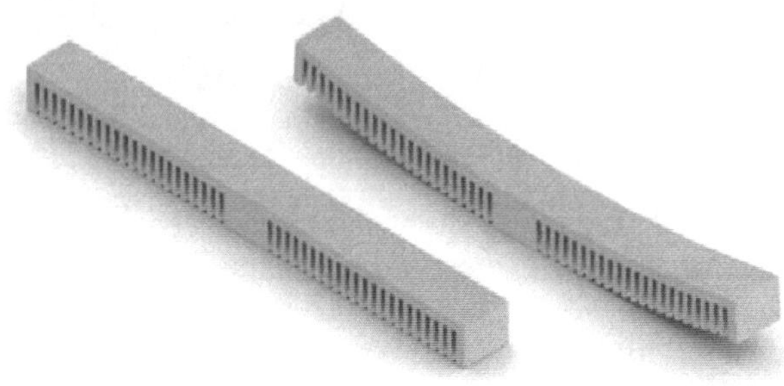

Figure 3. The calibration specimens in the neutral and deformed conditions.

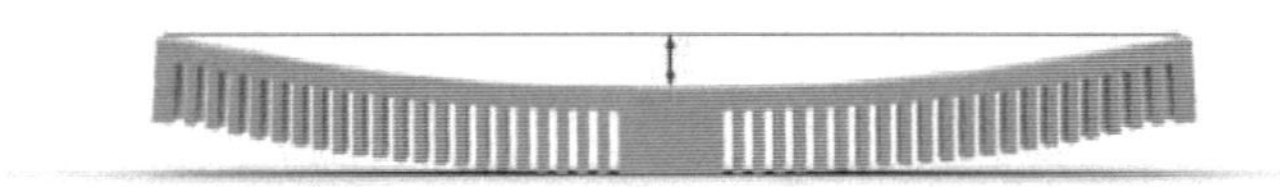

Figure 4. The analyzed value of the deflection.

Since the properties of the additive material depend on the history of the building process, it is recommended to consider the behavior of several calibration specimens, which are printed with different laser scanning strategies shown in Fig. 5.

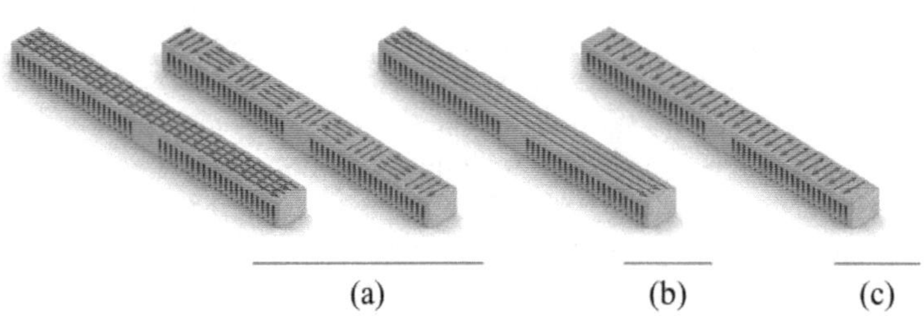

Figure 5. The scanning strategies: (a) balanced, (b) parallel and (c) perpendicular.

The "balanced" strategies are usually the default strategies that are used for the printing tasks, because they allow getting the most stable, precise and isotropic results.
The remaining "parallel" and "perpendicular" strategies are not balanced, because in these strategies one of the scanning directions dominates over another. The first specimen with the first scanning strategy shall have the highest level of deformation, and the specimen with the second strategy shall have the lowest deformation among all specimens. The unbalanced strategies usually lead to higher level of anisotropy and shape deviations in the printed parts.

5.3. Preparing the "digital twin" of the process

As it was mentioned, the process of calibration of the numerical algorithms of the software is started based on the input of the properties of the materials and based on the values of the deflection. During the calibration the system adjusts the parameters of the process in such a way that the resulting deflections of the virtually printed specimens equal the real experimental deflections. As the result, an object that might be called a "digital twin" of the process is generated. The term represents the set of characteristics and parameters, which allows the virtual printing of the parts.
The analyzed bracket is printed with AlSi10Mg aluminium alloy with the following mechanical properties:

- Young modulus – 76 GPa;
- Poisson ratio – 0.33;
- yield strength – 230 MPa.

The deflections of the calibration specimens equal:

- balanced strategy – 0.81 mm;
- parallel strategy – 1.08 mm;
- perpendicular strategy – 0.50 mm.

The layer thickness is 50 μm. The window of Amphyon showing the parameters of a "digital twin" is presented in Fig. 6.

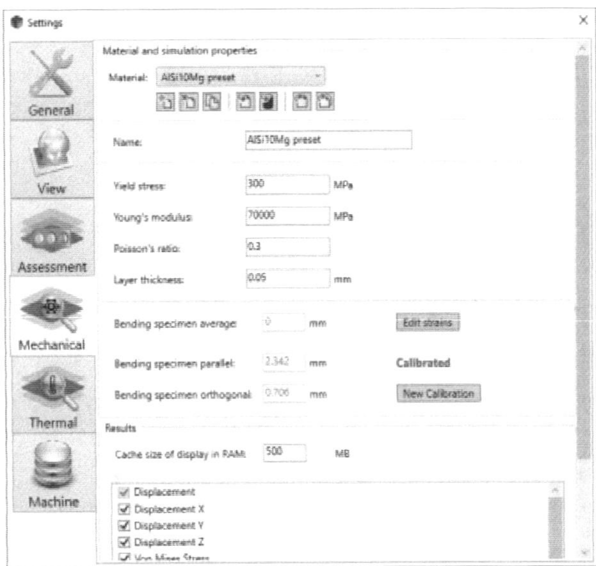

Figure 6. The window of Amphyon with the parameters of a "digital twin".

6. Generation and optimization of support structures

The research considers the printing of the analyzed bracket with two types of support structures – regular and optimized. Both types of supports are prepared in Amphyon.
The regular support structures are generated in Amphyon based on the value of the critical angle of 45 degrees with the standard settings. The resulting shape of the regular perforated support structures is shown in Fig. 7.

Figure 7. The analyzed bracket with regular support structures.

The optimized support structures are generated based on the results of the finite-element modeling of the printing process. The simulation starts with the geometry of the regions, which shall be filled with the support structures, shown in Fig. 8. The curve showing the variation of the deformation and volume of the support material is presented in Fig. 9. A series of fields illustrating the changes in the density of the support structures during the optimization process is shown in Fig. 10.

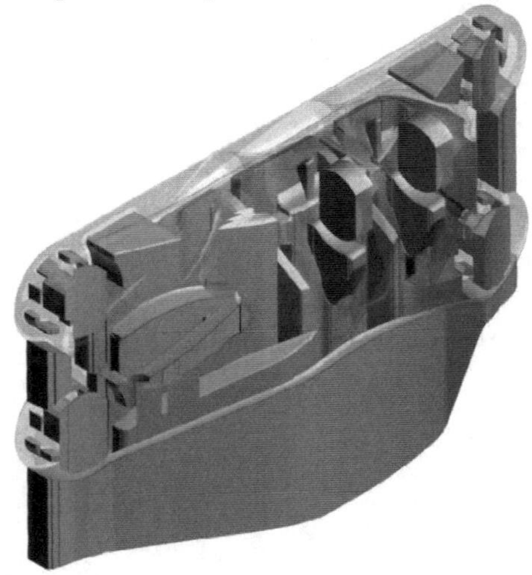

Figure 8. The regions that are to be filled with the support structures.

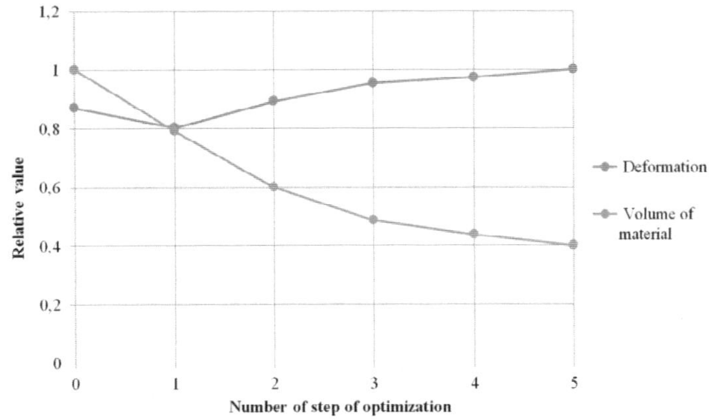

Figure 9. Relative values of the overall deformations and volume of support material during the support optimization.

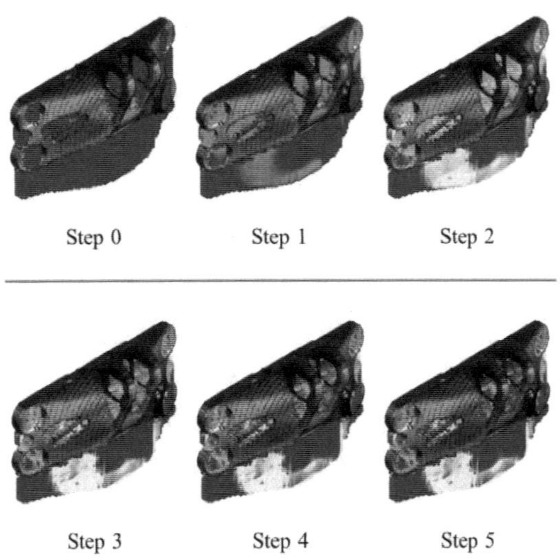

Figure 10. Changes in the density of the support structures during the support optimization.

7. Simulation of the additive manufacturing process

7.1. Finite-element meshing

The process of preparing the analysis model starts with the generation of the finite-element mesh on the main object and on the support structures. The characteristic size of the element in the build direction defines the thickness of the macro-layers.

Amphyon is capable of generating the mesh of two types: regular and adaptive. The difference between the types is that in case of adaptive mesh the elements become larger closer to the center of the printed object. Such approach allows reducing the number of equations in the system without any significant effect on the results.

Current study is based on the adaptive finite element mesh with the characteristic size of 1 mm both in the plane of the build platform and in Z direction. The generated mesh is presented in Fig. 11.

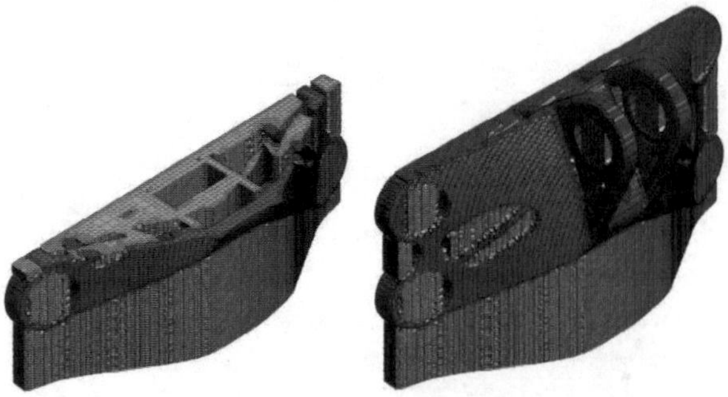

Figure 11. The finite element mesh for the bracket.

7.2. Simulation

The mesh generation is followed by the simulation itself. Before running the analysis it is required to set up several parameters, which define the type of the scanning strategy, which is used for printing the part, and the type of postprocessing or heat treatment that should be done.

In current research it is assumed that the additive manufacturing is done with the balanced scanning strategy, and that there is no heat treatment in order to maintain the integrity of the study of the effect of the part predeformation based on modeling.

After starting the simulation the system solves the task and shows the build process in the real time. The final results of the simulation are presented in Fig. 12. Several steps of the process are shown in Fig. 13.

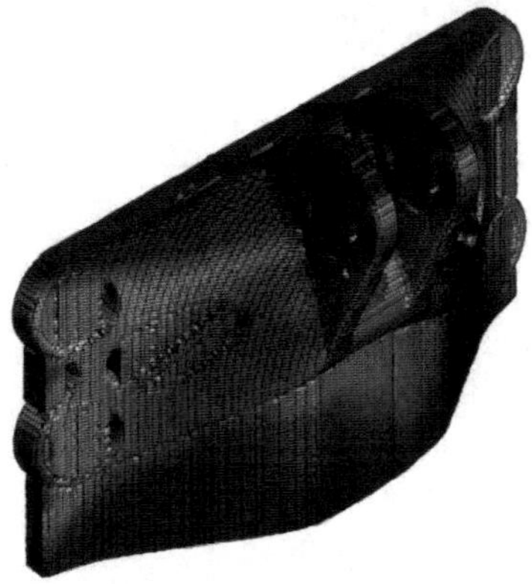

Figure 12. The final result of the simulation. The displacement field is shown.

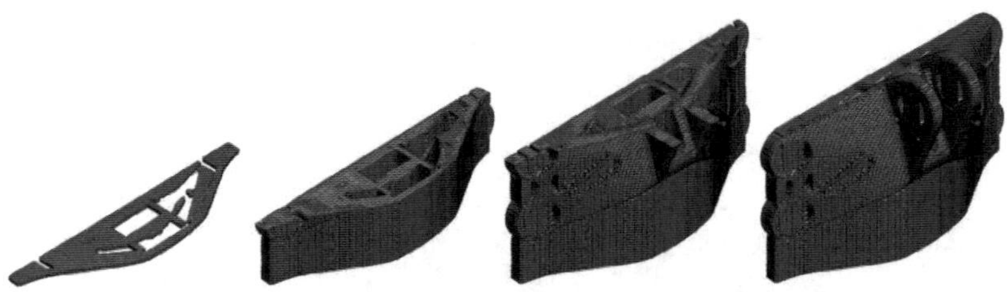

Figure 13. Several steps of the printing process.

7.3. Application of the results of the simulation

As it was discussed earlier in the document, the final goal of the additive simulation is to improve the geometry of the printed parts and stability of the printing process itself.

One of the possible types of result output from Amphyon are the deformed and predeformed shapes of the analyzed part. The deformed shape provides the information about the level of geometrical

IOP Conf. Series: Materials Science and Engineering **986** (2020) 012033 doi:10.1088/1757-899X/986/1/012033

deviations of the printed object. The predeformed shape in case of printing allows getting the original part with the nominal geometry.

8. Results

In order to estimate if the simulation of the 3D printing process has any significant result on the accuracy of the printed part, several brackets were printed:

- bracket based on the nominal geometry;
- bracket based on the predeformed geometry, which was obtained with the regular support structures;
- bracket based on the predeformed geometry, which was obtained with the optimized support structures.

The comparison between the real printed shapes and the nominal geometry of the bracket is shown in Fig. 14-16. The fields are the result of 3D scanning performed for the printed parts.

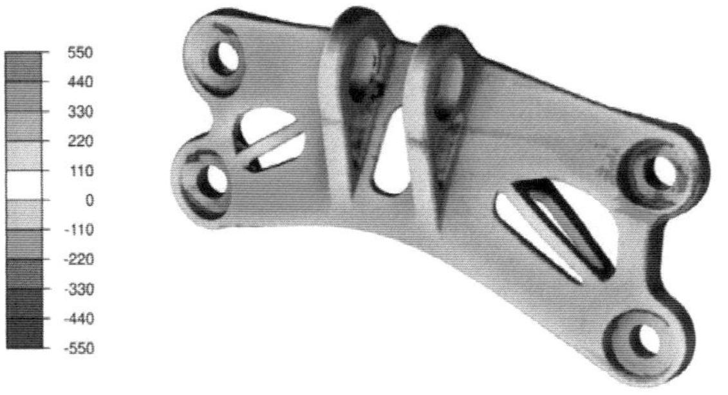

Figure 14. The field of geometrical deviations between the nominal geometry of the bracket and the shape of the part that is printed based on the nominal geometry of the bracket, μm.

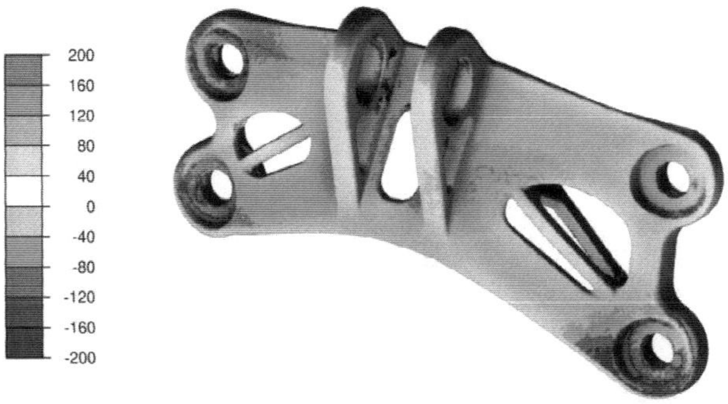

Figure 15. The field of geometrical deviations between the nominal geometry of the bracket and the shape of the part that is printed based on the predeformed geometry of the bracket with the standard supports, μm.

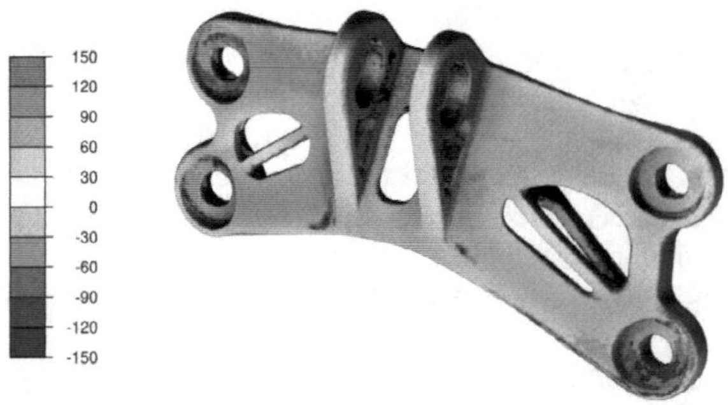

Figure 16. The field of geometrical deviations between the nominal geometry of the bracket and the shape of the part that is printed based on the predeformed geometry of the bracket with the optimized supports, μm.

9. Conclusion

The results of the simulation of the aluminium bracket additive manufacturing process are presented in the paper.

The comparison of the deviation fields obtained for the 3D printed brackets allows making a conclusion that the shape of the bracket, manufactured based on the simulation, has significantly lower values of the shape deviation from the nomimal one.

Besides, the shape obtained with use of the optimized support structures matches nominal one even better than the shape with the regular supports. Both these facts mean that the simulation of the additive processes makes it possible to improve the results of the printing.

From the practical point of view, the research confirms that including a 3D printing simulation as a step before manufacturing process should lead to a significant improvement of the quality of the prints, reduction of the production time and costs. This can be a crucial advantage in the fast-growing additive manufacturing market.

The outcomes of the research can be well applied for improving the production processes in the companies working with additive manufacturing in order to increase the stability of the processes and accuracy of the printed parts.

References

[1] Jing Zhang and Yeon-Gil Jung 2018 Additive Manufacturing : Materials, Processes, Quantifications and Applications (United Kingdom Butterworth-Heinemann) p 352.

[2] Frazier W E 2014 Metal additive manufacturing: a review *J. Mater. Eng. Perform.* (vol. 23, no. 6) pp 1917–1928

[3] Gouge, Michaleris 2017 Thermo-Mechanical Modeling of Additive Manufacturing *1st Edition. Oxford. Butterworth-Heinemann* p 294

[4] Michaleris 2014 Modeling Heat Transfer in Additive Manufacturing Processes. Finite Elements in Analysis and Design (18:51-60)

[5] Denlinger J Heigel and Michaleris P 2014 Residual Stress and Distortion Modeling of Electron Beam Direct Manufacturing Ti-6Al-4V *Journal of Engineering Manufacture*

[6] Martukanitz P, Michaleris T, Palmer T, DebRoy Z-K Liu, Otis R, Heo T W and Chen L-Q 2014 Towards an Integrated Computational System for Describing the Additive Manufacturing Process for Metallic Materials *Additive Manufacturing*

[7] Witherell S C, Feng T W, Simpson D B, Saint John P, Michaleris Z -K Liu, Chen L.-Q and Martukanitz R 2014 Toward Metamodels for Composable and Reusable Additive Manufacturing Process Models *Journal of Materials Processing Technology*

[8] Keller Neugebauer, Xu Ploshikhin 2013 Thermo-mechanical Simulation of Additive Layer Manufacturing of Titanium Aerospace structures *Proceedings of LightMAT*

[9] Neugebauer Keller, Ploshikhin Feuerhahn Koehler 2014 Multi Scale FEM Simulation for Distortion Calculation in Additive Manufacturing of Hardening Stainless Steel *Proceedings of International Workshop on Thermal Forming an Welding Distortion*
[10] Friedman 1975 Thermomechanical Analysis of the Welding Process Using the Finite Element Method *Journal of Pressure Vessel Technology* (3:206-13)
[11] Yang Knol, van Keulen Ayas 2018 A semi-analytical thermal modelling approach for selective laser melting *Additive Manufacturing* (2018;21:284-97)
[12] Mohanty S and Hattel J 2014 Cellular Scanning Strategy for Selective Laser Melting: Capturing Thermal Trends with a Low-Fidelity, Pseudo-Analytical Model *Mathematical Problems in Engineering* pp 1-14
[13] Bartsch Lange Gralow Emmelmann 2019 Novel approach to optimized support structures in laser beam melting by combining process simulation with topology optimization *Journal of Laser Applications* (2019;31(2))
[14] Strano Hao Everson Evans 2013 A new approach to the design and optimization of support structures in additive manufacturing *The International Journal of Advanced Manufacturing Technology* (66(9-12)) pp 1247-54
[15] Optimised bracket / 3D CAD Model Library / GrabCAD [electronic resource] 2013 *Stratasys* (Ltd., Rehovot, Israel.. Access mode: https://grabcad.com/library/optimised-bracket-1, free)

IOP Conf. Series: Materials Science and Engineering **986** (2020) 012034 doi:10.1088/1757-899X/986/1/012034

Methodology of constructing highly adequate models of additively manufactured materials

A I Borovkov[1], L B Maslov[1,2], K S Ivanov[1], E N Kovaleva[1], F D Tarasenko[1], M A Zhmaylo[1] and Barriere Thierry[3]

[1]Center of National Technological Initiative, Institute for Advanced Manufacturing Technologies, Peter the Great St. Petersburg Polytechnic University, 29 Politekhnicheskaya, St. Petersburg, 195251, Russia, E-mail address: tarasenko@compmechlab.ru
[2]Department of Theoretical and Applied Mechanics, Ivanovo State Power Engineering University, 34 Rabfakovskaya, Ivanovo, 153003, Russia
[3]Applied Mechanics Department, Institute "Franche-Comté Electronics Mechanics Thermal Science and Optics – Sciences and Technologies", 15B avenue des Montboucons, Besançon, France 25030

Abstract. In the this paper there is a description of the methodology of constructing highly adequate models of materials which are produced additively. The paper consists of introduction to the problem, demonstration of each step of the methodology and conclusions about the method and its applicability. The developed approach is presented on the two additively manufactured metal alloys. The modeling process is based on the orthotropic model of elasto-plastic material behavior. The list of material properties to be found was formulated according to proposed material model. These properties were found from several mechanical experiments. The developed mathematical models of the material studied were checked for compliance to experimental data with use of finite element simulations. Finite element modeling of a complicated 3d-printed lattice structure was performed and compared with the results of natural testing. In order to study the correlation between the material properties and manufacturing process the analysis of microstructure was done.

1. Introduction

This paper represents the methodology of material modeling for parts which are manufactured from metal powders with 3d-printing approaches. This methodology suggests a clear and easy way to construct such material models for further usage in finite elements simulations.

Additive manufacturing is one of the most rapidly developing advanced manufacturing technology [1]. The use of this technology removes a large number of traditional manufacturing barriers and obstacles [2]. That's why now it becomes possible to develop parts which producing was economically unreasonable or unfeasibly at all. Such manufacturing capabilities lead to significant changes in design approach. Now the goal of a part design is to implement the potential of additive manufacturing as efficiently as possible in order to get competitive products. This competitive product should consist of optimal decisions in each and every step of its lifecycle, from development to manufacturing.

One of these steps is quality of a material model. Utilized material model should match the high level of developing and manufacturing complicity and accuracy, and should be as close to reality as possible. That's why the main aim of this study is to develop a methodology for constructing highly adequate mathematical models of continuous materials, which are produced from metal powder with layer-by-layer technology, e.g. selective laser melting (SLM) or electron beam melting (EBM) [3, 4].

Performed analysis of papers about the same topic [5, 6, 7] showed that the comprehensive testing and analysis of additively produced materials is a live issue. But it is necessary to mention that in most cases only simple material models (like elastic isotropic ones) were used. Only few papers propose a hypothesis that the printed materials must be considered from the continuum mechanics point of view with using the anisotropic mathematical models [8, 9]. This all confirms the relevance of performed study.

So, the main aim of this study is the process of constructing material models based on the elastic and plastic anisotropic continuum media, which will match real life material behavior better than widely used elastic isotropic ones.

Onwards the developed methodology is performed for two well-known and widely used metallic materials, aluminum alloy AlSi10Mg and titanium alloy Ti6Al4V, and samples additively produced by using SLM machine for the aluminum alloy and EBM machine for the titanium alloy.

2. Methodology concept

First of all, the choice of material model type should be done. Chosen type of material model should be strongly connected with the purposes of model construction. For example, the model of materials used in thermonuclear reactors must include mechanical, thermodynamic and electromagnetic equations to describe all the possible processes. Talking about a way of usage additively manufactured parts, then their most common application are structures or construction parts. Thus, the mechanical behavior of the material is the main point of interest in this study.

Next step in the defining of a material model type is taking into account the influence of the manufacturing process on the material behavior. Here and later in this study, we are based on the assumption that additive manufacturing process strongly affects mechanical properties of a material.

The research of microstructure of considered materials was performed to figure out the type of its' behavior. On the below figures (1,2) obtained by using electronic microscopy we can observe the microstructure of AlSi10Mg and Ti6Al4V specimens.

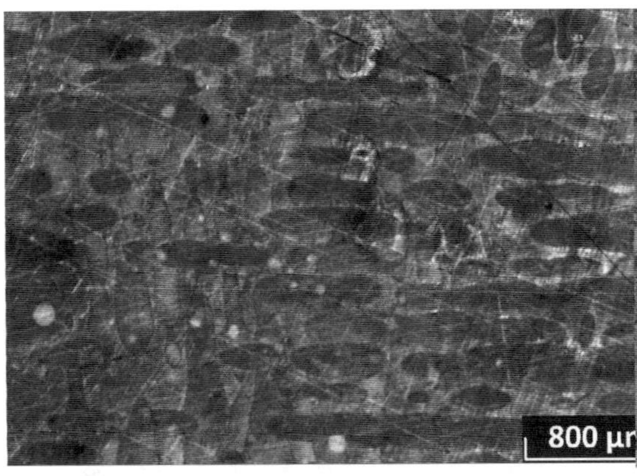

Figure 1. Microstructure of AlSi10Mg specimen.

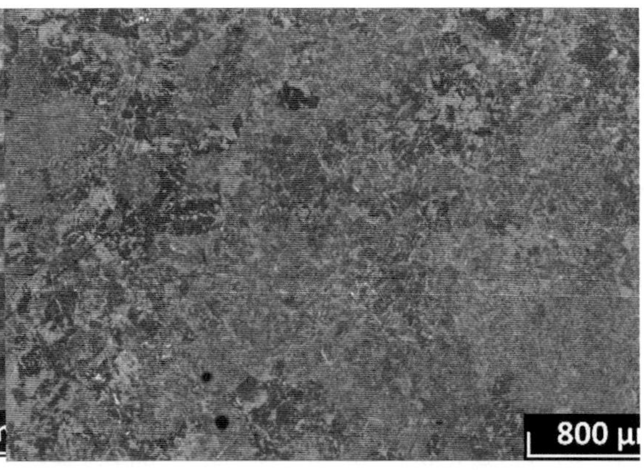

Figure 2. Microstructure of Ti6Al4V specimen.

On the pictures above there is a demonstration of microstructure in a cross-section of specimens made normal to the printing direction. Main conclusion that can be made here in terms of material behavior is that these structures might provide anisotropic properties of the material on macroscale. It is clearly visible on the Figure 1 where crystallites have pronounced elongation in two directions.

The microstructure on the above pictures is observed in a cross-section of each specimen made normal to the printing direction. On the Figure 1 it is clearly visible that crystallites have pronounced elongation in two directions. This fact leads to a conclusion that these structures might provide anisotropic properties of the material behavior on macroscale.

Characteristics of considered additive manufacturing processes and the conclusion made above allows to focus the methodology on the orthotropic model of elasto-plastic behavior.

Generally, the methodology proposed consists of several steps as follows:

- mathematical formulation of the material model based on the continuum mechanics approaches;
- identification of the equation material constants from the experiment;
- model verification.
- In this paper an additional step with the demonstration of applying constructed material model is performed.

3. Mathematical model of the material

Elasticity

In this study the orthotropic elasto-plastic model is constructed step-by-step. First of all, the elastic behavior is formulated by Hooke's law in tensor notation [10],

$$\boldsymbol{\sigma} = \mathrm{C} \cdot\cdot\, \boldsymbol{\varepsilon} \tag{1}$$

where $\boldsymbol{\sigma}$ – Cauchy stress tensor; C - elasticity tensor; $\boldsymbol{\varepsilon}$ – strain tensor.

The formula 1 can be represented in the inverse form in matrices notations and with use of the engineering constants:

$$\begin{Bmatrix} \varepsilon_{xx} \\ \varepsilon_{yy} \\ \varepsilon_{zz} \\ \gamma_{xy} \\ \gamma_{xz} \\ \gamma_{yz} \end{Bmatrix} = [S] \cdot \begin{Bmatrix} \sigma_{xx} \\ \sigma_{yy} \\ \sigma_{zz} \\ \sigma_{xy} \\ \sigma_{xz} \\ \sigma_{yz} \end{Bmatrix} \tag{2}$$

where $[S]$ – compliance tensor in matrix form:

$$[S] = \begin{pmatrix} \frac{1}{E_x} & -\frac{\mu_{xy}}{E_x} & -\frac{\mu_{xz}}{E_x} & & & \\ -\frac{\mu_{yx}}{E_y} & \frac{1}{E_y} & -\frac{\mu_{yz}}{E_y} & & 0 & \\ -\frac{\mu_{zx}}{E_z} & -\frac{\mu_{zy}}{E_z} & \frac{1}{E_z} & & & \\ & & & \frac{1}{G_{xy}} & 0 & 0 \\ & 0 & & 0 & \frac{1}{G_{xz}} & 0 \\ & & & 0 & 0 & \frac{1}{G_{yz}} \end{pmatrix} \tag{3}$$

The following notation was used in the formula above:

E_i – Young's modulus

G_{ij} – shear modulus

μ_{ij} – Poisson's ratio

We can observe twelve different material constants in the equation 3. However, there are three additional equation defining the relation between these constants. So, only nine of them are independent. Therefore, in order to fully describe elastic behavior of the material it is necessary to find these nine constants: three Young's moduli, three shear moduli, three Poisson's ratios

Plasticity

After defining the elastic behavior, the yield criterion should be formulated. The assumption of the orthotropy in material properties makes it impossible to utilize von Mises yield criterion as it can't describe differences in values of yield stress in proposed directions. Hill yield criterion is one of the most common way to describe the moment of transition to plasticity in case of orthotropic material. It can be represented in several ways. In this study the following equation was used [11]:

IOP Conf. Series: Materials Science and Engineering **986** (2020) 012034 doi:10.1088/1757-899X/986/1/012034

$$\sigma_{eq} = \sqrt{\frac{3}{2(H+F+G)}\left(H(\sigma_{xx}-\sigma_{yy})^2 + F\left(\sigma_{yy}-\sigma_{zz}\right)^2 + G(\sigma_{zz}-\sigma_{xx})^2 + 2N\sigma_{xy}^2 + 2M\sigma_{xz}^2 + 2L\sigma_{yz}^2\right)}$$

$$\sigma_{eq} \geq \sigma_y \qquad (4)$$

where σ_{eq} – equivalent stress; H, F, G, N, M, L – coefficients of material; σ_y – threshold (yield) stress. Six material constants H,F,G,N,M,L in the criterion (4) can be calculated from the values of yield stresses in each direction including shear ones. These six constants in the criterion allow to define the transition to plasticity by one specific value of threshold stress which will be compared to equivalent one calculated by the formula (4). According to this fact it is necessary to obtain six values of yield stress to define the yield criterion. These six values are three tension yield stresses and three shear yield stresses.

The last step in the process of defining the orthotropic elasto-plastic material model is description of the plastic material behavior. One of the possible ways here is to use stress-strain curve constructed from several specific points. Axes of this curve should be assigned as follows: abscissa is equivalent stress, calculated by the formula (4), and ordinate presents plastic strain. This choice should provide an alignment of stress-strain curves of any type and direction of loading. This means that chosen axes "scale" these curves and fit them to the one.

4. Testing and experimental data processing

Testing description

The list of material properties to obtain is formulated according to proposed material model. These properties are supposed to be found from several mechanical experiments: uniaxial tension in three directions and uniaxial torsion around three axes.

Tension experiments provide information about Young's moduli, Poisson's ratios and three yield curves. Specimens with circle cross-section for tension experiments are designed according to [12] and presented on Fig. 3.

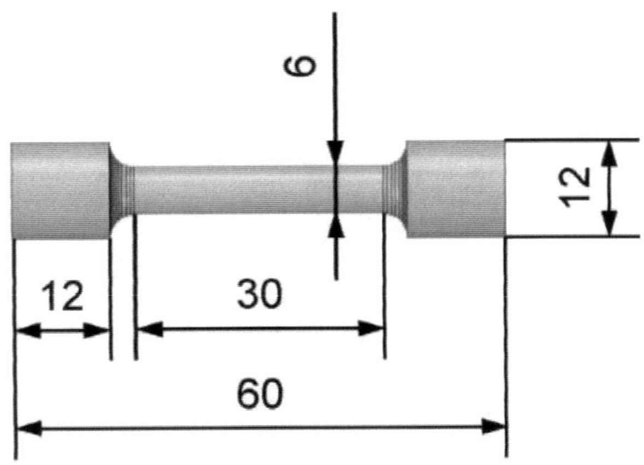

Figure 3. Uniaxial tension specimen (sizes provided in mm.

To carry out tension tests a test machine Zwick Z250 was used. Also the laser extensometer Zwick LaserXtens HP was used in order to measure transverse strain for Poisson's ratio estimation.

The torsion experiments give information about shear moduli and another three yield curves. Specimens with circle cross-section for torsion experiments are designed according to [13] and presented on Fig. 4.

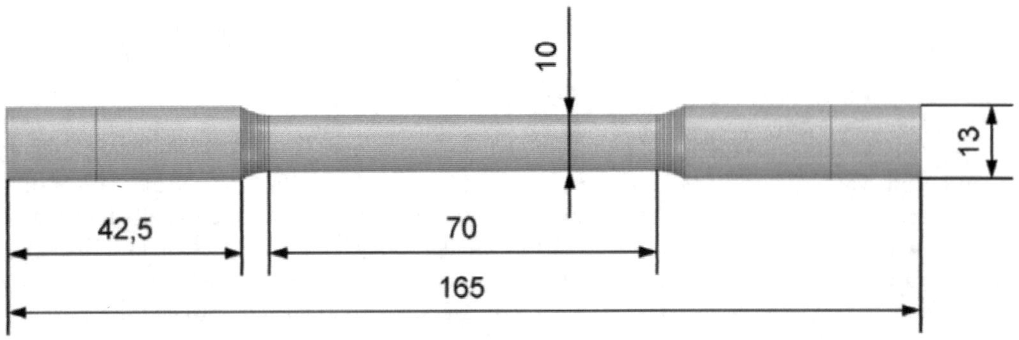

Figure 4. Uniaxial torsion specimen (sizes provided in mm).

A test machine Instron 8850 was used to carry out torsion tests.

Because of the assumption about orthotropic material behavior it is required to test each type of described loading in three directions. These directions should match the principal axes of anisotropy of a material. That's why it was decided to align the direction of loading of each specimen with the axes of the printing machine. The figure 5 is demonstrating the torsion specimens' orientation during printing process. Orientation of tension specimens was made in the same way.

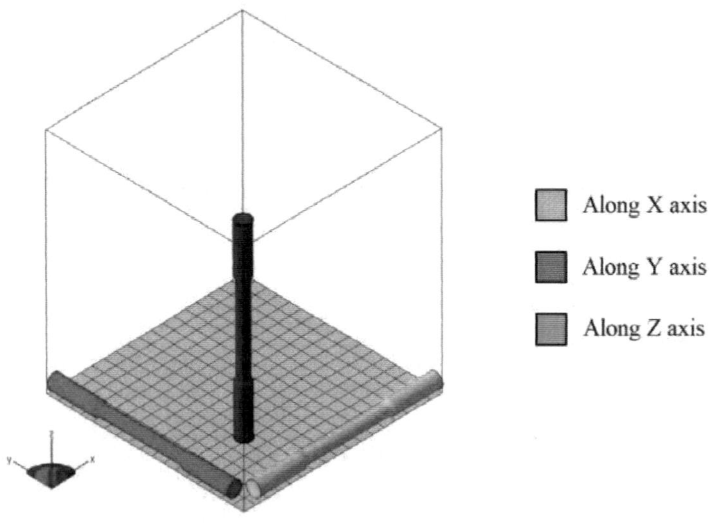

Figure 5. Torsion specimens` orientation.

To obtain values of desired material properties it is necessary to process the experimental data. The results of performed tests to process are stress-stain (load-displacement) curves.

In terms of minimizing the influence of random factors like manufacturing defects a set of three specimens for each orientation and type of loading is produced. Such specimens' amount makes it possible to perform a statistical processing of required values.

4.1. Experimental data processing

The processing of each coefficient of the model is based on techniques described in [12, 13, 14].

Young's modulus and tension yield stress of one direction obtaining is performed iteratively as follows. On a stress-strain curve a linear section is selected. Bounds of this section are determined based on the expected value of yield stress. Lower bound is 10% of this value, upper is 85%. This is needed to exclude all nonlinearities. Using the least squares method, a linear approximation of this section is done. A coefficient of this approximation is an applicant for the Young's modulus value. Afterwards plastic strain values are calculated by the subtraction of elastic strain from full ones. The stress value which corresponds to 0.2% plastic strain value is considered to be the yield stress value. If this value is equal to the expected yield stress value used in the beginning of the process, we consider that processing is completed and values are final. If not, we repeat the process with use the obtained yield stress value as new expected one. The described process is visually presented as a diagram on the Figure 6.

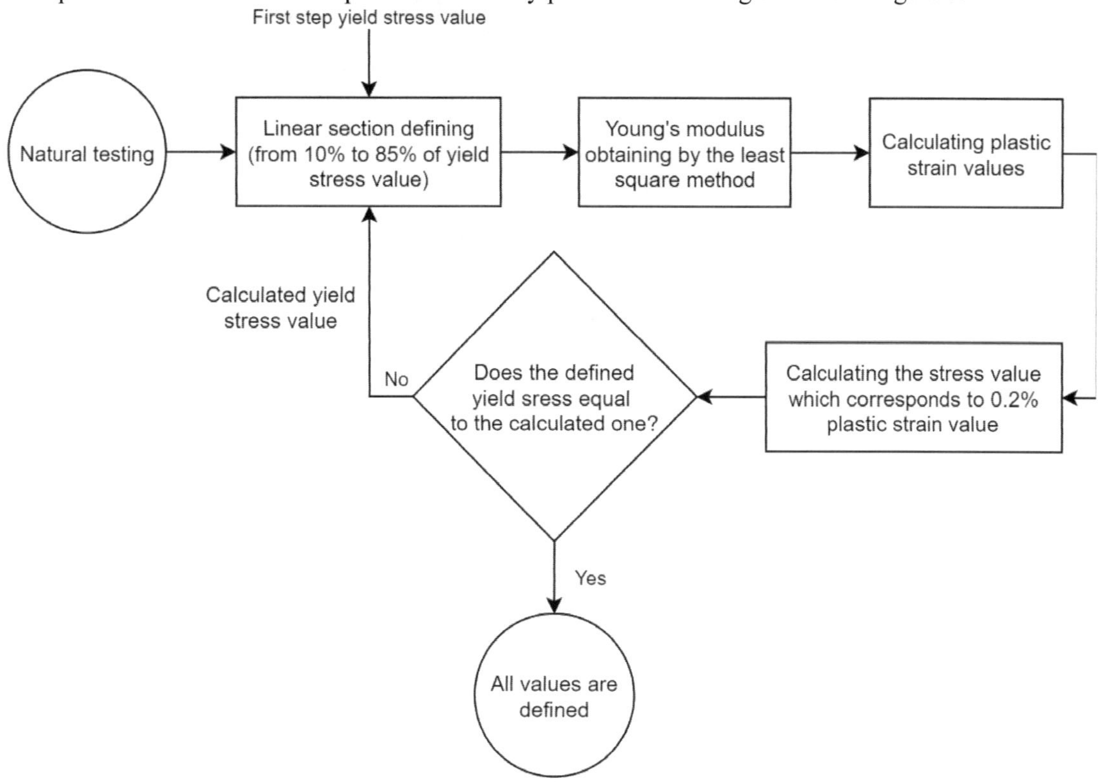

Figure 6. Tension test data processing diagram.

The process described above is performed for each tested specimen. So, each required value (Young's modulus and tension yield stress) is defined three times. This allows to calculate the final value as average from three available ones.

To estimate the value of the Poisson's ratio it is required to use results of measuring transverse strain values. The value of Poisson's ratio is calculated as a relation between the transverse strain to the axial strain measured in the linear section of stress-strain curve. As well as in the case of Young's modulus the final value of the Poisson's ratios in each direction is calculated as an average from three available ones.

All described processing techniques are relatively widely used and there is no effect on them from the proposed material model type. But in case of torsion problem it is impossible to easily use these methods, due to assumption about orthotropic material behavior. As it was mentioned before, obtaining the values of shear moduli is based on results of uniaxial torsion experiments. Its procedure is clear and easy in case of isotropic material model [13]. The value of shear modulus is calculated just as a ratio of shear stress to shear strain in a linear part of stress-strain curve. But if we take into account the assumption about orthotropic behavior the stress-strain state will be noticeably different. That's why in case of torsion problem it is necessary to perform an additional research.

Calculation submitted below are based on the anisotropic models of continuum mechanics [10].

First of all, we consider that we get the amount of values of torque and twist angle as a result of torsion tests. We assume the uniaxial torsion is performed about Z axis (Fig. 7).

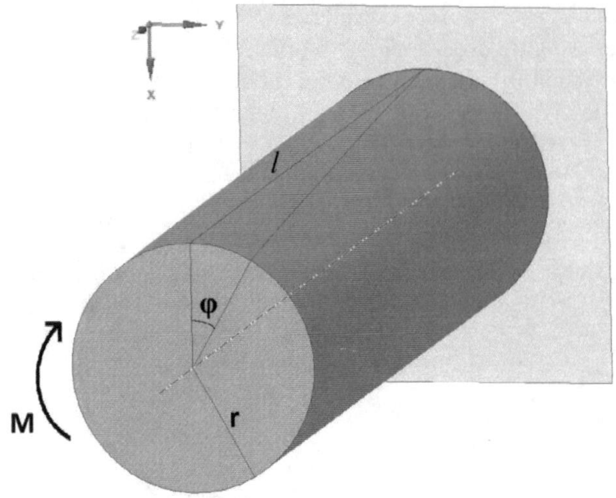

Figure 7. The statement of a torsion problem.

Let's assume the following relation between torque and twist angle:

$$\varphi = \frac{Ml}{C} \tag{5}$$

Second thing to mention is that there are two independent shear moduli which affect the stress-strain state during the uniaxial torsion along proposed axis. In case of torsion about Z axis these two moduli are G_xz and G_yz. This means that the relation between torque and twist angle includes both these moduli. We can describe the relation between torque and twist angle as torsional stiffness which can be calculated for circled cross-section as follows [10]

$$C = \frac{\pi r^4}{\left(\frac{1}{G_{xz}} + \frac{1}{G_{yz}}\right)} \tag{6}$$

From the above equation it is clear that it is impossible to get the values of two independent shear moduli from the results on one test. To resolve this problem all below calculations are performed.
We can obtain the following equation based on previous two

$$\varphi = \frac{Ml}{2J_p}\left(\frac{1}{G_{xz}} + \frac{1}{G_{yz}}\right) \tag{7}$$

And finally let's introduce B as follows

$$\frac{1}{B_z} = \frac{1}{2}\left(\frac{1}{G_{xz}} + \frac{1}{G_{yz}}\right) \tag{8}$$

Now the value of twist angle can be calculated in this way

$$\varphi = \frac{Ml}{J_p B_z} \tag{9}$$

These steps allow us to have the value which is possible to obtain from one experiment. This value is introduced to represent an average response of the material to the torsion loading in terms of two shear moduli. Main idea here is that it becomes possible to find the value of B with the procedure from [13] which relatively copies the procedure of obtaining Young's moduli. Now to get the value of each shear moduli we need obtaining same two values of B from experiments along X and Y axes. After this step we have three B values which are defined as follows

$$\begin{cases} \frac{2}{B_z} = \left(\frac{1}{G_{xz}} + \frac{1}{G_{yz}}\right) \\ \frac{2}{B_y} = \left(\frac{1}{G_{xy}} + \frac{1}{G_{yz}}\right) \\ \frac{2}{B_x} = \left(\frac{1}{G_{xz}} + \frac{1}{G_{xy}}\right) \end{cases} \quad (10)$$

This nonlinear system of equations has the following solution

$$\begin{cases} G_{xy} = \frac{B_x B_y B_z}{(-B_x B_y + B_x B_z + B_y B_z)} \\ G_{xz} = \frac{B_x B_y B_z}{(B_x B_y - B_x B_z + B_y B_z)} \\ G_{yz} = \frac{B_x B_y B_z}{(B_x B_y + B_x B_z - B_y B_z)} \end{cases} \quad (11)$$

Now all the required values of shear moduli are obtained.
Torsion yield stresses obtaining is also connected with the same difficulties caused by the problem explained above. All these difficulties can be resolved with calculations based on [10] and [11]. These analytical calculations are performed with a relatively same approach. The idea of these calculations is to obtain a full stress-strain state of a specimen up to the moment of transition to plasticity. Plastic strain values can be calculated with use of already processed shear moduli. The stress value at the moment when plastic strain reaches 0.2% is taken as shear yield stress.

4.2. Material properties obtained
Following Tables 1-3 contain several results of experimental data processing in all proposed directions.

Table 1. AlSi10Mg material properties.

Material property/ Direction	**X/XY**	**Y/YZ**	**Z/XZ**
Young's modulus [GPa]	82.5	76.7	76.8
Shear modulus [GPa]	23.8	27.1	21.4
Tension yield stress [MPa]	227.8	232.8	217.7
Torsion yield stress [MPa]	132.9	135.7	140.4
Poisson's ratio	0.32	0.32	0.33

Table 2. Ti6Al4V material properties.

Material property/ Direction	**X/XY**	**Y/YZ**	**Z/XZ**
Young's modulus [GPa]	124.2	121.9	124.1
Shear modulus [GPa]	37.5	41.8	42.0
Tension yield stress [MPa]	983.7	980.9	964.2
Torsion yield stress [MPa]	588.6	525.8	571.1
Poisson's ratio	0.26	0.25	0.26

Table 3. Parameters of Hill's plasticity models for AlSi10Mg and Ti6Al4V.

Parameter type	**F**	**G**	**H**	**L**	**M**	**N**
AlSi10Mg	0.517	0.559	0.424	1.444	1.293	1.384
Ti6Al4V	0.515	0.510	0.479	1.724	1.461	1.375

IOP Conf. Series: Materials Science and Engineering **986** (2020) 012034 doi:10.1088/1757-899X/986/1/012034

Values summarized in Tables 1-3 show that the materials studied have properties variations in different directions. This means that the assumption about anisotropic behavior was done correctly and that chosen model describes material closer to the real life than widely used elastic isotropic one.
These constants will be used in further finite element simulation and analysis of the obtained results.

5. Model verification

Models description
Two constructed material models were checked for compliance to experimental data. Calculations were performed in SIMULIA Abaqus for each test's direction. An orthotropic elasto-plastic material model was used. A short description of the modeling process is performed below.
On Figures 8-9 there are models which were used in FEA system to control the alignment of the obtained properties to real life ones. In the modeling process only the gauge section of a specimens is used in order to reduce computational complexity.

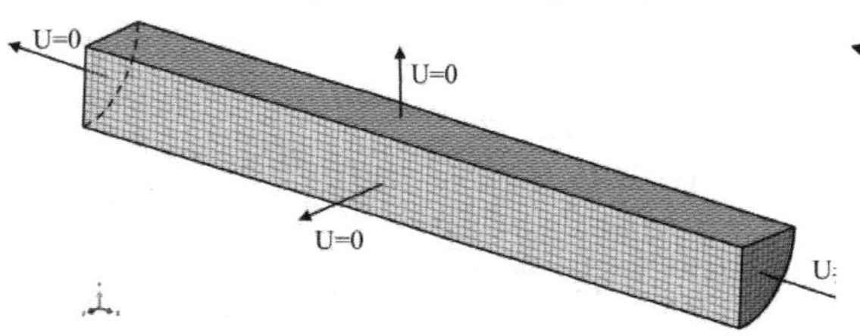

Figure 8. Tension model.

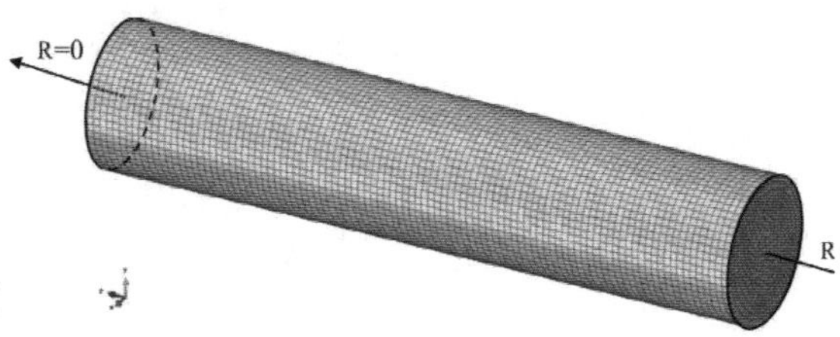

Figure 9. Torsion model.

Tension model includes 9600 linear hexa-elements.
Boundary conditions for tension modeling are as follows:
- On the flat side surfaces – symmetry conditions
- On left end surface – zero displacement along normal direction
- On the second opposite end surface – nonzero displacement along normal direction

Torsion model includes 44500 linear hexa-elements.
Boundary conditions for tension modeling are:
- On the one end surface – zero rotation along normal direction
- On the second opposite end surface – nonzero rotation along normal direction

As it seen in above description, we used kinematic boundary conditions. This displacement is applied in 100 steps with linear growth from 0.1% to 100% of its final value. Such way of defining boundary conditions allows to extract stress-strain curve in a resolution good enough to analyze it and compare with experimental data. The displacement is applied to the model through a reference point which is connected with end surface.

6. Results

On Fig. 10-13 there is a demonstration of results of finite elements simulation and experimental data. This allows to check the quality of constructed models. Such analysis was performed for each test done. Here there is an example of tests results along Y axis for two proposed materials.

IOP Conf. Series: Materials Science and Engineering **986** (2020) 012034 doi:10.1088/1757-899X/986/1/012034

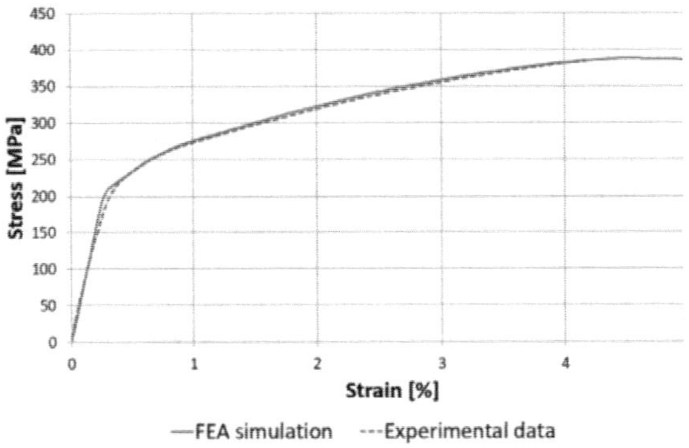

Figure 10. Tension model.

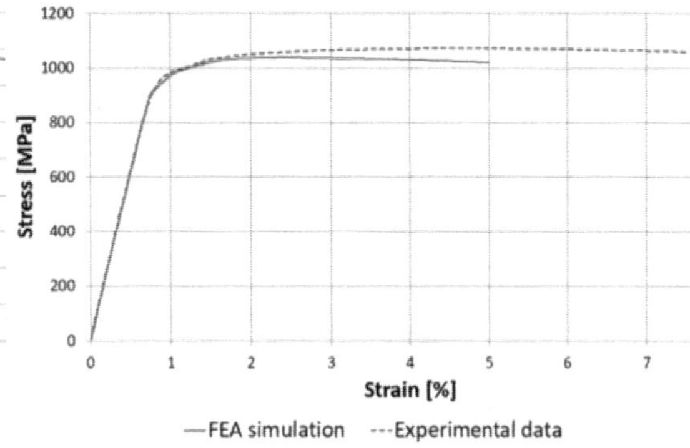

Figure 12. Torsion model.

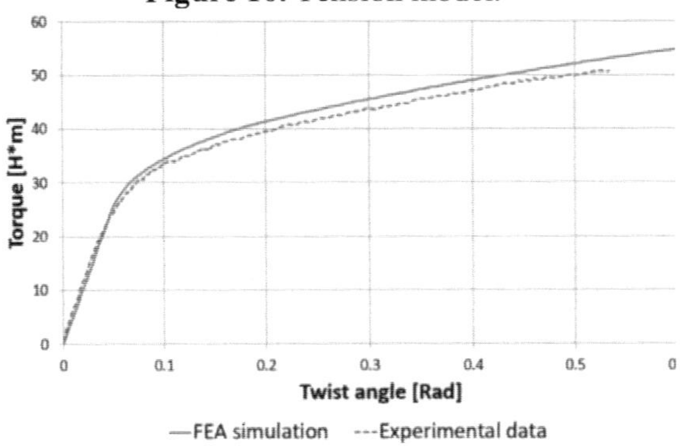

Figure 11. Torsion of AlSi10Mg specimen about Y axis.

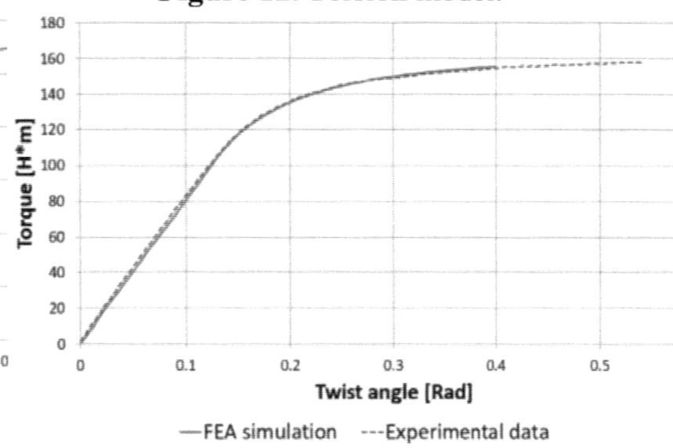

Figure 13. Torsion of Ti6Al4V specimen about Y axis.

7. Lattice structure FEA simulation

In this part there is an example of using a constructed material model in relatively complicated FEA simulation. This example is needed to demonstrate model applicability to structural analysis. 3d-printed lattice structure was chosen as a design for this research. This choice was made due to the fact that lattice structure is a well-known demonstration of additive manufacturing capabilities in terms of model complicity and further application possibilities.

Performed research was made for aluminum alloy printed lattice and consists of several steps: model preparation, natural testing, FEA simulation, result analysis. Lattices were tested to investigate their behavior in cases of uniaxial tension and compression. We decided to test both tension and compression as the lattice structure behavior may vary depending on the sign of the applied load [15].

7.1. Natural testing

As it was mentioned before, uniaxial tension and compression tests were carried out. Since structures have complex geometry, it is not possible to mount them in the test machine directly without any additional gear. So the specimens were adapted to the test machine by adding special blocks for installing. The shapes of the specimens and a basic cell are presented in Fig. 14.

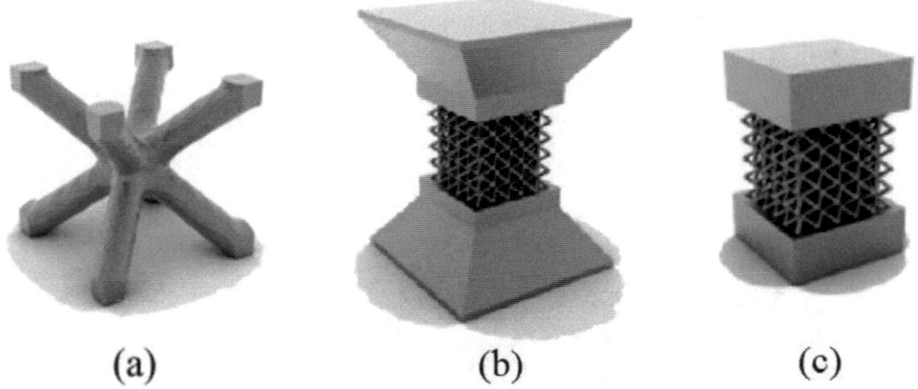

Figure 14. Shapes of (a) basic cell of specimen (b) tension specimen (c) compression specimen.

Tests were carried out with Instron 8850 Axial-Torsion System. The photographs of the specimens during tests are shown in Fig. 15.

Figure 15. Deformed shape of AlSi10Mg specimens.

7.2. Finite element modeling

FE modeling of uniaxial tension and compression was performed in SIMULIA Abaqus implicit solver with the same finite element models both for tension and compression. Constructed model is shown in Fig. 16.

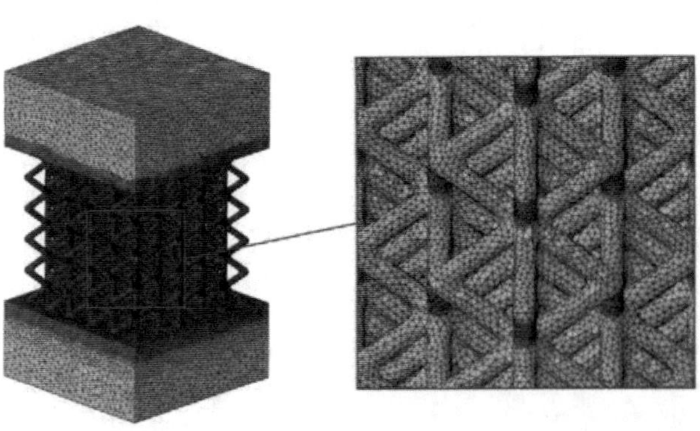

Figure 16. Finite element model for modeling of the experiments.

The characteristics of the finite-element model are as follows:
- element type – C3D4 (four-node linear tetrahedra);
- number of elements – 1 156 665;
- number of nodes – 273 733.

During the simulation the bottom face of the specimen is fully fixed, the top face has a prescribed axial displacement.
One run takes about five hours.

7.3. Results
On Fig. 17 – Fig. 18 there are curves obtained from the natural testing and FEA modelling. Axes of the curves are: abscissa is the axial reaction force and ordinate is the displacement of the top face.

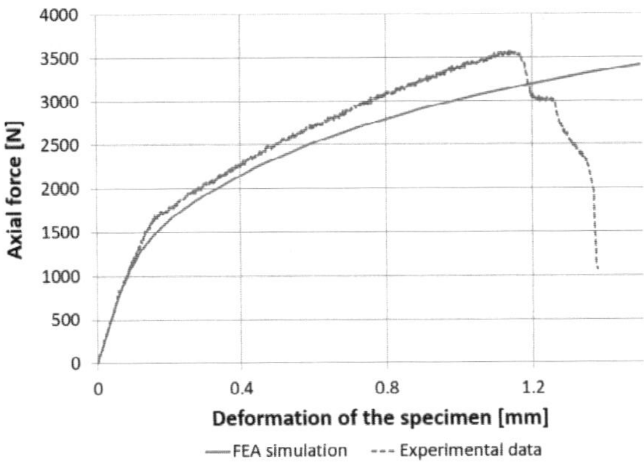

Figure 17. Force-deformation curve for tension.

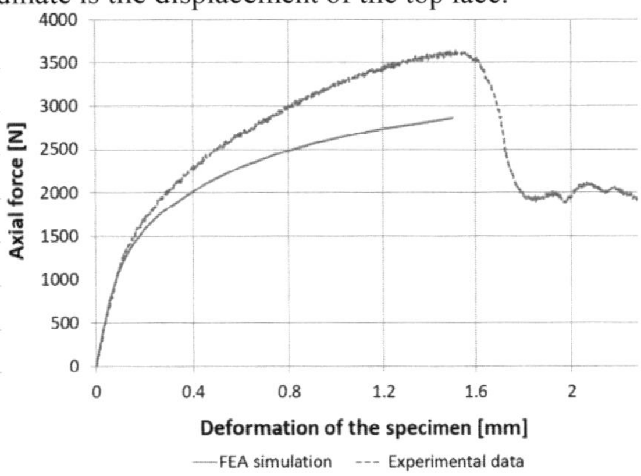

Figure 18. Force-deformation curve for compression.

Above figures show that the results of FEA simulation have good correlation in the linear range for both tension and compression cases. Also specimens' behavior at high level of deformation is described by the constucted model quite well. This model doesn't describe fracture region as if we didn't introduce fracture criteria into the model.

8. Discussion

In order to prove the assumption of orthotropic material behavior the analysis of material microstructure was performed. The analysis was based on several slice images obtained by using electronic microscopy in different cross-sections. Two of these slices were presented in this paper on Figures 1-2. From the analysis of all these slices several conclusions were made.

Despite the microstructure of researched materials closely matches their cast structure, it was discovered that AlSi10Mg crystallites have pronounced elongation in two orthogonal directions. This can be explained by the influence of the scanning strategy of a printing machine. This means that the path of laser or electron beam may define the shape of crystallites. Such irregularity in crystallite shape causes the differences in material properties in different directions. This can be observed in Table 1. Ti6Al4V structure is relatively more regular than AlSi10Mg one. This also can be observed in the deviations of material properties. Obtained Ti6Al4V constants are closer to each other in comparison to AlSi10Mg ones.

All these conclusions may prove the assumption that additive manufacturing process causes material properties change and that it is the reason of the material anisotropy.

As well there are some conclusions can be made based on the results of finite elements modeling for tested specimens and for lattice structures.

Main conclusion here is that constructed models describe elastic material behavior with high precision both in case of tested specimen and complicated nonlinear structure. This is seen by the coincidence of the linear parts of curves on all the performed the graphs. Moreover, constructed models provide the moment of transition to plasticity in the right moment in terms of stresses. This is confirmed by curves alignment in the early plasticity section of the graphs. One more point is that the models have noticeable deviations from experiments in the late plasticity region. But the global behavior still described quite well. Such deviations may be caused by the assumption of a single plastic curve which was made before. This assumption gave the possibility to properly and accurate describe transition to the plasticity and its early region, but to describe later regions some more assumptions need to be done.

9. Conclusions

All conclusions made before are mostly related to the performed research of material properties of additively produced parts, but not to the proposed methodology of the modeling of these materials which was the mail goal of the whole study. Despite all of those conclusions are relevant and may be used in further research, they all are still parts of the methodology, i.e. steps to be done during the material modeling. General conclusions which can be related to the methodology are presented below.

The proposed methodology contains a detailed description of a modeling process of metal materials obtained with layer-by-layer technology. Each step of this modeling process is provided with explanation and theoretical motivation. The example of applying this methodology is made for two additively produced metal alloys (AlSi10Mg and Ti6Al4V).

This methodology allows design and research engineers to easily master the submitted modeling process and know how to build a highly adequate model of additively produced materials for their needs. All the results obtained with the use of this methodology are strongly connected with the manufacturing process (printing machine and settings). This means that it is possible to obtain several different models of one metal material which was produced on different machines. This methodology may be a base for constructing even more complicated models which can contain extra properties for special purposes.

References

[1] Jing Zhang and Yeon-Gil Jung Additive 2018 Manufacturing : Materials, Processes, Quantifications and Applications *United Kingdom Butterworth-Heinemann* p 352

[2] Frazier W E 2014 Metal additive manufacturing: a review *J. Mater. Eng. Perform.* (vol. 23, no. 6) pp 1917–1928

[3] Polozov I, Sufiiarov V and Popovich A 2020 Investigation of Ti-6Al-4V alloy in situ manufactured using selective laser melting from elemental powder mixture *Solid State Phenomena*. (299:646-651)

[4] Hiemenz J 2007 Electron Beam Melting *Advanced Materials and Processes* (165(3)) pp 45-46

[5] Facchini L, Magalini E, Robotti P and Molinari A 2009 Microstructure and mechanical properties of Ti-6Al-4V produced by electron beam melting of pre-alloyed powders *Rapid Prototyp. J.* (15(3)) pp 171–78

[6] Rosenthal I, Stern A and Frage N 2014 Microstructure and mechanical properties of AlSi10Mg parts produced by the laser beam additive manufacturing (AM) technology *Metallogr. Microstruct. Anal* (3(6)) pp 448– 53

[7] Carroll BE, Palmer TA and Beese AM 2015 Anisotropic tensile behavior of Ti-6Al-4V components fabricated with directed energy deposition additive manufacturing *Acta Mater.* (87:309–20)

[8] Popovich A A, Sufiiarov V S, Borisov E V, Polozov I A, Masaylo D V and Grigoriev A V 2017 Anisotropy of mechanical properties of products manufactured using selective laser melting of powdered materials *Russ. J. Non-ferrous Metals* (58) pp 389–395

[9] Simonelli M, Tse Y Y and Tuck C 2014 Effect of the build orientation on the mechanical properties and fracture modes of SLM Ti–6Al–4V *Mater. Sci. Eng. A* (vol. 616) pp 1–11

[10] Lehnickii S G 1977 Continuum mechanics of anisotropic body *Moskow. Nauka* p 416

[11] Pisarenko G S and Mojarovskii N S 1987 Equations and boundary problems of the theory of plasticity and creep: a reference manual (Kiev. Naukova Dumka) p 496.

[12] USSR Standart 1497-84 Metals 2005 Tension tests methods (Moskow. Standartinform) p 24

[13] USSR Standart 3565-80 Metals 1980 Torsion tests methods (Moskow. Standarts publishing office) p 17

[14] Help V 2019 Instron Bluehill 2 Modulus calculations [electronic resource] *Illinois Work Tools, Inc., Norwood* (USA Access mode: https://www.instron.com, free)

[15] Ushijima, Cantwell, Mines, Tsopanos and Smith 2010 An investigation into the compressive properties of stainless steel micro-lattice structures *Journal of Sandwich Structures and Materials* (13) pp 303-29

Some aspects of custom-made 3d-printed hip joint implant structural simulation

L Maslov[1,2] a, b,*, M Zhmaylo[1], I Maslova[2], S Purkova[2], F Tarasenko[2], E Musaev[3], E Sushentsov[3] and T Barriere[4]

[1]Center of National Technological Initiative, Institute for Advanced Manufacturing Technologies, Peter the Great St. Petersburg Polytechnic University, 29 Politekhnicheskaya, St. Petersburg, 195251, Russia, E-mail address: leonid-maslov@mail.ru
[2]Department of Theoretical and Applied Mechanics, Ivanovo State Power Engineering University, 34 Rabfakovskaya, Ivanovo, 153003, Russia
[3]Federal State Budgetary Institution N.N. Blokhin National Medical Research Center of Oncology of the Ministry of Health of the Russian Federation, 23 Kashirskoye Highway, Moscow, Russia 115478
[4]Applied Mechanics Department, Institute "Franche-Comté Electronics Mechanics Thermal Science and Optics – Sciences and Technologies", 15B avenue des Montboucons, Besançon, France 25030

Abstract. Current article is devoted to the application of computer-aided engineering technologies to the virtual testing of customized endoprostheses of large joints. The research is performed for a representative case, where the personalized prosthesis of the hip joint is used as a part of the process of treatment of osteosarcoma, as conventional systems are not applicable. The study is highly relevant due to the intensive development of personalized medicine allowing for the patients' successful recovery and mobilization. However, since each of the customized implants is a unique part, a rapid, accurate and financially efficient methodology of performing the structural analysis of such components is required to avoid the real testing of each implant. The article describes the all steps of the structural simulation of the series of loading scenarios for the biomechanical system consisting of the pelvis and endoprosthesis. The process of preparing the finite element models of pelvic bones and components of the implant is also explained in detail. The stress-strain state of the pelvis and implant is investigated for a number of values of the pretension force of the screws. Based on the comparative analysis, a set of practical recommendations is formulated to reduce the stress in the bone tissue. The discussed steps are the elements of the methodology of the rapid finite element virtual testing of the customized implants that is currently developed by the research group.

1. Introduction

Every year, several million [1] of total hip arthroplasty surgeries are carried out in the world, and this number grows every year. In recent decades, such type of arthroplasty has become a standard procedure. The main reasons for the hip replacement are arthrosis and other types of degenerative diseases caused by the destruction of the cartilage of the hip joint.

Currently, due to the significant development of additive manufacturing technologies based on the metal powder compositions, the 3D printing of customized endoprostheses has become very common and popular. The key advantage of this approach is that such type of implant precisely fits the shape of the pelvis including the areas with defects.

The development of the market of customized orthopedic structures can be considered as the result of the recent improvements of the engineering software technologies. These technologies allow using the computer tomography data to generate the geometrical model of the bones, perform the surgery planning and design the customized implant.

The engineering simulation technologies such as finite element modeling make it possible to perform virtual tests of the designed implants. These tests are very important as the real testing of each implant will significantly increase the cost of such solution. However, the virtual testing or structural analysis itself leads to a number of specific technical issues.

The article presents the intermediate results of the development of the methodology of rapid virtual testing of customized implants. The process of preparing the finite element model, running the simulation and carrying out the analysis of the results is explained, and recommendations regarding the possible changes in the implant design are formulated.

2. Review of current studies

Numerical simulations based on the patient's computer tomography and the corresponding digital models of implants are widely used in modern orthopedics to analyze the stress-strain state of the prostheses of large joints [2]. This approach allows getting the stress and strain distribution in the implant and bone tissue and to assess their strength. Study [3] presents the results of the structural analysis of the hip reconstruction using rod-screw systems. The same method is used for simulation of the pelvic ring reconstruction with the double-barreled vascularized fibular free flap [4, 5] and for investigating the efficiency of the internal fixation of implants [6, 7].

In the considered articles, the finite element models are prepared on the basis of the geometry of the patient's skeleton elements obtained as a result of computer tomography. The implants are simulated as deformable three-dimensional solids. A vertical load is applied to the lumbar vertebrae, and the biomechanical system is analyzed in the standing position. Based on the prepared models, the strength of the system is assessed and the conclusions regarding the possible changes to the shape of the implant are made.

The paper [8] describes the results for a modular endoprosthesis that is used for the reconstruction of a half of the pelvis. A comparative analysis of the stress distribution is done between the normal pelvis and the pelvis with the implant considering three static positions: sitting, standing on two legs and standing on a leg of the injured side. The loads, areas of their application and boundary conditions are similar to those described in [3].

In clinical medicine, various reconstruction methods are used to treat osteosarcoma. The typical methods are arthrodesis, alloplasty, autoplasty and various types of prosthetics. The introduction of computer-aided design and engineering technologies, as well as additive manufacturing, opens up new opportunities in the reconstruction of the pelvic bones. Digital technologies make it easier to create geometrically accurate customized implants that can be used as an alternative to massive bone allografts, and can also be used in cases, where modular prostheses cannot be used [9]. The use of selective laser melting (SLM) and electron beam melting (EBM) technologies allows creating porous structures for better osseointegration [10, 11].

The practical results of implementation of the 3D printed implants in conventional and oncological orthopedy confirm that the patients with the individually designed implant show successful recovery and mobilization, which are the sequences of high anatomical accuracy of the structures, availability of various fixation methods and high manufacturing speed [9, 12, 13].

3. Object of the research

The article presents the process of preparing the finite element models and performing the structural analysis of the system containing the pelvis of a 14-year-old patient with the diagnosis of osteosarcoma of the left ilium.

The patient was admitted to the clinic of the Research Institute of Pediatric Oncology and Hematology of the National Medical Research Center of Oncology named after N.N. Blokhin with complaints of

IOP Conf. Series: Materials Science and Engineering **986** (2020) 012035 doi:10.1088/1757-899X/986/1/012035

pain in the left hip joint and limitation of mobility (Fig. 1). After the neoadjuvant chemotherapy, it was decided to perform the surgery with pelvic resection and reconstruction with a personalized implant.

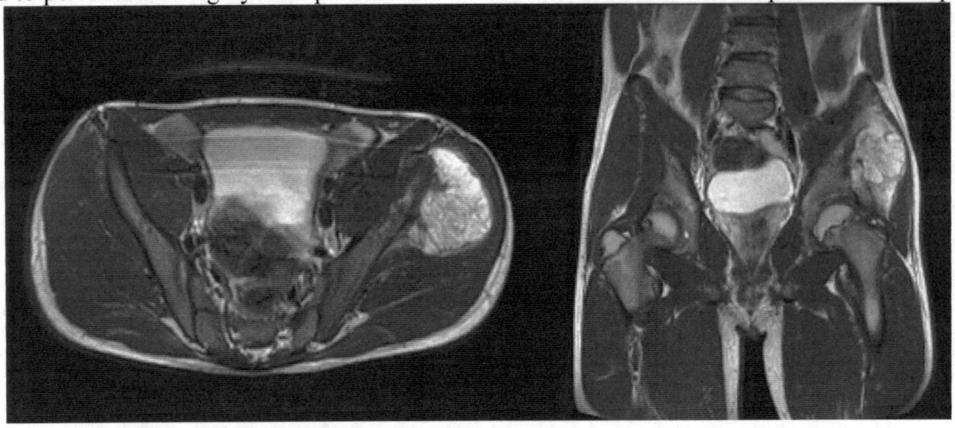

Figure 1. Magnetic resonance tomography image of the patient [14].

At the first stage of the design process of the endoprosthesis, the three-dimensional geometric models of the pelvic bones in STL format were prepared using the data obtained from the computer tomography study. On their basis, the specialists of the Research Center and their manufacturing partners determined the boundaries of the resection and developed the virtual model of the operation of osteosynthesis with the customized implant.

Using the anatomical data of the patient and the information about the surgery, the implant design was developed in such a way that it reconstructs the damaged regions of the bone tissue (Fig. 2). The implant includes the hemispherical acetabular cup and supporting flanges connecting the part with the ilium, ischium and pubic bone. The regions, where the implant contacts the bone tissue, were covered with a porous lattice layer to improve the osseointegration [11].

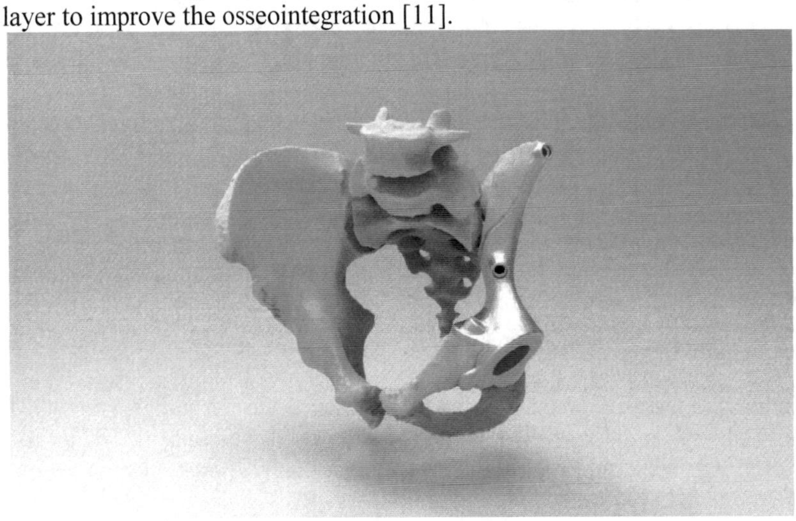

Figure 2. Location of the analyzed implant.

The part was printed with Ti6Al4V alloy powder with extra low interstitials (Fig. 3). After the printing, the implant was post-processed and examined for hidden defects.

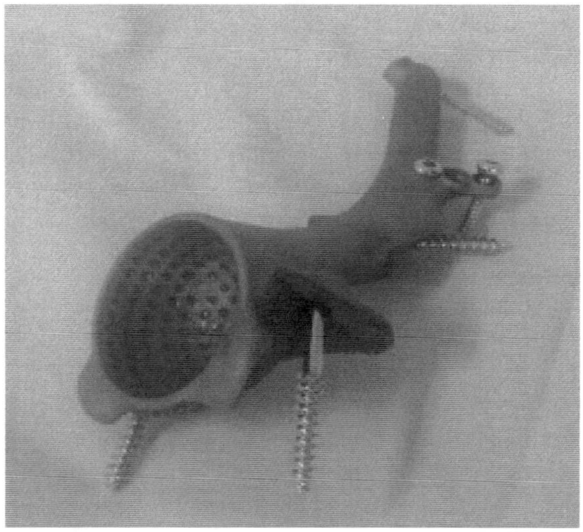

Figure 3. Additively manufactured customized implant [14].

4. Preparing the finite element model

The final goal of current project is to develop the methodology of rapid virtual testing of customized implants based on the detailed finite element analysis of the stress-strain state of the endoprosthesis and bone tissue. The first step of carrying out such simulation is the preparation of the finite element model of the object, which begins with preparing the geometric model.

Three-dimensional models of the pelvic bones generated on the basis of computer tomography data in STL-format consist of a large number of triangular surfaces and have a rather complex shape. At the first stage of the work with the model, the triangulated geometry of the pelvis is checked for intersections, missing elements and other defects. The mesh quality is fixed and homogenized (Fig. 4). While preparing the analysis model, it is taken into account that the bone has a double-layer structure and consists of the cortical and spongy layers with different properties. The outer cortical layer is formed by a stiffer and stronger material, while the inner spongy bone is much more flexible and weaker. The mesh near the screw holes is refined to increase the accuracy of the results. Based on the prepared surface mesh, a tetrahedral volume mesh is generated. The STL-mesh processing and the finite element mesh generation is performed with the special software tools.

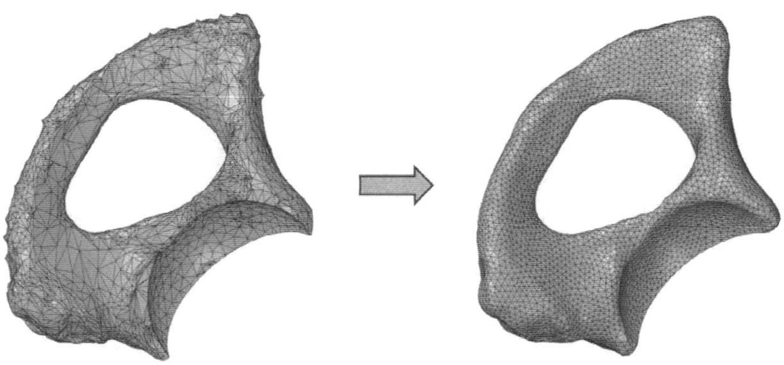

Figure 4. Results of the mesh processing from the geometrical triangular mesh to the finite element mesh.

The current study considers the presented biomechanical system in the standing position with the support from two legs. As the load case is symmetric, a half of the pelvis might be excluded from consideration if the geometry is also symmetric. Surely, the model of the pelvis after performing the resection and reconstruction is not symmetric, and the stress-strain state of the side with the prosthesis is not equal to the stress-strain state of the opposite side. However, since this research is devoted to the study of the stress-strain state of the damaged part of the pelvis, and the prosthesis allows reconstructing the shape of this part and the biomechanics of the hip joint quite well, it is assumed in the research that the opposite side of the pelvis can be replaced by the symmetry boundary conditions.

The developed finite element models were imported into SIMULIA Abaqus (Fig. 5). Such components as the polyethylene cup, femur implant and the simplified representation of the leg were added using the tools of Abaqus/CAE preprocessing system. The femoral component of the joint is modeled in the simplified way, because this part of the system is out of scope of the research.

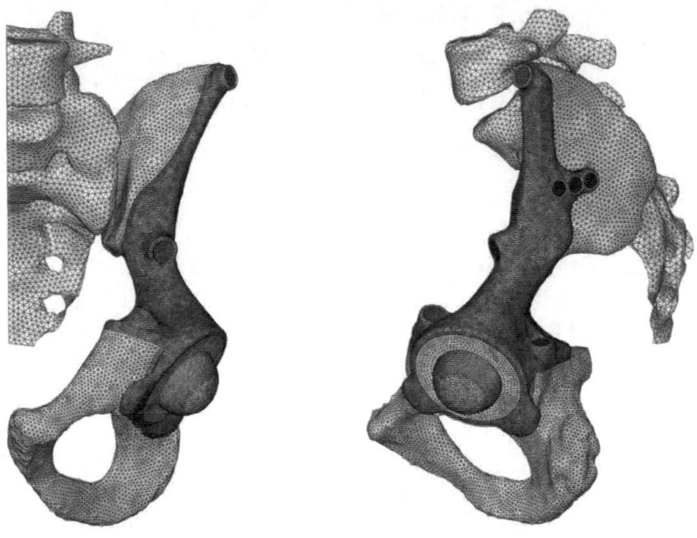

Figure 5. Finite element mesh of the pelvic, implant, acetabular cup and femur head.

The characteristics of the finite element model are:

- number of elements – 612 260;
- number of nodes – 172 997;
- number of degrees of freedom – 519 thousand.

The following element types are used:

- C3D10 – quadratic tetrahedral solid element (screws and polyethylene cup);
- C3D4 – linear tetrahedral solid element (implant, pelvic bones);
- S3 – triangular linear shell element (femur head);
- B31 – linear beam element (representation of the leg).

During the process of preparing the finite element model, additional kinematic constraints and contact interactions are created. The constraints are added to the model in order to simulate the real connections between the components of the biomechanical system. The contact interactions are specified to show the analysis system, in which regions in should control the relative positions of the parts in order to avoid their intersection.

5. Materials

The mechanical properties of spongy and cortical bone tissues were obtained experimentally in various studies and vary due to many factors. As a result of the analysis of the studies, the average values of the necessary parameters were chosen from [15]. The screws, implant and the femoral component of the implant are made of titanium alloy Ti6Al4V. The mechanical characteristics of this alloy are given

IOP Conf. Series: Materials Science and Engineering **986** (2020) 012035 doi:10.1088/1757-899X/986/1/012035

according to [16]. The polyethylene cup is made of ultrahigh molecular weight polyethylene that is characterized by high mechanical performance, low friction coefficient and biocompatibility [17].
Since the calculation takes into account the contact interaction with the nonlinear behavior, the effect of friction shall be taken into account. The friction forces are based on the coefficients from [18].
The mechanical properties of the materials are listed in Table 1, friction coefficients are listed in Table 2.

Table 1. Mechanical properties of the materials.

Material name	Density, kg/m3	Elastic modulus, MPa	Poisson ratio	Critical stress, MPa
Spongy bone	1 188	500	0.2	10
Cortical bone	1 470	10 000	0.3	160
Titanium alloy	4 410	110 000	0.3	1065
Polyethylene	930	27 000	0.35	27

Table 2. Friction coefficients for the pairs of materials.

First material	Second material	Coefficient of friction
Titanium alloy	Spongy bone	0.3
Titanium alloy	Cortical bone	0.2
Titanium alloy	Polyethylene	0.15

6. Loads and model setup

The problem described is solved in two steps. The first step is dedicated to the screw pretension, and the second one is used for applying the main loading.
At the first step the current research considers several values of the screw pretension force within the range between 100 N and 400 N. Such approach is implemented in order to estimate the optimal value of the force, which leads to the reliable connection between the bone and the implant in combination with the stress level within the allowed limits.
The second step is devoted to the modeling of the standing condition. In the basic statement of the task, the gravity force equal to 245 N (half of the patient's weight) should be applied to the center of gravity of the body located in the plane of symmetry of the human. But from the computational point of view, it is more convenient to fix the region of the sacrum, considering the pelvis to be immobilized, and apply the load at the end of the simplified model of the leg.
The following kinematic boundary conditions are applied:

- the top surface of the sacrum is fixed in two directions ($uy = uz = 0$);
- the mid-surface of the sacrum is fixed in normal direction ($ux = 0$).

The task is solved with Abaqus/Standard implicit solver. The solution is done with the assumption that the displacements are small. However, the task is solved in the nonlinear mode to make it possible to describe the contact behavior correctly.

7. Results

An important issue during the hip arthroplasty is the estimation of the proper screw pretension force between the implant and the bone. As it was mentioned earlier, this paper considers several values of the force. Particular attention is paid to the stresses in the bone tissue of the pelvis, since high pretension force can cause the bone destruction.
The stress distribution in the cortical and spongy layers of the pelvic bones for the cases of pretension forces of 100 N, 200 N, 300 N and 400 N are presented in Fig. 6 and Fig. 7. In each of the illustrations,

the upper limit of the legend is set equal to the value of the permissible stress in each of the materials. In the areas colored grey, the stresses exceed the allowable limits.

Neglecting the local hot-spots caused by the singularity of the stress fields, it is possible to conclude that the optimal value of the screw pretension force is in the range between 200 N and 300 N. The lower force of 100 N leads to the relatively low contact pressure between the components, and the higher force of 400 N causes the extreme stress level in the bone tissue.

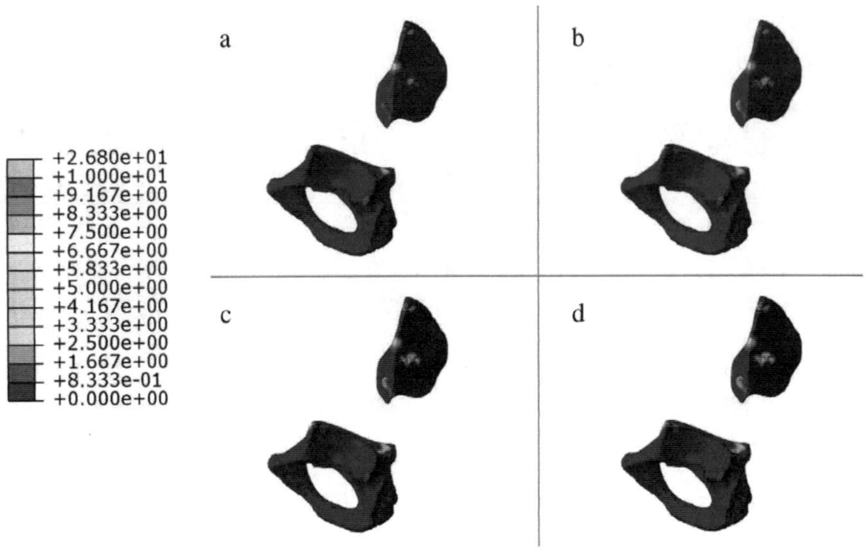

Figure 6. Stress field in the spongy bone for the screw pretension force of 100 N (a), 200 N (b), 300 N (c) and 400 N (d). Von Mises equivalent stress, MPa.

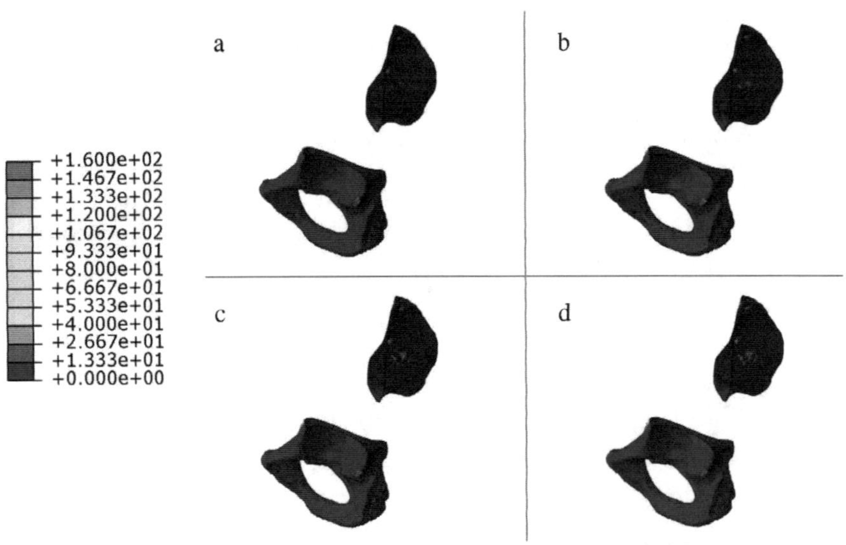

Figure 7. Stress field in the cortical bone for the screw pretension force of 100 N (a), 200 N (b), 300 N (c) and 400 N (d). Von Mises equivalent stress, MPa.

The following step representing the standing condition is based on the results of the first step for the values of pretension force equal to 200 N and 300 N.

IOP Conf. Series: Materials Science and Engineering **986** (2020) 012035 doi:10.1088/1757-899X/986/1/012035

The stress fields in the pelvis, implant, screws and polyethylene cup obtained from the static structural analysis and caused by the both screw pretensions and body weight are shown in Fig. 8 – Fig. 12.

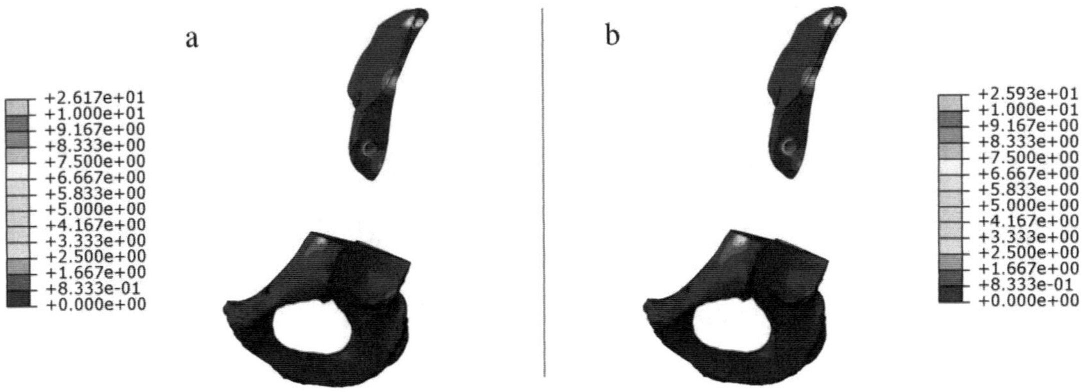

Figure 8. Stress field in the spongy bone for the screw pretension force of 200 N (a) and 300 N (b). Von Mises equivalent stress, MPa.

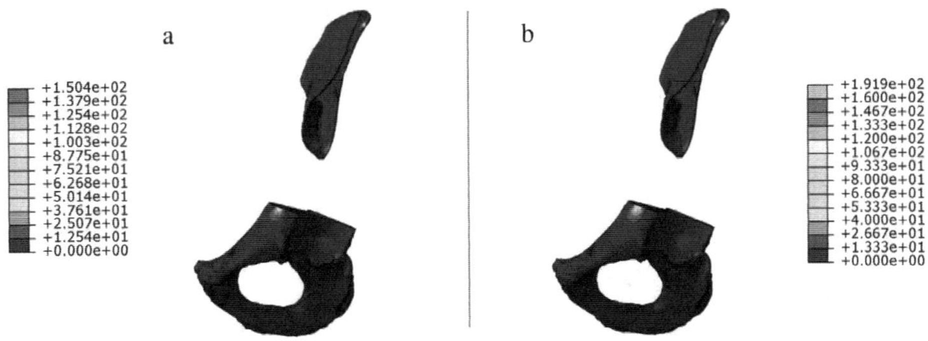

Figure 9. Stress field in the cortical bone for the screw pretension force of 200 N (a) and 300 N (b). Von Mises equivalent stress, MPa.

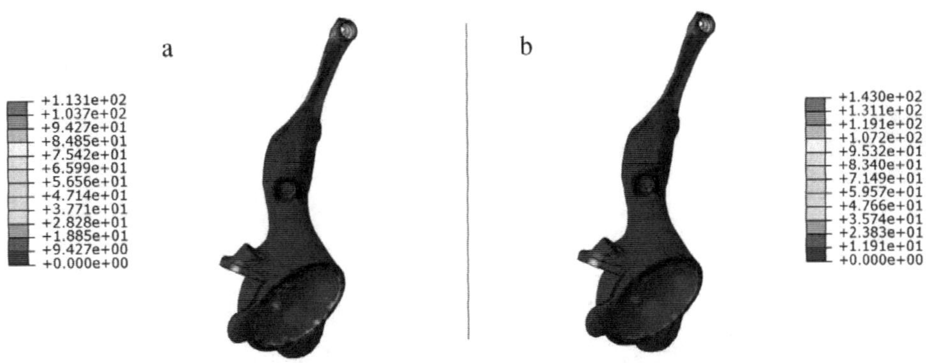

Figure 10. Stress field in the customized implant for the screw pretension force of 200 N (a) and 300 N (b). Von Mises equivalent stress, MPa.

IOP Conf. Series: Materials Science and Engineering **986** (2020) 012035 doi:10.1088/1757-899X/986/1/012035

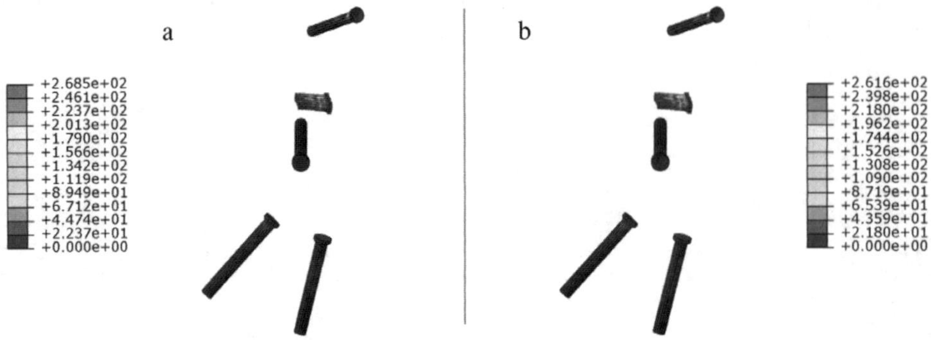

Figure 11. Stress field in the screws for the screw pretension force of 200 N (a) and 300 N (b). Von Mises equivalent stress, MPa.

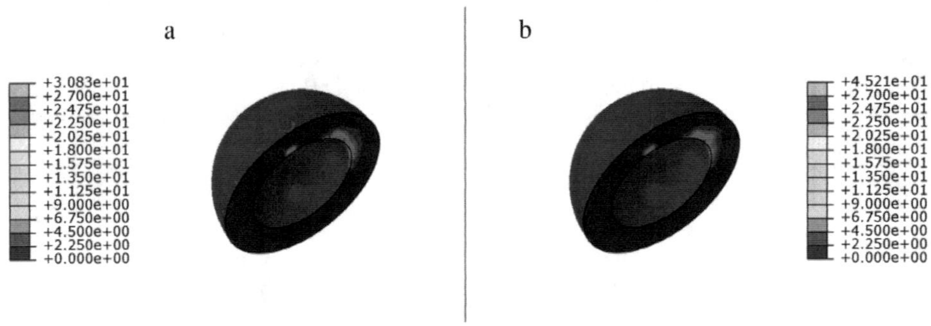

Figure 12. Stress field in the acetabular cup for the screw pretension force of 200 N (a) and 300 N (b). Von Mises equivalent stress, MPa.

The stress fields shown in Fig. 8 – Fig. 12 demonstrate that the critical value of the pretension force could be chosen equal to 300 N. For this value of the pretension, the stress levels in the implant and in the screws are well within the specified limits. The maximum stress values in the cortical and spongy bone formally exceed the allowable stress, but a more detailed analysis shows that the peak values take place mostly in the local hot-spots and should not cause any significant damage. But it is important to mention that the issue can be resolved by adjusting the positions of the screws in such a way that the screw holes do not form any thin bridges between them and do not approach the surface of the bone.

8. Conclusion

The article presents the basic information about the main steps of the development of the personalized hip joint prosthesis. The process of performing the finite element analysis of the pelvic implant is considered in detail as an important stage of the currently developed methodology for rapid virtual testing of individual endoprostheses of large joints.

The stress values, which were obtained from the analysis, show that the critical value of the pretension force equals 300 N. If this value is exceeded, then a high risk of destruction of the spongy layer occurs. The most dangerous are the areas, which are close to the screw holes, and regions with several screws placed close to each other or close to the surface of the bone.

Based on the simulation results, the following key conclusions and recommendations can be formulated to reduce the stress level in the pelvic bones with the endoprosthesis installed.

1. Some measures should be taken to reduce the stress level in the areas, where the screws intersect the surface of the bone. One of the possible solutions is to install additional flexible rings or gaskets

reducing the contact pressure. Such elements could be made of special biocompatible medical polymers [19].

2. The design of the implant could be improved to avoid such positioning of the screws, which leads to formation of thin bridges between the screw holes and intensive stress concentration due to the superposition of several stress concentrators.

The treatment of the patient was completed in April 2018 (Fig. 13). The observation period at the time of preparation of the article was 23 months with no indication of any local recurrence or progression of the underlying disease [14]. The functional assessment of the hip joint confirms the excellent result. There was no clinical or radiological indication of any instability at the time of the latest visit to the clinic. These results are in good agreement with the results of the structural assessment of the system performed with finite element analysis.

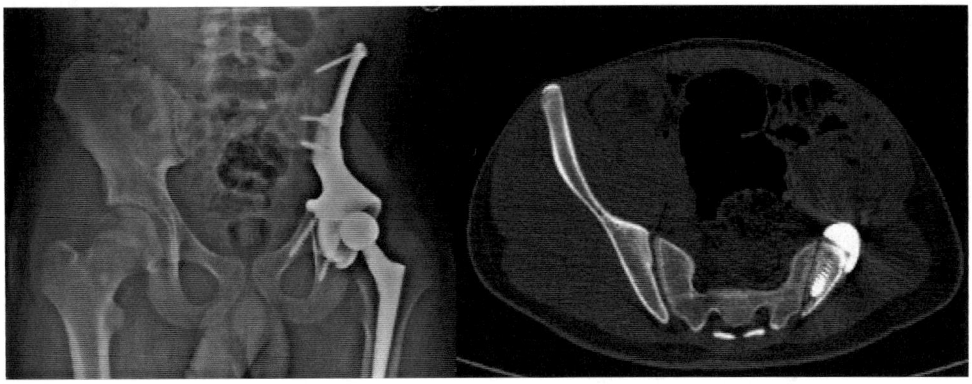

Figure 13. X-ray image and computer tomography study of the pation after the treatment completion [14].

References

[1] Kurtz S M, Ong, K L and Lau E et al. 2011. *International survey of primary and revision total knee replacement. International Orthopaedics* (SICOT) 35 pp 1783–1789

[2] Orthopaedic Biomechanics 2010 *Finite Element Analysis* pp 151–170

[3] Jia Y.-w, Cheng L.-m, Yu G.-r, Du C.-f, Yang Z.-y, Yu Y. and Ding Z.-q 2008. A finite element analysis of the pelvic reconstruction using fibular transplantation fixed with four different rod-screw systems after type I resection *Chinese Medical Journal* (Vol. 121) pp 321–326

[4] Sakuraba M, Kimata Y, Iida H, Beppu Y, Chuman H and Kawai A 2005 Pelvic ring reconstruction with the double-barreled vascularized fibular free flap *Plast Reconstr Surg*. (Vol. 116) pp 1340–1345

[5] Yu G, Zhang F, Zhou J, Chang S, Cheng L and Jia Y 2007 Microsurgical fibular flap for pelvic ring reconstruction after periacetabular tumor resection *J Reconstr Microsurg* (Vol. 23 pp 137–142

[6] Guo W, Tang S, Dong S and Li X 2006. Resection and reconstruction for tumors of iliac bone *Chin J Surg* (Chin) (Vol. 44) pp 813–816

[7] Yamamoto Y, Takeda N and Sugihara T 1997 Pelvic ring reconstruction with a vascularized bone flap of femur *Plast Reconstr Surg*. (Vol. 100) pp 415–417

[8] Zhou Y, Min L, Liu Y, Shi R, Zhang W, Zhang H, Duan H and Tu C 2013 Finite element analysis of the pelvis after modular hemipelvic endoprosthesis reconstruction *International Orthopaedics* (SICOT) (Vol. 37) pp 653–658

[9] Wong K C, Kumta S M, Geel N V and Demol J 2015 One-step reconstruction with a 3D-printed, biomechanically evaluated custom implant after complex pelvic tumor resection *Computer Aided Surgery* (Vol. 20, No. 1) pp 14–23

[10] Shah F A, Snis A and Matic A, Thomsen P and Palmquist A 2016 3D printed Ti6Al4V implant surface promotes bone maturation and retains a higher density of less aged osteocytes at the bone-implant interface *Acta Biomater* (30: 357-367)

[11] Maslov L B 2017 Mathematical model of bone regeneration in a porous implant *Mechanics of Composite Materials* (Vol. 53 № 3) pp 399–414

[12] Liang H, Ji T, Zhang Y, Wang Y and Guo W 2017 Reconstruction with 3D-printed pelvic endoprostheses after resection of a pelvic tumour *Bone Joint J.* (99-B: 267-275)

[13] Sushentsov E A, Musaev E R, Sofronov D I, Fedorova A V, Stepanova A M, Efimenko O S, Dzhenzhera G E and Aliev M D 2017 Custom-made endoprosthetics based on 3D-technology after resection of pelvic bones *Sarcomas of bones, soft tissues and skin tumors* (3) pp 14-20

[14] Sushentsov E A, Musaev E R, Maslov L B, Zhmaylo M A, Sofronov D I, Agaev D K, Dzampaev A Z, Romantsova O M, Fedorova A V and Aliev M D 2019 Computer simulation, 3D-printing and custom-made prosthetics in treatment of a patient with osteosarcoma of the pelvis *Sarcomas of bones, soft tissues and skin tumors* (4) pp 53-61.

[15] Borovkov A I, Maslov L B, Zhmaylo M A, Zelinskiy I A, Voinov I B, Keresten I A, Mamchits D V, Tikhilov R M, Kovalenko A N, Bilyk S S and Denisov A O 2018 *Russian Journal of Biomechanics* (Vol. 22, No. 4) pp 382–400

[16] ISO 5832-3-2014 2015 Implants for surgery. Metallic materials. Part 3: Wrought titanium 6-aluminium 4-vanadium alloy (Moscow, Standartinform Publ. Russian)

[17] ISO 5834-2-2014 2015 Implants for surgery. Ultra-high-molecular-weight polyethylene. Part 2: Moulded forms (Moscow, Standartinform Publ. Russian)

[18] Nedoma J, Stehlik J, Hlavacek I, Danek J, Dostalova T and Preckova P 2011 Mathematical and Computational Methods and Algotithms in Biomechanics of Human Skeletal Systems *A John Wiley & Sons* (Inc., Publication)

[19] Liksonov D, Barriere T, Boudeau N, Gelin J.-C and Maslov L 2004 Optimization of the elastic properties of an injection-molded composite hip implant, in XXIXème Congrès de la Société de Biomécanique SB-PARIS 2004, Paris, France, September 9-10, 2004 *Archives of Physiology and Biochemistry* (Vol. 112 (Suppl.)) p 43

IOP Conf. Series: Materials Science and Engineering **986** (2020) 012036 doi:10.1088/1757-899X/986/1/012036

Tibia and Fibula Stress Strain Research

A I Sudneva[1], A V Smirnov[1], M D Stepanov[1], N A Kharaldin[1], A V Stepanov[1] and A I Borovkov[2]

[1]Engineer of Automotive Departament SPbPU Engineering Center "Center of Computer-Aided Engineering" St. Petersburg, Russiae-mail: mstepanov@compmechlab.ru
[2]Ph.D., Professor, vice-rector for innovative projects of Peter the Great St. Polytechnic University of St.Petersburg St. Petersburg, Russia

Abstract. This article is devoted to the stress strain research of a biomechanical system formed by the tibia, fibula and muscles under external stress (axial compression). In this study an anatomically accurate geometric model of human bones was developed based on CT analysis data, which was the basis of the calculated finite element model. A numerical evaluation of the stress strain state was carried out using the Finite Element Method (FEM). The developed model of the tibia and fibula was validated based on the results of axial compression field tests. The analysis of mesh independence of finite element models of the tibia and fibula has been carried out.

1. Introduction

Detailed human bones modeling allows us to develop computational models of biomechanical systems to solve a number of problems in many areas related to ensuring human safety and security. This article describes the biomechanical system of the tibia and muscles, which in turn after a number of modifications can represent the lower limb of a person. This model can be used to model external mechanical effects at a wide range of speeds to solve various types of problems (safety of pedestrians and drivers, human safety under stroke, etc.). In this article the main focus is a damage modeling that occurs during axial compressive loading. The aim of this work is to create a detailed finite element model of the tibia and fibula, its physical and mechanical properties research, analyze the stress-strain state of the system, as well as the reaction force that occurs during loading in bones and evaluation of the model mesh independence.

2. Tibia and Fibula Anatomy

Each bone is a living, actively functioning and continually renewing organ. Blood vessels and nerves penetrating into the bone provide its interaction with the whole body, participation in metabolism, fulfillment of functions and necessary restructuring during growth, development and changing living conditions.

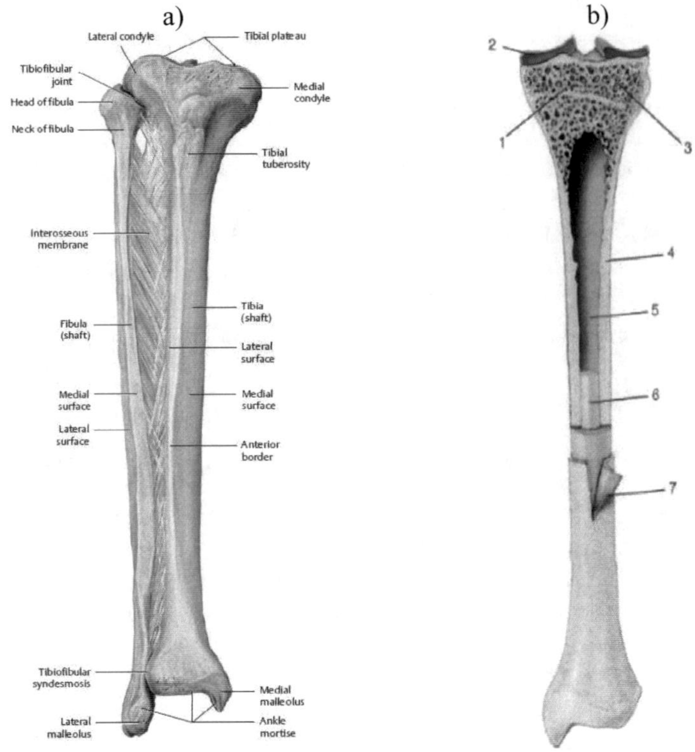

Figure 1. Structure of tibia and fibula a) External characteristics b) Internal characteristics (1-metaphysis; 2-articular cartilage; 3-spongy substance; 4-cortical bone; 5-cavitas medullaris; 6-bone marrow; 7-periosteum) [1].

The peculiarities of the bone's internal structure are due to its cortical bone and spongy substance. The cortical bone is in a dense layer located on the periphery of the bone. The spongy substance is under cortical bone and has a porous structure. The tibia is more massive and takes on the entire weight of the body [15]. An illustration the tibia and fibula which is shown in Fig. 1a and Fig. 1b is taken from the Atlas of Anatomy [1].

One of the goals of this article is the creation of a mathematical model with complex geometry and diversity. One of the first steps is to divide the model into 2 layers (cortical bone and spongy substance), giving them different properties, which will be discussed later. The difficulty is in determining the thickness of the cortical bone, since this value depends on the age of person, the geometric features of the bone, diseases and injuries. In the studied reference [3, 5, 9], the values range from 1 mm to 7 mm. This parameter greatly affects bone stiffness. Figure 2 shows the way the layer thickness is approximately calculated.

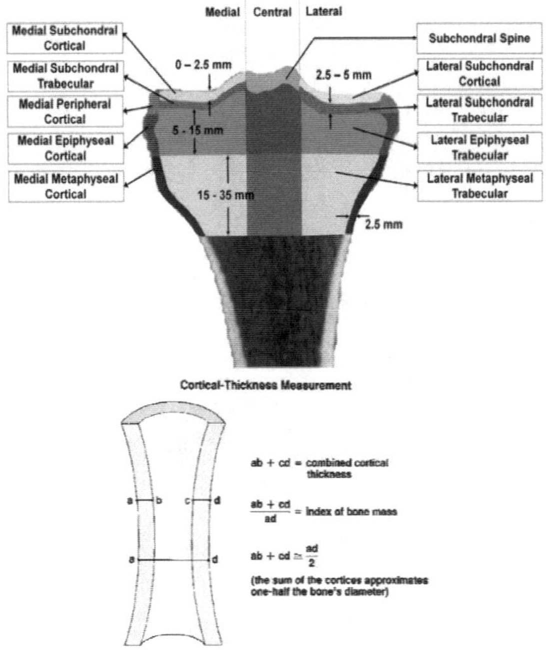

Figure 2. Cortical thickness measurement [9].

3. 3D Model creation

One of the goals of the research is the 3D model creation of tibia and fibula using CT data. In the process of preparing a three-dimensional model, we can distinguish two stages: a) Point clouds conversion from the DICOM to the STL b) Creation of STL solid-state geometry [6].

Point clouds conversion from the DICOM to the STL. The CT scanner mode of operation is based on the following simultaneous actions:

- continuous rotation of x-ray tube around human body that emits x-rays
- registration of this radiation with special detectors
- continuous movement of the bed with the person lying on it along the longitudinal axis.

X-rays are captured by detectors, then they are converting into electrical signals and transmitting to a computer. Information received from radiation detectors is processed using a CT scanner. As a result, a two-dimensional image of the human body cross section is created at different levels. (Fig.3a).

Computed tomography imaging pictures are a visualization of DICOM examination data. DICOM (Digital Imaging and Communications in Medicine) is an industry standard for creating, storing, transmitting and visualizing medical images and documents of examined patients.

Using the 3D Slicer an STL model of tibia and fibula was constructed (Fig. 4b) after processing DICOM files. Due to sound effects during tomography, the resulting model has a lot of imperfections and protruding surfaces, so it must be filtered. This was done by MeshLab.

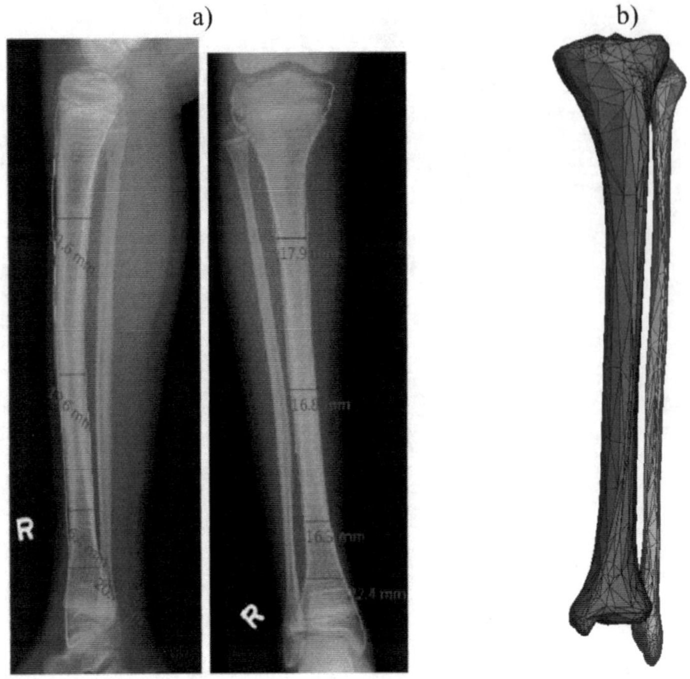

Figure 3. Image of bones a) MRI [2] b) STL model.

Creation of STL solid-state geometry. The STL format is used to store three-dimensional models of objects to use in rapid prototyping technologies. Information about the object is stored as a list of triangular faces that describe its surface, and their normals. To use the model for engineering calculations, it is necessary to build solid-state CAD (Computer-Aided Design) geometry, as this is required by engineering packages. Using the ANSYS SpaceClaim the described STL model was converted to a CAD model.

4. Physico-mechanical properties of bones

Bone is a solid body, one of the main properties of bone is strength. Bone strength is the ability to withstand external destructive forces. Quantitatively the strength is determined by the tensile strength and depends on the macro- and microscopic design, bony tissue composition. As for the macroscopic design, each bone has a specific shape that can carry the highest loads in a certain part of the skeleton. The strength of the bone is also significantly affected by bone structure. With decalcination the bone is easily bent, compressed and twisted, with an increase of calcium it becomes fragile. The bone strength of a healthy adult is greater than the strength of some construction materials, it is the same as of cast iron [12]. The analysis of scientific publications on the values of density, elastic modulus, Poisson's ratio, ultimate stresses and deformations showed a fairly wide scatter. Young's module of healthy spongy tissue along the fibers is 600–1050 MPa, across the fibers it is 375–600 MPa; tensile strength along fibers is 16–22 MPa, across fibers - 7–13 MPa. However, this data is reliable only if the conditional density of the material is sufficiently high. Experimental data on bone strength is available only for dried compact bones and relatively Young's module it has a fairly wide range of values ranging from 427 MPa to 22 GPa, and tensile strength range is 60–80 MPa. Moreover, in the experiment on a dried compact bone, the difference of the values across (4.2–9.9 GPa) and along (8.7–16.5 GPa) of bone trabecula was proved. These studies indicate anisotropy of bone tissue [13].
Two mathematical models can be distinguished which most accurately describe the movement of bones.

4.1. Model of an elastoplastic body with nonlinear hardening

The material is strained elastically until the yield point is reached. If this limit is exceeded under loading (load increase), the connection between stresses and strains will be nonlinear, i.e., the material will be strengthened nonlinearly (see Fig. 4). When unloading the material is linearly elastic and has permanent strains.

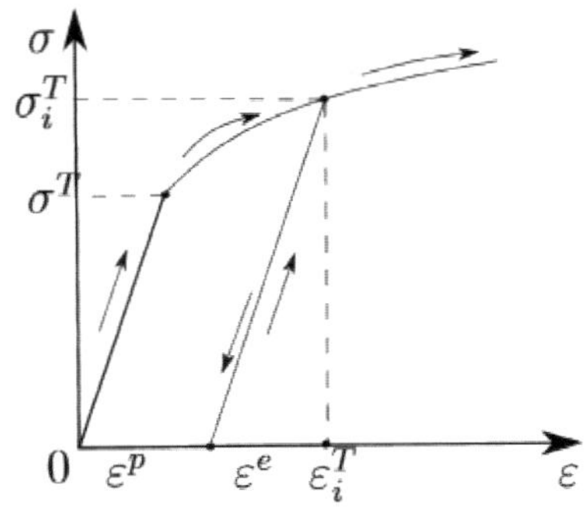

Figure 4. Model of an elastoplastic body with nonlinear hardening [17].

When loaded again the material will strain linearly-elastic (elastic modulus E) until the stress from which unloading begins is reached. This strain plays the role of a new yield strength (σiT) that exceeds the initial one, and as a result, we consider that the material has received plastic strain hardening [17]. The change in constant strains is described using an elastoplastic dependence for an instantaneous strain response. Equation (1) describes the connection between stress and strain in increments [11]:

$$\dot{\boldsymbol{\sigma}}^{\mathbf{0}}=\boldsymbol{D}^{e}\cdot\cdot(\dot{\boldsymbol{\varepsilon}}-\dot{\boldsymbol{\varepsilon}}^{p}) \quad (1)$$

where $\boldsymbol{D}^e$ is the instant elasticity tensor. The associated flow law (2) is used:

$$\dot{\varepsilon}^p = \dot{\lambda}\frac{\partial f}{\partial\sigma}, \quad (2)$$

where λ – plastic flow coefficient. The increase in stresses relatively to the strain above the elastic limit is taken into account when using the law of hardening (3), written in the form:

$$\sigma_T = f(\bar{\varepsilon}^p) \quad (3)$$

The plasticity criterion (4) is the Mises plasticity condition:

$$f(\sigma)=\sqrt{\frac{1}{2}[(\sigma_1-\sigma_2)^2+(\sigma_2-\sigma_3)^2+(\sigma_3-\sigma_1)^2]} = \sigma = \sigma_T \quad (4)$$

4.2. Viscoelastic body model

This model is described by equations (5) - (6), in which the dependence on the strain rate of the material is specified [14]. It is important to consider this parameter in calculations with high impact rates, dynamics problems (shock, explosion). The graphs in Fig. 5 [7] show the way the strain rate affects the stress-strain state.

$$\sigma_{ij} = \sigma_{ij}^0 + \int_0^t G(t-\tau)\frac{\partial\varepsilon_{ij}}{\partial\tau}d\tau;\ G(t) = \sum_{n=1}^N G_n e^{-\beta_n t}, \quad (5)$$

where σ_{ij}^0 are the components of the elastic stress tensor deviator, ε_ij are the components of the strain tensor.

The relaxation function is shown through the Prony serie

$$\int_0^t G(t-\tau)\frac{\partial\varepsilon_{ij}}{\partial\tau}d\tau = \sum_{n=1}^N G_n\int_0^t e^{-\beta_n(t-\tau)}\frac{\partial\varepsilon_{ij}}{\partial\tau}d\tau, \quad (6)$$

$\frac{\partial\varepsilon_{ij}}{\partial\tau}$ – engineering strain rate.

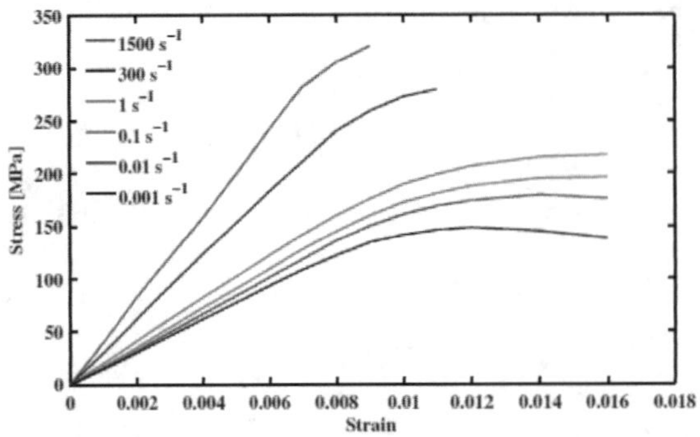

Figure 5. Stress – strain curves under compression for a cortical bone of thigh depending on the strain rate [7].

Studies show that by increasing strain rate, Young's modulus and tensile strength increase, and fracture strain decreases [8]. Figure 6 shows the dependence of Young's modulus on the strain rate on a logarithmic scale.

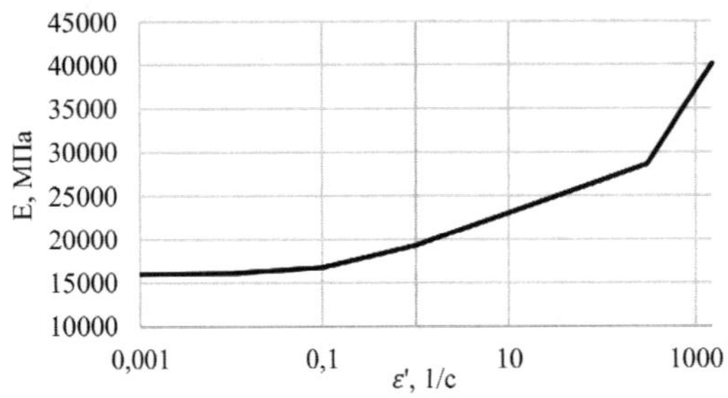

Figure 6. Dependence of Young's modulus on the strain rate.

5. Material characteristics selection

As the main mathematical model of materials used in the finite element model of the components of the biomechanical structure, a model of an elastic-plastic material with a fracture criterion based on the maximum value of effective plastic deformation is considered. The effect of plasticity should be taken into account in the model, since the accumulation of strain is significant, and it can also be an indicator of possible damage. In our article we do not consider the viscosity of a cortical bone, since it is known to be quasistatic.

As mentioned above, the scatter of these bone properties is great. We chose the following values (table 1 and Fig. 7-8.), since they are most often found in current sources [4, 10] when modeling the human lower limb. It should be noted that the bones behave differently when compressed and strained.

Table 1. Physico-mechanical properties of bones.

	Density, t / mm^3	Young's module, MPa	Poisson's ratio	Yield Strength, MPa	Fracture strain	Fracture stress, MPa
Tibia,cortical bone	2e-9	15 000	0.3	100	1.2%	157.2
Tibia,spongy substance	1.1e-9	445	0.3	-	13.4%	-
Fibula,cortical bone	2e-9	14 000	0.3	70	1.2%	127.2
Fibula,spongy substance	1.1e-9	292	0.3	-	13.4%	-

IOP Conf. Series: Materials Science and Engineering **986** (2020) 012036 doi:10.1088/1757-899X/986/1/012036

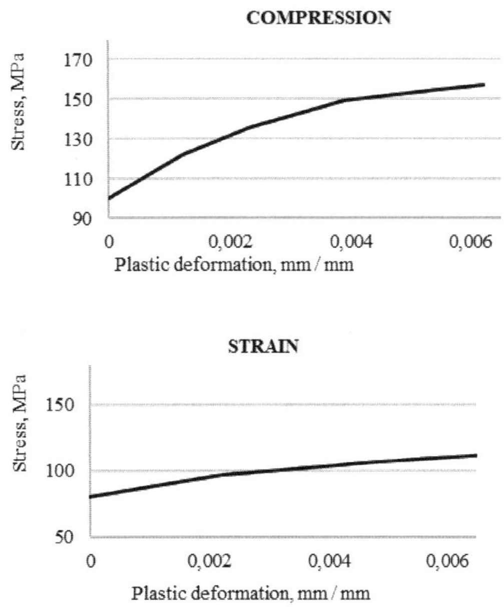

Figure 7. Hardening curves of compression and strain of Tibia.

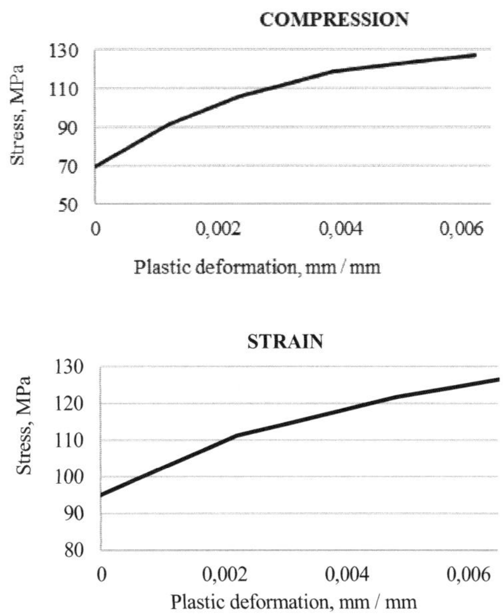

Figure 8. Hardening curves of compression and strain of Fibula.

6. Finite Element Method (FEM)

When designing vehicles, building constructions and technological processes in scientific research, computer software analysis systems (CAE – Computer – Aided Engineering) based on the finite element method (FEM) are widely used today. The finite element method (FEM) is a numerical method for solving differential equations with partial derivatives, as well as integral equations of applied physics. The method is widely used in mechanics tasks of a deformable solid body, heat transfer, hydrodynamics and electrodynamics. A detailed description of the method and the algorithms used is given in the reference list [16].

Advantages of the method:

- the possibility of splitting into finite elements of an area of any shape and complexity;
- the ability to calculate stress and strain fields in real details considering all their design features;
- excludes the variation that occurs during the field test (it is especially important when working with biological objects).

7. Finite element model

Figure 9 shows the finite element model of tibia and fibula, which consists of 30,000 elements of the solid type (spongy substance), 10,000 elements of the shell type (cortical bone). The size of the elements is 4 mm. The thickness of the tibia cortical bone varies from 1 mm to 3.2 mm, counting from the metaphysis, fibula - from 1 mm to 1.9 mm.

Figure 9. Finite element model of tibia and fibula.

8. Problem Statement

The problem statement (see Fig. 10a) [4] was selected as a task for validating the developed finite element model of bones, which allows to evaluate the movement of bones under axial compression. Tibia and Fibula are filled with epoxy resin above and below (E = 1000 MPa, p = 2.2E-9 t / mm ^ 3, v = 0.45). The upper part is fixed, while the lower part is affected by a steel plate with a constant speed of 2 mm / s. As a result, the validation of the developed finite element model of bones is carried out on the basis of a comparison of the contact force arising from the simulation with the results of field tests [4]. The finite element model of the described problem statement is presented in Fig. 10b.

Muscles, fat and other biological elements of the lower limb of a person are modeled using solid-state elements with averaged physical and mechanical parameters (E = 13.2 MPa, p = 1.3E-9E-9 t / mm ^ 3, v = 0.48).

IOP Conf. Series: Materials Science and Engineering **986** (2020) 012036 doi:10.1088/1757-899X/986/1/012036

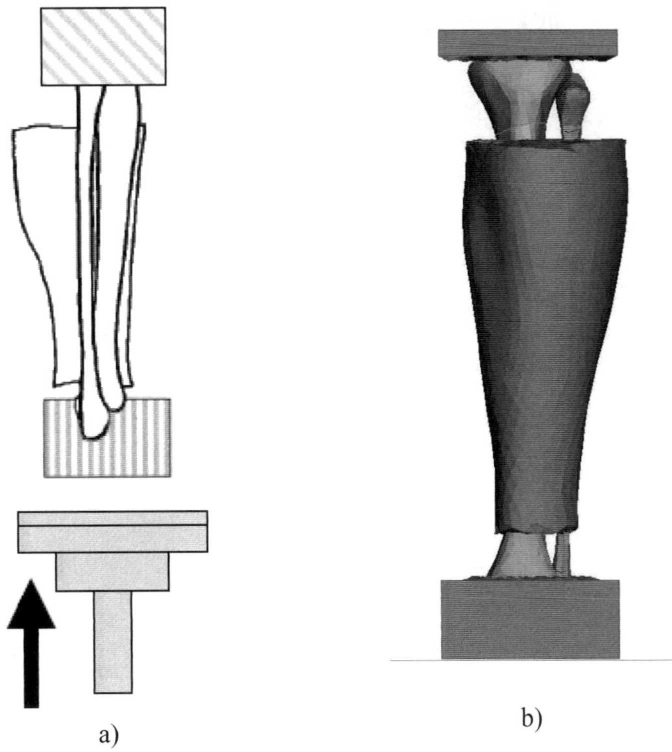

Figure 10. a) Problem statement б) Finite element model.

9. Results

The article [4] presents the results of a full-scale test for axial compression of tibia and fibula. Based on the sample, a corridor was created of the minimum allowable and maximum allowable values of the force arising in the bones (the contact force between the plate and the model). Fig. 11 shows a comparison of the obtained results of our model and experimental data.

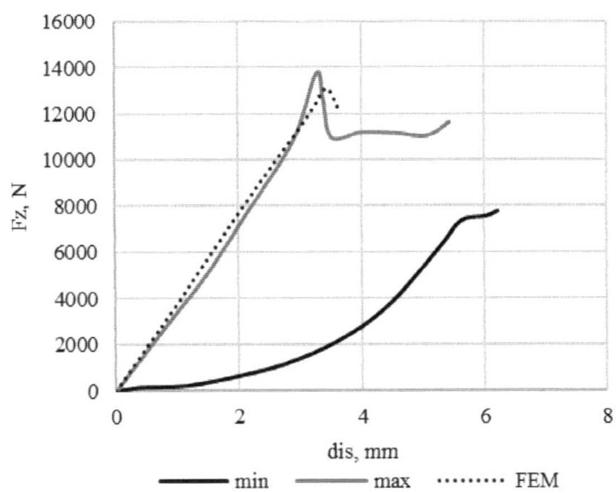

Figure 11. Comparison with the results of field tests.

An analysis of the results showed a good agreement with the results of field tests. The stiffness value of the entire system and the process of bone destruction in the developed model correspond to a biological sample.

When choosing the Finite element method as a numerical method for evaluating the stress-strain state, it is necessary to analyze the mesh independence. Thus, based on the developed finite element model of bones, models with element sizes of 3, 4, 8, 16, 24 mm were created.

In order to evaluate the mesh independence, we studied the dependence of the linear rigidity of the system on the number of elements in the model (NOE-Number of elements). The results of this analysis are shown in Fig. 12.

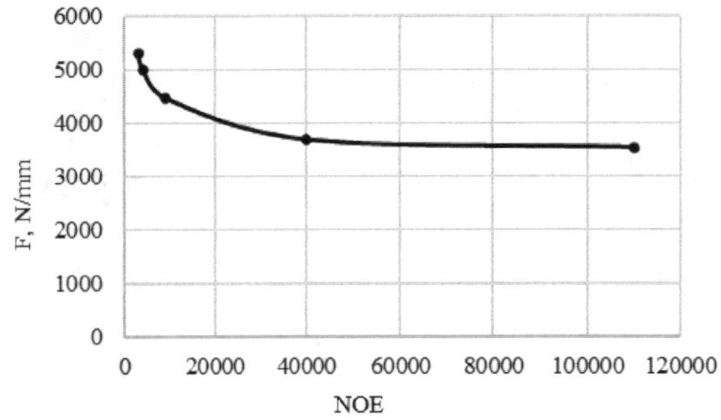

Figure 12. Mesh Independence Analysis.

It was found that the optimal element size is 4 mm (NOE = 40,000), since with a further element size decrease, the calculation time increases significantly with a slight increase in the accuracy of the results.

10. Conclusion

In this study we described the anatomy of the human tibia and fibula and created CAD and FEM models of bones, mathematical models were studied to describe biological tissues, physicomechanical properties of bones were selected, and a numerical analysis of bone compression was performed. The analysis of mesh independence was carried out. It is shown that the behavior of the developed model under external mechanical action is in good agreement with experimental data.

Thus, the biomechanical system described in this article as part of the tibia and muscle tissue describes the properties of the biological sample well and can be used to solve problems of external mechanical stress, for which the axial load (along the bone) is dominant. In future we plan to develop this model to increase the number of tasks to be solved.

References

[1] Knee & Leg Atlas of Anatomy ([Electronic resource] Access mode: https://doctorlib.info/medical/anatomy/27.html)

[2] Christiane Caouette 2016 Geometry reconstruction method for patient- specific finite element models for the assessment of tibia fracture risk in osteogenesis imperfecta *International Federation for Medical and Biological Engineering*

[3] Moeinoddin Mahmoudi 2017 Experimental investigation of compressive strength in diaphysis of human Tibia and Fibula bones *The 25th Annual International Conference on Mechanical Engineering ISME-2017.*

[4] Philippe Beillas and Paul C Begeman 2001 Lower Limb: Advanced FE Model and New Experimental Data *Stapp car crash journal*

[5] Sergio H Lüscher, Laura M Nocciolino and Nicolás Pilot 2019 Differences in the Cortical Structure of the Whole Fibula and Tibia Between Long-Distance Runners and Untrained Controls. Toward a Wider Conception of the Biomechanical Regulation of Cortical Bone Structure *Frontiers Endocrinology*

[6] Stepanov M D, Loboda O S, Novozhilov Y V and Vasilyev N V 2017 Finite element modelling of the mitral valve repair using an implantable leaflet plication clip *Materials Physics and Mechanics* (32(3)) pp 383-392

IOP Conf. Series: Materials Science and Engineering **986** (2020) 012036 doi:10.1088/1757-899X/986/1/012036

[7] Johnson T P M 2010 A viscoelastic, viscoplastic model of cortical bone valid at low and high strain rates *Acta Materialia Inc. Published by Elsevier Ltd* (vol. 6)
[8] Ulrich Hansen and Peter Zioupos 2008 The Effect of Strain Rate on the Mechanical Properties of Human Cortical Bone *Journal of Biomechanical Engineering* (Vol. 130)
[9] Varadaraju Magesh 2018 Finite element analysis of tibia bone by reverse engineering modelling approach *International Journal of Pure and Applied Mathematics* (Vol. 118, No.20) pp 839-846
[10] Yamada H 1970 Strength of Biological Materials *Williams and Wilkins* (Baltimore, MD)
[11] A.N. Natali, Issledovanie mekhanicheskogo povedeniya trabekulyarnoj kostnoj tkani s uchyotom vyazkouprugoplasticheskogo otklika/Rossijskij zhurnal biomekhaniki, 2009, tom 13, # 3.
[12] Mekhanicheskie svojstva kostnoj tkani (https://medbe.ru URL: https://medbe.ru/materials/kostnaya-i-khryashchevaya-tkan/mekhanicheskie-svoystva-kostnoy-tkani/)
[13] Osobennosti prochnostny`kh kharakteristik gubchatoj kosti pri zabolevaniyakh tazobedrennogo sustava (https://fundamental-research.ru URL: https://fundamental-research.ru/ru/article/view?id=31832)
[14] Rabotnov Yu N 1966 Polzuchest` e`lementov konstrukczij (M.: Nauka) p 753
[15] Samusev R P and Lipchenko V Ya 2000 Atlas anatomii cheloveka (M.: Izd-vo ONIKS: Al`yans V), p 320
[16] Segerlind L 1979 Primenenie metoda konechny`kh e`lementov (M.: Mir) p15
[17] Struzhanov V and Burmasheva N V 2019 Teoriya uprugosti: osnovny`e polozheniya: ucheb. posobie (M-vo nauki i vy`ssh. obra-zovaniya Ros. Federaczii, Ural. feder. un-t Ekaterinburg: Izd-vo Ural. un-ta) p 204

Tire tread optimization method to improve to push aside the water from the road contact patch

Y D Zhilin[1], N A Kharaldin[1], P S Cvetkov[1], A V Stepanov[1], M V Aleshin[1] and A I Borovkov[2]

[1]Engineer of Automotive Departament SPbPU Engineering Center "Center of Computer-Aided Engineering" St. Petersburg, Russia e-mail: zhilin.yad@gmail.com`
[2]Professor, vice-rector for innovative projects of Peter the Great St.Petersburg Polytechnic University, St. Petersburg, Russia

Abstract. The paper is devoted to the development of a method for optimizing the tread pattern to push aside the water from the road contact patch. This technique works with ready-made tread patterns of any type. Today, the methodology under consideration is relevant, since it prevents the occurrence of one of the most common causes of road accidents - aquaplaning. Currently, it is impossible to eliminate the risk of such an effect. It can only be reduced by slowing down or ensuring a uniform load on all wheels, trying to drive through the puddle with all four wheels.

1. Introduction

Aquaplaning is an effect that occurs when driving a car at high speed in wet weather. It can be observed at speeds ranging from 90 km / h. Aquaplaning occurs when the car slides on water and the feel between the tire and the road will be lost. A water wedge arising in front of the wheel raises the machine and deprives it of control. Most often, aquaplaning occurs in shallow puddles, in which the least water resistance to the wheel [2].

Automakers are struggling with the effect of aquaplaning by applying special patterns to the tread of the tire that pushes aside water from the contact area with the road and improves the grip, handling, and passability. Nevertheless, one needs to monitor the degree of tread wear-out, since the coefficient of the contact area between the worn-out tire and the asphalt is much less. Despite the huge selection of different tread patterns, universal tires that would be far ahead of the rest ones are almost impossible to create.

The method described below optimize absolutely any tread pattern. It can be used to calculate both a completely new tire and a worn one. In the future, there will be possible not to create the perfect tire, at least improve the performance indicators of any existing tires.

2. Problem statement

The case of a car colliding at high speed on a water volume simulating a puddle is considered in this paper. Calculations were performed on the basis of CML-Bench using commercial licenses LS-DYNA and computational resources of CNTI. The fluid volume is set using the SPH method.

This method - Smoothed Particle Hydrodynamic - simulates the volume of fluid in the form of pseudoparticles interacting with each other, connected by a general equation of state. Each pseudoparticle has its mass, speed, energy, and the so-called smoothing length, within which the core function "smooths" its properties. Thus, each of the particles can be calculated by summing the parameters of neighboring particles - within two smoothing lengths [5].

To simulate fluid flow using SPH in LS-DYNA, only a couple of points are different from conventional structural analysis. To ensure quasi-incompressibility, the Murnaghan equation of state can be used, allowing a reasonable time step in explicit calculations. Water is also modeled using the Grünazen or Mi-Grüneisen equations. In this problem, Murnagan equation was chosen because confirmed constants from experiments were found.

Murnagan's equation of state defines pressure as

$$p = k_0\left[\left(\frac{\rho}{\rho_0}\right)^{\gamma} - 1\right], \quad (1)$$

where γ and k_0 – constants.

For accurate modeling of fluid flows γ is often equal to 7, and k_0 is chosen so that

$$c_0 = \sqrt{\frac{\gamma k_0}{\rho_0}} \geq 10\upsilon_{max} \quad (2)$$

where υ_{max} – maximum expected fluid flow rate.

This provides low compressibility, given the relatively large time step size.

The task is divided into two main stages:

1) a preparatory stage for setting sph-elements, finding the optimization criteria and the evaluation of the model as a whole

2) parametric optimization to assess the performance of this technique

A. The first stage. Optimization preparation

The preparatory phase is carried out to determine and verify possible criteria for future optimization, as well as to configure the correct interaction of sph elements.

Due to the complexity of the geometrical shape of the parts, this problem cannot be solved by analytical methods; therefore, the calculation in LS-DYNA is carried out using the numerical finite element method. The design is divided into simple elements that connect at nodal points. To them, equations are written that describe their states, which add up to systems that work for the entire structure as a whole [1].

In order to reduce the necessary computing power, a reduced wheel model is considered instead of the whole car, taking into consideration the impact of the car on the wheel itself.

The test computational model is a car tire without a tread pattern, mounted on an aluminum rim (Fig. 1). It runs into a water volume created using SPH elements. The set speed of the wheel is 90 km/h.

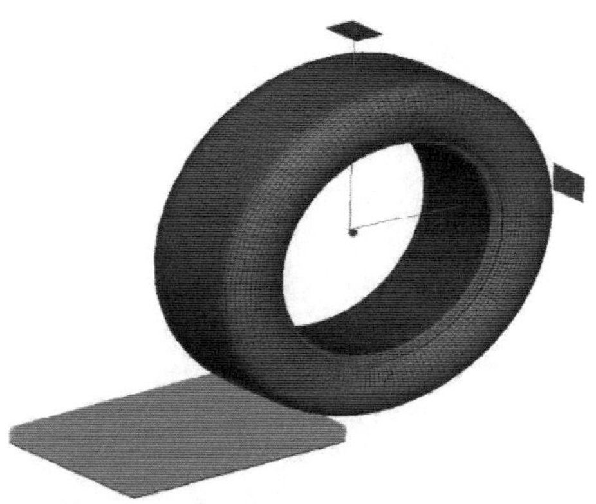

Figure 1. Test computational model.

At the center of mass of the tire, the concentrated mass of 300 kg is hung, simulating the effect of the machine on the wheel. This corresponds to the mass of the car per wheel with a total mass of 1200 kg and an ideal weight.

The tread consists of two layers: a rubber layer of SOLID type elements (ELFORM 3) and a simplified cord model, which is a homogenized layer of rubber and steel elements of the SHELL type (ELFORM 2). The sectional finite element tire model is shown in Fig. 2.

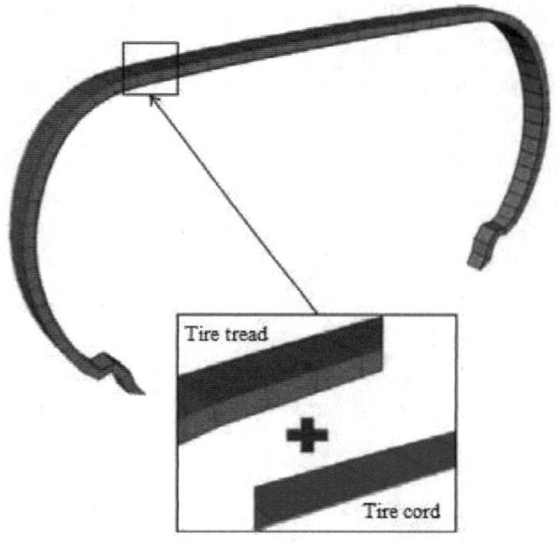

Figure 2. Finite element model of the tire.

The main properties of the materials of which the tire is made are presented in Table 1.

Table 1. Material properties.

Material	Young's module, MPa	Poisson's ratio	Density, kg/m3
Rubber	5	0,3	1500
Rubber+Steel	30000	0,3	7850

B. The second stage. Optimization

The optimized computational model is a tire with an arbitrary tread pattern (Fig. 3). In order to reduce computing power, one-fourth of the tire will be considered, which not influence the results. The statement of the problem with two different tire speeds - 90 and 110 km/h is examined that allows comparing optimization performance indicators.

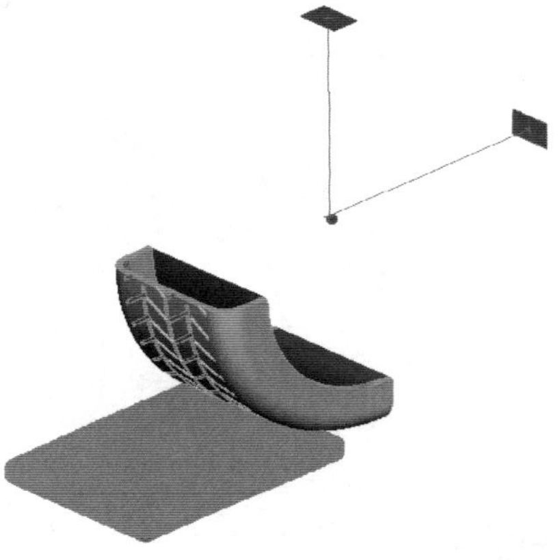

Figure 3. Optimized computational model.

The possible optimization criteria will be the resistance forces that are controlled by the system that simulates the car's suspension. The car body is imitated as two plates moving along the line of its movement and connecting to the center of mass of the wheel using control elements to remove the force.

At the point of their connection, there is a hinge working on torsion, so that the plates do not limit the rotation of the wheel.

Parametric optimization is carried out using the LS-OPT software package. This graphical tool allows the user to structure the design process, explore the design space, and calculate optimal projects according to specified constraints and goals.

The LS-OPT graphical tool interacts with LS-DYNA and provides an environment for setting optimization input, monitoring, and managing parallel modeling and optimization data after processing, as well as viewing multiple projects using LS - PREPOST.

The goal of optimization is to modify the tread pattern to a more effective one; the criterion is to reduce the water resistance force. The optimization parameter is the thickness and shape of the tread pattern grooves. Longitudinal grooves can vary in thickness from -3 to 3 mm, diagonal ones - from -2 to 2 mm. The extreme diagonal grooves change symmetrically, the middle ones change similarly.

Along the entire length of each diagonal groove, individually changing parameterizing points are set for the possibility of obtaining a curved shape (Fig. 4).

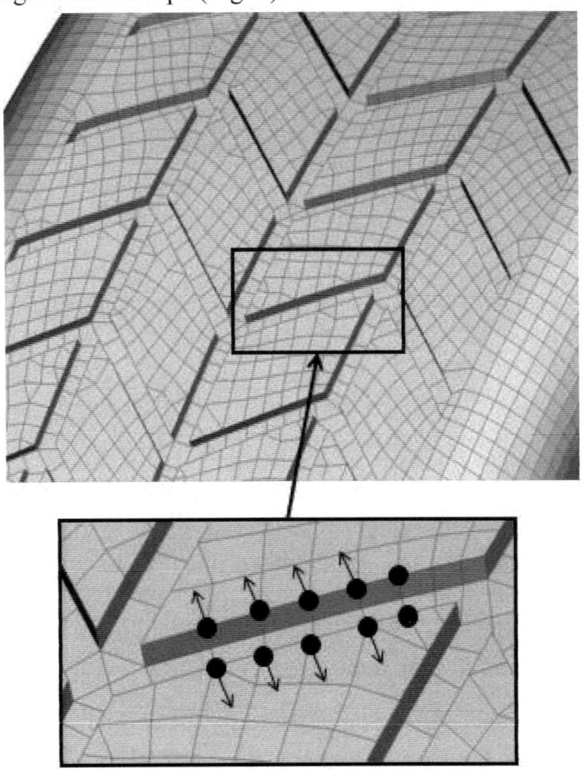

Figure 4. Parameters for changing diagonal grooves.

3. Results

A. The first stage. Optimization preparation

At the first stage, it was necessary to check the operability of sph particles as well as to find a suitable criterion for optimization. The estimated time is 0.015 s. At this point, the effect of aquaplaning is already observed. This time is sufficient to visualize the tire hits the puddle and analyze the behavior of SPH particles.

Figure 5 shows the visualization of how SPH particles work after the wheel hits them at different times. It can be noted that the particles interact with each other and act as a single unit, working effectively. The SPH method is suitable for modeling a puddle in the problem under study.

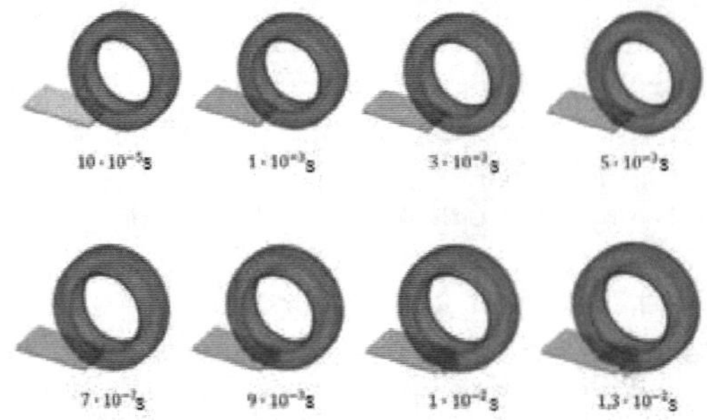

Figure 5. Visualization of particle motion.

At the first stage, the calculation resulted in finding the optimization criterion. The vertical force of water resistance – the upward force - was chosen as this criterion. Minimizing the force means more successful removal of water from the road contact patch. The graph of the upward force is shown in Fig. 6.

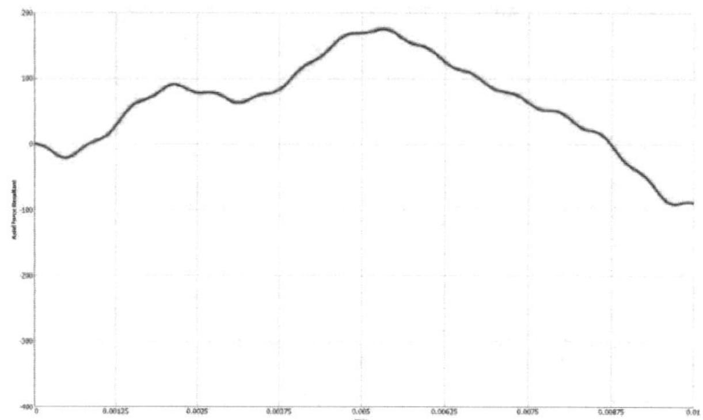

Figure 6. The upward force.

As can be seen on the chart, first, there is a small jump during hit the puddle, and then the force begins to increase rapidly, reaching its peak at the certain point. When the optimization is set, the force will be minimized directly in this time period.

B. The second stage. Optimization

The objective of the work was to develop the methodology for optimizing any tread pattern; therefore, a ready-made pattern was taken as the basis (Fig. 7). The symmetrical directional pattern was chosen because such ones are considered as the best way to prevent aquaplaning.

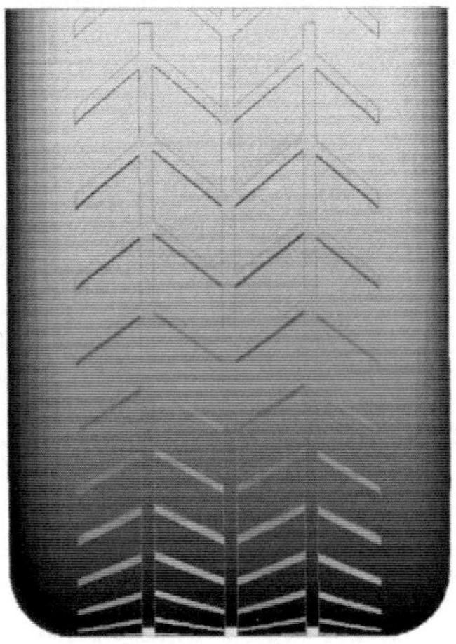

Figure 7. Initial tread pattern.

The optimization criterion is the upward force, which has to be minimized. The first optimization calculation was performed at the tire speed of 90 km/h.

It took a long time to go through the options for changing the original tread pattern due to the limited power of computer resources. Later, this method can be improved using more powerful computing technology, which would set a greater number of possible changes. The optimization process came together in 10 iterations, and the intermediate steps are shown in Fig. 8. The final tread pattern is shown in figure 9.

Figure 8. Intermediate steps to optimize the tread pattern at a speed of 90 km/h.

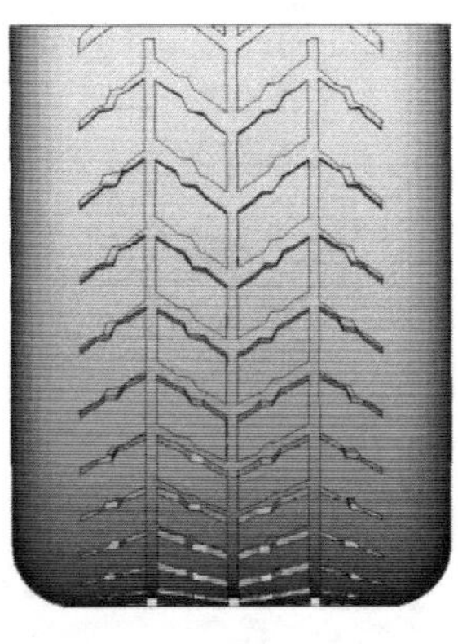

Figure 9. Tread pattern after optimization at a speed of 90 km/h.

At the same time, the force of pushing the tire out of the water decreased by 9%. The graph of changes in the indicators of this force depending on iterations is shown in Fig. 10.

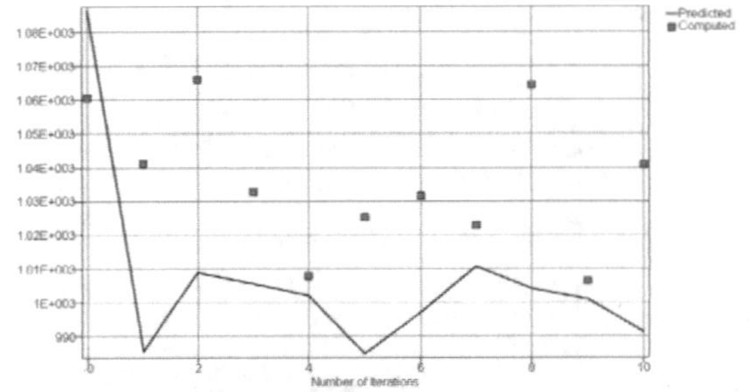

Figure 10. Optimization graph at a speed of 90 km/h.

The optimization process at the speed of 110 km/h converged in 10 iterations; the intermediate results are shown in Fig. 11. The final tread pattern is shown in Fig. 12.

Figure 11. Intermediate steps to optimize the tread pattern at a speed of 110 km/h.

Figure 12. Tread pattern after optimization at a speed of 110 km/h.

At the same time, the force of pushing the tire out of the water decreased by about 12%. The graph of changes in the indicators of this force depending on iterations is shown in Fig. 13.

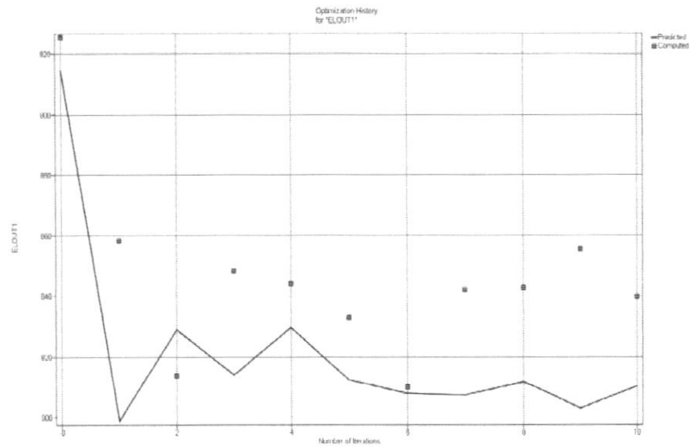

Figure 13. Optimization graph at a speed of 110 km/h.

The comparison of dependence graphs of the value of the upward force on the time before and after optimization is given below (Fig. 14). Before optimization, the maximum upward force was 916.8 N, and after that - 811.4. The decrease in strength indicators was 12%.

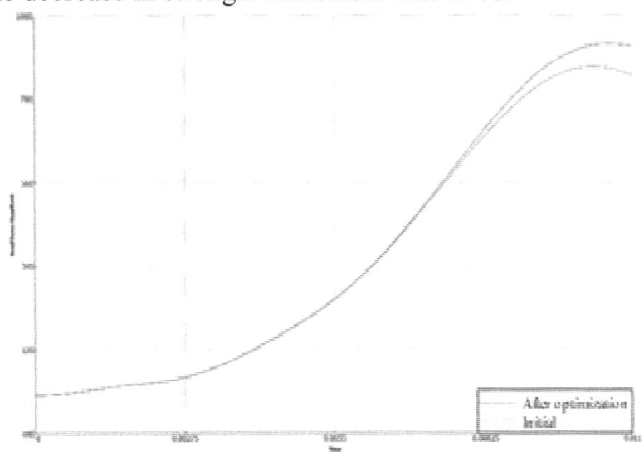

Figure 14. Comparison of upward force graphs.

4. Conclusion

This article has developed a method for parametric optimization of the tread pattern in order to reduce the risk of aquaplaning. This method is relevant, in view of the fact that today there is no ideal tread pattern that would exclude the possibility of such an effect. However, this method can be used to improve the efficiency of any existing tread pattern.

This method has been tested in the problem of driving a wheel over a volume of water at high speed, which creates a dangerous aquaplaning effect. Two design cases were considered, with a vehicle speed of 90 and 110 km / h. The force of water pushing was used as an optimization criterion. As a result of optimization, it decreased by 9 and 12%, respectively.

This technique can be improved, made more flexible and efficient. This requires more powerful computers that can be used to input more parameters into the optimization. Also, this technique can be adapted for other important calculations related to the cross-country ability of vehicles on various complex surfaces, which can be easily modeled using the sph method

References

[1]
[1] Agapov V P 2000 Finite element method in statics, dynamics and stability of spatial thin-walled reinforced structures *Moscow: Association of building universities* p 428
[2] Gorbachev M G 2009 The latest tutorial on safe driving *Publishing Ripol-Classic* p 384

[3] Prokhorenko D V 2015 Simulation of fluid using SPH *Mathematical and computer modeling: materials of the 3rd International Scientific Conference* (Omsk) pp 60-62
[4] Andre Haufe, Karl Schweizerhof and Paul DuBois 2013 Properties&Limits: Review of Shell Element Formulations *Developer Forum* p 35
[5] Campbell J C, Vignjevic R, Patel M and Milisavljevic S 2009 Simulation of Water Loading On Deformable Structures Using SPH *Tech Science Press* p 21
[6] Edouard Yreux Fluid Flow Modeling with SPH in LS-DYNA *15th International LS-DYNA Users Conference* p 6
[7] John O 2006 Halquist *LS-DYNA Theory Manual*
[8] LS-DYNA KEYWORD USER'S MANUAL *Material Models*
[9] LS-OPT Training Class Tutorial problems
[10] Tobias Erchart Reviewof Solid Element Formulations in LS-DYNA 2011 *LS-DYNA Forum* p 26
[11] Alekseev S, Tarasov A, Borovkov A, Aleshin M and Klyavin O Validation of Euroncap frontal impact of frame off-road vehicle: road traffic accident simulation p 11
[12] Borovkov A, Antonova O, Leontiev V, Mikhailov A and Nemov A 2019 Computational Mechanics of a Deformable Solid *Thermal conductivity and elasticity problems: tutorial - Polytech Press* p 162
[13] Davlechina A, Leoro Mendoza L, Klyavin O and Borovkov A 2018 Methodology of vehicle operational loads assessment during vehicle development process for further chassis and body-in-white strength and durability analysis Engineering Research and Development (IJMPERD) *International Journal of Mechanical and Production* (Vol. 8) pp 1415-1424
[14] Kukushkin K V 2018 Automobile as an ideal testing ground for the development ad testing of advanced production *ABS-auto* (No. 4) pp 38-43
[15] Tire aquaplaning: what is it and how to deal with it? *Behind the wheel*

Analyzing Causes of Damage to the Plunger Stem and Cylinder of a Sucker Rod Pump

D A Strekalovskaya, A D Davydov, A A Al'khimenko and D V Lyashenko

Peter the Great St. Petersburg Polytechnic University, St. Petersburg, Russia. darya.strek@gmail.com

Abstract. This paper presents the results of analyzing the reasons for damage to the plunger stem and cylinder of sucker rod pumps, which were operated in an environment with hydrogen sulfide. The chemical composition, mechanical properties, and hardness of the products were studied in order to assess how the material of the parts conforms to the technical documentation. Likewise, the reasons for damage were analyzed using an optical electron microscope. It was established that damage to the plunger stem happened along the first thread of the threaded connection of the rod inside the body of the adapter, while destruction to the cylinder was observed in the area of the threaded connection to the body of the pump's suction valve. Results showed that the main reason for damage was the combination of a corrosive H2S environment and mechanical stress, which led to sulfide corroded cracking under stress. Suggestions were put forward about extending the service period of the SRP parts and reducing the likelihood of their destruction.

1. Introduction

A significant number of oil wells in the trade are equipped with well sucker rod pumps (WSRP), whose number in oil drilling increases yearly. Besides the great advantages, these systems have drawbacks as well which are related to accidents of the deep pump equipment.

When operating the wells with sucker rod pump equipment, some of the main problems of damage are broken rods and mechanical friction of the parts, particularly the rods on the pump compressor tubing and the plunger on the cylinder during a long period of work in the wells. The wear and tear of the deep pump equipment also depends on the presence of formation mineralized water in the well products, which causes corrosion of the tubes, rods and parts and weakens their durability. Due to the causes mentioned above, broken and bent rods lead to accidents, while the leaks that increase in proportion to the increase in gaps between the SRP plunger and cylinder reduce the productivity of the pump after several months of work until the supply has completely stopped. Likewise, the occurrence of emergency situations can be related to the material of the parts not conforming to the requirements of technical documentation (TD), and to the improper selection of materials during operation, human factor, and so on.

The reliability of the sucker pump equipment, its functional possibilities and maintainability are determining factors for the efficient work of the sucker rod pumps.

Dealing with accidents at the place of extraction is a relevant and important problem and requires significant costs relating to stoppages, repairs, and eliminating consequences of the accidents. Reducing these costs is possible with accident prevention through timely repairs or replacing damaged parts. Therefore, papers aimed at increasing the operation efficiency of sucker rod pumps (SRP) are relevant.

Analyzing the damaged parts of SRP and identifying the reasons for the damage is necessary in order to recommend actions and protective measures focused on preventing and stopping accidents at the places of operation.
The goal of this paper was to identify the reasons for damage to the plunger stem and cylinder from two different SRP after being damaged during operation. In order to do this, mechanical tensile tests were studied, the material chemical composition of the parts and the measurements of hardness were analyzed. Breakages and damage to the parts were analyzed using an optical electron microscope.

2. Methodological Part

The object of research was Russian produced parts from two different sucker rod pumps after being damaged during operation: the plunger stem with the adapter from the pump and the cylinder with the valve from the pump. According to TD, it is known that the SRP products were operated in an environment containing H_2S and CO_2.
In order to confirm the conformity of the materials of the damaged parts with the TD, an analysis of the chemical composition and mechanical properties and hardness measurements was conducted.
The chemical composition of the SRP parts was analyzed using the method of atomic emission spectroscopy with the emission spectrometer "Iskroline-100" in accordance with GOST 54153-2010.
Testing of the mechanical properties and preparation of the samples were carried out according to GOST 1497-84, GOST 10006-80 and GOST 11701-84. The universal testing machine INSTRON 8801 (100 kN) was used with the software Instron Bluehill 2.6 for tensile testing.
Macro-hardness of the studied parts was measured based on the Rockwell method for measuring hardness on the hardness tester TP 5006 № 85 according to GOST 9013-59.
Additional study of the micro-hardness was carried out using the microscope attachment MICRO-DUROMAT 4000E in accordance with GOST 9450-76. The micro-hardness was assessed according to the recovered-print method on the optical microscope Reichert-Jung (×500) at ×500 magnification. Hardness was measured using Vickers hardness test with a load of 180g (HV0.18). The hardness was converted from the HV scale to the HRA scale according to the standard ASTM E 140 [15].
In order to do metallographic studies, the samples cut from damaged parts were pressed and their surfaces were sanded using abrasive paper with a grain of 120-1200 grid and polished with polishing cloths on the grinder-polisher BUEHLER ECOMET 4. In order to identify the microstructure of the metal, the samples were etched with a 3% nitric acid solution (HNO_3) in alcohol. The structure was studied using methods of optical metallography on the microscope Reichert-Jung with ×500 magnification.
The chemical composition of the inclusions was studied out using the scanning electron microscope SUPRA 55VP WDS with an X-ray spectral analysis attachment.

3. Results and Analysis

3.1. Researching Conformity of the Parts to Technical Documentation

SRP plunger stem

Figure 1 shows an SRP plunger stem with the adapter. As can be seen from the figure, the damage occurred along the first thread of the threaded connection of the stem inside the body of the adapter.

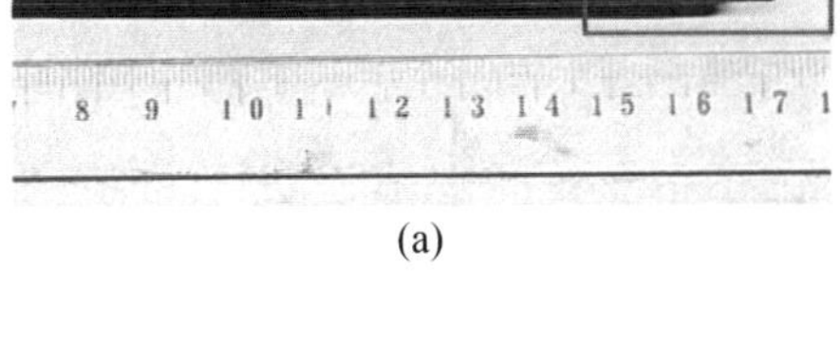

(a)

(b) (c)

Figure 1. SRP plunger stem with its adapter: (a) SRP plunger stem (the place of breakage is outlined in red); (b) stem adapter (side view, the place of breakage is marked in red); (c) stem adapter (view from above, the place of breakage is outlined in red).

The plunger stem part was studied according to the requirements of the technical documentation. Table 1 shows the requirements for the material of the plunger stem according to the TD and the results of mechanical tests of the sample material.

Table 1. Pivot table of results from mechanical testing of the plunger stem.

Parameter	Flow stress $\sigma_{0.2}$, MPa	Ultimate tensile strength σ_B, MPa	Hardness of material, HV
Test results	680	790	340
According to TD	> 586	> 793	-

According to the results from studying the mechanical properties of the stem metal, the value of the flow stress equaled 680 MPa, and the ultimate tensile strength equaled 790 MPa.

The hardness of the sample was 340 HV (34 HRC according to ASTM E140 [15]). According to the requirements of NACE MR0175 Part 2 for low-alloy steels used in hydrogen sulfide environments, the hardness of the material should equal less than 22 HRC; in rare exceptions up to 26-33 HRC [17].

Test results for the chemical composition of the stem material are given in Table 2. According to TD, the material of the plunger stem must meet the steel strength class "D" according to API Spec 11B [16], which includes the steel grades 40CrMnMoA, 30CrMoA and 38CrMnTi.

Table 2. Pivot table of results from testing the chemical composition of the plunger stem.

Parameter	Steel grade	C	Si	Mn	Cr	Mo	Cu	S	P
Results of chemical analysis		0.425	0.27	0.90	0.95	0.22	0.094	0.005	0.019
Steel of hardness class "D" according to API Spec 11B [16]	40CrMnMoA	0.37-0.42	0.17-0.40	0.6-0.9	0.9-1.2	0.15-0.25	Up to 0.3	0.035	0.035
	30CrMoA	0.26-0.33	0.17-0.37	0.4-0.7	0.8-1.1	0.15-0.25	Up to 0.3	0.025	0.025
	38CrMnTi	0.37-0.43	0.17-0.37	0.5-0.8	0.4-0.6	0.15-0.25	Up to 0.3	0.035	0.035

According to the chemical composition obtained, it was established that among steels of the hardness class "D", the material of the SRP stem corresponds to steel type 40CrMnMoA, meeting the requirements of the TD.

IOP Conf. Series: Materials Science and Engineering **986** (2020) 012038 doi:10.1088/1757-899X/986/1/012038

The material of the plunger stem meets the requirements of the technical documentation in terms of mechanical properties and chemical composition.

SRP Cylinder

The SRP cylinder with the body of the valve is presented in Figure 2.

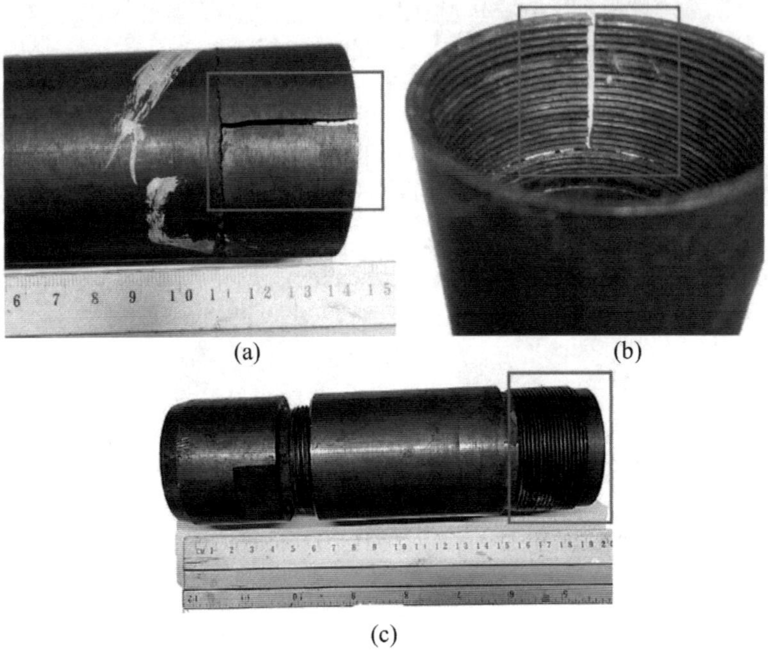

Figure 2. SRP cylinder and valve body: a) SRP cylinder, breakage area from the exterior; b) SRP cylinder, breakage area from the interior; c) SRP valve body (the area of attachment to the cylinder is outlined in red).

Study of the SRP cylinder part was carried out according to the requirements of the TD. Table 3 shows the requirements for the cylinder material according to the TD along with the results of mechanical tests of the sample material.

Table 3. Pivot table of results for mechanical tests of cylinder.

Parameter	Flow stress $\sigma_{0.2}$, MPa	Ultimate tensile strength σ_B, MPa	Presence of a nitrided coating	Material hardness
Test results	480	820	Hardness of coating: 1100 HV, at a depth of 0.127 mm – 800 HV	61.5 HRA [15]
According to TD	> 480	-	Hardness of coating: 870 HV, at a depth of 0.127 mm – 446 HV	55-62 HRA

According to the results of studying the mechanical properties of the cylinder metal, the flow stress value equaled 480 MPa, while the ultimate tensile strength equaled 820 MPa. The hardness of the sample was 61.5 HRA in accordance with ASTM E140 [15].

In order to monitor the quality of the material for the hardness parameter, it was of interest to check for the presence of a nitrided layer on the surface of the cylinder part. A section of the cylinder was studied

IOP Conf. Series: Materials Science and Engineering **986** (2020) 012038 doi:10.1088/1757-899X/986/1/012038

for hardness from the outer wall in order to identify the nitrided layer. For this, a lateral metallographic cut was made of the cylinder section. The results of the research are presented in Figure 3.

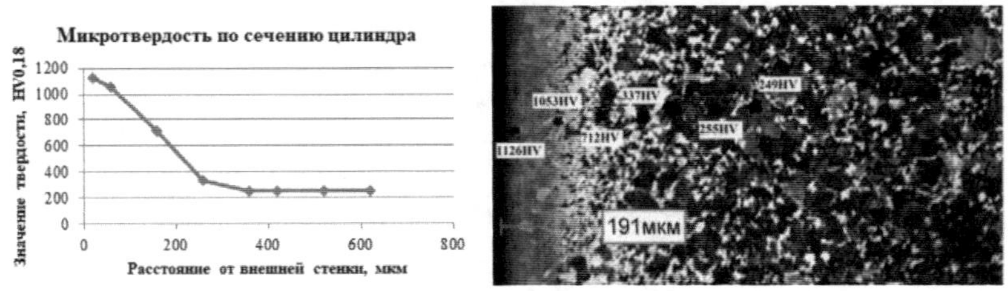

(a) (b)

Figure 3. Hardness values of the cylinder section from the outer wall: a) graph of dependency of hardness on distance; b) structure of the material surface laterally.

According to the results obtained, the thickness of the nitrided layer equaled around 200 μm, while the hardness of the main metal was between 240 and 260 HV0.18.

The results from studying the chemical composition of the cylinder material are presented in Table 4. According to the TD, the material of the cylinder should meet the steel grade 38Cr2MoAlA in accordance with API Spec 11B [16].

Table 4. Pivot table of results from studying the chemical composition of the cylinder.

Parameter	C	Si	Mn	Cr	Mo	Al	Cu	S	P
Results of chemical analysis	0.37	024	0.44	1.69	0.17	0.92	0.14	0.003	0.016
Chemical composition of the steel grade 38Cr2MoAlA according to API Spec 11B [16]	0.35-0.42	0.20-0.45	0.3-0.6	1.35-1.65	0.15-0.25	0.7-1.1	Up to 0.3	0.04	0.04

According to the results of studying the chemical composition of the cylinder material, it was established that the composition of the material meets the requirements of the TD and are of steel type 38Cr2MoAlA.

The cylinder material of the SRP valve meets the requirements of the technical documentation in terms of mechanical properties and chemical composition.

3.2. Analyzing the Reasons for Damage

SRP plunger stem

As it was previously established (from Figure 1), damage to the SRP plunger stem occurred along the first thread of the threaded connection of the stem within the body of the adapter. Impact marks of the counterparts appear on the stem guide and stem adapter during operation. During visual examination of the fracture, at a magnification up to 30 times, it was established that the damage was brittle in nature. In Figure 4, the red line marks the area the crack originated, the starting point of the first thread. The yellow line indicates the place of complete break.

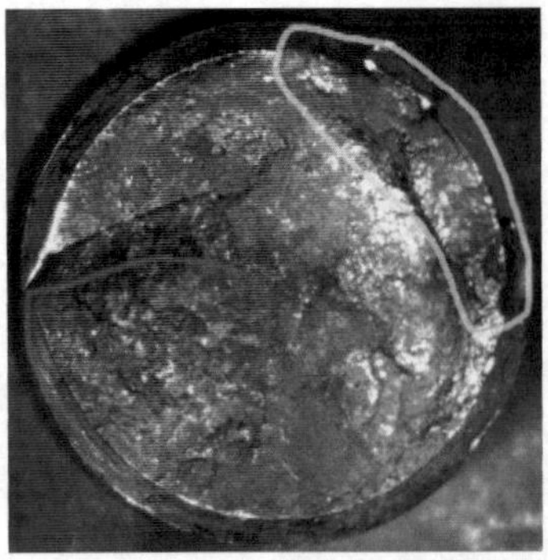

Figure 4. Surface of the fracture of the SRP plunger stem.

The structure of the sample material was studied on a lateral metallographic cut (Figure 5, a). It was determined that the structure of the material of the plunger stem was upper bainite.

The surface of the fracture and a chemical analysis of the corrosion products (Figure 4, highlighted in red) were studied using an electron microscope. The chemical composition of the corrosion products on the surface (Figure 5, b) is presented in Table 5.

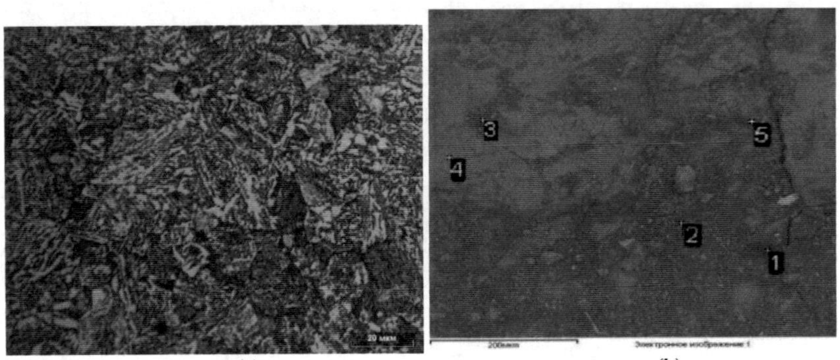

(a) (b)

Figure 5. Structure of the SRP plunger stem material magnified x500 (a) and surface of the fractured SRP plunger stem magnified x1000 (b).

Table 5. Chemical analysis results of surface of plunger stem fracture.

Spectrum	O	Al	Si	S	Cl	Fe
1	32.49	0.59	0.09	0.59	1.80	1.02
2	36.60	0.23	0.43	1.55	0.24	10.60
3	32.29	0.61	10.09	2.01	0.24	12.42
4	50.72	0.08	0.84	0.50	0.26	33.49
5	35.52	0.41	2.42	2.62	0.31	11.83

From the table it can be seen that sulfides and chlorides were discovered on the surface of the fracture in the area where the crack originated. This corresponds to the conditions of operation for the plunger stem part in an environment with hydrogen sulfide.

Judging by the corrosion products and the type of fracture (brittle), the reason for the damage to the stem could have been sulfide corrosion stress cracking (SCSC). An increased likelihood of damage is related to impact to the counterparts and a possible overload during operation.

SRP cylinder

As seen in Figure 2, the SRP cylinder has two cracks in the area of the threaded connection with the body of the pump's suction valve. Upon visual inspection of the fracture, at a magnification of up to 30 times, it was noticed that the damage to the lengthwise fracture was brittle in nature. In Figure 6, the broken fifth thread of the cylinder is marked with a red line.

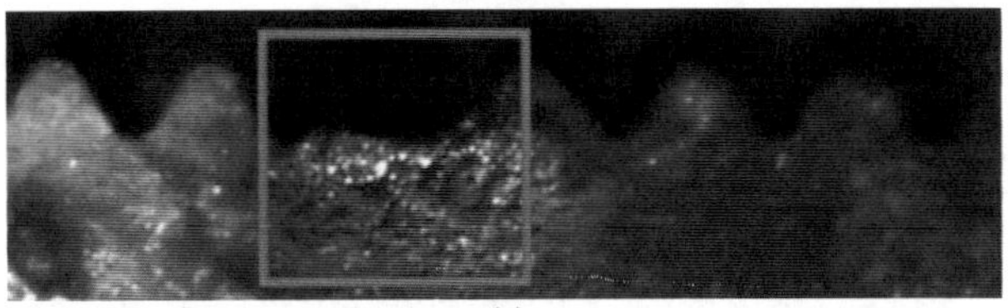

(a)

(b)

Figure 6. Macro images of lengthwise fracture of cylinder: a) lateral view; b) view from inside the tube, from the threaded side.

The structure of the sample material was studied on a lateral metallographic cut (Figure 7, a). When doing metallographic research, it was established that the material of the cylinder had a ferrite-pearlite structure, with a grain score of 8.5.

Figure 7, b presents the eighth thread in order to study the breakage area of the cylinder thread. In order to do this, a metallographic cut lengthwise to the cylinder thread was prepared. The depth of the cracks was established to be 100μm.

(a) (b)

Figure 7. Structure of the SRP cylinder material magnified to x500 (a) and cracks in the area between the 8th and 9th thread magnified to x100 (b).

A fracture spreading across the radius of the cylinder is presented in Figure 8.

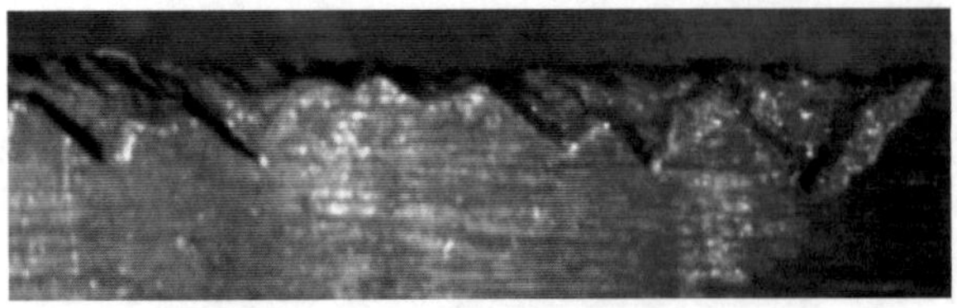

Figure 8. Radial fracture of cylinder.

According to the nature of the crack presented in Figure 8, it can be proposed that this radial crack is secondary, the origin and spread of which is a result of increased tensile loads (mechanical factor) on this section of threads after lengthwise damage to the cylinder.

The surface of the fracture and a chemical analysis of the corrosion products (Figure 6, highlighted in red) were studied using an electron microscope. The chemical composition of the corrosion products on the studied parts of the surface (Figure 9) is presented in Table 6.

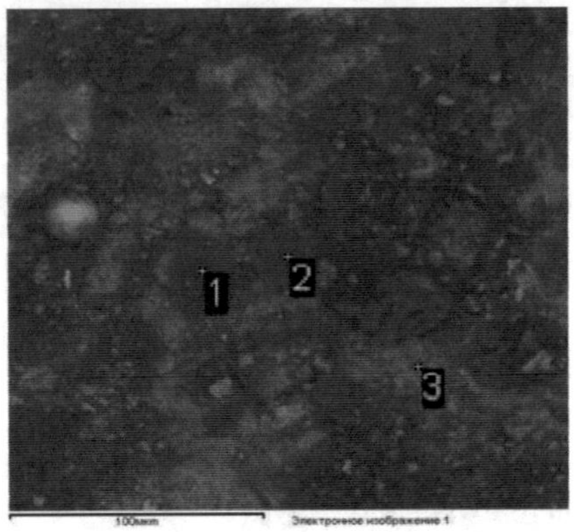

Figure 9. Studies of the surface of the SRP cylinder fracture, magnified to x1000.

Table 6. Chemical analysis results for the surface of the fractured cylinder.

Spectrum	O	Na	Si	S	Cl	Fe
1	21.14	0.18	-	1.38	0.23	4.12
2	5.02	-	-	5.73	1.96	16.59
3	2.85	-	0.39	0.87	0.36	74.64

Sulfides and chlorides were discovered on the surface of the fracture in the area where the crack originated. It is also known that the operational environment of the SRP cylinder was an environment containing H_2S. Thus, the overall high stress concentration in transition zones, i.e., mechanical impact, and the corrosive environment help to conclude that the main reason for damage to the material of the cylinder could be sulfide corrosion stress cracking (SCSC).

4. Discussion

Initially, the task was set to exclude the version that damage to the equipment was caused by not meeting the requirements of the technical documentation. All of the tests conducted showed that the materials of the SRP plunger stem and the SRP cylinder have the claimed indicators in accordance with the TD.

It is important to note that the TD for the material does not contain any requirements for the hardness of products. According to the research results, it was learned that at 33 HRC, the hardness of the parts exceeds the maximum acceptable values according to NACE MR0175 [17]. The hardness of the material is an important indicator capable of influencing the quality of the material, and a deviation of the hardness from recommended standard values can lead to damage to the material in aggressive environments and, as a result, to damage to the part. When used in H2S environments, it is necessary for the materials to be qualified according to the standard NACE MR0175, which contains a requirement for the maximum acceptable values of hardness. In the examined case, a deviation of the hardness values could have influenced the brittleness of the material and acted as one of the reasons for damage to the part.

Since the parts were used in an environment containing H2S and CO2, it was assumed that the main cause of damage was sulfide corrosion stress cracking. Study of the damaged surfaces and an analysis of the corrosion products showed that sulfides and chlorides were present in the fractured area. The joint effect of mechanical loads and a corrosive environment during operation likely led to damage to the SCSC mechanism. Similar incidents are described in the following works [].

In summary, it can be noted that steels of grades 40CrMnMoA and 38Cr2MoAlA are susceptible to sulfide corrosion damage in working environments containing H2S. Therefore, equipment from low-alloy steels is not recommended to use in hydrogen sulfide environments.

In order for material used in aggressive conditions such as H2S and CO2 to function, it is necessary to finalize the TD in terms of hardness, as well as to monitor that the specific material meets the given requirements. Lab tests can be conducted in order to be certain of the durability of the material.

For other protective measures of preventing such damage, it is recommended to lighten the load on the SRP, to use inhibitors to reduce the aggressiveness of the environment, to monitor the properties of the materials used or to use more corrosion resistant materials to manufacture SRP parts [3].

5. Conclusion

Foreign standards regulate the requirements for the hardness of materials used in hydrogen sulfide environments. However, such requirements are missing from technical documentation on the tested parts. Increased hardness leads to an increased tendency to sulfide corrosion cracking, and therefore it is recommended to take this parameter into account in documentation. Extending the service life of SRP parts and reducing the likelihood of damage is possible with a more detailed analysis of the anticipated conditions of operation and monitoring of the materials used.

References

[1] Sorokin V G, Volosnikova A V and Vyatkin S A 2012 Database of Steels and Alloys *Book on Demand* (640 [2])

[2] Yamaliev V U, Ishemguzhin I E and Latypov B M 2017 Assessing the frictional force of a plunger on the cylinder of sucker rod pump when designing a rod column *News of the Samar Research Center of the Russian Academy of Sciences* (V. 19(1-1)) pp 70-75

[3] Ponikarov S I, Karimov L F and Kalimullin N F 2017 Analysis of reasons for breaks to rod columns during operation of wells equipped with sucker pumps *Bulletin of Kazan Technological University* (V. 20(3)) pp 44-45

[4] Beketov S B and Kuchurin A E 2010 Features of using wells equipped with sucker pumps in fields with poorly cemented reservoirs *Mining information and analytical bulletin* (№ 12) pp 107-115

[5] Alkhimenko A 2019 Corrosion testing of experimental steels for oilfield pipelines *E3S Web of Conferences* (Vol. 121. EDP Sciences) pp 01001

[6] Ermakov B S, Alhimenko A A, Shaposhnikov N O, Tsvetkov A S and Shirokov A V 2020 Study of the crystallographic texture of pipe steel *Letters on Materials* (10.1) pp48-53

[7] Davydov A, Alekseeva E and Gaev A 2020 Specificity to the choice of materials for wellhead equipment *Materials today: Proceedings*

[8] Kovalev M, Lyashenko D, Kurakin M and Shakhmatov A 2019 Failure analysis of radial bearings used in the refining of petroleum products *E3S Web of Conferences* (Vol. 121. EDP Sciences) pp 03003

[9] Kostitsyna I, Shakhmatov A and Davydov A 2019 Study of corrosion behavior of carbon and low-alloy steels in CO2-containing environments *E3S Web of Conferences* (Vol. 121. EDP Sciences) pp 04006

[10] Laptev A and Bugay D 2010 Fighting against corrosion in oil and gas complex of Russia: Problems and the ways of their solving *European Corrosion Congress, EUROCORR-2010*

[11] Dvoinikov M V 2017 Research on technical and technological parameters of inclined drilling *Journal of Mining Institute* (Vol. 223) pp 86-92

[12] Litvinenko V S and Dvoynikov M V 2019 Monitoring and control of the drilling string and bottomhole motor work dynamics *Topical Issues of Rational Use of Natural Resources* (Vol. 2) pp 804-809

[13] Dvoynikov M V, Nutskova M V and Blinov P A 2020 Developments Made in the Field of Drilling Fluids by Saint Petersburg Mining University *International Journal of Engineering* (Vol. 33(4)) pp 702-711

[14] Moradi S S T, Nikolaev N I, Chudinova, I V and Martel A S 2018 Geomechanical study of well stability in high-pressure, high-temperature conditions *Geomechanics and Engineering* (Vol. 16(3)) pp 331-339

[15] ASTM E140 Standard Hardness Conversion Tables for Metals Relationship Among Brinell Hardness, Vickers Hardness, Rockwell Hardness, Superficial Hardness, Knoop Hardness, Scleroscope Hardness, and Leeb Hardness

[16] API SPEC 11B Specification for sucker rods (pony rods, polished rods, couplings and subcouplings) twenty-second edition

[17] NACE MR0175/ISO 15156 Petroleum and natural gas industries - Materials for use in H2S-containing environments in oil and gas production

Researching the operational properties of materials used in ECP

A A Alkhimenko, A V Shakhmatov and M L Shishkova

Peter the Great St. Petersburg Polytechnic University (SPbPU), Saint Petersburg, Russia, weatherford@onti.spbstu.ru

Abstract. The goal of this research was to determine the corrosion resistance of metal materials used in ECP constructions as well as to determine the degree of degradation of the metal properties of rubber compounds (RS-3, TER-10, AF-15) after exposure to a corrosive environment. In order to assess the impact of a corrosive environment on changes in the mechanical properties of rubber compounds used in ECP constructions, tests were carried out for rubber samples of the compounds RS-3 (TU 2512-028-46521402-2014), TER-10 (TU 2512-028-46521402-2014), AF-15 (TU 2512-028-46521402-2014) in their initial state and after testing in a corrosive environment in an autoclave. Tests on samples of structural metal materials (05Cr16Ni4, 40Cr13, 40Cr, 22MnAl, Nirezist) in an aqueous saline solution containing ions of chlorides. According to these tests at 240, 480 and 720-hour exposure in the autoclave, the steel 05Cr16Ni4 is passive in the given corrosive environment and has a corrosion rate of ≤ 0.01 mm / year. Tests of samples of four steels of grades 05Cr16Ni4, 40Cr13, 40XCr and 22MnAl in a solution of hydrochloric acid with pH = 0 over 6 hours at room temperature showed.

1. Introduction

Electric centrifugal pumps (ECP) are widely used in the oil industry as well as when developing mineral deposits by means of subterranean and above-ground methods [4, 5]. These pumps are operated in extreme well conditions: high temperatures and pressure, aggressive environment with a high concentration of chlorides, H_2S, CO_2, abrasive particles, excess gas - all of which leads to an intense deterioration of the working of the pump parts [1, 6]. New system standards of manufacturing processes, increased working temperatures, pressure, and rates lead to accelerated wear and tear both of individual parts and of the machines and mechanisms a whole. Combined with the need to automate production, the problem of increasing the durability of the machines becomes even more urgent [9]. Selecting optimal materials for aggressive working conditions is relevant due to the importance of solving the problem. In order to ensure an optimal service life of the SRP, it is necessary to pay special attention to the choice of materials which meet requirements for corrosion, abrasion and cavitation resistance [3]. As a rule, standard materials for pump parts (such as cast iron, bronze and low-carbon steel) have the lowest cost and are the easiest to replace. However, as a result of their low corrosion resistance, premature equipment failure occurs, which significantly increases the final cost, taking into account service and replacements [2, 7]. The goal of this research was to determine the corrosion resistance of metal materials used in ECP constructions as well as to determine the degree of degradation of the metal properties of rubber compounds (RS-3, TER-10, AF-15) after exposure to a corrosive environment.

2. Materials and method of experiment

For the metal materials of an ECP construction, samples of steel 22MnAl, samples of steel 40Cr in a heated treated condition of 22-32 HRC, and samples of Nirezist cast iron were selected. The corrosion resistance of the selected materials was assessed using an autoclave. Testing for resistance to general corrosion was carried out on flat polished samples with a size of 50×30×4 mm.

In order to assess the impact of a corrosive environment on changes in the mechanical properties of rubber compounds used in ECP constructions, tests were carried out for rubber samples of the compounds RS-3 (TU 2512-028-46521402-2014), TER-10 (TU 2512-028-46521402-2014), AF-15 (TU 2512-028-46521402-2014) in their initial state and after testing in a corrosive environment in an autoclave. Tensile testing of the rubbers was conducted on a modified universal testing machine, model R5. The rubber samples were rectangular strips 150 mm long, 20 mm wide, and 2.5 mm thick. The length of an unclamped stretchable part of the sample measured 100 mm.

Two environments were chosen as the corrosive environment for testing:

0) an aqueous saline solution containing dissolved hydrogen sulfide and carbon dioxide: H_2S - 1.25 g/l; CO_2 1.15 g/l; Cl^- - 75 g/l; HCO_3 - 1 g/l; Ca^{2+} - 9 g/l; (Na^++K^+) - 40 g/l. Tests were carried out at a temperature of 120 °C, pH=3.0 and a pressure of 12 atm in the autoclave. Testing lasted for 720 hours;

1) hydrochloric acid with pH=0 at room temperature. Testing lasted for 6 hours [8].

In order to analyze changes to the rate of corrosion of the studied materials in time, experiments of various time bases were conducted. For this, four grades of steel were chosen: 05Cr16Ni4, 40Cr13, 22MnAl, 40Cr. The samples were removed from the autoclave after 240, 480 and 720 hours.

3. Research results and discussion

The conducted autoclave tests on metal materials for overall corrosion in a aqueous saline solution containing dissolved hydrogen sulfide and carbon dioxide and heated to a temperature of 120 °C show that the average rate of corrosion for Nirezist material equaled 0.47 mm/year and for steels of grade 22MnAl and 40Cr – 0.13 mm/year and 0.37 mm/year, respectively. Thus, it can be seen that the steel of grade 22MnAl has higher corrosion resistance in comparison to 40Cr grade steel and especially in comparison with Nirezist cast iron.

As a result of tests on samples of steel grades 05Cr16Ni4, 40Cr13, 22MnAl, 40Cr at various time bases, the rate of corrosion was calculated for materials with consideration of duration of contact with the test environment (Figures 1, 2).

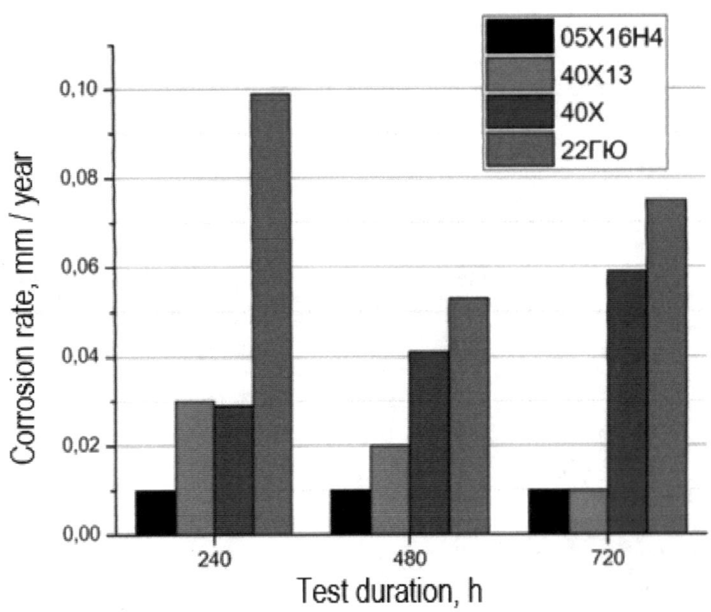

Figure 1. Changes in rate of steel corrosion.

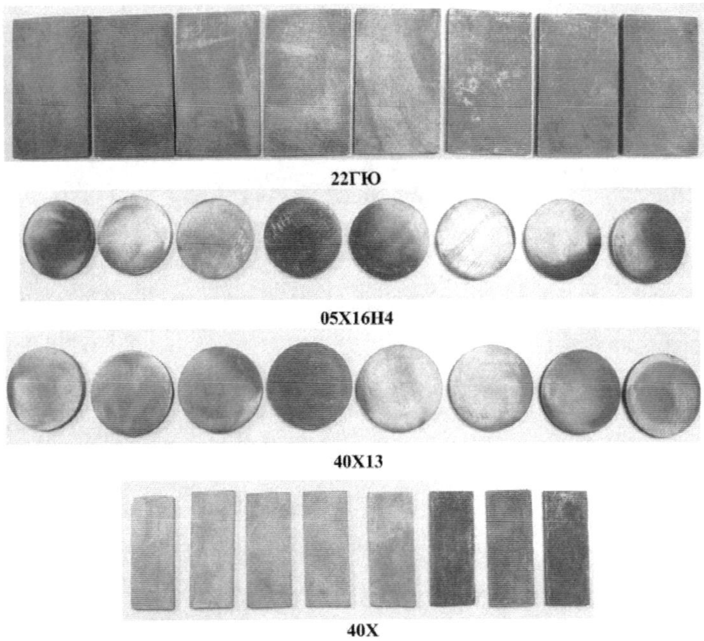

Figure 2. Samples after autoclave testing.

The tested steels significantly differ from each other in the rate of corrosion: the lowest corrosion rate, not dependent on the duration of testing, is seen with stainless steel 05Cr16Ni4, whose corrosion rate does not exceed 0.01 mm/year. A slightly higher but still relatively low corrosion rate was seen with stainless steel with 13% chrome, which also decreases while in a corrosive environment, in comparison with steel 05Cr16Ni4. The corrosion resistance of low-alloy steels 40Cr and 22MnAl is slightly higher. The corrosion rate of these steels fluctuates between 0.03 mm/year and 0.10 mm/year. At the same time, increasing the time of exposure in a corrosive environment does not lead to a reduced corrosion rate of these steels. This is due to the fact that corrosion products forming on the surface of the samples poorly protect the metal, leading only to a noticeable scattering of the data obtained on corrosion rate. This explains the discrepancy in corrosion rate values for the indicated steels, which were obtained during preliminary and subsequent tests with 720 hours of exposure.

It should also be noted that when conducting corrosion testing of structural metal materials which show no formation of a thick protective film (carbon and low-alloy steel) on the surface of the samples when interacting with a solution, the forming corrosion products poorly protect the metal from dissolving. Therefore, in such cases, a large scattering of data on corrosion rates is observed. In order to divide these materials more precisely according to corrosion rates, it is advisable to increase the number of test samples.

Study results of steel samples in hydrochloric acid showed that the average value of corrosion for the steel of grade 05Cr16Ni4 equaled 0.21 mm/year, for 40Cr13 steels – 2.43 mm/year, 40Cr – 1.63 mm/year, 22MnAl – 4.12 mm/year. Thus, it can be seen that the most resistant to a hydrochloric acid environment with pH = 0 turned out to be 05Cr16Ni4 grade stainless steel. The surface of the samples of this steel remained shiny, with no visible traces of corrosion. Samples of the steel of grade 40Cr13 experienced general corrosion. Likewise, individual small holes (pitting) was observed on the surface. Due to the pitting corrosion, the steel of grade 40Cr13 displayed a slightly greater inclination to corrosion than the low-alloy steel of grade 40Cr. The corrosion rate of the steel of grade 22MnAl was at a maximum, which can be explained by the presence of manganese as an alloying element, which is more thermodynamically active than iron and other elements included in the composition of the steels.

Figure 3 presents a change in the appearance of the sample surface of steel 22MnAl when in a hydrochloric acid with pH = 0 (exposure time of samples – 45 minutes).

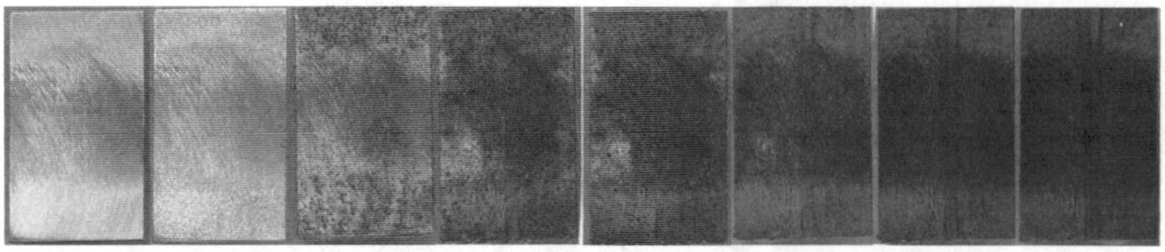

Figure 3. Change in appearance of sample 22MnAl over 6 hours.

Research on the degradation of the metal properties of the samples of the rubber compound TER-10 is given in Figure 4.

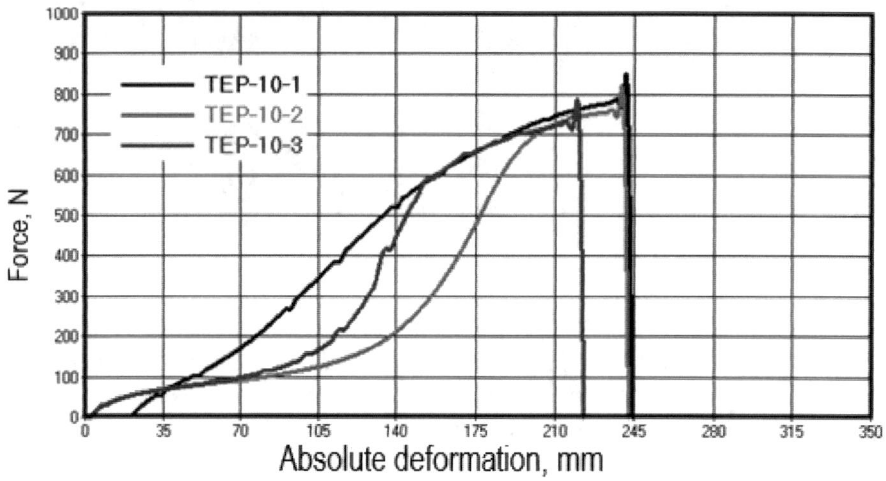

a)

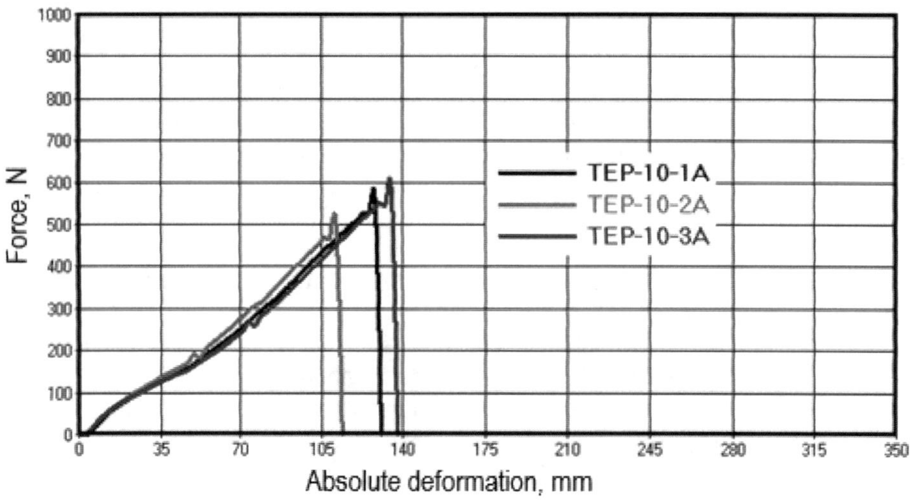

b)

Figure 4. Diagram of tension of TER-10 samples before (a) and after (b) exposure in an autoclave.

A diagram of the tension of RS-3 samples before and after exposure in an autoclave are given in Figure 5.

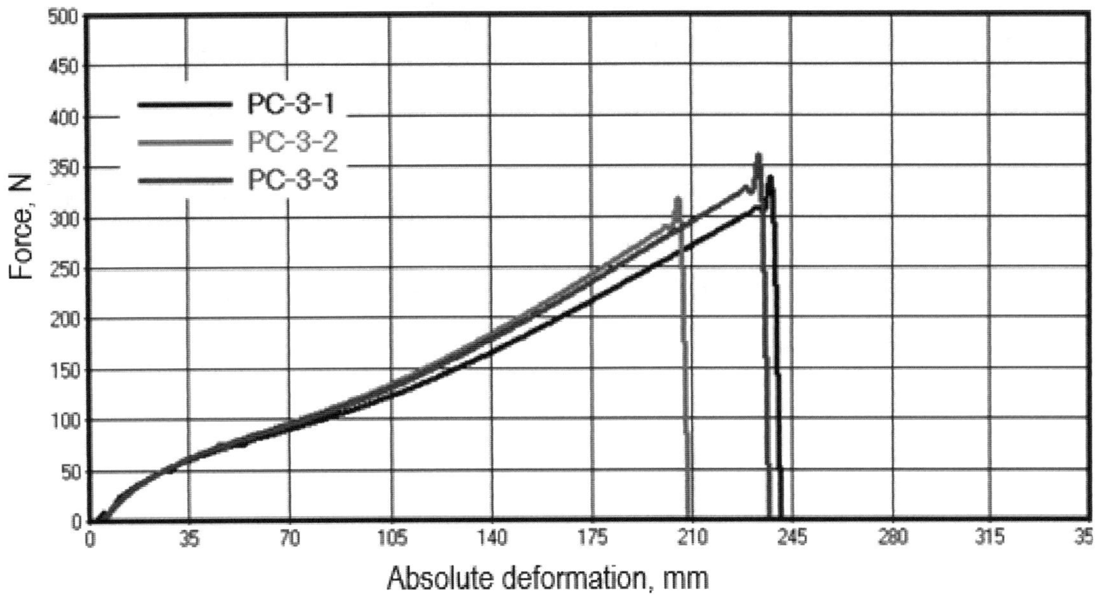

a)

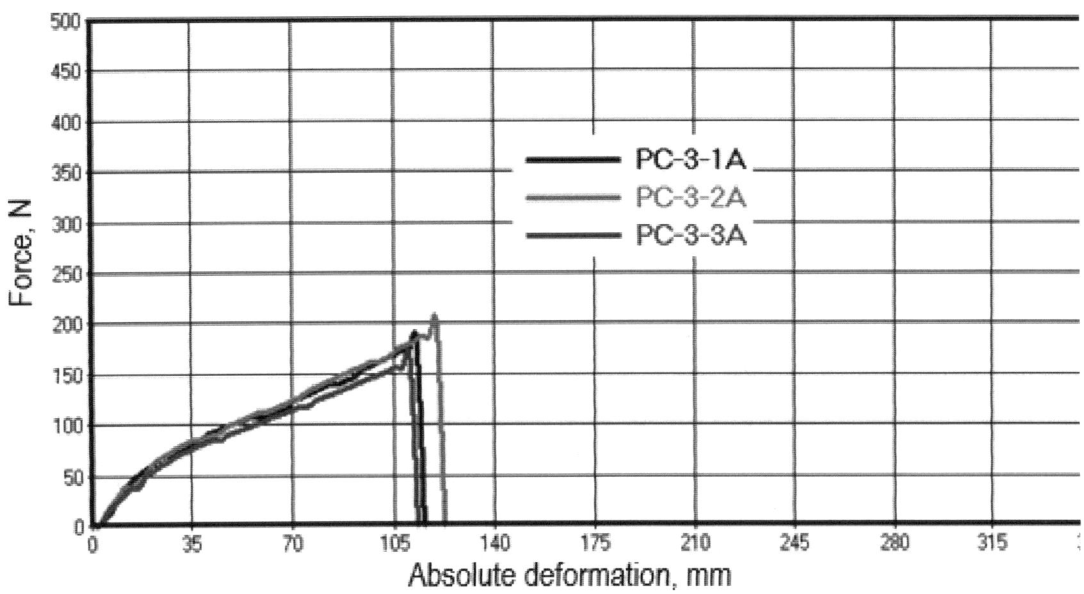

b)

Figure 5. Diagram of tension of RS-3 samples before (a) and after (b) exposure in autoclave.

The results of measuring the dependency of the load on the deformation for AF-15 samples before and after exposure in an autoclave are given in Figure 6.

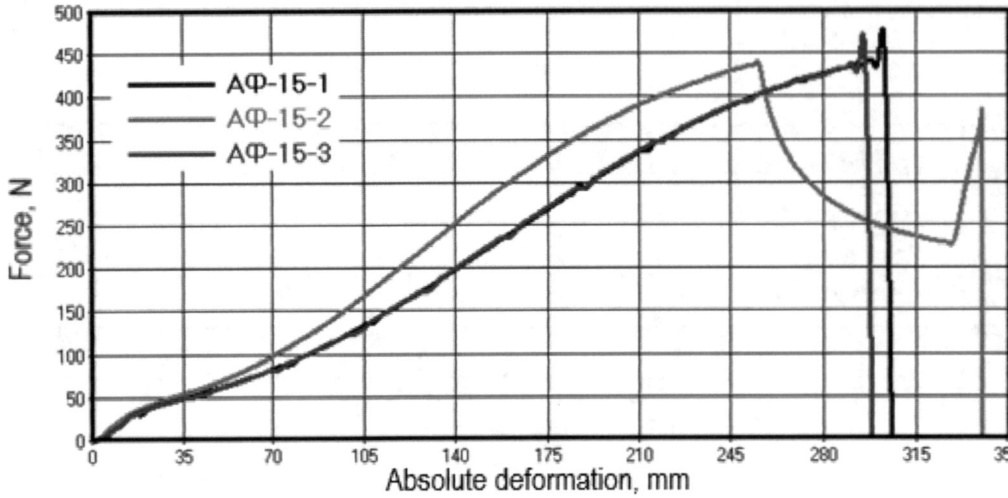

a)

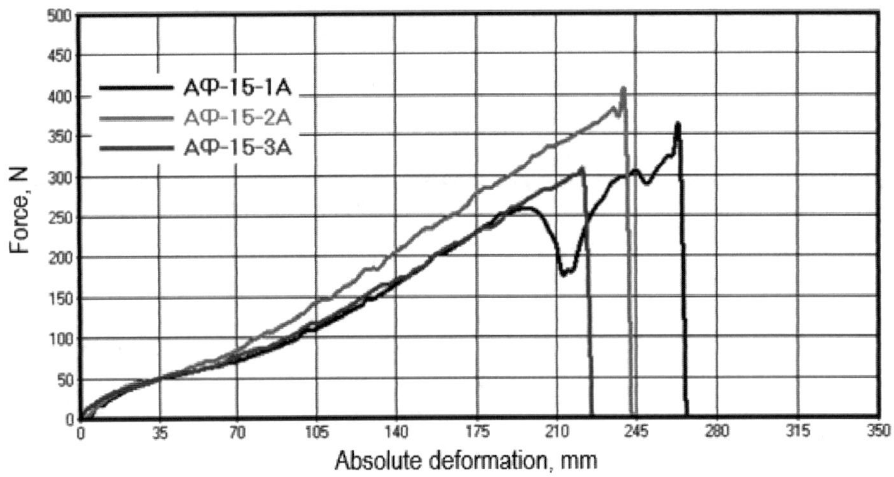

b)

Figure 6. Diagram of dependency on deformation for AF-15 samples before (a) and after (b) exposure in an autoclave.

Results of changes in the conditional tensile strength (GOST 270-75) and changes in the relative elongation during a rupture of samples of rubber compounds, grades RS-3, TER-10, AF-15, are given in Table 1.

Table 1. Mechanical properties of various grades of rubber compounds.

Grade of rubber compound	Mechanical properties during tension					
	Before testing		After testing		Degree of degradation, %	
	Force, N	Deformation, %	Force, N	Deformation, %	In force	In deformation

RS-3	337	128	190	18	44	86
TER-10	815	237	570	28	30	88
AF-15	427	215	365	148	14,5	31

An analysis of the obtained results shows that in terms of both strength and deformation, the rubber compound of grade AF-15 is subject to the least degradation under the impact of a corrosive environment. The conditional strength was reduced 14.5%, while the relative elongation during rupture reduced 31%. For rubber compounds of grades RS-3 and TER-10, strength dropped 44% and 30%, respectively, while the plasticity (relative elongation) reduced 86% and 88%.

4. Conclusion

Based on the research conducted and analysis of the obtained data, the following conclusions can be made:

1. Tests on samples of structural metal materials (05Cr16Ni4, 40Cr13, 40Cr, 22MnAl, Nirezist) in an aqueous saline solution, containing ions of chlorides, bicarbonate, calcium, sodium, potassium, as well as dissolved hydrogen sulfide and carbon dioxide at a temperature of 120°C and autoclave pressure of ~1.2 MPa showed:

1. After preliminary testing in the autoclave over 720 hours, all the studied materials (low-alloy steels of grade 22MnAl, 40Cr and Nirezist cast iron) displayed a tendency to general corrosion. The Nirezist cast iron has a maximum corrosion rate, which equaled around 0.47 mm/year. The corrosion rate of steels of grade 40Cr and 22MnAl was slightly lower than that of the cast iron, though noticeably different from each other (0.37 mm/year and 0.13 mm/year).

2. According to these tests at 240, 480 and 720-hour exposure in the autoclave, the steel 05Cr16Ni4 is passive in the given corrosive environment and has a corrosion rate of $\leq$ 0.01 mm/year. Likewise, a rather low corrosion rate is seen with stainless steel with 13% chrome, which reduces over time in a corrosive environment. Steels 40Cr and 22MnAl have a corrosion rate of 0.059 to 0.075 mm/year after exposure in the test environment over 720 hours. A reduced corrosion rate was not discovered for the indicated steels when the testing period was increased, which is evidence of the low protective ability of the corrosion products formed on the surface of the samples. For the same reason, a noticeable scattering of the corrosion rate data for low-alloy steels is observed, which makes it impossible to rank them according to corrosion resistance.

2. Tests of samples of four steels of grades 05Cr16Ni4, 40Cr13, 40XCr and 22MnAl in a solution of hydrochloric acid with pH=0 over 6 hours at room temperature showed:

2.1. The highest corrosion resistance was seen with the stainless steel of grade 05Cr16Ni4 with a corrosion rate of 0.21 mm/year.

2.2. Steel of grade 40Cr13 experienced general corrosion in a solution of hydrochloric acid. Likewise, individual small holes (pitting) was observed on the surface. For this reason, the corrosion rate of 40Cr13 grade steel was slightly higher than that of the low-alloy steel of grade 40Cr (2.43 mm/year and 1.63 mm/year, respectively).

2.3. The lowest resistance was seen with 22MnAl grade steel, which can be explained by the presence of manganese as an alloying element, which is more thermodynamically active than iron and other elements included in the steel's make-up. The corrosion rate of the 22MnAl grade steel equaled 4.12 mm/year.

3. Autoclave testing of three grades of rubber compounds (RS-3, TER-10, AF-15) in a corrosive environment (pH = 3, temperature 120°C, duration of testing 720 hours) showed the following:

3.1. In terms of both strength and deformation, the rubber compound of grade AF-15 was subject to the lowest degradation under the impact of a corrosive environment. The conditional strength reduced 14.5%, while the relative elongation during rupture reduced 31%.

3.2. For rubber compounds of grades RS-3 and TER-10, the strength dropped 44% and 30%, respectively, while the plasticity (relative elongation) reduced 86% and 88%, respectively.

References

IOP Conf. Series: Materials Science and Engineering **986** (2020) 012039 doi:10.1088/1757-899X/986/1/012039

[1] Alkhimenko A 2019 Several erosion test results of means of sand control
[2] Alzafin Y A Mourad A H I, Abou Zour M and Abuzeid O A 2007 A study on the failure of pump casings made of ductile Ni-resist cast irons used in desalination plants *Eng. Fail. Anal.* (V. 14 – № 7) pp 1294–1300
[3] Davydov A, Alekseeva E and Gaev A 2020 Specificity to the choice of materials for wellhead equipment *Mater. Today Proc.*
[4] Dvoynikov M, Syzrantsev V and Syzrantseva K 2017 Designing a high resistant, high-torque downhole drilling motor *Int. J. Eng. Trans. A Basics* (V. 30 № 10) pp 1615–1621
[5] Dvoynikov M V 2017 Research on technical and technological parameters of inclined drilling *J. Min. Inst.* (V. 223) pp 86–92
[6] Kostitsyna I, Shakhmatov A and Davydov A 2019 Study of corrosion behavior of carbon and low-alloy steels in CO2-containing environments *E3S Web Conf.* (V. 121) p 04006
[7] Sahoo T 2009 Making centrifugal pumps more reliable *World Pumps* (V. 2009 № 513) pp 32–36
[8] Smart N G, Hitchman M L, Ansell R O and Fortune J D 1994 A study of the electrochemical properties of NiResist in 3% sodium chloride solution *Corros. Sci.* (V. 36 № 9) pp 1473–1489
[9] Stewart M 2019 Centrifugal pumps (Elsevier) pp 61–309

IOP Conf. Series: Materials Science and Engineering **986** (2020) 012040 doi:10.1088/1757-899X/986/1/012040

Composite tube damage evaluation in aggressive oil-gas field

E L Alekseeva[1], M K Kurakin[1], M A Kovalev[1], A A Lapechenkov[1], M L Shishkova[1] and N A Devyaterikova[2]

[1]Peter the Great St.Petersburg Polytechnic University (SPbPU), St.Petersburg, Russia, alekseeva_el@spbstu.ru
[2]JSC Pervouralsk Novotrubny Plant, Pervouralsk, Russia

Abstract.
The purpose of this article is to lead a research program to assess the damaging of liners of a composite tube for operation in oil-gas fields containing carbon dioxide, including those with minor amounts of hydrogen sulfide at standard conditions and under conditions of elevated temperature and pressure. To assess the level of corrosion resistance of the composite tube liner, stress corrosion cracking tests were carried out, autoclave tests for general corrosion at elevated temperatures and pressures were also carried out, and electrochemical studies were carried out to assess the equilibrium corrosion potential and calculate corrosion rates analyzing Tafel curves.

1. Introduction
Steel tubing used in modern conditions of oil-gas fields suffer of severe wear as a result of the complex aggressive effects of recovered elements: hydrogen sulfide, carbon dioxide, oxygen, highly mineralized oilfield water, various types of bacteria, solid suspended particles and other aggressive components, included in the extracted products [1–3, 6–10]. In difficult production conditions the serviceability of tubing is several times lower than the declared service life of the pipe manufacturer. The number of premature tubing failures due to internal corrosion is significant (failures of more than 50%), which leads to a multiple increase of production cost [4]. Traditionally low alloyed steel pipes are used in the oilfield due to their low cost. However the low corrosion resistance of such pipelines dictates the need to improve their performance using various approaches. Due to their high cost, the use of high alloyed steel is not always economically reasonable. The use of inner covering, both non-metallic [12] and metallized [11], is widely used. The main disadvantages of non-metallic (polymer) coatings are the number of operational limitations (on the operating temperature and pressure in the system, resistance to chemically aggressive components, the contamination in the extracted product), as well as the lack of in-line inspection technology to predict their service life. [13]. The use of internal metallized coatings on tubing pipes, a relatively new and not yet very common technology in the Russian Federation - is an economically promising method of protection provided a thin coating is applied, however, the limitation of the use of this type of coating is connected with a high cost comparing to non-metallic (polymer) coatings, with the limited areas of application, as well as with the limited possibility of coating long pipes [5, 14].
One of the effective ways to multiply the operating life of tubing without a significant cost increase is the protection technology, where a stainless steel insert coating is used as an inner layer [15].

The purpose of this article is to lead a research program to assess the damaging of liners of a composite tube for operation in oil-gas fields containing carbon dioxide, including those with minor amounts of hydrogen sulfide at standard conditions and under conditions of elevated temperature and pressure.

2. Materials and experiment method

Samples for research were made of a composite tube pup joint. The inner part of the tubing (insert) is made of 15H13N2MB steel with a thickness of ≈ 1 mm. The material for the inner layer (60x1.0 mm size pipe) was gained by 4 times rolling of cold (warm) deformation from a rerolling material 108x8.0 mm from originally produced metal, followed by heat treatment.

To assess the level of corrosion resistance of the composite tube liner, stress corrosion cracking tests were carried out, autoclave tests for general corrosion at elevated temperatures and pressures were also carried out, and electrochemical studies were carried out to assess the equilibrium corrosion potential and calculate corrosion rates analyzing Tafel curves.

The cladding liability to stress cracking in H2S was assessed in accordance with the requirements of NACETM 0177, Method C. Sample surface roughness was achieved manually to get the required value of 0.81 Ra. Tests of cladding layer samples were carried out at loads of 60, 70 and 80% of the actual yield limit. Tests were carried out in brine A, NACETM 0177 standard. The brine was saturated with 100% H_2S. The concentration of H2S at the beginning of the tests was 2650 ppm, at the end of the tests it was 2570 ppm. The pH of the brine was 2.6 at the beginning of the test and 3.2 at the end of the test. The test lasted 720 hours.

The resistance of the composite tube liner to general corrosion at elevated pressure and temperature was evaluated by performing autoclave general corrosion tests. While testing GOST 9.905 and GOST 9.908 requirements (National state standards) were considered. Autoclave general corrosion tests were carried out on five samples with + 80 °C temperature, partial pressure CO_2 of 3,0 MPa and total pressure of 5,0 MPa. A water solution of 5% NaCl with 0.5% CH3COOH and a pH level in the range of 3.0 - 4.0 was chosen as a test brine. A pH level before test was 3,0, a pH level at the end of the test was 4,06. The test lasted 720 hours. As soon as the time is out, the samples were removed from the autoclave, cleaned of corrosion products, dried and weighed on an analytical scale SARTOGOSM LV 210-A with an accuracy of 10^{-4} g.

General corrosion rate assessment was carried out by gravimetric method according to the formula:

$$K = 1{,}129 * \frac{m_0 - m_1}{S * \tau}, mm/year \quad (1)$$

where 1,129 – coefficient for converting the dimension of the general corrosion rate of g/m^2 h into mm/year; m_0 – sample weight before testing, g; m_1 – sample weight after testing, g; S – sample surface area, m_2; τ – test duration, h. The coefficient 1.129 is calculated taking into account the density of steels of the martensitic class of the 13Cr type, taken equal to 7.76 g / sm^3, taking into account the correction factor equal to 0.989 in accordance with GOST R 53366.

To estimate the steel behavior in CO_2-saturated environment and to determine the theoretical corrosion rates and their subsequent comparison with the rates obtained by other methods in accordance with ASTMG3, G5, G59, G102 standards, electrochemical studies were carried out. To conduct electrochemical studies the surfaces of the composite tube liners samples with a working area of ~ 1 cm^2 were polished and grinded with a diamond suspension. Then the sample was soldered to the conductor in the form of a copper wire, the other parts of the sample were isolated with a heat-resistant paint coating and the surface was degreased. The tests were carried out using a VersaPrincetonAppliedResearch potentiostat supported with specialized software. The test brine was a deaerated 5% NaCl based on distilled water with a fixed pH = 3.0. The study consisted of putting the sample in a test environment, measuring the equilibrium corrosion potential ($E_{кор}$) for 55 min (3300 s) and carrying out subsequent linear polarization in the potential range from -100 to 100 mV with a sweep rate of 0.16 mV / s and getting a polarization curve. The theoretical corrosion rates were calculated graphically using polarization curves.

3. Research results and discussion

The results of assessing the tendency to cracking in a hydrogen sulfide-containing environment showed that all tested samples had a tendency to stress corrosion cracking in an H2S environment: at a load level of 60% of the actual yield strength, the samples did not fracture, however, they had longitudinal cracks, which is the evidence of the tendency to cracking under test conditions; at loads of 70 and 80% - the samples were completely destroyed (Figure 1).

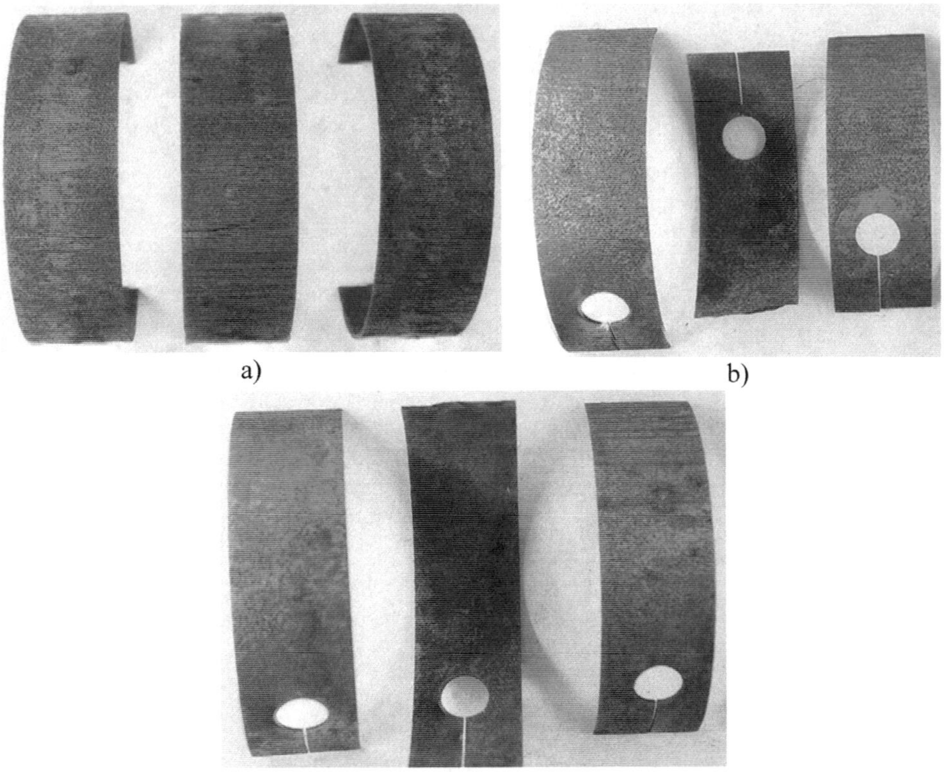

a) b)

c)

Figure 1. Samples after load testing: a) 60 %; b) 70 %; c) 80 %.

It was found that fracture cracks originated in the crosswise dimples appeared during pup joint production process. In addition to the main cracks, along which the destruction occurred, there are secondary cracks on the samples. Also, after testing, a large number of pitting with a diameter of ≈10 mm and metal blistering were found on the surface of all samples, which indicates local merging of chromium-nickel steel (Figure 2).

IOP Conf. Series: Materials Science and Engineering **986** (2020) 012040 doi:10.1088/1757-899X/986/1/012040

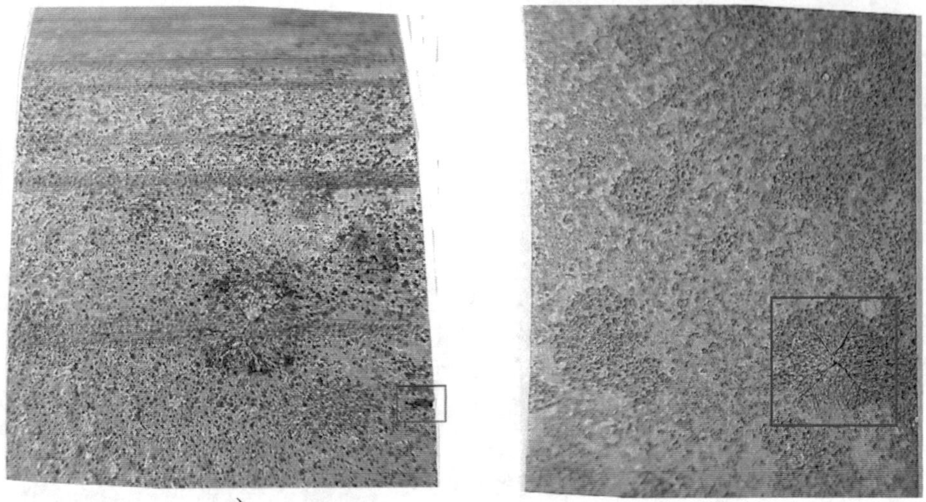

a) b)

Figure 2. Sample surface after testing: a) pitting, b) blistering.

Autoclave tests have shown that the average rate of general corrosion of the composite tube liner samples at increased pressure and temperature is 0.22 mm / year; the surface of samples were merged, and there were pitting with a diameter of 100 micron (Figure 3).

Figure 3. Surface sample after autoclave tests.

The results of measuring the corrosion potential and corrosion rates in a saturated CO2 brine, T = 60°C, pH ~ 3,0 showed that the corrosion potential of the steel is practically unchanged over time, which indicates the stability of the corrosion potential of the test material in this environment (Figure 4).

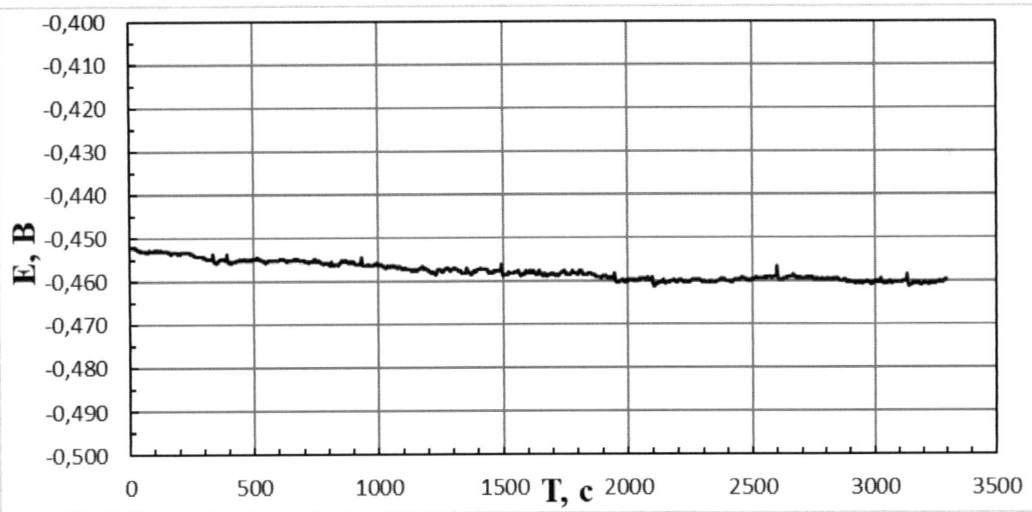

Figure 4. Corrosion potential of steel in a saturated CO2-brine, T=60oC, pH ~ 3,0.

During the electrochemical studies, the theoretical corrosion rate was calculated using Tafel curves. Research results showed the average theoretical corrosion rate obtained by the electrochemical method for steel 15H13N2MB under test conditions is 0.59 mm / year (Figure 5). So it has been shown that steel is actively merged under test conditions; there is no passivity.

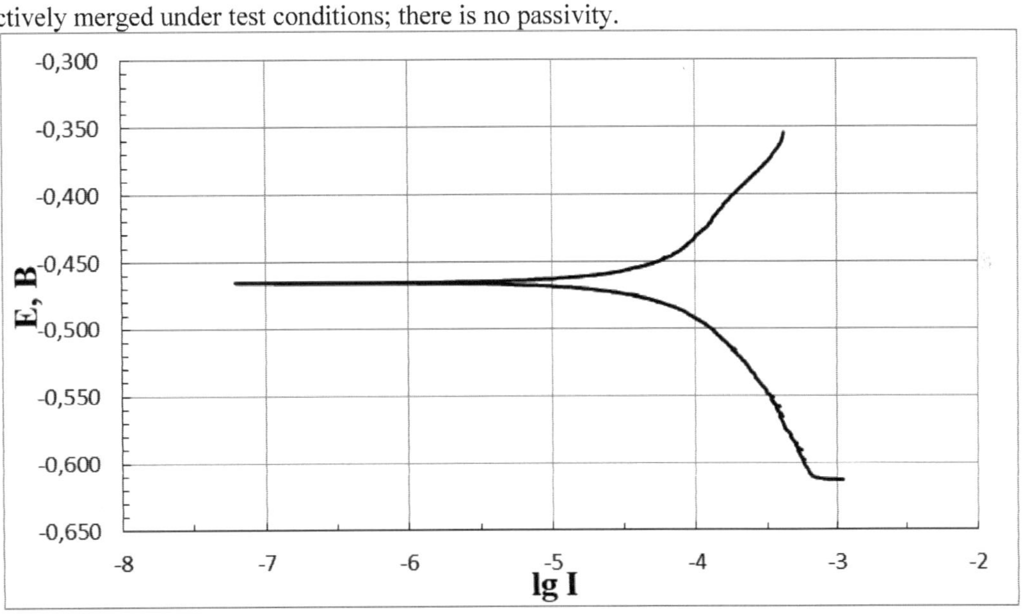

Figure 5. Polarization curve of the examined steel in a saturated CO_2-brine, T=60°C, pH ~ 3,0.

Stress corrosion cracking tests were carried out according to the most severe scenario environment aggressiveness wise (H2S concentration, pH level), moreover, the tendency to stress corrosion cracking was increased due to the crosswise dimples on the outer surface of the liner, which are the centers of crack initiation: in actual operating conditions, the outer part of the liner will not contact with an aggressive environment, therefore, its higher resistance to corrosion cracking can be expected.
Autoclave and stress corrosion cracking tests revealed pitting on the steel surface. It is probably connected with metal electrochemical contrast which needs further investigation. It was also shown that the susceptibility of chromium-nickel steel to pitting corrosion is significantly influenced by the quality of its surface: mechanical polishing reduces the tendency to pitting at normal temperatures, while electro

smoothing increases it. Preliminary metal passivation, for example (HNO_3 + $K_2Cr_2O_7$), increases the resistance of chromium-nickel steel to pitting corrosion.

The average general corrosion rate during autoclave tests in CO_2 saturated environment was 0.22 mm / year; under the condition of proportional steel merging, such corrosion rate data will provide 4 years of operation. However, we should take into account that under real conditions, the proportional steel merging is barely possible and with H2S, the rates of general corrosion can be higher.

In other words, it was found that this steel grade has limitations in corrosion resistance. It is necessary either to limit the operating conditions, or to increase the corrosion resistance of this steel grade by changing the quality, chemical composition, etc.

4. Conclusion

The comprehensive research revealed the level of corrosion resistance of the composite tubing liner.

The results show that the composite tube liner made of steel 15H13N2MB when tested in a brine with a H_2S2650ppm concentration, pH = 2.6, for 720 hours showed a tendency to stress corrosion cracking at a load level of 60-80% of the actual yield strength in H2S saturated environment.

The average general corrosion rate during autoclave tests in a CO2 saturated environment was 0.22 mm / year under the condition of proportional steel merging, such corrosion rate data will provide 4 years of operation. In electrochemical studies the theoretical corrosion rate in a 5% NaCl brine was 0.59 mm / year.

Also, during autoclave and stress corrosion cracking tests, the pitting on the steel surface was found. It was shown that the tendency of chromium-nickel steels to pitting corrosion has a significant effect on the quality of its surface.

References

[1] Akhiyarov R G, Laptev A B,. Movenko D A and, Belova N A 2016 The study of the ultralow corrosion resistance of tube steel for heat exchange equipment in oil refining (Russian) - OnePetro *Oil Ind. J.* (T. 2016, № 01 pp 118–121

[2] Alkhimenko A 2019 Several erosion test results of means of sand control

[3] Alkhimenko A 2019 Corrosion testing of experimental steels for oilfield pipelines

[4] Askari M 2019 A comprehensive review on internal corrosion and cracking of oil and gas pipelines J. *Nat. Gas Sci. Eng.* (T. 71) p 102971

[5] Daugherty R E, Zumbach M M and Sanders S F 2018 Golde T D 2018 Design challenges in electrodepositing metal-anionic clay nanocomposites: Synthesis, characterization, and corrosion resistance of nickel-LDH nanocomposite coatings *Surf. Coatings Technol.* (T. 349) pp 773–782

[6] Kondrat'Ev S Y and Shvetsov O V 2013 Effect of high-temperature heating on the structure and properties of aluminum alloys in the production of drill pipes *Met. Sci. Heat Treat.* (T. 55, № 3–40 pp 191–196

[7] Kostitsyna I, Shakhmatov A and Davydov A 2019 Study of corrosion behavior of carbon and low-alloy steels in CO_2-containing environments *E3S Web Conf.* (T. 121) p 04006

[8] Laptev A and Bugay D 2010 European Corrosion Congress (EUROCORR-2010)

[9] Latypov O R 201 Reduction of Salt Deposits on the Surface of Oilfield Equipment by Management of Electrochemical Parameters of the Medium *Chem. Pet. Eng.* (T. 51 № 7–8) pp 522–525

[10] Latypova D R, Latypov O R and Bugai D E 2018 Influence of electrode potential on the depth of pitting corrosion in the surface structures of placed steel *Nanotechnologies Constr. A Sci. Internet-Journal* (T. 10 № 3) pp 167–178

[11] Ponomarenko E P, Khokhlov V P, Rastorgueva V G, Levinskii N N and Porublev N I 1974 Increasing the erosion and corrosion resistance of steel by spray coating and vacuum diffusion metallization *Sov. Mater. Sci.* (T. 8 № 1) pp 114–115

[12] Umoren S A 2019 Protective polymeric films for industrial substrates: A critical review on past and recent applications with conducting polymers and polymer composites/nanocomposites *Prog. Mater. Sci.* (T. 104) pp 380–450

[13] Varley R J and Leong K H 2016 Polymer coatings for oilfield pipelines *Springer Ser. Mater. Sci.* (T. 233) pp 385–428
[14] Zamani Meymian M R, Ghaffarinejad A, Fazli R and Kosari Mehr A 2020 Fabrication and characterization of bimetallic nickel-molybdenum nano-coatings for mild steel corrosion protection in 3.5% NaCl solution *Colloids Surfaces A Physicochem. Eng. Asp.* (T. 593) p 124617
[15] Bogatov N A and Bogatov A A 2015 Development of corrosion-resistant pipe manufacturing technology. *Bulletin of the Manitogorsk State Technical University. G.I.Nosov* (T. 4 No. 52) pp 33–38

The Curing Simulation And Prediction Of Shape Distortion Of Thermoset Composite Reinforced With 3d Woven

M Kiauka, O Pudeleva, V, A Tamm and A Borovkov

Peter the Great St. Petersburg Polytechnic University, Polytechnicheskaya, 29, St. Petersburg, 195251, Russia, kiauka@compmechlab.ru

Abstract. The general modeling methodology for thermoset composite reinforced with 3D woven is described. In this study the focus is on the curing simulation и prediction of shape distortion for thermoset composites in ABAQUS with user subroutines. The procedure of homogenization and determination of effective properties of a thermoset composite reinforced with 3D woven for the rubbery and glassy state of the matrix is also described in detail.Depending on the thermoset polymer type, the different cure reaction formulations are implemented. The temperature expansion and chemical shrinkage tensors at each time step are set depending on the degree of cure and the glass transition temperature. Curing model with piecewise constant elastic modules is used. The model takes into account the shape distortion of the part due to chemical shrinkage and thermal expansion of the material.Curing simulation and prediction of shape distortion results for U-shaped part for Kamal-Sourour reaction type is presented. The effective properties of the composite reinforced with 3D woven for the rubbery and glassy states of the matrix are obtained.

1. Introduction

Polymer composite materials (CM) are widely used in various fields. Recently, there has been a significant increase in the research and development of composite reinforced with 3D woven preforms, in particular, to develop Gas Turbine Engine fan blades and housings (LEAP, GE9X engine blades) [1]. The 3D woven composites are highly resistant to delamination and impact due to reinforcement out-of-plane.

Nomenclature

CM	composite material
UD	unidirectional
UC	unit cell
$i,j = 1, 2, 3$	axis of the orthogonal coordinate system
E_i^*	effective Young's modulus
ν_{ij}^*	effective Poisson's ratios
γ_{ij}	engineering shear strain
G_{ij}^*	effective shear modules
< >	denotement of averaging

$<\sigma_{ij}^{(n)}>$	volume-average stresses in direction *ij* in the problem *n*
$<\varepsilon_{ij}^{(n)}>$	volume-average strains in direction *ij* in the problem *n*
u_i	displacement to direction *i*
h	initial length of the UC in the direction of shear
P	resulting force in the embedment
F	cross-sectional area
Δx	maximum displacement of the UC
l	initial length of the UC
ρ	density
C_p	specific heat capacity
T	temperature
k_{ij}	heat conductivity tensor components
∇	Laplace operator
q	internal heat generation rate
ρ_m	density of the matrix
H_T	total heat released by the matrix during the curing
V_f	fiber volume fraction
X	degree of cure
T_g	glass transition temperature
X_{gel}	degree of cure at gelation
T_{g0}	glass transition temperature of the uncured system
$T_{g\infty}$	glass transition temperature of the fully cured
λ	material constant
t	time
$K(T)$	Arrhenius-type relation
A	the pre-exponential factor
A_1, A_2	the pre-exponential factor for Kamal-Sourour model
m, n	constants for Kamal-Sourour model
E_a	apparent activation energy
$Ea1, Ea2$	apparent activation energy for Kamal-Sourour model
R	universal gas constant
ε_{ij}^{E}	non-mechanical strain components
ε_{ij}^{T}	thermal shrinkage strain components
ε_{ij}^{C}	chemical shrinkage strain components
α_{ij}	instantaneous coefficients of thermal expansion
$\alpha_{ij}^{l}, \alpha_{ij}^{r}, \alpha_{ij}^{g}$	linear coefficient of thermal expansion in the liquid, rubbery and glassy states
β_{ij}	instantaneous coefficients of chemical shrinkage
$\beta_{ij}^{l}, \beta_{ij}^{r}, \beta_{ij}^{g}$	linear coefficient of chemical shrinkage in the liquid, rubbery and glassy states
σ_{ij}	stress components
C_{ijkl}^{R}	components of the elastic modules tensor in the rubbery state
C_{ijkl}^{G}	components of the elastic modules tensor in the glassy state
A	heat transfer coefficient
q_t	heat flux from tool

There are many methods and some software for 2D textile composite manufacturing modelling. However, there is dearth of similar engineering methods and software for 3D textile composites. In particular, the literature describes the use of averaging techniques to obtain elastic properties, coefficients of linear thermal expansion and chemical shrinkage for UD fiber laminates based on micromechanical approach [2]–[4]. 3D woven composites with various textile architecture require the development of new averaging techniques.

Also, various manufacturing defects may appear during the manufacturing process, in particular, unplanned part shape distortion as a result of curing of the thermoset matrix.

The mechanical behavior of UD fiber laminates and 2D woven composites as a result of curing is well studied and presented in many publications.

Numerical implementation of the model for prediction of the degree of cure, the associated heat and temperature increase of undiluted polymer resin is presented in [5]. J. Zhang, Y. C. Xu and P. Huang studied the influence of temperature cure cycle have on the temperature and degree of cure gradient [6]. In [7], [8] numerical discretization methods of the kinetics equations for adhesives are described. J. Magnus Svanberg and J. Anders Holmberg an approach for prediction of shape distortion for different constitutive models is presented [9]. Modeling of residual strains for the pultrusion of composite beam is described in [10]. In [11] M.V. Kozlov et al. consider the main causes of shape distortion, conduct a comparative analysis of mathematical models of curing and offer mechanical contact between the part and the tool, which changes during the solution. M.V. Kozlov et al. using the constitutive relations of CHILE (cure hardening instantaneously linear elastic) and Svanberg models modeled the influence of tool on the shape distortion [12].

The overall goal of our work is development of modelling approach for manufacturing process of composite materials with 3D woven preforms and a thermosetting matrix. It includes integrated techniques and algorithms with wide possibilities for automation and solving optimization problems. Now we propose to use a complex of CAE solutions, including commercial packages ANSYS CFX (or ANSYS Fluent), ABAQUS and proprietary FORTRAN and Python codes.

In this paper a general modelling strategy for thermoset composite reinforced with 3D woven and in more detail results of the thermoset composite curing simulation and prediction of shape distortion are presented. The issue of homogenization and determination of the effective properties of CM with 3D fabric reinforcement is considered.

2. General approach description

Now our approach the manufacturing modelling of 3D woven composites includes the following main steps:

1) Creating of geometrical and finite element models of unit cell (UC) for a dry undeformed 3D woven.
2) Determination of the effective elastic properties of UC for a dry undeformed 3D woven.
3) Draping and compaction simulation of textile on a tool surface with using effective properties from step 2.
4) Dividing the textile into the finite number of regions where shear strain obtained in step 3 belongs to given range.
5) Shear strain averaging for regions obtained in step 4.
6) Application of average shear loading obtained in step 5 to the dry undeformed UC.
7) Determination of the permeability tensor components for deformed UC from step 6.
8) Modelling of textile preform impregnation for part with different permeability regions.
9) Homogenization and effective property's determination of the composite material for the rubbery and glassy matrix state for deformed UC obtained in step 6.
10) Transient heat transfer and cure kinetics modelling for part using effective elastic properties from step 9.
11) Residual stresses and strain determination for part, prediction of part shape distortion.

3. Homogenization procedure and effective properties determination

To determine the residual strains and stresses, it is necessary to obtain the effective properties of the 3D-woven composite material (Fig. 1) after impregnation for various states of the matrix during the curing cycle.

Figure 1. The finite-element model of the UC of the dry 3D woven.

Since the yarns of the reinforcing preform are composed of filaments, the effective properties of the threads for the rubbery and glassy states of the matrix are first calculated. In general, reinforcing fibers are not isotropic. Glass fiber is isotropic, carbon fiber is orthotropic. In this study, the filaments and matrix were assumed to be isotropic, with the same material properties in all directions. The UC for homogenization is shown in Figure 2. The volume fraction of filaments was taken to be 76% for both matrix states.

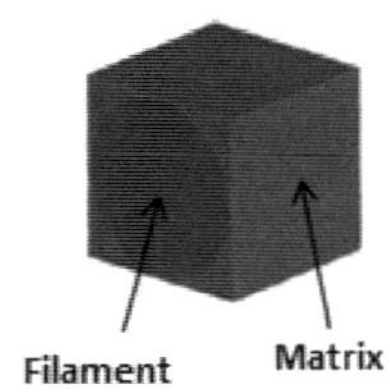

Figure 2. Finite-element model of the UC of the CM (filament and matrix).

The effective constitutive relations for an orthotropic heterogeneous medium in the orthogonal coordinate system $0X_1X_2X_3$ have the form [13]:

$$E_1^*\langle\varepsilon_{11}\rangle = \langle\sigma_{11}\rangle - \nu_{12}^*\langle\sigma_{22}\rangle - \nu_{13}^*\langle\sigma_{33}\rangle \tag{1}$$

$$E_3^*\langle\varepsilon_{33}\rangle = -\nu_{31}^*\langle\sigma_{11}\rangle - \nu_{32}^*\langle\sigma_{33}\rangle + \langle\sigma_{33}\rangle$$

$$G_{12}^*\langle\gamma_{12}\rangle = \langle\sigma_{12}\rangle;\;\; G_{23}^*\langle\gamma_{23}\rangle = \langle\sigma_{23}\rangle;\;\; G_{31}^*\langle\gamma_{31}\rangle = \langle\sigma_{31}\rangle$$

The equations (1) contain 12 effective elastic constants E_1^*, E_2^*, E_3^* – effective Young's modulues; $\nu_{12}^*, \nu_{23}^*, \nu_{31}^*, \nu_{21}^*, \nu_{32}^*, \nu_{13}^*$ – effective Poisson's ratios; $G_{12}^*, G_{23}^*, G_{31}^*$ – effective shear modules; effective Young's modules and effective Poisson's ratios are related by the equalities:

$$E_1^*\nu_{21}^* = E_2^*\nu_{12}^*;\;\; E_2^*\nu_{32}^* = E_3^*\nu_{23}^*;\;\; E_3^*\nu_{13}^* = E_1^*\nu_{31}^* \tag{2}$$

Therefore, the number of independent elastic constants for a macroscopically orthotropic heterogeneous medium is 9.

To determine the unknowns, six problems are solved in the finite element formulation for each state of the matrix. The statement of all tasks is shown in Figure 3.

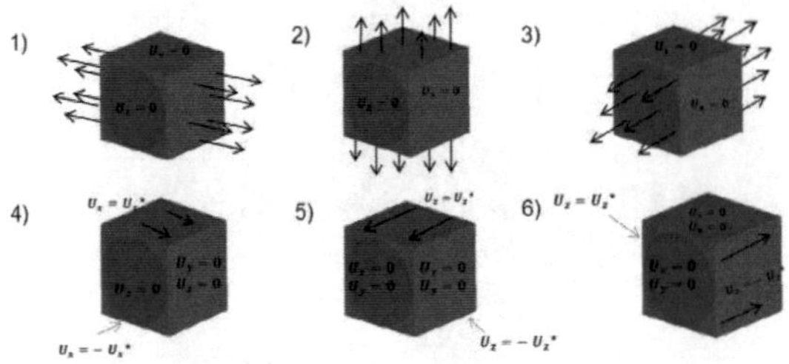

Figure 3. Applying boundary conditions to a UC .

In tasks 1-3 (Fig. 3), we calculate the average tensors of microstresses and microstrains, averaging the stresses and strains obtained at the nodes, respectively. Next, using the system of governing equations (1) and the relationship between Young's modules and Poisson's ratios (2), we find expressions for determining the effective Young's modules and Poisson's ratios (3):

$$E_1^* = \frac{\langle\sigma_{11}^{(1)}\rangle - \nu_{12}^* * \langle\sigma_{22}^{(1)}\rangle - \nu_{13}^* * \langle\sigma_{33}^{(1)}\rangle}{\langle\varepsilon_{11}^{(1)}\rangle} \tag{3}$$

$$E_2^* = \frac{-\nu_{21}^* * \langle\sigma_{11}^{(2)}\rangle + \langle\sigma_{22}^{(2)}\rangle - \nu_{23}^* * \langle\sigma_{33}^{(2)}\rangle}{\langle\varepsilon_{22}^{(2)}\rangle};$$

$$E_1^* = \frac{-\nu_{31}^* * \langle\sigma_{11}^{(3)}\rangle - \nu_{32}^* * \langle\sigma_{22}^{(3)}\rangle + \langle\sigma_{33}^{(3)}\rangle}{\langle\varepsilon_{33}^{(3)}\rangle};$$

$$\nu_{32}^* = \frac{\langle\sigma_{33}^{(2)}\rangle - \langle\sigma_{33}^{(1)}\rangle * \langle\sigma_{11}^{(2)}\rangle / \langle\sigma_{11}^{(1)}\rangle}{\langle\sigma_{22}^{(2)}\rangle - \langle\sigma_{22}^{(1)}\rangle * \langle\sigma_{11}^{(2)}\rangle / \langle\sigma_{11}^{(1)}\rangle};$$

$$\nu_{23}^* = \frac{\langle\sigma_{22}^{(3)}\rangle - \langle\sigma_{22}^{(1)}\rangle * \langle\sigma_{11}^{(3)}\rangle / \langle\sigma_{11}^{(1)}\rangle}{\langle\sigma_{33}^{(3)}\rangle - \langle\sigma_{33}^{(1)}\rangle * \langle\sigma_{11}^{(3)}\rangle / \langle\sigma_{11}^{(1)}\rangle};$$

$$\nu_{13}^* = \frac{\langle\sigma_{11}^{(3)}\rangle - \langle\sigma_{11}^{(2)}\rangle * \langle\sigma_{22}^{(3)}\rangle / \langle\sigma_{22}^{(2)}\rangle}{\langle\sigma_{33}^{(3)}\rangle - \langle\sigma_{33}^{(2)}\rangle * \langle\sigma_{22}^{(3)}\rangle / \langle\sigma_{22}^{(2)}\rangle};$$

$$\nu_{31}^* = \frac{\langle\sigma_{33}^{(1)}\rangle - \nu_{32}^* * \langle\sigma_{22}^{(1)}\rangle}{\langle\sigma_{11}^{(1)}\rangle};$$

$$\nu_{12}^* = \frac{\langle\sigma_{11}^{(2)}\rangle - \nu_{13}^* * \langle\sigma_{33}^{(2)}\rangle}{\langle\sigma_{22}^{(2)}\rangle};$$

$$\nu_{21}^* = \frac{\langle\sigma_{22}^{(1)}\rangle - \nu_{23}^* * \langle\sigma_{33}^{(1)}\rangle}{\langle\sigma_{11}^{(1)}\rangle},$$

where $<\sigma_{ij}^{(n)}>$ - the volume-average stresses in direction *ij* in the problem *n*, $<\varepsilon_{ij}^{(n)}>$ - the volume-average strains in direction *ij* in the problem *n*.

From tasks 4-6 (Fig. 3), effective shear modules of the CM are obtained. From the volume-average stresses and the defining equations (1) we obtain the following relations for the shear modules (4):

$$G_{12}^{*} = \frac{\langle \sigma_{12} \rangle}{\langle \gamma_{12} \rangle}, \quad \langle \gamma_{12} \rangle = \frac{\overset{\circ}{u_1}}{\left(\frac{h_2}{2} \right)} \tag{4}$$

$$G_{23}^{*} = \frac{\langle \sigma_{23} \rangle}{\langle \gamma_{23} \rangle}, \quad \langle \gamma_{23} \rangle = \frac{\overset{\circ}{u_3}}{\left(\frac{h_2}{2} \right)}$$

$$G_{31}^{*} = \frac{\langle \sigma_{31} \rangle}{\langle \gamma_{31} \rangle}, \quad \langle \gamma_{31} \rangle = \frac{\overset{\circ}{u_3}}{\left(\frac{h_2}{2} \right)}$$

where $\langle \sigma_{ij} \rangle$ - the volume-average stresses to direction *ij*; u – displacement to direction *i*; *h* – the initial length of the UC in the direction of shear.

To determine the effective properties of the UC of the 3D woven, nine simulations are carried out, similar to those performed for the UC of the filament, but with the three additional shear calculations due to the completely anisotropic behavior of the 3D woven CM.

Two different yarn materials are considered – with a rubbery and glassy matrix states. The yarn material is defined as an orthotropic material with the properties calculated above. The orientation of the filaments in the yarn for 3D woven composite is shown in Figure 4.

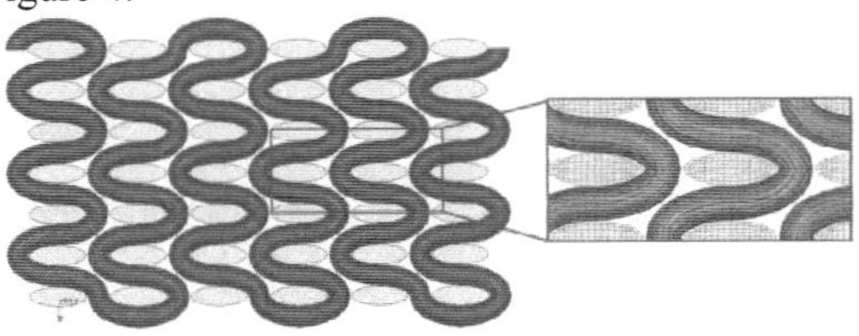

Figure 4. Orientation of the filaments in the yarn for 3D woven composite.

For a matrix between yarns, properties in rubbery and glassy states are specified.
The finite element model of a UC for 3D woven with a matrix is shown in Fig. 5.

Figure 5. The finite element model of the periodicity cell of the 3D-woven matrix composite material.

Tension in three directions was specified by displacement 0.1 mm to the face, while a restriction on six degrees of freedom was specified on the opposite side. Displacements were applied gradually, increasing according to a linear law.

The application of the boundary conditions to the UC of 3D woven CM and the deformed UC for tension along the x axis is shown in Fig. 6.

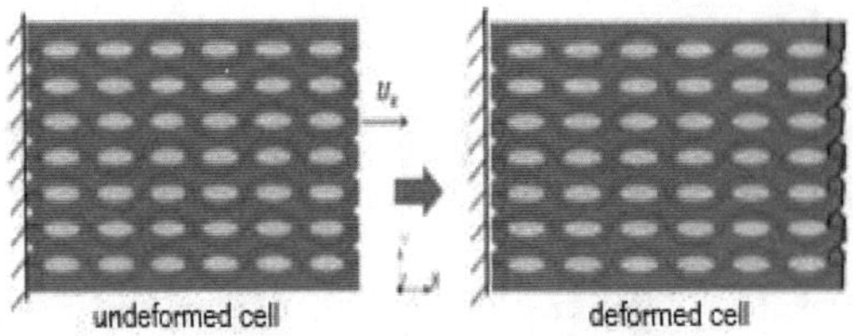

Figure 6. The application of the boundary conditions to the UC of 3D woven CM (along the x axis).

The effective Young's module in the X direction is defined as

$$E_x = \frac{\sigma_x}{\varepsilon_x} = \frac{P/F}{\Delta x/l}, \tag{5}$$

where *P* is the resulting force in the embedment (the resulting loads from all nodes are summed), *F* is the cross-sectional area, *Δx* is the maximum displacement of the UC, *l* is the initial length of the UC in the direction of load application. Similar calculations were performed for tension in the *Y* and *Z* directions. To determine the Poisson's ratios in tensile calculations in three directions, deformations in the longitudinal and transverse directions are calculated. Since the material is anisotropic, it will have six different Poisson's ratios in different directions, which are calculated as

$$\nu_{ij} = \frac{\varepsilon_{jj}}{\varepsilon_{ii}}. \tag{6}$$

where ε_{jj} – transverse strains, ε_{ii} - longitudinal strains.

Shear module are defined as

$$G_{xy} = \frac{\tau_{xy}}{\gamma_{xy}} = \frac{P/A}{\Delta x/l}. \tag{7}$$

The obtained effective properties of the 3D-woven CM for the rubbery and glassy states of the matrix are presented in table 1.

Table 1. Effective properties of the 3D-woven CM.

Property	Unit	Rubbery state	Glassy state
E_1*	MPa	7155,9	6112,8
E_2*	MPa	320,5	298,1
E_3*	MPa	70599,6	67120,6
ν_{12}*		0,081	0,066
ν_{23}*		0,006	0,004
ν_{13}*		0,098	0,040
ν_{21}*		0,384	0,255
ν_{32}*		0,015	0,008
ν_{31}*		0,015	0,008
G_{12}*	MPa	356,2	297,0
G_{23}*	MPa	1788,3	1679,6
G_{31}*	MPa	966,5	904,1
G_{21}*	MPa	1115,7	928,9
G_{32}*	MPa	213,2	208,3
G_{13}*	MPa	4100,8	3927,1

4. Cure kinetics simulation and shape distortion prediction

To verify the curing model, a 2D woven reinforced composite was considered. The choice of such a CM is explained by the fact that there are no known results in the literature on shape distortion for 3D reinforced composites. Such a statement of the problem made it possible to qualitatively assess the adequacy of the model and compare the simulation results with literature data. For 2D composites in the transverse direction, the coefficients of linear thermal expansion and chemical shrinkage are much larger than in the plane of the laminate. It is important to note that the developed technique allows predicting shape distortion for 3D reinforced composites with a thermosetting matrix. The difference between 2D and 3D reinforced composites is in the properties of the material after homogenization.

4.1. Main relationships for thermal behavior

The temperature distribution is defined by the heat equation and Fourier's heat law:

$$\rho\frac{\delta(C_pT)}{\delta t} = k_{i\,j}\nabla^2 T + q \tag{8}$$

where ρ is density, C_p is specific heat capacity, T is temperature, k_{ij} (i,j = 1, 2, 3) is the heat conductivity tensor components, ∇ denotes the Laplace operator, q is internal heat generation rate in a material as a result of an exothermic reaction.

Internal heat generation is defined as:

$$q = \rho_m H_T(1 - V_f)\frac{dX}{dt} \tag{9}$$

where ρ_m; H_T; V_f and dX/dt is density of the matrix, total heat released by the matrix during the curing reaction (energy per unit mass), fiber volume fraction, respectively, the chemical reaction rate.

In our algorithm depending on the thermoset polymer type, the following cure reaction formulations are implemented: nth order, Prout-Tompkins autocatalytic and Kamal-Sourour (Eq. 10-12, respectively) [14]. Which model to use can be determined using DSC (Differential scanning calorimetry).

$$\frac{d\alpha}{dt} = K(T)(1-\alpha)^n \tag{10}$$

$$\frac{d\alpha}{dt} = K(T)\alpha^m(1-\alpha)^n \tag{11}$$

$$\frac{d\alpha}{dt} = (K_1 + K_2\alpha^m)(1-\alpha)^n \tag{12}$$

$$K(T) = A\cdot\exp\left[\frac{-E_a}{RT}\right] \tag{13}$$

where $K(T)$ – an Arrhenius-type relation for the reaction temperature dependency; A is the pre-exponential factor; m, n – model constants; E_a is the apparent activation energy; R is the universal gas constant; T is the temperature.

4.2. Main relationships for mechanical behavior

During cure three states of the matrix are successively replaced – liquid, rubbery and glassy. The process of curing the resin is associated with the appearance of a non-mechanical strains [9].

Non-mechanical strains represent the sum of strains from thermal expansion $\varepsilon^T_{i\,j}$ and from chemical shrinkage $\varepsilon^C_{i\,j}$:

$$\varepsilon^E_{i\,j} = \varepsilon^T_{i\,j} + \varepsilon^C_{i\,j}. \tag{14}$$

Thermal strains are defined as:

$$\varepsilon^T_{i\,j} = \int_0^t \alpha_{i\,j}(T,X)\frac{\partial T}{\partial t'}dt', \tag{15}$$

where the instantaneous coefficients of thermal expansion $\alpha_{i\,j}$ depend on temperature and degree of cure as

$$\alpha_{i\,j} = \begin{cases} \alpha^{l}_{i\,j}, & X < X_{gel} \text{ и } T \geq T_g(X) \\ \alpha^{r}_{i\,j}, & X \geq X_{gel} \text{ и } T \geq T_g(X) \\ \alpha^{g}_{i\,j}, & T < T_g(X). \end{cases} \tag{16}$$

where $\alpha^{l}_{i\,j}$, $\alpha^{r}_{i\,j}$, $\alpha^{g}_{i\,j}$ – are linear coefficient of thermal expansion in the liquid, rubbery and glassy states, respectively; X – degree of cure; X_{gel} – degree of cure at gelation.

Chemical shrinkage is defined as

$$\varepsilon^{C}_{i\,j} = \int_0^t \beta_{i\,j}(T,X)\frac{\partial X}{\partial t'}dt', \tag{17}$$

where the instantaneous coefficients of chemical shrinkage $\beta_{i\,j}$ depend on temperature and degree of cure as,

$$\beta_{i\,j} = \begin{cases} \beta^{l}_{i\,j}, & X < X_{gel} \text{ и } T \geq T_g(X) \\ \beta^{r}_{i\,j}, & X \geq X_{gel} \text{ и } T \geq T_g(X) \\ \beta^{g}_{i\,j}, & T < T_g(X). \end{cases} \tag{18}$$

where $\beta^{l}_{i\,j}$, $\beta^{r}_{i\,j}$, $\beta^{g}_{i\,j}$ – the linear coefficient of chemical shrinkage in the liquid, rubbery and glassy states, respectively.

The dependence of the glass transition temperature T_g on the degree of cure α was defined using the Di Benedetto equation:

$$\frac{T_g - T_{g0}}{T_{g\infty} - T_{g0}} = \frac{\lambda X}{1-(1-\lambda)X}, \tag{19}$$

where T_{g0} and $T_{g\infty}$ are the glass transition temperature of the uncured (X=0), respectively, fully cured system (X=1); λ is a material constant.

In [9], [12] the main constitutive relations are given, which serve to describe the curing process of a composite material with a thermosetting matrix. These models describe the material in rubbery and glassy states.

In this paper, we use the defining relations of the elasticity model, in which the modules are piecewise constant and are written as:

$$\sigma_{i\,j} = \begin{cases} C^{R}_{i\,jkl}\varepsilon^{M}_{kl}, & T \geq T_g(X), \\ C^{G}_{i\,jkl}\varepsilon^{M}_{kl}, & T < T_g(X), \end{cases} \tag{20}$$

where $C^{R}_{i\,jkl}$ and $C^{G}_{i\,jkl}$ components of the elastic modules tensor in the rubbery and glassy states, respectively.

To be able to simulation using the finite element method, we use the direct Euler scheme

$$\varepsilon^{E}_{kl}(t + \Delta t) = \varepsilon^{E}_{kl}(t) + \Delta\varepsilon^{E}_{kl}, \tag{21}$$

where

$$\Delta\varepsilon^{E}_{kl} = \Delta\varepsilon^{T}_{kl} + \Delta\varepsilon^{C}_{kl}, \tag{22}$$

$$\Delta\varepsilon^{T}_{kl} = \alpha_{kl}\Delta T, \tag{23}$$

$$\Delta\varepsilon^{C}_{kl} = \beta_{kl}\Delta X, \tag{24}$$

where ΔT и ΔX – are the temperature, respectively, degree of cure increment over the time step. According to the article [9] the coefficients for linear shrinkage in the liquid state α_{kl} и β_{kl} are taken as zero.

4.3. Material properties

Resin properties were used from [3]. Table 2 shows the parameters for the kinetic equation (12) of the Kamal-Soro model.

Table 2. Cure kinetics parameters.

Property	Unit	Epoxy resin
m		0.51
n		1.49
A_1	s^{-1}	1528
A_2	s^{-1}	1.6
E_{a1}	kJ/mol	59.4
E_{a2}	kJ/mol	26.3
H_T	J/kg	198.9

Elastic properties, coefficients of thermal expansion and chemical shrinkage for 2D reinforced CM were also set based on the data of work [3].

4.4. Finite element simulation

Cure model has been implemented in ABAQUS with using user subroutines HETVAL, UEXPAN, UMAT. User subroutines are implemented in the FORTRAN programming language.

To check the model and FORTRAN code a test simulation for the U-shaped (60×40×20 mm) part was performed. Temperature cure cycle and boundary conditions shown in the Fig.7a and Fig.7b, respectively.

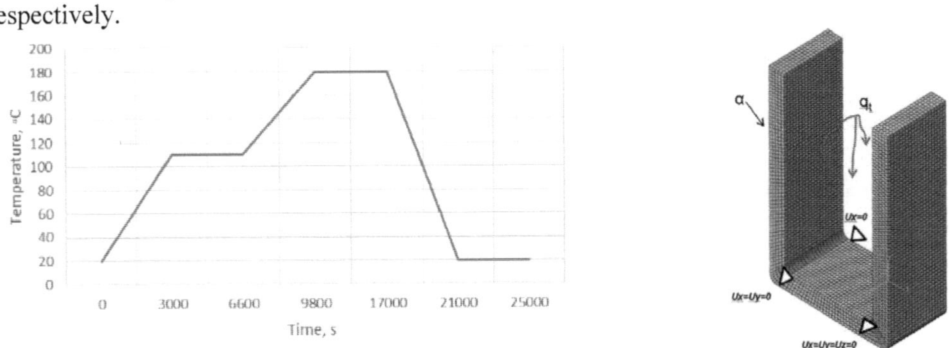

a) b)

Figure 7. (a) Temperature cure cycle; (b) Boundary conditions.

Boundary conditions include constraints, heat transfer coefficient α, heat flux from tool q_t. Constraints of the model simulates the location of the part on a tool (iso-static mounting).

In virtual tests a cure cycle in the furnace/autoclave and typical heat transfer coefficients on the part and tool surface is specified. Heat transfer coefficients on the surface of the part were taken equal 8 $W/(m^2 \cdot K)$.

Curing process is modeled using subroutines HETVAL according to equations (9), (12), (13). For this the cure kinetics equation was integrated by the improved Euler's method. The cure degree, cure rate and temperature at each integration point during a cure cycle are defined at each time step. The thermal conductivity tensor components of the composite, specific heat and density of the resin were defined by the functions of the process temperature and degree of cure.

The increment of temperature and shrinkage strains at each time step with using subroutines UEXPAN is specified according to equations (16), (18), (19) and (22)-(24).

The elastic properties of the orthotropic homogenized laminate with using subroutines UMAT are specified.
Simulation results presented in the Fig. 8-10 include thermal and mechanical effects. Residual strains are induced by chemical shrinkage during cure and thermal expansion during cooling.
The qualitative picture of the deformation corresponds to the experimentally observed decrease in the angle of L-shaped and U-shaped parts [3], [4], [11], [15]. Angular deviation in this work is 1,8 degrees.

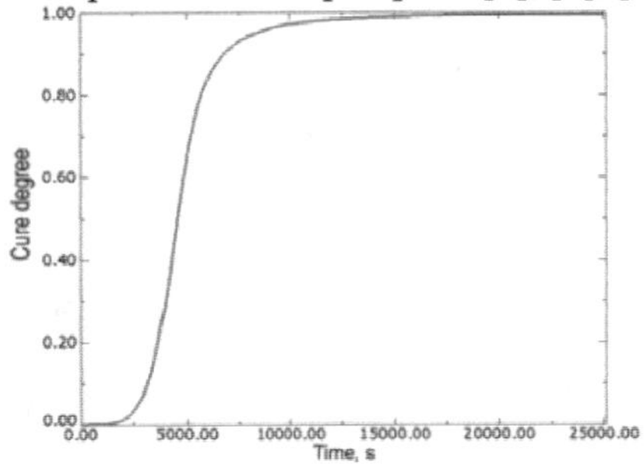

a)

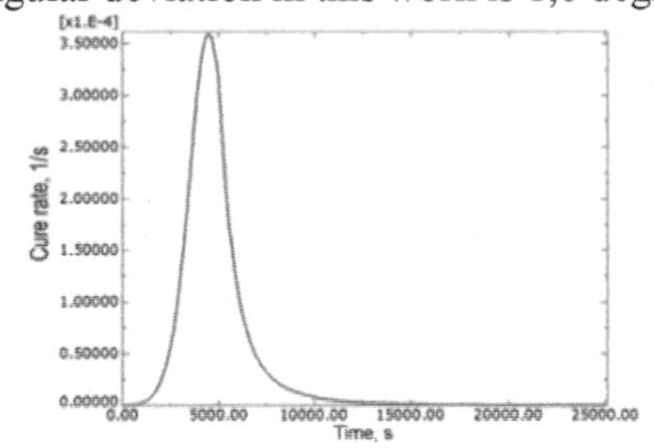

b)

Figure 8. Cure degree (a) and cure rate (b) at the laminate centre point.

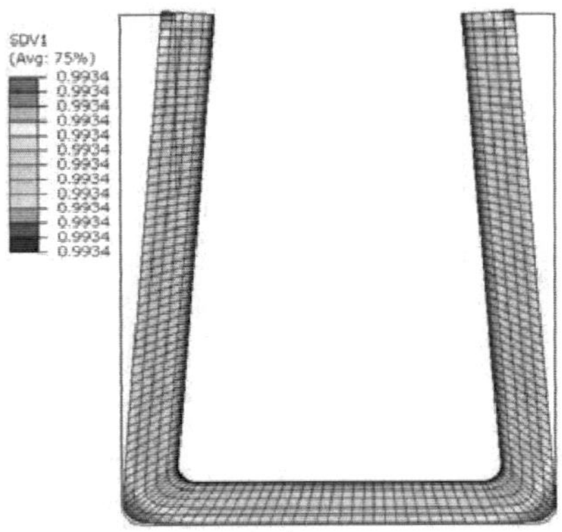

a)

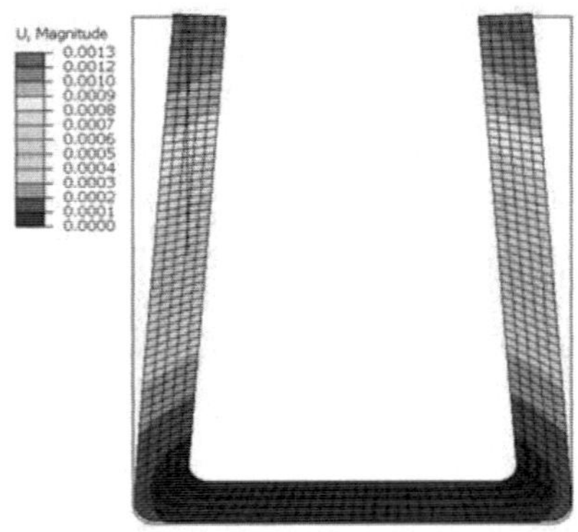

b)

Figure 9. (a) Cure degree after 17000 s; (b) Displacement (m) after 17000 s (the main contribution to shape distortion is chemical shrinkage).

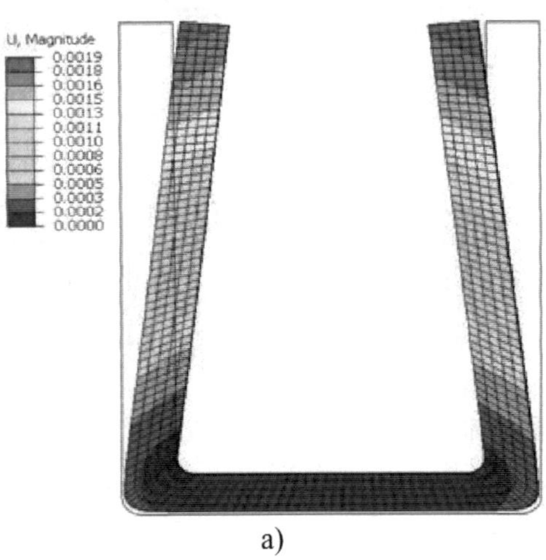

a)

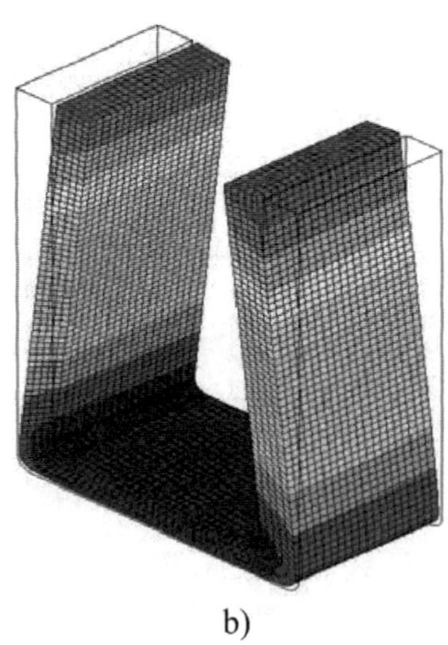

b)

Figure 10. Displacement (m) after 25000 s (the main contribution to shape distortion is thermal expansion).

5. Conclusions

General approach proposed in this paper permits to modelling the manufacturing process of thermoset composite reinforced with 3D woven. This approach can be used for both 3D textile and 2D textile composites.

A technique for homogenization and determining the effective elastic properties of CM with 3D woven reinforcement for the rubbery and glassy states of the matrix is proposed.

The general methodology for cure simulation and prediction of shape distortion is described. The technique was verified for laminate composites based on literature data.

In further development, it is supposed to clarify the actual distribution and values of the heat transfer coefficients on the part and tool surface by airflow simulation using CFD (ANSYS CFX, ANSYS Fluent). This may be important for products requiring increased manufacturing accuracy, such as reflectors for antennas and radio telescopes.

The literature describes relations for obtaining effective coefficients of linear thermal expansion and chemical shrinkage for CM with 2D woven reinforcement. These relations will not be valid when changing the architecture of weaving, in particular for CM with 3D woven reinforcement. In further development, a methodology will be developed for determining the effective coefficients of linear thermal expansion and chemical shrinkage for CM with 3D fabric reinforcement for different fiber volume fraction. Using these coefficients, an accurate calculation of residual strains can be performed using the proposed methodology for prediction of shape distortion.

In the future work, a tool will be added to the finite element model to more accurately take into account the effect of temperature expansion of the tool on shape distortion, mechanical and thermal contacts between the part and tool.

References

[1] Wadia A R 2017 Technologies for the next engine generation *29th Congr. Int. Counc. Aeronaut. Sci. ICAS*

[2] Dai J, Xi S and Li D 2019 Numerical analysis of curing residual stress and deformation in thermosetting composite laminates with comparison between different constitutive models *Materials (Basel)* (vol. 12, no. 4 doi: 10.3390/ma12040572)

[3] Wenani M, Andersen L, Hørlyk P and Nielsen M W 2013 Prediction of process induced shape distortions and residual stresses in large - With application to Wind Turbine Blades

[4] Svanberg J M 2002 Predictions of manufacturing induced shape distortions: high performance thermoset composites p. 131

[5] Heinze S and Echtermeyer A T 2018 A practical approach for data gathering for polymer cure simulations *Appl. Sci.* (vol. 8, no. 11 doi: 10.3390/app8112227) pp 1–31

[6] Zhang J, Xu Y C and Huang P 2009 Effect of cure cycle on curing process and hardness for epoxy resin *Express Polym. Lett.* (vol. 3, no. 9 doi: 10.3144/expresspolymlett.2009.67) pp 534–541

[7] Qin Z, Bin S, Simin W, Ling X and Sheng L 2009 Cure kinetics analysis and simulation of silicone adhesives *2009 Int. Conf. Electron. Packag. Technol. High Density Packag. August* (doi: 10.1109/ICEPT.2009.5270659) pp. 691–694

[8] Cassano A G, Stapleton S E, Hansen C J and Maiaru M Prediction of Cure Overheating in Thick Adhesive Bondlines for Wind Energy Applications pp 1–10

[9] Svanberg J M and Holmberg J A 2004 Prediction of shape distortions Part I. FE-implementation of a path dependent constitutive model *Compos. Part A Appl. Sci. Manuf.* (vol. 35, no. 6, doi: 10.1016/j.compositesa.2004.02.005) pp 711–721

[10] Shcheglov B A and Safonov A A 2005 Teoreticheskie osnovy i prikladnye zadachi tekhnologii kompozitov [Theoretical foundations and applied problems of composites technology] (Moscow: Lenand) p 112

[11] Kozlov M V, Sheshenin S V, Babkin A V, Kepman A V and Kudrin A M Modelirovanie formovaniya kompozitov na osnove termoreaktivnyh matric (no. 3) pp 11–17.

[12] Kozlov M V, Sheshenin S V, Makarenko I V and Belov D A 2016 Modeling the influence of tooling on the final shape of polymer composite parts *Comput. Contin. Mech.* (vol. 9, no. 2, doi: 10.7242/1999-6691/2016.9.2.13) pp 145–161

[13] Borovkov A I 1985 Effektivnye fiziko-mekhanicheskie svojstva voloknistyh kompozitov (M.: Izd-vo VINITI) p 113

[14] Menczel J D and Prime R B 2009 Thermal Analysis of Polymers *Fundamentals and Applications. 1st ed.* (New York: Wiley) Print.

[15] Benavente Miana M and Ruiz Martin E 2017 Viscoelastic distortion during manufacturing and post-curing of thermoset composites: characterization and modeling *Département de génie mécanique* p 122

Methods of improving service quality and increasing the reliability of locomotives registered in operational locomotive depot Agryz in the locomotive service depot Agryz-South

A T Kozlova

Federal State Budgetary Educational Institution of Higher Education «Kazan National Research Technical University named after A.N. Tupolev-KAI», e-mail: dashenka.fedagina@gmail.com, alsu_kozlova@mail.ru

Abstract.
The method for managing the technological processes of repairing diesel locomotives registered at the Agryz operational locomotive depot in the Agryz-Yuzhny service locomotive depot using network planning has been developed to monitor and adjust the performance of technological operations, taking indicators of technological preparation of repair into consideration. Using improved non-standard equipment in technological processes of current repairs of diesel locomotives with motor-axial rolling bearings allows mechanization of technological operations during the assembly of a wheel-motor unit to improve the quality of the technological process and reduces the time spent in repair. Assembly operations have a special place in mechanical engineering. It is at this stage that various connections of individual parts and assemblies arise and indicators of the reliability of a technical object are formed.

1. Introduction

The method for managing the technological processes of repairing diesel locomotives registered at the Agryz operational locomotive depot in the Agryz-Yuzhny service locomotive depot using network planning has been developed to monitor and adjust the performance of technological operations, taking indicators of technological preparation of repair into consideration. The proposed algorithm makes it possible to estimate the necessary transitional equipment, spare parts and materials to ensure high-quality performance of technological processes for repairing diesel locomotives of the 2TE25KM, TEP70 series and consumption rates of materials at unscheduled repairs, considering the likelihood of failures in operation. The developed method monitors and manages the quality of technological repair processes and assesses the efficiency of using mainline diesel locomotives based on predicting additional power losses in limiting nodes and units based on the results of the repair [1,5].

Using improved non-standard equipment in technological processes of current repairs of diesel locomotives with motor-axial rolling bearings allows mechanization of technological operations during the assembly of a wheel-motor unit to improve the quality of the technological process and reduces the time spent in repair.

2. Maintenance and repair as a system to maintain the technical condition and efficiency of using main diesel locomotive

Nowadays, the operational reliability of traction rolling stock becomes significant, which is directly impacted by the increase in speeds, in weight standards and the requirements for train safety.

The operational reliability of diesel locomotives is significantly influenced by timely and high-quality maintenance and repair, the quality of which, in turn, depends on the level of technological readiness of the locomotive repair production [2,4]. The problem of repair becomes especially urgent during the renewal of the locomotive fleet, when old machines are replaced by locomotives of new series with modern control systems, electrical equipment and new designs of the mechanical part and traction drives, such as TEP70BS and 2TE25KM. There are 29 diesel locomotives of the TEP70BS series and 23 diesel locomotives of the 2TE25KM series in the assigned fleet of the Agryz operational depot.

In general, the following main directions can be distinguished in the complex of issues of organizing high-quality repair of locomotives:

– availability of complete sets of design and technological documentation;
– technological equipment of workplaces and areas;
– staffing of workshops and departments with repair personnel of appropriate qualifications;
– material and technical supply;
– availability of the necessary production space.

All this determines the technological readiness of production. Technological preparation of the repair of railway rolling stock is a set of measures that ensure technological readiness to perform repairs on time with the specified technical and economic indicators at the required quality level. The requirements for the technological preparation of repairs are based on the Unified System for the Technological Preparation of Production and, in general, are aimed at solving the following problems:

– implementation of technological processes;
– design of technological equipment;
– tool preparation;
– providing the enterprise with means of transportation and storage;
– development of methods for technical control of technological operations and processes.

The design of technological processes takes a key position when preparing locomotive repair enterprises for the maintenance of new series locomotives, since the technical documentation prepared according to the GOST requirements contains comprehensive information about the necessary technological equipment, tools for performing operations, as well as the norms of time for their implementation. Thus, it is possible to assess the real need for resources (personnel, time, etc.), to determine the future layout of sites.

The continuity of the transportation process on railway transport makes it necessary to carry out maintenance and repair of rolling stock in the shortest possible time with minimal losses for the main technological process. The rolling stock of railways is subject to intensive exploitation, i.e. loads and wear, and therefore needs timely and high-quality maintenance and repairs to restore serviceability and performance.

The maintenance and repair system can be defined as a set of measures that allow the rolling stock to be kept in a given technical condition, providing a certain level of performance.

At present, the issue of choosing methods for organizing maintenance and repair of rolling stock, individual units, assemblies and parts have become relevant in addition to choosing a repair system [3].

In Russia, the operational locomotive depot is separated from the service locomotive depot. In the Agryz service locomotive depot, as in other regions of Russia in the field of mainline railway transport, a planned preventive system of maintenance and repair (M and R) of rolling stock has also been adopted, which is the main one in restoring the serviceability and operability of the locomotive fleet.

2.1 One of the advantages of service is an integrated approach to the organization of technological processes. Among the reasons for the transition to service maintenance, there is a lack of a reliable and effective supply of spare parts in the preventive maintenance system.

The Agryz-Yuzhny service locomotive depot has the opportunity to obtain spare parts directly from the locomotive manufacturer and trusted suppliers of parts and assemblies. This significantly reduces the level of prices, counterfeit goods and increases the efficiency of the maintenance and repair system.

Let us consider how the technical condition of the locomotive fleet is ensured in the service system. As of 01.01.2019, there are 23 units of diesel locomotives of the new 2TE25KM series manufactured in the conditions of the Bryansk Machine-Building Plant in 2019 in the registered (inventory) operational locomotive depot Agryz.

On average, five diesel locomotives are in the unexploited fleet per day (21.7% of the assigned depot fleet). This is caused by a large number of unscheduled repairs, overruns in scheduled repairs due to the lack of necessary spare parts for equipment and the unavailability of the Bryansk Machine Building Plant to provide a service locomotive depot Agryz-Yuzhny in the required amount for replacement during scheduled and unscheduled repairs [6-7].

At present, the interaction of service and operating enterprises in the Russian Railways system is evaluated on the basis of reliability and efficiency indicators, which are taken as the technical readiness coefficients, compliance with the standards for the duration of scheduled maintenance and repair, the reduced cost of scheduled maintenance and repair according to GOST R 56046- 2014.

The technical readiness coefficient (TRC) is the ratio of the time that railway equipment is in working condition to the total duration of operation in a given time interval, including all types of maintenance and current repair:

$$TRC = T_{WC}/(T_{WC} + T_{OT} + T_{pl}), \tag{1}$$

where T_{WC} is the total residence time of locomotives in working condition in the considered period of operation;

T_{OT} is the total time spent by locomotives in an inoperative state in connection with their repairs due to failures of established types in the considered period of operation;

T_{pl} is the total time spent by locomotives in an inoperative state in connection with their scheduled preventive maintenance and current depot repairs in the considered period of operation.

The coefficient of compliance with the standards for the duration of planned maintenance and repair (K_S) is the ratio of the standard duration of scheduled maintenance or repair of locomotives of a given series relative to the actual duration of scheduled maintenance or repair of these locomotives

$$K_S = T_{N.m(p)}/ T_{A.m(p)}, \tag{2}$$

where $T_{N.m(p)}$ is the normative duration of planned maintenance or maintenance of locomotives in this series;

$T_{A.m(p)}$ is the actual duration of scheduled maintenance or current repairs for the same type of locomotive.

The coefficient of the present value of planned maintenance and repair is the ratio of the actual cost of planned types of maintenance and repair for a certain calendar period of time to the cost of planned types of maintenance and repair for the same calendar period of time established by the regulatory documentation

$$S_{m(p)} = C_{a.m(p)} /C_{s.m(p)}, \tag{3}$$

where $C_{a.m(p)}$ is the actual cost of planned types of maintenance and repair for a certain calendar period;

$C_{s.m(p)}$ is the cost of scheduled maintenance and repair for the same calendar period established by the regulatory documentation.

1.3 Coefficient of output and technical condition of locomotives.

The locomotive productivity is assessed by quantitative and qualitative indicators.

The quantitative indicators of the locomotive fleet should include:

– mileage of locomotives in locomotive-kilometers;
– operating hours of locomotives in locomotive hours;
– the traffic volume in gross ton-kilometers.

Qualitative indicators expand the possibility of obtaining a more accurate assessment of the operation of locomotives and the locomotive complex as a whole. They can be roughly divided into three groups:

– by the time of using locomotives;
– by using locomotive power;
– by locomotive performance.

The power and productivity of locomotives are largely determined by their energy efficiency, which is taken as the absolute, specific or relative amount of consumption or losses of energy resources. The coefficient of performance (COP) is a value that characterizes the perfection of the processes of transformation, the transmission of energy, which is the ratio of useful energy to the supplied energy. This allows evaluating the energy efficiency of a locomotive. Thus, efficiency can be used as an indicator of locomotive utilization.

The quality indicators also include the total percentage of faulty locomotives, which reflects the technical condition and quality of repair and maintenance of locomotives. The indicator is expressed as a percentage measurement of the ratio of the front of locomotive repair to the fleet of locomotives at the disposal of the operational depot. It reflects the state of the repair base of the service depot and the degree of reliability of locomotives.

A decrease in the percentage of faulty locomotives is ensured by:

– improving the quality of performance and the technological processes of repair of locomotives;
– reduction of downtime for repairs;
– introduction of a diagnostic option for organizing service production and improving the professional culture of locomotive crews of the operational locomotive depot and employees of the service locomotive depot.

Thus, to ensure the reliable operation of locomotives and the efficiency of their use, it is necessary to maintain their performance and regulatory technical and energy indicators in operation. This is mainly determined by the quality of maintenance and current repair.

3. Locomotive failures in operation. indicators of reliability and productivity of mainline locomotive

To determine the main standard indicators (mileage rates, lists of work performed by type of repair, etc.) when performing service maintenance and current repair of traction rolling stock, it is necessary to establish and study the parameters that determine the patterns of wear, distribution of failures and the resource of parts and units of the locomotive.

The presence of these parameters would draw up a plan for maintenance and current repairs. Intervals between inspections and other maintenance and repair work are established for each part or assembly, depending on their reliability [8]. Besides, the amount of necessary spare parts and time standards for their repair and replacement are determined.

Thus, the most important task of any service locomotive depot is to monitor the locomotive fleet and technological support of repairs. The processing of big data for monitoring the reliability of locomotives is impossible without using mathematical and statistical methods that allow obtaining an actual objective assessment of the technical condition of the fleet based on the results of operational tests. Unsatisfactory provision of service locomotive depots with material resources, including linear equipment, and low qualification of repair personnel are some of the reasons for the increase in locomotive overtime for scheduled and long-term unscheduled repairs [10].

The duration of downtime of locomotives for maintenance or current repair directly depends on the volume and organization of work, on the reliability and maintainability of the repair facility.

For 12 months of 2019, according to the responsibility of SLD Agryz-Yuzhny, there admitted:

- Event - 6 cases (2018 - 4 cases), target - 5 cases (+ 16.6%).

1) 06.01.2019 TEP70 № 566 - Malfunction of the differential pressure gauge (malfunction in the plug connector - short circuit in the electrodes).

2) 03.11.2019 TEP70BS № 233 - Lack of diesel oil in the diesel crankcase due to its leakage through the technological hole of the diesel camshaft drive (69th group) thanks to unscrewing the technological plug.

3) 26.04.2019 TEP70BS № 278 - There is no automatic start of the diesel engine when controlled from both control cabins on semi-sets "A, B". When checking the control unit BU30T3-06M of the electronic speed controller ERCHM30T, the absence of signal transmission to the actuator was detected.

4) 10.06.2019 TEP70 No. 563 - Fracture of the fan wheel blade of the additional circuit as a result of the weakening of the blade landing on the hub in the fan wheel disk, followed by an imbalance of the

wheel, its wedging and the creation of increased oil pressure in the hydraulic system, which led to the squeezing out of the gasket in the bypass valve, and also a fracture of the blade along the old crack.

5) 22.09.2019 TEP70BS No. 072 - Self-excitation failure due to a malfunction of the power switch block BSK 27.T.245.08.00.001.1 No. 8067732 UOI of the MSU-TE system with the subsequent loss of the electrical circuit to the KVG-2 contactor.

6) 16.09.2019 TEP70BS No. 233 – A violation of the procedure for installing the copper-plated gas joint ring, as well as the moment of pulling the cylinder cover, which led to the breakdown of gases and the destruction of the rubber seals of the rings installed between the cylinder sleeve and the cylinder cover, followed by the flow of coolant through the control holes 8th cylinder.

Failures 1.2 categories - 53 cases (55 cases (3.6%) - 2018), the target - 44 cases (25%).

334 cases of unscheduled repairs of diesel locomotives registered in TCHE Agryz (+ 34% of the target indicator (249 cases) and -5.6% of those allowed by HP in the same period of 2018 - 354 cases of unscheduled repairs).

- by series of locomotives, unscheduled repairs registered in TCHE Agryz were distributed as follows: 2TE25KM - 9 cases (0 cases- 2018); TE10 v/i - 118 cases (186 cases - 2018); M62 v/i - 47 cases (37 cases - 2018); TEP70 in / and - 73 cases (82 cases - 2018); PME3 v/i - 87 cases (49 cases - 2018).

By the type of equipment, HP were distributed as follows: - diesel - 97 cases (128 cases -2018); auxiliary equipment - 35 cases (43 cases - 2018); malfunction of other equipment: Failure of the automated control system - 44 cases (70 cases - 2018), untimely withdrawal from TO-2 due to the fault of SLD - 9 cases (15 cases - 2018); - electrical equipment - 69 cases (46 cases - 2018); - TED –38 cases (14 cases - 2018); - braking equipment - 21 cases (15 cases - 2018); - fire alarm - 1 case (0 cases - 2018); - the crew unit (CP, TRP, KZP) - 13 cases (8 cases - 2018); - safety devices - 7 cases (15 cases - 2018).

Non-production losses (without downtime in anticipation of HP and on HP) for the repair of diesel locomotives amounted to 499.76 thousand rubles registered in TChE Agryz for 12 months of 2019:

- TE10 v/i - 432.9 thousand rubles; - M62 v/i - 15.94 thousand rubles; - TEP70 v/i - 18.466 thousand rubles; - ChMEZ - 32.455 thousand rubles).

Distributed by equipment:

- diesel - 221.8 thousand rubles; - auxiliary equipment - 126.76 thousand rubles; - braking equipment - 35.72 thousand rubles; - electrical equipment - 52.88 thousand rubles; - TED - 54.654; - crew equipment (KP, MOP, KZP, TRP) - 10,135 thousand rubles.

The number of unscheduled repairs of locomotives registered in TCHE Agryz by responsibility for 12 months of 2018/2019:

- SLD-59 - 110 cases (127 cases - 2018); - SLD-55 - 30 cases (30 cases - 2018); - SLD-57 - 24 cases (29 cases - 2018); - SLD-58 - 35 cases (45 cases - 2018); - SLD-61 - 30 cases (9 cases -2018); - other SLDs - 29 cases (32 cases - 2018); - Factories - 7 cases (1 case - 2018); - Russian Railways - 69 cases (81 cases - 2018).

The number of unscheduled repairs of locomotives registered in TCHE Agryz by category of malfunction for 12 months of 2018/2019:

- Production - 167 cases (205 cases -2018); - Degradation - 96 cases (64 cases - 2018); - Operational - 71 cases (85 cases - 2018).

The number of unscheduled repairs of locomotives registered in TCHE Agryz with distribution by blocks:

- Technology (group of the chief technologist) - 8 cases (22 cases - 2018); - Personnel - 66 cases (84 cases - 2018); - Spare parts and materials - 7 cases (3 cases - 2018); - Production (elimination of remarks in TU-152) - 72 cases (58 cases - 2018); - OTC - 14 cases (38 cases - 2018).

Downtime of diesel locomotives registered in TCHE Agryz for 12 months of 2018/2019 due to admitted HP (in hours):

- Awaiting for HP – 33986 h. (24462 h. – 2018);
- HP – 14241,34 h. (15926 h. – 2018).

Average idle time of diesel locomotives registered in TCHE Agryz for 12 months of 2018/2019 due to admitted HP (in hours):

- Awaiting for HP – 100,15 h. (89,12 h. – 2018);
- HP – 42,15 h. (48,11 h.–2018).

IOP Conf. Series: Materials Science and Engineering **986** (2020) 012042 doi:10.1088/1757-899X/986/1/012042

Downtime of diesel locomotives registered in TCHE Agryz for 12 months of 2018/2019, due to admitted unscheduled repairs, with distribution by series in thousand rubles:
TE10: - Costs for downtime pending – 1016 thousand rubles; - Costs for HP – 1313,8 thousand rubles;
M62: - Downtime costs – 15,5 thousand rubles; - Costs for HP – 8,21 thousand rubles;
TEP 70: - Downtime costs – 798,6 thousand rubles; - Costs for HP – 1339,6 thousand rubles;
ChMEZ: - Downtime costs – 510,6 thousand rubles; - Costs for HP – 366,6 thousand rubles.
Unscheduled repairs are mainly associated with the occurrence of system failures (traction motors, mechanical equipment, and electrical equipment). The main reason for the failures was poor quality maintenance or current repairs.
Traction motors, traction gears, wheelsets and electric devices can be identified as the main components and assemblies that limit using power and the operability of mainline diesel locomotives.

4. Technology management and quality of locomotive repairs

4.1 One of the tasks of the Agryz-Yuzhny service locomotive depot is to manage the technological processes of maintenance and routine repair of diesel locomotives. Their results are reducing the time for maintenance and repair of locomotives, also the number of over-cycle works, timely replenishment of inventories, ensuring the quality of technological processes [9]. Thanks to this, labor productivity is significantly increased; the cost of servicing diesel locomotives and their repairs is reduced. An important management function is planning and analysis of the availability of spare parts and materials for technological processes. Automation of technological process control increases the availability of locomotives and depot equipment.
One of the options for implementing production management of locomotive repair is network planning. Network planning is a method that is based on using the mathematical tools of graph theory and a systematic approach to display and algorithmization of complexes of interrelated works, actions or activities to achieve a clearly defined goal. Network planning determines, firstly, which works or operations are "critical" in their impact on the total calendar duration of the repair and, secondly, how to build the best plan for all work on this repair in order to meet the specified deadlines at a minimal cost.
The task of network planning is to graphically, visually and systematically display and optimize the sequence and interdependence of works, actions or activities that ensure the timely and systematic achievement of ultimate goals. For displaying and algorithmizing certain actions or situations, economic and mathematical models are used, which are usually called network models, the simplest ones of which are network graphs.
Analysis of the network schedule highlights those works, on which the total duration of the entire complex of works depends. The number of such jobs is small compared to the total number of jobs on the network. On average, the number of critical jobs, as a rule, does not exceed 10-15% of the total number of jobs in network models.
When using usual methods of planning a complex of works, it is not uncommon to reduce the duration of all or most of the works while reducing the total period of its implementation. This approach, as a rule, did not give the corresponding effect, since it required additional costs and slightly reduced the duration of the entire complex of works. Optimization of the network schedule reduces the duration of the critical path by carrying out a number of measures: varying the time indicators of events and works, using the aggregate-nodal repair method, redistributing resources between critical and non-critical works.
The feasibility of any event, the acceptability of the planned timing of its implementation, the effectiveness of the proposed organizational measures should be determined based on how they influence the corresponding indicators of the entire network model of locomotive repair.
After reducing the duration of work along the critical path, the network model is recalculated again to determine the sufficiency of the measures taken and to find out if new critical paths have appeared. If the new version of the network schedule did not ensure compliance with the established service intervals, then the whole procedure is repeated until a satisfactory result is obtained. In other words, bringing the network model in line with the established repair period for locomotives is carried out by consistently improving the original version of the plan with the required number of recalculations. At the same time,

IOP Conf. Series: Materials Science and Engineering **986** (2020) 012042 doi:10.1088/1757-899X/986/1/012042

to decrease the number of recalculations of the network model to its indicators, measures should be taken to reduce the work not only on the critical, but also on the subcritical paths [11].

One of the urgent tasks for any repair enterprise is to ensure technological readiness, which is influenced by scheduling, which efficiently uses the technological resources of the enterprise, ensures the continuity of the technological process and high quality of service.

Using network planning in the Agryz-Yuzhny service locomotive depot significantly simplifies the process of technological preparation of service production; make it transparent and easily reconfigured. Besides, network planning facilitates the depot's transition to a quality management system due to strictly prescribed time standards for performing technological operations, material consumption rates, processing modes and control methods.

This approach is proposed to be implemented on the basis of a model for quality management of technological processes of locomotives' repair using network planning within the automated control system (ACS) "Network Schedule", used to control and adjust the performance of technological operations.

When the diesel locomotive is placed for service, the supervisor of the production site begins to work with registering the payroll. After that, the master executes a request from the ACS "Network Schedule" about the locomotives set aside for service by the dispatcher. After processing the request, the ACS "Network Schedule" appoints locksmiths to work and issues problem areas for resources.

For each job (i,j) the index of the closeness of the job to the critical path D is estimated. A schedule is drawn up for the next job that has the largest value D_u and whose initial event has already entered the schedule. After the start and end of the next work (i,j), $C(i,j)$ is subtracted from the required value of the TPR C indicator. If $C^* \geq 0$, then the schedule (i,j) of the work is saved, C is taken equal to C^*, the start time of work according to the schedule t_u is taken equal to T, the end time of the work according to the schedule is $t_\kappa = T + t(i,j)$.

If $C^* < 0$, then the work schedule (i,j) is canceled. A job with the next D is selected, the initial event of which has already been included in the schedule and occurred by time T, for these jobs, steps 2 and 3 are repeated. The operation ends when all jobs whose previous event has already been included in the schedule are considered. The value of T is replaced by the nearest higher value from the set of values $t_\kappa(i,j)$. The work that ends by the time point T is determined, and the values of the TPR indicator for these works are taken into account in C.

The shop foreman adjusts the start of work according to the proposed ACS "Network Schedule" scheme, after which he starts the procedure for assigning orders and ordering materials and spare parts. The shop foreman enters the "Repair monitoring" subsystem and selects the desired locomotive from the list in the process of carrying out repairs. The system performs a request for data on completed orders, a request for data on parts and assemblies requiring repair in automated technical diagnostics systems, a request for data on repaired parts received at the warehouse. The received data is marked in the subsystem "Work with orders" and "Repair monitoring". In the case of lagging behind the work schedule, the start and end times of the work depended on them are automatically recalculated. When a diesel locomotive is handed over from repair, the shop foreman enters the "Start for repair" subsystem and selects the locomotive. The system performs a request for data on acceptance or remarks by the receiver and the issue of a diesel locomotive, respectively. After that, it goes to the subsystem "Work with orders".

When there are no comments, the shop foreman closes the orders, and the system sends data on the technological operations performed. This approach allows for operational control and management of technological processes servicing locomotives.

Using the proposed control technique, technological processes were optimized and network schedules for the repair of locomotives in the service locomotive depot were compiled.

Thus, for example, when performing repair in the amount of TR-2, the locomotive should be placed in the position of dismantling the roof equipment. Then the roofs over the diesel generator set are removed. Next, the diesel fuel equipment is dismantled and sent to the fuel equipment repair department for repair work. Then the exhaust manifolds are removed and transferred for repair to the department for the repair of diesel units. Then they dismantle the water pumps and transport them to the diesel unit repair department. After that, the turbocharger is removed and moved for repairs to the site for the repair of diesel units. After dismantling the turbocharger, proceed to the removal of the camshaft drive (group

69), in order to transfer it to the repair site for the repair of diesel units. After that, the cylinder kits are removed. Next, they remove the auto-brake equipment: the compressor and the driver's cranes, and send them to the auto-brake equipment repair department. In parallel, the electrical equipment of the locomotive and the refrigerator section is removed with their transfer to the electrical apparatus repair department and the section compartment, respectively. The rest of the work on the TR-2 cycle is performed on a diesel locomotive. After carrying out repair work on the dismantled units, they are installed in place in the reverse order of dismantling [12].

In this case, the cylinder kits are disassembled and sent for repair to the diesel unit repair section. Inspection and repair of pistons, connecting rods, cylinder covers and bushings are carried out, malfunctions are eliminated, after which the cylinder sets are assembled. The fuel equipment, oil and water systems of the diesel engine, the gas distribution mechanism, the block and the crankcase of the diesel engine are examined for oil and diesel fuel leaks. The hydraulic pushers of the diesel engine gas distribution mechanism in the diesel-unit compartment are repaired. In addition, the crankshaft and the diesel engine muffler, the gearbox and the coupling of the shafting drive of the exciter and the starter-generator are inspected; the coarse oil filters and the centrifugal filter, the water cooling system of the diesel engine are washed. The condition of pipelines, durites and flange connections of the oil and water systems of the diesel engine is checked. Also, in the diesel-aggregate department, repairs of the shaft line to the hydraulic pump gearbox, oil separator tank, controlled damper, water meter, cold and hot water pumps, camshaft drive, the cooling fans for traction motors are performed. The refrigerator sections and the air cleaner housing are inspected and cleaned. Maintenance of electrical equipment includes inspection and verification of the fastening of all electrical machines and devices, as well as lead wires, terminal strips and connections. The condition of collectors, slip rings, brushes and brush holders, insulators, bandages is checked, and parts are cleaned. In addition, a rectifier unit, fan wheel blades and refrigerator shafts are examined. The condition of the power and interlock contacts of electrical devices is checked. The alignment of the main generator, starter-generator, and exciter is checked, and the shaft lines are revised. The detected faults are eliminated. Also, relays, panels, voltage regulators are repaired [13].

4.2 Determination of the amount of materials and spare parts required to perform repair technological processes.

As mentioned earlier, maintenance and current repairs registered in operating locomotive depot Agryz are carried out on an outsourcing basis by the service locomotive depot Agryz-Yuzhny, LLC Lokotech-Service. At the same time, new approaches to the maintenance and repair system are introduced, associated with the growth of overhaul mileage. In order to ensure timely and high-quality repairs, it is necessary to create a revolving fund under these conditions, that is, the presence of a transitional stock of units and assemblies in a working technical condition in the depot. In this regard, the question arose about the quantitative and qualitative composition of this transitional stock. In addition, many locomotives of new series have been put into operation. For JSC "Russian Railways" and the service locomotive depot Agryz-Yuzhny, LLC "LocoTech-Service", there was an acute problem of accounting and ordering spare parts and materials to ensure the technological preparation of service maintenance of diesel locomotives.

Inventory rationing for material and technical resources. The stock rates are determined based on the maximum value of the current stock, the minimum stock rate and the transition stock rate.

However, the purpose of these norms is very difficult, especially for locomotives of new series, as well as when switching to new norms of overhaul mileage. Since for their use, it is necessary to know the planned average daily demand for materials, the interval between deliveries, the actual intervals, the planned volumes of deliveries, the size of the statistical sample for assessing the average deviation of the actual volume of deliveries from the planned one, the time of loading and unloading operations, the estimated time of cargo in transit, and other logistic parameters. It is possible to accurately estimate the required transition margin using the probabilistic approach. For this, the transitional margin is necessary to be determined, which consists of the technological stock Ftech, defined on the basis of the requirements of the technological process and the repair program, and the minimum stock Fst for replacing units during unscheduled repairs, as well as units that cannot be restored or which require restoration special conditions and time longer than the standard:

$$F_{ob}=F_{tech}+F_{st}. \tag{4}$$

The required number of spare units $F_{tech}=(t_a\pm\Delta t_a-t_b\ /\ t_p\pm\Delta t_p)+x_a)$,
where n_a is the number of similar units in the locomotive;
x_a is the number complementing the value F_{tech} to the nearest integer;
$t_a\pm\ \Delta t_a$ is the actual downtime of the unit for repair, taking into account the deviation Δt_a from the standardized one;
$t_p\pm\Delta t_p$ is the actual downtime of the locomotive, taking into account the deviation Δt_p from the standardized one; t_b is the period of time, during which the unit is absent in the locomotive according to the schedule of the repair process. If M_p locomotives are repaired at the same time, then

$$F_{tech}=M_p n(t_a\pm\Delta t_a-t_b/\ t_p\pm\Delta t_p)+x_a). \tag{5}$$

For the settlement period

$$M_p=M^2{}_p(t_p\pm\Delta t_p)\ /\ T^2{}_p, \tag{6}$$

where $M^2{}_p$ is the annual locomotive repair program;
$T^2{}_p$ is the time fund for the settlement period.
Then

$$F_{tech}=n(t_a\pm\Delta t_a-t_b\ /\ T^2{}_p)+xa)M^2{}_p. \tag{7}$$

Transition stock

$$F_{st}=an_aM_pt_b\ /\ T^2{}_p. \tag{8}$$

where a is the coefficient of the changeability of units (parts) that are necessary for unscheduled repairs and are not subject to the restoration process or require a long time to repair.
When determining the consumption rates of materials, the main initial information would be:
- drawings of parts (assembly units);
- specifications;
- technological documentation;
- standards of material consumption;
- standards for waste and losses.

One of the problems in determining the safety stock during the repair of new series locomotives is the data lack on the number of unscheduled repairs that occur during operation. However, the estimated level of unscheduled repairs can be estimated based on the adopted technical readiness factor (TRF). The initial data will be the operating fleet of locomotives, the estimated total residence time of locomotives in the operational state of T_{os}, for scheduled maintenance and repair (M and R) T_{pl}, as well as for unscheduled repairs of T_{ot} for the considered period of operation:

$$n=(t_{os}\cdot N\ /\ \mathrm{TRF}\ -t_{os}\cdot N-t_{pl}\cdot N)/t_{ot} \tag{9}$$

where n is the number of unscheduled repairs.

5. Forecasted power losses in assemblies and units by results of repairs and efficiency of using mainline locals

High-quality repair of diesel locomotives is the most important factor in the reliability and safety of the transportation process in railway transport. In connection with the restructuring of Russian Railways, the repair of traction rolling stock owned by the holding is carried out by repair plants and service locomotive depots. The main task of these enterprises is to ensure the required indicators of the technical readiness of the company's locomotive fleet at an economically justified level of financial costs. Technical, economic indicators, one of which is the energy efficiency (power use) of the repaired locomotive, characterize the quality of repair technological processes. When repair is performed in units and assemblies, additional power losses may appear due to the deviation of their technical parameters and characteristics from the factory values established in the design documentation, which reduces the efficiency of the locomotive. Additional power losses depend on the admissions set in the repair documentation for the technical parameters and characteristics of units and components. Generally, the larger the admissions, the more significant the additional power loss can be.

5.1 Technical parameters and characteristics of parts, components and assemblies after a repair, affecting power losses in diesel locomotives.

The procedure for performing maintenance and repair of parts and assembly units of locomotives is regulated by the Repair Rules, technological instructions, technological processes (maps), etc. Measurement data and values of technical parameters and characteristics of the main parts, assemblies and assemblies after a repair, replacement or modernization are recorded in the technical passport (assembly unit form) of the locomotive. To assess possible additional power losses in units and assemblies, we use only those technical parameters that are regulated by the current regulatory documents. Losses associated with operating modes are not taken into consideration. The energy efficiency of a locomotive is influenced by technical parameters measured during repairs, such as:

– radial clearance in axle box and motor-armature bearings;
– side clearance in the gear train;
– the difference in the diameters of the wheelset tires;
– frequency of rotation of the traction motor;
–contact gaps;
– failure of power contacts;
– pressing of power contacts (initial and final), etc.

To determine the increments of power losses in units and assemblies due to poor-quality repair, it is necessary to find an analytical or experimental relationship between the predicted additional power losses and the technical parameters measured during the repair.

5.2 Quality locomotive repair technological processes and energy efficiency. The time spent by locomotive repair is largely determined by the availability of technological equipment that ensures the mechanization of repair operations and repair quality. This is especially important when performing routine repairs in the amount of TR-3. The repair quality of technological processes also significantly influences the efficiency of locomotive use in operation, largely determining the degree of power utilization of the repaired locomotive. The research results show that one of the main reasons for the decrease in the efficiency of a diesel locomotive is the increase in power losses in units and assemblies. This is caused by the deviations of their technical indicators and characteristics from the nominal values obtained in the case of poor quality of the repair performed. At the same time, radial clearances in motor-axle bearings, general lateral clearances in gear drives, speed characteristics of wheel-motor units and diameters of wheel set tires have a significant effect on reducing losses and using the power of a locomotive. Thus, in order to reduce power losses and gain the energy efficiency of locomotives, it is necessary to ensure the quality of technological operations during the formation (assembly) of wheel-motor units and a selection of KMB for one locomotive with the same speed characteristics and diameters of wheel set tires. The fulfillment of these requirements is ensured by equipping locomotive repair enterprises with appropriate technological and diagnostic equipment.

One of the quality indicators of diesel locomotives is the quality of the connections. Deviations in the shape and relative position of parts from the standard values that occur at the assembly stage, as a rule, reduce the performance of the assembly unit. Thus, assembly operations have a special place in mechanical engineering. It is at this stage that various connections of individual parts and assemblies arise and indicators of the reliability of a technical object are formed.

For example, enhanced technological equipment improves the quality of technological operations for assembling a traction gear, installing a wheelset with motor-axial rolling bearings in the frame of a traction motor and forming a wheel-motor unit as a whole, and provides an increase in the level of mechanization of technological operations and reduces motor block.

It can be concluded that when performing repairs in assembly units, additional power losses may appear due to the deviation of their technical parameters and characteristics from the factory values established in the design documentation. This reduces the efficiency of the locomotive. Additional power losses depend on the admissions for technical parameters and characteristics of units and assemblies established in the repair documentation. Generally, the larger the admissions, the more significant the additional power loss can be.

References

[1] Davydov, A. D., Erokhina, O. O., Ryaboshuk, S. V., & Kovalev, P. V. (2020). Analysis of the causes of cracks in the production of ingots and forgings from austenitic stainless steel 08X18H10T (aisi 321) doi:10.4028/www.scientific.net/KEM.854.16

[2] Drobintsev, P., Voinov, N., Kotlyarova, L., Selin, I., & Aleksandrova, O. (2020). Optimization of technological processes at production sites based on digital modeling doi:10.1007/978-981-15-2341-0_75

[3] Kalinin, M. O., & Pavlenko, E. Y. (2015). Increasing the fault tolerance and availability of software defined networks using network equipment control based on multiobjective optimization by service quality parameters. Automatic Control and Computer Sciences, 49(8), 673-678. doi:10.3103/S014641161508026X

[4] Klochkov, Y. S., & Tveryakov, A. M. (2020). Approaches to the improvement of quality management methods. International Journal of Systems Assurance Engineering and Management, 11, 163-172. doi:10.1007/s13198-019-00939-x

[5] Abyzov, O. V., Galyshev, Y. V., Dobretsov, R. Y., Krasilnikov, A. A., Sidorov, A. A., & Uvakina, D. V. (2020). Stationary test complex for vehicle and tractor gearboxes. International Review of Mechanical Engineering, 14(2), 127-132. doi:10.15866/ireme.v14i2.18267

[6] Borisoglebskaya, L. N., Sergeev, S. M., Provotorova, E. N., & Zaslavskiy, A. A. (2020). Digital algorithms for supply chain automation of mechanical engineering production. Paper presented at the IOP Conference Series: Materials Science and Engineering, 862(4) doi:10.1088/1757-899X/862/4/042025

[7] Dobretsov, R. Y., Porshnev, G. P., Semenov, A. G., Bulakh, D. V., & Bondar, K. A. (2020). The estimation of main parameters of the power plant and electromechanical powertrain for the wheeled vehicle. International Review of Mechanical Engineering, 14(2), 139-145. doi:10.15866/ireme.v14i2.18262

[8] Kalyutik, A., Kiselev, V., & Rouzich, E. (2018). Impact of the underground metal construction cover layer on the cathode protection efficiency. Paper presented at the IOP Conference Series: Materials Science and Engineering, 463(2) doi:10.1088/1757-899X/463/2/022057

[9] Makarov, E. S., Cheglov, A. E., Gvozdev, A. E., Zhuravlev, G. M., Sergeev, N. N., Yusupov, V. S., . . . Breki, A. D. (2018). Power required in the plastic deformation of metallic powder materials. Steel in Translation, 48(9), 597-602. doi:10.3103/S0967091218090061

[10] Mamutov, V. S., & Mamutov, A. V. (2018). Numerical simulation of electrohydraulic processes. Paper presented at the MATEC Web of Conferences, 245 doi:10.1051/matecconf/201824509015

[11] Popovich, A. A. (2020). Additive technologies as breakthrough solutions for creating advanced functional materials. Metal Science and Heat Treatment, 62(1-2), 18-24. doi:10.1007/s11041-020-00507-2

[12] Provotorov, V. V., Danilevich, D. V., Fedotov, A. A., Sergeev, S. M., & Kravets, O. J. (2020). Digital management by supply networks in engineering. Paper presented at the IOP Conference Series: Materials Science and Engineering, 862(3) doi:10.1088/1757-899X/862/3/032024

[13] Tsvetkova, G., & Skotnikova, M. (2018). Engineering and research of wearability coating on the basis of high-strength steel. Paper presented at the Proceedings of 9th International Scientific Conference, BALTTRIB 2017 - Dedicated to 100th Anniversary of Restitution of Lithuania, 166-171. doi:10.15544/balttrib.2017.30

IOP Conf. Series: Materials Science and Engineering **986** (2020) 012043 doi:10.1088/1757-899X/986/1/012043

Electric Resistance Welding of Dissimilar Pipes

A D Davydov[1], Ya I Evstratikova[1], N O Shaposhnikov[1] and G G Klimov[2]

[1]National Technology Initiative Center for Advanced Manufacturing Technologies based on the Institute of Advanced Manufacturing Technologies of Peter the Great St. Petersburg Polytechnic University Polytechnicheskaya, 29, St.Petersburg, 195251, Russia, goy95.95@mail.ru
[2]Gottfried Wilhelm Leibniz Universität Hannover

Abstract. Numerous studies have shown that corrosion of the casing outer side is one of the factors that has a negative impact on the accident rate of the casing string and is the reason of loss of tightness and various failures and complications in the well. The article describes a method of electric resistance welding of dissimilar steel bonding, which allows to increase the corrosion resistance of casing strings in elevator sections with an increased exterior aggressiveness. The developed technology solutions for welding conditions are presented. Metallographic studies of welded joints and recommendations for technological welding conditions are given.

1. Introduction

Corrosion is a common cause of oil field equipment and flow string failures [6, 12]. Corrosion damage to casing pipes often occurs due to loss of cement integrity under the influence of groundwater. Despite many research works to improve the quality of casing cementing [9], damage cannot be avoided.

The choice of corrosion-resistant materials can reduce the possibility of corrosion damage, but it is not always economically justified [18]. Since the corrosive media at different levels of the casing string is different, it may technically and economically be feasible to setup the casing pipe from dissimilar materials, that means to use low-alloyed and corrosion-resistant pipes in one string [3, 10].

One of the main requirements for pipe steel is a combination of strength and plastic properties, high rate of impact elasticity, fatigue resistance and corrosion resistance. DSS provides a good combination of mechanical properties along with high corrosion resistance of these steels (table 1) [4, 15, 17]. Among austenite and ferrite corrosion-resistant steels, duplex steels (DSS) stand out for their higher strength, toughness, corrosion resistance and heat conductivity. [1].

Joining pipes made of dissimilar materials by electric resistance welding is one of the advance research directions, since it allows to stop using sophisticated thread connections. A proper connection is ensured at high current densities, minimal heating time, while the development of a coarse-grained structure is excluded and the decontamination and oxide phase and overheated metal disposal from the welding area is provided [7, 19]. However the process of electric resistance welding has difficulties with a decrease in the stress-strain properties of connections, especially plasticity [8], nonsymmetrical strain conditions, which complicate the substitution of oxide films from the joint and the development of high-quality connections [7], with the generation of hardening texture at a high rate cooling.

While crystallizing duplex steel weld pool there is a risk that the ferrite content will be about 65%, which reduces the toughness and corrosion resistance of the weld. Consequently while predicting the behavior of a welded joint, it is necessary to know about the distribution of ferrite and austenite over the entire section of the welded joint and in the heat-affected area. [14, 16].

Therefore, the object of the work is to study the possibility of building up a casing string from dissimilar steels (low-alloy and duplex steel), connected by electric resistance welding. For this purpose the structure of the material in the weld area as well as stress-strain properties and corrosion resistance were investigated.

2. Methods and materials

To get a weld joint we used duplex 2205 corrosion resistant duplex steel and 09G2S low alloyed steel (Table 1).

Table 1. S31803 (duplex 2205) [20] and 09G2S [11] steel chemical composition.

S31803(duplex 2205)		09G2S	
C, %	0,03	C, %	0,12
Cr, %	21,0 - 23,0	Cr, %	0,3
Ni, %	4,5 - 6,5	Ni, %	0,3
Mn, %	2	Mn, %	1,3-1,7
Si, %	1	Si, %	0,5-0,8
Mo, %	2,50 - 3,50	Cu, %	0,3
N, %	0,08 - 0,20	N, %	0,008
P, %	0,03	P, %	0,035
S, %	0,02	S, %	0,04
FPREN	31 - 38	As, %	0,08

For the study electric resistance welding was considered for five samples of pipes of duplex and low-alloy steel with different welding conditions (Table 2). In these welding conditions, the parts were compressed, followed by the electric current transmission, which led to the heating of the metal. As a result, interatomic bonds were set between the excited atoms located on the surfaces to be connected, that means a welded joint was formed.

Table 2. Welding conditions.

№	I, A·c	j, A/mm^2	t_w, s	U, V	p_{oc1}, bar	p_{oc2}, bar	Δoc, mm	IF, A·c
1	1845	3,33	5,58	24,1	122,2	120,2	4,95	53,3
2	1733	3,13	5,28	26,1	131,6	123,8	3,58	50,7
3	1948	3,52	5,77	25,1	127,9	123	4,86	54,8
4	1467	2,65	4,48	25,6	129,9	122	2,7	44,5
5	1701	3,07	5,28	28,6	136,6	124,8	4,07	50,7

Where j – current density, t_w – welding time, U – welding voltage, p_{oc1}- p_{oc2} – welding pressure, Δoc – welding value, IF – coil current [9].

The sample was a pipe with an external diameter of 48 mm, a wall thickness of 4 mm and a height of 90 to 100 mm (Figure 1). The distance from the current lead to the pipe joint was 50 mm.

After that a metallographic sample was cut in the longitudinal direction to study the heat-affected zone and the welded joint. The studies were carried out using an optical microscope (×500). Hardness measurements were performed by the Vickers method with a load of 180 g ($HV_{0,18}$).

IOP Conf. Series: Materials Science and Engineering **986** (2020) 012043 doi:10.1088/1757-899X/986/1/012043

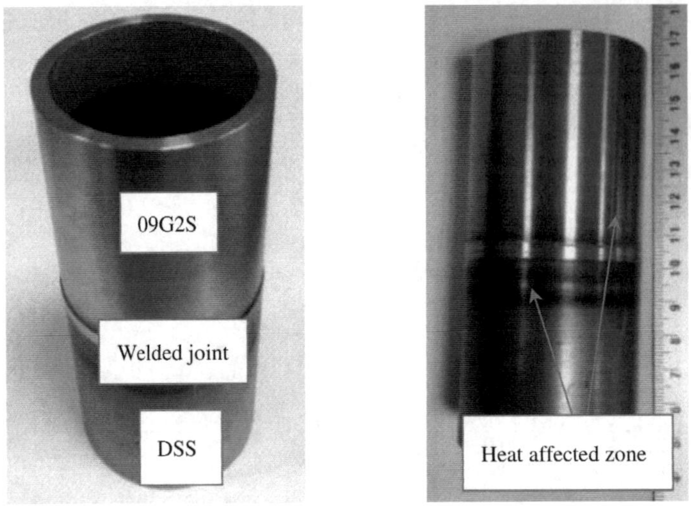

a b

Figure 1. Test item: a – general view, b – dimensions.

3. Research results

The microstructure in the welded joint is shown in Figure 2.

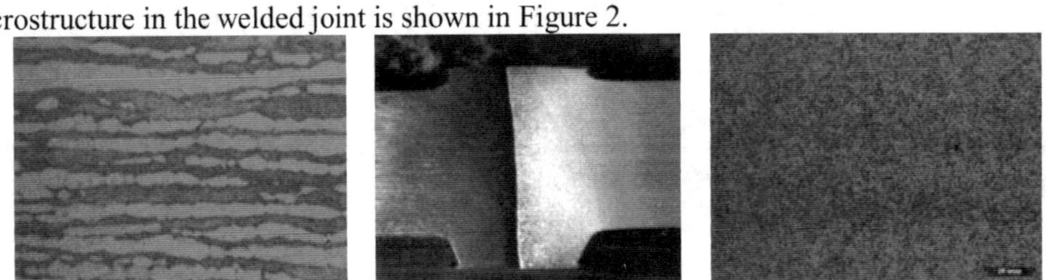

1) DSS microstructure 2) The welded joint macrophotograph 3) 09G2S microstructure

Figure 2. The metal structure of the welded joint of the sample №3.

During the compression of the parts towards the inner and outer walls of the pipe, the non-liquid metal spread out (Fig. 2). When the welded joint was formed, a line of separation of two metals with discontinuities appeared (Fig. 3).

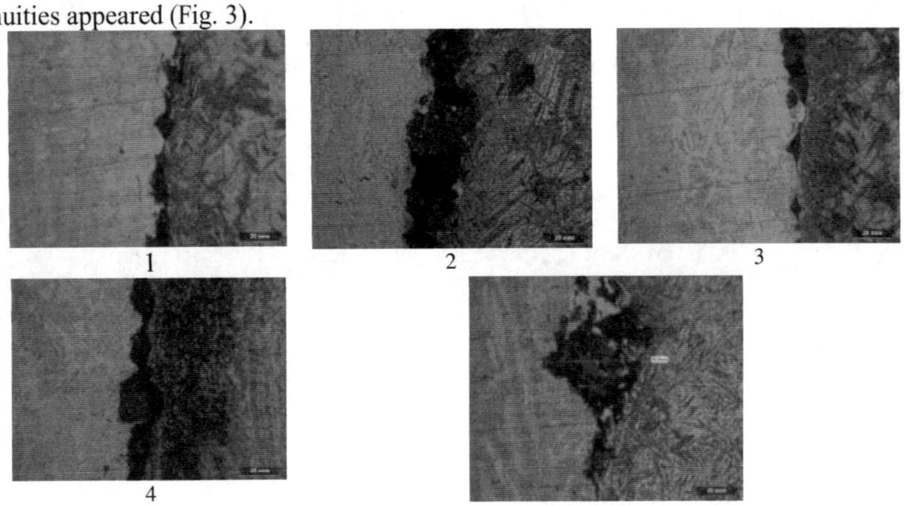

1 2 3 4 5

Figure 3. Discontinuities of different modes in the welded joint: 1-5, ×500.

For modes 2 and 4 discontinuities reach 20-25 microns, for mode 5 - 45 microns. Also, as we can see on Figure 4, there are drops of duplex steel metal on the inner surface of the pipe, which is the result of heating above the solidus during welding. Also some part of duplex steel flowed into low-alloyed steel, which also confirms the heating which exceeds the duplex steel solidus temperature.

Figure 4. Dropping metal duplex steel side (mode 5).

Low alloyed steel pipe grain in the welded joint is a martensite (Fig. 5, area 1), average hardness by mode was approximately 600 HV0,18. In the heat-affected area near the welded joint (Fig. 5, area 2), we have the banding of the low-alloy steel grain in martensite strips form, which are characterized by reduced hardness values), alternating with bainitic ones.

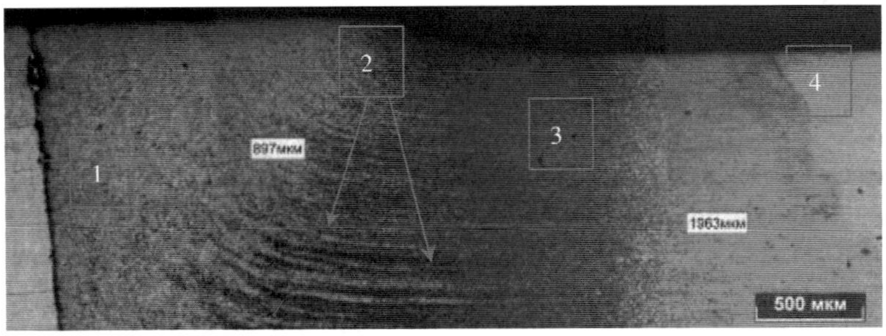

Figure 5. Panorama of the heat-affected zone of the welded joint low-alloyed steel side (mode 5), ×25.

Then there is the area 3, which has a bainite grain (Figure 6, area 3), the average hardness of the area by mode is about 390 $HV_{0,18}$. The base metal grain is sorbite (Fig. 6, area 4).

Area 3 grain

Area 4 grain

Figure 6. Low-alloyed steel grain in the welded joint, ×500.

The grain of a part of a duplex steel pipe in the welded joint area is a mixture of ferrite and austenite in an approximate ratio of 50/50 banded morphology (Fig. 7, dark areas are austenite, light areas are ferrite). Average hardness of area 3 (Fig. 7) for all welding conditions was ≈ 290 $HV_{0,18}$. In the heat-affected zone close to the welded joint (area 1), there is no banding of the grain, the average hardness of the area for all welding conditions is about 320 $HV_{0,18}$.

Microstructure close to welded joint

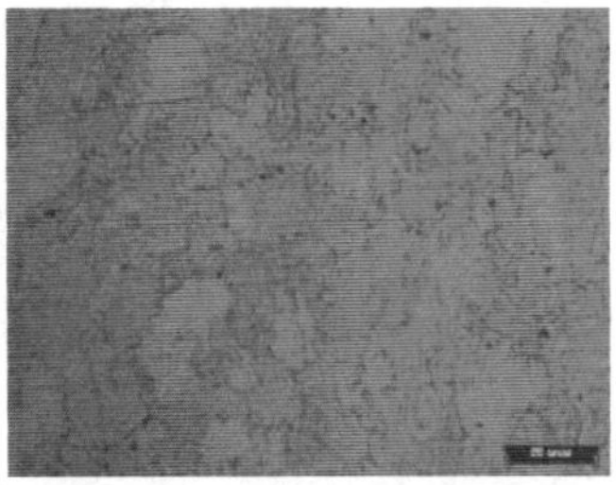

Base metal microstructure

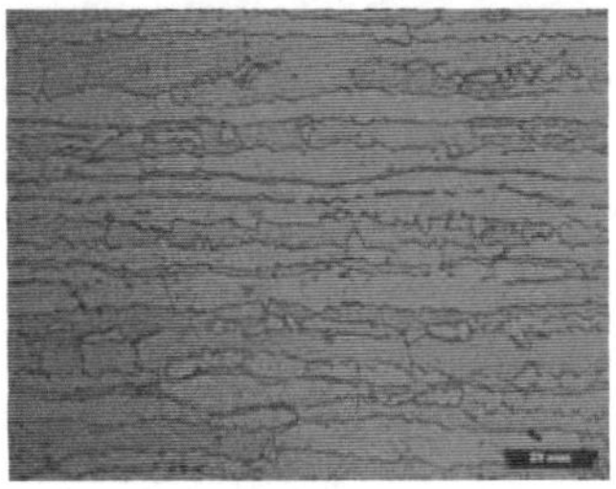

Figure 7. Duplex steel grain close to welded joint, ×500.

To accurate determine the grain in different areas of welded joints, as well as to determine the dimensions of the heat-affected area at the weld boundary, the hardness of the welded joints was measured by metallographic study. The results of measuring the hardness of welded joints, carried out after welding, including the heat-affected area and welded joint metal are shown in Fig. 8.

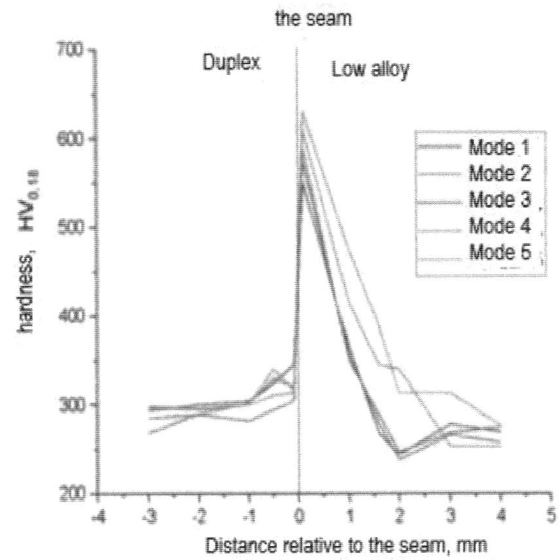

Figure 8. Distribution of hardness over the pipe section in the joint area.

According to NACE MR0175 the maximum permitted hardness value for the welded area must be 250 HV. These hardness values are unacceptable for H_2S media and the requirement is not met [20].

4. Discussion

Metallographic studies have shown that there are discontinuities in the joints, which affects the mechanical properties of the construction. There are areas of increased hardness in the welded joint area, which can also adversely affect the corrosion resistance of the pipe under voltage [20].

Hardening grains have also been found in the welded joint, which can adversely affect resistance to corrosion cracking, especially in a hydrogen sulfide environment.

For this welding method, there are low specific pressures, which are responsible for plastic deformation, which is the main activating factor in a welded joint development in the solid phase. Heating is also an activating factor, but its main function is to reduce the resistance to deformation and, as a result, to lighten plastic deformation. Although with a temperature increase, the concentration of vacancies increases, which intensifies the diffusion, which is an accompanying process, but rather important in the welded joint development in the solid phase. Due to relatively low specific pressures and increased contact resistances in comparison with classical electric resistance welding, local melting and drops splashing in the joint are observed (Figure 4) The use of heating to certain temperatures reduces the deformation resistance of the metal and forces diffusion processes. According to this the temperature increase contributes to all those processes that can provide a reliable welded joint, and the correct technology will ensure that hardening grains and discontinuities are avoided [2, 5].

So the welding technology requires modification of the welding conditions. The optimal parameters for the welding conditions are: higher welding current density and lower voltages [5, 13].

5. Conclusion

We have studied the result of electric resistance welding of dissimilar pipes - duplex and low-alloyed steel. The investigated technology allows us to use corrosion-resistant casing pipes in the well lift in those areas where their use is economically reasonable, but the technology has its own requirements and limitations, therefore the welding conditions should be carefully selected. To improve the welding, a higher welding current density, about 20-60 A / mm2, and lower voltages of 1.5-5 V are required.

In the material grain in the area of the welded joint on the low-alloyed steel side, hardening grains of the martensite and tempered martensite type with high values of hardness were noted, which is unacceptable in a hydrogen sulfide environment and can significantly reduce the corrosion resistance.

IOP Conf. Series: Materials Science and Engineering **986** (2020) 012043 doi:10.1088/1757-899X/986/1/012043

This technology requires further study for choosing the optimal welding conditions.

References

[1] Kall`yari I, Breda M, Frigo M 2014 Izotermicheskoe starenie e`konomno legirovannoj dupleksnoj nerzhaveyushhej stali *Vestnik Magnitogorskogo gosudarstvennogo tekhnicheskogo universiteta im. G. I. Nosova* (№ 4) pp 44-53

[2] Karakozov E` S Soedinenie metallov v tverdoj faze

[3] Blinov P A, Dvoynikov M V, Sergeevich K M and Rustamovna A E 2017 Influence of mud filtrate on the stress distribution in the row zone of the well *International Journal of Applied Engineering Research* (12(15) ISSN 0973- 4562) pp 5214-5217

[4] Ermakov B S and Shaposhnikov N.O Effect of Production Factors on Main Oil Pipeline Pipe Metal Property Formation *Metallurgist* (62(7-8) DOI: 10.1007/s11015-018-0718-7)

[5] Kuchuk-Yaczenko S I and Lebedev V K 1976 Kontaktnaya sty`kovaya svarka neprery`vny`m oplavleniem (Kiev: Nauk. Dumka) p 216

[6] Blinov P A and Dvoynikov M V 2018 Rheological and filtration parameters of the polymer salt drilling fluids based on xanthan gum *Journal of Engineering and Applied Sciences* (13(14) DOI: 10.36478/jeasci.2018.5661.5664) pp 5661-5664

[7] Kuchuk-Yaczenko S I, Nakonechny`j A A, Zyakhor I V, Chernobaj S V and Zavertanny`j M S Tekhnologiya i oborudovanie dlya sty`kovoj svarki soprotivleniem detalej bol`shogo secheniya iz raznorodny`j stalej (IE`S im. E.O. Patona)

[8] Kunshin A A, Dvoynikov M V and Blinov P A 2019 Topology and dynamic characteristics advancements of liner casing attachments for horizontal wells completion *Youth Technical Sessions Proceedings: VI Youth Forum of the World Petroleum Council - Future Leaders Forum (WPF 2019), June 23-28, 2019, Saint Petersburg, Russian Federation* (ISBN 9780367346683) pp 376-381

[9] Chudinova I V, Nikolaev N I and Petrov A A 2019 Design of domestic compositions of drilling fluids for drilling wells in shales *Youth Technical Sessions Proceedings: VI Youth Forum of the World Petroleum Council - Future Leaders Forum (WPF 2019), June 23-28, 2019, Saint Petersburg, Russian Federation* (DOI: 10.1201/9780429327070) pp 371-375

[10] Davydov A, Alekseeva E and Gaev A Specificity to the choice of materials for wellhead equipment *Materials Today: Proceedings*

[11] GOST 19281-2014

[12] Nutskova M V, Kupavykh K S, Sidorov D A and Lundaev V A 2019 Research of oil-based drilling fluids to improve the quality of wells completion *IOP Conference Series: Materials Science and Engineering* (Vol. 666, No. 1, 012065, DOI: 10.1088/1757-899X/666/1/012065)

[13] Krasulin Yu L and Nazarov G V 1976 Mikrosvarka davleniem *Metallurgiya* p 160

[14] Zubchenko A S and Fedorov A V Kompleksny`e issledovaniya svarivaemosti i korrozionnoj stojkosti austenitnoferritnoj (dupleksnoj) stali saf 2507

[15] Alkhimenko A 2019 Corrosion testing of experimental steels for oilfield pipelines *E3S Web of Conferences* (121, 01001, DOI: 10.1051/e3sconf/201912101001)

[16] Kovalev M, Alkhimenko A and Shakhmatov A 2020 Electrochemical studies of welded joints corrosion resistance made from stainless steels *Materials today: proceedings* (DOI: 10.1016/j.matpr.2020.01.034)

[17] Andrey Lapechenkov, Aleksandr Fedorov, Ekaterina Alekseeva, Lyudmila Galata and Aleksey Piskarev 2020 Comparative analisys of corrosion resistance of S31200 duplex stainless steel and its analogue *Materials today: proceedings February 2020* (DOI: 10.1016/j.matpr.2019.12.174)

[18] Laptev A and Bugay D 2010 Fighting against corrosion in oil and gas complex of Russia: Problems and the ways of their solving *European Corrosion Congress* (EUROCORR 2010)

[19] Sakhaczkij G P 1963 Issledovanie kontaktnoj sty`kovoj svarki oplavleniem i soprotivleniem (№ 10) pp 26–32

[20] NACE MR0175/ISO 15156-1

IOP Conf. Series: Materials Science and Engineering **986** (2020) 012044 doi:10.1088/1757-899X/986/1/012044

Analysis of Technical Aspects of Choosing Air Conditioners and Temperature Parameter Control for Developing Air Conditioning Systems for Premises of Various Area at Fishery Enterprises

V V Kiryuha

Department of Electric Power Engineering and Automation, Far Eastern State Technical Fisheries University, ul. Lugovaya 52b, Vladivostok, 690087 Russia E-mail: vkiryuha@list.ru

Abstract. This article considers the theoretical and practical aspects of optimal construction of air conditioning systems for production enterprises of various area. The importance of this task is determined by the requirements on providing a comfortable work environment for maintenance personnel. The data on air conditioner operation, the theoretical surveys, and the factors that influence the temperature conditions and distribution of air streams are used to propose a number of recommendations for optimizing the construction of noise-proof conditioning systems. In addition, some guidelines are proposed for upgrading remote control systems for split conditioning systems.

1. Introduction

One of the key aspects of occupational safety is the maintenance of optimal air parameters in production premises. These parameters are temperature, humidity, and dust content in the air. Currently, the most favorable temperature in terms of comfort and on medical indications is provided by means of air conditioning equipment.

Air conditioning is the setting and automatic maintenance or adjustment of all or individual indoor parameters, for example, a certain level of temperature to provide optimal conditions for people's general state and managing operating procedures [8].

Air conditioning is provided by a set of technical means referred to as air conditioning system (ACS). This system consists of air entrainment and treatment components: some of them (filters, heat exchangers, air humidifiers, air driers) serve to endow the air with certain parameters, move (fans) and distribute the air; others include cold and heat supply, automation, and remote control devices. ACS at large public and production buildings are usually serviced using integral automated control systems.

2. Materials and methods

The application of ACS at production premises has a number of peculiar features. Production premises, especially the ones used for fish processing, have unevenly distributed air streams. This is why, a situation occurs in which zones with uncomfortable conditions are formed, including too low or too elevated temperatures, intensive cold air streams, etc. This issue can be partially tackled by choosing a suitable air conditioning system: a central ACS, a comfortable ACS, a reheat air conditioning plant, etc. In its respect, the choice is made on the basis of classifying conditioning systems and analyzing factors that influence the distribution of heat in a concrete premise.

Before getting down to the classification of conditioning systems, one should note that there is no conventional classification of ACS, which is due to the large number of wiring circuits, engineering and functional specifications that depend not only on the functionality of the systems themselves but also on their target facilities (conditioned premises) [11].

Modern conditioning systems can be classified by the following characteristics: [16]

-main purpose (target facility) (relief and process ACS);

-position of the conditioner relative to the conditioned premise (central and local ACS);

-availability of an in-system (embedded in the conditioner configuration) source of cold and heat (autonomous and non-autonomous ACS);

-action principle (direct-flow, recirculating, and combined ACS);

-number of serviced premises (local zones).

A classification according to other parameters is also possible.

In addition to the above classifications, there are diverse conditioning systems for specialized operating procedures, including ACS with meteorological parameters varied in time according to a certain program.

Relief ACS are intended to generate and automatically maintain air temperatures, relative air humidity, air purity, and air flow rates that meet optimal sanitary and hygienic requirements on residential, public, and administration and amenity buildings or premises.

Process ACS are intended to provide air parameters compliant with production requirements to the full. The process conditioning of premises with people is provided, taking account of the sanitary and hygienic requirements on the state of the air environment.

Central ACS are supplied from outside with cold (with cold water or coolant), heat (with hot water, steam, or electricity), and electric power for actuating electric motors of fans, pumps, etc.

Central ACS are mounted outside service premises and designed to condition one such premise, several of its zones, or a lot of separate premises.

Central ACS are equipped with central non-autonomous conditioners made according to basic (standard) equipment layout drawings and their modifications.

Local ACS are developed on the basis of (non-) autonomous conditioners installed right in serviced premises. Local ACS are easy to install and mount.

Autonomous ACS, for example, split-system ACS, are supplied from the outside only with electric power.

These conditioners have in-built compression chillers, usually run on Freon-22.

Autonomous systems cool and dry the air, for which purpose the fan blows the recirculating air through chiller evaporators used as surface air-coolers. In the midseason and in winter these systems can heat the air with the help of electric heaters or by reversing the chiller's operation in the so-called heat pump cycle.

Split-system conditioners can be considered the most basic option for decentralized maintenance of target indoor temperatures.

Air conditioning systems for production premises have several various configurations the most popular of which are single or multi-unit split-systems, central conditioning, chiller-fancoil water cooling systems, or multizonal systems.

The most expedient solution for small production premises of up to 200 sq. m in area are generic single-unit or multi-unit split-systems. The standard capacity of these conditioners from top specialized manufacturers allows maintaining optimal indoor air temperatures, usually, from 18 to 30^{0}C, without forming low-temperature zones right near internal split conditioner units. Standard internal units can be placed so as to ensure air exchange across the entire air mass of the premise.

As an example, let us consider the specs of the LG LSK2460HL split conditioner: the cooling/heating rate is 24,000 WTE/h, the air circulation rate is 16/42 cubic meters per minute, the consumed capacity is ~ 2.7 kW. Four such conditioners mounted in the opposite corners of a premise of 250 sq. m in area and four to five meters in height allow generating optimal temperature conditions (adjustable within a reasonable temperature range) and optimal air exchange conditions.

It is also quite justified to mount simple split-systems in facilities of 300 to 500 sq. m in total area. Although this approach does have obvious drawbacks, it has some obvious advantages as well, notably, a relatively low cost, a high work delivery speed, and the option of stage-by-stage equipment installation. However, basic split-systems do not allow organizing centralized management. Central or multizone conditioning is a sensible solution for facilities of more than 500 sq. m in area. Conventional central conditioning involves supplying cool air from one ventilation plant to all the premises in the building along a network of air conduits. This approach has a number of advantages, including the synchronous provision of conditioning and ventilation and the possibility to save energy resources by using a regenetative heat exchanger in the plant. However, this method is not always optimal because the high air consumption necessary for compensating heat emissions requires using a system of heavy-gauge air conduits, which is not always possible. In addition, the air conduits must be insulated from heat to reduce heat losses and avoid the risk of condensate formation. The main issue with central conditioning is that it implies using low-heat capacity air as heat carrier. Another weakness of this method is that it requires using an automatic air consumption control system in every premise to precisely maintain the temperature in each single premise. This makes it way more complicated to manage the equipment, increases the system's overall cost, and results in tough requirements on its maintenance. A modification of the described method is the VAV (variable air volume) system. Unlike conventional central conditioners, these plants allow regulating air supply to different premises. However, the enumerated flaws make it difficult to use these systems in premises of more than 500 to 1 000 sq. m. in area.

The issue with conditioning these premises is solved using an intermediary coolant of high heat capacity, for example, water (in chiller-fancoil systems). Cold water is treated in the chiller and then supplied to the fancoils installed in the premises. Other coolants it is possible to use instead of water are freon and variable refrigerant flow (VRF) conditioning systems. Essentially, these are multi-systems with expanded capabilities. The outer unit (compressor-condenser unit) is either air-cooled or water-cooled.

These systems have such advantages as a large number of internal units connected to the outer unit, the possibility to form an integral system with common control, and the capability to run internal units simultaneously for cooling and heating.

Multizonal conditioners are used most often at such facilities as production premises divided in separate workshops and sections, and also production premises on different storeys of the building. In this case, the outer unit can be installed in a convenient location far away from workspaces.

However, the broad capabilities mean certain difficulties with designing and choosing the assemblies, considering the need to maintain an optimal temperature and set the conditions for its adjustment. It is pointed above that VRF systems allow running the internal units in one such system for both, cooling and heating. In this case the air streams of various temperatures in one production premise are distributed among uninsulated areas placed close to each other.

To choose conditioners and their capacity for ACS in design, it is proposed to follow the procedure described below. This express method is used mainly to develop ACS based on simple-design climate solutions such as split-system conditioners and also window-type and single-unit conditioners.

To select a conditioner of required cold output, it is necessary to calculate the heat supplied to the premise with solar radiation, lighting, people, office automation equipment, etc.

The components of the main heat leakages to the premise are enumerated below.

1)Heat leakages produced by the difference between the indoor and the outdoor air temperature as well as from the influence of solar radiation are found as:

$$Q_1 = V \cdot q_{sp}, \tag{1}$$

where $V = S \cdot h$; S is the premise area; h is the premise height; q_{sp} is the specific heat load; it is taken that the heat leakage for the premise without sunlight is 30 to 35 W/m^3, the average heat leakage rate is 35 W/m^3, and the heat leakage to the premise with a large glazed area on the sunny side is 35-40 W/m^3.

2)Heat leakages Q_2 produced by the office automation equipment used inside the premise.

The average value of Q_2 is 300 W per full-packaged computer (or 30 % of equipment capacity).

3)Heat leakages Q_3 produced by the people in the premise.
It is usually taken for calculations that the heat leakage rate is 100 W per capita in the office and 100 to 300 W per capita in restaurants and premises, where people take to physical labour;

$$Q=Q_1+Q_2+Q_3. \tag{2}$$

The calculated heat leakages are added to 20 % of unaccounted heat leakages so that

$$Q_{ttl}=(Q_1+Q_2+Q_3)\cdot 1.2\ \text{W}. \tag{3}$$

If the premise has additional heat-emitting equipment (electric cookers, production equipment, etc.), the given calculation must take account of the respective load as well.
The main user-relevant indicator of comfort and indoor air quality is indoor temperature.
The so called optimal indoor air temperature used in the design of air conditioning systems ranges from 20 to 25^{0}C in residential, public, and administration and amenity buildings, depending on the time of the year and the speed of indoor air streams. According to GOST 30494-26, the optimal microclimate parameters are those values of microclimate indicators that ensure the sensation of comfort and normal body thermal status at the minimal stress of thermal control mechanisms for at least 80 % of the people indoors at regular and consistent exposure.
The design engineer uses this range to choose the design indoor air temperature against which he determines the indoor thermal balance in the warm and the cold part of the year as well as in the midseason. Moreover, the maximal and the minimal indoor air temperature are often chosen from the range of optimal temperatures in the warm and the cold period, respectively, for the purpose of power supply.
This choice of design temperatures is absolutely adequate from the standpoint of the existing reference documents and broadly used in the design of conditioning systems. In case of multizonal systems (VRF, VAV, chiller-fancoil system, sometimes, split-systems), however, this choice makes it impossible to sustain the optimal air parameters in serviced premises.
This is caused by the fact that these systems allow the user to individually choose indoor temperature values necessary for a given premise. The range of necessary indoor temperatures is fairly broad; in most systems, it extends from 18 to 30^{0}C. The temperature is chosen at random, which depends on individual thermal control peculiarities. The chosen temperature often falls out of the range of standardized optimal indoor air parameters, which is conditioned by the very definition of optimal microclimate parameters that are optimal only for 80 % of the people. The statutory indicators calculated for the average person can be used only in designing large premises for lots of people, for example, cinemas, conference rooms, and sports centres. Multizonal conditioning systems providing comfort to particular individuals must be designed, taking account of their subjective characteristics.
The procedure of attaining comfortable specifications is described in work [1].

3. Results

It is usually accepted in practical calculations of air stream distribution that jet speeds and temperatures drop according to the regularities set for free jets (Fig. 1). In multizonal conditioning, however, the interaction of jets that will redistribute air streams and temperatures is inevitable [13].
The jet's behavior will depend, first of all, on the correlation of forces the most critical of which for the task in question will be inertia, viscosity, gravity, and pressure. Surface tension, elastic strains, and unsteady motion forces can be ignored because our case covers real stationary gas streams. The relative change in air volume and, therefore, density, will be considered negligibly small because conditioners in on mode have a narrow range of temperature variations.
Then the main criteria of similarity that characterize relations of applied forces will include
1)Reynolds criterion that reflects the inertia-viscosity ratio as

$$Re=\frac{ul}{\nu}, \tag{4}$$

where u and l are the characteristic speed and size, ν is the kinematic viscosity factor;

2)Froude density coefficient (reverse to Archimedes criterion) that characterizes the inertia-buoyancy force ratio as

$$Fr_\rho = \frac{u^2\rho}{gl(\rho_o - \rho)}, \tag{5}$$

where ρ and ρ_0 are the jet and indoor air density, g is the acceleration of gravity;
3)Eulerian number found as

$$Eu = \frac{p}{\rho u^2}, \tag{6}$$

where p is the pressure in the fluid.
According to the experimental investigation results presented in the paper by V. A. Bakharev and V. I. Troyanovskii, the air chute sizes that exclude the influence of side walls on jet spread must meet the following ratio:

$$\frac{\sqrt{\omega}}{d_0} \geq 71.5, \tag{7}$$

where ω is the passage area of the air stream indoors.
Usually, the criterial equation in the theory of similarity is recorded as

$$Eu = f(Re, Fr_\rho), \tag{8}$$

which allow formulating the research task as to determine the influence of friction, gravity, and inertia on pressure variations in the air stream.
Let us distinguish the influence domains of each of the indicated criteria and respective forces.
According to the jet theory, turbulent jets are self-similar, beginning from Re ≥ 200. In this work's tasks self-similarity means the independence of the process from Re numbers and, therefore, from viscosity forces.
The significance of the Fr_ρ criterion has no unequivocal definition. It is known from (Abracham, 1972) that at $Fr_\rho \geq 20$ buoyancy forces stop affecting the jet spread and the streams can be treated as turbulent nonbuoyant. At 20>Fr>4 the jet spread is influenced by both, gravity and inertia forces; at Fr< 4 the dominant forces are buoyancy forces and the jets go down or up, depending on the sign of the difference between the jet and the ambient temperature.
Abracham's investigations have allowed finding critical Froude's air stream numbers; once these are exceeded, the jets become self-similar relative to buoyancy forces, whereas inertia forces begin to exert the dominant influence that determines the regularities in the jet's speed drops and geometrical parameters.
In multizonal conditioning air streams can spread as jets in an unbounded space (Fig. 1) or as jets with circulating zones (Fig. 2). These zones begin to form when the values of the Bakharev-Troyanovskii criterion are exceeded.
The pressure for free jets spreading in a space with infinitely remote boundaries is taken constant, and the equation of motion for the jet spreading along the *x* axis (Fig. 1) is recorded as Eq. 8

$$u\frac{\partial u}{\partial x} + v\frac{\partial u}{\partial y} = \pm 2c^2x^2\frac{\partial u}{\partial y}\frac{\partial^2 u}{\partial y^2}, \tag{9}$$

where ν is the kinematic viscosity factor, *c* is the coefficient of turbulent mixing in Prandtl's free turbulence theory.

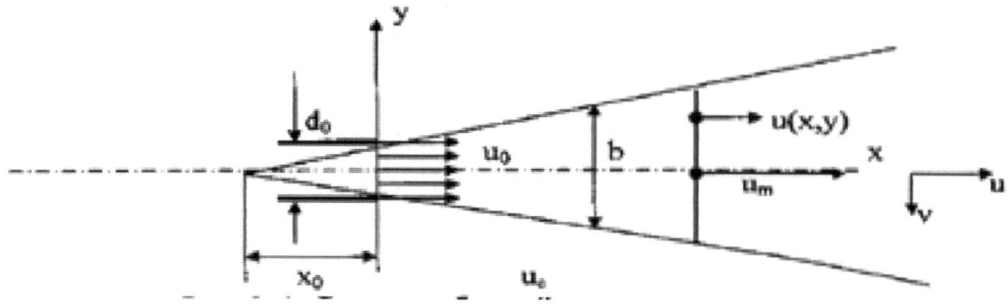

Figure 1. Free jet configuration.

Prandtl's, Taylor's, and Kármán's half-empirical turbulence theories have been used to develop self-similarity solutions that allow attaining the results close to the test data for pairings of half-infinite flat parallel streams, turbulent free jets, and tracks at a large distance from streamlined bodies.
The drop rate of the maximal speed along the axis, jet width, initial section length, and speed and temperature profiles for such jets are determined according to regularities for free boundary layers.
The successes in describing free jets and tracks as part of half-empirical solutions apply, to a lesser extent, to more complex solutions such as the system of interacting jets with circulating zones (Fig. 2).
Jets with circulating zones form when the stream expands in a bounded space. The indicated limitations can be created by the walls on the sides and butt ends of jets and by the neighbouring jets that create low pressure gradients the value of which found on the basis of D. Bernoulli's equation corresponds to stream speed values.

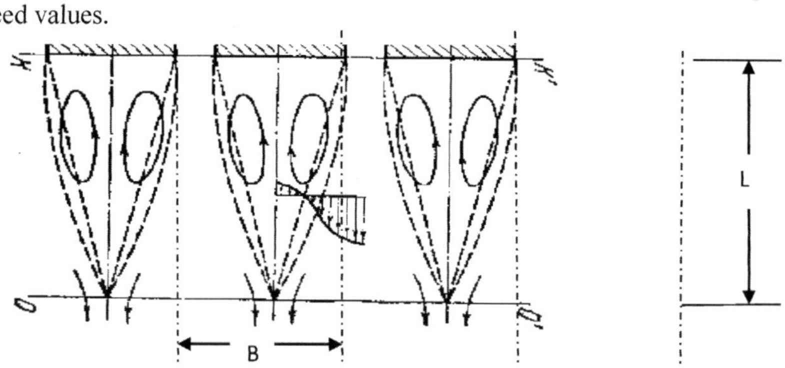

Figure 2. Interaction pattern and speed profile of air jets forming zones of circulation.

Thus proper correlations for calculating speed and temperature profiles and impurity concentrations must be proposed for each of the highlighted jet spread regions (nonbuoyant turbulent jet Fr>20, buoyant turbulent jet Fr<4, and intermediate jet 20<Fr>4). It is assumed at a first approximation that these profiles coincide.
As shown above, turbulent nonbuoyant air jets are calculated using the results of solving free turbulence equations. The results of the theoretical and experimental investigations conducted by C. J. Chen and W. Rodi in 1980) [14] can be used for buoyant and mixed jets without zones of circulation.
Work [Bolshakov, Konstantinov, Popov et al., 1984] presents a list of relations for calculating self-similar jet flows, turbulent buoyant jets in a stationary, an unstratified, and a linearly stratified medium, and also turbulent buoyant round jets in a stationary unstratified medium, taking account of the influence of the envelope. In this paper the initial jet pulse is usually ignored at $Fr_\rho \rightarrow 0$ and the flow is referred to as blast (plume in foreign literary sources). At $Fr_\rho \rightarrow \infty$ buoyancy effects are ignored and the flow is referred to as jet. In addition, the jet flow ($x` < 0.50$), the intermediate ($0.5 < x` < 5$), and the plume zone ($x` > 5$) are distinguished in the positive-buoyancy round jet, where

$$x' = \frac{x+x_0}{d_0}\left(\frac{\rho_o}{\rho_e}\right)\frac{v_o}{\sqrt{\frac{\rho_e-\rho_o}{\rho_o}gd_0}}, \tag{10}$$

istinguished in the positive-buoyancy round jet, where and x_0 is the distance to the jet pole.
The successes in describing the spread of various jets, jets with zones of circulation included, were achieved by using model tasks based on the joint solution of equations of motion, equation of turbulent energy balance k, and balance equation of turbulent energy dissipation ε.
For the needs of practice, however, half-empirical correlations are used most often; according to them, circulating zone length $\boldsymbol{L}$ and distance to whirlpool centre $\boldsymbol{L_ц}$ (Fig. 2) are found as

$$L = 3B,\ L_ц = 2B. \tag{11}$$

The distribution of all of the parameters along two thirds of the whirpool zone length (to the whirlpool centre) corresponds to the distribution in a free jet; then the air stream sharply expands to the limiting surfaces or to the line of contact with a neighbouring jet. The speed profile in this part of the jet also remains universal but the profile becomes more slanting.
In view of the foregoing it can be concluded that the use of multizonal systems to condition large production premises makes it necessary to rightly choose the design air temperature. The approach to solving this problem is exposed in work [2].
Considering that the air space of any premise can be presented as a closed heterogenous environment, the existence of air streams with various temperatures will lead to the heat exchange that can be tentatively described as

$$P = \xi S(Q_2 - Q_1) + mc\,\frac{d\,(Q_2 - Q_1)}{dt}, \tag{12}$$

where P is the absorbed (released) capacity;
ξ is the heat transfer coefficient;
S is the heat exchange surface;
$Q_2 : Q_1 :$ m are the temperature and mass parameters of the environment.
The heat exchange will affect the steady-state temperature, especially at a significant difference in the steady-state temperatures of inner units of a multizonal system. The aspects of controlling these parameters are exposed in work [3].
The temperature control issue matters in engineering solutions applied in the design of conditioning systems. They can be controlled in an optimal manner when the true temperature in a certain point of the internal ecosystem is known. In addition, the actuating devices of the conditioning system are tied with the data from thermal sensors installed in the system's devices.
The device within can be represented as a closed heterogeneous environment.
The temperature control in heterogeneous closed environments can be the parameter to which one can confine the diverse processes of thermal change in both, various engineering procedures and macrosystems, for example, the control over the air temperature in air-conditioned amenity premises.
To make the right choice of a circuitry and design solution of a temperature control system, optimal for a concrete engineering procedure or macrosystem, it is necessary to distinguish the main criteria of forming requirements on controlling devices. It is expedient to include in these requirements the range of measured temperatures, accuracy of control, delayed action, and thermal pollution level of a measured object.
The range of measured temperatures can be found as

$$\Delta\Theta = \Theta_2 - \Theta_1,, \tag{13}$$

where Θ_2 is the highest control temperature;
Θ_1 is the lowest possible control temperature.
The range of measured temperatures must be taken into account when choosing thermal elements of sensors for control instruments. The accuracy of control is found as

$$\sigma = \frac{\theta_2 - \theta_1}{\theta_1} \cdot 100\%, \tag{14}$$

where Θ_2 is the measured temperature;

IOP Conf. Series: Materials Science and Engineering **986** (2020) 012044 doi:10.1088/1757-899X/986/1/012044

Θ_1 is the true temperature.

The accuracy of control is affected most of all by circuitry and design solutions chosen for control devices [4, 5].

Heat retention has the strongest influence on the accuracy of control, especially for express control and intermittent control with short time spans. The need for solving the heat retention problem arises in two cases, In the first one the sensor with mass *m* and specific heat capacity *C* is located in a medium with constant temperature Θ_1 and sets into action at instant t_1. The sensor in on mode releases capacity *p* (active sensor), and its temperature $\Theta_2(t)$ begins to rise. When the balance with the medium is reached, the temperature stops changing.

The process can be tentatively described as

$$P = \xi S(\Theta_2 - \Theta_1) + mc\,\frac{d(\Theta_2-\Theta_1)}{dt}, \tag{15}$$

where S is the heat exchange surface (sensor surface area plus lead-in parts area);

ξ is the heat transfer coefficient.

This equation corresponds to the aperiodic link equation with the time constant found as

$$T = \frac{mc}{\xi S}. \tag{16}$$

In the second case the sensor with *m*, heat capacity *c,* and temperature Θ_2 at instant *t* is placed in a controlled medium with temperature Θ_1. As a result of the heat exchange with the medium, the sensor temperature begins to tend to Θ_2. This process is described as

$$\Theta_1 = \Theta_2 + \frac{mc}{\xi S}\cdot\frac{d\Theta_2}{dt}. \tag{17}$$

The experimental curve of the transition process in the heat converter is similar to the plot of the transition process in the aperiodic link. The difference between the curve and the aperiodic link plot is that in initial section ($t_1 - t_2$) of the curve subregular mode related to the distribution of heat inside the sensor body is observed plus the thermal gradients correspondent to the unidirectional flow of heat takes place are established.The subregular mode section curvature depends on the uniformity of the sensor's structure and can be changed on intention by using a sensor with a high non-uniformity of masses. The regular temperature setting mode is observed in section ($t_2 - t_3$), whereas a steady-state thermal equilibrium occurs after t_3.

The irregular mode takes fairly little time at a small nonuniformity of masses in the sensor, and the thermal lag design for the sensor can be conducted as for the aperiodic link. The thermal time constant is determined by the full heat capacity of the converter and the conditions of the heat exchange between the converter and the ambient medium; this is why, one and the same converter has different time constants, depending on heat-exchange conditions. To calculate the time constant, it is necessary to find the overall heat capacity of the elements included in the sensor. Approximate calculations can be conducted proceeding from the average heat capacity of metals within 400 to 600 J/kg·K, the heat capacity of inorganic insulation materials (mica, china) of 800 to 1 000 J/kg·K, and the heat capacity of organic materials (resin-dipped fabric laminate, SRB paper laminate) of 1 200 to 1 400 J/kg·K.

The heat transfer factor depends on the medium, sensor surface treatment, and convection characteristics of the control medium. Persistent variables differ in very broad limits. Commercial heat meters have a time constant of three to six minutes. To shorten this period, it is expedient to use p-h thermal sensors because these are much lighter and, therefore, have a smaller heat retention.

The general analysis of the choice of temperature-sensitive sensors for designing circuits for heat meters meant to control the temperature in closed heterogeneous environments must rely on choosing the main criteria of assessing a particular kind of sensors for applicability. Some of these criteria are the range of temperatures, required accuracy of control, delayed action, allowed thermal pollution level. It is also expedient to assess sensors of various kinds against the chosen criteria by means of computer programs.

The above exposed considerations applied to macrosystems, in particular, to split air conditioning systems used in closed premises, allows optimizing the conditioner management system and cut energy expenditures.

4. Discussion

As a rule, modern conditioning systems have remote control consoles (RCC) that allow the operator to change the system's running modes by directly manipulating the operating controls on this console [12, 17].

The RCC of a split conditioning system has an on button, an off button, buttons for different running modes (cooling, heating, drying), buttons for changing the fan's blow angle and intensity, and also temperature-setting buttons. The RCC electronic circuit converts preset parameters to a coded signal transferred with an infrared beam to the inner unit receiver. After a certain temperature is set using the RCC, this mode sets up within a certain period of time, when the air blown by the internal unit fan reaches the preset temperature, because the temperature sensor (thermal resistor) is installed right in the internal unit radiator. In this case, the temperature in the zone, where the operator with the RCC is found, will differ from the preset temperature. For the diagram of the control based on temperature parameters see Fig.3.

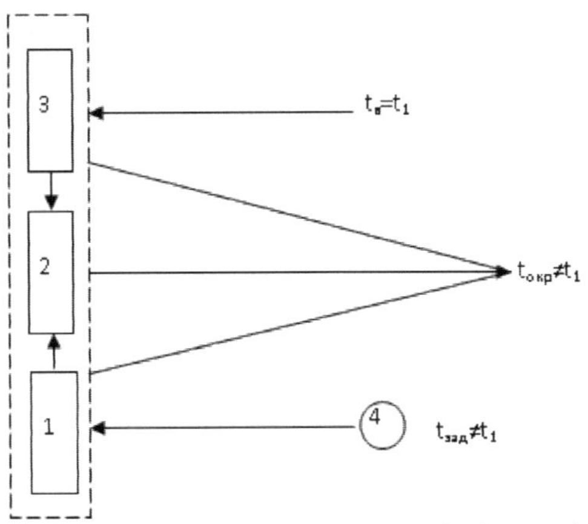

Figure 3. Control based on temperature parameters: 1 is the receiver of control signals; 2 is the control circuit; 3 is the temperature sensor; 4 is the RCC with temperature-setting buttons.

The difference between the temperature in the zone, where the operator with RCC is found, and the preset temperature depends on the conditioner's internal unit, the air stream direction angle, the existence of eddy zones (walls or items on the way of the air stream), etc. That said, the actual temperature may differ from the preset one by several degrees, which does not allow creating comfortable conditions for the personnel in the operator's zone. The difference between the actual and the preset temperature is illustrated by the plot in Fig. 4.

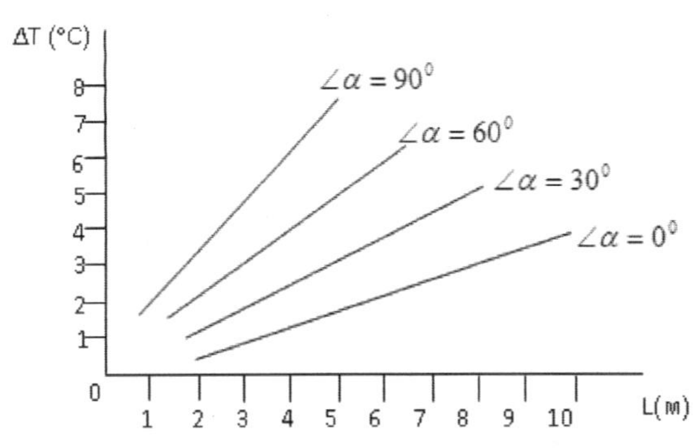

Figure 4. Changes in the difference between the actual and the preset temperature, depending on the distance and the stream angle.

This plot is valid only at a certain difference between the actual and the preset temperature. The greater is this difference, the more pronounced is the nonlinearity in the temperature rise along the plot's curves. Sometimes, it is expedient to use an RCC with a corrective thermal resistor connected by the control signal to the main thermal resistor in the internal unit. This configuration is fairly easy to implement by introducing minor modifications to the RCC because the basic RCC package has a temperature sensor for informing the operator about the ambient temperature.

This upgrade may allow improving labour conditions for people who are found in the conditioned zone and use the RCC.

5. Conclusion

The above exposed considerations are an attempt to determine an approach to solving the problem of optimizing the choice of conditioning systems for production premises. The aspects of calculating air streams inside large premises have been considered. The influence of thermal error with modifications in remote control system has been analyzed. The recommendations for upgrading remote control systems are provided.

When designing conditioning systems for large production premises, it is expedient to make a preliminary mathematical modeling by probabilistic techniques for identifying internal temperature fields and their distribution indoors. This will allow making a more optimal choice of type and capacity of conditioning systems being designed.

References

[1] Burtsev S I and Tsvetkov Yu I 2002 Thermal and Gas Comfort, Taking Account of Individual Human Features [Teplovoy i gazovyy komfort s uchetom individual'nykh osobennostey cheloveka]. *Energy efficient thermal technologies (Teploenergoeffektivnye tekhnologii)*

[2] Brukh S L 2002 Probabilistic Technique of Selecting Design Internal Air Temperature for Designing Multilink Conditioning Systems [Veroyatnostnyy metod vybora raschetnoy temperatury vnutrennego vozdukha pri proyektirovanii mnogozvennykh sistem konditsionirovaniya]. *Arctic SNiP (Arkticheskii SNiP)* (No. 4)

[3] Kiryukha V V 2001 Controlling Temperature Parameters in Closed Heterogeneous Environments [Kontrol' temperaturnykh parametrov v neodnorodnykh zamknutykh sredakh] *Academic papers of Far Eastern State Technical Fisheries University (Nauchnyye trudy Dal'rybvtuza)* (Issue 14)

[4] Kiryukha V V and Mazur N F Optimizing Air Temperature Parameters in Large Production Premises [Optimizatsiya temperaturnykh parametrov vozdushnoy sredy proizvodstvennykh pomeshcheniy bol'shoy ploshchadi] *in Collection of Papers of the Conference Primorskie Zori-2003 of Pacific Academy of Ecology, Man, and Nature Protection Sciences (TANEB), branch of International Academy of Ecology, Man and Nature Protection Sciences (MANEB)* (Vladivostok)

[5] Kiryukha V V and Mazur N F 2004 Aspects of Aerial Environment Control [Voprosy kontrolya i upravleniya kachestvom vozdushnoy sredy] *Belye Nochi: Saint-Petersburg*

[6] Kiryukha V V and Mazur N F 2005 Optimizing Selection of Coolants for Conditioning Systems, Taking Account of Environmental Safety Requirements [Optimizatsiya vybora khladagenta v sistemakh konditsionirovaniya s uchetom trebovaniy ekologicheskoy bezopasnosti] *in Collection of Papers of the Conference Primorskie Zori-2005 of Pacific Academy of Ecology, Man, and Nature Protection Sciences (TANEB), branch of International Academy of Ecology, Man and Nature Protection Sciences (MANEB)* (Vladivostok)

[7] Kiryukha V V and Mazur N F 2007 Choosing Filters for Eliminating Molecular Pollution from Air in Conditioning Systems [Vybor fil'trov dlya ochistki vozdukha ot molekulyarnogo zagryazneniya v sistemakh konditsionirovaniya] *in Collection of Papers of the Conference Primorskie Zori-2007 of Pacific Academy of Ecology, Man, and Nature Protection Sciences*

(TANEB), branch of International Academy of Ecology, Man and Nature Protection Sciences (MANEB) (Vladivostok)

[8] Kiryukha V V and Mazur N F 2009 Improving Labour Conditions and Environmental Safety in Air Conditioning System Repairs [Uluchsheniye usloviy truda i ekologicheskoy bezopasnosti pri remonte sistem konditsionirovaniya vozdukha] *in Collection of Papers of the Conference Primorskie Zori-2009 of Pacific Academy of Ecology, Man, and Nature Protection Sciences (TANEB), branch of International Academy of Ecology, Man and Nature Protection Sciences (MANEB)* (Vladivostok)

[9] Zemlyanaya N V and Kiryukha V V 2012 Distribution of Air Streams in Multizonal Air Conditioning [Raspredeleniye vozdushnykh potokov pri mul'tizonal'nom konditsionirovanii vozdukha] *in Collection of Papers of International Scientific and Technical Conference "Urgent Problems of the World Ocean Biological Resources Development"* (Vladivostok)

[10] Kiryukha V V 2013 Analyzing Technical Aspects of Temperature Control for Developing Air Conditioning Systems of Various Area at Fish-Processing Enterprises [Analiz tekhnicheskikh voprosov kontrolya temperaturnykh parametrov pri razrabotke sistem konditsionirovaniya vozdukha dlya pomeshcheniy razlichnoy ploshchadi na ryboobrabatyvayushchikh predpriyatiyakh] *Academic papers of Far Eastern State Technical Fisheries University (Nauchnyye trudy Dal'rybvtuza)* (Vol, 28)

[11] Venco S 2016 New Engineering Solutions for Conditioning Large Premises *Heating, Ventilation, Air-Conditioning, Heat Supply and Building Thermal Physics (AVOK)* (No 2)

[12] Kiryukha V V 2009 Air Medium Temperature Optimization for Choosing and Using Conditioning Systems at Large Production Premises [Optimizatsiya temperaturnykh parametrov vozdushnoy sredy pri vybore i ekspluatatsii sistem konditsionirovaniya proizvodstvennykh pomeshcheniy bol'shoy ploshchadi] *Academic papers of Far Eastern State Technical Fisheries University (Nauchnyye trudy Dal'rybvtuza)* (Vol 21)

[13] Balandina L Ya 2006 Peculiarities of Distributing Warm Air in Premises with Air Heating Systems [Osobennosti razdachi teplogo vozdukha v pomeshcheniyakh s sistemami vozdushnogo otopleniya]. *Arctic SNiP (Arkticheskii SNiP)* (No 19) pp 10-14

[14] Balandina L Ya and Moshkova L P 2005 Modern Aspects of Using Air Dispensers in Building and Renovating Premises in Various States [Osobennosti primeneniya vozdukhoraspredeliteley pri sovremennom stroitel'stve rekonstruktsii pomeshcheniy razlichnogo sostoyaniya] *Arctic SNiP (Arkticheskii SNiP)* (No. 18) pp 10-14

[15] Apigkhin A G 2004 Conditions for Thermal Dynamic Optimization of Heat Transfer by Intercoling Systems [Usloviya termodinamicheskoy optimizatsii peredachi teploty sistemami s promezhutochnym teploobmenom] *Arctic SNiP (Arkticheskii SNiP)* (No 16) pp 21-24

[16] Sotnikov A 2005 Modern SIV and VS for Buildings as Design Facilities for Construction and Operation [Sovremennyye SIV i SV zdaniy kak ob"yekty proyektirovaniya stroitel'stva i ekspluatatsii] *Civil Engineering (Stroitel'naya inzheneriya)* (No 10) pp 11-14

[17] Kiryukha V V 2006 Temperature Measurement [Izmereniye temperatury] *Ch. P in Instrument Transducers in Automation Systems. Study Guide [Izmeritel'nyye preobrazovateli v sistemakh avtomatiki. Uchebnoye posobiye]* (Vladivostok) pp 216-235

[18] Podobed M Yu and Soroka V V 2018 Adaptive Management of Air Conditioning System by Generalized Temperature Mode Criterion [Adaptivnoye upravleniye sistemoy konditsionirovaniya vozdukha po obobshchayushchemu kriteriyu temperaturnogo rezhima] *Proceedings of BSTU Series 3* (No 1) pp 67-71

IOP Conf. Series: Materials Science and Engineering **986** (2020) 012045 doi:10.1088/1757-899X/986/1/012045

Determination of the velocity coefficient of a turbine nozzle diaphragm with partial blading of the runner

A. A. Kryukov[1], G. L. Rakov[2], S. V. Chekhranov[1], R.R. Simashov[1]

[1] Far Eastern State Technical Fisheries University, 690087, Vladivostok, ul. Lugovaya 52b, aleksey902@mail.ru
[2] Peter the Great St.Petersburg Polytechnic University, St.Petersburg, Russia

Abstract. A method for determining the velocity coefficient of a nozzle diaphragm for a turbine with partial blading of the runner using the ANSYS software package is presented in the paper. The velocity coefficient of the nozzle diaphragm along the active arc length is distributed and plotted. The averaged values of the velocity coefficient of the nozzle diaphragm of the turbine with partial blading of the runner are determined. The dependences of the speed coefficients of the turbine nozzle diaphragm with partial blading of the runner in the range of πt 1,5÷2,5 and the degree of partiality in the range of 0,059÷1 are presented. The approximation resulted in an empirical dependence that takes into account the influence of the degree of partiality and the Mach number on velocity coefficients of the nozzle diaphragm with partial blading of the runner. The empirical dependence φ=f(M1t; ε) obtained in this work is essential for modeling variable modes and multi-mode optimization of low-flow turbines.

1. Introduction

Low-flow turbines (LFT) are reliable, productive, small-sized drives of various aggregates and devices. The experience of production and widespread usage of low-flow turbine stages determines the need to improve efficiency by simplifying and reducing the cost of manufacturing technology for blades and the stage as a whole. Improving the efficiency of low-flow turbines requires solving the problem of aerodynamic improvement of the flow part and calculating the optimal geometry and operating modes of low-flow turbines.

Today, theoretical and experimental material has been accumulated that allows us to summarize and formulate the general principles of the approach to the design of partial turbines. Turbines with partial blading of the runner occupy a special place. This turbine has a number of properties not previously seen in other LFT designs. In this regard, there is a need to study the processes occurring in the flow part of partial LFT with partial blading of the runner [1].

2. The goal of the work

The work's purpose is to determine the velocity coefficient of the turbine nozzle diaphragm with partial blading of the runner by numerical simulation using the ANSYS CFX software package [2]. This will allow to consider each nozzle passage and determine its gas-dynamic characteristics. The total velocity coefficient of the nozzle diaphragm averaged over the arc is possible to determine basing on the determination of the coefficient φ in each nozzle passage.

The influence of the extreme channels does not allow to properly determine the value of the total velocity coefficient of the nozzle diaphragm by the usual calculation method due to a large number of influencing parameters and factors. Therefore, it is viable to create a model for determining the coefficient φ by approximating the experimental data obtained on a simulation bench using the ANSYS CFX software package. The empirical dependence should take into account the influence of the degree of partiality and the Mach number, and the possibility of its using in modeling in multi-mode optimization of low-flow turbines.

3. Formulation of the problem

A feature of the low-flow turbine with partial blading of the runner shown in Figure 1 is the design of the runner. When the nozzle diaphragm with blades distributed along the arc of 360o is presented, the blades are located only on a part of the circle of the runner, forming an active arc, and the solid body of the metal disk occupies the inactive arc.

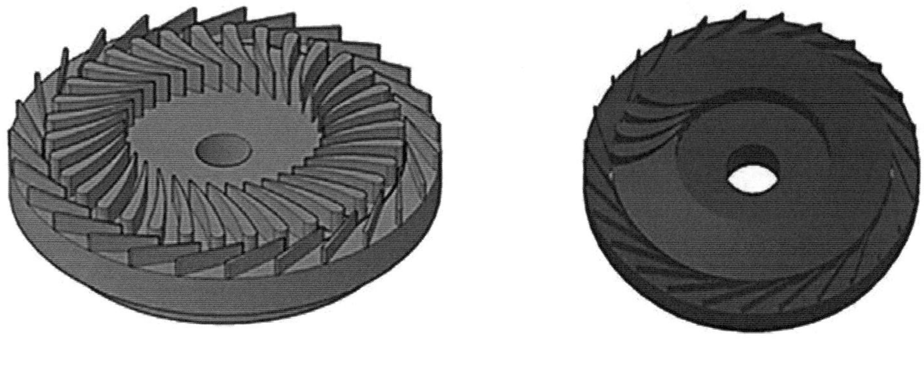

Figure 1. Examples of centripetal turbines: all-wheel drive and partial runner blading.

In such a stage, the working fluid is supplied to the nozzle passages around the entire circumference of the nozzle device (ND), as in conventional turbines with a full admission. The flow is similar to the flow in the nozzles of traditional turbines in the nozzle passages located opposite the active arc of the runner. Moreover, the flow is limited by the minimum dimensions of axial and radial clearances in nozzles located opposite the solid disk of the runner. Consequently, the movement of the working fluid in these nozzles will be determined by its flow rate and the direction of leakage through the gaps. The more commensurable the area of the nozzle exit section and the transverse annular gap between the crowns, the more intensive the flow leaving the nozzles in the gaps or eject an inactive medium from them. Clearances provide constant working fluid leakage in the inactive arc runner. These circumstances should be taken into account when assessing the value of the suction of the working medium from inactive to the active arc region.

The following initial parameters are known in a three-dimensional gas-dynamic calculation [3]: working fluid pressure (air) at the inlet to the nozzle diaphragm P_0*= 0,15 ÷ 0,25 MPa, working fluid temperature at the inlet to the nozzle T_0*=293 K, the pressure of the working fluid at the exit of the wheel P_2=0,1 MPa, the gas constant of the working fluid R=287 J/kg ·K, the isentropic index of the working fluid k=1,4, the rotational speed of the runner 30 000 min^{-1}. The diameter of the runner D=50 mm, the degree of radiality - 0, 55, the nozzle diaphragm has 27 blades. The main geometric parameters: the angle of exit of the flow from the nozzle diaphragm α_1=16,31°, the angle of entry of the flow into the runner β_1=90°, the angle of exit of the flow from the runner β_2= 42°. The most suitable turbulence model for the selected model is SST (Shear Stress Transport). The SST model is a set of turbulence models: the free flow is calculated by the k-epsilon equation and in the region near the walls by the k-omega equation.

The calculation model is a combination of flowing parts of the nozzle diaphragm, runner, radial and axial clearances with labyrinth seals. The runner has a different number of blades, depending on the degree of partiality - one, three, six, thirteen, and thirty-four blades. The computational grid generated for the turbine stage under study complies with all the main recommendations for ensuring the required reliability of the calculation results [4]. A thickening of the grid was created near the streamlined surfaces. This is necessary to increase the quality of the calculation of parameters in the boundary layer. The solution method is the Transient Rotor Stator unsteady task.

4. Results

Three-dimensional calculations of turbine stages with varying degrees of partiality made it possible to determine velocity fields at rotation frequencies of 30,000 min^{-1}. In order to average the velocity coefficient of the nozzle diaphragm φ, 10 positions of the runner were examined by the angle of rotation relative to the nozzle diaphragm when the wheel rotates counterclockwise to the left, Fig. 1. The value of the total velocity coefficient of the nozzle diaphragm φ at a specified time moment is determined by the averaged value of the velocity coefficients in each nozzle passage of the active arc. Averaging the total velocity coefficient of the nozzle diaphragm over 10 characteristic time instants determines the velocity coefficient of the nozzle diaphragm φ. Table 1 shows the values of the velocity coefficients of the nozzle passages and the nozzle diaphragm of the turbine with six blades on the runner. Parameters were averaged using the areaAve method (over the area). When the extreme left channel is opened up to 50%, five channels of the nozzle device are involved in the active arc. As one opens the extreme left channel, the right one will close. The value of the velocity coefficient of the middle nozzle passage is relatively constant. The value of the velocity coefficients of the extreme nozzle passages varies depending on the degree of openness. The total velocity coefficients of the nozzle diaphragm are determined at each time moment.

Table 1. The main results of a three-dimensional gas-dynamic calculation with six blades on the runner.

Nozzle passage number, N		The degree of openness of the extreme left channel of the active arc, F									
		10%	20%	30%	40%	50%	60%	70%	80%	90%	100%
1	P_1, MPa	0,13	0,13	0,13	0,13	0,13	0,13	0,13	0,13	0,13	0,13
	c_1, m/s	58,9	57,23	56,63	58,83	67,13	85,06	107,65	130,05	184,26	193,02
	φ	0,224	0,217	0,216	0,225	0,257	0,326	0,412	0,498	0,706	0,733
	ζ	0,950	0,953	0,953	0,950	0,934	0,894	0,830	0,752	0,502	0,463
2	P_1, MPa	0,13	0,13	0,13	0,13	0,13	0,13	0,13	0,13	0,13	0,13
	c_1, m/s	209,2	225,86	238,39	242,45	237,86	232,35	223,88	221,07	229,83	237,48
	φ	0,795	0,858	0,909	0,926	0,861	0,890	0,857	0,847	0,880	0,902
	ζ	0,368	0,264	0,173	0,143	0,259	0,208	0,265	0,283	0,225	0,186
3	P_1, MPa	0,13	0,13	0,13	0,13	0,13	0,13	0,13	0,13	0,13	0,13
	c_1, m/s	237,7	228,25	220,31	218,66	215,59	218,21	231,1	245,48	238,38	233,81
	φ	0,903	0,867	0,840	0,835	0,826	0,836	0,885	0,940	0,913	0,888
	ζ	0,185	0,248	0,294	0,303	0,318	0,301	0,217	0,116	0,166	0,211
4	P_1, MPa	0,13	0,13	0,13	0,13	0,13	0,13	0,13	0,13	0,13	0,13
	c_1, m/s	231,4	232,12	225,86	225,90	238,12	250,10	250,34	240,55	231,83	234,95
	φ	0,879	0,882	0,861	0,862	0,912	0,958	0,959	0,921	0,888	0,892
	ζ	0,227	0,223	0,258	0,256	0,168	0,082	0,081	0,151	0,212	0,204
5	P_1, MPa	0,13	0,13	0,13	0,13	0,13	0,13	0,13	0,13	0,13	0,13
	c_1, m/s	227,3	225,74	237,68	249,17	249,76	241,04	232	226,49	223,49	221,83
	φ	0,864	0,857	0,907	0,951	0,957	0,923	0,889	0,867	0,856	0,850
	ζ	0,254	0,265	0,178	0,095	0,085	0,148	0,210	0,247	0,267	0,278
6	P_1, MPa	0,13	0,13	0,13	0,13	0,13	0,13	0,13	0,13	0,13	0,13
	c_1, m/s	228,7	238,49	239,35	232,24	221,82	206,86	210,45	213,72	216,82	224,57
	φ	0,869	0,906	0,913	0,887	0,850	0,792	0,806	0,819	0,830	0,853
	ζ	0,245	0,179	0,167	0,214	0,278	0,372	0,350	0,330	0,310	0,272
7	P_1, MPa	0,13	0,13	0,13	0,13	0,13					
	c_1, m/s	230,9	234,84	231,45	217,71	195,26					
	φ	0,877	0,892	0,883	0,831	0,7478					
	ζ	0,230	0,204	0,221	0,309	0,4406					

Figures 2–5 show graphs of the breakdown of the velocity coefficient of the nozzle diaphragm along the length of the active arc of the turbine with partial blading of the runner. The graph is divided into zones: the extreme left channel, the core of the flow and the extreme right channel. The red graph shows the value of the coefficient φ in the nozzle passage. The value of the coefficient in the flow core is relatively constant and varies depending on the position of the inlet lump of the runner blades relative to the nozzle passages. As the extreme left channel opens, the coefficient φ increases reaching a maximum at 100% opening. When the extreme right channel is closed (narrowing the passage section of the channel) to the size of the nozzle passage neck a_2, the coefficient φ first increases. When the extreme right channel is closed further, the coefficient φ decreases, since the expanded gas in the oblique section begins to pass through a narrower section than the neck of the nozzle passage a_2 that leads to a decrease in speed. The green line shows the average value of the total velocity coefficient of the nozzle diaphragm along the arc. The characteristic features in the extreme nozzle passages are the same in all stages of the turbine with partial runner blading. The difference in the geometry of the curve of the extreme right channel indicates a different relationship between the degree of openness of the extreme left and right channels with different runner blading.
Figures 6 ... 8 show the velocity fields of turbine stages with one, three blades and a three-dimensional model of a turbine stage with current lines.

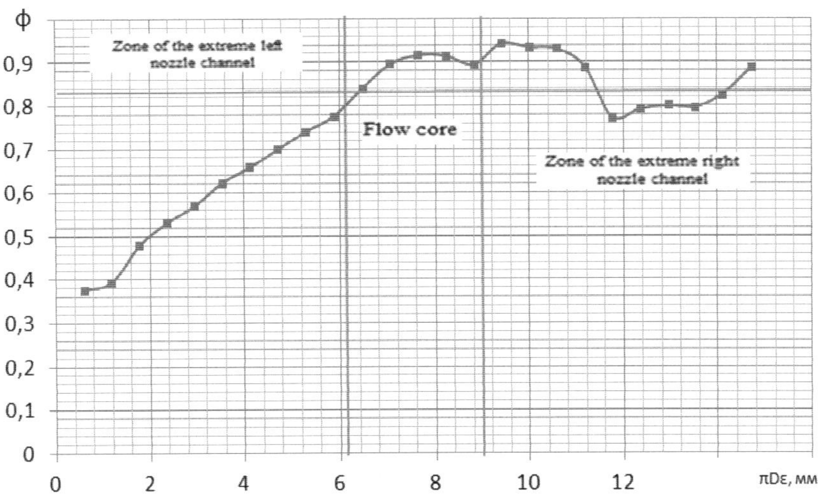

Figure 2. Dependence of the velocity coefficient of the nozzle diaphragm along the length of the active arc of the turbine with partial blading of the runner with one blade on the runner.

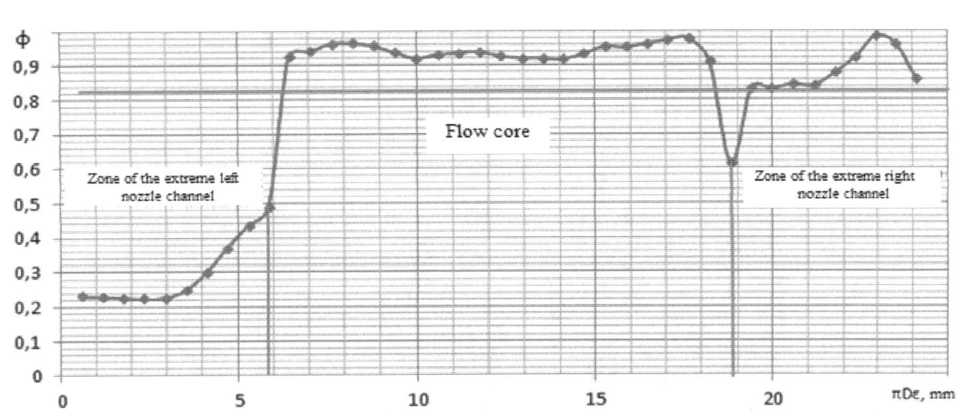

Figure 3. Dependence of the velocity coefficient of the nozzle diaphragm along the length of the active arc of the turbine with partial blading of the runner with three blades on the runner.

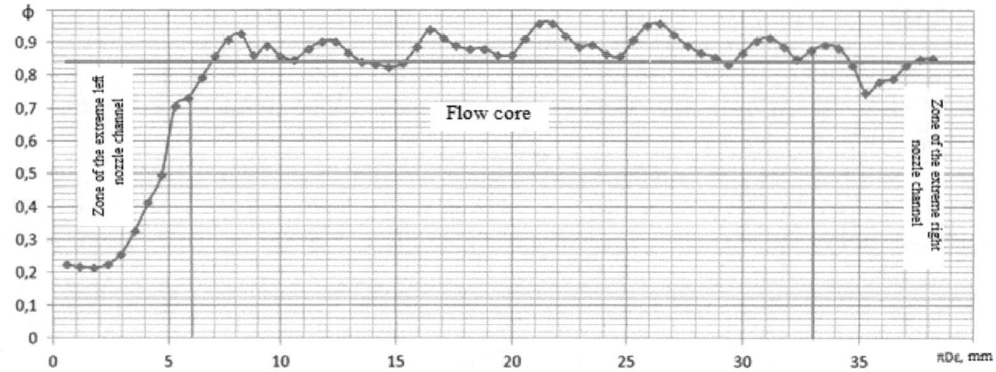

Figure 4. Dependence of the velocity coefficient of the runner along the length of the active arc of the turbine with partial blading of the runner with six blades on the runner.

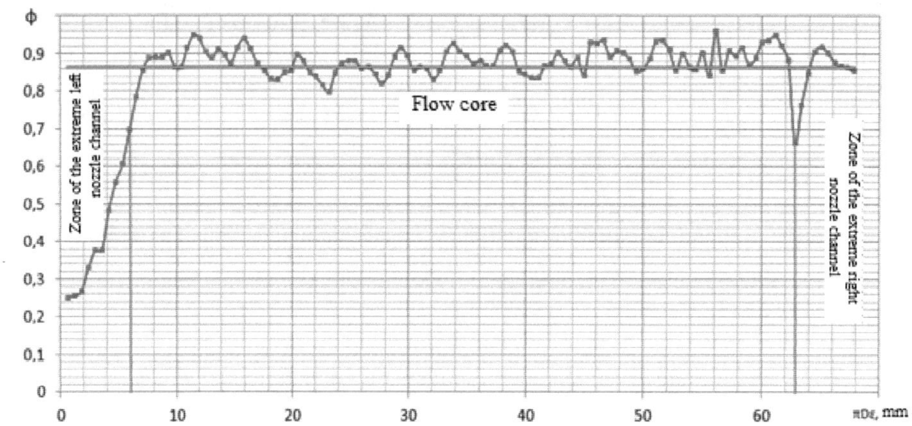

Figure 5. Dependence of the velocity coefficient of the runner along the length of the active arc of the turbine with partial blading of the runner with thirteen blades on the runner.

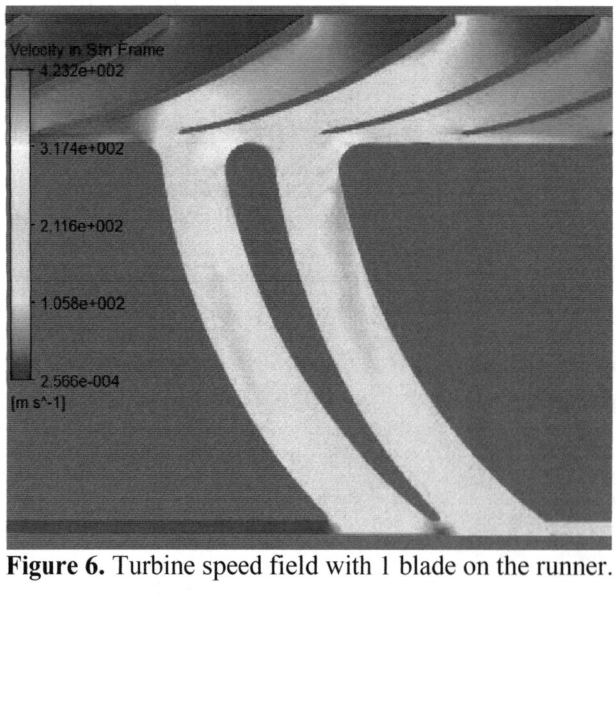

Figure 6. Turbine speed field with 1 blade on the runner.

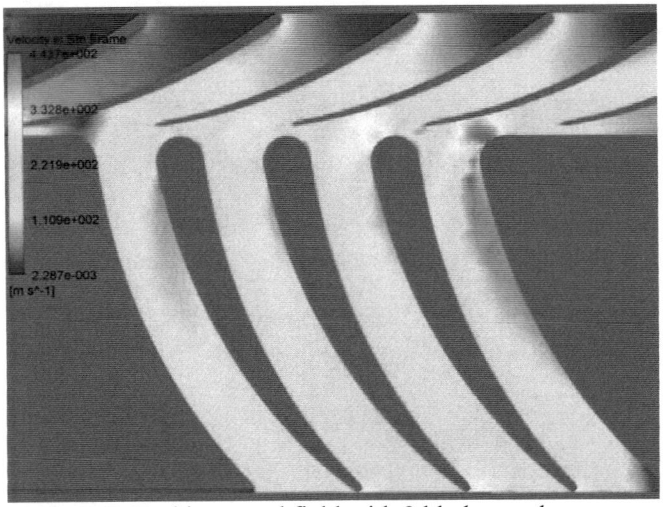

Figure 7. Turbine speed field with 3 blades on the runner.

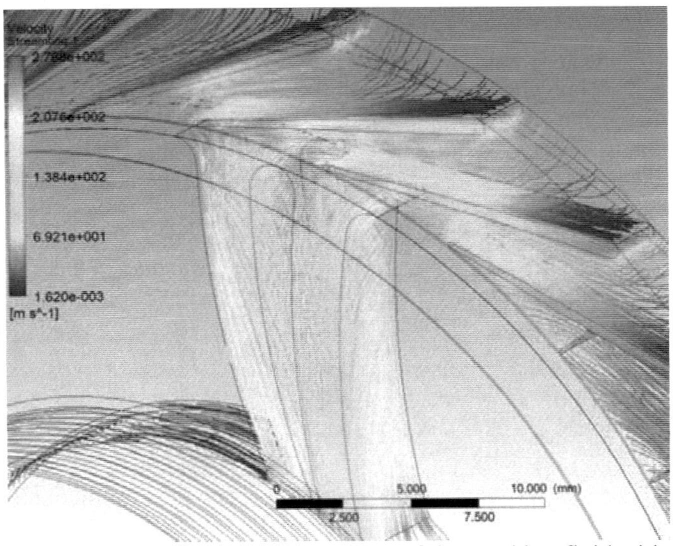

Figure 8. Three-dimensional model of the flow of the working fluid with Streamline.

The value of the velocity coefficient of the nozzle diaphragm is also determined by the variation range of π t from 1.5 to 2.5 to determine the dependence on the Mach number. Figure 9 presents graphs of the dependence of the velocity coefficient of the nozzle diaphragm on the Mach number for various degrees of partiality. The value of the velocity coefficient of the nozzle diaphragm grows after increasing in the degree of partiality. The maximum value of the nozzle diaphragm velocity coefficient is achieved in the range of the Mach number 0.55 ÷ 0.60. In this area, the optimal value is located, where the value of profile losses is minimal. This is characterized by the presence of local supersonic zones on the convex surface of the profile and shock waves. [5].

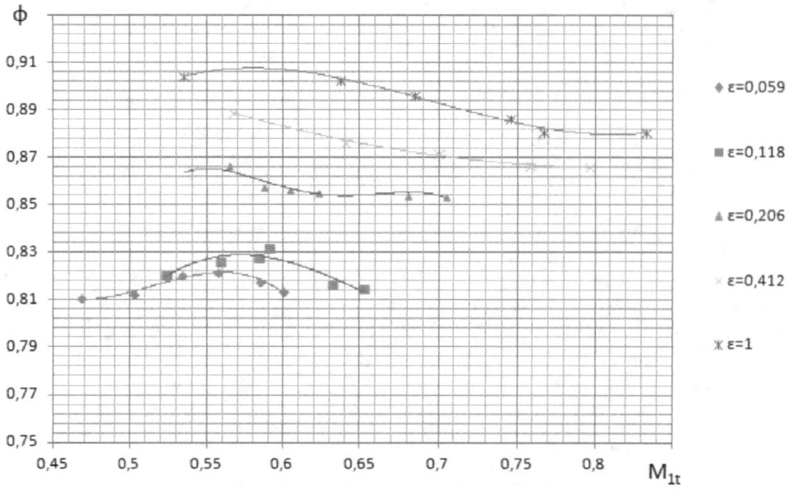

Figure 9. Dependence of the velocity coefficient of the nozzle diaphragm on the Mach number and the degree of partiality.

The mathematical model for determining the coefficient φ depending on the main determining factors ε and M_{1t} can be expressed: $\varphi = \varphi(\varepsilon, M_{1t})$. To obtain this dependence, the values of the coefficient φ are used in the range of π_t from 1,5 to 2,5 during the numerical experiment.

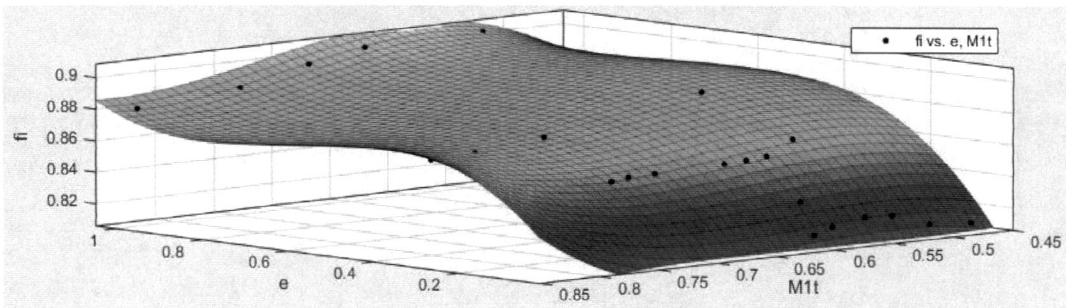

Figure 10. Dependence of the nozzle diaphragm velocity coefficient on the Mach number and degree of partiality.

The two-parameter dependence $\varphi = \varphi(\varepsilon, M_{1t})$ shown in Figure 10 is approximated by a cubic polynomial:

$$\varphi = \varphi(\varepsilon, M_{1t}) = 0{,}1502 + 0{,}2148 \cdot M_{1t} + 1{,}506 \cdot \varepsilon - 2{,}23 \times {M_{1t}}^2 - 2{,}035 \cdot \varepsilon \cdot M_{1t} - 1{,}486 \cdot \varepsilon^2 + 0{,}6226 \cdot M_{1t}^3 + 1{,}042 \cdot M_{1t}^2 \cdot \varepsilon + +0{,}5765 \cdot M_{1t} \cdot \varepsilon^2 + 0{,}6007 \cdot \varepsilon^3 \quad (1)$$

A comparison of the data obtained during the numerical experiment with the calculation by formula (1) showed that the model presented in the work is adequate for determining the coefficient φ since the discrepancy between the calculation and the numerical experiment does not exceed 2.5% in a wide range of the degree of partiality and mode parameter. Therefore, there is every reason to consider the created two-parameter dependence $\varphi = \varphi(\varepsilon, M_{1t})$ adequate to recommend it for practical use in modeling variable modes and multi-mode optimization of low-speed turbines.

5. Conclusion

Analysis of the results of three-dimensional modeling leads to the following conclusions:

- to determine the total velocity coefficient of the nozzle diaphragm, it is necessary to take into account the values of the velocity coefficient of each nozzle channel, as well as the influence of the extreme channels;

IOP Conf. Series: Materials Science and Engineering **986** (2020) 012045 doi:10.1088/1757-899X/986/1/012045

- averaging of the total velocity coefficient of the nozzle diaphragm at various degrees of partiality has to be carried out by the characteristic positions of the runner by the angle of rotation relative to the nozzle diaphragm;
- the obtained values of the velocity coefficient of the nozzle diaphragm at various degrees of partiality in the range of πt from 1,5 to 2,5 are approximated by a cubic polynomial (1);
- A comparison of experimental data with the calculation according to formula (1) showed that the difference lies within the experimental error and does not exceed its value by more than 2.5%. This result can be considered satisfactory.

References

[1] Chekhranov S V 2015 An experimental study of radial turbines with partial blading of the runner *Transport business in Russia* (№ 6) pp 280-284

[2] Rakov G L, Zabelin N A, Rakov G L, Rassokhin V A, Sebelev A A and Smirnov M V 2013 Investigation of flow characteristics in low-flow turbine stages of the design of LPI *Scientific and technical statements of St. Petersburg State Polytechnical University* (№1 (166)) pp 45 – 53

[3] Kryukov A A 2018 Analysis of using the ANSYS CFX software package for a low-flow turbine model with partial runner blackening *Actual problems of the development of biological resources of the World Ocean. Materials of the V International scientific and technical conference - Vladivostok: FESTFU "Dalrybvtuz"* pp 263-267

[4] Khitrykh D 2014 Recommendations for using the ANSYS Meshing *ANSYS Advantage mesh preprocessor. Russian edition 2014* (No. 20 M.: CADFEM CIS) pp 34-43

[5] Kholshchevnikov K V Theory and calculation of aircraft blade machines (M.: Mechanical Engineering) p 614

IOP Conf. Series: Materials Science and Engineering **986** (2020) 012046 doi:10.1088/1757-899X/986/1/012046

The Actualized Complex of Indicators of Quality Monitoring of New Cars in Operation

V N Kozlovsky[1], D V Aydarov[1], M V Shakursky[1] and V A Kozhevnikov[2]

[1]Samara State Technical University, Samara, Russia, kozlovskiy-76@mail.ru
[2]National Technology Initiative Center for Advanced Manufacturing Technologies based on the Institute of Advanced Manufacturing Technologies of Peter the Great St. Petersburg Polytechnic University Polytechnicheskaya, 29, St.Petersburg, 195251, Russia

Abstract. The article devotes to the analysis and development of a modern analytical complex of quantitative indicators of the quality of new cars in operation.

1. Introduction

High-tech engineering, in general, and the automotive industry, in particular, are one of the key sectors of the global economy. It is here that a significant share of growing points is formed in the field of innovation, including in the field of competitiveness, quality, modernization, and development of quality management systems (QMS).

The analysis of the existing quality management systems (QMS) of national automakers at present, in comparison with the QMS landscapes of world leaders, showed the lack of comprehensiveness in solving analytical problems related to measuring product quality in operation, revealing key quality problems. It turns out that in practice, all of the most important tasks listed above, as before, are essentially solved in stages and separately, within the established rigid functional management system at enterprises. Therefore, it is obvious that the prospects for the development of the QMS at national car assembly plants are directly related to a comprehensive, within a single system, consideration of the tasks of monitoring, analysis, and solving problems of product quality.[1-3]

In this regard, the complexity of solving problems in the field of quality should be determined by such an important, fundamental aspect of the new management system as the possibility of a comparative analysis of the achievements in the field of engineering management of a particular automobile assembly plant with the corresponding achievements of other enterprises in the competitive market. This is exactly what can be called benchmarking in technical management.

2. Key research findings

One of the key tasks of transferring the QMS of domestic car assembly enterprises to a new level of development is the implementation of a new measuring system for assessing the quality of products at the after-sales stage of the life cycle. To solve this problem is required critical analysis and development of a solution in the field of creating a new group of indices that most adequately reflect aspects of the measurement process for quality assessment of automotive equipment in operation. In terms of relevance for all participants in the process, there are the manufacturer and consumer of the product, and from the point of view ensuring the consistency and adequacy of technical benchmarking of relevant products and processes. [4]

The analysis results of the existing activities of national automakers in the field of monitoring the quality of cars in operation highlight a number of important aspects reflecting the current practice. Considering the quantitative quality indicators and tools of the corresponding analysis used at Russian car assembly plants, it should be noted that they fit quite adequately into the traditional QMS, in which the quality assessment of their products was carried out practically without comparison with the quality of competitors' products. Moreover, this is the historical paradigm of the development of technical management at enterprises, when monopolists produced both cars and trucks. In addition, the denominator was historically the guarantee fleet, to which absolute quality indicators of cars in operation were given to determine the given values. In general, such a compromise option was justified until some time. [5-7]

Let us consider the problematic aspects of monitoring the quality of vehicles in operation based on the above aspect of the warranty fleet, using an example (Fig. 1). [8]

a)

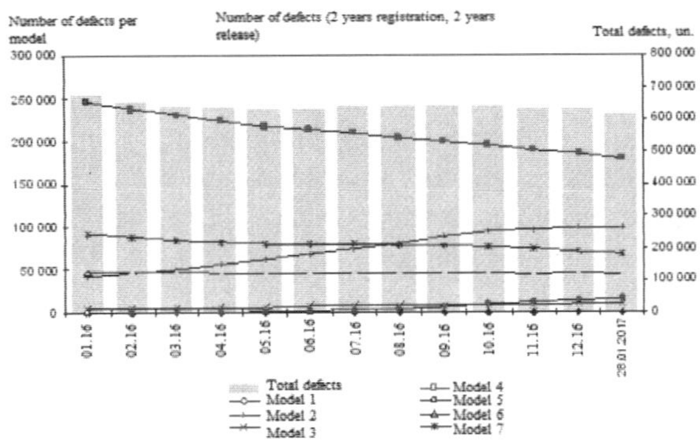

b)

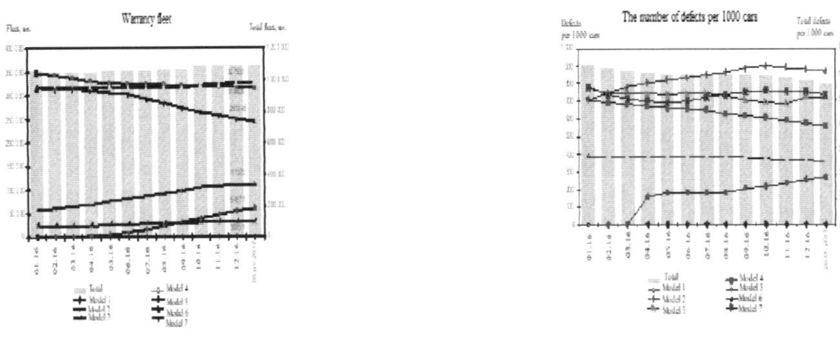

c)

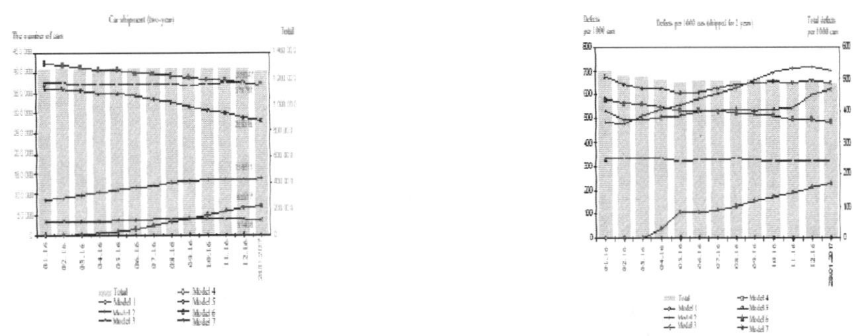

d)

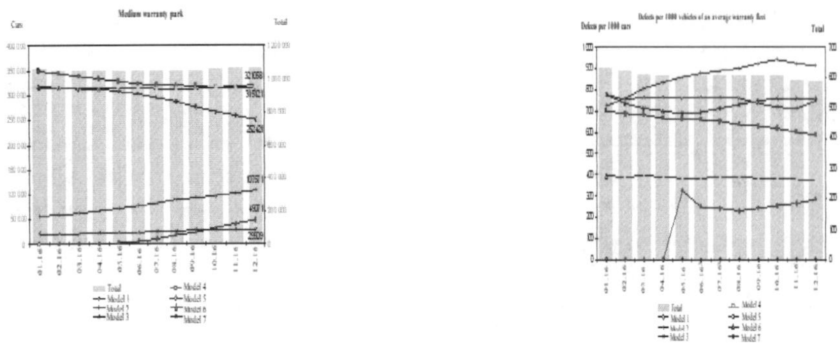

Figure 1. The specified defect level for various methods of calculating the volume of the warranty fleet of cars.

Figure 1a shows the curves reflecting the total (for the entire model range) and private (for each model) number of defects for vehicles of one of the leading national automakers according to the production date, taking into account the two-year operation (warranty period). Figures 1b - d (right) show curves reflecting the change in the car warranty fleet calculated by specific methods in the same analyzed period: b - by the number of cars registered in the specialized information system of car dealers; c - by the number of vehicles shipped within two years; d - the average warranty fleet, determined taking into account the date of release of cars, and a two-year stay in the park. Figures 1 b - d (left) present curves showing the corresponding reduced to the volume of the warranty fleet (according to the methods under consideration), the level of defective cars in the warranty period of operation. Corresponding recounts can also be made for cost indices for the elimination of defects in automotive equipment in operation.
The data analysis (Figure 1) shows that there are a number of methods for assessing the level of the guarantee fleet that are not correlated with each other. Because of the above variety of unrelated methods for calculating the warranty fleet, it turns out that the use of the amount of the warranty fleet, as the denominator does not create the prerequisites for obtaining a reliable and adequate result of analyzing the quality of cars in operation. [9-11]
The next aspect, which logically fits into the justification block for the need to develop a new group of quality indicators, is the modern requirement of international standards. For example, ISO 9001: 2015 is related to the analysis of changes in defectiveness indicators both in the warranty and in the after-warranty period of automotive equipment operation that is exactly impossible to do based on the ideology and understanding of the performance of the guarantee fleet. At present, an aspect should be highlighted that reflects the level of stability in sales of automotive equipment. Under a smooth change in the warranty fleet, the corresponding level of defectiveness behaves adequately. However, in the case of sharp changes in sales, the given indicator begins to behave unpredictably. For example, production volumes are sharply reduced under crisis conditions, a paradoxical situation arises that is associated with an unreasonably high level of product quality that is monitored. The denominator (warranty fleet) behaves more statically because of a significant decline in the level of defectiveness in new cars due to a decrease in output and sales (numerator). Finally, two final aspects of the justification block for the need to solve the problem under consideration are presented. This is an aspect of the reliability of the car warranty base. The problem is the lack of technical ability to fully account for all cars that are in the warranty period of operation due to there are entire systems of corporate clients in which, for example, the cars can be preserved for a rather long time. At the same time, it is natural that in dealer systems they do not go through accounting during repair work, and in general corporate accounting, cars that are under warranty at the car assembly plant, they are usually presented. [12-15]
The last aspect of the justification block is the relevance of the used indices for monitoring the quality of products in operation based on the need to solve the technical benchmarking problem. Let us consider the study results of consumer satisfaction indexes for car quality in operation (Fig. 2). [16]

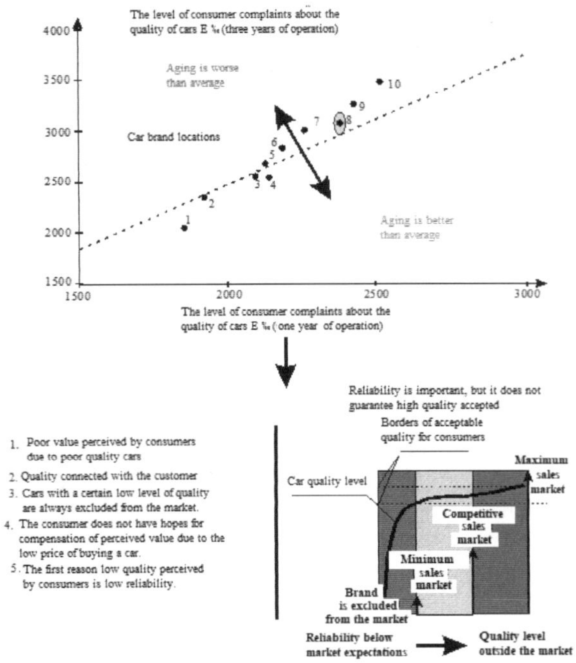

Figure 2. Consumer automotive reliability assessment.

The upper diagram (Figure 2) presents an assessment of the dependence of consumer complaints in the first year and the first three years of operation, reduced to 1000 (E ‰), for ten well-known automotive brands operating in the Russian market. On the basis of averaging indicators, a trend line was obtained dividing the plane into two areas: “aging is worse than average” and “aging is better than average”. The analysis of the diagram clearly defines the consumer’s preferable level of defectiveness for the first four brands working in the field of “aging is better than average”. It is obvious (lower part of Figure 2) that reliability is the foundation of an overall quality assessment. High reliability is important to consumers. The consumer understands by a high level of reliability the absence of problems that disable automotive equipment or significantly reduce the safety of its operation. In practice, such problems are called blocking, since they block the possibility of full use of products. At the same time, the deterioration in reliability is the first of the reasons for the low value of product quality perceived by consumers and leads to the exclusion of the brand from the competitive market. At the same time, a high level of reliability is not a guarantee of a high consumer assessment of product satisfaction. [17]

3. Results

Comprehensive results based on the analysis are presented in Figure 3, and they consist of the need to develop and implement the whole system of indexes for monitoring the quality of cars in operation that adequately reflect consumer opinion on product quality.

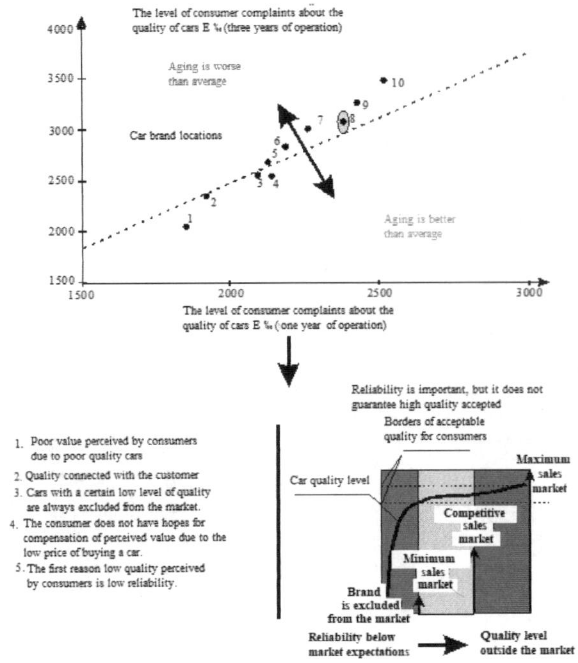

Figure 3. Comprehensive findings on consumer preferences in the field of car quality.

Based on the analysis of the complex conclusions (Figure 3), it is quite obvious that the system of indicators of car quality should reflect the level of defectiveness, reliability, operating costs, and the level of consumer complaints. It is also obvious that the solution to the problem of ensuring a comparative assessment of the quality indicators of own products and competitors' products should be based on using the methodology of international practice for monitoring car quality, thereby approaching advanced quality measurement standards. Prerequisites are created to increase the level of adequacy of the relevant processes of national automakers to modern industry requirements. Moreover, the reason is quite simple. How the indicators presented above can be used when benchmarking quality with cars of other manufacturers. This is difficult to do because there is no work in the system of indicators already generally accepted in the global automotive industry. Today, in order to compare the level of quality of own products with the quality of competitors, it is necessary to do complex autonomous recounting, since in corporate specialized information systems the indicators of a new monitoring level are not standardized and put into practice. That is why new indicators are needed that are well broadcast in relation to the indicators of the leaders in the global automotive industry. However, it is also obvious that the implementation of a system of new quality indicators for automotive equipment in operation should be carried out taking into account existing industry practice and some specifics of the domestic auto industry.

The goal of this study is to improve the process of monitoring the quality of cars at the after-sales stage of the life cycle by developing and improving the methods and tools of the system for monitoring the quality of cars in use.

The objectives of the study cover the following issues: development of a system of indicators for assessing the quality of automobiles in operation, taking into account international practice of the monitoring process, and industry-specific processes and products of national automakers.

Figure 4 presents the developed scientific and practical concept of the process of monitoring the quality of cars in operation in the form of a pyramid, adequately meeting today's challenges. Based on the obtained complex conclusions (Figure 3), the solution to the problem of monitoring quality should be carried out according to the main levels of the pyramid: basic quality; durability, and attributes of product quality; inconsistencies with consumer expectations; brand identification in terms of quality. Moreover, as mentioned above, ensuring a high level of reliability is very important, but it does not guarantee a high level of assessment of the quality of automobiles perceived by consumers (customer satisfaction).

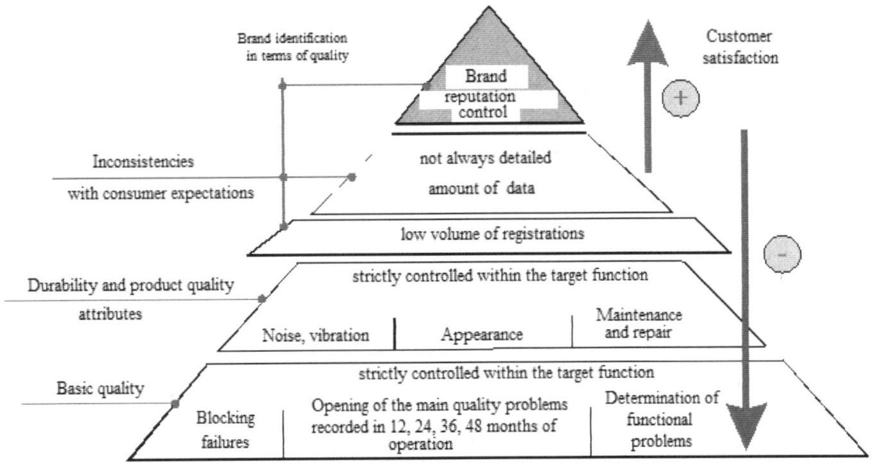

Figure 4. Scientific and practical concept of the process of monitoring the quality of cars in operation.

Turning to the international practice of monitoring the quality of vehicles in operation, the quantitative index GMF (guaranty method factory) is necessary to be highlighted, which is currently a kind of reference point for organizing the methodology of the corresponding measurement process for many enterprises in the automotive industry, including companies such as GM, Renault, Nissan, etc. This index is the level of defective cars in operation by month, group of months of operation (K ‰), per 1000 cars:

$$K_i \text{‰} = \frac{\sum_{i=1}^{M} Q_j}{N} \times 1000, \qquad (1)$$

where Q_j — number of defects, breakdowns, failures for each of the j- th system, cars in general, for i months of operation; N — the total number of cars having the corresponding (analyzed) date of production (release) and i – months of operation.

The international index is not burdened by the base of the warranty fleet and is calculated only on the basis of the number of defects on vehicles with a specific release date and having a certain operating limit (3, 6, 12, 24, 48 months). An important aspect of putting the system of quantitative indicators for monitoring vehicle quality into the operation of domestic enterprises to the ideology of the K ‰ index is that international marketing institutions, such as ADAC, JD Power, Ipsos, MICA, etc., that conduct market research in the field of quality, also use appropriate methodology. Currently, taking into account the needs of the car industry in such research, this aspect is very important to ensure the sustainability of the process of strategic positioning of car brands in a competitive market.

Considering domestic work practices, several important problems are highlighted that impede the direct application of the GMF indicator.

Firstly, the principle of pushing out a production facility still works in domestic production systems. Western competitors have the principle of pulling, that is, custom-made assembly of cars. If the GMF indicator is directly applicable, then the date of the cars' release is needed to consider in the database. In this case, this cannot be done, as there will be a significant distortion of the result, due to a significant amount of production can be stored for some time on the sites before the sale (Figure 5).

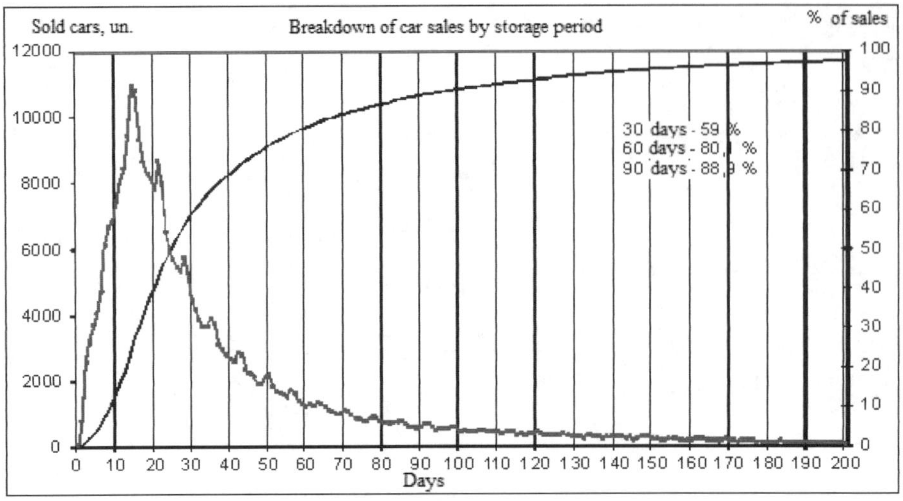

Figure 5. Breakdown of car sales.

It turns out that for justification of the indicator, the calculation base needs to be complicated by translating the quality assessment from the date of production to the date of cars' sale, that is, and the calculation needs to be carried out only for sold cars at the studied production date. At the same time, a certain time limit will be fixed determining the acceptability or admissibility of the quality assessment. The quality assessment of cars, for example, after the first month of operation, with an adequacy level of 80%, is obtained only after 60 days from the date of release of the cars based on the data obtained on car sales (Figure 5). The adequacy of the indicator corresponding to the level 90% is reached only after 90 days from the date of issue. This aspect is well traced using Figure 6.

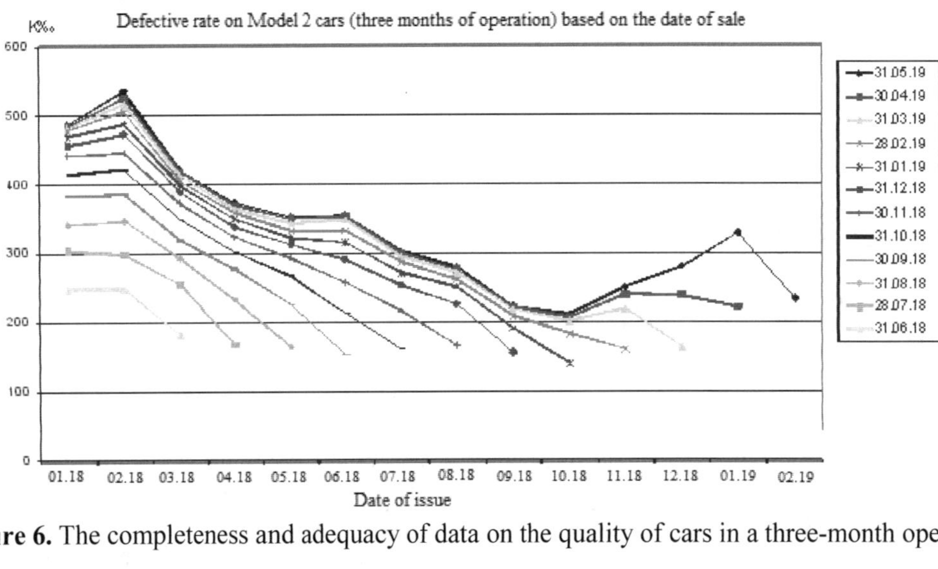

Figure 6. The completeness and adequacy of data on the quality of cars in a three-month operation.

Figure 6 presents the results of a series of recalculations of the quality indicator K ‰ for the vehicles of the product in question (Model 2), one of the largest automakers in Russia. The calculation was carried out in dynamics, starting from the first point of June 31, 2018, for cars produced in January 2018. The process of data accumulation, to the required completeness, is traced by the points of conversion of the considered quality indicator, performed during the next 11 reporting periods (months of analysis). The graphs show that for a three-month operation, the achievement of the level of adequacy of the indicator K ‰ corresponding to 90% can be obtained only after 6 months of data analysis. Approaching 100% adequacy K ‰ can be obtained after 12 months of continuous analysis of vehicle quality.

For competitors, "Western" car assembly plants, issues of product quality assessment become relevant after at least 3 months of operation. For example, Renault company targets are built on a period of at least 6 months of operation, while Nissan has at least 12, and preferably 24 months of operation. Due to the specifics, it is very important to provide quality analysis in the first weeks of operation.

Based on the foregoing, it turns out that a promising basic quality indicator that provides a solution to a number of important tasks in the existing system for monitoring the quality of automotive products have a disadvantage associated with an insufficient level of efficiency. This is unacceptable for the practice of domestic car assembly enterprises. This raises the problem associated with the development, in addition to the main indicator, of an operational index providing the ability to measure the quality of cars from the first days of operation.

The operational index of the level of alarm signals from the guarantee (ASG) is proposed as an additional statistical indicator.

Practice shows that most often bursts of defects, due to the insufficient level of assembly quality of automobiles or the production of automotive components, appear in the first few months of operation. It turns out that ASG is a set of indicators reflecting the number of failures for each of the positions of the defect codifier during the first weeks of vehicle operation. Figure 7 shows the alarm diagram. An alarm is declared in case of a sharp increase in the number of defects in the first weeks of operation of vehicles with a given release date, with a 3-fold excess of the acceptable level. In this case, the permissible and boundary levels are set by the standards taking into consideration data from the guarantee for previous periods, and targets for the quality of cars in use. Thus, the levels are determined by averaging the operational indicators for failures, for each of the positions, in the period of acceptable stability.

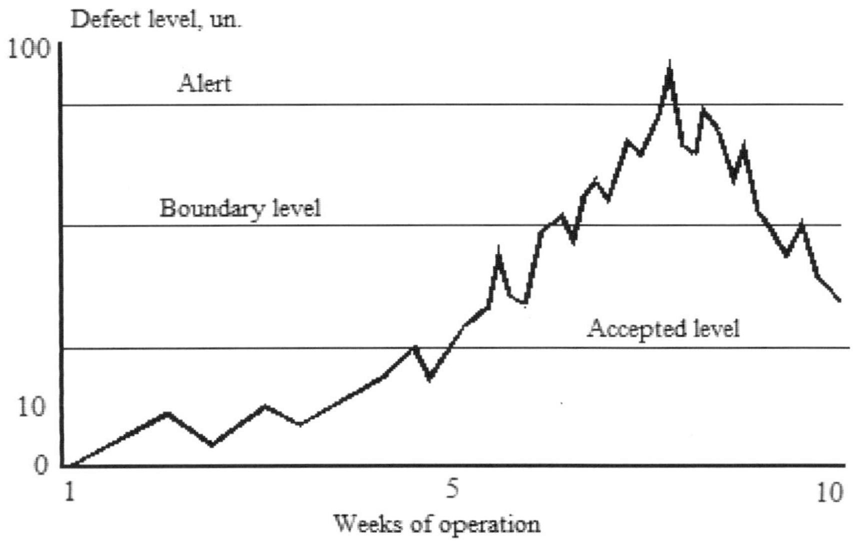

Figure 7. Implementation of the first approach to determining the alarm in the warranty operation of cars.

ASG is assessed according to the number of defects registered for the reporting week (according to the date of defects registration) relative to the average value for the previous 10 weeks. The formula for its calculation is as follows:

$$ASG^{n-mis} = \frac{D^{n-mis}_{rep.week}}{D^{n-mis}_{av}} \geq 3, \qquad (2)$$

where ASG^{n-mis} – warranty defective alarm indicator; $D^{n-mis}_{rep.week}$ – the number of defects for the reporting week, calculated by the date of defect registration (Monday to Sunday); D^{n-mis}_{av} – the average number of defects over the previous 10 weeks; $n-mis$ – period of warranty operation of the car, on which defects are identified. The following periods 0-3 months, 3-6 months, 6-12 months are considered.

Another important aspect that determines the specifics of the domestic automotive industry is the availability of a pre-sale preparation system, with a whole range of tools to ensure that vehicles are ready for use. Many of the competitors have a pre-sale concept no longer. In accordance with state and own regulations, this is still an important process. That is, the transition to a system of new quality monitoring indicators should be accompanied by the development of an index characterizing the quality of products at the pre-sale stage.

Proceeding from the above, let us present the new most comprehensive and appropriate set of indicators to monitor vehicle operation, which determines progressive currently methodology for measuring the quality of the process (Figure 4) and taking into account the specifics of the domestic industry.

The level of defects in the month's group of analysis for 1000 MIS cars sold (K ‰) is calculated by the formula:

$$K\text{‰}_B(MIS) = 1000 \times \sum_{i=0}^{i=MIS} k_i\,, \tag{3}$$

where k_i – the coefficient reflecting the ratio of the number of failures to the number of cars sold for each of the months included in the analysis group; MIS – month in service, the number of months between the date the vehicle is placed under warranty.

The "zero" month in the guarantee (0 mis) starts from the moment the vehicle is placed under the guarantee and lasts until the end of the current calendar month. From the beginning of the next calendar month - 1 mis, etc.

The number of blocking, leading to the inability to operate the car failures per 1000 cars according to the group of months of the analysis MIS (R ‰) is carried out according to the formula:

$$\text{R}\text{‰}_B(MIS) = 1000 \times \sum_{i=0}^{i=MIS} r_i\,, \tag{4}$$

where r_i – the coefficient reflecting the ratio of the number of blocking failures to the number of cars sold for each of the months included in the analysis group; MIS – month in service, the number of months between the date of sale of the device and the date of opening of registration of documents for repair.

The maintainability index Z ‰ determines the level of costs for the elimination of failures in vehicles with fixed service life.

The indicator of the level of costs for eliminating operational failures per 1000 cars according to the group of months of analysis cv (Z ‰) is calculated according to the formula:

$$\text{Z}\text{‰}_B(MIS) = 1000 \times \sum_{i=0}^{i=MIS} z_i\,, \tag{5}$$

where z_i – the coefficient reflecting the ratio of costs to eliminate failures to the number of cars sold for each of the months included in the analysis group; MIS – month in service, the number of months between the sale date of the device and the opening date of documents registration for repair.

The failure rate detected at pre-sale preparation (PSP) P is the ratio of the number of failures detected at the PSP stage, as well as all the failures detected during a car drive during maintenance before the PSP with a mark in the service book to the number of vehicles that have passed PSP (Np) having a specific release date reduced to 1000 cars (P ‰):

$$P\textperthousand_i = 1000 \times \frac{\sum_{i=1}^{n} Y}{N_P}, \quad (6)$$

where Y – the number of failures on cars of a known month of production identified in pre-sale preparation.

4. Conclusion

The study resulted in identifying key problems from the practice of organizing a monitoring system of the quality of new cars during warranty operation that is typical for domestic automakers. A new set of quantitative indicators of the quality of the cars in operation is substantiated and proposed, which have no previously identified shortcomings. The proposed complex has been successfully implemented at a number of domestic enterprises of automakers.

References

[1] Klochkov Y, Gazizulina A and Golovin N 2016 Assessment of organization development speed based on the analysis of standards efficiency *Proceedings - 2nd International Symposium on Stochastic Models in Reliability Engineering, Life Science, and Operations Management, SMRLO 2016* (№ 7433168 DOI: 10.1109/SMRLO.2016.93) pp 530-532

[2] Klochkov Y, Klochkova E, Antipova O, Kiyatkina E, Vasilieva I and Knyazkina E 2016 Model of database design in the conditions of limited resources *5th International Conference on Reliability, Infocom Technologies and Optimization, ICRITO 2016: Trends and Future Directions* (№ 7784927 DOI: 10.1109/ICRITO.2016.7784927) pp 64-66

[3] Kozlovsky V, Klochkov Y, Ostapenko M, Ushanova N and Antipov D 2016 Conformity assessment of car quality databases exemplified on the case of the Russian car manufacturer 'Auto VAZ' *5th International Conference on Reliability, Infocom Technologies and Optimization, ICRITO 2016: Trends and Future Directions* (7784925 DOI: 10.1109/ICRITO.2016.7784925) pp 57-60

[4] Klochkov Y, Gazizulina A, Golovin N, Glushkova A and Selezneva Z 2017 Information model-based forecasting of technological process state *International Conference on Infocom Technologies and Unmanned Systems: Trends and Future Directions, ICTUS 2017* (2018-January DOI: 10.1109/ICTUS.2017.8286099) pp 709-712

[5] Klochkov Y 2018 Conflicts between quality management methods *International Conference on Infocom Technologies and Unmanned Systems: Trends and Future Directions, ICTUS 2017* (DOI: 10.1109/ICTUS.2017.8285970) pp 34-36

[6] Gluhov V, Leventsov V, Radaev A and Nikolaevskiy N 2018 Analytical Modeling of Development and Implementation of Telecommunication Technologies *Lecture Notes in Computer Science (including subseries Lecture Notes in Artificial Intelligence and Lecture Notes in Bioinformatics)* (11118 LNCS DOI: 10.1007/978-3-030-01168-0_39) pp 428-440

[7] Shirokova S V, Bolsunovskaya M V, Gintciak A M and Zharova M S 2018 The Substantiation of the Project Management System for Company by Means of the Functional Completeness Criteria in Innovative Company *Proceedings of the 3rd International Conference Ergo-2018: Human Factors in Complex Technical Systems and Environments, Ergo 2018* (№ 8443942 DOI: 10.1109/ERGO.2018.8443942) pp 40-44

[8] Kalinin M O and Pavlenko E Y 2015 Increasing the fault tolerance and availability of software defined networks using network equipment control based on multiobjective optimization by service quality parameters *Automatic Control and Computer Sciences* (49 (8) DOI: 10.3103/S014641161508026X) pp 673-678

[9] Bolsunovskaya M, Shirokova S, Loginova A and Uspenskij M 2019 The development and application of non-standard approach to the management of a pilot project *IOP Conference Series: Materials Science and Engineering* (497 (1) DOI: 10.1088/1757-899X/497/1/012024)

[10] Iskoskov M O, Kargina E V and Klochkov Y 2019 Innovative. Model of the Organization of a Marketing System at Engineering Enterprises *Proceedings - 2019 Amity International*

Conference on Artificial Intelligence, AICAI 2019 (№ 8701397 DOI: 10.1109/AICAI.2019.8701397) pp 992-997

[11] Drobintsev P D, Kotlyarov V P, Letichevsky A A and Selin I A 2017 Industrial software verification and testing technology *CEUR Workshop Proceedings* (№1989) pp 221-229

[12] Ognev R A, Zhukovskii E V and Zegzhda D P 2019 Clustering of Malicious Executable Files Based on the Sequence Analysis of System Calls *Automatic Control and Computer Sciences* (53 (8) DOI: 10.3103/S0146411619080212) pp 1045-1055

[13] Papirovskiy A, Alekseev S, Tarasov A and Borovkov A 2019 Development of numerical methods and virtual tests for cyclic durability by finite element method *Materials Physics and Mechanics* (42 (6) DOI: 10.18720/MPM.4262019_17) pp 859-869

[14] Pyayt A L, Shevchenko D V, Kozionov A P, Mokhov I I, Lang B and Krzhizhanovskaya V V 2015 P.M.A. Sloot, Combining data-driven methods with finite element analysis for flood early warning systems *Procedia Computer Science* (51 (1) DOI: 10.1016/j.procs.2015.05.404) pp 2347-2356

[15] Glukhov V, Klochkov Y, Gazizulina A, Krasulina O and Vlasova N 2018 Evaluating quality of adhesive joint of missile radomes *International Conference on Infocom Technologies and Unmanned Systems: Trends and Future Directions, ICTUS 2017* (DOI: 10.1109/ICTUS.2017.8286096) pp 694-697

[16] Alkhimenko A 2019 Corrosion testing of experimental steels for oilfield pipelines *E3S Web of Conferences* (121 № 01001 DOI: 10.1051/e3sconf/201912101001)

[17] Grashchenko A S, Kukushkin S A and Osipov A V 2019 Strength and structural properties of AlN films grown on SiC/Si substrates synthesized by atomic substitution *Journal of Physics: Conference Series* (1410 (1) № 012244 DOI: 10.1088/1742-6596/1410/1/012244)

[18] Zhukov D O, Zaltcman A G and Khvatova T Y 2019 Forecasting Changes in States in Social Networks and Sentiment Security Using the Principles of Percolation Theory and Stochastic Dynamics *Proceedings of the 2019 IEEE International Conference Quality Management, Transport and Information Security, Information Technologies IT and QM and IS 2019* (№ 8928295 DOI: 10.1109/ITQMIS.2019.8928295) pp 149-153

IOP Conf. Series: Materials Science and Engineering **986** (2020) 012047 doi:10.1088/1757-899X/986/1/012047

Research of New Cars Quality

V. N. Kozlovsky[1], D. V. Aydarov[1], A. P. Novikova[1] and P.S. Cvetkov[2]

[1]Samara State Technical University, Samara, Russia, kozlovskiy-76@mail.ru
[2]National Technology Initiative Center for Advanced Manufacturing Technologies based on the Institute of Advanced Manufacturing Technologies of Peter the Great St. Petersburg Polytechnic University Polytechnicheskaya, 29, St.Petersburg, 195251, Russia

Abstract. The article deals with the study results of the quality of new cars in the production control process in the delivery condition to the consumer. The research was conducted on the basis of resource tests and using information-statistical analysis. The quality assessment of new cars and their role in customer satisfaction is given.

1. Introduction

Currently, ensuring the quality of the car and its operational efficiency most significantly depends on the relevant indicators of the electrical complex. Moreover, this issue is important given the prospects for the development of new types of vehicles, such as electric vehicles and cars with a combined power plant.

Quality is known to lay in the design and is ensured in production. The processes of measuring the quality of new cars have to be implemented at different stages of the life cycle. This should be covered by design tools and the production system and, of course, by car operation. [1, 2]

The goal of the research is to actualize the problem of ensuring the quality of automotive equipment based on a comprehensive study of the data array reflecting defects of new cars that are identified at the stage of production output quality control. Besides, the information-statistical analysis of the data array on car system failures during the warranty period of operation is carried out in the article, highlighting the most important systems in terms of operational quality. As a result of the study, an analysis and comparison of production control data and warranty operation data was conducted to assess the equivalence of the weight of the detected defects of new cars at the stages of the life cycle.[3]

2. Key research findings

As a research base, data selection is defined as the results of industrial output quality control, as well as data from the warranty operation of one of the most popular passenger car models made in our country. The study period covered the entire 2017 and 6 months of 2018. In addition, the data of an independent expert laboratory were used in the research process, which conducted the study of car shipping of the considered brand as an analog of production control and during operation, as an analog of the warranty period. In conclusion, summarized data on the results of the resource tests of cars of the investigated brand are presented. [4, 5]

As a research tool, a modified Pareto diagram is used, which reflects the accumulated level of defects according to statistical data for specific vehicle systems. The expert method based on information - statistical analysis of new car quality problems is also used. [6]

Let us consider the study results of the quality of cars according to the data from the output quality control department of the product and from the enterprises of a branded car service to the parent company. They reflect the level of defects in new vehicles before a sale and during warranty service (Figures 1 - 4). According to the workshop, the largest number of cars' defects of the considered brand for a year and a half of analyzing (from January 2017 to June 2018) is manifested on the systems "Car Interior" (about 34%) and "Car Body" (about 26%). These systems make up the bulk of the cost of fixing defects. The electrical equipment system takes third place both in defects (23.8%) and in costs (20.2%) and, therefore, is the most defective one of the functional systems. [7-10]

It should be noted that not all cars with one defect are fixed during the inspection in the workshop. Because of this, the actual values of defectiveness and advertise ability of the electrical equipment system may be higher.

According to the warranty operation (the results of the elimination for the same year and a half), the majority of vehicle defects are manifested in the electrical equipment system of 17.5%. The electrical equipment system takes the second place and makes up 17.8% of the total costs of the guarantee in terms of warranty costs (the first place is occupied by the engine system, which accounts for 17.9% of the costs).[11-15]

Thus, the analysis of defects and their causes revealed in the output inspection workshop, as well as during the period of warranty operation, allows to single out the most significant problems of car quality using informatization: body assembly defects, defects in the internal combustion engine, problems with the quality of electrical equipment. It is these problematic aspects of the quality of automotive equipment that currently determine the list of the most pressing problems. Their solution can significantly reduce the cost of eliminating defects both within production and in warranty, as well as the quality of products. [16-21]

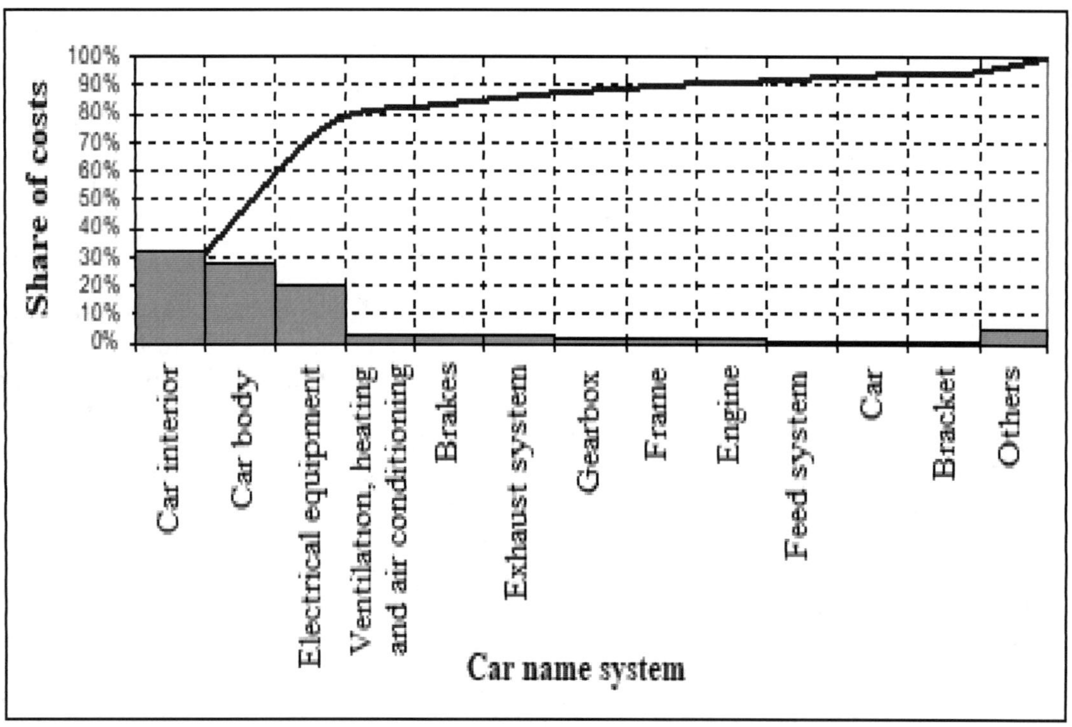

Figure 1. Costs broken down by systems in the output control workshop.

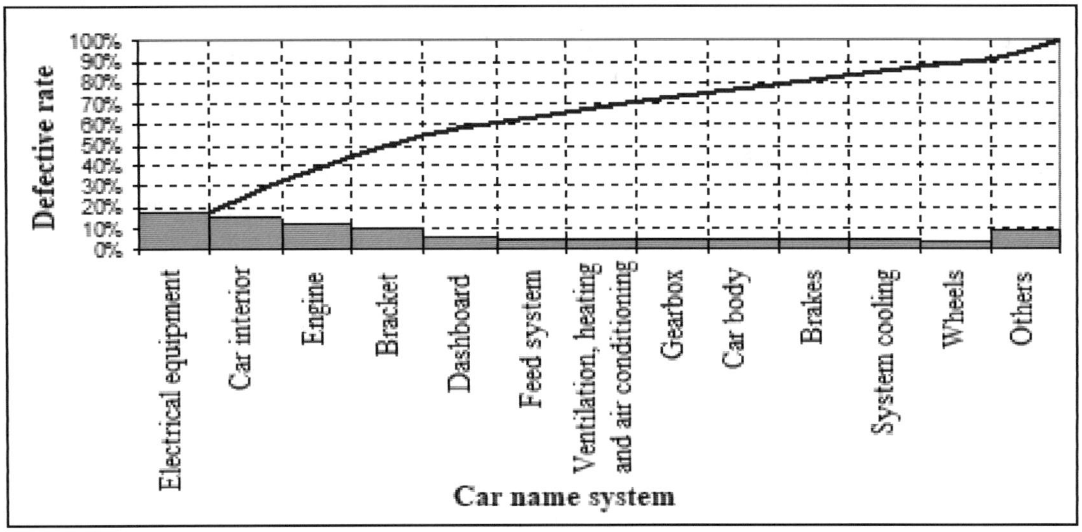

Figure 2. Defects broken down by systems in the warranty.

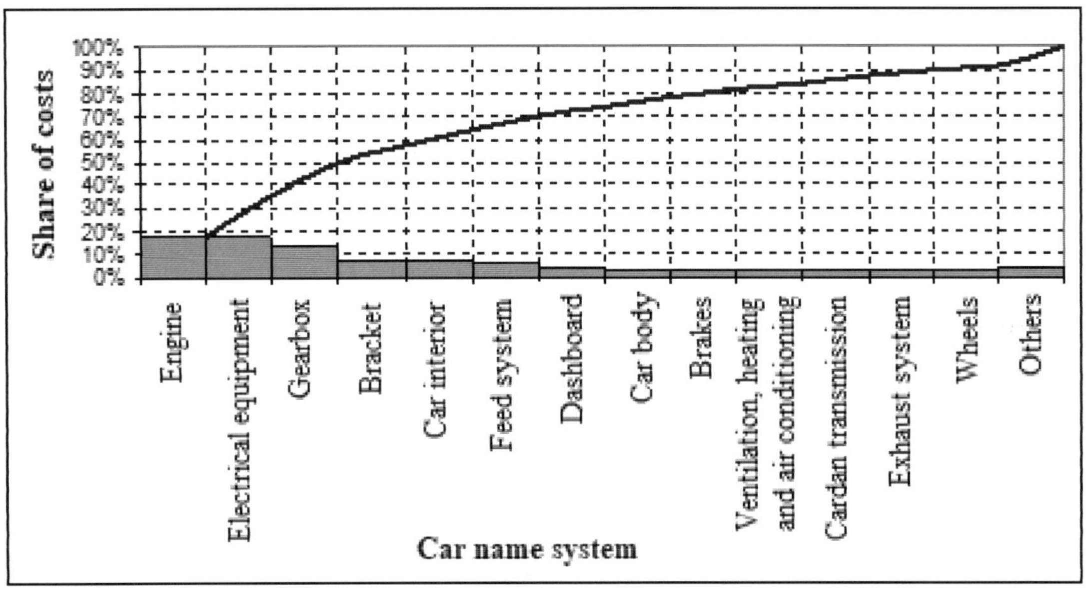

Figure 3. Costs broken down by systems in the warranty.

3. Results

Let us turn to the analysis of the results of independent tests of automotive vehicles obtained in an international certified laboratory. In summary, separate problems during control tests of the study six brands of cars and their visualization are presented in Table 1. Tests were conducted in several areas of expert assessment based on information - statistical analysis of quality: quality of car shipping; quality according to the results of testing the components and assemblies of cars for durability; risk assessment of vehicle operating safety.

There is an analysis of the quality of car shipping in columns 1, 2, Table 1. The evaluation resulted in the following problems were identified: the requirements for fastening the fuel supply hose (problem 1, Table 1) to the internal combustion engine (ICE) were not met; for this reason, a fuel leak occurred; mismatch of the braking characteristics of the rear wheel lock during sudden braking; broken wire connector to the headlamp lighting system; indicators on the dashboard do not work. There are the

IOP Conf. Series: Materials Science and Engineering **986** (2020) 012047 doi:10.1088/1757-899X/986/1/012047

problems that can reduce consumer satisfaction in the delivery condition: noise and vibration of the internal combustion engine; gear shifting is difficult (problem 2, Table 1); poor quality fastening of upholstery panels; difficulty opening / closing the tailgate (problem 3, Table 1). Considering and summarizing these problems, the main reason for the non-compliance with the quality requirements for car shipping can be revealed. This is the same reason shown earlier, when analyzing the quality of production results, and associated with low quality of car assembly.

Quality according to the results of testing units and assemblies for durability is in columns 3, 4, Table 1. The following problems were identified in cars quality during road tests on public roads within 5000 km: transmission failure in the car: the driver could not engage the second and third gear of the gearbox (gearbox) after stopping; loss of the initial tightening torque of the front left drive shaft nut (problem 1, Table 1); failure of the fuel pump connector (problem 2, Table 1); the front left cap of the wheel disc is disconnected (problem 3, Table 1); failure of the front left power window; high engine oil consumption; incorrect connection of the engine temperature sensor; leaking rear windshields.

Risk assessment of vehicle operation safety is in columns 5, 6 Table 1. The following problems were identified during the tests: harmful contact of the throttle valve control cable with the engine fan cover (problem 1, Table 1); harmful contact of the brake tube with the fuel hose (problem 2, Table 1); harmful contacts of cooling hoses with several connections (problem 3, Table 1); harmful contact of the spark plug wiring with the engine cover bracket (problem 4, Table 1).

Table 1. Results of testing new Russian-made cars.

In delivery condition		The results of durability tests		Safety risk assessment	
Problems	Visualization	Problems	Visualization	Problems	Visualization
1	2	3	4	5	6
1.		1.		1.	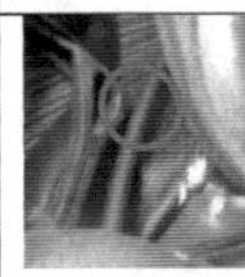
2.	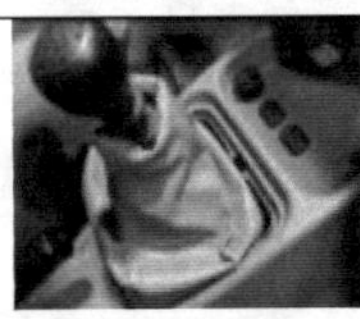	2.		2.	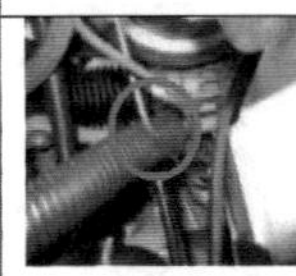
3.		3.		3.	
		4.		4.	
		5.			

Summarizing the results of the resource tests of six cars, special attention should be paid to quality problems, the reasons for which are the quality of cars assembly. Most of the identified problems related to issues of executive quality formed at the assembly lines of automobile manufacturers.

In conclusion, we turn to the results of operational tests of 10 cars of this brand on a range of up to 240 thousand kilometers on public roads.

The ranking of the data obtained during the operational tests of automotive equipment for the main classification reasons for failures showed that:

- about 40% of the identified problems were related to the assembly quality of units and cars;
- about 30% of the problems were connected with the insufficient structural quality of brackets, tubes, hoses and other elements, various automobile systems;
- up to 20% of problems were related to failures of elements of the electrical system;
- about 10% of problems were due to the presence of harmful contacts of various elements and systems of cars.

Thus, the results of operational tests and other studies discussed above reveal the most significant problems of vehicle quality. They are the insufficient quality of assembly operations on conveyors, the presence of quality problems of individual units and assemblies (ICE, gearbox), as well as quality problems of elements vehicle electrical systems. Moreover, the importance of automobile electrical equipment in terms of quality and operational efficiency would steadily increase. Data on the level of defectiveness and the costs of their elimination clearly show that the electrical system confidently occupies one of the two leading places of anti-rating. At the same time, the results of expert assessments based on information - statistical analysis and resource tests of automobiles clearly show the importance of ensuring the quality of production and assembly of electrical components.

Let us go to the issue of assessing the quality of new cars on the part of the consumer environment. The analysis of the correlation between individual estimates that determine the unit elements of ensuring the quality of new cars allows us to draw some conclusions regarding the initial, after-production quality of new cars in the delivery condition. According to the obtained data, the consumer is more susceptible to changes in satisfaction: the correlation coefficient is 0.65 in case of a change in the quality of the finished product (car). The consumer is suggested to be most sensitive to the primary quality of new cars, and if it is high, then satisfaction is relevant. Ensuring the quality of service (correlation coefficient 0.61) and the cost of ownership (0.6) are also important. A sensitive relationship is observed between satisfaction with service and cost of ownership (0.56); between satisfaction with service and car quality (0.51).

In other words, it turned out that it is important for the consumer: the primary high quality of the purchased car, the acceptable perceived value of quality (cost of ownership), and quality of service. Moreover, all this should be already provided at the start, that is, when the car is bought. The consumer is not very sensitive to the fact that the possible quality problems that arise during the operation will be resolved quickly, whether at the design or production level or in the service. This does not increase loyalty and satisfaction.

Conclusion

The comprehensive study showed that in the current production conditions, the level of defectiveness, as well as the predicted level of relevant indicators for the component base of the electrical equipment system, could not be estimated. Unfortunately, it is necessary to use primarily data from the period of operation to assess the quality of electrical equipment.

At the same time, based on a cross-analysis of data on defects detected in the processes of output quality control, warranty operation, expert assessment based on information - statistical analysis and life tests, a complete picture is formed. It allows highlighting the electrical system as a key one in terms of ensuring the quality of new cars. In this regard, there is a need for the development of analytical and hardware methods and tools for assessing and predicting the quality of electrical equipment of cars at the production stage. This requires the development of appropriate diagnostic production systems that solve the problems of both assessing and predicting the operational efficiency of new cars, at least within the warranty period.

References

[1] Klochkov Y, Gazizulina A, Ostapenko M., Eskina E and Vlasova N 2016 Classifiers of nonconformities in norms and requirements *5th International Conference on Reliability, Infocom Technologies and Optimization, ICRITO 2016 Trends and Future Directions* (№ 7784933. DOI: 10.1109/ICRITO.2016.7784933) pp 96-99

[2] Kozlovsky V, Klochkov Y, Ostapenko M, Ushanova N and Antipov D 2016 Conformity assessment of car quality databases exemplified on the case of the Russian car manufacturer 'Auto VAZ'. 2016 *5th International Conference on Reliability, Infocom Technologies and Optimization, ICRITO 2016: Trends and Future Directions* (№ 7784925. DOI: 10.1109/ICRITO.2016.7784925) pp 57-60

[3] Glukhov V, Turichin G, Klimova-Korsmik O, Zemlyakov E and Babkin K 2016 Quality management of metal products prepared by high-speed direct laser deposition technology *Key Engineering Materials* (684 DOI: 10.4028/www.scientific.net/KEM.684.461) pp 461-467

[4] Gravit M, Dmitriev I and Ishkov A 2017 Quality control of fireproof coatings for reinforced concrete structures *IOP Conference Series: Earth and Environmental Science* (90 (1), № 012226, DOI: 10.1088/1755-1315/90/1/012226)

[5] Klochkov Y, Klochkova E, Kiyatkina E, Skripnuk D and Aydarov D 2018 Development of methods for business modeling *International Conference on Infocom Technologies and Unmanned Systems: Trends and Future Directions, ICTUS 2017* (DOI: 10.1109/ICTUS.2017.8286034) pp 366-369

[6] Zharkovskii A A and Pospelov A Y 2015 Use of 3D Methods for Flow Analysis, Prediction of Characteristics, and Optimization of the Shape of Settings of Hydraulic Turbines *Power Technology and Engineering* (49 (1). DOI: 10.1007/s10749-015-0568-4) pp 27-32

[7] . Iskoskov M, Smagina A and Gazizulina A 2019 Comprehensive Approach to Tax Analysis as a Procedure for Management Control of Tax Liabilities of Commercial Organizations of the Automotive Industry in Russia *Proceedings - 2019 Amity International Conference on Artificial Intelligence, AICAI 2019* (№ 8701271. DOI: 10.1109/AICAI.2019.8701271) pp 1015-1021

[8] Rostova O, Shirokova S and Sokolitsyna N 2019 Management of project for automation of investment control at industrial enterprise *IOP Conference Series: Materials Science and Engineering* (497 (1) № 012017 DOI: 10.1088/1757-899X/497/1/012017)

[9] Shirokova S V, Bolsunovskaya M V, Gintciak A M and Zharova M S 2018 The Substantiation of the Project Management System for Company by Means of the Functional Completeness Criteria in Innovative Company *Proceedings of the 3rd International Conference Ergo-2018: Human Factors in Complex Technical Systems and Environments, Ergo 2018* (№ 8443942. DOI: 10.1109/ERGO.2018.8443942) pp 40-44

[10] Kalinin M O and Pavlenko E Y 2015 Increasing the fault tolerance and availability of software defined networks using network equipment control based on multiobjective optimization by service quality parameters *Automatic Control and Computer Sciences* (49 (8) DOI: 10.3103/S014641161508026X) pp 673-678

[11] Bolsunovskaya M, Shirokova S, Loginova A and Uspenskij M 2019 The development and application of non-standard approach to the management of a pilot project *IOP Conference Series: Materials Science and Engineering* (497 (1) № 012024 DOI: 10.1088/1757-899X/497/1/012024)

[12] Pyayt A L, Shevchenko D V, Kozionov A P, Mokhov I I, Lang B and Krzhizhanovskaya V V 2015 P.M.A. Sloot, Combining data-driven methods with finite element analysis for flood early warning systems *Procedia Computer Science* (51 (1) DOI: 10.1016/j.procs.2015.05.404) pp 2347-2356

[13] Iskoskov M O, Kargina E V and Klochkov Y 2019 Innovative Model of the Organization of a Marketing System at Engineering Enterprises *Proceedings - 2019 Amity International Conference on Artificial Intelligence, AICAI 2019* (№ 8701397 DOI: 10.1109/AICAI.2019.8701397) pp 992-997

[14] Glukhov V, Klochkov Y, Gazizulina A, Krasulina O and Vlasova N 2018 Evaluating quality of adhesive joint of missile radomes *2017 International Conference on Infocom Technologies and Unmanned Systems: Trends and Future Directions, ICTUS 2017* (DOI: 10.1109/ICTUS.2017.8286096) pp 694-697

[15] Alkhimenko A 2019 Corrosion testing of experimental steels for oilfield pipelines *E3S Web of Conferences, 121* (№ 01001 DOI: 10.1051/e3sconf/201912101001)

[16] Drobintsev P D, Kotlyarov V P, Letichevsky A A and Selin I A 2017 Industrial software verification and testing technology *CEUR Workshop Proceedings, 1989* pp 221-229

[17] Grashchenko A S, Kukushkin S A and Osipov A V 2019 Strength and structural properties of AlN films grown on SiC/Si substrates synthesized by atomic substitution *Journal of Physics: Conference Series* (1410 (1), № 012244, DOI: 10.1088/1742-6596/1410/1/012244)

[18] Zhukov D O, Zaltcman A G and Khvatova T Y 2019 Forecasting Changes in States in Social Networks and Sentiment Security Using the Principles of Percolation Theory and Stochastic Dynamics *Proceedings of the 2019 IEEE International Conference Quality Management, Transport and Information Security, Information Technologies IT and QM and IS 2019* (№ 8928295, DOI: 10.1109/ITQMIS.2019.8928295) pp 149-153

[19] Papirovskiy A, Alekseev S, Tarasov A and Borovkov A 2019 Development of numerical methods and virtual tests for cyclic durability by finite element method *Materials Physics and Mechanics* (42 (6). DOI: 10.18720/MPM.4262019_17) pp 859-869

[20] Garanin D A, Lukashevich N S, Salkutsan S V and Svirina AA 2017 Formalization of stochastic restrictions in risk assessment models of investment projects *Proceedings of the 2017 International Conference "Quality Management, Transport and Information Security, Information Technologies", IT and QM and IS 2017* (№ 8085875. DOI: 10.1109/ITMQIS.2017.8085875) pp 521-524

[21] Yakovis L, Sporyagin K and Chechurin L 2016 Control with Constraints for Multivariable Objects with Time Delay *IFAC-PapersOnLine* (49 (12), DOI: 10.1016/j.ifacol.2016.07.572) pp 996-1001

Performance analysis of electrical systems of vehicles under the external electromagnetic exposure

V. N. Kozlovsky[1], P. A. Nikolaev[2], A. S. Podgorny[1], Y. V. Brachunova[1] and N.K. Efimov-Soini[3]

[1] Samara State Technical University, Samara, Russia, kozlovskiy-76@mail.ru
[2] JSC "AVTOVAZ", Togliatti, Russia
[3] National Technology Initiative Center for Advanced Manufacturing Technologies based on the Institute of Advanced Manufacturing Technologies of Peter the Great St. Petersburg Polytechnic University Polytechnicheskaya, 29, St.Petersburg, 195251, Russia

Abstract. The study results of malfunctions of electrical systems of modern vehicles due to external electromagnetic exposure are presented in the paper.

Introduction

The collection and analysis of statistical data on malfunctions of the onboard electrical complex due to electromagnetic impacts is a complex problem. First of all, this is caused by the complexity of the unambiguous interpretation of the data on operational failures of the complex and the classification of these failures due to external electromagnetic disturbances. That is why it is necessary to determine problematic issues in the operation of automotive electrical systems in order to provide a research base. To solve this kind of problem, given the high importance of the problem of electromagnetic compatibility (EMC), not only leading car manufacturers, but also component manufacturers have test centers that allow them to identify and solve problems before they appear in operation. [1, 2].

Extensive experience in the collection of statistical data on malfunctions and understanding the physics of the interaction of electrical systems of vehicles with electromagnetic exposure is obtained in the process of testing. First, it modifies the product to meet the requirements. Secondly, it imposes the existing experience on the problem that manifested itself in operation and, with a certain probability, to give a competent answer about the effect of an electromagnetic field (EMF).

A high-energy effect is required for the product not to fail due to an electromagnetic field (Figure 1) [3, 4]. The functioning of systems will be disrupted earlier at lower levels of interference. Failures with low energy impact are possible, but they do not occur because of direct impact. This mainly happens because of the transition to the open state of transistors and the subsequent thermal destruction of the elements.

Let us present the obtained results of the study of maloperation of electrical systems of vehicles due to external electromagnetic influences.

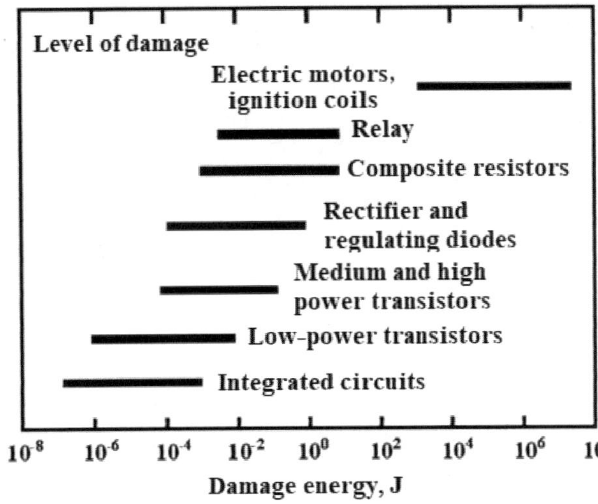

Figure 1. Damage energy of electrical systems vehicle elements.

Electronic engine control system (EECS). Cases of data corruption from various auxiliary sensors are typical, which do not critically influence the engine control process. These are, for example, temperature, phase and detonation sensors. They belong to functional class B [5].

Failures associated with changing the speed or stopping the vehicle are characterized by classes C, D and E. They mainly occur through the sensor channels: crankshaft position, airflow, throttle position, accelerator pedal, as well as measuring the current in the primary circuit of the ignition coil [5 - 7]. Class E problems are less common, where electromagnetic interference acting on the power supply circuits of the engine control system controller destroys the injector drivers, and also leads to the "freezing" of the microprocessor, so that it can be resumed after the operation of turning off/turning on the power from the battery [5, 6]. The immunity characteristics of an electronic engine control system, obtained from a standard EMC proof test, are shown in Figure 2.

Electromechanical power steering (EMPS). EMC problems typical for EMPS are reduced to the following: spontaneous rotation due to the appearance of a parasitic compensating moment and the inability of EMPS to create the required compensating moment. The vehicle spontaneously turns the steering wheel to the side in operation, or it is very difficult to control it while driving [6 - 8]. In general, class C malfunctions occur when information from the torque sensor is distorted, and class D - with erroneous current diagnostics of the EMPS state (Figure 3).

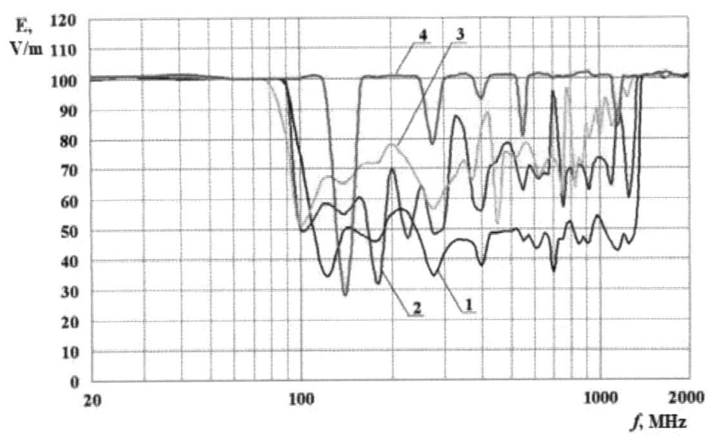

Figure 2. EECS immunity characteristics: 1 – class B; 2 – class C; 3 – class D; 4 – class E.

IOP Conf. Series: Materials Science and Engineering **986** (2020) 012048 doi:10.1088/1757-899X/986/1/012048

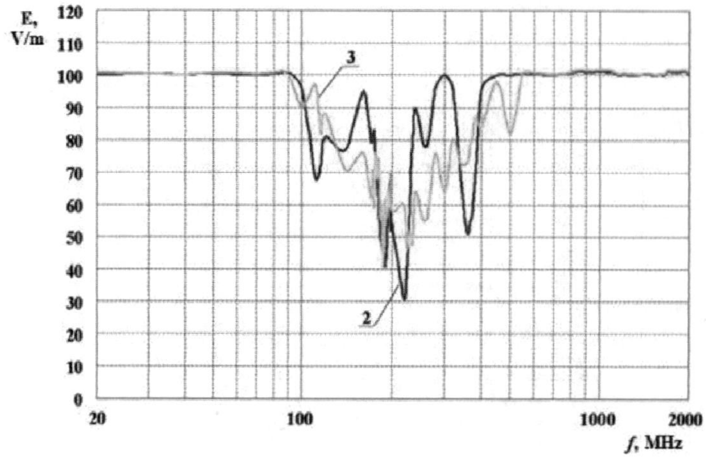

Figure 3. EMPS immunity characteristics: 2 – class C; 3 – class D.

Air conditioning and heating systems [5]. These systems are characterized by malfunctioning associated with a change in the speed and temperature of the air entering the passenger compartment in relation to a given value (class B) and the termination of the system - class C (Figure 4).
Power supply systems. Voltage generator failures are common. Depending on the type of the regulator, there are cases of voltage decreases and increases up to 21 V [9]. If the deviation from the nominal voltage limits is no more than 5%, then this is considered class B, and if more - class C (Figure 5).
Increasing voltage is more critical than decreasing, since electrical systems of vehicles can fail..

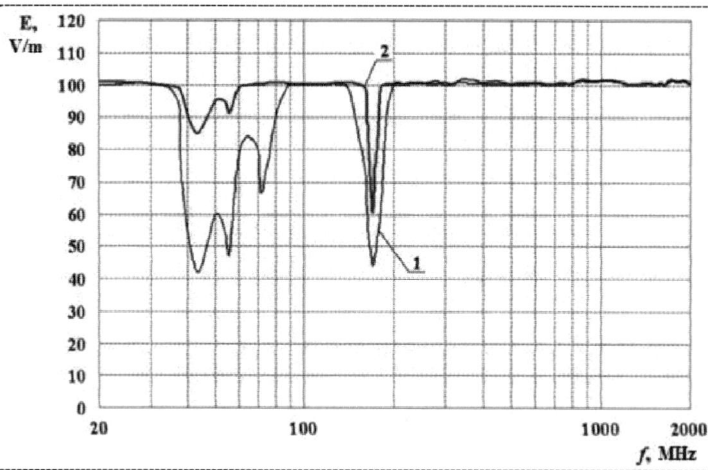

Figure 4. Immunity of air conditioning and heating systems: 1 - class B; 2 - class C.

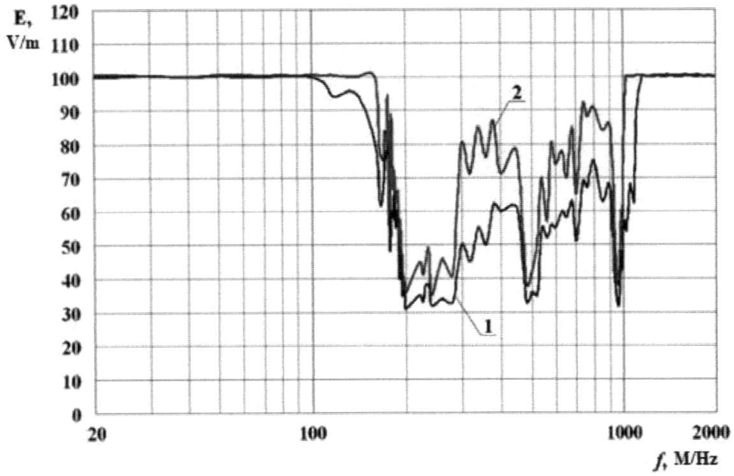

Figure 5. Voltage generator immunity characteristics: 1 - class B; 2 - class C.

Safe parking systems [5, 12-14]. Violations of the system's performance were revealed in the form of: a change in the tone of the sound, distortion of the video image on the display, as well as a decrease in the detection range of an obstacle (class B); loss/appearance of an acoustic signal in the "presence of an obstacle"/"absence of an obstacle" mode and distortion of the video image from the rearview camera (class C); a complete cessation of functions (class D) (Figure 6).

Lighting systems. The lighting system is an example where the exposure of an electromagnetic field on one system leads to malfunctions of another one. So increasing the voltage from the generator, the luminous flux grows under the influence of interference, and the lamps of headlights and lanterns burn out. The direct exposure of the electromagnetic field is observed in the form of turning off the fog lamps and stopping the operation of direction indicators (Figure 7).

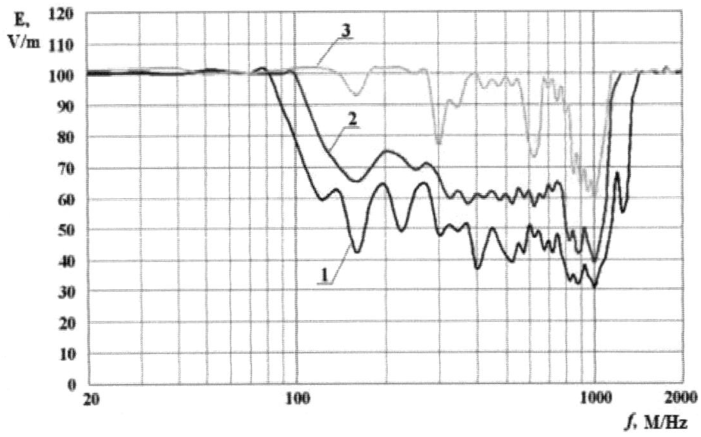

Figure 6. Immunity of the safe parking system: 1 - class B; 2 - class C; 3 - class.

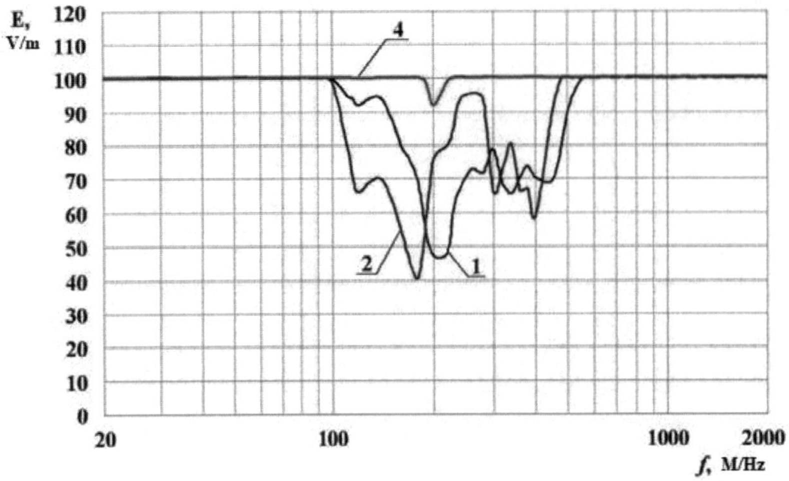

Figure 7. Immunity of the lighting system; 1 - class B; 2 - class C; 4 - class E.

Information display systems. There are malfunctions of the following nature: slight distortions of the image on the screen and the brightness of symbolic alarms, as well as deviations of instrument readings (class B); strong distortions of the image on the display, changes in the brightness of symbolic signaling devices, significant deviations of instrument readings, spontaneous reboot of the system (class C); "freezing" of individual elements or the entire system, and resetting the set settings, for example, time (class D); failure of elements, and information erasure from the EEPROM (class E) (Figure 8).

Multimedia systems [5]. These systems have malfunctions of the following nature: slight distortion of sound (class B); strong distortion of sound, it is impossible to select a multimedia file during EMF (class C); "freezing" of the multimedia system (class D) and failure of radioelements (class E) (Figure 9).

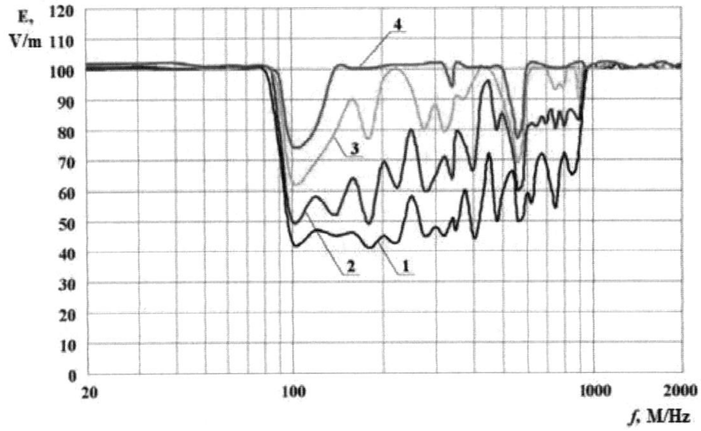

Figure 8. Immunity of the information display system: 1 - class B; 2 - class C; 3 - class D; 4 - class E.

IOP Conf. Series: Materials Science and Engineering **986** (2020) 012048 doi:10.1088/1757-899X/986/1/012048

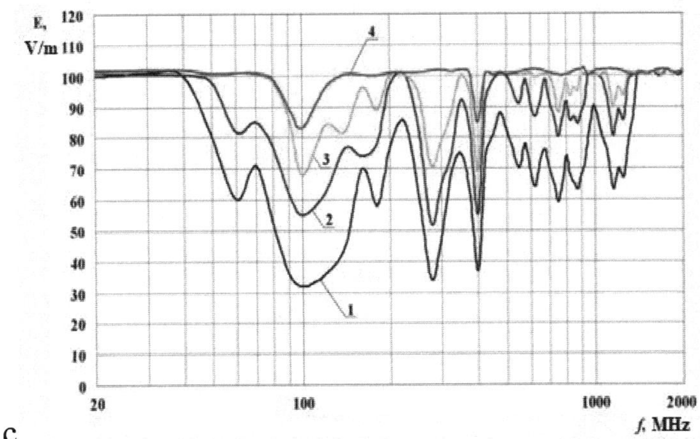

c

Figure 9. Immunity of the multimedia system: 1 - class B; 2 - class C; 3 - class D; 4 - class E.

Security systems [8]. Failures of the automobile anti-theft system (ATS) are expressed in the form of a prohibition to start the engine during the period of the EMF action (Figure 10). "Cruise control" [5]. Functional disorders are revealed in the form of a spontaneous change in the set speed without returning to the initial value, as well as the inability to activate the function (Figure 11).

Wiper systems. There are malfunctions of the following character (Figure 12): a small, less than 10%, decrease in the speed of the wipers (class B) and a decrease in the speed of the wipers by more than 10% of the nominal one, as well as their complete stop (class C).

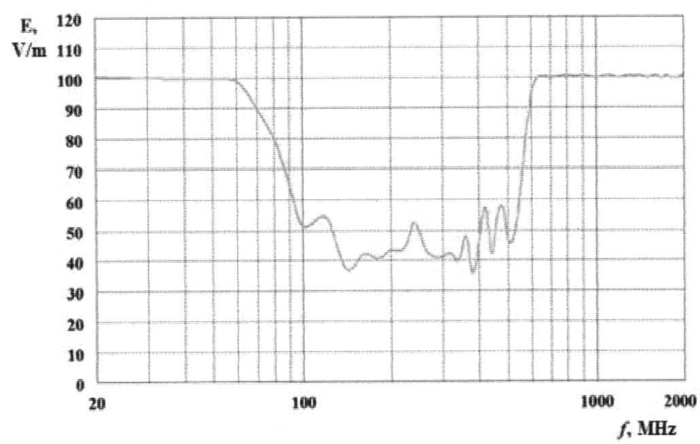

Figure 10. Anti-theft system class D immunity characteristics.

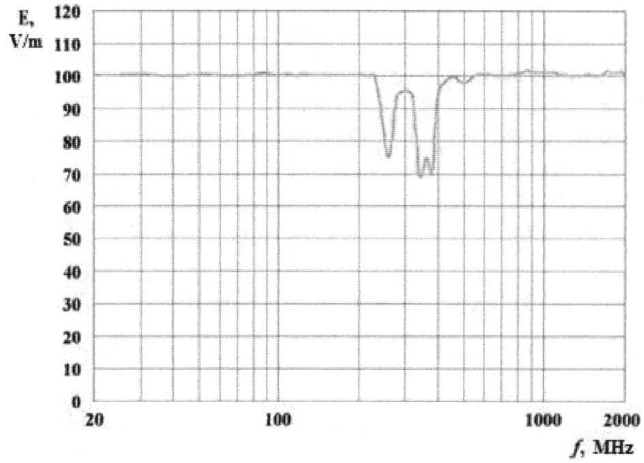

Figure 11. Cruise control class D immunity characteristics.

Navigation system. Radio channels are the least resistant to immunity. During the action of EMF with a strength of 1 V/m in the frequency range from 800 MHz to 2 GHz, the system does not see satellites (Figure 13).

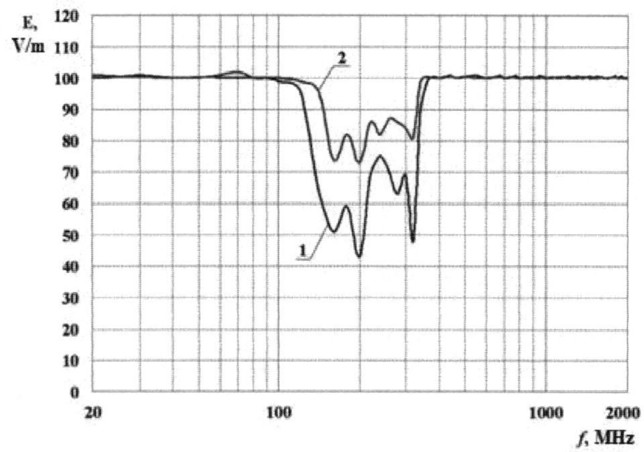

Figure 12. Wiper system immunity characteristics: 1 - class B; 2 - class C.

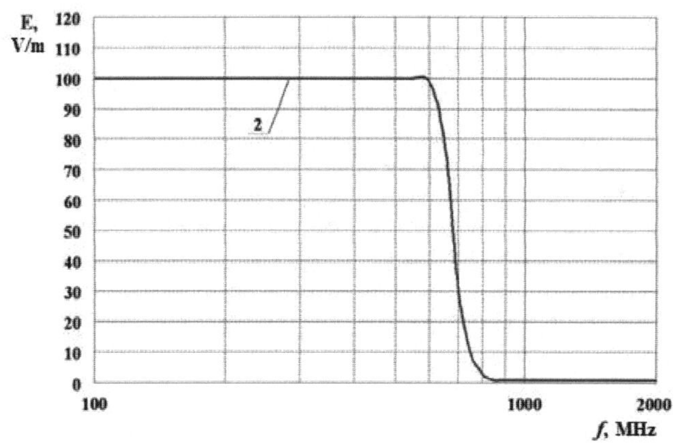

Figure 13. Immunity characteristics of the navigation system (2 - class C).

IOP Conf. Series: Materials Science and Engineering **986** (2020) 012048 doi:10.1088/1757-899X/986/1/012048

Transmission control system. There is the following type of malfunctions caused the EMF exposure: a decrease in the speed of the car due to spontaneous switching to a lower gear and a failure in the chain of the position sensor of the gear selection drive. In the latter case, the movement of the vehicle would be possible only in the gear that was included before the moment of malfunction.

Malfunctions of the transmission control system are presented at frequencies from 22 to 60 MHz and EMF levels of 34 V/m.

Electric vehicles. The following problems are observed (Figure 14): a decrease in the traction battery charge and an increase in the current consumption of the traction asynchronous electric motor (class C) are displayed on the instrument cluster; "freezing" of the electronic engine management system, which is eliminated by disconnecting/supplying power from the battery (class E).

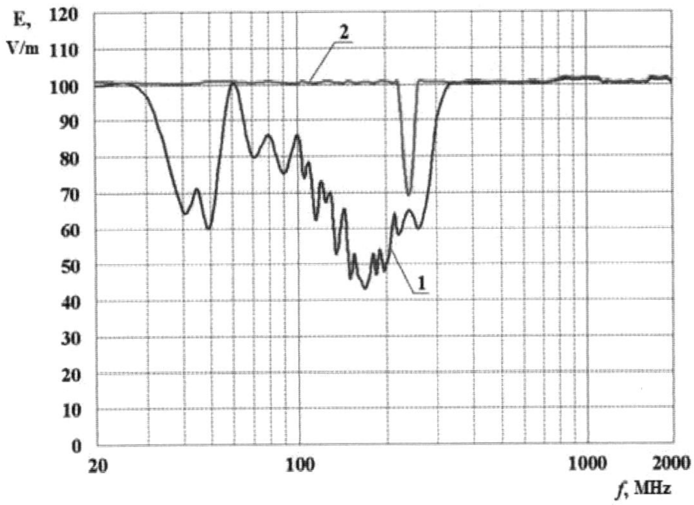

Figure 14. Immunity characteristics of the electric car: 1 - class C; 2 - class E.

CAN bus. CAN-controllers may send incorrect data «Error frame» under the EMF exposure, or are useful signals are distorted in the tire themselves. This causes disconnection from the bus of electrical systems with advanced diagnostics. For example, the anti-lock braking system is disconnected from the CAN bus when the counter of broken frames is overflowed, so that signals from the wheel speed sensors are not received by EECS. EECS goes into emergency control mode because of that. This type of malfunctioning was revealed at frequencies above 110 MHz and voltages of 60 V/m.

The foreign literature also contains information about the immunity of the vehicle. According to [10, 11], vehicle malfunctions were observed at frequencies above 1.5 GHz and a level of 500 V/m, expressed in stopping the engine, and failures of electrical systems occurred at 15 kV/m at 1.5 GHz and 24 kV/m at 2.86 GHz. The engine control, relay, instrument cluster, car anti-theft system and external view video camera were damaged. The author showed that it is possible to disrupt the trajectory of the vehicle from a distance of 500 m and stop it, disabling it at a distance of 15 m.

Conclusion

The analysis shows that the electrical systems of vehicles are most vulnerable to electromagnetic radiation at frequencies above 20 MHz. A malfunction when low-energy EMF exposure is most likely in the range from 100 MHz to 1.5 GHz. These data correlate to a certain extent with the evaluation of immunity for information structures. Based on this, one of the most important tasks is to improve the existing methods and develop new ones for testing electrical systems of the vehicle for resistance to external electromagnetic radiation.

References

[1] 2010 Ensuring the quality and reliability of the electrical equipment system of cars *Dissertation abstract for the degree of Doctor of Technical Sciences* (Moscow state car-road university Tolyatti)

[2] Stroganov V I and Kozlovsky V N 2014 Modeling of systems of electric vehicles and vehicles with a combined power plant in the design and production processes (monograph Moscow) p 264

[3] Balyuk N V Kechiev L N and Stepanov P V 2008 Powerful electromagnetic pulse: impact on electronic equipment and protection method (LLC "Group IDT") p 478

[4] Williams T 2004 EMC for Product Developers (M.: Publishing house "Technologies") p 508

[5] Nikolaev P A and Kechiev L N 2015 Electromagnetic compatibility of vehicles (M.: Grifon) (EMC Library) p 424

[6] Nikolaev P A and Nikolaev A D 2014 Immunity of cars to electromagnetic exposure *Technologies of electromagnetic compatibility* (№4 (51) pp 72-76

[7] Nikolaev P A 2011 Electromagnetic compatibility of modern automobile ignition systems (monograph, Tolyatti: VRSUS) p 224

[8] Nikolaev P A, Konyukhov N E and Chervyakov S A 2006 Analysis of possible reasons for the failure of electronics and electrical equipment of the car in electromagnetic terrorism *Infocommunication technologies. Appendix "Methods and means of ensuring security in infocommunications"* pp 234-245

[9] Nikolaev P A and Nikolaev A D 2011 Influence of the electromagnetic field on the car voltage generator *Technologies of electromagnetic compatibility* (№1 (36)) pp 39-44

[10] Backstrom M 1999 HPM Testing of a car: a representative example of the susceptibility of civil systems *Supplement to Proc. of the 13th Int. Zurich Symp. on EMC. Zurich, Switzerland* (February 16–18) pp189–190

[11] Backstrom M 1995 Testing of a car's immunity against intense electromagnetic radiation *FOA Report* (FOA-R-00099-3.2-SE)

[12] Alekseev, A., Maksimov, A., & Tarasov, A. (2020). CFD investigations of the effect of rotating wheels, ride height and wheelhouse geometry on the drag coefficient of electric vehicle. International Journal of Mechanics, 14, 130-134. doi:10.46300/9104.2020.14.17

[13] Aleshin, M., Gavrilova, L., Goryainov, I., & Melnikov, A. (2020). Dem generation based on commercial uav photogrammetry data. Paper presented at the Engineering for Rural Development, , 19 1750-1756. doi:10.22616/ERDev2020.19.TF461

[14] Maksimov, A., Igoshina, D., Petrov, R., & Klyavin, O. (2020). Optimization of the electric vehicle hvac duct system based on gradient method. International Journal of Mechanics, 14, 135-140. doi:10.46300/9104.2020.14.18

IOP Conf. Series: Materials Science and Engineering **986** (2020) 012049 doi:10.1088/1757-899X/986/1/012049

Electromagnetic compatibility of the on-board electrical complex as a key factor in ensuring the operational safety of vehicles

V. N. Kozlovsky[1], P. A. Nikolaev[2], A. S. Podgorny[1], A. V. Kritsky[1] and M.A. Zhmaylo[3]

[1] Samara State Technical University, Samara, Russia, kozlovskiy-76@mail.ru
[2] JSC "AVTOVAZ", Togliatty, Russia
[3] National Technology Initiative Center for Advanced Manufacturing Technologies based on the Institute of Advanced Manufacturing Technologies of Peter the Great St. Petersburg Polytechnic University Polytechnicheskaya, 29, St.Petersburg, 195251, Russia

Abstract. The issues of functional safety of the on-board electrical complex of vehicles under exposure of electromagnetic interference are considered in this paper. Besides, the modern practice of organizing the process of assessing the immunity of electrical systems of cars is shown.

Introduction
The vehicle is a source of increased danger because of their mass operation. Despite the measures taken, a large number of people constantly die and receive injuries of varying severity in car accidents. Because of accidents, technical damage is significant. Undoubtedly, a malfunction of the on-board electrical complex of the vehicle due to electromagnetic impacts is introduced into these statistics [1, 2].

In the context of the problem of immunity of road transport to electromagnetic interference (EMI), one proceeds from the concept of its functional safety, based on the standards [3-9].

The functional safety of electrical systems, as receptors, is associated with their response to the acting EMI. Failures occur when interfering signals are induced in electrical circuits that exceed the sensitivity thresholds. Summarizing practical experience, the following classification can be made.

Disturbance of functions related to driving vehicles; with the protection of the driver, passengers and/or other road users; disorienting the driver and/or road users; with the functioning of data buses, which, in the event of a malfunction, can adversely impact the readings of devices displaying information.

For the first point, this can be obtained, for example, by deteriorating or changing the performance of the engine control system, brakes, active control, transmission, suspension, etc. For the second point, examples are the malfunctioning of the airbag system, as well as child restraints. For the third one, cases of creating optical and acoustic interference are evident, such as the incorrect operation of direction indicators, sidelights, brake lights, light signaling devices of the emergency system, incorrect operation of the anti-theft device and sound signal. For the fourth point, this may be a malfunction of the odometer, speedometer or navigation device.

The most critical consequences occur due to failures and failures of electrical systems that determine the safety of the vehicle. Functional safety is important not only for a single device, but for the entire

IOP Conf. Series: Materials Science and Engineering **986** (2020) 012049 doi:10.1088/1757-899X/986/1/012049

on-board complex as a whole, because within modern architecture, almost all electrical equipment is interconnected. EMI failure in the operation of one unit may lead to a malfunction of the other.

The functional safety of the vehicle is indirectly related to the human factor. The behavior of vehicles on the road, especially those in private use, in case of malfunctioning of electrical systems due to EMI depends on the proper operation of the vehicle and the behavioral characteristics of the driver.

A broad contingent of different people drives a vehicle, because of its massive availability. Therefore, there is a possibility that the driver's response to a failure of electrical systems due to its specific features would be inadequate. The degree of behavior during movement is determined by the psychophysical state, professional qualities: medical indications, age factor and temperament.

The statistics have several examples: drivers who are used to being guided by the sound signal of safe parking allow hitting an obstacle when driving in reverse when the system fails under EMI or the driver loses control of the road due to his distraction from control to the radio that is disconnected under EMI exposure.

Main part

A certain category of drivers does not pay the necessary attention to the maintenance of vehicles, operating them in disrepair. Such treatment leads to a decrease in the immunity of electrical systems (Figures 1, 2) [10].

To ensure functional safety, electrical systems must have a certain margin of immunity that has to be guaranteed to be confirmed taking into account a given probability during testing.

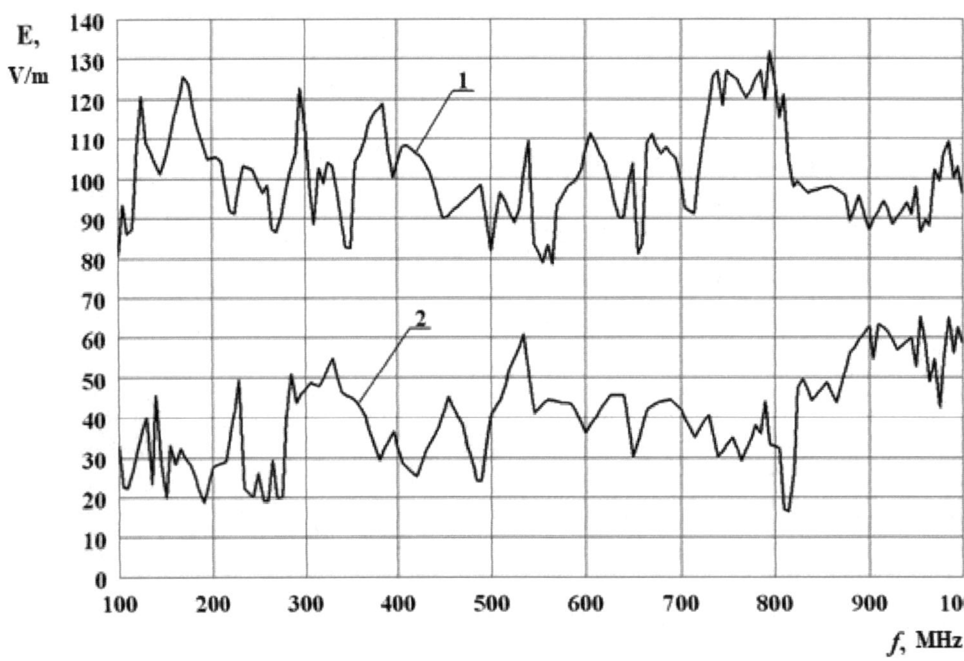

Figure 1. The characteristics of the immunity of the instrument cluster of cars: 1 - Russian brand; 2 - foreign brand.

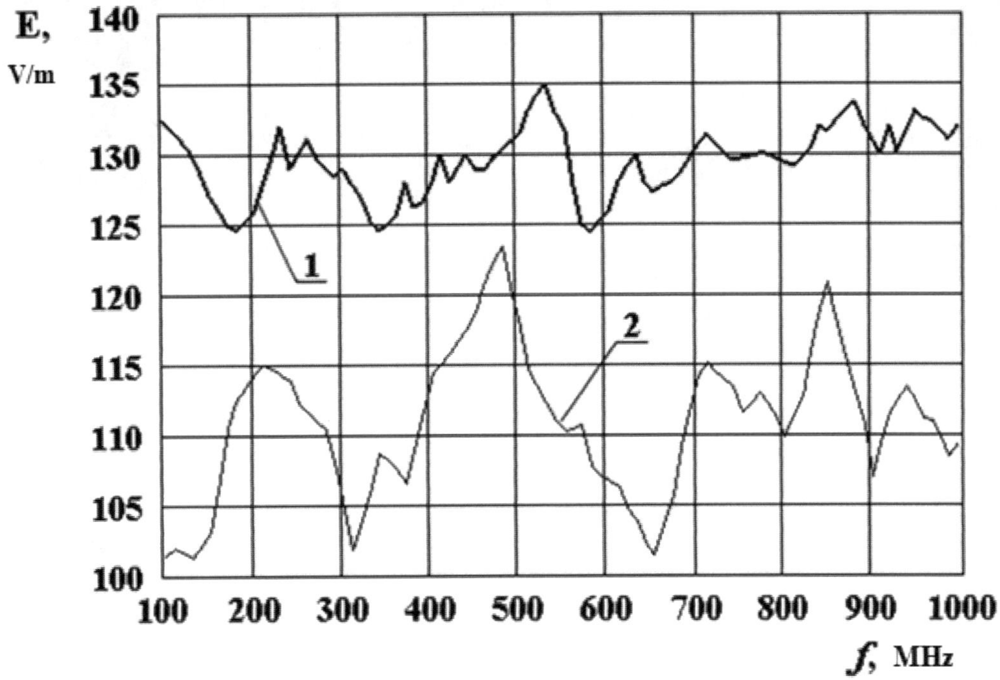

Figure 2. The immunity characteristics of the Russian-made automobile anti-theft system: 1 - 10 km; 2 - 140 thousand km.

There exist test methods prescribed in national and international standards for the civil automotive industry. They make it possible to directly or indirectly assess the immunity of the vehicle's electrical systems to external electromagnetic influences.

The indirect method is the volume current injection. It is based on a non-contact indication of electromagnetic interference in electrical circuits using a current transformer [11, 12, 13, 14]. This method is mainly used when testing individual units and systems. The frequency range of its applicability is limited to 400 MHz.

Other methods are based on direct exposure to electromagnetic radiation. Generalizing them, it can be distinguished tests using radiating antennas and in stripline systems. Tests in stripline systems [14, 15] are based on the electromagnetic effect of a plane-transverse wave distributed between two parallel metal plates. Unshielded stripline systems and shielded TEM-cameras have become widespread [16, 17, 18] (Figure 3). The frequency range of field forming systems depends on their size. The upper range is limited by the frequency at which higher types of electromagnetic waves appear. The lower limit starts at 0 Hz. A typical dependence is the larger the system, the lower the upper operating frequency. In general, field-forming systems intended for the vehicle operate in the range up to 20 MHz, and up to 300 MHz - for products. The main disadvantage is the impossibility of testing at higher frequencies.

Tests using radiating antennas are carried out in the frequency range above 20 MHz in anechoic shielded chambers (ASC) [19, 20]. Open areas are an alternative to ASC, but their use is minimized due to the dependence of the results because of weather conditions. The versatility of the ASC allows it to be used both for testing products and for the vehicle.

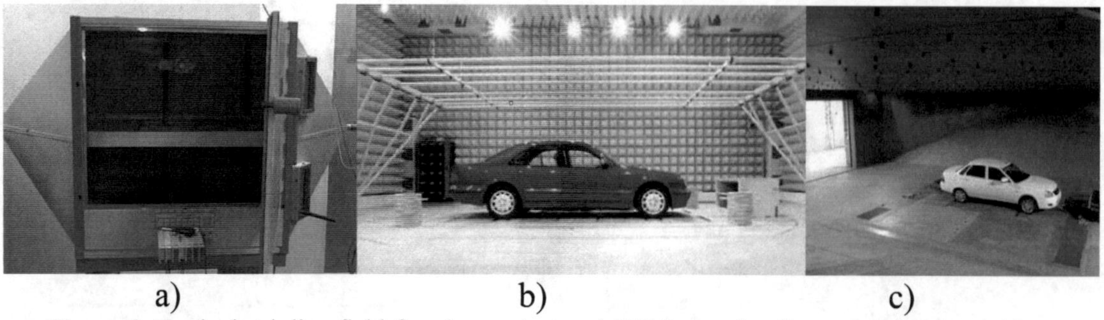

Figure 3. Typical stripline field-forming systems: a) TEM-chamber for testing products; b) unshielded asymmetrical strip line; c) TEM camera for testing the vehicle.

According to the requirements for civilian vehicles, the exposure is produced by antennas with vertical polarization. When testing products up to 1 GHz, the phase center of the antenna should be opposite the center of the wire harness, and an antenna operating above 1 GHz should be positioned opposite the product. During the vehicle's certification [11-13], tests are performed in the range from 20 MHz to 2 GHz. The frequency tuning step in terms of electromagnetic radiation is chosen linear or logarithmic steps (Table 1) [21]. The analysis shows that the larger the step, the wider the frequency range that is not captured during testing. As a result, during the tests, one can skip frequencies at which low noise immunity.

Table 1. Step frequency.

Frequency range, MHz	0,01–0,1	0,1–1	1–10	10–200	200–400	400–1000	1000–2000
Linear step, MHz	0,01	0,1	1	5	10	20	40
Logarithmic step, %	10	10	10	5	5	2	2

For speeding up the test, it is recommended to act on the object only at a few resonant frequencies in [22].

The regulated impacts have the following parameters (Figure 4):

– amplitude modulation 1 kHz with modulation factor 0.8 in the frequency range 20-800 MHz;

– pulse modulation with a duration of 577 μs and a period of 4600 μs in the range from 0.8 to 2 GHz MHz

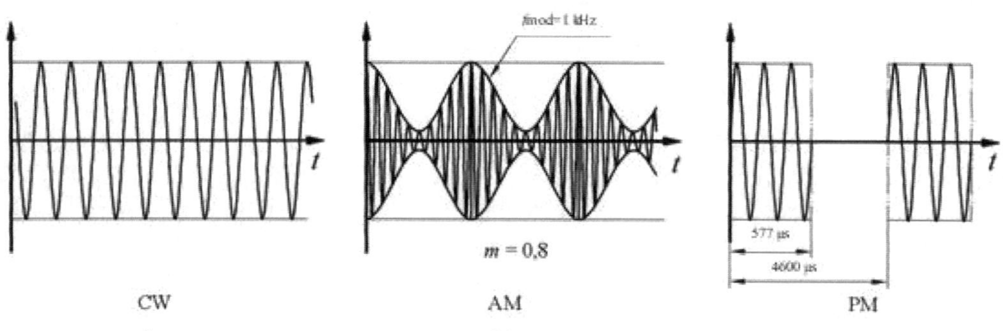

Figure 4. Standard test signals: a) unmodulated harmonic signal; b) amplitude-modulated signal; c) pulse-modulated signal.

The minimum exposure time at a given frequency is 2 s. In the testing process, the vehicle moves on the stand at a constant speed of 50 km/h, and if it is slow-moving, then the speed is set to 25 km/h.
The distance from the car to the radiating antenna is chosen so that the entire front part of the vehicle falls into the main lobe of the radiation pattern of the radiating antenna [23].
Before testing, the vehicle is installed with its front part to the fixed antenna so that its center line is at the control point (line) [23]. If the electrical systems are mainly located in the back or on some side of the vehicle, then the vehicle is oriented with the corresponding part to the radiating antenna [11-13].
In tests (Figure 5), a passenger car was tested for resistance to microwave fields. It was irradiated from the front and obliquely from the rear towards the left side, because in the latter case, the electrical engine management systems were located in the passenger compartment in front of the metal shield separating the passenger compartment and the engine compartment [24].

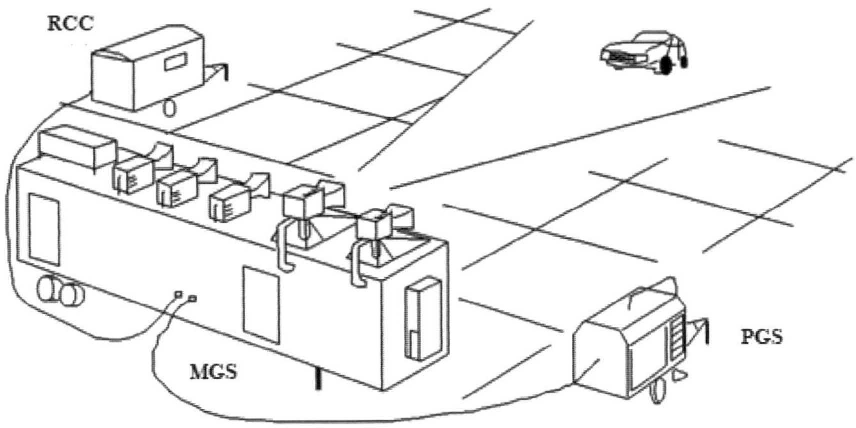

Figure 5. Microwave device for testing the vehicle: MGS - microwave generation system; PGS - power generation system; RCC - remote control cabin.

Leading car manufacturers are additionally testing horizontal polarization at 1 MHz to improve the quality of their products, applying unmodulated harmonic effects as well. Special tests, taking into consideration the peculiarities of the operation of these systems, are developed to check the special functions of electrical systems. [25].
An alternative test for susceptibility is testing the vehicle to the fields generated by on-board transmitters [26, 27]. There are two types of tests, characterized by the location of the antenna in relation to the body. In one test, an antenna is mounted externally to the body of a vehicle and power is supplied from an external source or an on-board radio. The tests are carried out at multiple antenna locations. In another test, a handheld transceiver with a built-in antenna is positioned in the interior of the vehicle. The program includes tests at several locations of the airborne transceiver when located in three mutually perpendicular planes. Recommended modulations: Telegraphy, AM, SSB, FM, TDMA / FDMA, Tetra: π/DQPSK and GMSK with modulation parameters AM 1 KHz with 80% depth; CW and PM 18 Hz with duty ratio 50%.
Tests for resistance to fields of power frequency are divided into two types: the exposure of the electric and magnetic components. In the first case, testing is carried out using a strip line. In the second one, Helmholtz coils are used [28, 31].
Most tests for external EMI, the spectrum of which does not exceed 20 MHz, such as remote lightning impact, by the fields of the contact network of railways and power lines, are carried out in strip systems. [29, 32].
Special purpose technical means are additionally tested for direct and close lightning strikes (Figure 6) [30, 33-34].
All considered tests involve one or several fixed positions relative to the field-generating system, which is insufficient in considering the vehicle as a complex antenna system. This approach does not assess the immunity of vehicles from those directions where slots and holes form an effective antenna at a

specific frequency. A comprehensive, uniform effect is achieved in a reverberation chamber. However, they are mainly used for product testing. There are size limitations for testing full-size vehicles. The larger the camera, the higher its lower frequency range. The typical test frequency is 200 MHz

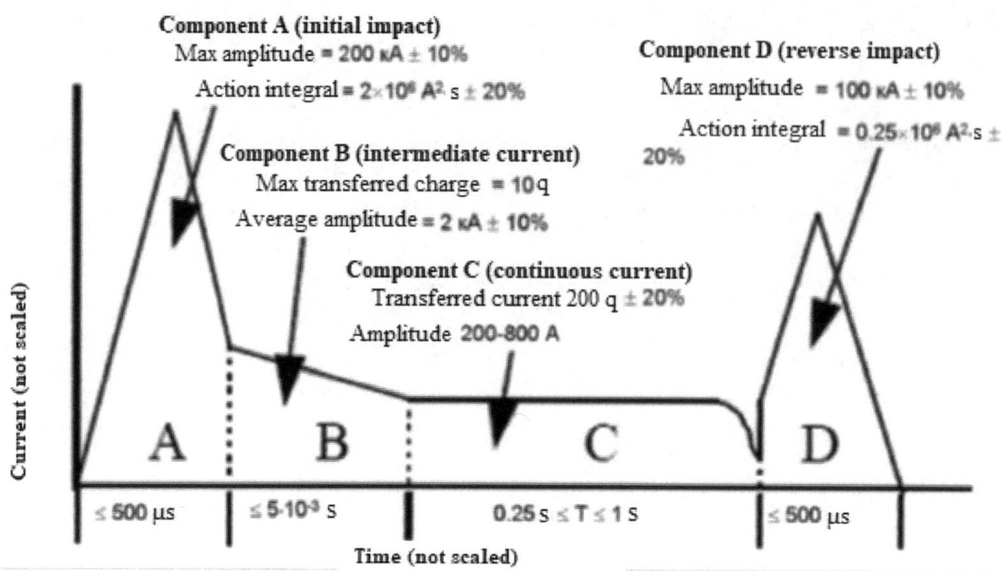

Figure 6. Test impulse of a direct lightning strike.

Conclusion

The analysis of the considered methods shows that the existing requirements for frequency tuning step, positioning relative to the field generating system, to the parameters of impulse influences, as well as for the operating modes during testing, are not sufficient for a full assessment of the noise immunity of the vehicle. This imposes certain risks in the process of their operation.

Thus, the future development of the above methods is directly related to the solution of problems associated with the development of relevant techniques and tools. That solves the problem of determining the required frequency tuning step, taking into consideration the uncertainty factor in the positioning of the vehicle relative to the field-forming system, as well as providing a set of operating modes for the on-board electrical system of the vehicle during control testing.

References

[1] Kozlovsky V N 2010 Ensuring the quality and reliability of the electrical equipment system of cars *Dissertation abstract for the degree of Doctor of Technical Sciences* (Moscow automobile and road construction state technical university Tolyatti)

[2] Stroganov V I and Kozlovsky V N 2014 Modeling of systems of electric vehicles and vehicles with a combined power plant in the design and production processes: monograph (Moscow) p 264

[3] GOST R IEC 61508-1-2012 2013 Functional safety of electrical, electronic, programmable electronic systems related to safety. Part 1. General requirements. (Standartinform) p 53

[4] GOST R IEC 61508-2-2012 2013 Functional safety of electrical, electronic, programmable electronic systems related to safety. Part 2. Requirements for systems (Standartinform) p 87

[5] GOST R IEC 61508-3-2012 2013 Functional safety of electrical, electronic, programmable electronic systems related to safety. Part 3. Requirements for software (Standartinform) p 133

[6] GOST R IEC 61508-4-2012 2013 Functional safety of electrical, electronic, programmable electronic systems related to safety. Part 4. Terms and definitions (Standartinform) p 45

[7] GOST R IEC 61508-5-2012 2013 Functional safety of electrical, electronic, programmable electronic systems related to safety. Part 5. Recommendations for the application of methods for determining safety integrity levels (Standartinform) p 62
[8] GOST R IEC 61508-6-2012 2013 Functional safety of electrical, electronic, programmable electronic systems related to safety. Part 6. Guidance on the application of GOST R IEC 61508-2 and GOST R IEC 61508-3 (Standartinform) p 60
[9] GOST R IEC 61508-7-2012 2013 Functional safety of electrical, electronic, programmable electronic systems related to safety. Part 7. Methods and tools (Standartinform) p 177
[10] Moroz S M, Nikolaev P A and Nikolaev A D 2012 The concept of increasing the electromagnetic safety of vehicles in service *Electronics and electrical equipment of transport* (№2-3) pp 51-55
[11] Regulation №10. Uniform provisions concerning the approval of vehicles with regard to electromagnetic compatibility (Appendix 9, Revision 3, UNECE) 2008
[12] Regulation №10. Uniform provisions concerning the approval of vehicles with regard to electromagnetic compatibility (Appendix 9, Revision 4, UNECE) 2012
[13] Regulation №10. Uniform provisions concerning the approval of vehicles with regard to electromagnetic compatibility. (Appendix 9, Revision 5, UNECE) 2014
[14] ISO 11451-2 2005 Road vehicles. Vehicle test methods for electrical disturbances from narrowband radiated electromagnetic energy. Part 2: Off-vehicle radiation sources
[15] ISO 11452 2005 Road vehicles - Component test methods for electrical disturbances from narrowband radiated electromagnetic energy
[16] Electromagnetic susceptibility measurements of vehicle components using TEM cells (14 kHz – 200 MHz), SAE, 1984, June) pp 130–136
[17] MIL-STD-464A 2002 US Department of Defense Interface Standard. Electromagnetic Environmental Effects for Systems (USA DoD 19.12.2002) p 121
[18] Open field whole-vehicle radiated susceptibility 10 KHz – 18 GHz, Electric field (SAE, 1981, June) pp 115-125
[19] Kovneristy Yu K, Lazarev I Yu and Ravaev A A 1982 Materials that absorb microwave radiation. (Moscow: Nauka) p165
[20] Mitsmaher M Yu and Torgovanov V A 1982 Microwave anechoic chamber (M.: Radio and communication) p 129
[21] ISO 11451-1 2005 Road vehicles. Vehicle test methods for electrical disturbances from narrowband radiated electromagnetic energy. Part 1: General principles and Terminology
[22] Komyagin S I 2015 Electromagnetic resistance of unmanned aerial vehicles. (M.: KRASANDR) p 432
[23] ISO 11451-2 2005 Road vehicles. Vehicle test methods for electrical disturbances from narrowband radiated electromagnetic energy. Part 2: Off-vehicle radiation sources
[24] Microwave test facility *Ericsson Saab Avionics, Electromagnetic Technology Division* (Linkoping, Sweden) pp 581-588
[25] Nikolaev P A and Kechiev L N 2015 Electromagnetic compatibility of vehicles (Ed. L.N. Kechiev, M.: Grifon) p 424
[26] ISO 11451-3 2007 Road vehicles. Vehicle test methods for electrical disturbances from narrowband radiated electromagnetic energy. Part 3: On-board transmitter simulation. Second edition
[27] SAE J 551/12:1996 Vehicle Electromagnetic Immunity-On-Board Transmitter Simulation
[28] Knyazev A D, Kechiev L N and Petrov B V 1989 Design of radio-electronic and electronic-computing equipment taking into account electromagnetic compatibility (M.: Radio and communication) p 224
[29] Davydov A A, Plygach V A and Chibisov Yu F 2010 Electromagnetic factors of natural and man-made origin and methods of their reproduction during testing of weapons and military equipment *EMC technologies* (№1 (32)) pp 38-48
[30] MIL-STD-461F US Department of Defense Interface Standard. Requirements for the control of electromagnetic interference characteristics of subsystems and equipment (10.12.2007) p 255

[31] Alekseev, A., Maksimov, A., & Tarasov, A. (2020). CFD investigations of the effect of rotating wheels, ride height and wheelhouse geometry on the drag coefficient of electric vehicle. International Journal of Mechanics, 14, 130-134. doi:10.46300/9104.2020.14.17

[32] Aleshin, M., Gavrilova, L., Goryainov, I., & Melnikov, A. (2020). Dem generation based on commercial uav photogrammetry data. Paper presented at the Engineering for Rural Development, , 19 1750-1756. doi:10.22616/ERDev2020.19.TF461

[33] Maksimov, A., Igoshina, D., Petrov, R., & Klyavin, O. (2020). Optimization of the electric vehicle hvac duct system based on gradient method. International Journal of Mechanics, 14, 135-140. doi:10.46300/9104.2020.14.18

[34] Panyukov, D., Kozlovsky, V., & Klochkov, Y. (2020). Development and research FMEA expert team model. International Journal of Reliability, Quality and Safety Engineering, doi:10.1142/S021853932040015X

Actualization of the problem of developing an electronic control system for the oxygen sensor heater of an automobile internal combustion engine

V V Debelov[1], V N Kozlovsky[2], D V Aydarov[2] and P.S Cvetkov[3]

[1] The Central research and development automobileand engine institute NAMI
2, Avtomotornaya street, Moscow, 125438, Russia, Moscow
[2]Samara State Technical University, Samara, Russia, kozlovskiy-76@mail.ru
[3]National Technology Initiative Center for Advanced Manufacturing Technologies based on the Institute of Advanced Manufacturing Technologies of Peter the Great St. Petersburg Polytechnic University Polytechnicheskaya, 29, St.Petersburg, 195251, Russia

Abstract. The presented work is devoted to the problem of developing an electronic control system for the oxygen sensor heater of an automobile internal combustion engine.

Introduction

A catalytic converter of exhaust gases plays an important role in reducing the emission of harmful substances into the atmosphere when considering a system for neutralizing harmful emissions from vehicles. Exhaust gases, passing through a special grid, having platinum and rhodium spraying to reduce NOx, and platinum and palladium spraying to reduce CO, transform from aggressive toxic gases into relatively harmless and neutral gases.

In order for the neutralizer to effectively clean the exhaust gases entering it, the content of each component in them must fit into rather narrow frames corresponding to the combustion in the cylinders of a stoichiometric working mixture of fuel and air, which is characterized by the so-called excess air ratio λ. If λ is more than 1, 0, the mixture is lean, if λ is less than 1, 0, the mixture is rich. If the mixture contains such an amount of air that is required for complete combustion of the fuel, the mixture is called stoichiometric (theoretically ideal) - this is the range of λ values near 1.0. In order for the combustion efficiency to be high, fluctuations in the composition relative to the optimal one should not exceed 1% relative to λ=1. The stoichiometric ratio for air and gasoline has a value of about 14,4…14,7 / 1, where λ = (air)/(fuel). Such a composition of the air-fuel mixture provides the lowest content of toxic substances in the exhaust gases and their effective "afterburning" in the neutralizer. The greatest efficiency at 100% open throttle valve of a gasoline engine is achieved at λ =1,1-1,3. The maximum power is provided when λ =0,85-0,9 (Figure 1).

Accurate dosing of the mixture composition and maintenance of its value is required for the neutralizer to effectively perform its function of reducing toxic emissions. This problem is most effectively solved by using an engine control system that provides a stoichiometric ratio by controlling the fuel injectors, the ignition timing spark, and the throttle valve opening angle. But this task of achieving the stoichiometric composition would be difficult to solve without using an oxygen sensor, which is an important feedback element in the fuel mixture injection system. That allows maintaining the stoichiometric composition at steady-state engine operating conditions with an accuracy of at least ±1%.

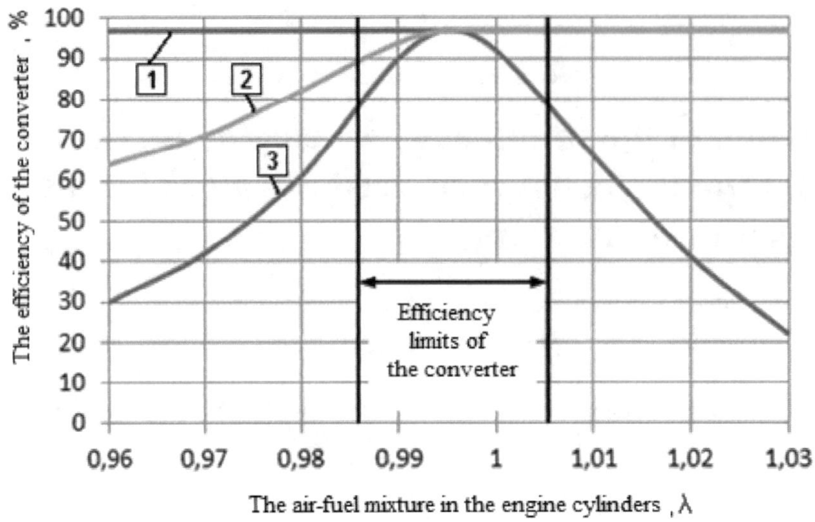

Figure 1. Efficiency dependence of the converter on the composition of the air-fuel mixture in the engine cylinders, where the curve: 1- Nitrogen oxides NO_X, 2- Unburned hydrocarbons C_3H_3, 3- Carbon monoxide CO.

Neutralization system equipped with two oxygen sensors is widely used to obtain the best effect on exhaust gas cleaning. The first oxygen sensor is installed above the vehicle catalyst closer to the engine. This sensor allows the controller of the engine control system to determine the oxygen content in exhaust gases. The engine controller fine-tunes the air-fuel ratio.

The development of electrical engine control systems using a lambda probe is an important scientific and technical task. Its solution enables achieving the best results in the management of mixture formation that leads to a decrease in toxic emissions into the environment. At the same time, the problems of ensuring the reliability of the sensor readings, which can be observed on an aged or partially poisoned probe, also remain important. The issue of protection against burnout of the heating coil of the sensitive element is one of the priority tasks when working with a lambda probe, since the quality of measurement and regulation of the mixture composition depends on its continuous operation and integrity.

Problem

The problem of developing mathematical and simulation models is solved in the paper that provides the most efficient control of the oxygen sensor to obtain the best indicators for controlling the stoichiometric composition.

An oxygen sensor is an electronic device that assesses the amount of unburned fuel or the ratio of oxygen in exhaust gases. The sensor signal is used by the electronic engine management system to maintain an optimal fuel ratio.

The oxygen sensor contributes to the determination of λ in the exhaust gases and generates a corresponding signal by changing the output voltage of the sensor, converts the composition of the mixture into a voltage value λ. Besides, it transfers the exhaust gases through the protective tube to the sensitive element, isolates the inside of the sensor from environmental influences, isolates the inside sensor against the influence of exhaust gases.

The main purpose of the sensor is two-stage control of the mixture composition with feedback and diagnostics of the catalytic converter and as well as diagnostics of the state of the sensor itself in on-board diagnostics systems in systems with two sensors (Figure 2).

Figure 2. Planar oxygen sensor (lambda probe).

Planar oxygen sensors have become very popular in Russian-made vehicles since 2015. They use the Nernst principle in their work, equipped with integrated heaters and a pumped sensing element. These sensors make it possible to accurately determine the composition of the mixture for λ values close to unity. The main manufacturers of lambda probes for the domestic automotive industry are BOSCH and DELPHI.

The effective operation of the sensor is possible at temperatures not lower than 650-900°C. Therefore, modern sensors are equipped with an electric heating element for quick warm-up after starting the engine, which is a ceramic rod containing a sensitive element with an incandescent spiral inside.

All elements of the oxygen sensor are made of heat-resistant materials, since its operating temperature can reach 950 °C. Electric power cables are equipped with heat-resistant insulation. Because of the oxygen sensor can generate an electrical signal only at high temperatures 300-350 ° C and above, the sensors are often installed in the exhaust manifold next to the neutralizer [6, 7].

Since the effective operation of the oxygen sensor directly depends on how much exhaust gas enters it, the installation is carried out close to the catalytic converter before and after it. Because of the operation of the sensor directly depends on the temperature of its sensitive element, it is required to constantly maintain a constant temperature of the sensor, regardless of what temperature of the exhaust gases is currently entering it. For this, the sensor is equipped with a special heating element, which is controlled by the motor controller and ensures the stability of its temperature regime.

A special heating element, which is part of the oxygen sensor, is always controlled from the moment the engine starts until it stops completely. The time should be minimal for the sensor to reach the operating temperature. It is possible to install a lambda probe at a distance from the exhaust manifold, next to the catalyst.

The place of installation of the oxygen sensor is selected in such a way as to exclude the ingress of condensing moisture into it, which can lead to the destruction of the sensitive element when it warms up [8].

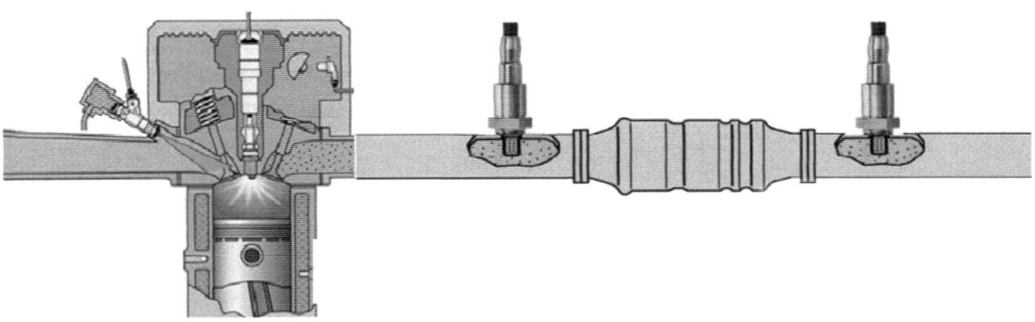

Figure 3. A lambda probe near the catalytic converter.

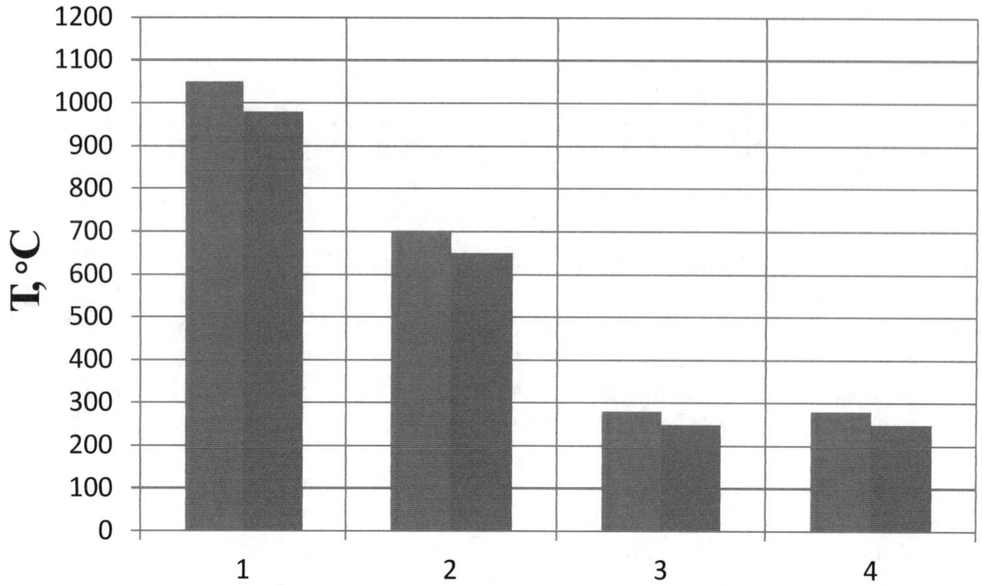

Figure 4. Limiting characteristics of modern planar-type lambda sensors used in passenger cars of domestic production maximum values and at long-term exposure to temperature: 1 – Exhaust gas temperature, 2 – Temperature of the lambda probe housing at the junction with the exhaust manifold, 3 – Temperature of the seal at the junction of the harness to the sensor, 4 – Temperature of cable and protective sleeve.

The ceramics begin to conduct oxygen ions when the probe is in operation, upon reaching a temperature of 300–350 ° C [9-11]. The maximum operating temperature is limited depending on the type of sensor to prevent overheating that inevitably leads to damage. In the event of damage to the lambda probe heater, the engine management system goes into emergency mode without correcting the air content in the mixture.

Conclusion

Thus, the most effective way to solve the problem of reducing harmful emissions into the atmosphere from road transport, as before, is to create and improve the corresponding control system with an oxygen sensor. When such a system is developed, it is necessary to strictly adhere to the requirements for the temperature conditions of the sensor. Hence a highly efficient electronic control system for the oxygen sensor heater is needed to develop.

References

[1] Kamenev V F and Pugachev I O 2016 Domestic control systems for diesel engines with battery-type fuel equipment and a complex anti-toxic system *Transport on alternative fuels* (№ 1 (49) pp 56-62

[2] Kamenev V F and Shcheglov P A 2016 Principles of constructing a mathematical model of a modern low-emission diesel engine for transport and technological means and agricultural machines *Tractors and agricultural machines* (№1) pp 3-8

[3] Debelov V V, Kozlovsky V N, Pyanov M A and Stroganov V I 2015 Modeling and development of an electrical engineering complex for vehicle control in START AND STOP *Truck modes* (№5) pp 15-20

[4] Kozlovsky V N and Debelov V V 2015 Electrical control complex for a car engine *Monograph Samara* p 160

[5] Debelov V V, Kozlovsky V N, Pyanov M A and Stroganov V I 2015 The concept of creating an electrical and electronic control system "START-STOP" of a passenger car *Electronics and electrical equipment of transport* (№ 1) pp 2-6

[6] Erokhov V I 2013 Features of the oxygen sensor of the control system of transport engines *Transport on alternative fuel* (№6 (36)) pp 17-26
[7] Erokhov V I 2011 Gasoline engine injection systems (design, calculation, diagnostics) *college textbook* (Hot line) p 567
[8] Solomatin N S 2013 Testing of components, assemblies and systems of the car *Monograph* (Tolyatti) p 220
[9] Yutt V E 2009 Electrical equipment of cars. Textbook for students of automotive specialties of higher educational institutions (Moscow Textbook for higher educational institutions. Specialty (4th ed., Revised and enlarged))
[10] Karunin A L, Bakhmutov S V and Selifonov V V 2007 Hybrid Cars - The Pillar Road to Economical and Environmentally Friendly Transport *Journal of Automotive Engineers* (№3 (44)) pp 38-45
[11] Chechurin, L. S., Wits, W. W., & Bakker, H. M. (2015). Invention software support by integrating function and mathematical modeling. Chemical Engineering Research and Design, 103, 32-39. doi:10.1016/j.cherd.2015.07.013

IOP Conf. Series: Materials Science and Engineering **986** (2020) 012051 doi:10.1088/1757-899X/986/1/012051

Improve the integrity testing process based on QFD, FMEA and TRIZ

Maria V Vysotskaya[1]
[1]Graduate student of institute of the aircraft equipment of the Samara national research university named after the academician S.P. Korolev; Samara
Manya_93@mail.ru

Abstract. In today's aircraft industry – competitiveness in the domestic and foreign markets, is the main criterion for the development of quality, a high level of which is achieved through methods and approaches: QFD, FMEA, TRIZ. There is a possibility of application of these methods at tests of production, for the purpose to meet expectations of consumers with the most optimum combination of characteristics of production.

The design process is often associated with a compromise of conflicting parameters, which means increased costs and loss of quality. Innovation in the production process can create great value. The rapid development of modern technologies and the commercial environment suggests that the design of production processes requires the application of a systematic method of improvement. The proposed method combines a number of quality tools to support the process of modern development of aviation. In order to implement the acceptance improvement, it is necessary, first, to analyze the product testing process in accordance with the influencing parameters and use QFD (quality function deployment) to obtain the basic requirements for the testing process. Secondly, to find the contradictions of the testing process and analyze them comprehensively. To creatively solve the revealed contradictions, apply the Russian theory of inventive problem solving (TRIZ). When using this technique increases both the efficiency and maneuverability of the use of quality tools to solve technological problems. The effectiveness of the proposed method of improvement is illustrated and confirmed by the example of the process of testing the tightness of the hose of the life support system of the product "Bion - M".

1. Introduction

Aircraft belong to complex technical systems, failures of which often lead to serious consequences.

Production of any production is based on stability of technological processes which confirmation serve static and dynamic tests of production with the maximum approach to real operating conditions. Static tests to a greater extent characterize the properties of the material, and dynamic – the validity of the selected parameters of structures and technologies.

Operational testing plays an important role in the production life cycle, as it provides an opportunity to confirm compliance with the requirements that should be assessed in conditions as close as possible to the actual operation.

The main methods of conformity assessment are measurement, testing and control, which must be carried out in accredited laboratories with certified testing equipment.

It is advisable to improve the quality of products to use modern tools and methods of quality management, such as: QFD, FMEA, TRIZ.

IOP Conf. Series: Materials Science and Engineering **986** (2020) 012051 doi:10.1088/1757-899X/986/1/012051

FMEA – "Analysis of the types and consequences of potential nonconformities" assesses the risks and possible damage caused by design and process nonconformities at an early stage in the design and construction of the finished product or components. Is divided into: PFMEA to identify potential inconsistencies of the process and DFMEA to identify potential inconsistencies of the design [4].

QFD – "quality function deployment Method", provides high quality of the created products at each stage of the life cycle with a guarantee of obtaining the final result that meets the requirements and expectations of the consumer [5].

TRIZ – "Theory of inventive problem solving", used to identify and use the laws, laws and trends in the development of technical systems. TRIZ is designed to organize the creative potential of the individual so as to promote self-development and the search for solutions to creative problems in various fields. The main task of TRIZ is to propose an algorithm that allows to find the most suitable variant without searching for infinite variants of solutions to the problem, discarding the less qualitative ones. Or, to put it more simply, TRIZ allows you to solve the inventive problem so that the output to get the highest efficiency. [1]

According to modern research in the aerospace industry, the test process plays an important role between design development and production. However, the testing process has limitations or contradictions and requires a compromise between many factors. The rapidly changing modern market makes us strive for high competitiveness and develop production processes with the use of systematic improvement techniques to ensure stable product quality. A number of quality tools, methods and approaches are used to improve test design processes. Innovation processes usually mean a change in test technology that includes new designs, processes, the use of modern equipment and test models. This article describes the design strategy of production processes on the example of combining modern quality management tools: QFD, FMEA and TRIZ [2].

QFD and FMEA can help to identify the requirements to test the system and determine actions to address the shortcomings of the system. The TRIZ method is used to improve the testing process and to solve the conflict parameters of the process system. Since the design process at an early stage plays a crucial role, which determines the degree of compliance of the product with the declared requirements of the consumer, it is extremely important to build a systematic approach to the design of the test process. The implementation of a compromise strategy in the design of test processes helps to recognize and manage conflicting process parameters and find creative solutions to the problem [6].

TRIZ is a theory that can effectively resolve technological conflicts or contradictions with the help of original ideas. In the design of the test process and the use of TRIZ analysis of the contradictions of the process, which are the source of conflict moments of the test environment, is inevitable. However, if you solve the problem comprehensively, especially in a complex system of production processes, it is necessary to learn about all the requirements, not only about the demands of consumers, but also the requirements for the test system. And QFD (quality function deployment) can be an effective tool for requirements analysis.

In accordance with the above, this article proposes a strategy for the improvement process. QFD, FMEA and TRIZ are used as an integrated methodology to select the appropriate solution for emerging problems in the testing process. First, QFD is used to transform process requirements into technical specifications. Meanwhile, many technical contradictions can be found, given the complexity of the system of production processes. In this case, TRIZ is used, for example, standard methods of eliminating technical contradictions, which are one of the effective tools to break down the complex system of the design process into principles and achieve several alternative modern solutions. Common design parameters for selecting a suitable process system include productivity, safety, environment, quality, flexibility and cost. If you cannot achieve a solution to the problems, you must apply TRIZ again. An example of the improvement process strategy is shown in figure 1 [3].

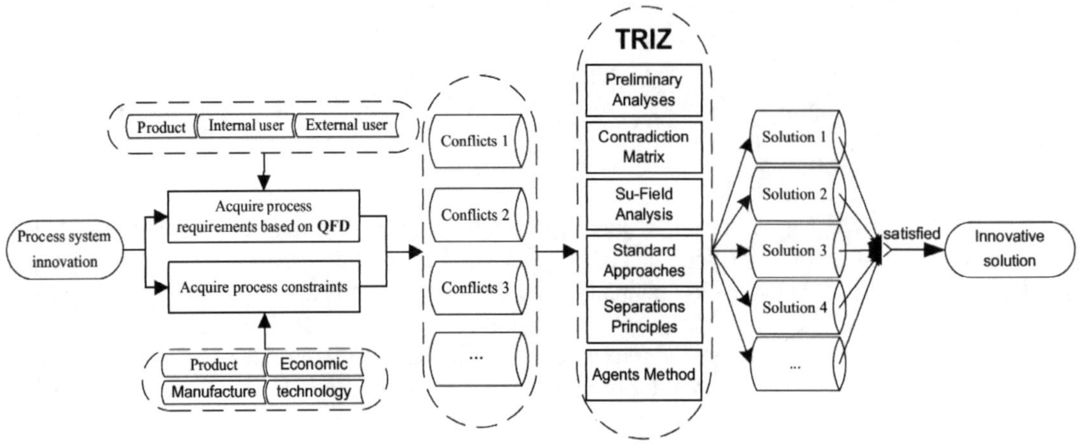

Figure 1. Improvement process strategy

2. Object of research

The strategy of the improvement process is applied on the example of testing the tightness of the hose of the life support system of the product 12XM. Tests are carried out on the pneumatic panel VK 70209-398A. The block diagram of the test bench is shown in the figure.

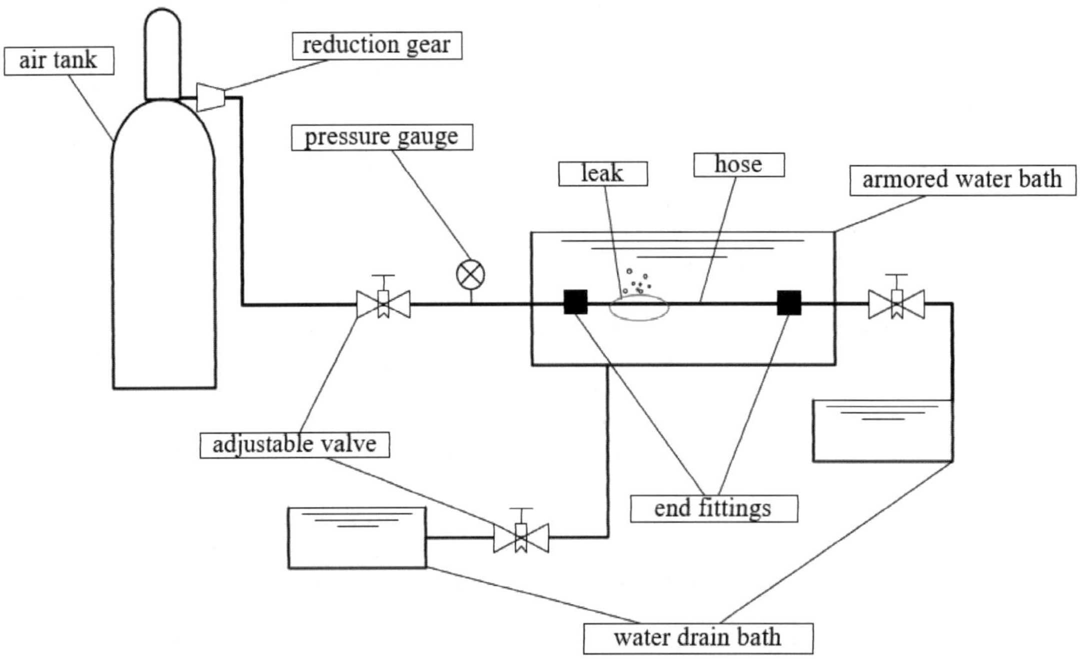

Figure 2. Structural diagram of the test bench

The object of the test is the hose of the life support system of the product 12 KSM presented in figure 3. Hose construction consists of seal (pin & sleeve) and the sleeves 2K7-30, supplied by THAT 38.0051515-92. The elements of the sealing hose is exposed to the greatest stress, as it affects them the most free weight of the sleeve 2K7-30. The hose is installed on the outer surface of the spacecraft on the cable mast and serves to supply oxygen into the spacecraft from oxygen cylinders.

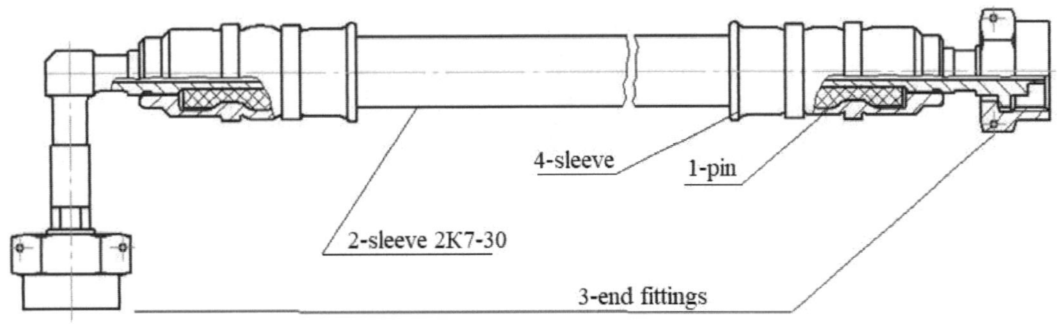

Figure 3. Hose, life support system

Based on the test procedure program, there are test requirements.

1. Before each test:
   - ✓ inspect hoses for mechanical damage and contamination;
   - ✓ in the case of the presence of moisture on the hoses to keep them dry.
2. After each test, the hoses must be completely drained and kept plugged. Drying is carried out by blowing air until complete removal of moisture.
3. When installing the hoses in a device for testing the tightness of coupling nuts of the joints to produce 1.0 to 1.5 facets of focus, are not allowed twisting of the hoses around the axis. Seal the joints with gaskets.
4. Test the hose for tightness with compressed air pressure (3.45±0.15) MPa [(35±1.5) kgf/cm2] for 5 minutes under water with a preliminary exposure of 10 min. No tightness is not allowed (no air bubbles from the hose in the water and pressure drop on the pressure gauge).
5. The necessary measuring instruments used in the tests must be verified and the test equipment must be certified.

3. Application of the QFD method to identify the basic requirements and process parameters «leak test»

The term QFD is a method that allows to deploy the customer requirements into performance characteristics, performance characteristics - into part characteristics, part characteristics into manufacturing process and manufacturing process - into control tools [6, 7]. QFD should be used since the conceptual design stage, because the decisions made at this stage provides about 70% of the project's success with minimal costs [8, 9].

Figure 1 shows the structure of the main QFD tool - the House of Quality, which received such a name for its shape.

Requirements for the process of testing for tightness and contradictions of parameters exist simultaneously, which usually affects the appearance of many technological conflicts between the accuracy of observations and the speed of testing, the presence of air bubbles and the lack of pressure, etc. it is Necessary to assess the requirements and contradictions of the test process, this requires to find out the controversial points of the process without missing anything. The first thing to do is to find out the exact parameters of the requirements, not a General description. QFD (quality function deployment) can be an effective tool for this. Secondly, the limitations of the process system, which are related to costs, technologies and resources, etc., must be fully taken into account.

The purpose of the deployment of quality functions (QualityFunctionDeployment - QFD) is to ensure the quality of the created products at each stage of the life cycle, which would guarantee the final result that meets the requirements and expectations of the consumer [7].

Technology QFD includes a 4-phase (level): quality plan product (phase 1), the quality of the product components (phase 2), quality plan process (phase 3), the quality plan of operations (phase 4), each of which used a matrix diagram is a special kind of quality. Quality house is an element of quality functions deployment technology (QualityFunctionDeployment - QFD) [8].

To identify the quality requirements of the process and operations, we need the third and fourth level of QFD methodology. For QFD the third level, we need requirements to test the system and steps of the process of "testing for leaks".

Requirements for the test system according to the degree of failure are divided into three main types:

1. Anticipatory:
   - ✓ ergonomics of testing.
2. To health:
   - ✓ the presence of the user in the workplace;
   - ✓ high performance (maximum number of tests per day);
   - ✓ the safety of good products;
3. Obligatory:
   - ✓ quality assurance seal of rubber hoses to end fittings
   - ✓ impact of loads provided by operational factors,
   - ✓ reproducible taking into account the capabilities of the test equipment;
   - ✓ evaluation of the reliability of the test object according to the test results;
   - ✓ security testing;

The stages of the process "test for leaks»:

1. Install the hoses in the equipment and connect to the pneumatic console;
   To drown out the hoses;
2. Give pressure to hoses with compressed air P=0,5 kgf/cm2;
   Place the hoses in an armored bath water;
3. To increase the pressure in the hoses to 36 kgf/cm2;
   Withstand pressure for at least 10 minutes;
4. Check for the tightness of the sealing space and the wall of the hose within 5 minutes. Leakage is not allowed (Signs of leakage are air bubbles in the chromatic solution or pressure drop on the pressure gauge);
5. To relieve the pressure to 0.5 kgf/cm2;
6. Remove hoses from armored bath;
   Bleed the pressure to 0 kg/cm2.;
7. Disconnect the hoses from the pneumatic console;
   Remove plugs from hoses and tape;
8. Blow the hose with compressed air P=10 kgf/cm2 until the moisture is completely removed;
9. To reflect the result of the test;
10. To establish technological plugs in the hose.

Figure 4 shows the third level QFD.

As a result of the QFD of the third level, the key operations and parameters of the "leakproofness test" process are determined»:

1) Withstand pressure for at least 10 minutes (14.5%)
2) Reflect the result of the test in the sheet of test marks (11.5%)
3) Install the hoses in the equipment and connect to the air gun (10,6%)

For QFD the fourth level, we need the operation to the process "test hose for leaks".

Auxiliary requirements for the "leak test" for the hose object of the life support system of the product 12 KSM are presented in the second section "Object of study". Figure 5 shows the QFD of the fourth level.

Based on the matrix diagram, QFD allows you to understand the key operations that affect product performance, product quality and process parameters of the "leak test".

Row #	Max Relationship Value in Row	Relative Weight	Weight / Importance	Quality Characteristics (a.k.a. "Functional Requirements" or "Hows") / Demanded Quality (a.k.a. "Customer Requirements" or "Whats")	Install the hoses in the equipment and connect to the pneumatic panel	Plug the hoses	Give the pressure in the hoses with compressed air 0.5 kgf / cm2	Place the hoses in an armored water bath.	Increase the pressure in the hoses to 36 kgf / cm2	Hold under pressure for at least 10 minutes	Check tightness of the place of seal and wall - 5 minutes	Relieve pressure to 0.5 kgf / cm2	Remove hoses from armored bath	Relieve pressure to 0 kgf / cm2	Disconnect the hoses from the pneumatic panel	Remove plugs and cassette from hoses	Blow out the hose with compressed air 10 kgf / cm2 to remove moisture	Reflect the test result	Install process plugs on the hose
1	9	18,5	10,0	Checking the quality of sealing rubber hoses in the end fittings under the influence of loads	O					⊙									
2	9	18,5	10,0	Evaluation of the reliability of the test object according to test results		O			⊙	⊙	⊙			⊙			⊙	⊙	
3	9	18,5	10,0	Safety test	⊙										⊙	⊙			⊙
4	9	14,8	8,0	Safety of tested tests	▲	▲	⊙			⊙	O	⊙		⊙			⊙	⊙	
5	9	13,0	7,0	The presence of instructions in the workplace	⊙	▲	⊙			O	O	⊙							▲
6	9	11,1	6,0	High performance (maximum number of tests per day)				O										⊙	
7	9	5,6	3,0	Ergonomic testing	O								O		⊙				▲
				Target or Limit Value			0.5 kgf / cm2	armored bath with water	up to 36 kgf / cm2	at least 10 minutes	5 minutes	up to 0.5 kgf / cm2		0 kgf / cm2			purge with a pressure of 10 kgf / cm2	test mark sheet	
				Difficulty (0=Easy to Accomplish, 10=Extremely Difficult)															
				Max Relationship Value in Column	9	3	9	3	9	9	9	9	3	9	9	9	9	9	9
				Weight / Importance	370,4	83,3	250,0	33,3	166,7	505,6	250,0	250,0	16,7	300,0	216,7	166,7	300,0	400,0	185,2
				Relative Weight	10,6	2,4	7,2	1,0	4,8	14,5	7,2	7,2	0,5	8,6	6,2	4,8	8,6	11,4	5,3

Figure 4. Third level QFD for "leak test»

Row #	Max Relationship Value in Row	Relative Weight	Weight / Importance	Quality Characteristics (a.k.a. "Hows") / Demanded Quality (a.k.a. "Whats")	Inspect hoses for mechanical damage and impurities	If there is moisture on the hoses, make them dry	After each type of test, the hoses must be completely drained	Hoses should be kept completely plugged	Air drying is carried out to completely remove moisture	To make a cape of nuts on 1,0-1,5 sides from an emphasis	Test with compressed air pressure of 3.45 ± 0.15 MPa (35 ± 1.5 kgf / cm2)	Test for 5 minutes under water at 10 minutes pre-exposure	Means of measurement are verified, test equipment certified
1	9	10,6	370,4	Install the hoses in the equipment and connect to the pneumatic panel	O	▲		▲			O		Θ
2	9	2,4	83,3	Plug the hoses				Θ		O			
3	3	7,2	250,0	Give the pressure in the hoses with compressed air 0.5 kgf / cm2	O						O		
4	9	1,0	33,3	Place the hoses in an armored water bath.	Θ								
5	9	4,8	166,7	Increase the pressure in the hoses to 36 kgf / cm2							Θ		
6	9	14,5	505,6	Hold under pressure for at least 10 minutes								Θ	
7	9	7,2	250,0	Check tightness of the place of seal and wall - 5 minutes	▲							Θ	
8	9	7,2	250,0	Relieve pressure to 0.5 kgf / cm2							Θ		
9		0,5	16,7	Remove hoses from armored bath									
10	3	8,6	300,0	Relieve pressure to 0 kgf / cm2			O		O		▲		
11	1	6,2	216,7	Disconnect the hoses from the pneumatic panel									▲
12		4,8	166,7	Remove plugs and cassette from hoses									
13	9	8,6	300,0	Blow out the hose with compressed air 10 kgf / cm2 to remove moisture		▲	Θ		Θ				
14	9	11,4	400,0	Reflect the test result					Θ		O	O	Θ
15		5,3	185,2	Install process plugs on the hose									
				Target or Limit Value									
				Difficulty (0=Easy to Accomplish, 10=Extremely Difficult)									
				Max Relationship Value in Column	9	1	9	9	9	3	9	9	9
				Weight / Importance	69,0	19,2	103,0	32,1	206,0	7,2	203,5	228,9	204,6
				Relative Weight	6,4	1,8	9,6	3,0	19,2	0,7	19,0	21,3	19,1

Figure 5. QFD fourth level for the process of verification of impermeability of the hose

IOP Conf. Series: Materials Science and Engineering **986** (2020) 012051 doi:10.1088/1757-899X/986/1/012051

4. The application of TRIZ

There are many controversial points in the testing process, so it is advisable to use TRIZ at an early stage of design. One of the prerequisites of TRIZ is that there are objective laws of development and functioning of systems, based on which it is possible to build inventive solutions. In other words, many technical, production, economic and social systems are developing according to the same rules and principles. Soviet scientist and inventor Heinrich Saulovich Altshuller discovered them by studying the patent Fund and analyzing the ways of development and improvement of technology for a long time. A list of 40 standard methods of eliminating technical contradictions has been created – a kind of desktop guide for the inventor, but a special kind of guide: the inventor should consider it as the basis for which it is necessary to replenish the new technical and patent publications. TRIZ is a human-oriented systematic methodology for solving inventive tasks based on knowledge. It can solve technical problems and offer innovative solutions using a knowledge base.

To study the similar features of other inventions for the use of TRIZ and determine the patent purity of the studied object of technology: a stand for checking products for leaks, a patent search was conducted and materials were selected for further analysis:

1) Stand tightness control of filter elements and filter packages – utility model №74617.
2) Stand for testing the tightness of the transformer housing – utility model №106368.
3) Stand for checking the tightness of radiators before and after repair – utility model №172904.
4) Stand for checking the tightness of gas valves of the boiler – utility model №99849.
5) Walls for leak test – the invention №2583880.

Stand for testing air hoses – utility model №38 463 is selected as an analogue for the use of TRIZ, the stand refers to devices for monitoring and testing of brake systems and is designed for testing brake hoses with air. Currently, the company is known to stand the test of hoses, which hoses the consolidation is done manually, the lifting and lowering of the frames with handheld hoses.

The disadvantage of the known stand is the use of manual labor, that is, manual hose clamping and lowering of the framework.

The purpose of the proposed utility model is to increase the performance of testing hoses by automating the testing process by using a pneumatic actuator and a sealing unit.

In this article TRIZ is applied in relation to the process of testing the hose of the life support system of the product 12KSM, the purpose of which is to check the quality of sealing of rubber hoses in the end fittings by the loads provided by the operational factors reproduced taking into account the capabilities of the test equipment, as well as to assess the reliability of the test object according to the test results.

Risk analysis by FMEA [4] allowed to determine a number of contradictions of the test process, for example, such as: the dependence of the pressure in the test bench, affecting the test result, i.e. if the pressure in the system is up to 36 kgf/cm2, it is possible defective hose is allowed to work, at a pressure above 36 kgf/cm2 there is the appearance of microcracks in the hose.

On the basis of the table of application methods of solving technical contradictions the selected factors: No. 11 Pressure and No. 27 Reliability. To solve these contradictions, standard methods of their elimination are proposed, such as: №19 "Principle of periodic action", which consists in the transition from continuous to periodic (impulse) and №35 "Change of physical and chemical parameters of the object", which allows to change the concentration or consistency.

On the basis of the above methods, measures have been developed to improve the stand for tightness control:

✓ Air supply will be carried out according to a certain law (pulse mode) short-term supply to the system is not constant, but variable, with a certain frequency and amplitude of oscillations.

✓ Adding a chemical reagent to the test environment of the bath to change the physical and chemical properties of the water, in order to improve the visualization of the detection of non-tightness of the hoses.

IOP Conf. Series: Materials Science and Engineering **986** (2020) 012051 doi:10.1088/1757-899X/986/1/012051

5. Conclusion

1. Operational testing plays an important role in the production life cycle, as it provides an opportunity to confirm compliance with the requirements that should be assessed in conditions as close as possible to the actual operation.
2. Modern methods and approaches are applied and investigated: QFD, FMEA, TRIZ, for improvement of process of tests on tightness.
3. QFD fourth level for the process of verification of impermeability of the hose allowed us to identify key activities that could affect efficiency of production.
4. Risk analysis by FMEA revealed a number of contradictions in the testing process, for which TRIZ was used.
5. A review of patents for existing analogues of the test bench to test products for leaks, as well as the application of the theory of inventive problem solving allowed to come to solutions of technical contradictions.
6. Measures to improve the test stand are defined.

References

[1] Duflou, J. R., & D'hondt, J. (2011). Applying TRIZ for systematic manufacturing process innovation: the single point incremental forming case.Procedia Engineering, 9, 528-537.

[2] Li, T. S., & Huang, H. H. (2009). Applying TRIZ and Fuzzy AHP to develop innovative design for automated manufacturing systems. Expert systems with applications, 36(4), 8302-8312.

[3] Wang, C., Zhao, W., Wang, J., Li, X., Zhang, K., Guo X., & Zeng J. (2015). The Optimizing Strategy of Systematic Process Innovation based on QFD TRIZ and AHP, Applied Mathematics & Information Sciences, No. 3, 1593-1604.

[4] Vysotskaya M.V., Khryashcheva D.V., Orlova S.I., Tsyman VN, Russkikh I.V., Shurshev M.A. Improving the process of leak testing with modern quality management methods. Quality and life. 2018. No. 3 (19). Pp. 16-24.

[5] Dmitriev A Ya and Mitroshkina T A Improving the efficiency of aviation products design based on international standards and robust approaches 2019 IOP Conference Series: Materials Science and Engineering.

[6] Dmitriev A Ya, Vashukov Yu A and Mitroshkina T A 2016 Robust design and technological preparation of the manufacturing of aviation equipment (Samara: Samara University) p 76 [In Rus].

[7] Dmitriev A Ya and Mitroshkina T A The ontological model and the hybrid expert system for products and processes quality identification involving the approach based on system analysis and quality function deployment 2016 ITM Web of Conferences 6 02005.

[8] Dmitriev A, Mitroshkina T and Rogachev G Structural and parametric analysis of robust design quality of complex technical systems 2017 ITM Web of Conferences 10 01001.

Numerical and experimental investigation of the main parameters of a small gas turbine engine

A S Tikhonov, A I Borovkov and A Yu Tamm

Center of National Technological Initiative, Institute for Advanced Manufacturing Technologies, Peter the Great St. Petersburg Polytechnic University, 29 Politekhnicheskaya, St. Petersburg, 195251, Russia, tikhonov@compmechlab.com

Abstract. A numerical and experimental study of the main parameters of a small gas turbine engine is conducted in this paper. An experimental bench was developed for determining both the integral parameters (thrust and fuel consumption) of the gas turbine engine and the local parameters (temperatures and pressures) of its gas-air path. Besides, full-scale gas-turbine engine tests Jetcat P200-RX were done with and without gas-air path equipment. The existing small gas turbine engine sample was digitized and the gas flow in its gas-air path was simulated using the CFD method in a three-dimensional setting in ANSYS CFX. A comparative analysis of the results of full-scale tests and CFD modeling was carried out.

1. Introduction

The interest in the development of unmanned aerial vehicles (UAV) has grown significantly over the past decade. This is caused by performing various military and civilian missions using UAV, for example, photo and video reconnaissance, reconnaissance of chemical and radiation conditions, as well as using false targets to complicate the air situation.

The key point in UAV development is small gas turbine engines (SGTE). SGTE currently has a wide scope. SGTE are used as auxiliary or power plants of light aircraft, cruise missiles, helicopters, as well as power plants of land and water vehicles. In addition, they are used as drives of electric generators, and as sources of compressed air. Nowadays, finite-element supercomputer modeling tools for the basic physical processes that occur in the SGTE (hydro and gas dynamics, heat transfer, combustion, etc.) are actively used. This can significantly accelerate and reduce the cost of the design process for new SGTEs. At the same time, there is still a problem of insufficient verification of design models at both the engine level and its components.

The analysis and selection of the object of study were conducted for SGTE research. The main engine selection parameters were:

– market availability of finished small gas turbine engines in Russia;
– relativity low cost;
– dimensions allowing easy transport and equipment;
– universal scope;
– availability of scientific and technical research materials, which develops a methodology for designing and calculating a small gas turbine engine.

Based on the above parameters, a small single-shaft Jetcat P200-RX turbojet engine was adopted as an analog by the Customer. The appearance of this engine is shown in Figure 1.

Figure 1. The appearance of the small single-shaft turbojet Jetcat P200-RX.

The goal of this work is to determine the main parameters of the Jetcat P200-RX SGTE and its components using experimental research, as well as mathematical modeling methods.

On the one hand, there is a sufficiently large data on this engine, but on the other hand, the incomplete set of experimental data.

A large number of direct measurements of the thermal state of the stator parts of this engine is considered in the dissertation [1]. However, the parameters of the gas-air path are not fully measured, and those that are measured are given in dimensionless relative values, which makes them inaccessible for analysis.

Reverse engineering of this engine was performed, and the aerodynamic characteristics of its centrifugal compressor were studied in detail in [2] and [3] works. At the same time, there are no experimental and numerical data on the hot part in the engine. Similar data on experimental studies of the compressor parameters of this engine are given in [4] and [5].

The literature analysis shows that the experimentally measured characteristics of the gas flow for the engine under consideration are still insufficient in an open press, as well as insufficient data on a comprehensive comparison of the measured and obtained by CFD modeling of the main and local parameters of the engine. This makes the work relevant.

The main specifications of the Jetcat P200-RX engine, as declared by the manufacturer [6], are shown in Table 1.

Table 1. Specifications of the Jetcat P200-RX engine.

Brand	JetCat
Model	P200-RX
Idle speed, rpm	33000
Maximum, rpm	112000
Thrust at idle, N	9
Thrust maxRpm, N	230
Exhaust gas temperature, °C	750
Pressure ratio	3.7
Mass flow, kg / s	0.45
Exhaust gas velocity (km/h)	1840
Power, kW	58.8
Consumption Full load (ml/min)	730
Consumption idle (ml/min)	129

Repair interval, h	25
Weight, g	2560
Dimensions of the diameter (mm)	132
Length (with starter), mm	355

2. Description of the experimental setup

x`On the basis of the "Turbine" department of St. Petersburg Polytechnic University, an experimental bench has been made for a full-scale engine test, which is shown in Figure 2.

The bench consists of the main blocks: the research object, the support frame, the flow measurement system, the fuel supply system, the JetCat P200-RX gas turbine engine, the fuel combustion products exhaust pipe, the plant control system and the experimental data collection.

The object of the study is installed in the block supports and is attached with clamps using M8 bolts. Asbestos tapes provide thermal insulation of the support system from the engine housing. Both supports are attached to the base with four M8 bolts.

The support block allows the engine to move along its axis on radial ball bearings along two guides, fastened with M8 bolts to the foundation plate. Bearing blocks are installed in pairs in two housings attached to the base with four M8 bolts.

The fuel system provides the engine with the necessary fuel consumption and measuring its consumption. Fuel is supplied by a gear pump TN under pressure through a filter. The system pressure is set and maintained automatically by a regulating solenoid valve installed on the bypass line of the fuel system.

The control system for the installation and the collection of experimental data consists of a remote control panel, a personal computer, a jet engine thrust measurement system (JTMS).

The JetCat P200-RX engine was equipped with appropriate thermal and pneumatic combs to measure temperatures and pressures that characterize the working process in the flow part (Figures 3-5).

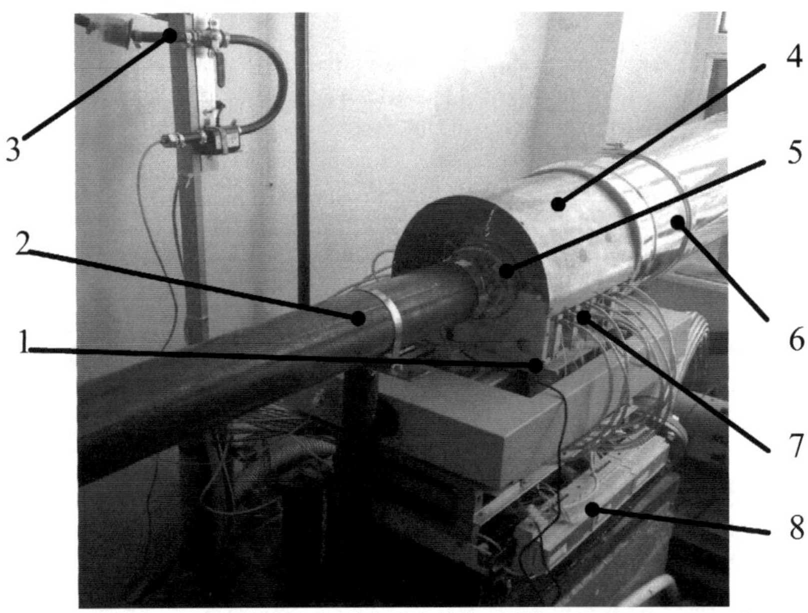

1 – support frame; 2 – air supply pipe; 3 – fuel supply systems; 4 – protective casing; 5 – investigated object; 6 – exhaust pipe for fuel combustion products; 7 – pressure measuring unit; 8 – temperature measuring unit

Figure 2. The experimental bench.

IOP Conf. Series: Materials Science and Engineering **986** (2020) 012052 doi:10.1088/1757-899X/986/1/012052

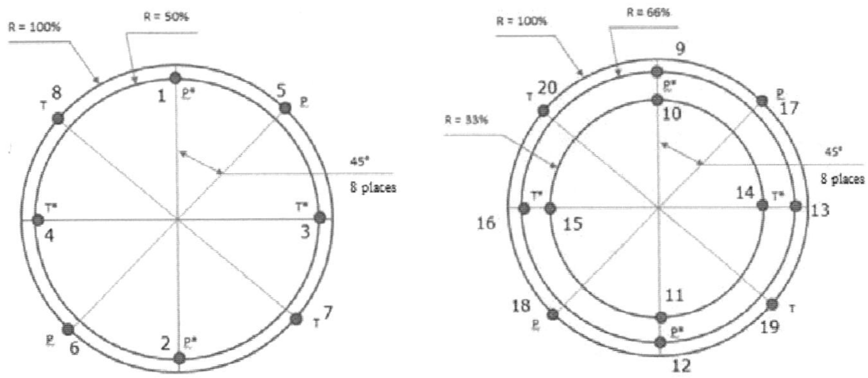

Figure 3. Section A (left, behind the compressor) and B (right, behind the compressor). View from the nozzle side.

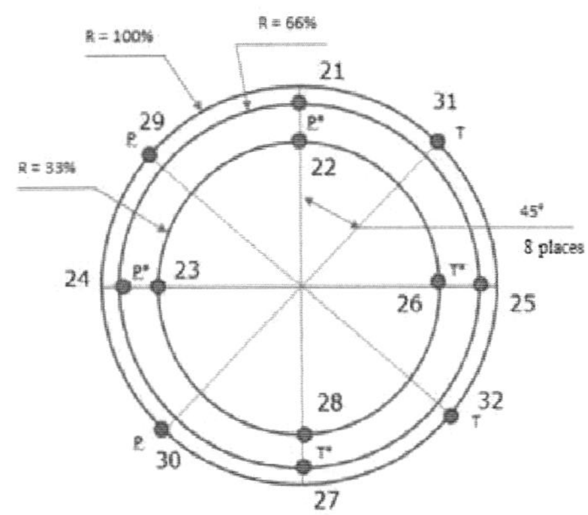

Figure 4. Section B (behind the turbine). View from the nozzle side.

Figure 5. The general view of the equipped SGTE.

IOP Conf. Series: Materials Science and Engineering **986** (2020) 012052 doi:10.1088/1757-899X/986/1/012052

The remote control panel (Figure 6) is intended for monitoring, registering, displaying parameters and automatically regulating the bench. Circuit breaker, fuses, terminal block for connecting temperature and pressure sensors and emergency sirens are installed in the panel.

Figure 6. Remote control panel connected to a personal computer.

The design of the experimental setup allows the engine to move on the support frame along the axis of the engine (Figure 7).

Figure 7. Jet engine thrust measurement system.
1 – support frame; 2 – force sensor frequency

A frequency force sensor is used to measure the amount of jet thrust. Calibration of JTMS was carried out using a traction calibration system by applying a known axial load and determining the frequency of oscillations of the force sensor. Five measurements were done with increasing load in the entire range of changes in traction force and five measurements with decreasing load. After the experiment, the calibration was repeated.
Characteristics of bench measuring devices are given in Table 2.

Table 2. Characteristics of measuring devices used at the experimental bench.

Equipment	Main characteristics
Thermometer (measuring the ambient temperature)	Measurement limit: +283…+323 K The limits of permissible error: 0.5 K
Barometer (environmental pressure measurement)	Measurement limit: 98655… 103990 Pa The limits of permissible error: no more than 100 Pa
Manometer (fuel pressure measurement)	Measurement limit: 1 … 10 atm The limits of permissible error: 3 %
Thermometer (fuel temperature measured at the outlet of the fuel system)	Measurement limit: +283…+373 K The limits of permissible error: 0.5
Load cell with MVA8 analog input module to load cell	Measurement limit: 0…30 kg The limits of permissible error ±1%
Temperature measuring unit (in sections) TRM38+ MVA8	limit of the basic reduced error (BRE) up to 0.5% temperature ranges +278…+1373 K nature of the medium being measured: air, fuel combustion products
Pressure measuring unit (in sections) AIR20	limit of the basic reduced error (BRE) up to 0,5 % static pressure limits: 10…400000 Pa temperature ranges +283…1373 K nature of the medium being measured: air, fuel combustion products

Engine tests were performed in 2 stages:
1. First, the engine was tested without equipment, its main parameters were measured;
2. Then the engine was equipped and the tests were repeated.

At the stage of preliminary preparation of the experiment, a fuel mixture in the ratio of 5% oil and 95% kerosene was prepared in the fuel tank. A visual inspection of the cleanliness of the fuel supply line was performed, as well as checking the charge level of the battery of the gas turbine engine control unit.
The main unit with the engine was placed in the protective box; visual observations were conducted through the armored glass of the protective box. To ensure safety, carbon dioxide fire extinguishers were located near the front door of the box and at the operator's position.
Then, the engine was started and waited for when the automatic start-up mode of the gas turbine engine is stabilized at low gas, and engine control would be transferred from the electronic unit to the engine control lever.
After starting and switching the gas turbine engine to idle gas mode (corresponding to an engine rotor speed of approximately 33,000 rpm), measurements were made in the following sequence:
- the frequency mode of rotor rotation using the engine control lever was not regulated and the engine was heated for at least 1 minute to stabilize the parameters in this mode;
- engine performance was measured for 30 seconds.

Measurements of modes in the range from 60,000 to 105,000 rpm were performed in a similar sequence.

3. The mathematical model of the engine

Numerical simulation of the aerodynamic parameters of the engine on the SGTE virtual test bench by the resources of the Polytechnic supercomputer center was performed using the CML-Bench digital platform and ANSYS CFX 2019R3 software.
The numerical model of the engine was based on geometric models obtained by detailed three-dimensional scanning of the real engine. The numerical models of the turbine flow section and the SGTE

nozzle were scaled taking into consideration the engine heating process based on the assessment of its thermal state [1]. Figure 8 shows the constructed computational domain; the entire engine was simulated.

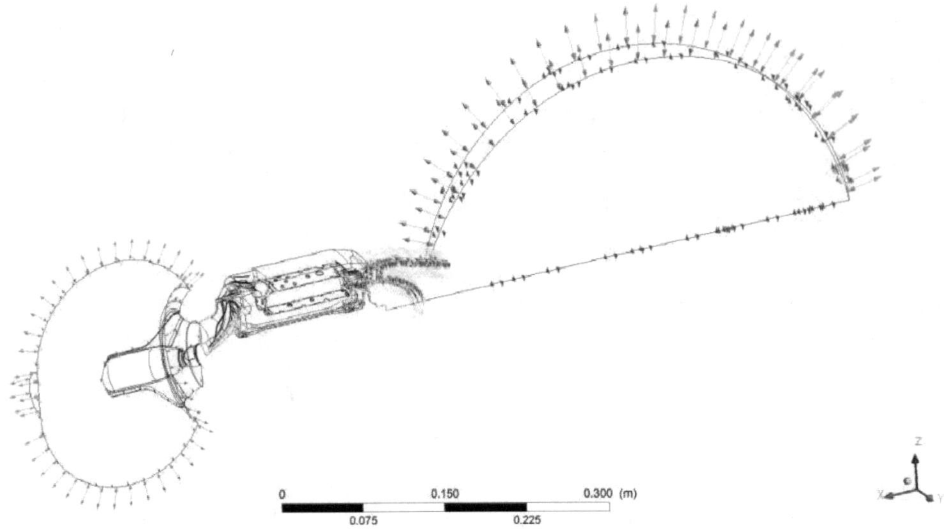

Figure 8. The estimated domain of SGTE.

The estimated SGTE domain modeled with the following features:

- the sector of the SGTE input device was modeled in 180 ° setting under the condition of symmetry;
- two compressor blades (blade and splitter), instead of 12 ones were modeled under a periodicity condition;
- one compressor diffuser blade, instead of 17 ones was modeled under a periodicity condition;
- two straightening compressor blades, instead of 34 ones were modeled under a periodicity condition;
- the combustor sector modeled directed 30 ° under a periodicity condition;
- one nozzle blade of the turbine, instead of 22 ones was simulated under a periodicity condition;
- one turbine blade, instead of 29 ones was simulated under a periodicity condition;
- the sector of the nozzle chamber was modeled at 12.414 ° under the condition of periodicity;
- the radial clearance on the compressor working blades was taken equal to 0.2 mm, turbines - 0.45 mm.

The SGTE grid model has a dimension of about 7.5 million nodes and 22.5 million elements. The type of grid model is mixed; the model consists of tetrahedral elements in the main volume and prismatic elements of the wall zone with a triangular base. The wall layer is divided into 10 layers. The minimum size of the final element is 0.1 mm; the maximum one is 1 mm.

According to [7], the BSL turbulence model with a turbulent Prandtl number equal to 0.2 was chosen as the turbulence model according to [8]. The equation for total energy includes accounting for the work of friction forces.

Kerosene was specified as a dispersed liquid phase to simulate combustion processes and a particle-droplet transfer model was used to describe the motion path, based on the equation of motion of Lagrange particles taking into account the influence of the carrier gas mixture flow (flow resistance force, pressure gradient force). The possibility of crushing droplets, as a result of which the droplets break up into smaller particles under certain conditions, was taken into account in the model.

The phase transition of particles from the liquid state to the gas state was determined using a liquid evaporation model, which uses different heat and mass transfer ratios depending on the boiling point. The boiling point was determined using the Antoine equation.

The theory of combustion of a pre-mixed mixture (diffusion combustion) was used to simulate a chemical reaction in which fuel and oxidizer enter the reaction zone separately, in contrast to pre-mixed systems in which the reactants are mixed at the molecular level before combustion. To describe diffusion combustion, such an approach was used in which the instantaneous thermochemical state of the liquid (concentration of the components of the substance, temperature, and density) is unambiguously associated with a passive scalar quantity - a mixing variable. This approach solves only one transport equation for the mixing variable regardless of the number of chemical components.

The interaction of the chemical kinetics of the diffusion flame with the turbulent flow field was described using the model of micro laminar flame. There, the turbulent flame is considered as a combination of thin, one-dimensional, laminar flame structures embedded in the field of the turbulent flow. The statistical method of the probability density function was used to couple the chemical kinetics of the laminar flame with the changing field of the turbulent flow.

A simplified spectral equation of radiation transfer for isotropic radiation intensity without scattering was used to simulate radiation.

To simulate the combustion process, a chemical kinetic combustion scheme was used for aviation kerosene (with a 100% content of n-C10H22) with the complete formation of nitrogen oxides. The following stoichiometric ratio was used: n-C10H22+14O2→(combustion product) for the mixing variable. This kinetic combustion scheme contains 1127 chemical reactions for 127 components. Kerosene properties were taken according to the reference data [9].

The real air model is used as the working medium in the compressor, i.e. the values of heat capacity, thermal conductivity and viscosity for calculations are selected for air [10,11].

The mixture of air and kerosene combustion products at the outlet of the combustion chamber was selected as the working medium.

Between the domains, the Stage (MixingPlane) interfaces were used while keeping the velocity vector. Unbalances of the results of the equations being solved, as well as the values of the main parameters of the engine components, were monitored to assess the stability and convergence of the calculation. The convergence of the calculation was considered achieved when the parameters were established, and the fluctuations in the unbalance took on an unchanged character (with the amplitude of the oscillations reaching 0.1%).

The time step was taken equal to 10-4 s, the calculation scheme is High Resolution.

The boundary conditions were the atmospheric parameters (pressure and temperature) in the box, the rotor speed of the turbocompressor, and the fuel consumption measured during the SGTE test.

4. The results of comparing the simulation and experimental data

For getting a general idea of the mathematical model of the engine, Figure 9 shows the calculated field of the total gas temperature in the middle section of the engine at the maximum thrust mode.

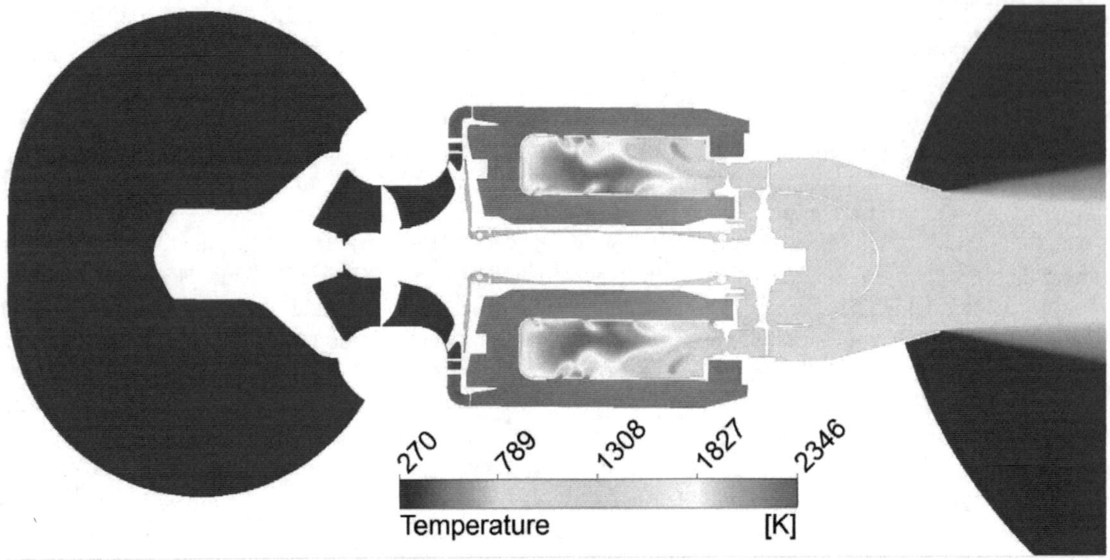

Figure 9. Field of full gas temperature in the middle section of the gas turbine engine at the maximum thrust mode.

Figures 10-16 show the main results of comparing mathematical modeling data with experimental data. The results of mathematical modeling are designated as VTB, i.e. virtual test bench.

All figures show the extrapolation of graphs up to 112,000 rpm (maximum). Extrapolation was performed by constructing a trend line (polynomial of the 3rd degree) and is shown on the graphs as dashed lines.

Figure 10 compares the measured and simulated jet thrust of the engine. The simulation results are in agreement with the test results of the unequipped SGTE in the range of rotor rotation frequencies from 60 to 105 thousand rpm. The equipped SGTE shows 10 ... 12% less thrust than the unequipped one, which indicates the negative impact of the preparation of the compressor and turbine on the integral parameters of the engine.

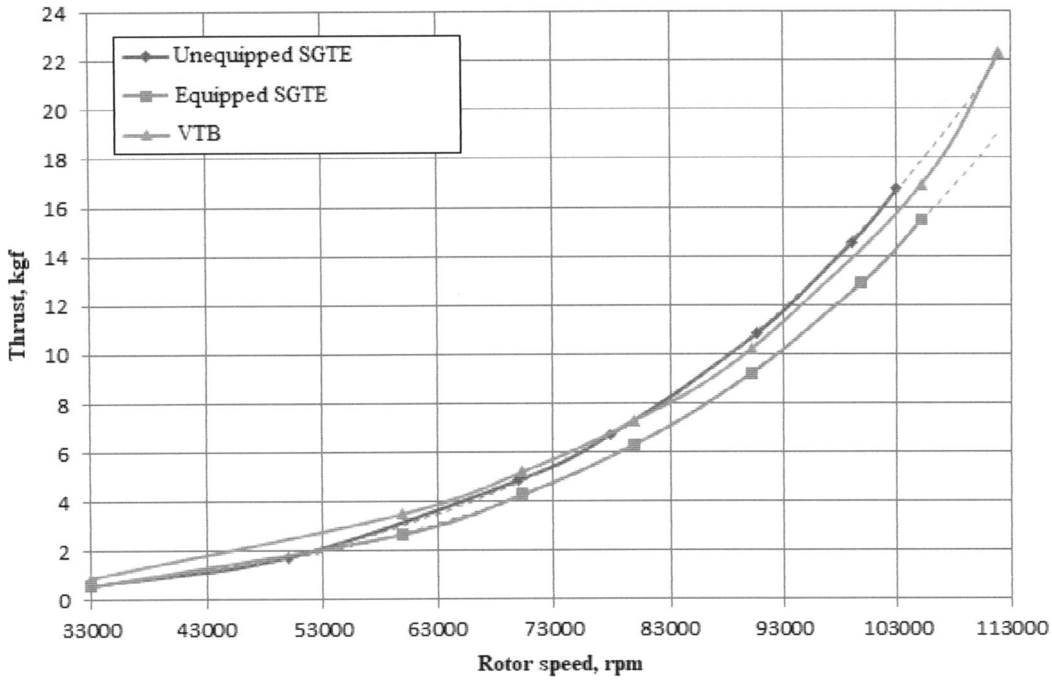

Figure 10. Dependence of engine thrust on rotor speed.

Figure 11 presents a comparison of the fuel consumption of the equipped and unequipped engines. They differ slightly in fuel consumption, the unequipped engine has a fuel consumption of 3% more than the equipped one. The fuel consumption of the equipped engine was used as a boundary condition for mathematical modeling, except the point 112000 rpm, where the fuel consumption declared by the manufacturer was used.

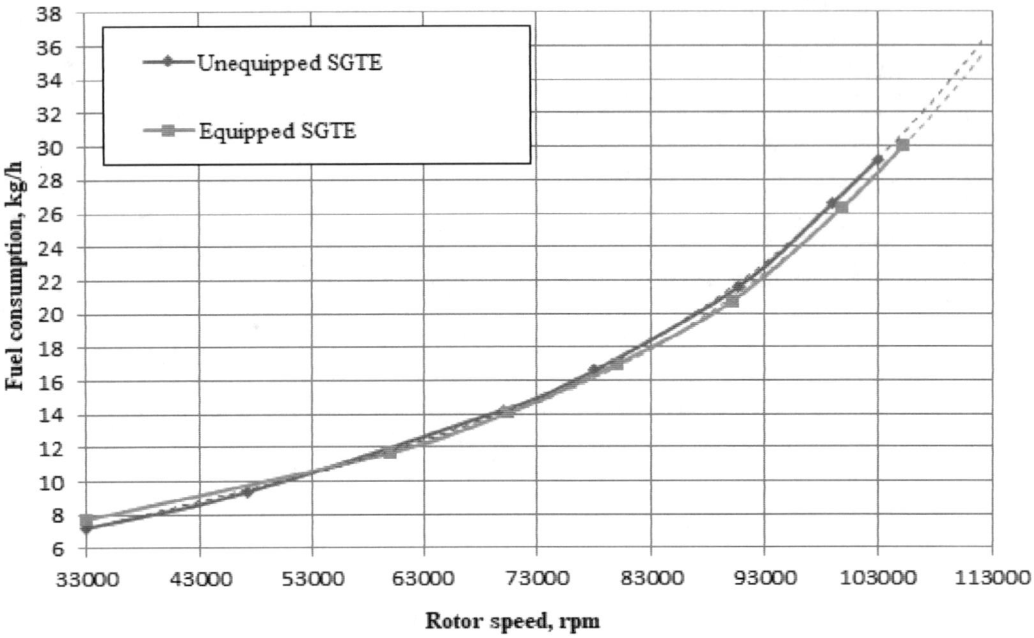

Figure 11. Dependence of fuel consumption on rotor speed.

Figure 12 presents a comparison of the measured and simulated degree of pressure increase in the compressor. The estimated P*k is higher than the measured one by about 8%, which can be caused by the following reasons:

The modeling error of the compressor;

Blockage of the compressor flow section by equipment;

Non-measurement of the total pressure of the sensors caused by a non-zero angle of attack of the flow on them.

Given that the height of the channel at the outlet of the compressor straightening device is extremely small (about 5 mm), and its unevenness can be significant. This indicates more about the difficulties of measuring the total pressure field (measured using only two sensors) in small flow parts of turbomachines, than about modeling errors. Additionally, this conclusion can be confirmed by the measured total pressure behind the compressor practically do not differ from the measured total pressure in front of the combustion chamber (the difference is less than 2%, while the total pressure loss on the combustion chamber in such a design should be about 5 ... 7%). This also indicates an underestimation of the measured total pressure behind the compressor.

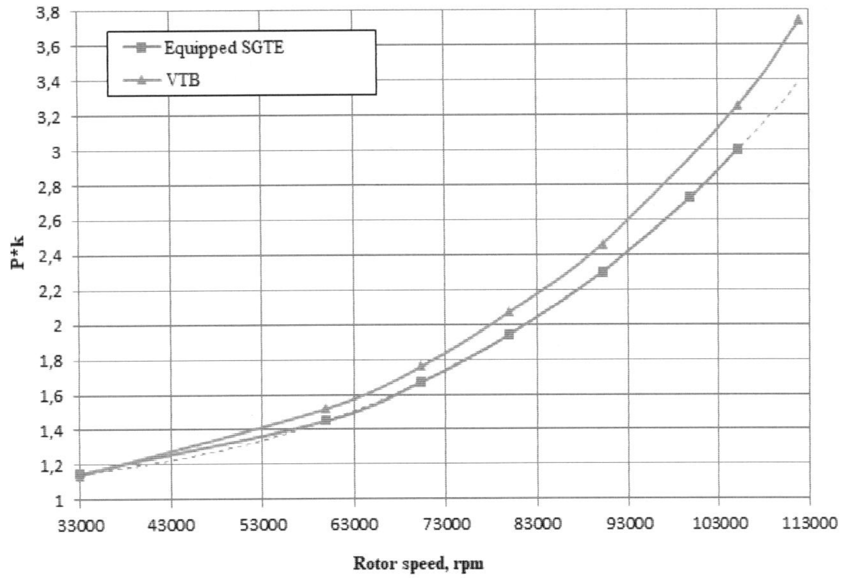

Figure 12. Dependence of compressor pressure increase on rotor speed.

Figure 13 shows a comparison of the measured and simulated air temperature behind the compressor. The results of the simulation and experiment are quite close, the difference is about 1 % (about 4...5 degrees K) except the "Low gas" mode (33000 rpm), where the measured temperature exceeds the calculated one by 7 degrees K (the difference is about 2%). Taking into account that the temperature unevenness behind the compressor is extremely small, the accuracy of measurements and modeling can be considered high.

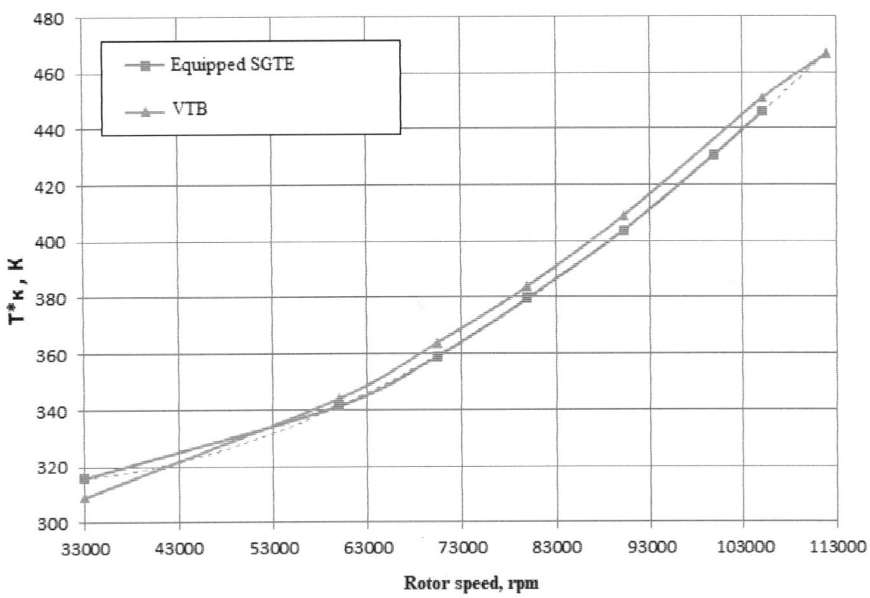

Figure 13. Dependence of the air temperature behind the compressor on the rotor speed.

Figure 14 presents a comparison of the measured and simulated degree of pressure drop behind the turbine. As soon as the channel height becomes acceptable (at least 10 mm), and the number of total pressure sensors increases (from 2 to 4 behind the turbine), the measurement results become more reliable, and the difference between simulation and experiment does not exceed 1 %.

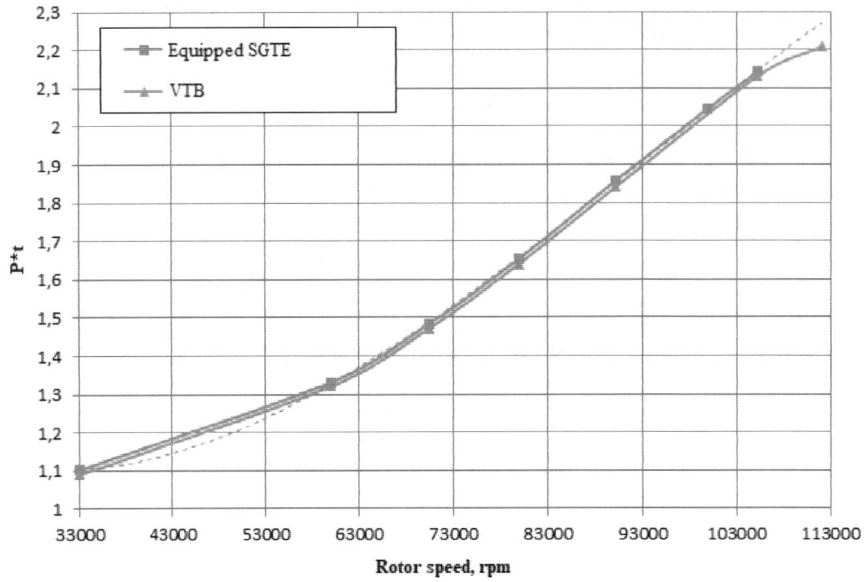

Figure 14. Dependence of decrease in turbine pressure on the rotor speed.

Figures 15 and 16 show a comparison of measured and simulated gas temperatures before and behind the turbine. The measured total gas temperature in front of the turbine is incorrect, since it exceeds the measured total gas temperature behind the turbine that cannot be. The comparison of the measured and simulated temperature behind the turbine shows that the calculated temperature is about 15% higher than the measured temperature, which can be caused by the following reasons:

- The modeling error of the combustion chamber;
- Failure or incorrect operation of part of the temperature sensors before and behind the turbine during test runs, when the fuel does not burn in the combustion chamber and burns out in the turbine and nozzle;
- It should be noted that the temperatures in the "hot part" of the turbine engine were measured by chromel-alumel thermocouples, which could easily be damaged during numerous experimental engines start due to temperature overshoot in the turbine and nozzle. The results of this experiment show that it is highly desirable to use substantially more expensive platinum-rhodium thermocouples.

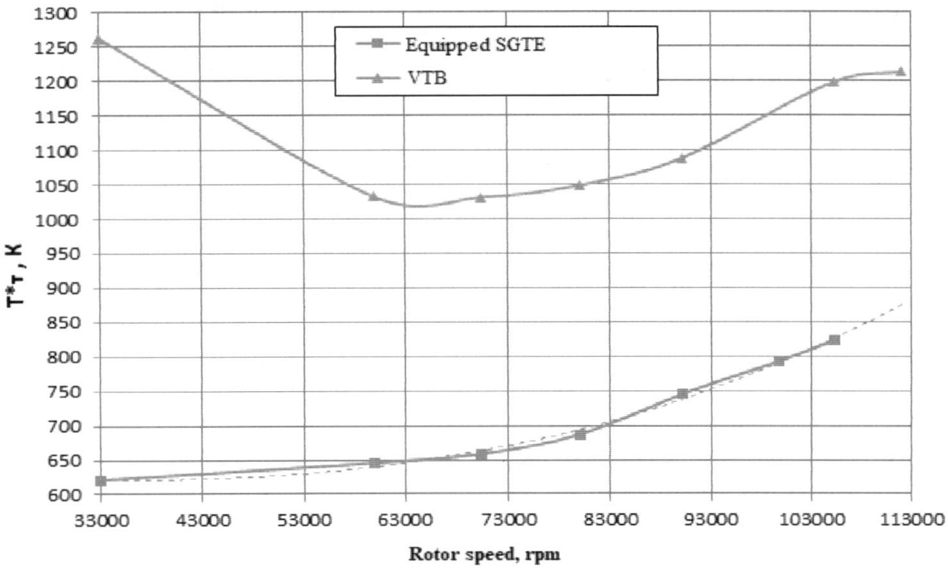

Figure 15. Dependence of the gas temperature in front of the turbine on the rotor speed.

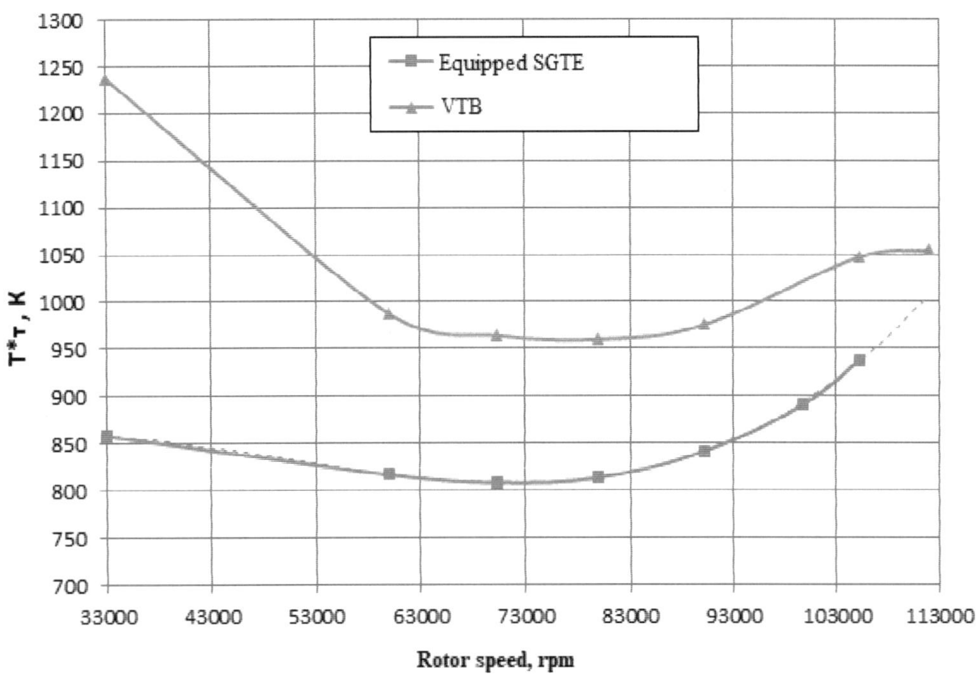

Figure 16. Dependence of the gas temperature behind the turbine on the rotor speed.

All results at the maximum engine operating mode (112,000 rpm), including extrapolation of the experimental curves, as well as calculated data and manufacturer data are summarized in Table 3. The following conclusions can be drawn from the table:

- the computation data on the degree of pressure increase in the compressor are in agreement with the manufacturer's data, which indirectly confirms the thesis about the shortage of the total air pressure behind the compressor;

- the levels of the calculated and measured gas temperatures at the turbine outlet, taking into account the measurement error and the data declared by the manufacturer, can be considered satisfactory
- extrapolation of the fuel consumption measurement curves in the experiment is in agreement with the manufacturer's data, taking into account the measurement accuracy;
- the difference in thrust measurement / calculation compared to the manufacturer's data was about -5 ... 6%. Given the growth of the calculated thrust curve when changing the operation mode of SGTE from 105,000 to 112,000 rpm (see Figure 10), it can be noted that the forecast curves for the experimentally measured engine thrusts are likely not correct and for the maximum mode the measured thrust of the unequipped engine would be close to that declared by the manufacturer.

Table 3. Comparison claimed/pilot/current characteristics of the Jetcat P200-RX engine at maximum thrust.

Parameter	Manufacturer data	Unequipped engine	Equipped engine	VTB
Full thrust, kgf	23.45	~22	~19	22.28
Outflow gas temperature, K	1023	-	~1010 ± 34	1054
Degree of pressure increase	3.7	-	~3.4 ± 0.1	3.74
Fuel consumption, kg/h	34.2	~36 ± 4	~35 ± 4	34.2

5. Conclusion

The work, performed on the numerical and experimental study of the parameters of the working path of SGTE, results in the following main conclusions:

- Under conditions where the main prerequisites for the experimental measurement of parameters are favorable (sufficient channel heights, a sufficient number of sensors, the absence of extreme temperatures and high irregularities), a comparison of the experimental and simulation results are in agreement (within 1 ... 2%) that is evident from the comparison total air temperature behind the compressor, differential pressure of the gas on the turbine and engine thrust;
- Measurement of parameters in difficult conditions (small channel heights, large non-uniformity of parameters, temperature "spikes", and a small number of sensors) leads to significant uncertainties in determining the parameters of the SGTE path and the inability to separate modeling errors and experimental errors. It is seen from the results of gas temperature measurements behind the combustion chamber and the turbine;
- Despite the incomplete success of the tests (which attributed to their difficult conditions because of the rather small overall dimensions of the research object), it can be confidently said that modern modeling techniques make it possible to reliably predict the key parameters of the main SGTE components. Besides, virtual tests allow partially replacing the full-scale ones under conditions when the experimental measurement of the necessary parameters is difficult.

References

[1] Schilling F U Untersuchengen zum Betriebsverhalten von Mikrogasturbinen *Institut fur Luftfahrtantriebe der Universitat Stuttgart*

[2] Tamer S Fathy Aly M Elzahaby and Mohamed K Khalil 2018 2018 Micro TJE centrifugal compressor performance prediction Engineering of Science and Military Technologies (Volume (2), Issue (4), Oct 2018)

[3] Tamer S Fathy, Aly M Elzahaby, Mohamed K Khalil 2018 Reverse Engineering of Micro Turbojet Centrifugal Compressor Design Based on Parametric CAD Data *International Journal of Scientific & Engineering Research* (Volume 9, Issue 1, January 2018)

[4] Roberto Capata 2015 Experimental Tests of the Operating Conditions of a Micro Gas Turbine Device *Journal of Energy and Power Engineering* (9)

[5] Nicholas D Grannan and John Hoke 2017 Trends in JetCAT Microturbojet-Compressor Efficiency *AIAA SciTech Forum, 9 - 13 January* (Grapevine, Texas)

[6] https://www.jetcat.de/en/productdetails/produkte/jetcat/produkte/hobby/Engines/p200_rx

[7] Samokhvalov N Yu, Khairulin V T, Tikhonov A S 2017 Verification of settlement three-dimensional models of the formation of secondary vortex flows in the interscapular channels of turbines of gas turbine engines according to the test profile T106 *PNRPU Mechanics Bulletin* Aerospace Engineering №49

[8] Aerospace engineering and technology 2013 №8 ([Electronic resource] Access mode: https://www.khai.edu/csp/nauchportal/Arhiv/AKTT/2013/AKTT813/Petelchi.pdf - On the issue of increasing precision RANS simulation of film cooling)

[9] Dubovkin N F and Malanicheva V G 1985 Reference "Physico-chemical and operational properties of jet fuels." (Moscow)

[10] Vargaftik N B 1972 A reference book on the thermophysical properties of gases and liquids *Nauka Publishing House* ([Electronic resource] Access mode: http://bookre.org/reader?file=476288&pg=1 –Reference book on the thermophysical properties of gases and liquids Moscow)

[11] NASA 2002 Glenn Coefficients for calculating thermodynamic properties of individual species, (Washington, NASA/TP 2002-211556).

IOP Conf. Series: Materials Science and Engineering **986** (2020) 012053 doi:10.1088/1757-899X/986/1/012053

Development of a method for multidisciplinary parametric optimization of vehicle hood reinforcement

T V Filina, A N Leontev, N A Kharaldin, M V Aleshin, O I Klyavin and A I Borovkov

Engineer of Automotive Departament Engineering Center «Center of Computer-Aided Engineering» Peter the Great St.Petersburg Polytechnic University St. Petersburg, 195251, Russia. filina_tv@spbstu.ru

Abstract. The goal of this paper is to develop and test a multidisciplinary optimization technique for designing structures that are subjected to both static and dynamic loads, using the digital twin of a vehicle hood as an example. This method reduces the weight of the hood reinforcement and increases the probability of pedestrian survival when hitting the head. This is due to the implementation of technological and layout constraints, as well as the requirements for stiffness, frequency characteristics and hood denting.

1. Introduction

Currently, when structures are developed, the engineer is faced with the need to find a compromise between the technical requirements for the product, layout and technological limitations, requirements for the weight and cost of the final product.

Designing using Digital Twin technology and conducting Digital Certification avoids the long-term costly testing of prototypes and significantly reduces time-to-market, which gives a serious competitive advantage in global high-tech markets ([1], [2], [3], [4], [5], [6]).

A particular challenge is the design of structures that provide at the same time sufficient levels of stiffness under static loading and energy absorption during impacts both from the viewpoint of the opposite of these requirements and from the viewpoint of multidisciplinarity. For a hood, stiffness is calculated in a linear static formulation, denting - in a non-linear quasistatic one, and energy absorption - in a non-linear dynamic one.

In practice, to simulate the full range of structure operating conditions is often necessary to use several systems of finite element (FE) analysis, which means the Digital Twin contains several FE models for one structure for different types of calculations. This complicates the process of fine-tuning the design to the targets, because to account for the impact of structural changes in the design, one needs to repeat these changes in each of the models.

The principal novelty proposed in this work approach is the use at early finishing of the design to target of cross-platform parametric optimization methods, which allows combining design cases of different types and at the same time to consider the impact of design changes on all its features (unlike the existing method [7]) and to build into the model a number of technological limitations.

Implementation of the virtual tests described in the work was carried out on the basis of the platform CML-Bench^{TM} (digital platform for the development of digital twins and an activity management system in the field of digital design, mathematical modeling and computer engineering, development of the engineering center CEC SPbPU) using computational resources of CNTI.

2. Pedestrian protection

According to NHTSA statistics ([8]), despite the total number of traffic fatalities decreases, more than 6,000 pedestrians die in accidents each year in the United States alone (Tab. 1).

Table 1. Statistics of pedestrian deaths in road accidents in the united states.

Year	Died, people	
	Total	*Including pedestrians*
2017	37 473	6 075
2018	36 560	6 283

In Russia, the number of deaths per 1,000,000 population is 2.5 times higher than the average for European countries (Fig. 1). The high death rate is affected by both poor road conditions and using old cars with low quality pedestrian protection systems. However, certification rules regarding pedestrian protection are being tightened in Russia year by year, approaching European standards.
Virtual tests of the hood for pedestrian protection are carried out on average at 150-180 points for each structure (Fig. 2).

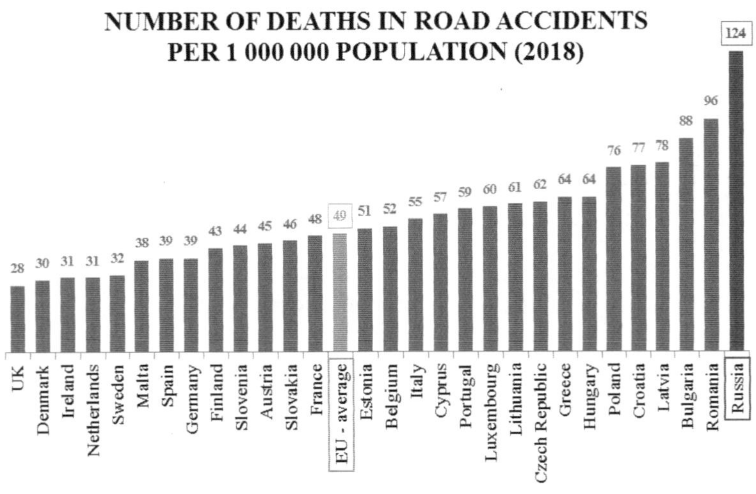

Figure 1. Number of people died in road accidents per 1 million population in Europe and Russia in 2018 ([2], [3]).

When a car collides with a pedestrian, as a rule, the vehicle bumper first contacts lower leg and thigh, and then the head hits the hood area. Leg injuries are painful and can lead to long-term disability. However, head injuries are the most dangerous, which are often fatal ([11]). Therefore, it is so important to predict and prevent overloads that a person's head experiences when hitting a car. The decisive factor determining the degree of head injury is the structure of the hood, since it should slow down the head, preventing critical overloads and excluding the possibility of contact of the head with massive elements of the underhood space ([12]).
The hood structure is tested for pedestrian protection according to the rules of the EuroNCAP rating test in this paper ([13], [14]).
The success of these tests is evaluated using HIC (Head Injury Criteria), which is based on the Wayne State Tolerance Curve ([15]) and displays the likely assessment of head injuries on impact.

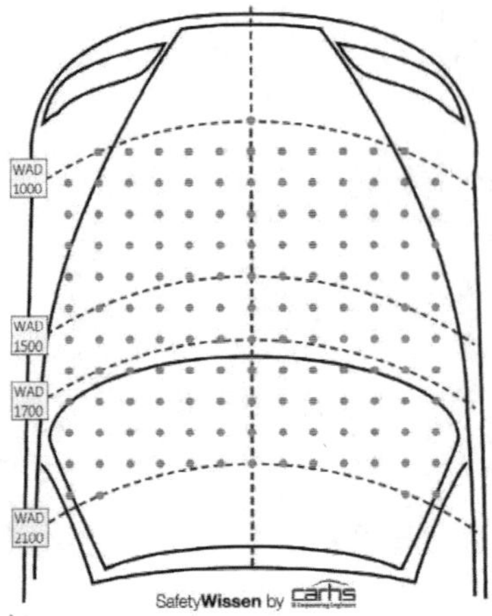

Figure 2. Determination of the hood test points for pedestrian protection when hitting a head according to EuroNCAP rules ([4]).

$$HIC_{15} = \max_{t_0}\left\{\Delta t\left(\frac{1}{\Delta t}\int_{t_0}^{t_0+\Delta t} a(t)dt\right)^{2.5}\right\} \quad (1)$$

Tab. II presents the rules for setting EuroNCAP rating points for the hood point where the test took place, depending on the value of the criteria.

Table 2. EuroNCAP rating points depending on the value of HIC15 at the point ([4]).

$HIC_{15} < 650$	1.00 Point
$650 \leq HIC_{15} < 1000$	0.75 Points
$1000 \leq HIC_{15} < 1350$	0.50 Points
$1350 \leq HIC_{15} < 1700$	0.25 Points
$1700 \leq HIC_{15}$	0.00 Points

The percentage of the maximum number of points in the workspace was selected as the evaluation criteria, because in this task, virtual tests are not performed for all hood points

$$S = \frac{\sum_i Point_i}{Number\ of\ grids} \quad (2)$$

Achieving the main goal - reducing the total criteria of HIC15 (increasing the S rating) when testing the hood for pedestrian protection - has to satisfy both the targetbook (requirements for stiffness, eigenfrequency and denting) and the requirement to reduce the weight of the structure while fulfilling technical limitations (technological and layout ones).

3. Smart digital twin of the vehicle hood

The developed highly adequate model (Digital Twin) of the vehicle hood contains:

a) Full-scale geometric (CAD) and finite element (CAE) models of the hood - Fig. 3a,

b) Information about all supports and joints (spot and arc welding, glue, bolts, hinges, rebound buffers, sealing) both between the hood parts and with adjacent assemblies (car body) and their necessary characteristics - Fig. 3b,

c) Complete data on the materials necessary for the assigned tasks: the elastic-plastic characteristics of steels, taking into account the dependence on the deformation rate for impacts and the fracture criteria, is an example for steel DX56D in Fig. 3c, as well as the assortment of material-thickness combinations available for the production of the hood by cold stamping (this information was provided by the customer and was later used to parameterize the model),

d) Information on operating conditions (Fig. 3d) and a matrix of targets for each type of load - Tab. III,

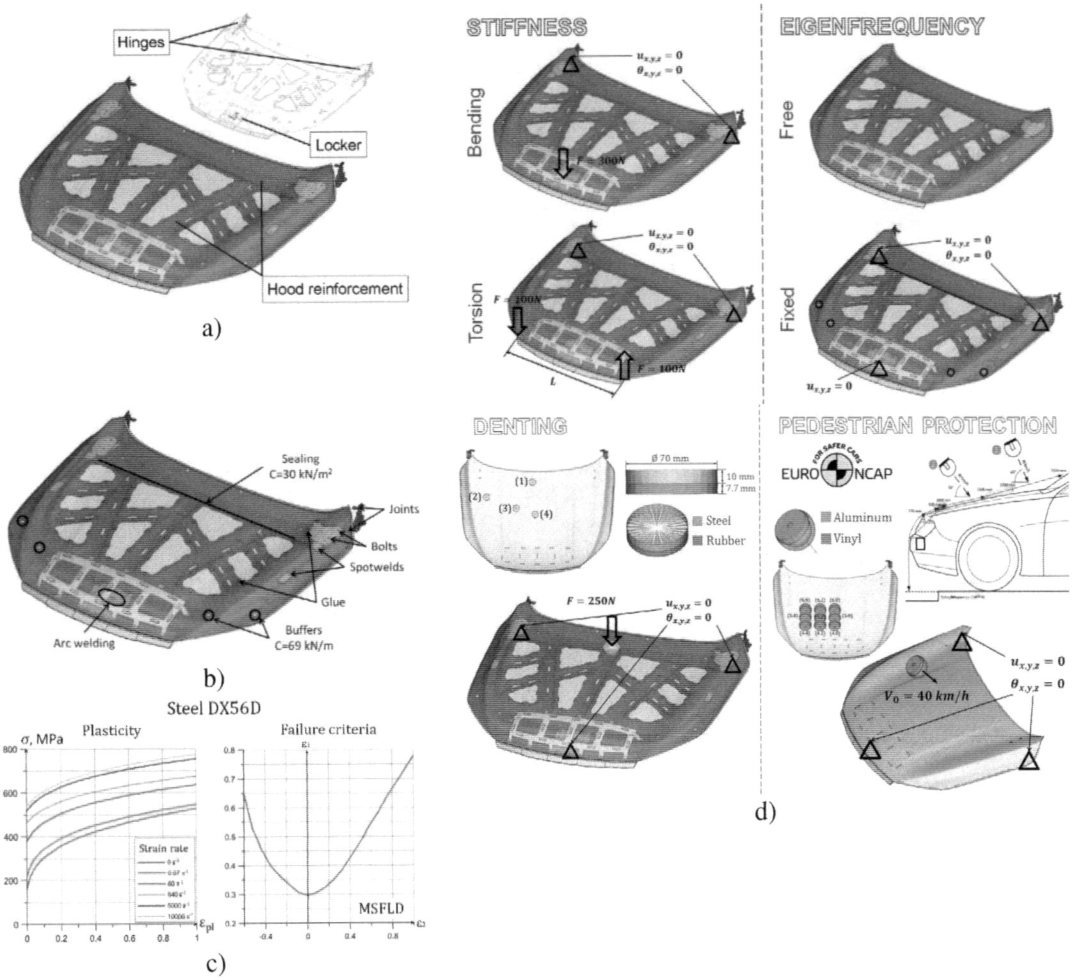

Figure 3. The digital twin of the car vehicle: a - FE model of the hood, b - interfaces between parts and the car body, c - steel deformation curves DX56D, d - computational cases corresponding to the target matrix.

In this work, the operating modes of the hood are two tests for static stiffness (bending and torsion), two modal analyses (in the free and fixed state), a test for denting at four points, and a test for pedestrian protection at nine points. The last two tests were performed for the positions of the acting indenters/impactors located in the right (in the direction of the car) half of the hood, because the hood structure is symmetrical with respect to the central plane. Tests for pedestrian protection and denting at other points as part of the optimization are not carried out, because a change in the structure of the reinforcement in the central part slightly influences the test result in other parts of the design.

e) Information on layout and technological limitations.

The choice of materials and manufacturing technology, as a rule, is dictated by the desire to minimize the cost of the final product. In the case when the original design was made, the selection criteria may be to minimize investment in an existing process. In this work, the technological limitations on the hood are the inability to use lighter materials (aluminum alloys) and other manufacturing technologies, except for cold sheet stamping.

Layout constraints prohibit changes to the exterior styling surface of the hood, as well as changes in the interface between the hood and the car body (i.e. in the areas of loops, rebound buffers, and sealing).

Thus, the central part of the hood reinforcement is the working area for making changes to the structure that meets all the constraints (Fig. 4).

Table 3. Matrix of target indicators of the vehicle hood.

Loadcase		FE system	Controlled parameter	Units	Target
Static	Bending	Optistruct	Bending stiffness	kN/m	≥ 15
	Torsion		Torsional stiffness	kNm/rad	≥ 30
Modal	Free		First nonzero eigenfrequency	Hz	≥ 35
	Fixed		First eigenfrequency	Hz	≥ 26
Denting		LS-Dyna	Intrusion	mm	≤ 6
Pedestrian Protection			HIC_{15}	-	< 650

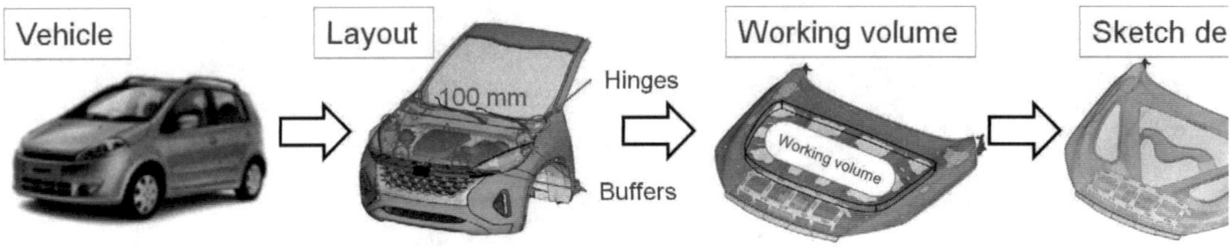

Figure 4. The working capacity for optimization and the sketch design of the hood.

4. Sketch design and parametrization

The goal of parametric multidisciplinary hood optimization is to minimize the total pedestrian head injury criteria for all loadcases of the pedestrian protection test (maximize the S rating). In the adopted formulation, the target values from the targetbook were set as limitations on the responses from each loadcase.

The parametrization of the sketch FE model of the hood reinforcement is shown in Fig. 5. The thickness and material of the reinforcement can be changed within a discrete set of values corresponding to the available material-thickness combinations within the selected manufacturing technology. The shape parameters can vary in a continuous range of values determined by geometry feasiblity (intersections inside the part and with other structural elements are unacceptable); reinforcement design changes are applied symmetrically (left and right).

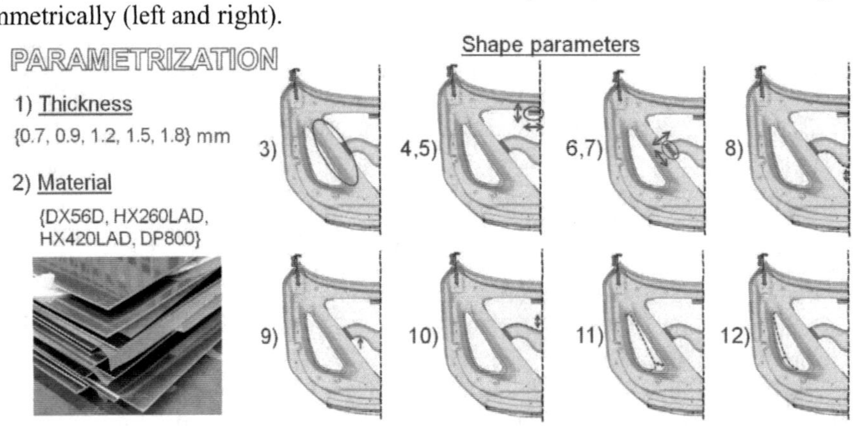

Figure 5. Parametrization of the FE model of the hood for multidisciplinary optimization.

IOP Conf. Series: Materials Science and Engineering **986** (2020) 012053 doi:10.1088/1757-899X/986/1/012053

The ARSM (Adaptive Response Surface Method) algorithm was used for optimization, which works only with single-purpose tasks, but is the most effective for solving linear and nonlinear problems with a mixed type of variables (discrete and continuous) at the same time.

The optimization problem was formulated in the Altair HyperStudyTM, software system, which provides the user with a convenient environment for parameterizing models of various FE analysis systems (in this case, Optistruct and LS-DYNA). This program allows to achieve the identity of changes made to the design by pairing the parameters of different FE models within each other one Digital Twin, determining responses (controlled parameters) for each loadcase and setting goals on them, and also has a wide range of parametric optimization algorithms and tools for post-processing optimization results.

5. Result of the optimization

The parametric multidisciplinary optimization of the hood resulted in a set of model parameter values that ensured a minimum of the total criterion for a head injury during pedestrian protection tests.

Fig. 6 shows a comparison of sketch and optimized structures of the hood reinforcement. The comparison of characteristics of these designs with the targets is presented in Tab. 4.

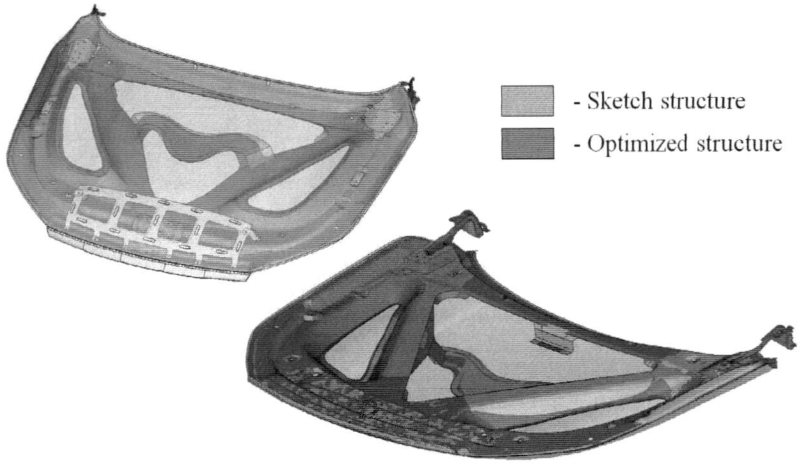

Figure 6. Comparison of the geometry of the sketch and optimized hood reinforcement.

Table 4. Comparison of characteristics of initial and optimized hood structures with targets.

Loadcase		Controlled parameter	Units	Hood structure		Target
				Initial	*Optimized*	
Static	Bending	Bending stiffness	kN/m	12.9	17	≥ 15
	Torsion	Torsional stiffness	kNm/rad	18	36.3	≥ 30
Modal	Free	First nonzero eigenfrequency	Hz	28.4	38.7	≥ 35
	Fixed	First eigenfrequency	Hz	24.8	26.6	≥ 26
Denting	(1)	Intrusion	mm	6.7	5.9	≤ 6
	(2)			5.5	4.3	
	(3)			3.4	5.4	
	(4)			4.4	3.8	
Pedestrian Protection	4.0	HIC_{15}	-	1 112	618	≤ 650
	4.2			1 096	668	
	4.4			993	959	
	5.0			1 035	490	
	5.2			895	624	
	5.4			969	873	
	6.0			917	477	
	6.2			857	583	
	6.4			940	799	
		Mass	kg	22.1	21.5	-

6. Conclusion

The resulting design of the central part of the hood using parametric multidisciplinary optimization satisfies the layout and technological limitations, meets the requirements for static stiffness, eigenfrequency, and denting. The weight of the optimized structure is 600 grams less than the original, while the rating of the optimized part of the hood according to the criteria of injury (S) increased from 68% to 87%. This increases the probability of a person's survival when hitting his or her head against the central part of the hood.

When structures subject to dynamic influences are designed, this technique reduces the laboriousness of the design and obtains effective structures that go beyond the bounds of the intuition of the designer. Besides, this approach can be extended to other classes of problems.

References

[1] Borovkov A I and Ryabov Yu A 2019 Digital Twins: definition, approaches and methods of development *Digital transformation of the economy and industry: proceedings of a scientific-practical conference with foreign participation, June 2019* (ed. A. V. Babkin, SPb.: POLYTECH-PRESS (in Russian))

[2] Borovkov A I, Gamzikova A A, Kukushkin K V and Ryabov Yu A 2019 Digital Twins in the high-tech industry. Brief report (September 2019 SPb.: SPbPU publishing (in Russian))

[3] Borovkov A I and Ryabov Yu A 2019 Definition, development and application of Digital Twins: the approach of the NTI Competence Center «New manufacturing technologies» *Digital substation* (№ 12 (in Russian))

[4] Factories of Future 2019 *TechNet NTI infrastructure center* (in Russian) https://technet-nti.ru/article/fabriki-buducshego)

[5] Borovkov A I Maruseva V M and Ryabov Yu A 2018 «Smart» Digital Twins are the basis of a new paradigm of digital design and modeling of globally competitive products of a new generation» *Springboard to success* (№ 13 (in Russian))

[6] Borovkov A I, Ryabov Yu A and Maruseva V M 2017 A new paradigm of digital design and modeling of globally competitive products of a new generation» *Book: «Digital production: methods, ecosystems, technologies: working report of the corporate training Department of the Moscow school of management SKOLKOVO»* (ch.2 (in Russian)

[7] Xiaomin Zeng and Xiongqi Peng 2014 Optimization Design of Bonnet Inner Based on Pedestrian Head Protection and Stiffness Requirements *13th International LS-DYNA Users Conference*
[8] 2018 fatal motor vehicle crashes: Overview» (Traffic Safety Facts Research Note. Report No. DOT HS 812 826) 2019 *National Center for Statistics and Analysis* (Washington, DC: National Highway Traffic Safety Administration)
[9] «2018 road safety statistics: what is behind the figures?» 2019 *European Comission - Fact Sheet* (Brussels)
[10] The number of people who died in traffic accidents per 100 thousand people 2019 Federal State Statistics Service ((in Russian) https://www.gks.ru/)
[11] https://gufo.me/dict/medical_encyclopedia/Черепно-мозговая_травма (in Russian)
[12] Kalliske I and Friesen F 2001 Improvements to pedestrian protection as exemplified on a standard-sized car *17th international technical conference on the Enhanced Safety of Vehicles* (Amsterdam, Netherlands)
[13] https://www.carhs.de/en/, Safety Companion 2020
[14] EuroNCAP «PEDESTRIAN TESTING PROTOCOL» 2017 (v 8.4)
[15] Voronov A S, Kaligin N N 2013 Determining the moment of an accident and assessing the probability of life-threatening injuries (Moscow, Polzunovsky Almanac #1(in Russian))

IOP Conf. Series: Materials Science and Engineering **986** (2020) 012054 doi:10.1088/1757-899X/986/1/012054

The study of the fatigue strength characteristics of welded joints in car components

D V Nikolaev, O I Klyavin, A V Tarasov, M V Aleshin, N A Kharaldin and A I Borovkov

National Technology Initiative Center for Advanced Manufacturing Technologies, Engineering Center "Center of Computer-Aided Engineering" Peter the Great St.Petersburg Polytechnic University Polytechnicheskaya, 29, St. Petersburg, 195251, Russia nikolaev@compmechlab.com (corresponding author), klyavin@compmechlab.com, tarasov@compmechlab.com, aleshin@compmechlab.com, kharaldin@compmechlab.com

Abstract. The computational study of the fatigue life of car body components and its spot welds was carried out using two methods in this article. Besides, there analyzes the impact of spot welds failure on the stiffness characteristics of the car and a comparison of the sites of breakdown that arose during computer modeling using a digital twin with the sites of a breakdown in a real test. The study resulted in a significant influence of fatigue life on the stiffness characteristics of the car body. This noticeable effect is explained by most of the failed spot welds are located in the areas of the front and rear suspension fastenings, experiencing significant stress both during the car operation and during bench tests for bending and torsional stiffness of the body.

1. Introduction

The geometric parameters of the body structure and the quality of technological and welded processes determine the reliability of the vehicle design, as well as its service life. At the early stages of the car development, under a large number of changes made to the design, engineers do not have a full set of information about the car's wheel loads and their impact on its performance. This leads to the potential need for a constant recalculation of the target characteristics, including fatigue strength of car components.

Most of the changes and the impossibility of studying their influence using real prototypes can be compensated by creating a digital twin of the car [1, [2]. Moreover, a simplified assessment technique for rapid evaluation of fatigue life can be used, with complete or partial absence of data on wheel loads. According to this method, the loads applied to the front axle are calculated based on the design characteristics of the car, which in most cases are known and unchanged from the very beginning of the development process: the full curb weight of the car and the width of its track, and the passage of various geometrically shaped obstacles is modeled by a test bench in the form of applying loads with different frequency and amplitude. This technique has significantly lower accuracy compared to a full virtual fatigue test and compared to the site or full bench tests. However, at the early stages of the car development, critical areas of the structure are quickly identified and compare the impact of changes, introduced into the design in terms of fatigue life, evaluating their effectiveness.

At later stages of development, in order to understand not only the qualitative but also the quantitative characteristics of fatigue life, it is possible to test a digital twin of a car at a virtual testing ground, using real loads coming to the body during various types of operation and road surface [3, [4].
In this paper, two approaches to the study of the fatigue life of a car body were considered. These approaches can significantly reduce time and increase the speed of development of the vehicle design, allowing to abandon some of the expensive and resource-intensive real tests.

2. Determination of fatigue life of a car

The durability of a structure can be calculated using two methods that are radically different in approach way. The development of defects can be simulated, taking into account the boundary conditions in a particular structural element, as well as the rebuilding of the mesh in the model with a gradual increase in the defect size [5]. In addition, one can evaluate the degree of damage to the material in the car components without explicitly considering defects and their growth [6]. Since the goal of this work is to evaluate the fatigue properties and not the sequence of defects development, a method with damage assessment was chosen as providing all the results necessary for the development and, besides, significantly faster in terms of computing resources. To analyze the fatigue life of a structure, initial data are required. As initial data, there needs information on the elastic and fatigue properties of structural materials, including the SN curve. A model is needed that describes the geometry of the structure with detailed modeling of welded joints, as well as information on the points of attachment of the suspension to the car body, namely, the stress-strain state of the structure under the action of loads applied at the corresponding interface points [7, [8].
Static strength analysis of the car body structure under the action of unit loads applied to the interface points of the front suspension is carried out based on the available initial data. Subsequently, based on the analysis data, areas of the structure can be identified that bear heavy loads and require special attention when fatigue life is analyzed. Besides, it seems possible to identify structural components that are likely to cause residual deformations during operation. Such areas should be the subject of low-cycle fatigue studies. When describing the process of multi-cycle fatigue, the SN curve is used, which explains the ability of a material to resist multi-cycle fatigue [9]. The SN curve describes the dependence of the peak voltage in the cycle on the total number of cycles to the complete failure of the element at the maximum voltage level, which is constant during the entire process of loading the element. For various materials, the SN curve has a different form in the area of multi-cycle fatigue.

A. Model Problem

Before solving the main problem, the model problem was considered to adjust the parameters of the solver and assess the adequacy of the approximation of the fatigue properties of materials according to known mechanical and physical properties. To solve the model problem, the design shown in Fig. 1, consisting of two pipes welded together from steel of the FE510 grade, was considered, the characteristics of which are presented in Table I, one of which was applied along the edges with a harmonic law force Wohler A 1858 Report on test of the Royal Niederschleesich-Märkkischen Railway made with apparatus for the measurement of the bending and torsion of railway axles in service (in German) Zeitsch.Bauwesen (N8) pp 642-651
[10]. In total, several tests were performed at different values of the amplitude of the applied load.

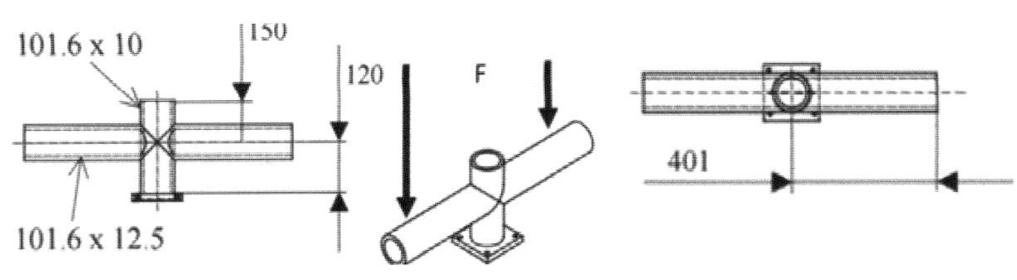

Figure 1. Dimensions of the test specimen

IOP Conf. Series: Materials Science and Engineering **986** (2020) 012054 doi:10.1088/1757-899X/986/1/012054

Table 1. Characteristics of the material FE510

Material	Fe510
Yield strength	345 MPa
Strength limit	500 MPa
Fatigue limit	250 MPa
The number of fatigue limit cycles	1.0e+5
Elastic modulus	210000 MPa

To solve the model problem, a finite element model was constructed that takes into account the elastic properties of the material. The weld between the two pipes was modeled using shell finite elements. The appearance of the model is shown in [11].

Using the MSC Nastran software package, the problem was solved in a linear formulation for studying the stress-strain state under the action of a load of 1 kN. The stresses arising in the structure when such a load is applied are shown in [12].

Using the obtained stress-strain state and the fatigue properties of the material, the life expectancy of the structure under the action of periodic loads at three different loading amplitudes: 80 kN, 60 kN, and 40 kN were obtained.

The comparison of the number of cycles before fracture obtained in a numerical solution and during the test are given in Table II.

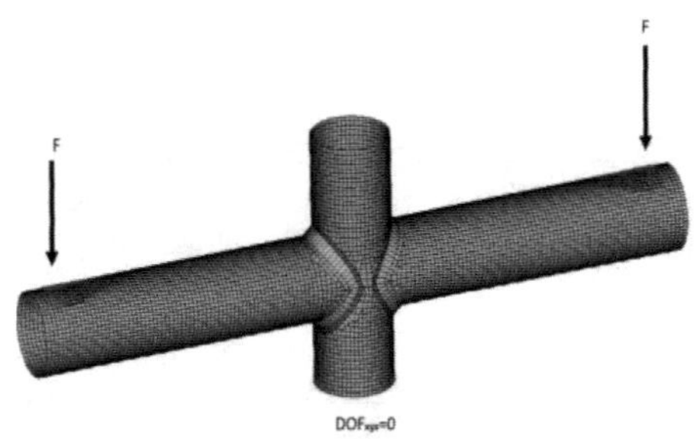

Figure 2. Finite element model of the design.

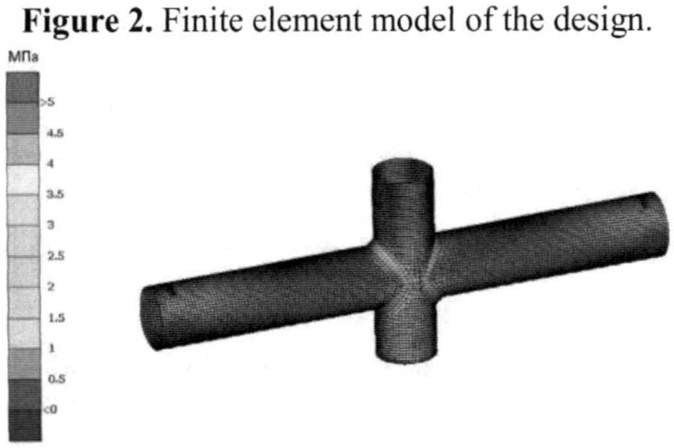

Figure 3. Stresses in the tubular structure.

Table 2. The comparison of numerical modeling with experimental data

Load, kN	Numerical solution, number of repeats	Test solution, number of repeats	Difference %
80	37500	35000	7
60	86000	80000	7
40	315000	300000	5

The difference between the results of numerical modeling and experimental data was less than 7 percent. Thus, the calculation parameters and the general approach to solving the model problem can be used further under assessing the fatigue life of a car body, and the fatigue properties of materials can be approximated based on physical and mechanical properties.

B. The computational model of the car

To conduct a numerical assessment of fatigue life, a digital twin of the car under study was developed, which includes the following structural components that can influence the parameters under study: body, front suspension frame, dashboard reinforcement, front bumper reinforcement, rear bumper reinforcement, car front, roof, windshield and rear window. The calculated model of the car body is shown in [13].

Figure 4. The computational model of the car body.

The presented body model was prepared for static calculation in the MSC Nastran finite element analysis package and calculation of fatigue life in the MSC Patran/Fatigue package and contains:

- 1034640 shell elements
- 67234 volumetric elements
- 644 absolutely rigid connections (finite elements RBE2)
- 77603 interpolation links (finite elements RBE3)
- 28 beam elements

Shell elements are used to model structural parts of the car body, for example for stamped metal parts, as well as for windshields and rear windows.

Volumetric elements are used in the task of spot welds, adhesive joints, and cast parts of the car. Elements of an absolutely rigid connection are used to specify bolted connections between parts, since when an absolutely rigid connection is used, the relative location of the nodes included in it does not change, they move as a whole. For the simulation of bolted joints, the properties of an absolutely rigid connection are set so that to transfer only translational degrees of freedom of the nodes, but not rotational ones, providing the connection model more adequate to the real object. In addition, around bolted joints, an additional element area is added to an absolutely rigid connection for reducing the edge effect.

The disadvantage of absolutely rigid connections is the stiffness they introduce into the model. If in the case of bolted joints this modeling method is quite physical, then for modeling spot welds where relative movement of different sections of the joint is possible, modeling using an absolutely rigid connection can introduce a discrepancy in compared to reality [11].

To simulate spot weld, a volumetric finite element is used, connected to the mesh using elements with interpolation coupling.

Interpolation link distributes balanced the load of the control node between the nodes managed, while not introducing additional non-physical rigidity into the connection model itself, resulting in a positive effect on the accuracy of the model.

Elements of the beam type are used in the model to specify bolted connections in the front and rear axle boxes. Due to these bolted connections experience the highest loads when connecting the subframe to the body, they need to be set using deformable elements, and not an absolutely rigid connection.

By analogy with spot welds, the connections to the parts to be glued were also modeled using interpolation links. Modeling of welded lines of linear arc welding was carried out using shell elements. The materials used in the car body and their distribution are shown in 0. In the future, the developed digital twin of the car body can be used both for assessing fatigue life by a simplified method and for using a virtual test site.

Figure 5. Breakdown of materials in the car body model

C. Determination of durability by stress level

To explain multi-cycle fatigue, it is generally accepted to describe durability as a function of the characteristics of a stress cycle. The fatigue curve is called the Weller curve, and the approach in foreign literature is called the SN approach, by the name of the abscissa and ordinates on the Weller curve (Stress vs. Number of cycles) [12].

Moreover, the Weller curve has a different characteristic form for different materials and is described by the corresponding dependence proposed by Baskvin.

Therefore, the fatigue curve equation of materials with a physical fatigue limit has the form:

$$N = \begin{cases} \dfrac{\sigma_R^m N_G}{\sigma_{max}^m}, & \sigma_{max} \geq \sigma_R \\ \infty, & \sigma_{max} < \sigma_R \end{cases} \quad (1)$$

The equation of the fatigue curve of materials with two inclined sections in the fatigue curve takes the form presented in formula 2.

$$N = \begin{cases} \dfrac{\sigma_{RN}^m N_G}{\sigma_{max}^m}, & \sigma_{max} \geq \sigma_R \\ \dfrac{\sigma_{RN}^{m_1} N_G}{\sigma_{max}^{m_1}}, & \sigma_{max} < \sigma_R \end{cases} \quad (2)$$

If the fatigue curve does not have an inflection when the number of cycles is increased, the Baskvin dependence takes the form presented in formula 3:

$$N = \frac{\sigma_{RN}^m N_b}{\sigma_{max}^m} \quad (3)$$

where m – the slope of the first section of the fatigue curve, m_1 – the slope of the second section of the fatigue curve, N_G – the abscissa of the inflection point of the fatigue curve, σ_{RN} – fatigue limit based on N_G, and N – the number of cycles to failure. [13]

The above relations describe the SN curve only for a given asymmetry coefficient of the loading cycle. Thus, a complete description of the behavior of the material requires a set of SN curves obtained for different values of the asymmetry coefficient.

This approach may be inconvenient because when studying the structure fatigue resistance loading cycles can occur with a wide range of asymmetry coefficient values, in addition, the SN curve is an experimental curve and an experiment is required for each asymmetry coefficient value. To simplify this approach, most of the experimental data are presented for a symmetric loading cycle, after which the loading history is reduced to a symmetrical stress cycle with respect to damage using relation 4, constructed on the basis of the Goodman dependence, after which relations 1 - 3 are used to directly evaluate the durability.

$$\sigma_{a\text{equiv}} = \begin{cases} \dfrac{\sigma_a^{pr}}{1-\frac{\sigma_m^{pr}}{\sigma_v}}, & \sigma_m^{\text{pr}} \geq 0 \\ 0, & \sigma_m^{\text{pr}} < 0 \end{cases} \quad (4)$$

D. Material properties used in the car body

Since the S-N method based on the stress amplitude is used to calculate fatigue life, to set the properties of the material requires knowledge of the S-N curve characterizing the number of cycles to failure that the test material specimen bears at a given stress amplitude [12].

High-cycle fatigue breakdown can be predicted with high accuracy using the S-N method, which is ideal for analyzing the durability of structural elements of a car body under operational loads. However, using this technique, it is impossible to reliably predict low-cycle fatigue failure caused, for example, by suspension breakdown. So, during the breakdown of the suspension, the stresses in the body structure can reach the yield strength and complete failure of the structure is possible after several loading cycles. Table III shows the elastic and fatigue characteristics of the materials used in the car body model. Used fatigue properties of materials were obtained on the basis of physical and mechanical properties and the relationships between them and fatigue properties.

At the later stages of designing a car, it makes sense to use the fatigue properties obtained using fatigue tests of material samples [14]. This will level out the inaccuracies in the results caused by the discrepancy between the used, approximated properties of materials with real brands of metals and batches that are used in the manufacture of the body.

Table 3. Material properties

Material	DX56D/10 X	HX260LAD	HX420LAD	HCT780X	ALU
Yield strength	210 MPa	227 MPa	481.6 MPa	500 MPa	130.2 MPa
Strength limit	270 MPa	372 MPa	598.8 MPa	800 MPa	200 MPa
Fatigue limit	121 MPa	155.5 MPa	307.2 MPa	360 MPa	60 MPa
The number of fatigue limit cycles	2.0e+6	7.54e+5	2.06e+5	2.0e+6	1.0e+7
Elastic modulus	210000 MPa	210000 MPa	210000 MPa	210000 MPa	70000 MPa

E. A simplified method for assessing fatigue life

Evaluation of fatigue life by a simplified method is based on the results of static strength analysis under the action of unit loads applied to the attachment points of the front suspension struts.

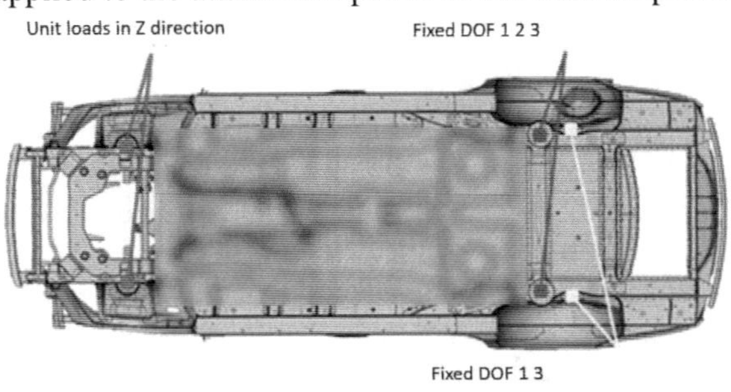

Figure 6. Boundary conditions and loads in static analysis

The translational degrees of freedom of the car body are fixed in the attachment points of the rear suspension according to 0. An analysis of the stress-strain state in two design cases of application of a single static loading is carried out: alternately, at the points of attachment of the left and right pillars of the front suspension to the body, a unit force is applied in the direction of the z-axis.

When analyzing the durability, the force along the z axis calculated by the formula below is applied to the left and right struts of the front suspension with a phase difference π according to the harmonic law:

$$P = \frac{R_{st} * B_f}{4 * L} \quad (5)$$

where: R_st - the static load on the front axle, B_f – front wheel track equal to 1700 mm, L – the meter mounting lever. (4)

$$R_{st} = 0.49 * P_{calc} * g = 2588\ kg \quad (6)$$

The loading scheme used in the analysis of durability by a simplified method consists of four stages and, taking into account the meter-mounting lever, is presented in Table IV.

Table 4. Loading by the simplified method

Loading stage	Amplitude loading, N	Frequency of loading cycles, Hz	Number of loading periods
1	0.4*P=2125	5	50000
2	0.8*P=4251	4	100000
3	1.2*P=6377	4	100000
4	1.6*P=8503	3	50000

The sequence of test loads presented in Table IV reflects the full cycle of operation of the vehicle under study, corresponding to 200,000 kilometers. Thus, all fatigue breakdown occurring in the first test cycle is undesirable.

When conducting a virtual durability test, the fatigue module of the MSC Patran software package calculates the stresses that occur in the structure under the action of forces applied to the interface points, superposing the results of the stress-strain state under the action of unit loads. This method, using a superposition of stress-strain states, can be applied in cases where a linear-elastic formulation of the problem is used. In this case, this is so, since we are talking about multi-cycle fatigue and the effects that lead to stresses significantly lower than the yield strength of the materials used.

Based on these data, and taking into account the fatigue properties of the material, it becomes possible to indicate the areas where failure is most likely to occur due to cyclic exposure.

F. Advanced fatigue life assessment methodology

Fatigue life using an advanced method accurately estimates the fatigue life of a body structure and welded joints in comparison with the simplified method.

As in a simplified technique, stress-strain states from the action of unit forces and moments are used as input data. However, if using the simplified methodology, unit loads were applied only to the attachment points of the front suspension struts, and then in the advanced methodology, all interface points are exposed to single impacts through which loads are applied to the body during vehicle operation. So in the advanced methodology for assessing durability, interface points are used:

- Front and rear connection points of the upper arm of the front suspension with the support of the front pillar of the car;
- The point of connection of the front suspension strut with the front support cup of the car body;
- Front and rear connection points of the silent block with the rear subframe bolt;
- The connection point of the rear suspension strut with the rear support cup of the car body;
- The attachment point of the anti-roll bar to the subframe;
- The front mounting point of the lower front suspension arm to the subframe;
- The front point of attachment of the lower arm of the front suspension to the subframe;
- Steering rack mounting point;

- The point of contact with the support of the transfer gearbox (TGB).

For all the indicated points, including symmetrical points, load profiles were used that correspond to the real road tests of the prototype car for an advanced assessment of the fatigue life of welded joints of a digital twin of a car using a virtual test site. These load profiles correspond to ten different types of maneuvers for five different types of pavement. This type of loading can be considered forced with an equivalence coefficient equal to 4.36. Thus, this loading can be considered forced, since 1 kilometer passed by this method corresponds to 4.36 kilometers of actual use.

A complete list of loading profiles and their parameters is presented in table V.

Given the equivalence coefficient, approximately 9.2 repeats of the loading cycle shown in Table V are required to evaluate fatigue life after 2,000,000 kilometers.

G. Methodology for assessing changes in body stiffness

Body stiffness was chosen as a characteristic used to assess the consumer properties of the car. This value influences both the driving performance of the car and its further deterioration [15].

Spot welds failed at this stage of fatigue loading were sequentially removed to assess the dynamics of changes in body stiffness from a finite element model of a car. After that, torsional and bending stiffness were evaluated. To calculate torsional stiffness, the digital twin of the car was fixed in the areas of attachment of shock absorbers of the rear suspension, and multidirectional force was applied along the z-axis to the right and left supports of the shock absorbers of the front suspension.

Table 5. Types of roads for fatigue life calculation

Road type	Load type	The required duration, km	One cycle length, km
High-speed mountain road	Acceleration; Braking; Double rebuild	3700	0,74
Smooth pavement		350	0,07
Profiled pavement		350	0,07
Access road	Sinus load; Hit a curb; Hit a curb in reverse; Exit from the curb during braking; Exit from the curb with one wheel during braking;	600	0,12
Total		5000	

A diagram of the application of forces and boundary conditions for assessing torsional stiffness is presented in 0, where the prohibition of translational degrees of freedom along the x and y axes is introduced at points 1 and 2, and the ban on movement over all translational degrees of freedom is introduced at points 3 and 4.

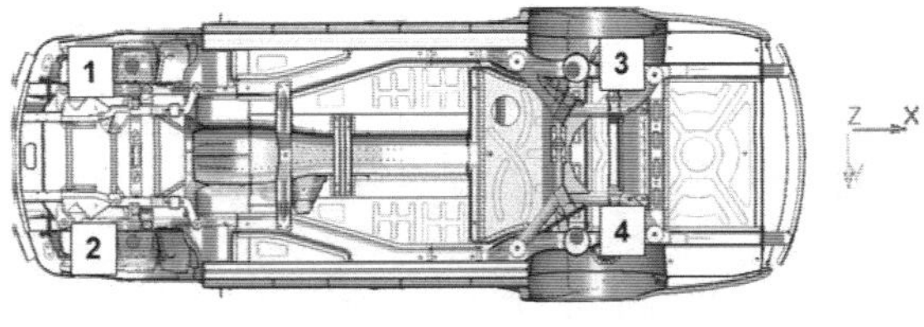

Figure 7. The calculation of the torsional stiffness of the car body.

Torsional stiffness is evaluated according to formula 7:

$$c=\frac{2Fl}{\operatorname{arctg}(\frac{w_1-w_2}{2l})} \quad (7)$$

where F – applied force, l – half the distance between points 1 and 2, w1 and w2 – vertical displacements at points 1 and 2 of the application of forces, respectively.

Similarly, the body flexural rigidity is estimated: fixation is carried out according to the translational degrees of freedom at points 1, 2, 3, 4, and at the attachment points of the seats 5, 6, 7, 8, loading along the z axis is performed. The layout of loading points and boundary conditions is shown in [5].

The value of bending stiffness is estimated by the formula 8:

$$c=\frac{4F}{\langle w\rangle} \quad (8)$$

where F – force intensity applied at points 5, 6, 7, 8; and <w> - average vertical displacement at these points.

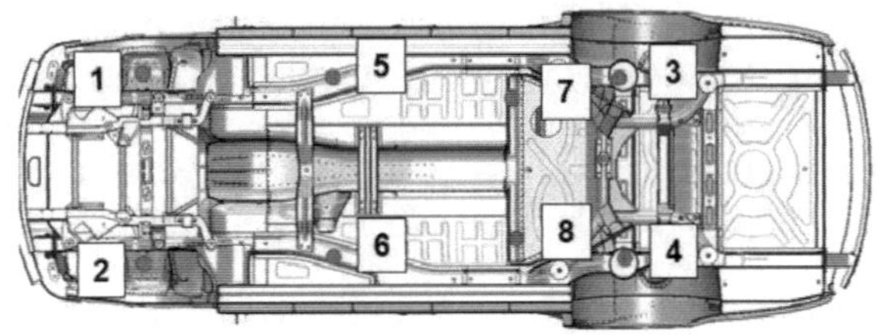

Figure 8. The calculation of the bending stiffness of the car body.

3. Tests results of fatigue life of the car body

A. Car body test results using a simplified technique

The implementation of the virtual tests described in the work was carried out on the basis of the platform CML-BenchTM (digital platform for the development of digital twins and an activity management system in the field of digital design, mathematical modeling and computer engineering, development of the engineering center CEC SPbPU). The calculations were performed using the supercomputer of the center "Polytechnic" SPbPU.

According to the results of a calculation check of the fatigue life of body structural elements using a simplified method, the integrity of most parts is maintained. However, minor fatigue breakdown zones are observed in motor shield reinforcements.

The analysis results showed from the entire cloud of spot welds of the digital twin of the car, shown in Fig. 9 and containing 9281 points, the main sites of the breakdown of the spot welds were in the area of attachment of the front and rear struts of the shock absorber and the area of reinforcements of the motor shield. In total, during the evaluation of the operation period, 175 spot welds were failed, which corresponds to less than two percent of their total number.

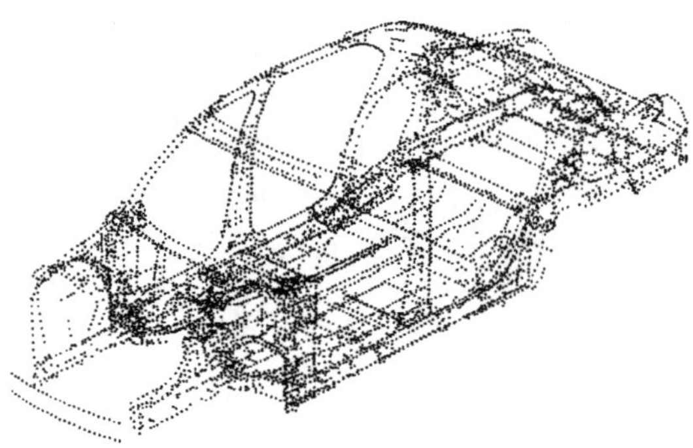

Figure 9. Cloud of all spot welds of the car.

[15] shows the main sites of the breakdown of welded joints are located around the most highly loaded areas, which are used to apply loads when calculating by a simplified method.

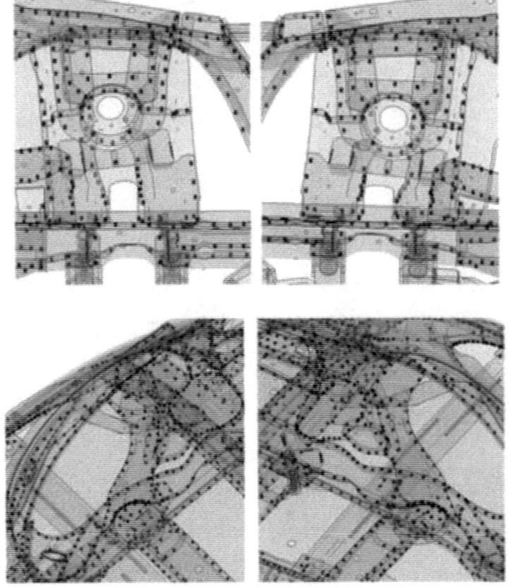

Figure 10. The main site of the breakdown of weld points in the areas of attachment of the front and rear shock absorbers using a simplified modeling technique.

In addition, the sites of the breakdown of spot welds were located in areas that were used to set the boundary conditions: rear shock absorber supports.
Additionally, there is a breakdown of the welded joints in the parts connecting the motor shield with the fastening areas of the front shock absorbers. The image of failure of spot welds in this area is shown in [12].

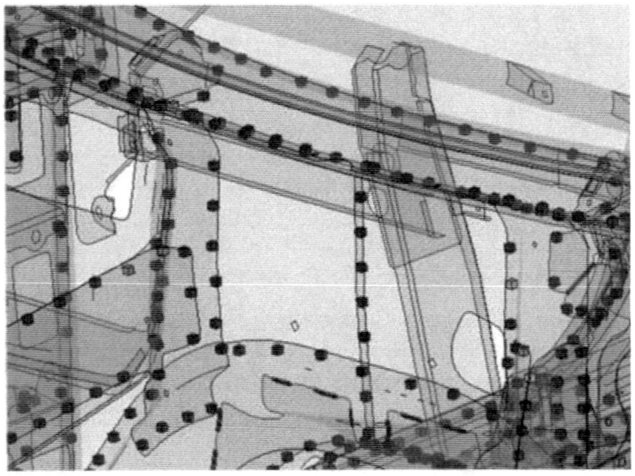

Figure 11. Fatigue breakdown of spot welds in the area of the motor shield.
Expanded body test results

Compared to the results of the simplified method, the advanced method shows more areas of failure of the body parts themselves, rather than welded joints. So, the failure is observed in not only the areas of application of loads and boundary conditions (the area of attachment of the front and rear struts of shock absorbers, subframes, suspension arms) but also relatively lightly loaded parts of the car, such as the bracket for attaching a high-voltage battery. However, the main site of the breakdown is still in the areas where the struts are mounted. What is shown in [9].

IOP Conf. Series: Materials Science and Engineering **986** (2020) 012054 doi:10.1088/1757-899X/986/1/012054

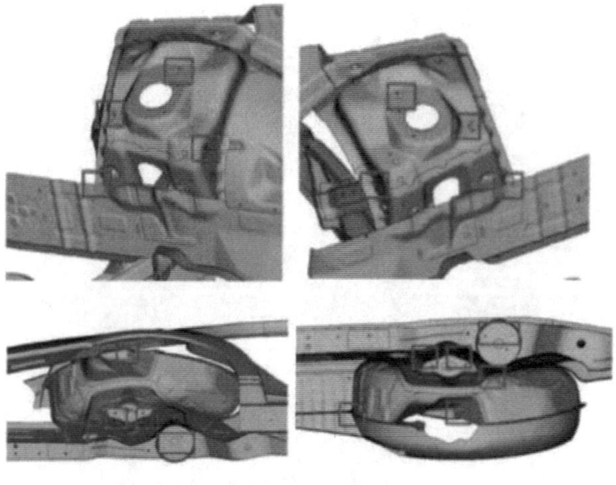

Figure 12. The main site of the breakdown of spot welds in the areas of attachment of front and rear shock absorbers in the advanced modeling method.

In addition, single sites of the breakdown were identified in the front and rear spars.

B. Assessment of the drop in body stiffness during fatigue breakdown of welded joints

A comparison of the torsional and bending stiffness of the body was made for the entire production cycle, with a step equal to 10% of the cycle duration. Figure 13 shows a comparison of the stresses arising in the car body during the bending stiffness test before and after passing the fatigue test and Fig. 14 a similar comparison is presented for the body torsional stiffness test.

As can be seen from the presented comparison, the fatigue fracture of weld points during the first operating cycle does not significantly influence the location of the centers of high stresses during testing for bending and torsional stiffness. However, it significantly influences the magnitude of stresses: for a body that has not undergone fatigue, the maximum stresses in a bending stiffness test are 110 MPa, 200 MPa - in torsional stiffness, and after a durability test, the maximum stresses are 120 MPa for bending testing and 350 MPa for torsional, thus showing the growth of 10% and 75%, respectively.

There is a significant increase in stress values, especially in the areas of fastening of the front struts of shock absorbers that experience significant loads during operation. This can lead to accelerated fatigue breakdown of the car body structure during further operation.

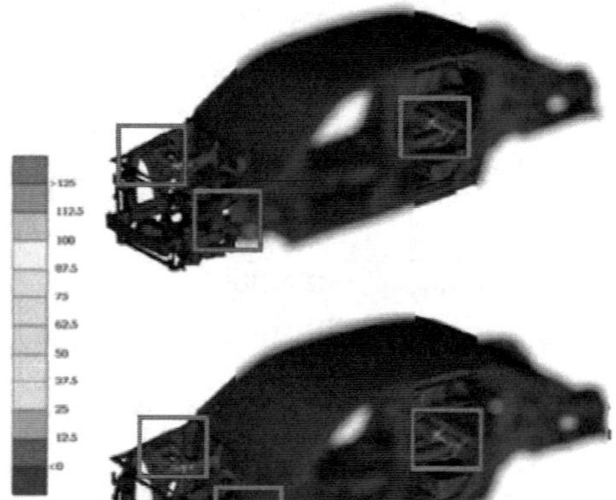

Figure 13. Comparison of stresses arising in the body when bending stiffness is tested.

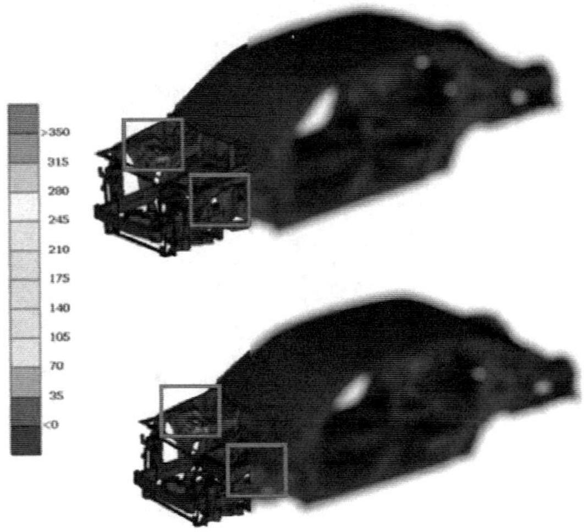

Figure 14. Comparison of stresses arising in the body when torsional stiffness is tested.

The maximum displacement along the z axis during the bending stiffness test was 0.77 mm for a body without a fatigue breakdown of welded joints and 1 mm for a car body after the first operating cycle.

[6] presents a comparison of movements along the z axis that occur in a car body when bending stiffness is tested, before and after passing the test for fatigue life.

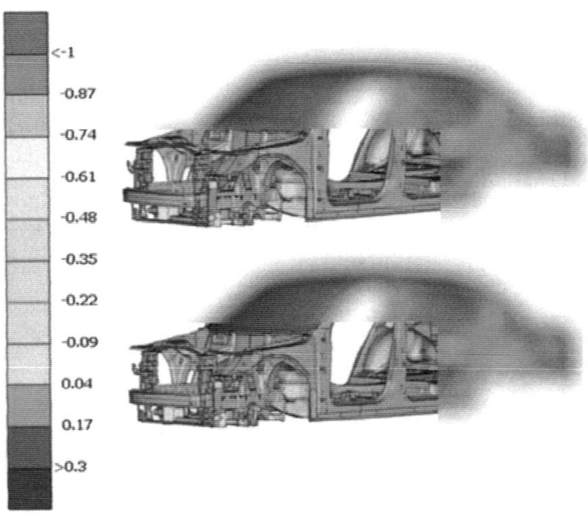

Figure 15. Comparison of movements along the z axis during the test for bending stiffness of the body.

The maximum displacement along the z axis during the torsional stiffness test was 1.25 mm for a body without fatigue breakdown of welded joints and 2.6 mm for a car body after the first operating cycle. [4] shows a comparison of the displacements along the z axis that occur in a car body during a torsional stiffness test before and after passing the fatigue test.

IOP Conf. Series: Materials Science and Engineering **986** (2020) 012054 doi:10.1088/1757-899X/986/1/012054

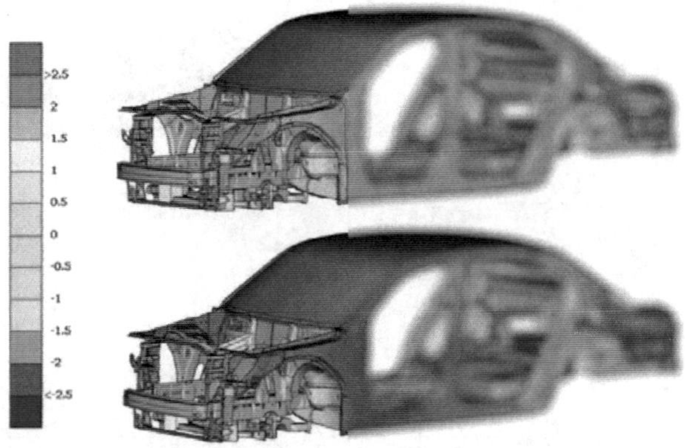

Figure 16. Comparison of movements along the z axis when torsional stiffness of the body is tested.

Based on the results of a numerical calculation of the bending and torsional stiffness of the body, graphs of changes in these parameters were constructed taking into account the fatigue breakdown of the spot welds of the body. These graphs are shown in [1] in comparison with the fracture schedule of spot welds.

In total, the bending stiffness of the car body decreased by 30 percent during one operation cycle: from 11,600 N/mm to 8,100 N/mm. Torsional stiffness decreased by more than 45 percent: from 26,700 Nm/deg to 14,600 Nm/deg.

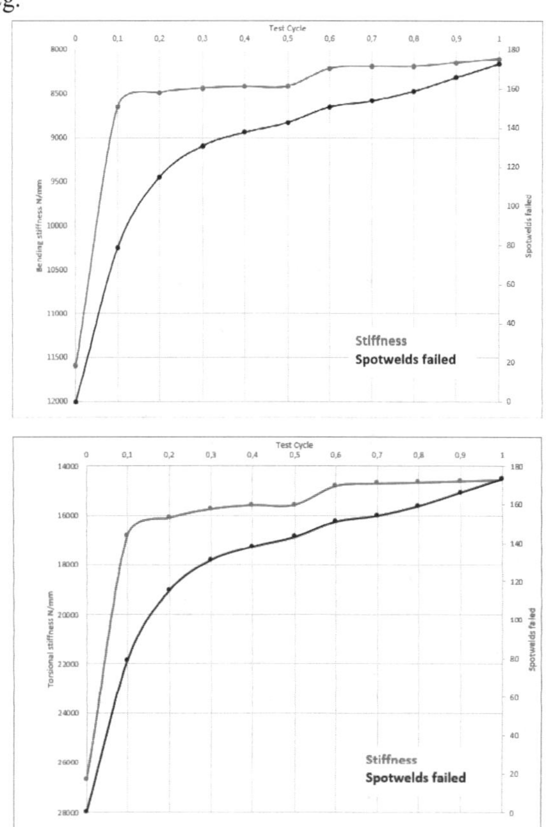

Figure 17. Change in body stiffness during the failure of spot welds.

IOP Conf. Series: Materials Science and Engineering **986** (2020) 012054 doi:10.1088/1757-899X/986/1/012054

C. Comparison of the areas of failure obtained using simplified and advanced techniques with real-life tests

The results obtained in the previous paragraphs showed the main site of fatigue breakdown according to the assessment of the digital twin of the car are:

- In the area of fastening the front shock absorbers;
- In the area of the motor shield;
- In the area of mounting the rear shock absorbers.

Resource tests conducted for a real prototype of the car under development allow comparing the results obtained for the digital twin using a simplified technique and a virtual test site and the results obtained in reality.

This comparison enables to assess the applicability of the used methods for further development and subsequent projects.

During the inspection of the prototype body, the failure of the body structure elements and welded points were found in the following areas of the structure:

- Area of fastening the front shock absorbers;
- Spot welds of the motor shield reinforcement.

A comparison of the areas of actually occurring and predicted fatigue breakdown is shown in Fig. 18 [2].

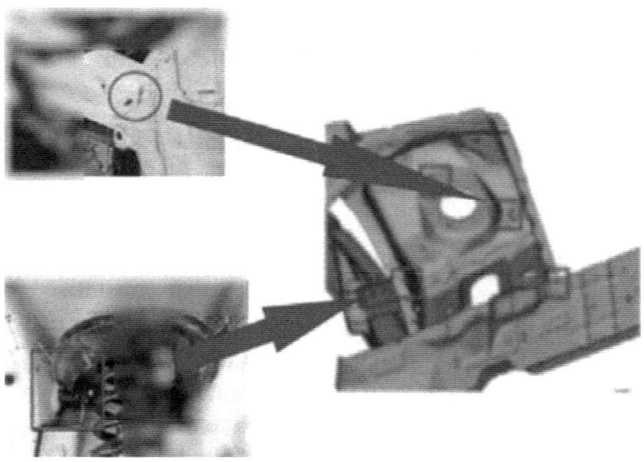

Figure 18. Comparison of fatigue breakdown areas.

At the same time, the area of fatigue breakdown was not found in the supports of the rear shock absorbers during real tests.

4. Conclusion

Various approaches to solving the problem of assessing the fatigue life of the car body and its welded joints were considered in this paper.

Despite the conservatism in the results obtained using these techniques, the main site of fatigue breakdown coincide with those occurring during a real test of a car. At the same time, the considered methods can significantly reduce the cost and speed up the process of developing the car in comparison with real tests, especially in the initial stages. This helps to assess the location of potentially problematic areas of the body in terms of fatigue life and the decrease in consumer characteristics of the car during its operation.

The obtained results suggest the prospect of using these methods for assessing fatigue life in car development.

References

[1] Borovkov A I, Gamzikova A A, Kukushkin K V, Ryabov Yu A 2019 Digital twins in high-tech industries (SpB, SPBSTU Publ) p 62

[2] Borovkov A I, Ryabov Yu A 2019 Digital twins: definition, approaches and development methods *Digital Transformation of Economics and Industry: Proceedings of a Scientific and Practical Conference with Foreign Participation* p 780

[3] Davlechina A, Leoro, Mendoza L, Klyavin O and Borovkov A 2018 Methodology of vehicle operational loads assessment during vehicle development process for further chassis and body-in-white strength and durability analysis Engineering Research and Development (IJMPERD) *International Journal of Mechanical and Production* (Vol. 8) pp 1415–1424

[4] Borovkov A I, Burdakov S F, Klyavin O, Melnikov M P, Mikhailov A A and Nemov A S 2012 *Computer-Aided Engineering* (SpB, SPBSTU Publ) p 93

[5] Morozov E M and Nikishkov G P 1980 Finite element method in fracture mechanics (M .: Nauka) p 256

[6] Hanq D A, Walters A J and Beuth J L 2000 Development of an object-oriented fatigue tool *Engineering with computers* (V. 16) pp 131 – 144

[7] Nikulin M V 2017 Digital design technique for car body design for a given resource *SPBSTU Science week*

[8] Kwang-Jae Son, Young-Soo Yang, Ki-Sup Choi and Sung-Kyu Cho 2003 Fatigue Strength Evaluation on Resistance Spot Welds of the Vehicle Body *Mechanics Based Design of Structures and Machines* (31:1) pp 79-92

[9] Wohler A 1858 Report on test of the Royal Niederschleesich-Märkkischen Railway made with apparatus for the measurement of the bending and torsion of railway axles in service (in German) Zeitsch.Bauwesen (N8) pp 642-651

[10] Atzori B, Manara G and Meneghetti G 2015 Experimental Analysis of the Fatigue Strength of a Tubular Welded Joint adopted in a Roller Coaster Structure *Padova : University of Padova*

[11] MSC Nastran 2013 Advanced Linear Analysis using MSC Nastran.

[12] Berendeev N N 2012 Methods for solving fatigue problems in the Ansys Workbench package *Nizhny Novgorod*

[13] Morozov E A 2015 The method of analysis of the fatigue life of electronic equipment on the basis of mathematical modeling of thermal strength processes (Tomsk)

[14] Zuev L B and Danilov V I 2013 The physical basis of the strength of materials *Dolgoprudny: Intelligence* p 376

[15] Patrikeev A, Tarasov A, Borovkov A, Aleshin M and Klyavin O 2017 NVH analysis of offroad vehicle frame. Evaluation of mutual influence of body-frame system components *Materials Physics and Mechanics* (no. 34) pp 70–75.

IOP Conf. Series: Materials Science and Engineering **986** (2020) 012055 doi:10.1088/1757-899X/986/1/012055

Study of the vibration state (natural frequency spectrum) of the rotor of a microturbine unit with a capacity of 100 kW

V Rassokhin[2], A Sukhanov[2], A Smetankin[1], V Barskov[2], Yu Matveev[2] and O Nikiforova[1]

[1]National Technology Initiative Center for Advanced Manufacturing Technologies based on the Institute of Advanced Manufacturing Technologies of Peter the Great St. Petersburg Polytechnic University Polytechnicheskaya, 29, St.Petersburg, 195251, Russia, v-rassokhin@yandex.ru
[2]Higher school of power engineering, Institute of power engineering Peter the Great St. Peersburg Polytechnic University, Saint Petersburg, Polytechnicheskaya, 29, St.Petersburg, 195251, Russia

Abstract. The vibrational state of the rotor of a turbine installation, as is known, determines the level of reliability and the resource of its operation. In this article, by the vibrational state of the rotor we understand the vibrational state of all its elements, including the vibrational state of the impellers of the turbine stages. An acceptable level of vibrational state of the rotor impellers (RI) of the turbine stages is provided by various factors, among which there is the most important one - the presence of vibrational detuning between the natural frequencies of the RI and the frequencies of the disturbing flow effect, which are usually multiples of the rotor speed [1]. From what has been said, there is a substantial need for an accurate determination of the eigenfrequencies of turbine impellers. The accuracy of determining the frequencies is achieved by comparing the results of the calculations with the results of the experimental determination of the indicated natural frequencies.

1. Introduction
In the production of power plants, the issue of ensuring the reliability of the impellers is solved in several stages: Stage 1 - calculation of the natural frequencies of individual blades and impellers; Stage 2 - direct measurement of the natural frequencies of single working blades (WB); Stage 3 - direct measurement of the natural frequencies of stationary impellers with typed impellers; Stage 4 - experimental determination of the natural frequencies of the assembled impellers at an operating speed of rotation on the Campbell machine. (Stage 5 - the possible measurement of the frequency spectrum of the vibration of the supports of the assembled rotor at the operating speed of rotation during its high-speed balancing in the vacuum chamber of the Schenk machine). The main feature of the design of the rotors of microturbine plants is the use of impellers made at the same time with the blades, and also often at the same time with the installation shaft [2-10]. Considering the indicated design features, it is advisable to conduct vibrational studies of the impellers of microturbines directly on the assembled rotor, bypassing the preliminary stages of vibrational studies indicated above.

2. Methods
To perform the indicated vibrational studies of microturbine rotors, it is necessary to use highly sensitive measuring equipment with an extended frequency range up to 30 kHz and higher. An additional very

important point for microturbines is the ability to perform non-contact vibration measurements [11-15]. In this work, we consider laser scanning Doppler vibrometer from Politek PSV-400 (LDV) [16] as such equipment. The specified vibrometer meets the above requirements.

The frequency measurement in the device is based on the use of the Doppler effect. The scheme for measuring the vibration frequency using the indicated device is shown in fig. 1 and explained below.

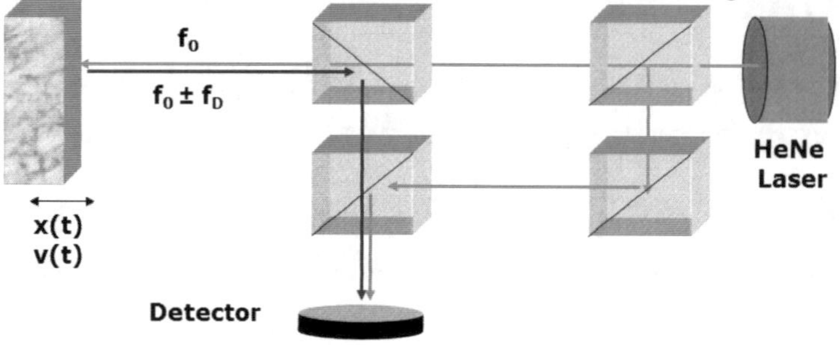

Figure 1. Scheme for measuring the frequency of a vibrating surface using a laser Doppler vibrometer [11].

A beam from a helium-neon laser, characterized by a frequency f0, passes through a system of optical prisms and falls on a vibrating surface, indicated in fig. 1 by x(t). The reflected beam (in fig. 1 it is a blue beam) is characterized by some frequency shift f0 ± fD; the amount of displacement is determined by the magnitude of the vibration velocity of the reflecting surface v(t). The detector perceives a total beam, consisting of a reference (red) and reflected (blue) beam. The total beam is characterized by beating, the value of which is determined by the frequency offset of the reference and reflected beam. The detector perceives the total beam as a system of concentric light and dark spots shown in fig. 2.

Figure 2. The optical form of the interaction of direct and reflected rays at the detector of a laser vibrometer [16].

When using a PSV-500-H laser vibrometer, the maximum vibration frequency of the studied structures can be from 0Hz to100 kHz with surface vibration velocities from 10-5 m/s to 10 m/s [16]. The indicated parameters provide overlapping of the values of frequencies and vibration velocities of structural elements encountered in turbine construction.

Given the positive experience of Polytec in the field of frequency measurements on the rotors of small turbomachines, we attempted to measure the natural frequencies of the rotor of a gas turbine plant with a capacity of 100 kW. The overall dimensions of the rotor of the GTU 100 kW unit are shown in fig. 3. The number of blades on the impeller of the Z2 turbine is 17.

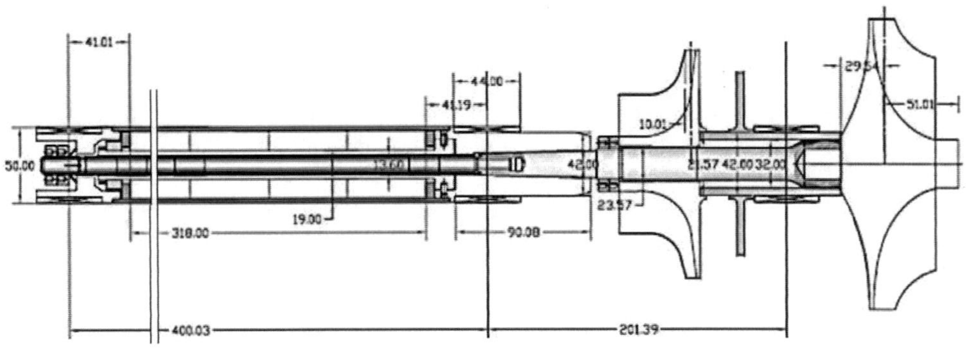

Figure 3. Overall dimensions of gas turbines with a capacity of 100 kW.

3. Results and discussion

Some results of vibration studies of a rotor of GTU 100 kW, obtained using a PSV-400 scanning laser meter, are presented in fig. 4 and 5. Fig. 4 shows the visualization of the first knotless form of the rotor mounted on the mounting supports. The impellers of the compressor and turbine are involved in the oscillations in the indicated form as massive nondeformable bodies.

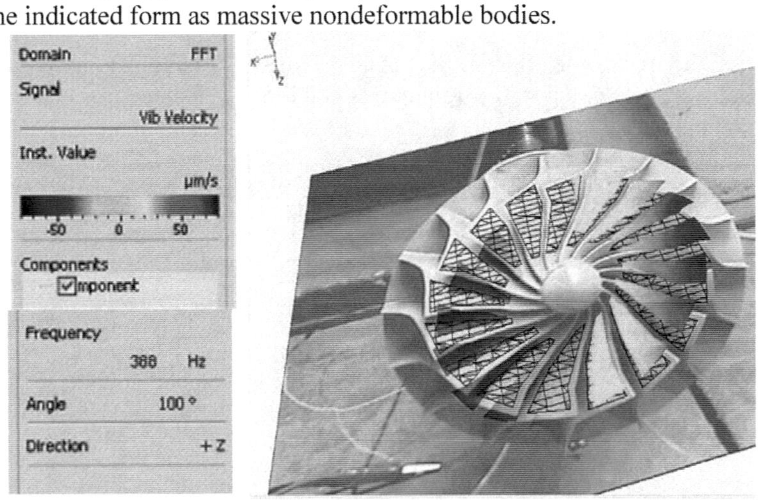

Figure 4. The oscillation form of the rotor of GTU 100 kW, corresponding to a frequency of 388 Hz. The maximum vibration amplitude is 36 µm / s. (view of the rotor from the side of the turbine impeller).

As a result of the measurements, a spectrum of natural vibrations of the rotor of GTU 100 kW was also obtained. Fig. 5 shows the low-frequency part of this spectrum. The spectrum contains discrete components corresponding to the lower rotor form considered above (388 Hz), as well as discrete components corresponding to the lowest forms of natural vibrations of the turbine impeller (1900 … 2400 Hz).

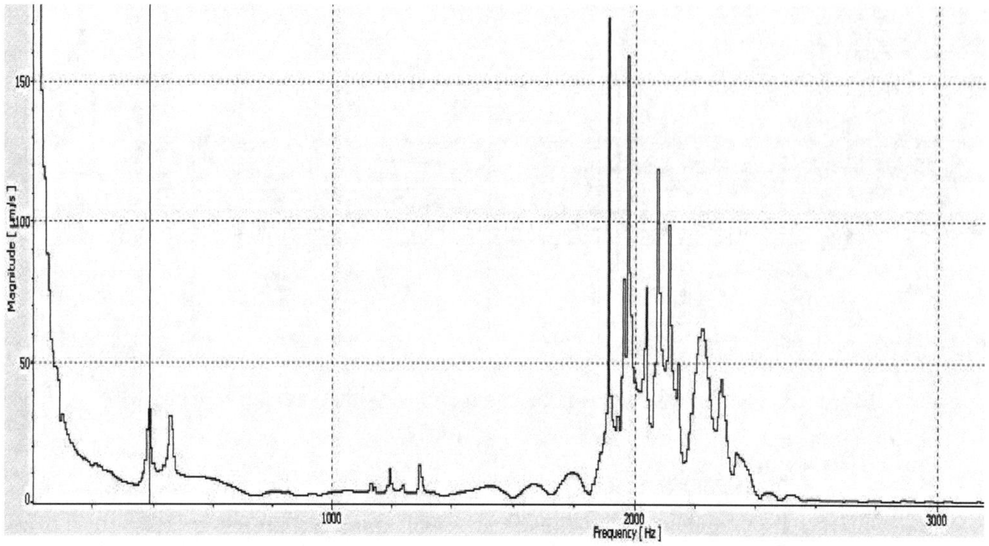

Figure 5. Vibration spectrum of GTU 100 kW rotor.

To clarify the structure of the spectrum in the range 1900 ... 2400 Hz, it is necessary to refer to the analysis of the results of calculations of the frequencies and forms of the centrifugal turbine of the turbine under consideration in the GTU 100 kW performed in the ANSYS program [17-21]. It can be seen from the vibration diagram shown in Fig. 6 that the frequency range of the first in-phase form of the RI corresponds to the indicated frequency range; a general view of this form is shown in Fig. 7.

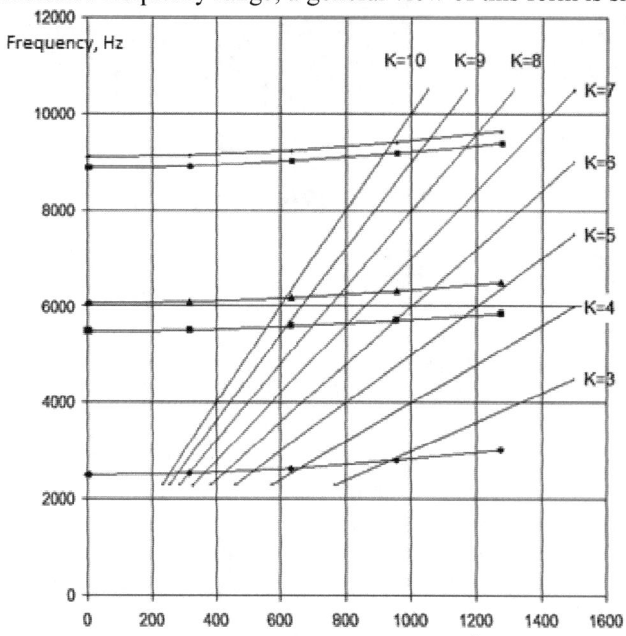

Figure 6. Vibration diagram of the RI turbine turbine 100 kW.

IOP Conf. Series: Materials Science and Engineering **986** (2020) 012055 doi:10.1088/1757-899X/986/1/012055

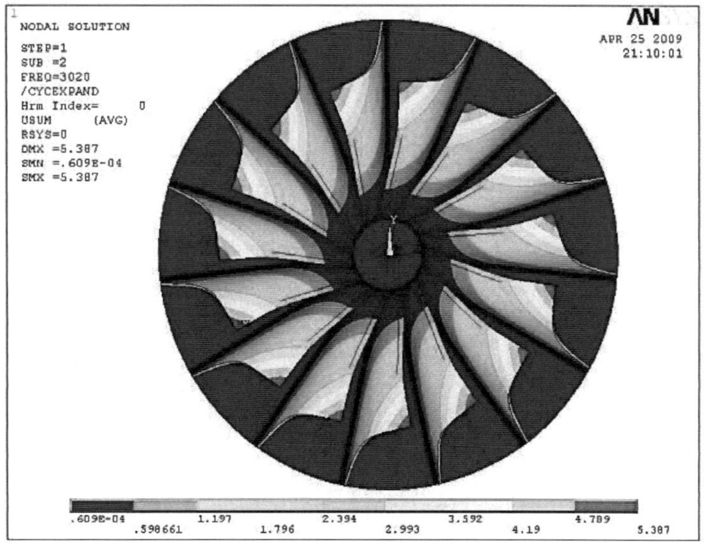

Figure 7. The low-frequency common-mode form of the RC (3020 Hz), obtained by calculation in the ANSYS program at a rotation speed of 1300 rps.

Let us consider the possible reasons for the presence of several discrete components close in frequency in the frequency range under consideration of 1900 ... 2400 Hz.

The first reason may be the presence of several similar in frequency forms of the RI on the basis of one form of the working blade (in this case, knotless). The second reason may be the presence of some slight different frequencies of the adjacent working blades of the RI [22]. Given the high accuracy of manufacturing the RI on a CNC machine, we will consider the first reason more significant. Let's consider it in more detail. As the calculations in ANSYS showed, the number of similar types of impellers of a centripetal turbine is determined by the number of working blades Z2. With an even number of working blades, the number of such forms will be equal to Z2 / 2 and these forms will be distinguished by the number of nodal diameters [1]. The lowest-frequency form will be a form with one nodal diameter, the higher-frequency form will be a form with two nodal diameters, etc. The highest frequency of the indicated homogeneous forms will be the form with common-mode vibrations of all working blades (a similar form is shown in fig. 7). A visual representation of these uniform forms with a different number of nodal diameters is given by the frequency map of a single-type RI of centripetal turbine with the number of blades Z2 = 18 (see fig. 8).

Here, the shape number is plotted along the horizontal axis, and each number formally corresponds to two numbers (oscillations along the X and Y axes). The eigenfrequencies of the considered RI are plotted along the vertical axis. The frequencies shown on the map can be grouped by type of blade shape. A conventional view of the shapes of the blades is shown on the left. Each scapular form forms the basis of a group of RI forms with a different number of nodal diameters, where this number is m = 0, 1, 2, ... Z2 / 2. The indicated forms of the RI with a different number of nodal diameters are conventionally drawn in the middle of the figure.

Thus, each RI form with in-phase oscillations of the blades corresponds to a group of lower-frequency forms that differ from the in-phase form by the presence of nodal diameters from 1 to Z2 / 2 (with an odd number of blades this number is Z2 / 2 + 1).

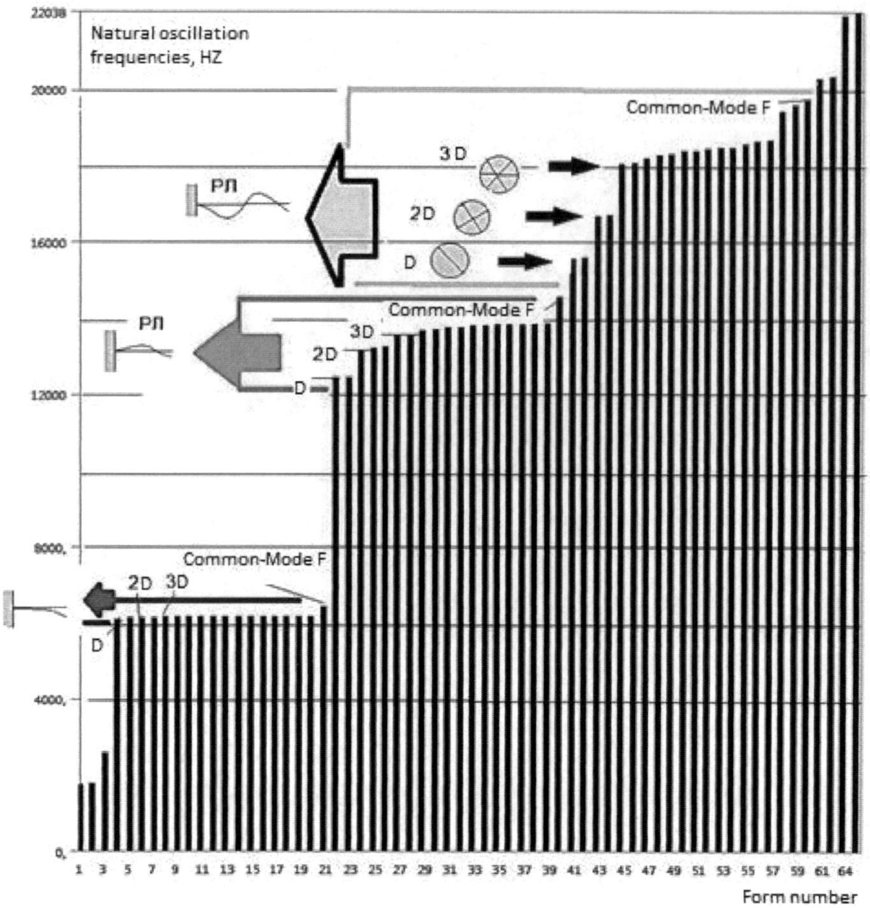

Figure 8. The frequency map of the impeller of a centripetal turbine.

Considering the fact that the number of rotor blades on the impeller of the considered GTU 100 kW is 17 (Z2 = 17), there should be nine corresponding fan forms of the turbine RI. As can be seen from the spectrum shown in fig. 5, the frequency range of 1900 ... 2400 Hz corresponds to nine discrete components. Thus, the number of predicted and available on the spectrum of discrete components coincided.

4. Conclusions

The obtained experimental and calculated results [23-26], as well as an attempt to jointly analyze them, are of scientific and methodological significance, since they realize the desire to obtain mutual confirmation of the obtained values of the natural frequencies of the RI. This approach will guarantee reliable vibrational detuning of the GTU 100 kW rotor and its main elements, and therefore ensure the reliability of the GTU. The materials presented in the article indicate that the laser scanning method allows one to determine with high accuracy the natural frequencies and shapes of the impellers of low-consumption gas turbine plants.

References

[1] Kostyuk A G Dynamics and strength of turbomachines. Publishing House MPEI (2007)
[2] Rassokhin V A, Rakov G L, Sukhanov A I, Kondratyev V et al. 2009 Development of a turbocharger of a small-sized gas turbine unit with a capacity of 100 kW with an internal efficiency of at least 30%. SPbSTU *Research report*

[3] Besedin S N, Rassokhin V A and Okunev E I 2010 *Scientific and Technical Journal of St. Petersburg State Polytechnical University* pp 57-61
[4] Barskov V V 2013 *Scientific and Technical Ved SPbPU* (4-1 (183)) pp 244-249
[5] Barskov V V, Rassokhin V A, Besedin S N and Osipov A V 2017 *Bulletin of the Bryansk State Technical University* (8 (61)) pp 25-33
[6] Barskov V V 2017 Thesis of the candidate of technical sciences
[7] Barskov V V, Rassokhin V A, Besedin S N and Osipov A V 2015 *Bulletin of the Bryansk State Technical University* (3 (47)) pp 6-14
[8] Barskov V V 2018 *Jour of Adv Research in Dyn & Control Sys* (10 (06)) pp 384-392
[9] Besedin S N 2009 *Scientific and Technical Sciences SPbSPU* (4-1 (89)) pp 153-166
[10] Besedin S N, Zabelin N A, Olennikov S Yu, Polishchuk V G, Rassokhin V.A, Rakov G L and Fokin G A 2009 *Innovac politics and inventors* pp 58-61
[11] Analysis Of Dynamical Properties of a 700 Kw Turbine Rotor Designed to Operate in an ORC Installation. Diagnostyka 2016 (Vol.17, № 2, BRENKACZ, ZYWICA, BOGULICZ. The Szewalski Institute of Fluid-Flow Machinery, Polish Academy of Sciences Centre of Mechanics of Machines, Department of Turbine Dynamics and Diagnostics Gen. J. Fiszera 14 St. 80-231 Gdańsk, Poland, lbrenkacz@imp.gda.pl.) pp 17-23
[12] Ge Li-juan, Zhang Chun-hui, Hao Min and Zhang Yong 2013 Vibration Analysis of the Steam Turbine Shafting caused by Steam Flow *TELKOMNIKA* (Vol. 11, No. 8, August 2013, Inner Mongolia Agricultural University, 306 Zhaowuda Road. Hohhot, Inner Mongolia. P.R.China, 010018, e-ISSN: 2087-278X) pp 4422-4432
[13] Banakh L, Nikiforov A and Panovko G 2008 LATERAL TURBINE AND GENERATOR VIBRATIONS. REPORT 2016:294. VIBRATIONS IN NUCLEAR APPLICATIONS. Vibroimpact Motion of Rotor taking into account Friction at the Contact *Proc. 6th EUROMECH Nonlinear Dynamics Conference (ENOC 2008)* (Saint Petersburg (Russia)).
[14] Bachschmid N, Pennacchi P and Tanzi E 2009 Cracked Rotating Shafts: Typical Behaviors, Modeling and Diagnosis *IUTAM Symposium on Emerging Trends in Rotor Dynamics Proceedings of the IUTAM Symposium on Emerging Trends in Rotor Dynamics, held in New Delhi, India, March 23 – March 26* (Department of Mechanical Engineering, Politecnico di Milano, Via La Masa 1, 20156 Milano, Italy. e-mail: nicolo.bachschmid@polimi.it) pp 441-454
[15] Tejas H Patel and Ashish K Darpe T H Patel 2009 Application of Full Spectrum Analysis for Rotor Fault Diagnosis *IUTAM Symposium on Emerging Trends in Rotor Dynamics Proceedings of the IUTAM Symposium on Emerging Trends in Rotor Dynamics, held in New Delhi, India, March 23 – March 26* (Department of Mechanical Engineering, University of Alberta, Edmonton, Alberta T6G 2G8, Canada A.K. Darpe. Department of Mechanical Engineering, Indian Institute of Technology, New Delhi 110 016, India. e-mail: akdarpe@mech.iitd.ac.in) pp 535-545
[16] Polytec. Laser Doppler Vibrometer (LDV) http://www.polytec.com
[17] Barskov V V, Rassokhin V A, Besedin S N and Tulubensky E V 2016 *Bulletin of the Bryansk State Technical University* (1 (49)) pp 9-15
[18] Besedin S N, Rassokhin V A, Fokin G A and Rakov G L 2010 *II All-Russian scientific and practical conf Actual probl Mashinostro* pp 284-289
[19] Besedin S N, Matveev Yu V, Mironov V V, Rassokhin V A and Rakov S G 2010 *XXXVIII week of science SPbSPU* pp 57-59
[20] Besedin S N and Fokin G A 2010 Gazoturb technol, (1 (82) pp 10-13
[21] Barskov V V and Besedin S N 2013 Copyright Certificate No. 132195
[22] Keith W 2008 *Jones. Journal of Turbomachinery* p 130
[23] Barskov V 2017 *Int Jour of Adv Biotech and Research* (8 (4)) pp 1708-1715
[24] Besedin S N, Zabelin N A, Matveev Yu V and Rassokhin V A 2010 *High intellect technol and innovations in education and science* (1) pp 262-263
[25] Rassokhin V A 2004 *Trudy SPbSPU* p 491
[26] Barskov V V and Besedin S N 2018 Ned Science SPbPU pp 153 - 154

IOP Conf. Series: Materials Science and Engineering **986** (2020) 012056 doi:10.1088/1757-899X/986/1/012056

Numerical simulation of solid data carrier grinding process

G V Monakhovskiy, N A Kharaldin, M V Aleshin and O I Klyavin

Center of National Technological Initiative, Institute for Advanced Manufacturing Technologies, Peter the Great St. Petersburg Polytechnic University, 29 Politekhnicheskaya, St. Petersburg, 195251, Russia, monahovskij_gv@spbstu.ru

Abstract. In this article we consider a general approach and method to assess the solid data carrier destruction in a shredder using grinding rotors with teeth. Numerical simulation of the grinding process was carried out in the LS-DYNA by discrete element method and finite element method (FEM-DEM). Evaluation and comparison with a real grinding process was based on a phenomenological approach.

1. Introduction

One of the most effective means of physical destruction of various data carrier (paper, memory cards, CDs, etc.) is crushing and grinding them into small fractions in specialized shredders. Traditionally, engineering uses analytical methods [1], [7], [8], [10], [12], calculation of crushing parameters, calculation of kinematics and strength. The complexity and engineering time, as well as the required number of prototypes and full-scale tests to develop engineering to the performance goals, depends on the requirements and targets of the design. In turn, the use of digital design and digital twins methods [14], [15], which use numerical methods for evaluating the prototype design at an early stage, will reduce the time spent on engineering, optimization and testing and, therefore, will reduce financial costs.

The purpose of this work is to develop and test a numerical modeling method of a grinding process of a solid data carrier (CD-disk) using crushing roller with cutting teeth installed in the prototype of shredder. Another goal of this work is to demonstrate the possibilities of combined application of the finite element and discrete element method for evaluating the grinding process of solid data carriers.

The implementation of the virtual tests described in the work was carried out on the basis of the platform CML-BenchTM (digital platform for the development of digital twins and an activity management system in the field of digital design, mathematical modeling and computer engineering, development of the engineering center CEC SPbPU) using commercial licenses LS-DYNA and computational resources of CNTI.

2. Finite element model description

The shredder considered in this work belongs to the cascade type of rotary crushers [16] and has three rows of cutting rollers. The simulation was carried out for the second level of the rollers, which should provide data carrier grinding to a 2 mm fraction. The developed finite element model of the shredder contains the following:

a) Geometric (CAD) (Fig. 1) and finite element (CAE) plastic shredder block models (Fig. 2);

The object of finite element modeling is a plastic grinding block, as well as a sample of a 14 мм×14 мм polycarbonate CD disk with thickness of 1.2 mm, located directly between the cutting teeth of the grinding block rotors (Fig. 3). Within this study three modifications of this finite element model have been developed:

- with cutting rotors made of absolutely hard material and an average element size of 3 mm (Fig. 2, a),
- - with three rows of cutting teeth made of strain sensitive material and an average finite element size of 1.5 mm (Fig. 2, b),
- one of the rotors is modeled entirely, the second rotor is made in the form of three rows of cutting teeth, an ejector is added under one of the cutting rollers (Fig. 2, c). An average size of finite element on rollers is 1.5 мм. A strain sensitive material is used.

Configuration №1 of the finite element model was used to estimate the size of the produced fractions of the shredded object because of the longer calculation time. The main purpose of the grinding modeling in configurations №2 and №3 was to estimate the stress-strain rate of the rotors and ejector. To study the ejector operation, which cleans the space between the rows of rotor teeth, a set of "glued" CD fragments was created, simulating a disk fragment stuck in the space between the rows of teeth. For

effect of fragments to the rotors, and an artificial obstacle was created to keep these fragments at the ejector.

b) Information about all connectors;

Bolted connection are modeled using absolutely rigid constraints, bearings - using absolutely rigid cylindrical joints.

c) The complete materials data required for the assigned task used in the models is shown in Table 1:

Table 1. Elastoplastic properties of materials.

Material	E, MPa	υ	ρ, kg/m^3	$\sigma_{0.2}$, MPa	σ_B, MPa
HVG Steel	$1.9 \cdot 10^5$	0.3	7850	445	790
Steel 45	$2 \cdot 10^5$	0.3	7826	245	470
D16 aluminium	$7.2 \cdot 10^4$	0.34	2800	360	460

d) Operation mode data.

The rotation speed of the grinding and cutting rollers is 25 rpm.

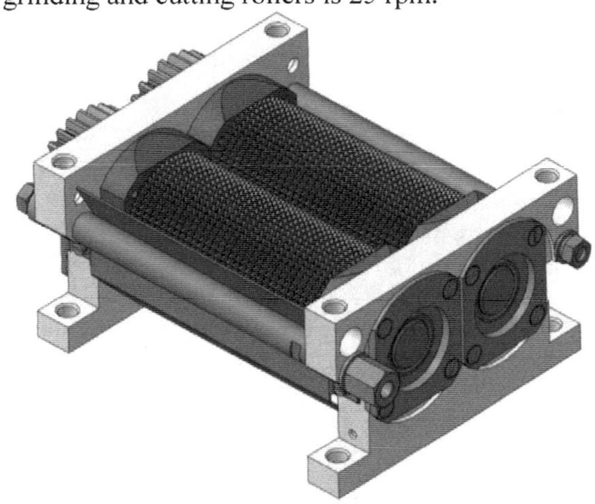

Figure 1. CAD model of the shredding block.

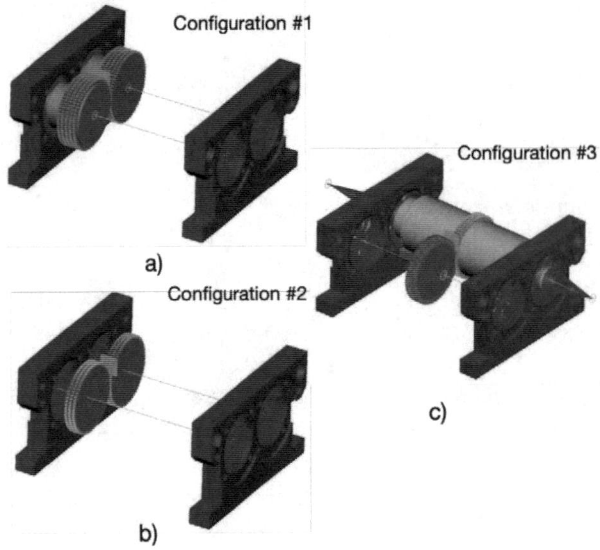

Figure 2. FE-models of the shredding block shown in 3 configurations.

Figure 3. CD-disk specimen, containing discrete elements.

3. Numerical simulation of shredding process

The study of the grinding process is a numerical simulation of the process of crushing a sample of a solid data carrier - a CD-disk. To simulate the destruction of a CD-disk sample the discrete element method was chosen which works in conjunction with the finite element method in the LS-DYNA. The discrete element method is the most optimal and convenient in the grinding process simulation, which is confirmed by numerous studies and tracking the size of fragmented particles.

The discrete element method is widely used in various industries and science for numerical modeling of soils [2], [9] and other flowing medium, crushing of rocks [3], and business process modeling [4].

The material settings [11] describing the CD sample, as well as the particle size are shown in Fig. 4. These parameters were optimized on the basis of numerous design estimates using a phenomenological approach.

```
*CONTROL_DISCRETE_ELEMENT
$#   ndamp     tdamp      fric     fricr     normk    sheark       cap     mxnsc
       0.7       0.9       0.3       0.2      0.01     0.003         0         0

*DEFINE_DE_BOND
$#     sid     stype    bdform
       200         3         1
$#     pbn       pbs     pbn_s     pbs_s       sfa     alpha         -    maxgap
       0.8       0.8      70.0      20.0       1.0       0.0       0.0       0.4
```

Figure 4. Discrete elements settings in LS-DYNA input file.

At the beginning, the CD sample is directly above the cutting teeth of the rollers. The cutting rollers have an initial rotation speed of 25 rpm, which corresponds to the operating mode of the shredder prototype (II.d). Thus, under the gravitation the CD sample falls under the cutting teeth and is shredded by rotors.

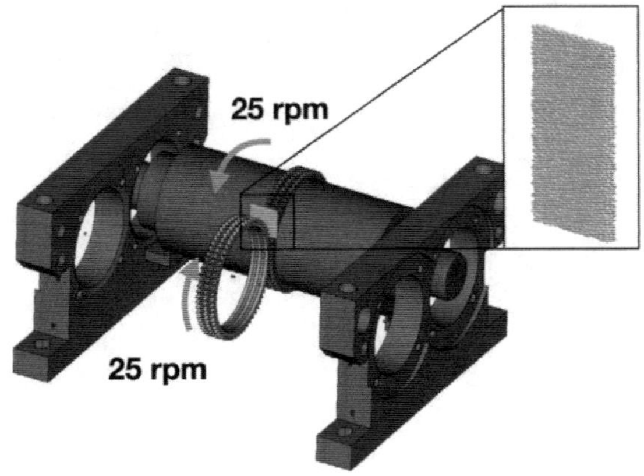

Figure 5. Graphical Interpretation of the problem formulation.

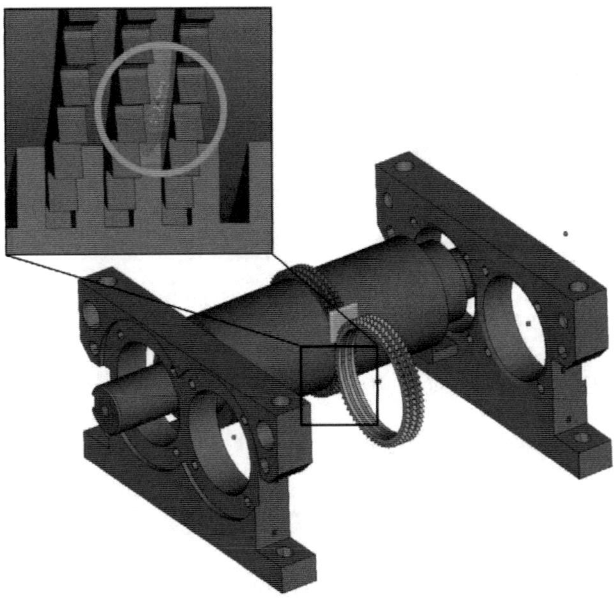

Figure 6. FE-model of the shredding block, configuration №3.

To carry out numerical modeling of the grinding process, an explicit integration method (Explicitdynamics) was used, presented in the finite element software package LS-DYNA.

4. Results

Within this study we estimated not only the nature of destruction and the size of the breakaway fragments of the CD-disk sample, but also the stress-strain state of crushing rotors teeth and the ejector.

Fig. 7 and Fig. 8 shows storyboards of the destruction process of a CD-disk sample in three configurations of the finite element model.

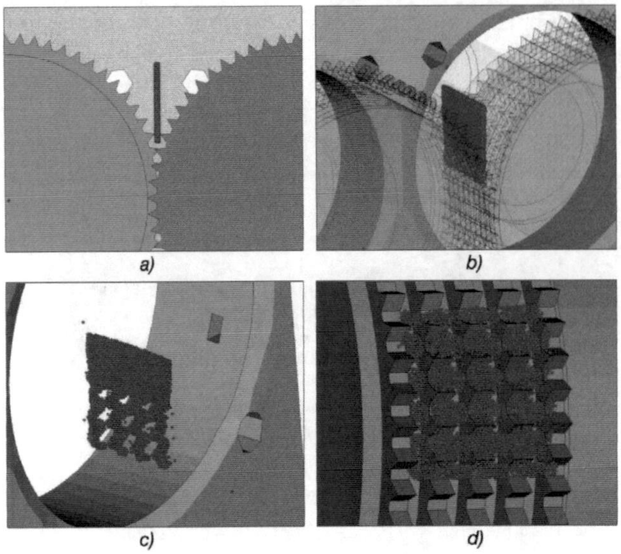

Figure 7. Shredding process of CD-disk specimen, configuration №1.

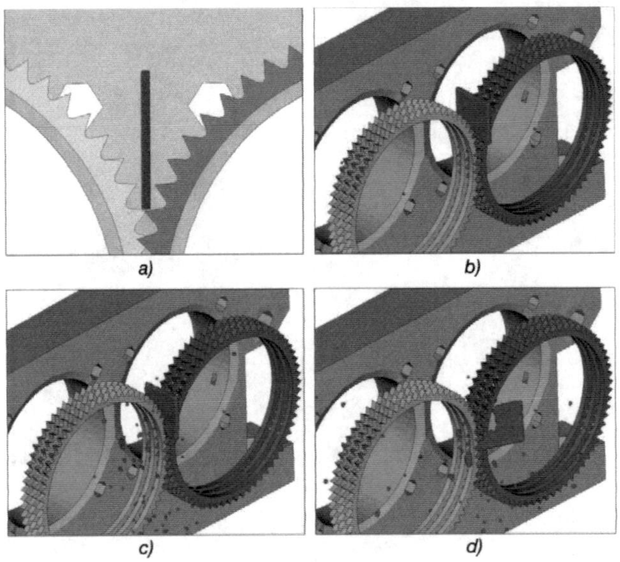

Figure 8. Shredding process of CD-disk specimen, configuration №2.

Fig. 9 and Fig. 10 shows the fields of equivalent stresses in the teeth of crushing rotors and parts of the ejector arising during grinding.

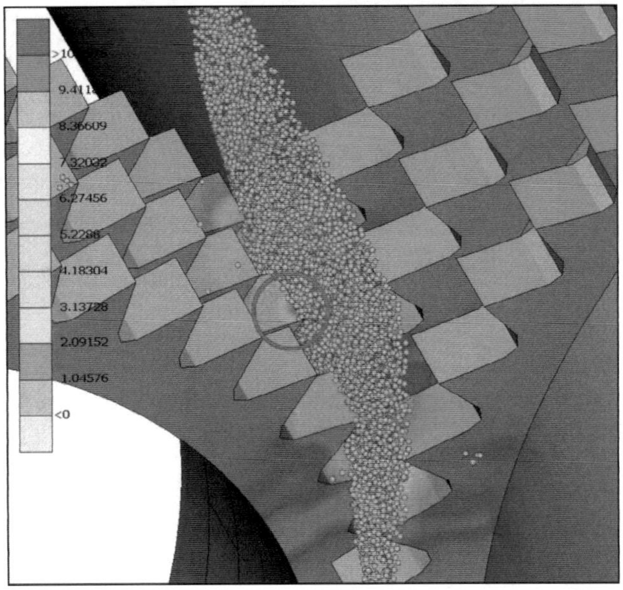

Figure 9. Equivalent stress field in rotors,MPa.

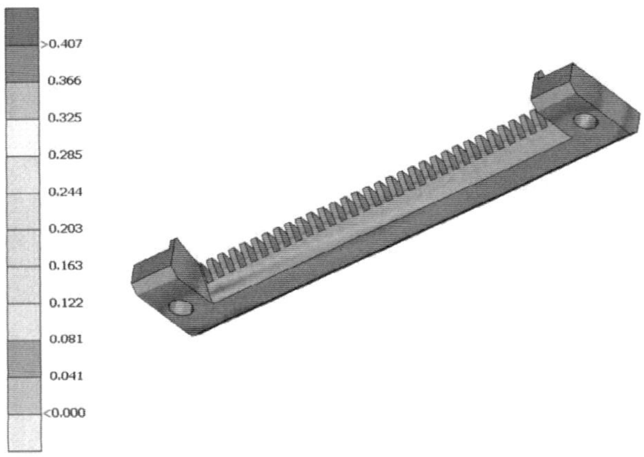

Figure 10. Equivalent stress filed in wiper, MPa.

5. Discussion

Numerical modeling of grinding showed that the size of the data carrier fragments after processing is 1.7-1.9 mm, which corresponds to the characteristics of the second level of grinding rotors. The maximum values of the equivalent stresses in the grinding rollers teeth are 10.4 MPa and they appear at the moment of direct interaction of the tooth with CD-disk sample, which were not yet grinded. (Fig. 9). The resulting stresses do not exceed the yield point of the tooth material (II.c), therefore, they are in the elastic deformation zone.

The maximum equivalent stresses arising in the part of the ejector do not exceed 0.5 MPa, which in turn also indicates the absence of plastic deformations. The maximum nodal deformations observed in the ejector are 4×10^{-4} mm, which excludes the possibility of rotor jamming due to the contact of the teeth with the ejector comb.

The results of the study demonstrate the effectiveness of the application of the developed method and in particular the use of finite element and discrete element modeling of the grinding process. This approach makes it possible to effectively assess material grinding by measuring its size, number of fragments, stress fields and internal deformations of the working parts of the product. The implementation of this method in future will allow to reduce the design and testing time of products. In comparison with scientific studies [5], [6], [13], which describe the application of a similar approach in the numerical modeling of the process of crushing rocks, this work considers the process of grinding samples of solid data carrier. This fact is the uniqueness of this study and its practical value.

In this work the material of which the CD was made was not validated, and the process of fragment expansion after grinding was not analyzed due to the lack of some of initial data. What we also have to analyze in this work will be a full-sized validation of the mathematical model of the material (parameters of particles and their interaction with each other) of the shredded object, the transition to a single universal finite element model of the grinder, as well as the full simulation of the grinding process from the beginning till the end.

6. Conclusion

The developed method demonstrates the possibility of practical application in describing grinding processes of objects. Furthermore, using a verification process in order to get more realistic and valid model can improve the quality of overall development process.

References

[1] Austin L G and Klimpel R R 1964 The theory of grinding *Industrial & Engineering Chemistry* (56(11)) pp 18-29

[2] Asaf Z, Rubinstein D and Shmulevich I 2007 Determination of discrete element model parameters for soil tillage *Soil & Tillage Research* (92)

[3] Barrios Gabriel K P and Tavares L M 2016 A preliminary model of high pressure roll grinding using discrete element method and multi-body dynamics coupling *InternationalJournal of Mineral Processing*

[4] Bing Du, CangcaiZhai, Guojiang Dong and Jiang Bi 2017 FEM-DEM coupling analysis for solin granule medium forming new technology *Journal of Materials Processing technology*

[5] Cleary P W and Sinnot M D 2014 Simulation of particle flows and breakage in crushers using DEM: part 1 – compression crushers *Miner. Eng.*

[6] Du Bing, Zhao Cangcai, Dong Guojiang and Bi Jiang 1961 FEM-DEM coupling analysis for solid granule medium forming new technology *Journal of Materials Processing Technology*

[7] Bond F C 1961 Crushing and Grinding Calculations *Trans. Am. Inst. Min. Eng* (Allis-Chalmers, Milwaukee,Wisc)

[8] Bond F C 1952 The Third theory of communition *Trans. Am. Inst. Min. Eng.*

[9] Jiang M J, Yu H.-S and Harris D 2005 Discrete element modeling of deep penetration in granular soils *International Journal for Numerical and Analytical Methods in Geomechanics*

[10] Kick F 1883 Dinglers Polytech

[11] LS-DYNA® 2017 *Keyword User's Manual* (Volume I R10.0) p 2065

[12] Rittinger R P 1857 von, "Reprocessing science" (Berlin)

[13] Yu Nagata, Yuki Tsunazawa, KoujiTsukada, Yuichi Yagauchi, Yosuke Ebisu, Kohei Mitsuhashi and Chiharu Tokoro 2020 Effect of the roll stud diameter in the capacity of a high-pressure grinding roll using the discrete element method *Minerals Engineering*

[14] Borovkov A I and Ryabov Y A 2019 Determening, development and application of digital twins: Approaches of Competence center of NTI SpBSTU New manufacturing technologies *Digital substation* (№12 (in Russian))

[15] Borovkov A I, Ryabov Y A, Kukushkin K V, Maruseva V M and Kulemin V Y 2018 Digital twins and digital transofration of military industrial enterprises *Defence technics* (in Russian)

[16] Borhsev V Ya 2004 Material grinding equipment: crushers and mills *TGU* (in Russian) p 75

BIW structure development in accordance with passive safety requirements using several optimization methods

O Fedotova, A Korostelkin, K Zatylkin, V Gorunov, A Efimov

National Technology Initiative Center for Advanced Manufacturing Technologies based on the Institute of Advanced Manufacturing Technologies of Peter the Great St. Petersburg Polytechnic University Polytechnicheskaya, 29, St.Petersburg, 195251, Russia, fedotova@compmechlab.com

Abstract. The article is focused on the BIW characteristics and behavior optimization in terms of EURONCAP rules. Variable parameters: part's shape, material, thickness. With multi-disciplinary, topological and topography optimization the most acceptable BIW behavior during virtual crash-tests has been achieved. All design improvements were implemented in accordance with stamping and assembly capabilities, restrictions on the use of each type of material. These requirements are especially important because updated design has been developed for the final stage and supposed to be frozen.

1. Introduction

Approximately 1.35 million people die in road crashes every year [1]. Factors which contribute road crashes can hardly be changed in short-time perspective (e.g. poor road infrastructure and management, unenforced traffic laws, etc.) however the death probability can be decreased by improving the vehicle behavior in road crash [2]. Typical road crash is a head-on collision between two vehicles at relatively high speed [3].

In the full-scale test, the car is driven at 64km/h and with 40 percent overlap into a deformable barrier which represents the oncoming vehicle. The test replicates a crash between two cars of the same weight, both travelling at a speed of 50km/h [3].

In this paper authors consider mass decreasing for sedan vehicle while crash-safety requirements are satisfied (C-NCAP). The peculiarity is that the Customer desired to unify BIW structure for gasoline and electric types of vehicle so during design development the features of the both vehicle types have to be considered. Optimization process includes multi-disciplinary optimization (MDO) for dynamic tests and forces redirection, parametric optimization for NVH parameters improvement and topological optimization for strength and stiffness values increase. Also, local dynamic stiffnesses and frequency response analysis have to be considered during design development [4]

In addition to the frontal-crash test side-crash test has been considered. This test describes a collision on the cross-road which also can be included in road crash statistics.

One more restriction is a percentage of high-strength steel implementation. Replacement normal steel with high-strength is a typical method of certain areas reinforcement [5].

Mainstream technique of vehicle improvement in terms of crash-safety is preventing deformations behind the firewall and keeping the living space size [6]. This goal can be reached by redirecting forces and energies between the loadpathes so that energy could be mainly absorbed in the front part of the vehicle, before the firewall [2].

IOP Conf. Series: Materials Science and Engineering **986** (2020) 012057 doi:10.1088/1757-899X/986/1/012057

The main innovation of proposed method is that topology, MDO and multivariate optimization have been consistently simulated on the basis of CML-Bench platform for developing digital twins and for control of digital design, mathematical simulations and computer engineering (developed by the Computer Engineering Center, CompMechLab, St. Petersburg Polytechnic University).
In this research full-scale model of vehicle has been tested and optimized in accordance with customer's requirements [7].

2. Optimization

All improvements were based on CAE-model provided by customer. First of all, the initial model has been calculated. Expert assessment of the vehicle behavior figured out that shotgun is too high to provide the force distribution properly and the subframe carries over not enough force so that all the force distribution can go through the longitudinal (middle loadpath). But in current case crash-boxes don't collapse properly, there is no bending in longitudinals and lack of deformations in the front area, all these factors involve energy and deformation transfer to the living space which supposed to be a safe area without significant decrease.

The goal for the design of the front part of the vehicle is to evenly spread forces between three loadpathes: upper (shotgun), middle (longitudinal) and lower (subframe) because it's the best way to improve the vehicle's behavior in crash collision (Figure 1).

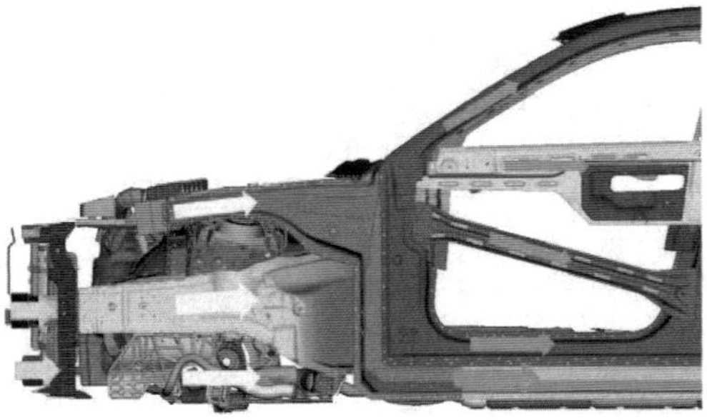

Figure 1. even distribution of forces

2.1. Multi-Disciplinary optimization

For the behavior improvement marked parts considered in several optimization types (Figure 2). Front parts of the vehicle play an important role in vehicle behavior during head-on collision. Shape of the longitudinals and inner reinforcements, stampings and inner ribs in crash-boxes define the middle loadpath and determine force distribution and ways of the energy absorption in the lower front part. Middle loadpath has to be balanced in terms of stiffness in such a way that it would be stiff enough to prevent deformations behind the firewall and soft enough to absorb significant part of the energy to avoid energy distribution on the middle and rear vehicle's parts.

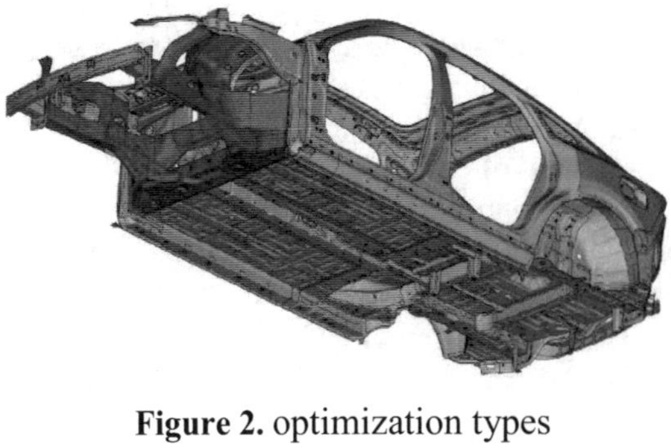

Figure 2. optimization types

In CAE-engineering there is growing interest in multi-disciplinary optimization method which allows to consider different parameters in one optimization process. This method includes several optimization loops of virtual tests, after each loop variable parts are modified in accordance with specified limits to bring optimized value closer to the target and new iteration of test launches [8], [9]. Current case includes CAE-model for crash and frequency response analysis CAE-model (Figure 3).

Figure 3. CAE-model

Front part numbers match for both models, thickness of each part set as variable parameter in optimization task. For current task Response Surface Optimization Method can be used. This method is based on the development of a mathematical model approximating the response of the objective function withing the range of design variables [11]. Then optimization algorithm uses this surface. About 22 variants have been considered in each iteration, the best one has been chosen. At the end of each iteration variable limits were clarified and the range of possible variable values decreased.
So, parts thicknesses were optimization variables. Mass decreasing is a mainstream optimization goal. In current case optimization constrains were: frequency, b-pillar acceleration, steering column displacement, acceleration pedal displacement, brake pedal displacement, door frame deformation.
Virtual crash-test (ODB), frequency response analysis, local dynamic stiffness analysis for the front longitudinals are presented in each iteration. For computational time saving ODB model was reduced to front part while rear part was replaced with mass (Figure 4).

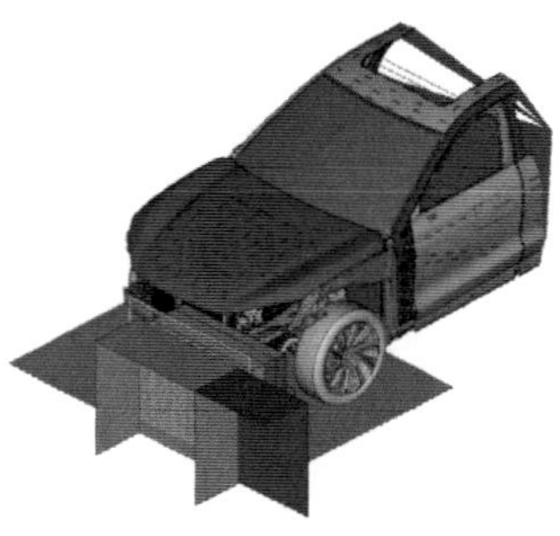

Figure 4. Virtual crash-test (ODB)

As result left B-pillar acceleration value reaches target due to next improvements: thickness of longitudinal and its reinforcement increased and material has been replaced with more durable to provide energy absorption increase and forces redistribution; thickness of a-pillar's reinforcements increased to minimize a-pillar intrusion, keep living space and decrease door frame deformation; rocker's thicknesses have been increased to improve behavior during side crash-test and keep frequency response values acceptable (Figure 5, Figure 6).

IOP Conf. Series: Materials Science and Engineering **986** (2020) 012057 doi:10.1088/1757-899X/986/1/012057

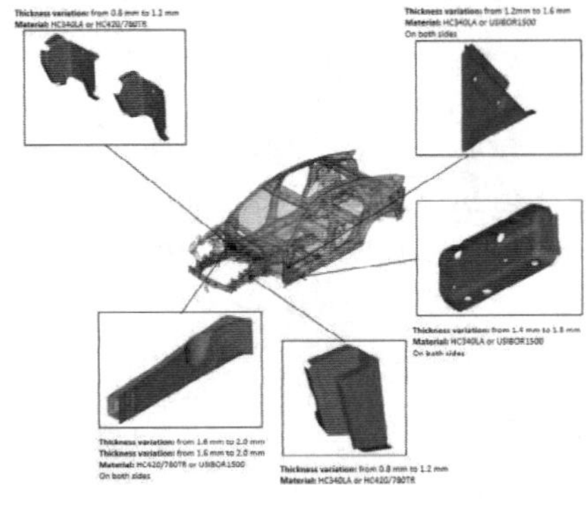

Figure 5. acceleration left B-pillar

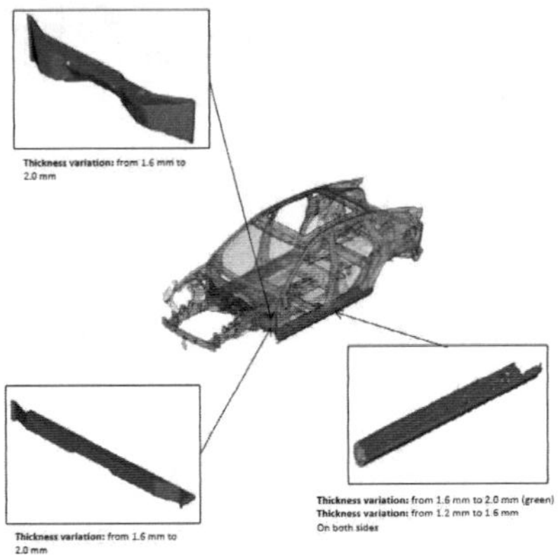

Figure 6. acceleration left B-pillar

To clarify the shape of crashbox reduced task has been calculated with dynamic optimization method [6]. Here only crashbox and frontbeam on the driver's side are left, the other side has been replaced with symmetrical condition, the rest of the vehicle has been replaced with mass and inertia characteristics. Front crash-test has been calculated in optimization process. This optimization supposed to help with the investigation of the best shape and inner structure of crashbox and frontbeam [10].

Optimization goal in this case is energy absorbing increasing. Inner crashbox and frontbeam ribs position and thickness and stampings distribution are design variables for current optimization. As a constrain dashboard intrusion has been considered, it is a fine marker of the energy absorption efficiency (Figure 7).

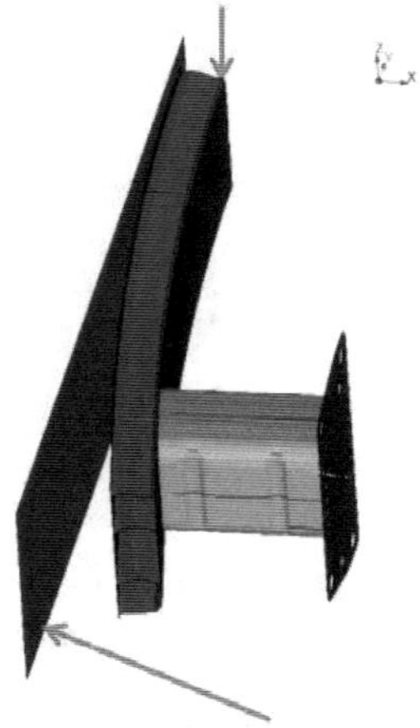

Figure 7. dynamic optimization method

During optimization process mass of specified area decreased by 0.3 kg.

2.2. Topology optimization

Topology optimization aimed to help with determination of extra material areas and define where the holes can be created. This method allows to decrease mass of the BIW and keep stiffness values and frequency response analysis values in target. In current case optimization constrain were mass and the first frequency value, while compliance minimization was an objective. Variable parts of the BIW are shown on the figure below (Figure 8)

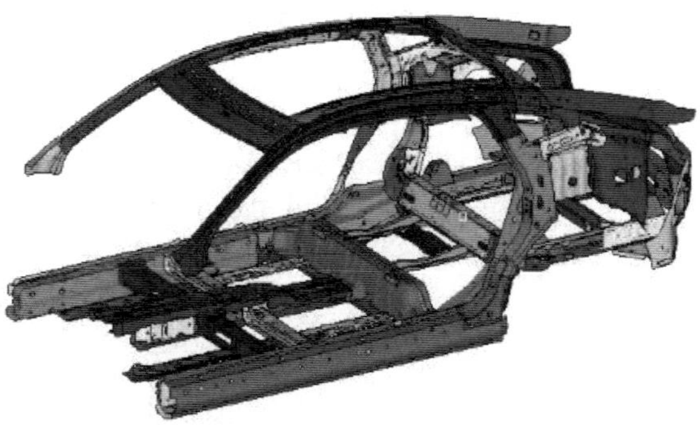

Figure 8. Various parts of BIW

During topology optimization areas with extra material have been identified. Blue areas define not necessary material areas, in red areas all current material should be kept (Figure 9).

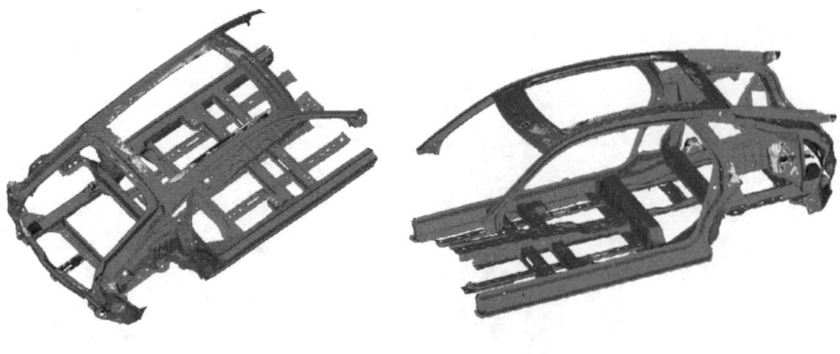

Figure 9. Topology optimization

Based on topology optimization results cutouts in BIW parts have been created.

2.3. Parametrical optimization

One more significant marker of BIW structure quality is global stiffness, certain level of stiffness characteristics for the whole BIW helps with acceptable behavior during crash-test achievement. Global stiffness increase provides proper behavior for the rear and middle parts of the vehicle. Parametrical optimization helps with thickness redistribution for chosen parts (Figure 10)

Figure 10. Parametrical optimization

Optimization objective is mass minimization. Global stiffness target values set as constrains, thickness of chosen parts is optimization variable. As result most of involved parts thickness decreased by 0.2 mm while global stiffness values are still in target. Mass decreased by 20 kg (~4% of BIW's mass).

2.4. Multivariant optimization

The next step is multivariant optimization. This process takes into account production restrictions and expert assessment which can hardly be considered in previous optimization methods. Front parts have been redesigned and reinforcements redistributed during several loops of multivariant optimization. After each iteration expert assessment of the vehicle behavior helps to figure out which of proposed changes have to be implemented into design. Multivariant optimization in current case aimed to improve vehicle behavior during side crash. This test consists of the vehicle and mobile barrier, crash collision speed 50 kmph.

Huge influence on the deformation and force distribution in side crash-test has b-pillar area. Initiate creasing in the bottom part of b-pillar is a mainstream practice to minimize risks during side-crash collision.

B-pillar reinforcement strain is one of the most important markers, so focus of the multivariant optimization is on b-pillar strain decrease and crease in the bottom of the b-pillar provision. There are several methods to improve b-pillar behavior: modification of rocker, b-pillar reinforcements and roof

rail parts, spotweld points rearrangement. In current case the best results in terms of side crash have been achieved due to the following proposals:
- divide b-pillar reinforcement into two parts
- increase p-pillar reinforcement bead height
- replace b-pillar reinforcement material with high strength steel
- inner rocker reinforcement shape and thickness change
- b-pillar to roof rail overlap
- spotweld points rearranged in roof rail area

All the proposals implementation helped to achieve significant improvement of the vehicle behavior. Overlap to b-pillar and strengthen with spotweld points in the roof rail area allow to avoid local buckling; bead height increase and division b-pillar reinforcement into two parts connected with spotwelds helped with decreasing b-pillar intrusion on the passenger's head level; rocker reinforcements modification and lower part in the b-pillar reinforcement increase helped to organize strong b-pillar to rocker joint and have controlled deformation in the bottom area.
The result structure is shown on the picture below (Figure 11)

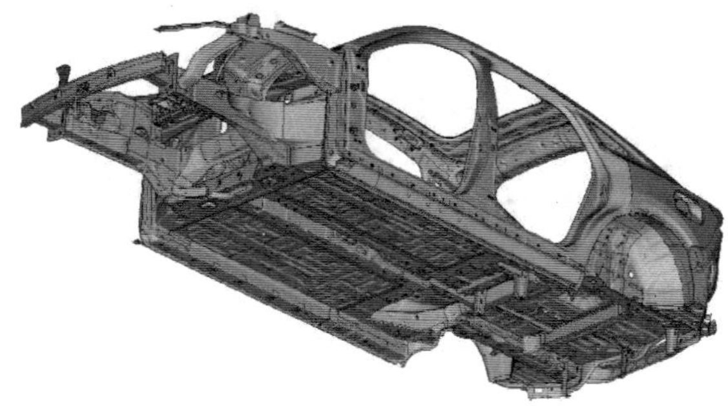

Figure 11. Multivariant optimization

3. NVH

Important part of vehicle's design development is tracking other key parameters during BIW optimization in terms of crash safety. Previously only frequencies and global stiffness have been considered in optimization process. NVH (Noise, Vibration, and Harshness) performance is significant for quality level of the vehicle. During the driving of a car, the main sources of vibration noise include: vibration noise generated by engine excitation, vibration noise generated by the interaction of automobile tires and the road surface, and vibration noise generated by friction between the air flow and the vehicle body. After the various vibration noises are superimposed, they are transmitted to the entire body, causing the vibration of the body structure, radiating the noise to the interior of the vehicle, and making the passengers in the vehicle feel uncomfortable [12]. During vibrations transfer it is important to avoid resonance, so frequency values of the BIW, engine, closures, package and all other units have to be far enough apart. Mainstream practice is BIW frequency values and global stiffness increase, which has been considered in parametrical optimization and MDO method. High values of these parameters will help with further NVH performance improvement such as IPI (Inertance point interface), LDS (Local dynamic stiffness), VTF (Vibration transfer function), NTF (Noise transfer function).
IPI (Inertance point interface) and LDS (Local dynamic stiffness) performance shows vehicle response for harmonic load applied to each interface point. Displacement and acceleration are evaluated and stiffness value can be estimated in comparison with target value.
VTF (Vibration transfer function) aimed to evaluate vibration response from sources of noise to parts the driver has contact with (steering wheel, seat mounting points).
NTF (Noise transfer function) characterizes which part of the noise can reach the driver's ear. Vibrations from engine, road and airflow is transferred through the BIW and closure panels and create pressure fluctuations which generate noise inside the vehicle living space.
All the proposals provided during global stiffness and frequency response optimization helped to improve IPI and LDS performance. NTF performance problem has been solved with applying extra

damping material layer. VTF and NTF performance is very important in terms of comfort for the driver and passengers.

Calculations of the NVH performance were based on the trim-body model. All vehicle units except BIW were replaced with masses in the mass center of corresponding unit.

As result four improved variants of the BIW structure have been considered. The main proposal allowed to reduce 8 kg while mass reduction was from 20 to 27 kg in alternative proposals. Some compromises in terms of crash and NVH performance took place for the alternative proposals and the Customer could set priority level for each requirement. Changed parts for each proposal marked on the picture below (Figure 12). Red color – parts changed in the main proposal, green and red – parts changed in the first alternative proposal, blue, green and red – parts changed in the second alternative proposal, all colored parts – changes in the third alternative proposal.

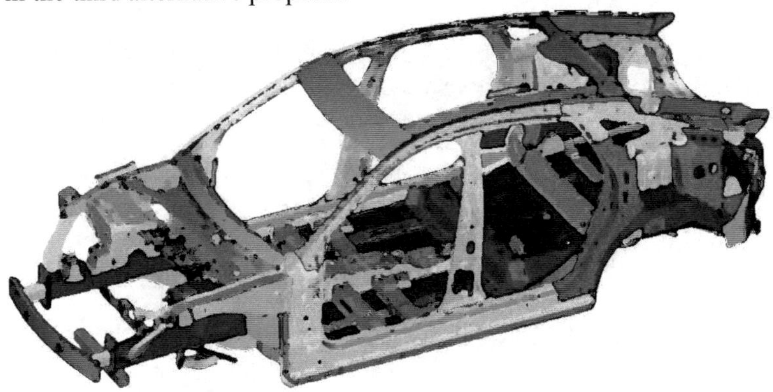

Figure 12. Calculations of the NVH

4. Conclusion

During optimization process vehicle mass has been reduces by 12 kg, crash-safety requirements have been satisfied (C-NCAP) and frequency response values kept in target. Shapes of parts have been changed during multivariant optimization, thickness and material redistribution provided by MDO and topology optimization. Improvements implementation took into account the ease of stamping, manufacturing cost and capabilities. Most of design proposals have been realized in the final vehicle design.

References

[1] Global status report on road safety 2018, ISBN 978-92-4-156568-4, World Health Organization 2018
[2] Vehicles A A, Korostelkin O I, Klyavin, M V, Alechin Wang Guodong, Wang Suifeng and Liu Jingyi Optimization of Frame Mass in Crash Testing of Off-Road (DOI: 10.3103/S1068798X19120116)
[3] Alekseev S, Tarasov A Validation of Euroncap Frontal Impact of Frame Off-Road Vehicle – Road Traffic Accident Simulation (Peter the Great St.Petersburg Polytechnic University, Russia)
[4] Shahram Azadi, Mohammad Azadi, Farshad Zahedi NVH Analysis and Improvement of a Vehicle Body Structure Using DOE Method (DOI: 10.1007/s12206-009-0909-z)
[5] Cornette D, Hourman T, Hudin O, Laurent J P High Strength Steels for Automotive Safety Parts (DOI: 10.4271/2001-01-0078)
[6] Yijung Chen, Derek Board, Omar Faruque, Cortney Stancato MMLV: Crash Safety Performance (DOI: 10.4271/2015-01-1614)
[7] C-NCAP Management Regulation 2018, China Automotive Technology and Research Center
[8] Ren-Jye Yang, Liya Gu, Cao Chi Tho, Jaroslaw Sobieszczanski-Sobieski Multidisciplinary Design Optimization of a Full Vehicle with High Performance Computing
[9] Rebecka Domeij Bäckryd Multidisciplinary Design Optimization of Automotive Structures

[10] Qing-fen Li, Yan-jie Liu, Hai-dou Wang, Sheng-yuan Yan Finite Element Analysis and Shape Optimization of Automotive Crash-Box Subjected to Low Velocity Impact
[11] Allen Sheldon, Edward Helwig, Yong-Bae Cho 2011 Investigation and Application of Multi-Disciplinary Optimization for Automotive Body-in-White Development *8 th European LS-DYNA Users Conference, Strasbourg - May 2011*
[12] LIU Xing Research on Improvement Methods of Automotive NVH Performance (Sichuan Aerospace Vocational College, Chengdu 610100, China)

On shale shaker's structural optimization and performance improvement

A A Korostelkin, A A Filintsev, V V Novokshonov and A V Nikitin

National Technology Initiative Center for Advanced Manufacturing Technologies based on the Institute of Advanced Manufacturing Technologies of Peter the Great St. Petersburg Polytechnic University Polytechnicheskaya, 29, St.Petersburg, 195251, Russia, korostelkin@compmechlab.com

Abstract. Recent trends, predictions [1] and overall assessments [2] in extractive industries show there will be changes in many aspects of business, varying from geopolitical issues up to the optimization of working processes and machines. In severe conditions of drilling stations usually based quite a way off to cities and are difficult to access, it is important for all the equipment and machines to operate without failure and provide forecasted level of durability, because unexpected shutdown may lead to halting of production and economic losses. This article is focused on investigation of shale shaker primary characteristics and its behavior during operational loads. In the result of the work an optimization has been performed on the basis of kinematic, strength and NVH performance. Special emphasis was put on the value of vibrational acceleration which reflects an overall possibility of a shale shaker to sift drilling mud. It is a widespread practice to implement optimization tools into different fields. The methods of usage topology optimization can follow the same ideas as it is done in automotive [3] or agricultural industries [4]. Advantage of this paper based on intense cooperation between manufacturing department of the customer and CAE engineers which is the basis of a good design process [5] and overall approach to successful development [6], that includes manufacturing restrictions and that allowed to prove sustainability of the models through validation phases, in the beginning of the project for the initial structure and in the middle of the project for several newly assembled prototypes. So the results of the work have been confirmed by real experiments. Another strong point was a possibility to obtain a feedback right from drilling stations about the condition of optimized shale shaker after it started to work under common operational loads. Constant monitoring, predicting systems and other methods to evaluate overall condition of the machines under operational loads or drilling fluids [7] during the extraction process are important measures in order to prevent unexpected failures. Process and result of this work can be assembled in a step-by-step method for simplified evaluation of shale shakers with vibrators of different amounts and powers. Such method allowed to obtain an overall glimpse on main characteristics, performance, to locate durability weak zones and shows potential for possible structural optimization. The new approach has been used in this research by using the CML-Bench platform, which is a system for managing the simulation driven design projects field of digital design and modeling [8]. This system allows tracking and observing all necessary variables, simulation history and mutual influence of different parameters on different structural changes. Such a method for design engineering allows easily carrying out multidisciplinary optimization of the objects and always being aware of the performance state of the whole system [9].

IOP Conf. Series: Materials Science and Engineering **986** (2020) 012058 doi:10.1088/1757-899X/986/1/012058

1. Introduction

Shale shakers belong to the components of the mud handling system and are used to separate large solid particles from drilling fluid. On high level main parts of the shale shaker are vibrator, screen basket with angling mechanism, feeder and base [10].

An interfusion of muck, drilling fluid, ground and other solid particles come from the borehole to a collection pan called feeder and then transferred to the shale shaker screening area which is essentially responsible for shaking and sifting [11]. Such behavior is caused by the work of vibrators which provide necessary load to the shaker bed.

One of the most important values which characterize an intensity of shaking is called vibrational acceleration and measure in G's. It is assumed higher values of G's provide more effective process of sifting.

Some investigations in the area of performance improvement are based on using different kinds of screen, separators and types of drilling liquid [12]. This article is focused on structural changes of the shale shaker bed.

In most cases two vibrators with linear motion is used, however some machines have an additional option of adding a third motor providing elliptical motion, so an overall motion of the basket tends to be more circular and softer. This kind of motion always decreases an overall value of G and is used depending on the consistence of the fluids from the drill or customer specific requirements.

The main goal of this paper was to increase maximal vibrational acceleration of the shale shaker while keeping structural performance, design feasibility and cost at the same level. Using different amount of vibrators or different engine powers is an obvious option to increase the result, so in addition a structural optimization and mass reduction of the shale shaker vibrating bed has been performed in order to increase G. Though shale shaker is a relatively easy to be controlled in terms of drive system [13], in this investigation it was important to suggest sustainable from the performance and possible from manufacturing point of view structural changes of the shale shaker.

A layout of a conventional shale shaker is depicted in the figure 1.

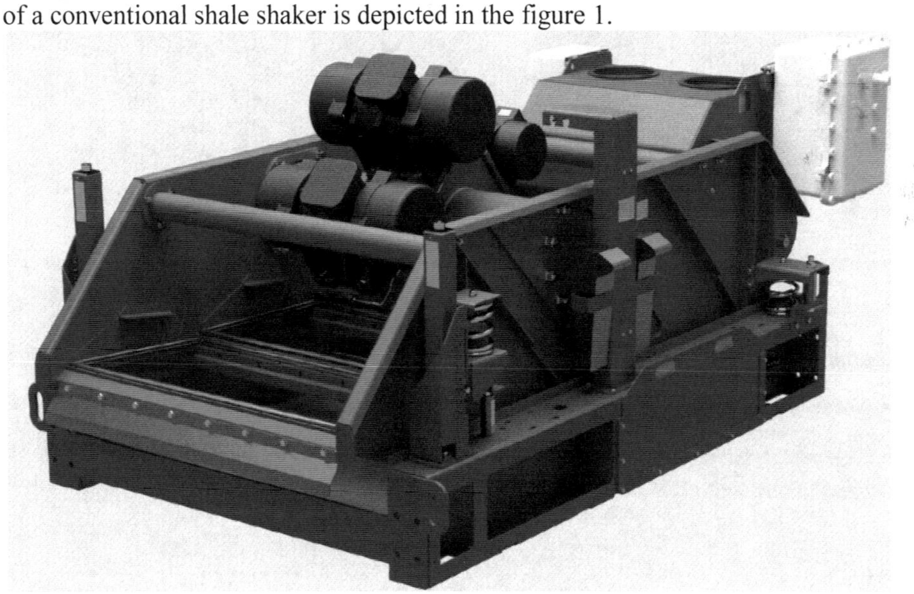

Figure 1. Shale shaker layout.

2. Project guidelines

In order to make an overall estimation of the structure the following steps has been proposed and performed by the authors for the optimization process:

1) Creation of MBS model on the basis of actual CAD-Data, vibrator properties and mass distribution of all relevant components

2) Simulation of idle motion operational state of the shale shaker. Vibrational acceleration and displacement estimation.
3) Validation on the basis of real tests and model confirmation
4) Obtaining maximum load values from fixed shale shaker base to the spring mounting points and from vibrators to the vibrator support
5) Static strength evaluation using maximum load values from MBS model
6) Durability evaluation using load history from MBS model. Weak areas determination
7) Global stiffness and frequency evaluation
8) Structural changes of the shaking bed

All necessary steps have been repeated after every optimization loop in order to track the influence of structural changes or vibrator parameters on overall behavior of the shale shaker.
Following software has been used during optimization process:
1) MBS analysis in MSC Adams
2) NVH, strength and durability in MSC Nastran

First step was to create an MSB model of the shale shaker and perform kinematic simulation of the idle motion [14] . MBS model of the shale shaker with three vibrators is shown in the figure 2.

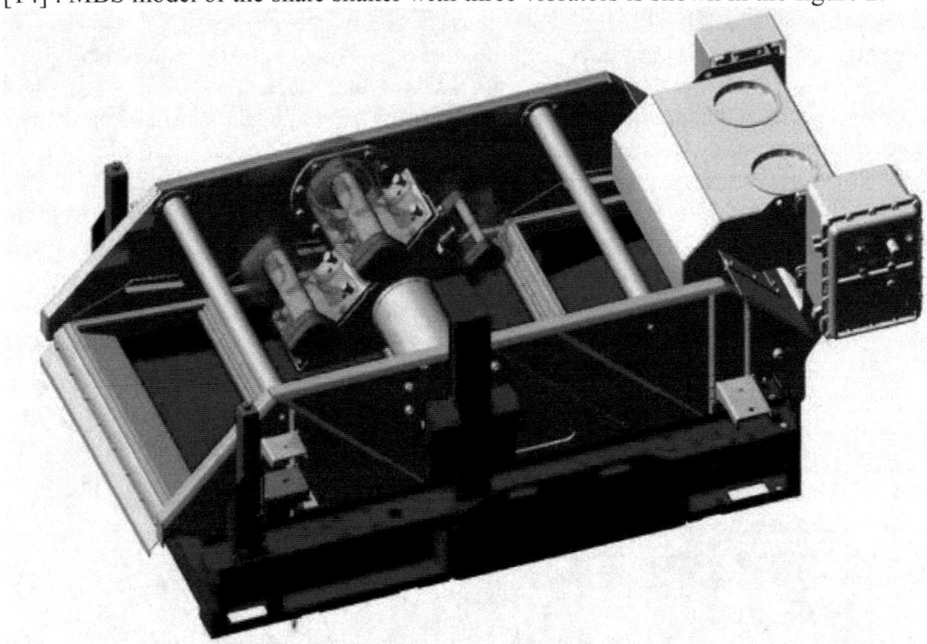

Figure 2. MBS Model.

In order to ensure model sufficiency, a validation has been performed on the basis of initial shale shaker structure. Comparison of the results between initial digital model and real test are shown in the table 1.

Table 1. Initial model validation.

	Linear operating mode		Linear-elliptical operating mode	
	Maximal vibrational acceleration, G	Maximal vibrational displacement, mm	Maximal vibrational acceleration, G	Maximal vibrational displacement, mm
Difference, %	4.1	0	1.89	2.59

Initial MBS-Model showed high correlation between real tests in terms of measurement of vibrational acceleration and displacement, both on linear and elliptical operating modes and could be regarded as sustainable.

3. Overall performance and optimization

After that a series of calculations have been performed on the basis of validated kinematic analysis. Using results from MBS analysis, loads have been applied to the spring mounting points and vibrators mounting points to estimate static strength, stiffness, frequencies and durability.

Boundary conditions for corresponding loadcases are shown below. For static strength load points with corresponding maximal forces are marked with red dots.

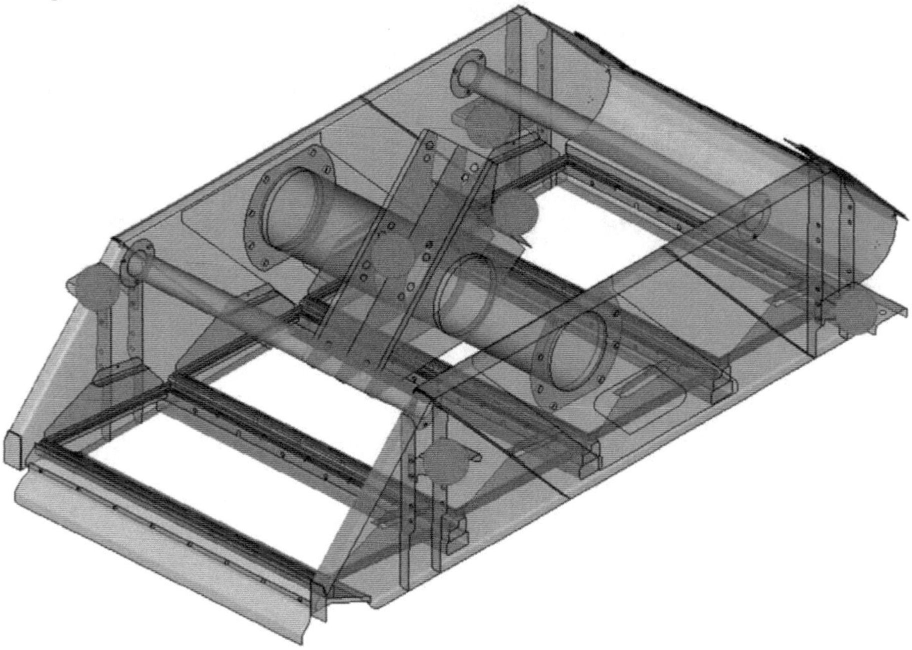

Static strength state of the shale shaker showed areas with.

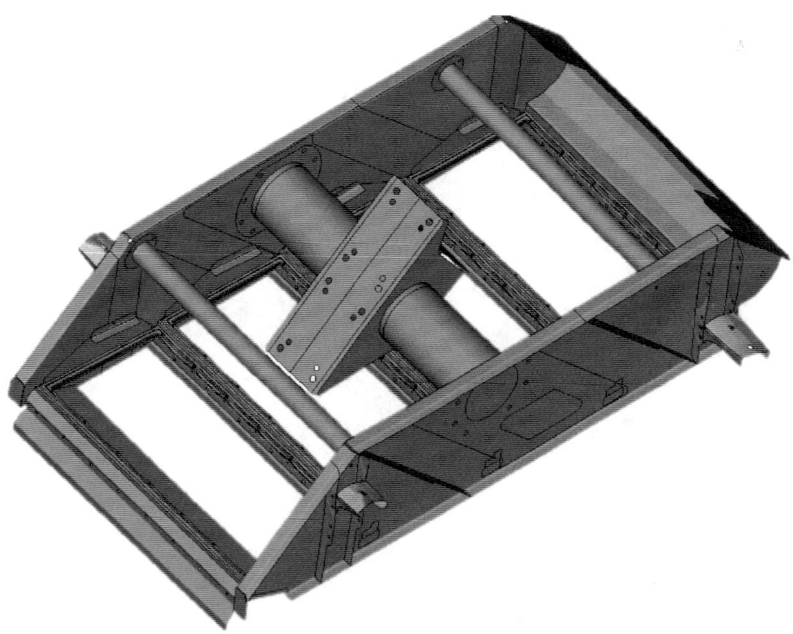

Figure 3. Static strength.

For torsional and bending stiffness corresponding boundary conditions and loads have been applied in the following way.

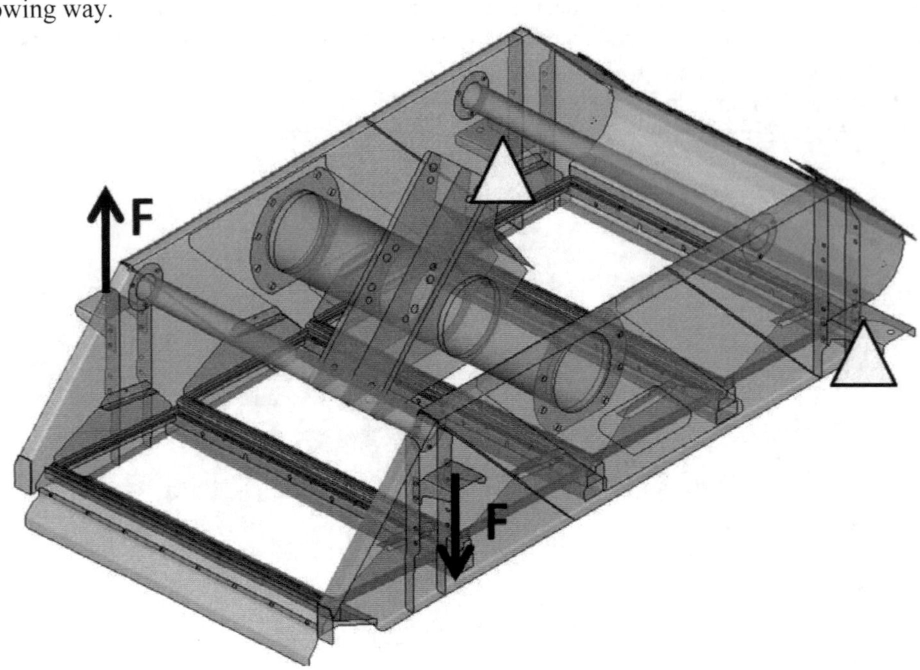

Figure 4. Torsional stiffness.

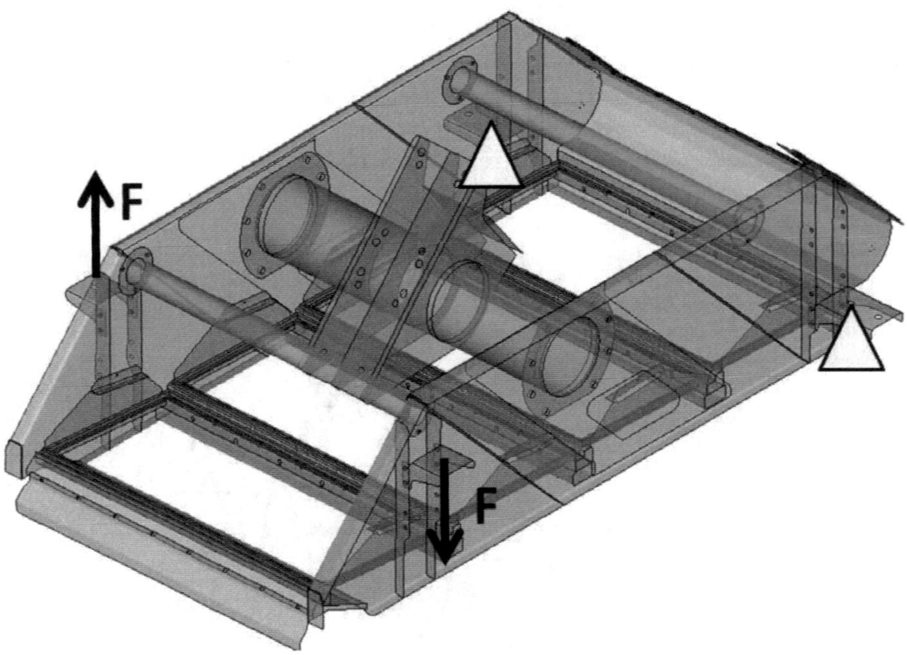

Figure 5. Bending stiffness.

All above mentioned loadcases have been calculated along with modal and kinematic analysis for initial and optimized structures with different types of vibrators. As result of the work, three variants of optimized shale shakers vibrating beds have been proposed to the customer. Optimized variants contained new set of thicknesses and shapes of main transversal and longitudinal parts.

4. Optimized structures

Optimized structure №1 contained general proposals based on parametrical optimization of parts of shale shaker, didn't require any changes from manufacturing point of view and didn't increase complexity of assembly.
Several prototypes of this proposal have been assembled and tested by the customer during the project phase for another round of model validation. Results of comparison are depicted in the table 2 below.

Table 2. Optimized structure №1 model validation.

Difference between virtual results and real tests	Linear operating mode	
	Maximal vibrational acceleration, G	Maximal vibrational displacement, mm
Difference with prototype 1	2.12%	0.44%
Difference with prototype 2	3.06%	1.76%

Results of validation between virtual model and real tests showed high correlation and even better convergence in comparison for initial model validation.
Optimized structure #2 was based on the variant №1 and contained additional proposals based on topology optimization of parts of shale shaker. This proposal didn't require any changes from manufacturing point of view, however at some extent increased complexity of assembly and working operations. This proposal has been accepted by the customer for as a modification for mass production. Result of one of the proposals for cut outs on the motor mount is depicted in the c 6. Red lines represent main load paths to bear the load from the motors.

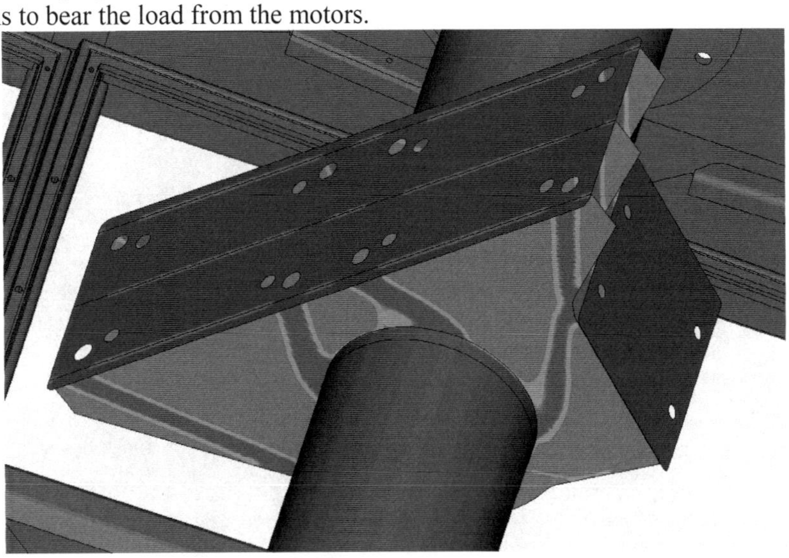

Figure 6. Motor mount topology optimization results.

Optimized structure №3 contained significant structural changes, required changes from manufacturing point of view, however at some extent increased complexity of assembly and working operations. This proposal has been also accepted by the customer for theoretical possibility of production upon updating manufacturing line.
An overview comparison table with the results of initial structure is depicted in the table 3.

Table 3. Results of initial structure is depicted.

Comparison with initial structure	Kinematic		Stiffness		Modal	Durability	Weight
	Maximal vibrational acceleration, G	Maximal vibrational displacement, mm	Torsional	Bending	First natural frequency, Hz	Life cycles	Mass, kg
Optimized structure №1	+ 4.3% (**+0.35G**)	+4.5%	+69%	+8.5%	-4%	No failure	**-10%**
Optimized structure №2	+ 12.9% (**+1.05G**)	+15%	+76%	-1.5%	-2.2%	No failure	**-16.6%**
Optimized structure №3	+ 24.8% (**+2.02G**)	+25%	+116%	+0.3	-3.3%	No failure	**-19.2%**

5. Additional calculations

In order to get a more realistic influence of a load coming to the motor from drillhole liquid and particles an explicit process this behavior has been calculated, based on the data of an average particles dimension and density which usually travel through the shaking bed. Result of simulation process is depicted in the figure 7.

Figure 7. Explicit process of particles distribution.

A series of additional calculations have been performed to locate weak zones in terms of durability. To estimate influence of material thinning or its properties degradation, assembly imperfections, different ways of vibrator mounts positioning and several versions of mobile modifications for shale shakers.

One of customer existing and operating shale shaker modifications suffered from fractures in the area vibrator mounts. After performing kinematic simulation and fatigue estimation most dangerous zones have been observed, which have been located at the same areas as in the figures provided from the customer. Result of simulation and comparison with real photos are depicted in the figure 8.

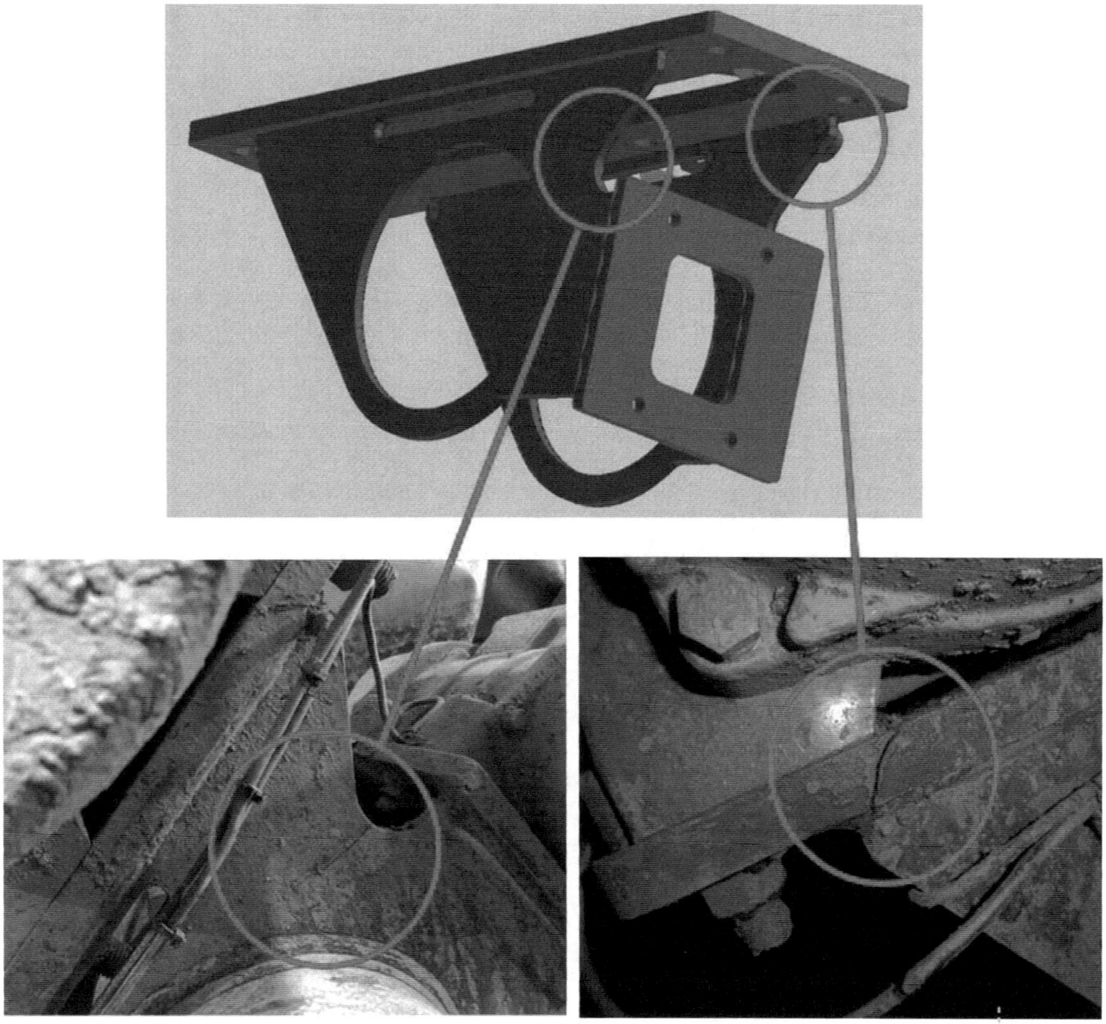

Figure 8. Result of simulation.

A simple coupling from kinematic analysis to fatigue calculation allowed to observed zones, tend to have stress concentrations.

6. Conclusion

As a result of the work more than 100 modifications have been investigated for all abovementioned loadcases. An optimized structure showed higher G value while keeping the rest of performance at required level.

A series of step-by-step calculations allowed to assemble results of the work into a method for performance and durability estimation of shale shakers of different modifications.

References

[1] Littleboy A, Keenan J, Ordens C M, Shaw A, Tang R H, Verrier B and Hodge R A 2019 A sustainable future for mining by 2030? Insights from an expert focus group *The Extractive Industries and Society*

[2] António Mateus and Luís Martins 2019 Challenges and opportunities for a successful mining industry in the future

[3] Korostelkin A A, Klyavin O I and Aleshin M V et al 2019 Optimization of Frame Mass in Crash Testing of Off-Road Vehicles *Russ. Engin. Res.* (39) pp 1021–1028
[4] Alexander Leontev, Mikhail Aleshin, Oleg Klyavin, Dmitrii Lebedev, Aleksey Okunev, Ani Petkova and Aleksey Borovkov 2018 Multidisciplinary Optimization in Design of Tractor Cab Frame *International Journal of Mechanical and Production Engineering Research and Development (IJMPERD)*
[5] https://hbr.org/1988/07/manufacturing-by-design
[6] Ashton J E 1993 Relationship Between Design for Manufacturing, a Responsive Manufacturing Approach, and Continuous Improvement. In: Haug E.J. (eds) Concurrent Engineering: Tools and Technologies for Mechanical System Design
[7] Magalhães S, Calçada L A, Scheid C, Paraíso E, Moraes G, Valim E, and Marques de Sá C H 2020 Field Results of a Real Time Drilling Fluid Monitoring System *SPE Latin American and Caribbean Petroleum Engineering Conference*
[8] https://fea.ru/article/cml-bench [Electronic resource]
[9] Davidenko L M 2019 Texnologicheskie aspekty` sistemnoj modernizacii e`konomiki *Bol`shaya Evraziya: Razvitie, bezopasnost`, sotrudnichestvo* (№2-2)
[10] Merrill J, and Robinson L 2005 Shale Shakers *Drilling Fluids Processing Handbook*
[11] American Association of Drilling Engineers's. Shale Shaker Screens 1999 *Shale Shaker and Drilling Fluids Systems* (doi:10.1016/b978-088415948-3/50008-3) pp 120–138
[12] Golenko A V and Yamaliyev V U 2016 MAIN DIRECTIONS FOR SHALE SHAKERS IMPROVEMENT
[13] Atroshchenko V A, D'yachenko R A, Kichkar' I Yu and Kichkar Yu E 2019 CONTROL OF A SHALE-SHAKER ELECTRICAL DRIVE SYSTEM *Chemical and Petroleum Engineering* (Vol. 55, Nos. 5-6, September, 2019 (Russian Original Nos. 5-6, May–June, 2019))
[14] Zhu W, Wang H and Dong L 2012 Dynamics Analysis of a Linear Shale Shaker in Considering Forces of Solids *Measuring Technology and Mechatronics Automation in Electrical Engineering* (doi:10.1007/978-1-4614-2185-6_26) pp 213–220

The method of RTM process modeling using porous medium

N Efimov-Soini, M Kiauka, A Borovkov

National Technology Initiative Center for Advanced Manufacturing Technologies based on the Institute of Advanced Manufacturing Technologies of Peter the Great St. Petersburg Polytechnic University Polytechnicheskaya, 29, St.Petersburg, 195251, Russia, efimovs_nk@spbstu.ru

Abstract. This paper proposes a method to simulate the Resin Transfer Molding process by using Ansys Fluent software. The offered approach permits to calculate impregnation time, to predict dry zones after impregnation and proposes a method to calculate porosity of the final product to predict the mechanical properties. The method is illustrated in the plate resin impregnation case.

1. Introduction

Resin Transfer Molding (RTM) is a manufacturing process in which a liquid resin is injected into a closed mold pre-loaded with a porous fibrous preform, producing complex composite parts with good surface finishing. Resin flow is a critical step in the process. One of the greatest difficulties in applying RTM and LRTM processes is related to mold filling, i.e. to guarantee that the fibrous reinforcement is completely impregnated by the resin inside the mold. Besides, in order to manufacture quality composites through RTM, it is necessary to minimize the void content inside them. The presence of voids is harmful to predict the mechanical properties of the parts, such as the decreasing of shear, compression, impact and fatigue. There are different approaches to simulate this process. On the one hand there are some special commercial software e.g. PAM Composite (ESI Group), in particular module PAM-RTM. The proposed software permits to simulate composite manufacturing process, but this software doesn't work with 3D composite materials. On the other hand, the offered software simulates air voids by using empirical formulas.

The mechanical characteristics of the obtained material with a given distribution and void size can be estimated by using the homogenization method. To do this, you can use any commercial software package for modeling the mechanics of a deformable solid or special software e.g. Digimat (e-Xstream/MSC Software Corp.).

In this paper the manufacturing process is descripted as a part of the unique framework. The last one takes into account the product life-cycle. It means the manufacturing process results are used in stress analysis. This approach develops the concept of Virtual Manufacturing. It means that for fiber reinforced polymers (FRPs) the process from modeling of the textile reinforcement material behavior on tool, impregnation and curing. All these processes are associated with defects such as air voids, incomplete polymerization, residual stress and unplanned shape distortion., It is the result of decrease in strength properties and deterioration of product characteristics. Product defects, acquired during the manufacturing process, can be taken into account in the simulation process. The proposed approach can be used to the conceptual design of the product and permits to decrease the final product cost. It is radically importantб,because 75% of the product cost is based on the results of the conceptual design stage [1].

In this work, the numerical study of the resin flow in RTM applications was performed by employing a general Computational Fluid Dynamics software which does not have a specific RTM module, making necessitate to use the Volume of Fluid method for the filling problem solution. The porous medium model is used for reinforcement modeling. These models are implemented in the ANSYS Fluent commercial code. Examples were presented and compared with experimental and numerical results showing the validity and effectiveness of the present study, with maximum difference among these solutions of around 5%. On the other hand the air voids development permits to analyze the strength properties of the product. In the contrast of existed approaches, offered method takes into account the manufacturing process to creat the conceptual design of the product. It means we can predict the strength properties of the product in the early development stage.

Voids in composite products are known to seriously impair the mechanical properties of the substance. For example, according to the aviation standard ASTM, a product with a porosity >2% should be rejected [2]. Therefore, the mechanism analysis of pore formation and ways to reduce porosity is of great importance for improving the composite parts quality.

The paper remainder is structured as follow: in Section 2 states-of-the-art is presented, Section 3 is devoted to method description, and finally section 4 present the conclusions.

2. State-of-the-art

In last decennaries composite materials have come into position of significant and beneficial high-productivity material solution in different areas like automobile and city transport, aviation and rocket and space technology shipbuilding, bridge building, oil refining industry and other.

The necessity of improving performance, reliability and quality of products led to strictering requirements which have to meet the composites industry. The resin transfer molding (RTM) is a complicated production process. A preformed reinforcement of glass or carbon fiber is placed in a closed mold and a viscous resin is injected into the mold. The intricacy of the process occurring in the production of composites does a trial and error approach effect less. Partially, the problems can be solved by designing the process and integrating the simulation into the design cycle. Accordingly, performing numerical optimization is perspectively. For distinguish and improve optimum settlement parameters it is necessary to elaborate methods relating predictive simulation with numerical optimization way. G. Struzziero et al.[3] reviewed numerical and experimental results concerning the optimization techniques and methodologies implemented in literature to address the optimization of thermoset composite manufacturing processes.

M. Hattabi et al. [4] proposed a procedure of simulation the flow in the LCM (Liquid Composites Molding) processes by finite difference discretization in a curvilinear coordinate system adapted to the shape of the saturated zone. They simulated two-dimensional mold filling of Newtonian fluid through the porous reinforcement under an isothermal condition. They proposed the concept of the capillary number to explain the variations of the permeability obtained for pressure values lower than 0.25 Bar.

In the RTM process the mold contains three constituents including fibres, resin and gas. Due to its similarity to a porous medium, the process can be modeled within the Theory of Porous Media (TPM), where the simulation of the moving resin front generates an inherent difficulty. Commonly used two approaches are: the moving resin front is either approximated roughly, such that an interface thickness of multiple finite elements arises, or respectively by a sharp interface with zero-thickness which requires a special numerical treatment. Dammann C. et al. [5]applied regularized sharp interface theory as a branch of phase-field models which go back to the classical Ginzburg-Landau equation. In this way, the sharp discontinuities between different phases, i.e. resin and gas, are approximated by smooth transitions of suitable order parameters, of which one is the resin fraction. Thus, the sharp interface topology as well as the surface energy of the interface are smeared out over a region proportional to a chosen regularization length scale. As an additional advantage, the phase-field method facilitates a thermodynamic treatment of phase interfaces rendering it more physically consistent in combination with the TPM, which also works within a thermodynamically sound framework. To this end, a coupled finite-element strategy was presented accounting for both, a Ginzburg-Landau- and Cahn-Hilliard-type regularization scheme for the interface. Two boundary value problems were formulated in the context of the TPM. They consider a multiphase-system, represented by deformable fibers interacting with the resin saturating the pores that was coupled to one of both regularization schemes. Accordingly, two numerical examples demonstrate the capabilities of both regularized formulations by simulations of RTM process.

Liércio A. Isoldi et al. [6] performed numerical studied of RTM by employing a general Computational Fluid Dynamics software which does not have a specific RTM module, made it necessary to use the Volume of Fluid method for the filling problem solution. J. da S. Porto et al. [7] investigated a numerical model developed in the FLUENT package to study the resin flow behavior in the LRTM process.

The mold filling process during resin-transfer molding has been simulated numerically by Moon Koo Kang et. al [8] by using a modified control volume finite-element method (CVFEM) along with a fixed-grid method to handle problems associated with the moving resin front. The fixed grid was refined in an adaptive manner, by dividing the flow-front elements into two regions using the estimated flow front. Imaginary new nodes were placed at the intersections between the original element border and the temporary flow front. By using the mass conservation of resin, the pressures at these newly added `imaginary' nodes were expressed in terms of the pressure at the old `real' nodes. Through this elimination, the imaginary nodes do not affect the size of the global matrices and thus do not increase the computation time. The proposed method, referred to as the FINE method, yielded smoother flow fronts and reduced the error in the pressure at the flow front that plagued the conventional fixed-grid methods. The solution accuracy was considerably higher than that of the conventional method for the same number of nodal points and elements, without a significant increase in the computation time.

Anita Zade et al. [9] had presented the isothermal mould filling simulation using Ansys Fluent for LCM process. 2-D rectangular geometry was used for analyzing the effect of process and raw material parameters on filling time. Sensitivity analysis was done using velocity and viscosity of injection resin and orthotropic permeability of reinforcement.

In full 3D analyses the major concern is computer time for calculations in RTM flow simulations. F.Trochu et al.[10] proposed to simulate 3D composite shells with multi-layer reinforcements with several enhancements. The first one is connected with appropriate domain discretization. The second consists of using prismatic finite elements instead of tetrahedrons to reduce the number of degrees of freedom in 3D analyses. The third improvement concerns the different levels of coupling that can be implemented between in-plane and transverse calculations.

Modeling the inhomogeneous microstructures of fibrous tows is important for analyzing the process of resin transfer molding because dual-scale pores in a preform can lead to void formation. Shigeki Yashiro et al.[11] developed a microscopic flow analysis method to predict the impregnation of fiber bundles. The moving particle semi-implicit method was adopted to model the microstructure of a fiber bundle explicitly and interparticle potential force was introduced into the numerical model to take account for the capillary effect. The predicted process of impregnation and void formation agreed with empirical observations. The developed approach was applied to predict the relationship between the modified capillary number and void content to identify the optimal molding conditions to reduce micro voids.

3. Method description and case study

The proposed method permits to calculate the impregnation time for the composite material. This approach uses model, based in Ansys Fluent. This software is based on the Finite Volume Method (FVM) and to solve this problem the Volume of Fluid (VOF) model are used. Proposed by Hirt and Nichols in 1981 [12], the VOF model solves the fluid dynamic and heat transfer problems of two or more immiscible fluids. Through VOF, it is possible to identify the position of the interface among different phases of fluid. In addition, the fluid phases are well separated and the volume of a phase cannot be occupied by another one. The computational modeling of the RTM process in both two-dimensional and dimensional geometries has already been validated by Porto et al. [13].

In this approach two earlier phases are defined – air and resin. The plate is presented as a porous medium with porosity 0.5. This method is based on [13] . The proposed approach is used in the set of cases based on [14] . The set of cases are presented in the figures 1.

As the input the offered approach uses results of periodicity cell impregnation. The information about air traps [15] is transferred to model in the Ansys Fluent. These air traps are included to the model as an additional fluid domain with air.

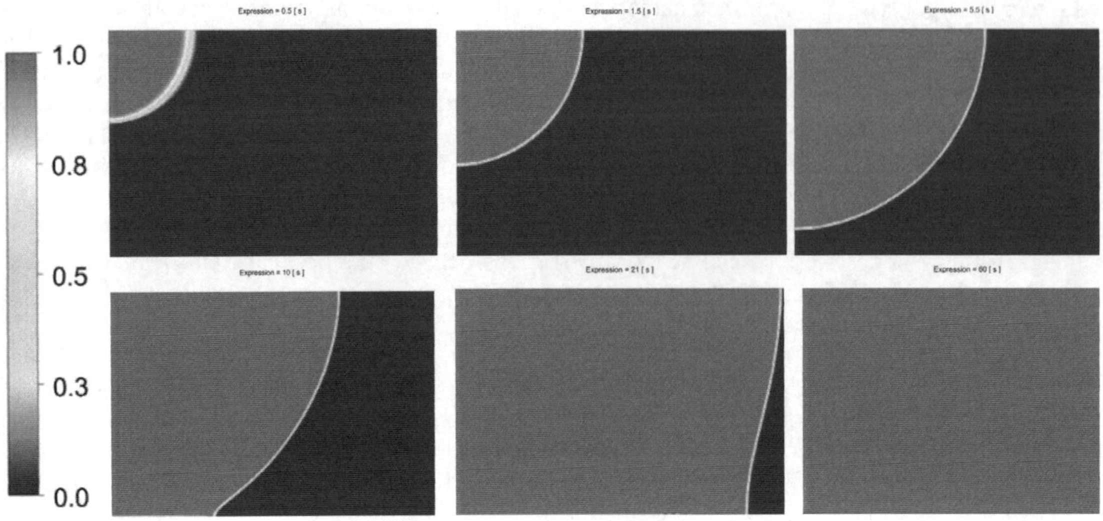

Figure 1. Resin impregnation.

Set of three cases are completed in this paper – diagonal impregnation on diagonal and lengthwise direction. The last one is calculated for one and two outputs. In all cases 200 mm length, 300 mm and 5 mm thickness plate is used. Permeability is isotropic. The additional domain with air is added to emulate the air traps. There is a set of equally spaced elements in this domain. The 1 mm3 size of these elements are chosen.

It is known that the heterogeneous flow of the binder is due to the heterogeneous microstructure of the wavy preform and the air is trapped along the flow front of the binder during the impregnation process. In the microstructure of such textile forms, you can find two types of pores with very different sizes: micropores (inside fiber bundles), which occupy a tiny space between individual fibers of the filler, and micropores, which are the empty space between individual threads.

Thus, this method permits to analyze the porosity of the final composite material by means analysis of the created voids. We take into account only macrovoids created in air traps without account sources of porosity [16].

Most studies [17]–[20] describe semi-empirical models for air void prediction. These approaches calculate the size and distribution of the air voids using experimental data and semi-empirical formulas. It is experimentally established that air void formation correlates with the capillary number:

$$Ca = \frac{\mu \cdot u}{m \cdot \gamma \cdot \cos Q} \qquad (1)$$

where: μ is the resin viscosity; u is the resin velocity; m is the porosity of the textile preform; γ is the interfacial tension between resin and air; Q is the contact angle.

The PAM-RTM software implements the determination of the air voids size depending on the capillary number according to formula 1.

According to experimental observations and the formula 1 there is an optimal resin flow velocity at which the number of macro-and micro-voids will be minimal [16]. It is about 0.002-0.003 m/s. At a low flow velocity of the resin, the air voids between the yarns (macro-voids) prevail, at a high flow velocity of the resin – inside the yarns (micro-voids).

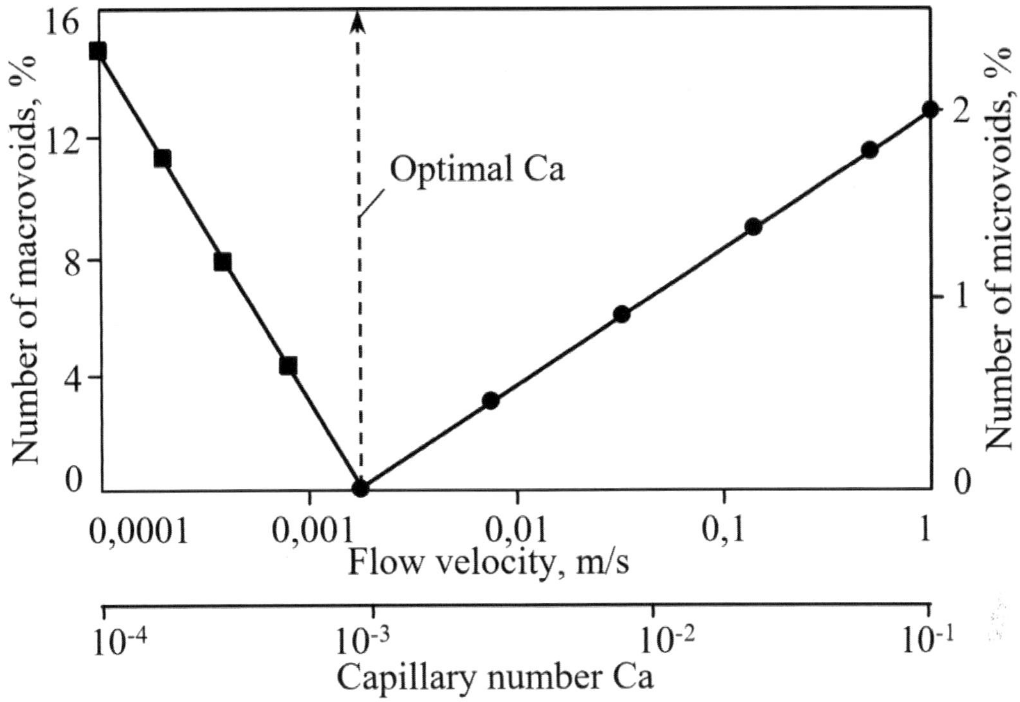

Figure 2. Dependence of macro- (rectangles) and micro-voids (cirles) on resin flow rate [15].

In this paper, we offer only to model the macro-void formation between individual yarns. The micro-void formation inside the yarns is not considered. The flow rate resin at each point of plate allows you to obtain the fraction of voids using the graph in Fig. 2. The voids are modeled with additional fluid domain with air. The distribution is done hand using the velocity and Fig. 2.

Proposed method is a part of the perspective model of the composite components conceptual design development presented in the Fig.3. In perspective this model will takes into account the manufacturing stage permit to develop the composite component.

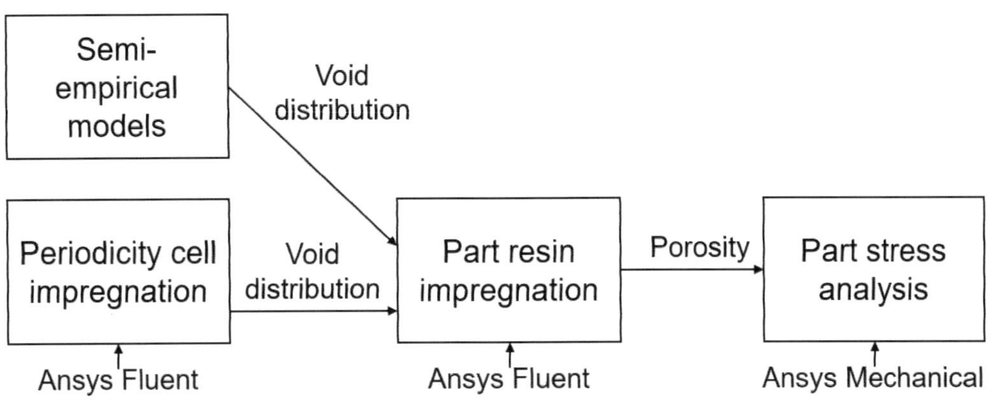

Figure 3. Method diagram.

4. Conclusions

This method can be used in industry and science. At further research we will use multiphysical approach – using of thermal and CFD approaches, also we guess to add automatic procedure of macro-void distribution by means Fortran or Python software. On the other hand, this approach permits to simulate all production process in one unique framework.

Size and distribution of air voids will be obtained when modeling the unsteady flow through the unit cell of textile depending on geometry of unit cell and impregnation direction at each point of the dry textile.

Acknowledgement

This work was financial supported by Ministry of Education and Science of the Russian Federation as part of the Project (unique identifier XXX).

References

[1] Ullman D 2010 *The mechanical design process 4th edition* (McGraw-Hill)

[2] *ASTM D2734 - Void content of reinforced Plastic*

[3] Struzziero G, Teuwen J J E and Skordos A A 2019 Numerical optimisation of thermoset composites manufacturing processes: A review *Composites Part A: Applied Science and Manufacturing* (vol. 124. Elsevier Ltd, Sep. doi: 10.1016/j.compositesa.2019.105499)

[4] Hattabi M, Echaabi J and Bensalah M O 2008 Numerical and experimental analysis of the resin transfer molding process *Korea Aust. Rheol. J.*

[5] Dammann C and Mahnken R 2019 Simulation of a resin transfer molding process using a phase field approach within the theory of porous media *Compos. Part A Appl. Sci. Manuf.* (vol. 120 May doi: 10.1016/j.compositesa.2019.02.022) pp 147–160

[6] Isoldi L A, Oliveira C P, Rocha L A O, Souza J A and Amico S C 2012 Three-dimensional numerical modeling of RTM and LRTM processes *J. Brazilian Soc. Mech. Sci. Eng.* (doi: 10.1590/S1678-58782012000200001)

[7] Porto J D S, Letzow M, Dos Santos E D, Amico S C, Souza J A and Isoldi L A 2012 COMPUTATIONAL MODELING OF RTM AND LRTM PROCESSES APPLIED TO COMPLEX GEOMETRIES *Rev. Eng. Térmica* (doi: 10.5380/reterm.v11i1-2.62007)

[8] Kang M K and Il Lee W 1999 A flow-front refinement technique for the numerical simulation of the resin-transfer molding process *Compos. Sci. Technol.* (doi: 10.1016/s0266-3538(99)00029-9)

[9] Zade A and Kuppusamy R. R P 2019 Mould fill time sensitivity analysis using isothermal mould filling simulations for applications in liquid composite moulding processes *Mater. Today Proc.* (Nov. doi: 10.1016/j.matpr.2019.09.212)

[10] Trochu F, Ruiz E, Achim V and Soukane S 2006 Advanced numerical simulation of liquid composite molding for process analysis and optimization *Composites Part A: Applied Science and Manufacturing* (doi: 10.1016/j.compositesa.2005.06.003)

[11] Yashiro S, Nakashima D, Oya Y, Okabe T and Matsuzaki R 2019 Particle simulation of dual-scale flow in resin transfer molding for process analysis *Compos. Part A Appl. Sci. Manuf.* (doi: 10.1016/j.compositesa.2019.03.038)

[12] Hirt C W and Nichols B D 1981 Volume of fluid (VOF) method for the dynamics of free boundaries *J. Comput. Phys.* (vol. 39, no. 1, , Jan. 1981, doi: 10.1016/0021-9991(81)90145-5) pp 201–225

[13] Porto J D S, Letzow M, Dos Santos E D, Amico S C, Souza J A and Isoldi L A 2012 Computational Modeling of Rtm and Lrtm Processes Applied To Complex Geometries *Rev. Eng. Térmica* (vol. 11, no. 1–2, doi: 10.5380/reterm.v11i1-2.62007) p 93

[14] Coulter J P and Güçeri S I 1989 Resin impregnation during composites manufacturing: Theory and experimentation *Compos. Sci. Technol.* (vol. 35, no. 4, doi: 10.1016/0266-3538(89)90055-9) pp 317–330

[15] Matsuzaki R, Naito M, Seto D, Todoroki A and Mizutani Y 2016 Analytical prediction of void distribution and a minimumvoid angle in anisotropic fabrics for radial injection resin transfer

molding *Express Polym. Lett.* (vol. 10, no. 10, doi: 10.3144/expresspolymlett.2016.80) pp 860–872

[16] Dushin M I, Khrulkov A V, and. Karavaev R Y 2015 Parameters that influence the formation of porosity in the products made of polymer composite materials (PCM) produced by out-of-autoclave methods *Proc. VIAM* (vol. 0, no. 2, (doi: 10.18577/2307-6046-2015-0-2-10-10) pp 10–10

[17] Patel N and Lee L J 1996 Modeling of void formation and removal in liquid composite molding. Part II: Model development and implementation *Polym. Compos.* (vol. 17, no. 1, doi: 10.1002/pc.10595) pp 104–114

[18] Greszczuk L B 1967 Effect of voids on strength properties of filamentary composite *22nd Annual Meeting of the Reinforced Plastics Division of the Society of the Plastics Industry*

[19] Chen Y T, Davis H T and Macosko C W 1995 1995 Wetting of fiber mats for composites manufacturing: I. Visualization experiments *AIChE J.* (vol. 41, no. 10, doi: 10.1002/aic.690411009) pp 2261–2273

[20] Rohatgi V, Patel N and James Lee L 1996 Experimental investigation of flow-induced microvoids during impregnation of unidirectional stitched fiberglass mat *Polym. Compos.* (vol. 17, no. 2, doi: 10.1002/pc.1060) pp 161–170

IOP Conf. Series: Materials Science and Engineering **986** (2020) 012060 doi:10.1088/1757-899X/986/1/012060

Methods for identifying risks and opportunities in industrial organizations

A T Kozlova[1], D A Stepanova[1] and O I Klyavin[2]

[1]Federal State Budgetary Educational Institution of Higher Education «Kazan National Research Technical University named after A.N. Tupolev-KAI», e-mail: alsu_kozlova@mail.ru, dashenka.fedagina@gmail.com,
[2]National Technology Initiative Center for Advanced Manufacturing Technologies based on the Institute of Advanced Manufacturing Technologies of Peter the Great St. Petersburg Polytechnic University Polytechnicheskaya, 29, St.Petersburg, 195251, Russia

Abstract. Many factors negatively influence the Russian economy in modern conditions. They are the instability of industrial relations, the growth of inflation, the rise in the cost of loans, a significant decrease in the state support for the sectors of the economy and the functioning of the organization, a decrease in real incomes of the population, and many others. The chosen topic is relevant because many enterprises and organizations are unable to adapt to market conditions and make informed decisions under uncertainty. This is one of the reasons for the decline in their activities. This article is devoted to modern methods of managing the efficiency of an organization, proves the importance of considering the risks of the external and internal environment, as well as a methodology for determining the risks and opportunities for organizing the real sector of the economy is presented.

1. Introduction

There are many factors that negatively impact on the state of the Russian economy in modern conditions: the instability of industrial relations, growth of inflation, and the rise in the cost of loans, significant reduction in the state support for the sectors of the economy and the functioning of the organization, a decrease in real incomes of the population, and many others. Many enterprises and organizations are not able to adapt to market conditions and make informed decisions in conditions of uncertainty were one of the reasons for the decline in their activities to the complete bankruptcy of individual enterprises.

Continuous improvement of the organization's work is the key to survival in the modern market. Regardless of industry and location, organizations manage their performance. This involves a series of repetitive activities, namely the development of organizational goals and assesses progress in achieving them. At the same time, the heads of organizations rely on the ratio of economic results and used resources. In the modern world, the criterion for assessing efficiency is the satisfaction of stakeholders - internal and external stakeholders.

The available and most popular ways to manage organizational performance are:

1. process approach and elimination of cross-functional inconsistency;
2. risk-oriented thinking in strategic and operational planning, including analysis and processing of statistical data;
3. definition and continuous measurement of targets, the appointment of responsible persons, analysis and assessment of non-conformities.

All activities of an organization involve risk. Organizations manage risk by identifying it, analyzing it, and then assessing whether the risk will be modified by exposure to meet established risk criteria. Throughout this process, they share information and consult with stakeholders, observe and analyze risk and management actions that modify the risk to ensure that further risk mitigation is no longer required [2].

2. Material and methods

Industrial enterprises of the military-industrial complex occupy a special place in the structure of the national economy. The increased attention of the state to this industry is explained by the desire to create an effective innovative resource on the basis of the scientific, technical, and production and technological potential of enterprises.

When the quality management system is planned, the organization determines the risks and opportunities, considering the factors of the external and internal environment and the requirements of interested parties for:

a) ensuring confidence that QMS achieves the intended results;

b) increase the desired impact of opportunities;

c) preventing or reducing the unwanted impact of risks;

d) achieving improvement.

The organization analyzes the risks of problems expressed in non-fulfillment of requirements, failure to achieve planned results in all processes and activities of QMS, and takes actions to prevent risks or reduce their possible unwanted impact.

At this stage, an analysis of the influence of the external environment on the strategic goals of the enterprise and the QMS policy can be carried out. When the external environment is examined, it is recommended to use PESTLE-analysis (Figure 1) - studying the positive and negative influence of the following factors: P – political, E – economic, S – social, T – technological, L – legal, E –environmental [5].

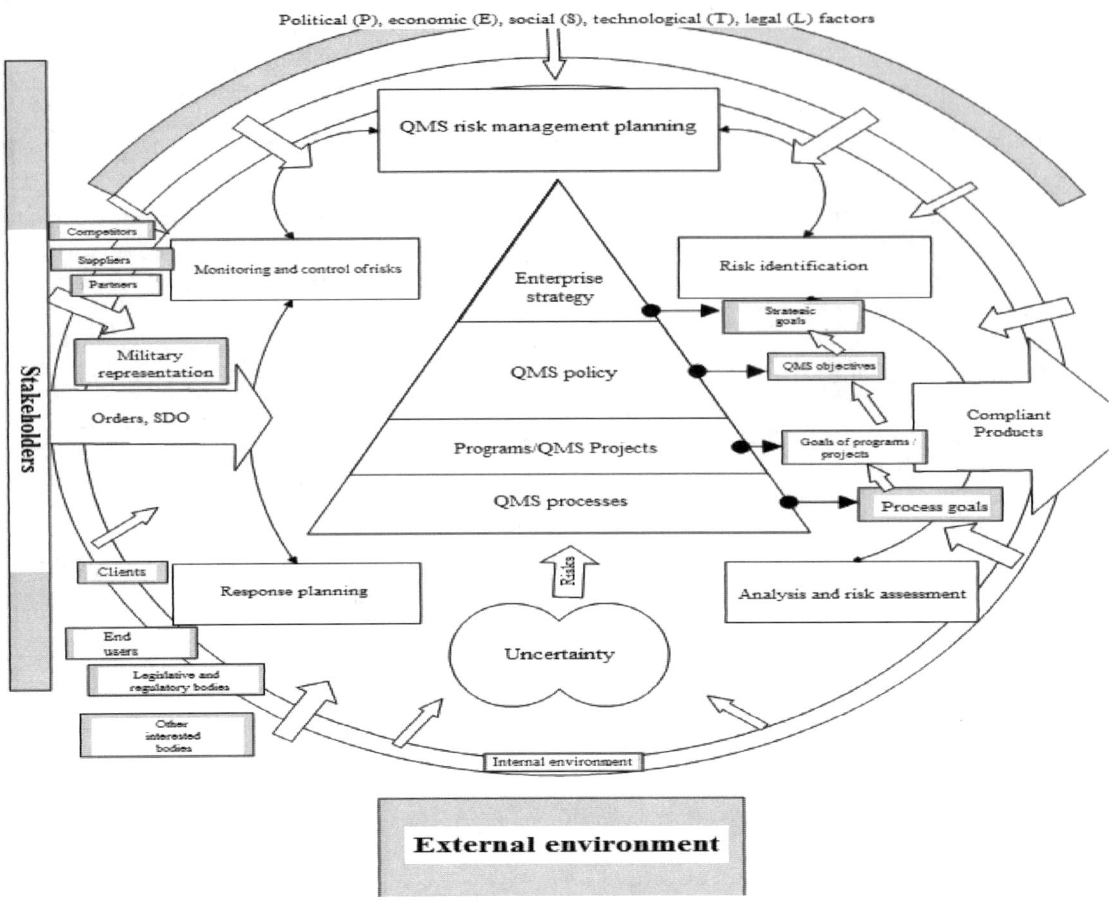

Figure 1. Typical risk management model in an industrial enterprise.

Risk management in an organization without fail includes the following steps:
- Risk identification;
- Risk description;
- Search for the sources of the identified risks;
- Search for the causes of the identified risks;
- Identification of chains of consequences of the detected risks;
- Ranking chains of consequences in order to identify the most significant events;
- Selection of the most significant risks;
- Choice of a response method for each significant risk;
- Planning of preventive actions for each selected method;
- Analysis of the adequacy of planned actions;
- Implementation and control of planned activities (actions);
- Analysis and evaluation of the effectiveness of the measures taken.

Table 1 shows the classification of risks used for an industrial enterprise.

Table 1. Industrial enterprise risk classification.

Risk	Main reasons	Results
Risks in procurement activities	-working with an unscrupulous supplier -procurement of goods at inflated prices -refusal of planned suppliers from drawing up supply contracts	-failure to meet the planned production targets for finished products, disruption of the state assignment;

		-increased costs for the purchase of raw materials and materials for the production of finished products; -non-recognition of the organization's expenses in full due to inspections by fiscal authorities
Risks in production activities	-insufficient design and technological preparation of production -irregular operation of production -insufficient qualification of personnel -insufficient level of production automation	-failure to meet the planned production targets for finished products, disruption of the state assignment; -decrease in the quality of manufactured products; -disqualification of an official
Risks in innovation	-the need to improve the competitiveness of production -the need for the development of civilian products -the wrong choice of innovative projects -insufficient innovative activity of personnel	-reduction of the income from the project (the enterprise as a whole); -failure to meet the planned indicators of the project (the enterprise as a whole)
Financial management risks	-lack of highly liquid operating assets - ineffective accounting system (management, accounting) - the complexity of obtaining credit resources	-loss of operating assets placed on bank accounts with revoked licenses - “freezing” of funds in separate accounts due to the complexity of justifying spending; -increase in interest rates on lending; -failure to meet the planned production targets for finished products, disruption of the state assignment; -penalties
HR risks	-insufficient qualification of personnel -lack of staff	-non-fulfillment of planned indicators of production of finished products, disruption of the state assignment

Sources of information for identifying risks can be results of the analysis of the functioning of the QMS processes; supplier assessment results; internal and external audit reports; analysis of nonconforming products and ones returned by consumers and complaints from consumers; information about violations of technological discipline; analysis and assessment of customer and employees of the organization satisfaction; minutes of meetings of the Quality Council, meetings of scientific and technical councils.
Actions that can be used to implement risk response methods include:
Correction is the action taken to eliminate the inconsistency that has arisen.
Corrective actions are used to eliminate the causes of existing nonconformities of products, processes, QMS procedures, and prevent these nonconformities in the future. The organization's improvement process using corrective actions includes:

1) setting specific goals;
2) control (assessment) of the implementation of the set goals;
3) accumulation of data based on the results of control (assessment);
4) identification of existing inconsistencies;
5) identification (registration) of non-conformities;
6) analysis of inconsistencies, including consumer claims, and identification of their causes;
7) elimination of detected inconsistencies (if necessary);
8) assessment of the need for action to avoid recurrence of nonconformities;
9) development of corrective actions to prevent the causes of nonconformities;
10) implementation of corrective actions;
11) control of the implementation of corrective actions and registration of their results;
12) analysis of the effectiveness of the taken corrective actions.

Preventive actions are used to eliminate the causes of potential nonconformities of products, processes, QMS procedures, which may entail nonconformities of various natures. For using preventive actions, "weak", "insufficiently reliable", "potentially dangerous" and similar areas of already ongoing activities are identified.

The order and methods for taking preventive actions may include:

- using appropriate sources of information to identify, analyze and eliminate potential causes of nonconformities;
- analysis of the risks of problems with the quality of military products at all stages of its life cycle and the development of measures to reduce these risks;
- the procedure for initiating (submitting and registering proposals) preventive actions;
- providing relevant information about the actions taken to analyze their effects on the part of management.

The risk management process should be consistent with the culture, processes, structure, and strategy of the organization. An internal situation (context) is something within the organization that can influence how the organization will manage risk. Table 2 shows a typical generalized process model of risk management in an industrial enterprise.

Table 2. Stages of the risk management process at the industrial organization.

Input	Process algorithm	Output
1.1. The strategic goals of the enterprise 1.2. QMS policy, QMS goals 1.3. Objectives of quality projects and programs 1.4. Goals of QMS processes 1.5. Report on risk management in the enterprise	1. Risk management planning in QMS	1.1. Risk management plan for the year
2. Risk management plan	2. Risk identification	2. Risk register
3.1. Risk register	3. Analysis and assessment of risks	3.1. Risk register 3.2. Risk map 3.3. Summary of the risk map
4.1. Risk register 4.2. Risk map	4. Response planning	4.1. Response plan 4.2. Risk register
5.1. Response plan 5.2. Risk register	5. Monitoring and control of risks	5.1. Report on the results of response actions 5.2. Problem registry

		5.3. Risk register 5.4. Risk Management Implementation Report 5.5. Suggestions for improving the risk management process

3. Conclusion

The analysis of the experience of the best organizations that have implemented a risk-based approach would show that their success is due to the mandatory personal participation of decision-makers in risk analysis.

It should not be forgotten about the involvement of all stakeholders in the development of targets. This allows to create an open and trusting environment within and around the company. This approach would help build a cross-responsibility system for meeting targets in end-to-end processes and increase customer satisfaction.

References

[1] GOST R ISO 9000-2015 2015 Quality management systems. Fundamentals and vocabulary (M .: Standartinform) p 50
[2] GOST R ISO 31000-2010 201 Risk management. Principles and Guidelines Introduction (M .: Standartinform 1) p 21
[3] GOST R ISO 9001-2015 2015 Quality management systems. Requirements (M .: Standartinform) p 24
[4] GOST R ISO / IEC 31010-2011 2012 Risk management. Risk assessment methods (M .: Standartinform) p 70
[5] Yashin N S and Andreeva T A 2017 Understanding the context of the enterprise within the quality management system *Bulletin SSTU* (№3 (67)) pp 115-124

IOP Conf. Series: Materials Science and Engineering **986** (2020) 012061 doi:10.1088/1757-899X/986/1/012061

Review of the current state of research of microturbine units

N Myakishev[2], M Laptev[2], A Pekarskii[2], V Barskov[1], V Rassokhin[1], V Chernikov[1], E Semakina[1], V Yadikin[1], Y Matveev[2] and A Smetankin[1]

[1]National Technology Initiative Center for Advanced Manufacturing Technologies based on the Institute of Advanced Manufacturing Technologies of Peter the Great St. Petersburg Polytechnic University Polytechnicheskaya, 29, St.Petersburg, 195251, Russia, v-rassokhin@yandex.ru
[2]Higher school of power engineering, Institute of power engineering Peter the Great St. Peersburg Polytechnic University, Saint Petersburg, Polytechnicheskaya, 29, St.Petersburg, 195251, Russia

Abstract. The article provides an overview of the current state of micro-turbine units with an external supply of heat. In particular, there have been described various approaches of companies to the creation of this type of units, priority directions of micro-turbines design have been identified, too.

1. Introduction

Micro-turbine units are regarded to be a new type of micro-turbines which are used to solve applied problems of energy generation. At cogeneration mode, this type of units is able to solve the problems of heat supply for various objects. This type of units accepts wide variety of fuel. Micro-turbines are also used in residential areas, gas platforms and villages [1].

It should be considered that the energy saving system is now in crisis because the development of the generation capacity does not keep up with the increased consumption, therefore micro-turbine is an effective solution for integrated energy security [2]. All things above considered, this topic can be regarded as relevant [3].

The system of centralized energy supply in Russia lives through the crisis, i.e. the equipment wears out and the development of generation capacity does not keep up with the increased consumption. Decentralized forms of energy supply of cities, towns and villages are a very real prospect in the emerged crisis. Marketing research, analysis of technical and operational characteristics showed that micro-turbine is suitable for mini-thermal power stations projects [4].

Micro-turbine units are used for:

- electricity and heat supply of construction projects in a city center or its suburbs, resort areas;
- own needs of boiler rooms;
- enterprises, industrial and telecommunication facilities and other facilities interested in independent energy supply;
- hospital complexes;
- sports facility. These are mainly swimming pools and water parks, by which both electricity and heat are in demand [5].

In Russia, serious works upon this type of units has not been carried out yet. The main development in this sphere was performed by foreign specialists, namely by the engineers from Capstone (the USA), Institute of Engineering Thermophysics, Chinese Academy of Sciences (China), MTT Micro Turbine

Technology BV (the Netherlands), Delta Motorsport (the UK), all of them have achieved significant results. All the companies held their studies in different directions and sought for their own way of solving the problem, so finally they got the results which are described further.

2. Overview of research and installations of foreign companies

2.1. Capstone Corporation

Capstone Corporation chose their own way, the main idea of which was the improvement of the thermodynamic model of Capstone C30 micro-gas turbine. The methodology was based on the genetic optimization algorithm where variable solutions and targets are set depending on the available experimental data. The results of the studied case underline the ability of the method to identify some inconsistencies in the experimental data and lead to a consistent thermodynamic reconstruction of the behavior of the micro-turbine [6].

As a result, there have been found out the lack of constructions on the micro-turbine that let achieve declared total efficiency value and effective capacity. For this purpose, a new methodology was accepted, which defined the optimal set of thermodynamic input data using the Euclidean norm to estimate the preferred solution.

Another direction, which improved rapidly, was devoted to the development of the method of rapid identification to obtain simplified dynamic models that were tested using the Capstone C30 micro-turbine [7]. For this purpose the model of the second order was identified through the adoption of the theory of sensitivity which included Lyapunov function that updates parameters of the model every time the physical system is changed. This characteristic can be effectively accepted to implement real models in simulations in the time domain, giving accurate results even if some internal parameters change. Several simulations and experimental tests were conducted in order to study the effectiveness of the proposed methodology in obtaining a simplified dynamic model.

The developed methodology was tested with the help of several simulations and experimental tests conducted at the Capstone C30 micro-turbine. The results showed that the suggested methodology can be characterized by reliability, self-adaptability, high accuracy, stability and good convergence rate.

2.2. Institute of Engineering Thermophysics, Chinese Academy of Science

At the Institute of Engineering Thermophysics at Chinese Academy of Sciences, a computational model of heat exchange and differential pressure was established to optimize the design of the ring heat exchanger in a micro-turbine with a transversely wavy primary surface [8]. The genetic algorithm method is used to solve the problem of optimization of a ring heat exchanger with several constructive variables.

Thus, the optimal approach of designing such type of a heat exchanger is formed. The reliability of the computational model is verified by comparing the results of calculations with experimental data of heat exchanger prototypes C30 and C65 (Capstone Turbine Corporation). Compared with the original design, the optimal design leads to a significant reduction in pressure loss if all the limitations are met, which proves that the formed optimal design approach is effective for the ring heat exchanger. Replacing the optimization goal, the approach to optimal design presented in this article can be used in a single-purpose optimization for other goals (compactness, heat transfer area, mass, volume, efficiency, etc.) and even in a multi-objective optimization in accordance with the specific requirements of various micro-turbines [9].

2.3. Institute of Engineering Thermophysics, Chinese Academy of Science

At this university the studies on the use of fast pyrolysis of bio-oil (FPBO) were held [10]. The studies were carried out on a small scale of non-regenerated micro-gas turbine, created in a special test unit [11]. The system includes an injection line and a modified combustion chamber to burn the effectively high volume parts of FPBO/ethanol. The effect from the increased volume of the combustion chamber increased the quality of combustion of the reference diesel oil and pure ethanol in relation to exhaust emissions while maintaining the same fuel consumption of the original configuration [12]. By increasing the volume fraction of FPBO [13], there was an increase of CO emissions in the fuel mixture, probably

IOP Conf. Series: Materials Science and Engineering **986** (2020) 012061 doi:10.1088/1757-899X/986/1/012061

caused by larger droplets derived from more viscous fuel, as well as increase of NOX emissions, which probably happened due to fuel-related nitrogen [14].

With the proposed modifications being taken into account and the FPBO/ethanol mixture accepted as fuel, the overall electrical efficiency of the engine was higher than that of the reference diesel fuel. The final launch with 100% FPBO feed showed unstable combustion with the presence of carbon deposits in the hot parts of the system, which showed that the current configuration requires further changes to achieve the goal. Some fundamental principles of the introduction.

Fundamental principles were developed for the further integration of improved solutions for the micro-gas turbine (MGT) for the delivery of viscous, acidic and aqueous fuel.

2.4. Inha University

Scientists from the Korean Inha University conducted a study to identify the most preferred cycle for the heat recovery system. For their experiment they took the Rankine cycle (ORC) and transcritical CO_2 cycle (tCO_2) [15], which were applied to the micro-gas turbine (MGT), subsequently a comparative analysis of productivity was made, too.

Non-constructive characteristics of two cycles with MGT (micro-gas turbine) load change were analyzed, too. According to the calculations, the output power of ORC [16] was higher than in the tCO_2 cycle when MGT was running at full load. At the same time, the change in the output power during the change in the MGT load is greater in ORC. tCO_2 output power becomes higher when MGT is running under partial load, namely below 75%. Consequently, CO_2 is suitable for a heat recovery system that uses MGT exhaust gas in places where MGT must operate at low loads for many hours of its operation, while ORC is more suitable in applications where MGT can run at almost full load.

The results were as follows:

- by capturing heat from MGT operating at the target point, the additional power produced by using ORC is about 5.5% higher than that of tCO_2. It is connected with the fact that in this case the heat that can be extracted from the HRU is bigger;
- when the MGT works at partial load, the change of the temperature in the inlet of the cold side of HRU is bigger in the ORC, if the MGT load is changed. Consequently, HRU tCO_2 can recover more heat than HRU ORC when MGT is running at a partial load below 75%;
- the appropriate system should be selected according to the operating conditions. If MGT must work at low loads for many hours of operation, tCO_2 is more suitable as a bottom cycle. ORC is more suitable for applications where the system can run at almost full load for the biggest part of its working time [17].

3. The review of the Russian studies in the sphere of micro-turbines

3.1. Peter the Great St. Petersburg Polytechnic University (MGTU-100)

The specialists from SPbSTU made the analysis, the main goal of which was finding methods of increasing the coefficient of efficiency for the cycle of a a simple thermal scheme. For this purpose four schemes were analyzed, which were suitable for this type of micro-turbine units, and, as a result, their advantages and disadvantages were revealed [18].

At the end of the study the a thermal scheme of a two-shaft small-size gas-turbine with a compressor drive from a separate electric motor was chosen, as it possesses a simple design and high reliability. Some other very important advantages of this scheme include the drive of the compressor from a separate electric motor, with the help of which the increase of the coefficient of efficiency of MGTU-100 (micro-gas turbine 100) without using the recuperative air heater, and the absence of "surge" zones due to the variable speed of the compressor impeller.

3.2. Combined cycle gas compressor station "Severnaya", «Gazprom transgaz St. Petersburg»

On the basis of «Gazprom transgaz St. Petersburg», which is one of the biggest gas companies in Russia, the concept of the turbine-generators construction was formed, which use the energy of compressed natural gas for own needs of the “Gazprom transgas”. This concept let creating environmentally friendly sources of generating electricity without burning additional fuel [19]. Therefore the steam turbine unit

"MTU-500" was made, which worked on the basis of the ORC cycle. The main task was the creating a steam turbine on the basis of already existing prototypes. Apart from this, the emphasis was put on the increase of the coefficient of efficiency. While designing a unit, one also examined several variants of thermal schemes, where advantages and disadvantages of each scheme were revealed [20].

According to the results of the preliminary calculation, a scheme of a three-circuit power source based on the MTU-500 ORC was chosen, which is the most promising at the present considering both increasing efficiency through the use of residual heat for thermal needs, safety and easy operation.

4. Conclusion

To sum up, one may say that each company described above has chosen its own way of solving the problem touched upon in the article. For instance, Capstone in its study put the emphasis on the increase of coefficient of efficiency and useful power, while Chinese scientists paid their attention to the optimization of the ring heat exchanger. Scientists from University of Florence decided to create a turbine which works on bio-oil. As a result, the increase of efficiency coefficient of the engine compared to the reference diesel engine was achieved The scientists from the Korean Inha University compared the organic Rankine cycle and the transcritical CO2 cycle, which were used in a micro-gas turbine, and afterwards the a comparative analysis of the productivity was made. The specialists from SPbSTU aimed to the increase of efficiency coefficient of MGTU-100 based on the analysis of thermal schemes. What about the company "Gazprom transgaz St. Petersburg", there has been chosen the most promising scheme of a three-loop energy source.

References

[1] IntechOpen 2013 (https://www.intechopen.com/books/progress-in-gas-turbine-performance/micro-gas-turbine-engine-a-review [Electronic resource] [website])
[2] Min Jae Kima, Jeong Ho Kima and Tong Seop Kimb 2018 The effects of internal leakage on the performance of a micro gas turbine *Applied Energy* p 212
[3] Kanareykin A 2012 Big role "small" energy *Power and industry of Russia* (20)
[4] Mikheev N 2013 Microturbines - an innovative technology for creating nuclear power *NETWORK&BUSINESS* (5)
[5] Microturbine is an effective solution for integrated energy security [Electronic resource] (electrosystems: [website]. URL: http://www.electrosystems.ru/experience.php (date accessed: 12.12.2018))
[6] Alfredo Gimelli R S 2017 Thermodynamic model validation of Capstone C30 micro gas turbine *Energy procedia* (126)
[7] Cagnano A and Tuglie E De 2018 On-line identification of simplified dynamic models: Simulations andexperimental tests on the Capstone C30 microturbine *Electric Power Systems Research* (157)
[8] Jun Cai, Xiulan Huai and Wenxuan Xi 2018 An optimal design approach for the annular involute-profile cross wavy primary surface recuperator in microturbine and an application case study *Energy* (153)
[9] Aiguo LIU and Yiwu WENG 2009 Effects of Lower Heat Value Fuel on the Operations of Micro-Gas Turbine *Scientific Research* (1)
[10] Hossain A K 2013 Philip Andrew Davies, Pyrolysis liquids and gases as alternative fuels ininternal combustion engines - A review *Renewable and Sustainable Energy Reviews* (21)
[11] Poritosh Roy and Goretty Dias 2017 Prospects for pyrolysis technologies in the bioenergy sector: A review *Renewable and Sustainable Energy Reviews* (77)
[12] ResearchGate 2009 (https://www.researchgate.net/publication/285676129_System_identification_and_performance_improvement_to_a_micro_gas_turbine_applying_biogas [Electronic resource] [website])
[13] Woudstra T, Schoenmakers L, Oldenbroek V, Thallam Thattai A, Aravind P V 2016 Experimental model validation and thermodynamic assessment on high percentage (up to 70%) biomass co-gasification at the 253 MWe integrated gasification combined cycle power plant *Applied energy* (168)

[14] Marco Buffi, Alessandro Cappelletti, Andrea Maria Rizzo, Francesco Martelli and David Chiaramonti 2018 Combustion of fast pyrolysis bio-oil and blends in a micro gas turbine *Biomass and Bioenergy* (115)

[15] Xurong Wang and Yiping Dai 2016 Exergoeconomic analysis of utilizing the transcritical CO_2 cycle and the ORC for a recompression supercritical CO_2 cycle waste heat recovery: A comparative study *Applied energy* (170)

[16] Joon Hee Lee and Tong Seop Kim 2006 Analysis of Design and Part Load Performance of Micro Gas Turbine/ Organic Rankine Cycle Combined Systems *Journal of Mechanical Science and Technology* 20

[17] Suk Young Yoon Min Jae Kim, In Seop Kim and Tong Seop Kim 2017 Comparison of micro gas turbine heat recovery systems using ORC and trans-critical CO_2 cycle focusing on off-design performance *Energy procedia* (129)

[18] Fokin G A 2015 Autonomous sources of electric and thermal energy for main gas pipelines and gas distribution stations *Monograph* (Moscow: FIZMATLIT)

[19] Fokin G A Methodology of creation of the Autonomous sources of electric energy using energy of the compressed natural gas for own needs of gas transportation system of Russia *St. Petersburg: Peter the Great St. Petersburg Polytechnic University*

[20] Zabelin N, Lykov A V, Rassokhin V A, Sivokon N V and G A Fokin 2013 Steam-gas plant compressor station "Severnaya" *Science and technology in the gas industry* (20)

IOP Conf. Series: Materials Science and Engineering **986** (2020) 012062 doi:10.1088/1757-899X/986/1/012062

Increment of steam turbine blades service life by means of erosion-resistant plasma coating deposition with meso-ordered structure

V E Bogdanovich and M G Giorbelidze

Samara National Research University academician S.P. Koroleva, Samara, Russia, bogdanovich@ssau.ru

Abstract. Main reasons and wear mechanism of steam turbine blades have been examined. Efficient recovery method of steam turbine blades allowing to increase their service life and enhance greatly service life of power units have been suggested. Materials to deposit coating layers have also been selected and deposition technology have been developed. Structure, porosity, micro-hardness, adhesion strength, erosion resistance and heat resistance of deposited coating have been researched. Suggested method can be used both for deposition of a coating on new turbine blades and for recovery of worn surfaces without overhaul of a turbine due to applying of a mobile compact plasma unit of in-house design. Composition of the coating featuring meso-arranged structure during deposition process greatly increases most operational performances of the whole 'coating – substrate' system.

1. Introduction

Main reason of exhaust of service life of power units of most heat-power stations is connected with erosion and contact wear of different surfaces of a turbine because of high loads, high rotation rates and impact of fast steam gas flows including a lot of moisture drop phase at its last stages [1,10,12-14]. In recent two decades, it is noted that while performing scheduled repair works, there is a persistent tendency of parts wear increase of steam turbines, especially leading edges of blades. It is stipulated for alternate operational modes of heat-power stations because of necessity of more economy mode of their operation. In spite of considerable experience of designing and operation of moist-steam turbines in Russia and abroad and numerous theoretical and experimental research, the problem of efficient protection of blades of a low-pressure cylinder at the turbine last stages from considerable erosion during operation comparable with technological lifespan is apart from its solution.

Solution of the problem is carried out due to constructively - technical measures when designing, development and operation of steam turbines to decrease specific content, dispersity and normal velocity component of the drop phase hitting blade surface. Research of physics and dynamics of development, movement and hit of drop phase on the surface, research of mechanisms of initiation and development of surface erosion, erosion resistance of structural materials and development of methods of surface hardening along with methods of repair and recovery of structural components after erosive damage are also considered.

2. Problem solving method

Method of performance restoration of power units assumes essential economic expenditures because of dismantling and transportation of units when replacing old items that depleted their service life with new ones [1,2]. This, in turn, results in additional economic wastes because of lengthy downtime of equipment up to finishing of overhaul. To recover surfaces of worn parts of power units there exist different methods such as vacuum ion-plasma deposition, high-speed deposition and others. However, most of these methods are ineffective or their application is not admissible in heat-power station conditions [3-7].

Specialists of Research institute of Technology and quality problems of Samara National Research University named after academician S.P. Korolyov have discovered an opportunity to recover and to increase essentially service life of power unit parts by means of plasma deposition method [1]. Besides, the process of surface hardening and recovering of sizes of blades, shafts and other parts can be carried out directly at heat-power stations without total unit dismantling.

It is generally known, that blades of the last stages of steam turbines of heat-power stations operating in moist steam environment are subjected to erosive wear as a result of impact and cavitation action of the environment, especially the surface of the leading edge [8].

At the surface of the leading edge of newly manufactured blades there are soldered hard-alloy stellite plates providing protection from erosive wear. After long-term operation of blades, in the spots of plate locations erosive wear arises. They look like cavities, superficial and deep grooves, breakoff of stellite plates and turbine blades damage are also possible (Figure 1).

To repair leading edges of such blades it is usually necessary to dismantle the turbine and soldering of new stellite plates. It is exceptionally labour intensive process, as it requires matching and soldering of a great number of such plates to each blade maintaining given temperature mode when soldering.

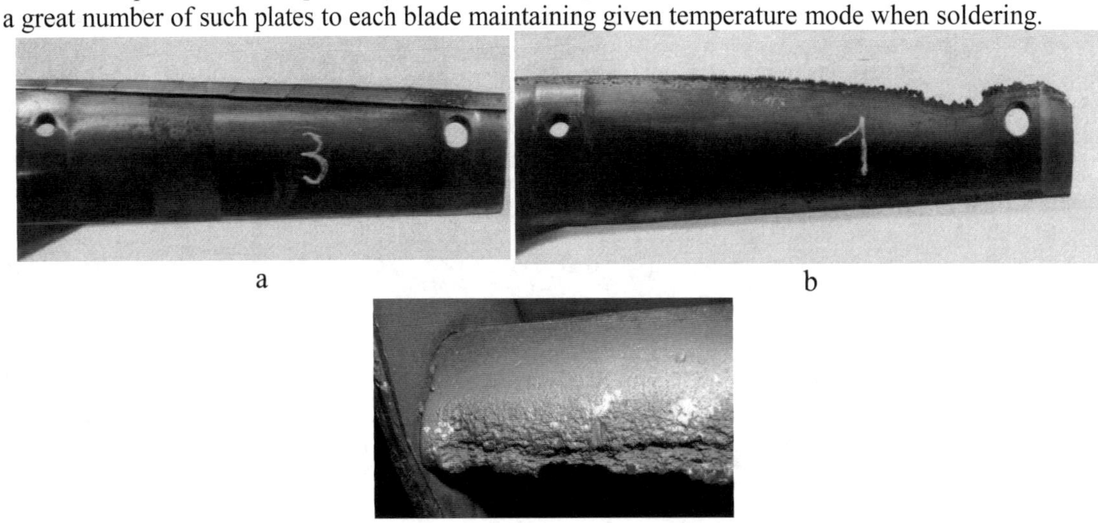

a b

c

Figure 1. Turbine rotor blade of a heat-power station: a – with a stellite plate; b – after stellite plate loss; c – view of damages and erosion of a blade after operation without coating.

Authors of this article have substantiated and proved in practice more than once necessity of application on recoverable items plasma coatings protecting expensive parts of units from wear and destruction arising as a result of deep damage in structural material. It should be noted that it is prospectless to find ideally resistant coating material in view of economic inexpediency and technical impossibility.

The task of the coating is to protect structural material over the time of overhaul period of unit operation when permissible wear of the coating is taking place. Technology's task is to provide coating recovery without considerable thermal influence on structural material happening during welding, overlaying welding, soldering, etc., moreover, less expensive cost of works in comparison with the cost of replacement of a set of used parts with new ones. A protective coating is to have substantially smaller wear rate then the stellite protection used, and coating thickness is to provide operating time much greater than existing stellite protection of 2 mm thickness.

Analysis of available research data on mechanisms of erosive fatigue damage of homogeneous materials, stellite and plasma coatings, as well as carried out theoretical and experimental research allowed to explain significant difference of their wear mechanisms. One of the essential results of such research is the fact that obtaining coatings with the structure having location of intercrystalline boundaries mainly along the outer surface of the item, it greatly increases majority of performance characteristics of the whole 'coating-substrate' system [1]. It has been found that such a structure called meso-arranged according to the authors [9], could be obtained by means of plasma gas-thermal deposition from most powder materials due to the choice of deposition mode [9].

3. Problem solving method

To choose the composition of erosion resistant coating for blades protection of steam turbines, nickel – aluminum, nickel – cobalt – chromium – aluminum – yttrium, titanium – nickel, chromium carbide – nickel systems and their combinations were researched. The subjects of inquiry were the structure, porosity, phase composition, micro-hardness, adhesion strength, heat resistance and erosion resistance of deposited coatings.

Metallographic research were carried out using cross sections of coatings. Coating samples were filled in special rings, and treated by the buffing and grinding machine Gripo 1V. Microstructure research was carried out using special equipment complex for metallographic analysis including a metallographic microscope and software to provide image analysis based on the computer Image Expert Pro 3.0 Professional. Application of this software allows to determine porosity of deposited layers. Research of phase composition of the coating were carried out by means of scanning electron microscope TESCAN Vega SB.

Micro-hardness of phase terms of coatings was determined under indentation load 0.2 and 0.49 N at 100 points for a length of 10 mm every other (80-100) μm. According to measurement results, values distribution histogram of layer micro-hardness in coordinates 'micro-hardness – number of prints' in percent were built. Measurement results and well known research data on micro-hardness of phase terms allows to evaluate percentage and phase distribution in the coating, as well as the conservation degree of a carbide phase under coating deposition, which can partly recover under isothermal particle deposition by a plasma jet.

Research of adhesion strength were carried out applying glue method according to State Standard 9.304 – 87 'Gas-thermal coatings. General requirements and inspection methods'. To carry out research for heat resistance special equipment was used. Qualitative method was used. The samples were heated with after air-cooling, and arising stresses and strains were not evaluated. Number of cycles before peeling or destruction of the coating were determined. The sample with coating was secured in shunts. Heating was provided by a current transformer up to the temperature 1000 °C. The temperature was measured by means of a potentiometer and a thermocouple. Potentiometer output was connected to the automatic control unit, which maintained sample temperature within the prescribed limits by means of a contactor. Heating rate was 10 °C per second, it was adjusted by the autotransformer, and number of cycles was recorded by the counter. Tests were carried out in automatic mode keeping the prescribed temperature for 20 seconds. Checking took place every 20 cycles up to destruction or peeling of the coating.

Tests for erosion resistance carried out at special testbed allowing to simulate conditions of interaction of drops of liquid with the surface of structural material and protective coatings samples. When researching erosive wear process of the material, the value of mean erosive wear of the material or coating depending on specific weight of drops per a unit of eroded surface was measured.

Research for choosing coating composition was carried out using samples. For each of researched materials the mode was determined using mathematical modelling. In this mode powder particles of deposited materials of narrow fractions are accelerated by plasma jet during coating layer formation, they are heated up to the temperature when only particle surfaces are melted, but the deposited particle itself is deformed in a cluster of a meso-arranged coating. Meso-arranged coatings have layer structure consisting of separate disk-shaped clusters oriented along the surface of protected basic material, vertical and horizontal boundaries, as well as micro-cavities located within vertical boundaries between adjacent clusters. Thickness of a separate disk-shaped cluster is from 2 to 10 micrometers μm, it is several times less than characteristic horizontal dimension. Boundaries of such structure are limited by the surface of other crystallites in a vertical direction, as a result, closed porosity type is developed that prevents penetration of chemically active components to the basic material. As mentioned before, we have established that the coatings with meso-arranged structure having intercrystalline boundaries mainly along the outer surface of an item have higher performance characteristics.

Research of the coating structure deposited under strict conformity of the deposition mode of a disperse fraction of the deposited material and improvement of the powder feed system has shown that we are able to form cluster meso-arranged structure from powders of all researched materials.

Results of testing for adhesion strength by glue method have confirmed that cohesive strength of coatings reaches (36 – 40) MPa, and it exceeds adhesive strength. Coating porosity is (2 – 5) %, heat resistance of coatings exceeds 200 cycles at the mode (20 – 1000) °C. Determining factor for choosing composition of deposited coating were the test results on erosion resistance. Research of erosion resistance of plasma coatings has shown that the coating with composition Ni-Co-Cr-Al-Y + CrC-Ni in conditions of the test by the drops $d_k = 400$ μm at speed $V_k = 25$ 0 m/s has provided better protection than stellite. Under drop load $5 \cdot 10^4$ kg/m^2 coating (worn about 30%) has shown erosion rate 1.4 times as less than stellite.

For deposition of the given coating, two materials were used: powder Ni-Co-Cr-Al-Y fraction (60 – 80) μm and powder CrC-Ni fraction (40 – 60) μm. Chemical composition of the powder: Ni – basis, Co – 21,3 %, Cr – 17,4 %, Al – 14,1 %, Y – 0,34 %. Powder deposition Ni-Co-Cr-Al-Y was carried out at the design mode: current strength 380 A, voltage (50 – 52) V, deposition distance 120 mm, argon consumption 35 l/min, hydrogen consumption 3.5 l/min. Thickness of the deposited layer was (250-300) μm.

Powder CrC-Ni was deposited at the mode: current strength 420A, voltage (52-54) V, deposition distance 120 mm, argon consumption 35 l/min, hydrogen consumption 3.5 l/min. Thickness of the deposited layer was (150-170) μm. Chemical composition of the powder: Cr – 70 %, C – 9,7 %, Ni – 20 %, Si – 0,3 %.

Structure of the deposited meso-arranged coating Ni-Co-Cr-Al-Y + CrC-Ni is shown in Figure 2, table 1 displays main parameters of its layers. The coating has homogeneous, dense structure; peelings, cavities, cracks and other defects have not been discovered.

Figure 2. Coating microstructure Ni-Co-Cr-Al-Y + CrC-Ni (X200).

Table 1. Research of microstructure and micro-hardness of the coating.

layer	Thickness, μm	Micro-hardness, GPa	Porosity, %
Bonding layer (powder of the system Ni- Co-Cr-Al -Y)	250 – 300	4.5 – 5.5	2.5
Working layer (powder of the system CrC-Ni)	150 – 170	9.5 – 10.0	3.5

To repair, harden and recover turbine blades of heat-power stations without their dismantling we have developed mobile compact unit and deposition technology. The unit is fitted with plasmatrons [11], meter regulators, manipulators of in-house design. It is also fitted with computer control system for deposition process, which allows to deposit homogenous coating on the surfaces of complex geometry (blades) without labor-intensive dismantling of power turbines and other similar assemblies. Mobile unit consists of separate assemblies. It is delivered to a heat-power station in a truck body and mounted at the turbine hall during a shift. It should be noted that the units of such type are not manufactured at present. Development of stationary plasma rooms at locations of large-dimension assemblies are not economically profitable.

During deposition turbine rotor is set on special support-bearings by a beam crane of the machine shop, deposition area is enclosed by polyathylene foil shields to eliminate opportunity of contamination of neighboring workshop objects. Examination and preparation for deposition of worn blades of the last rotor stages assembled are to be done. Deposition is performed sequentially along the worn surface of each blade using special manipulator; travel from the deposited blade to the following one is done by means of the rotor movement (Figure 3).

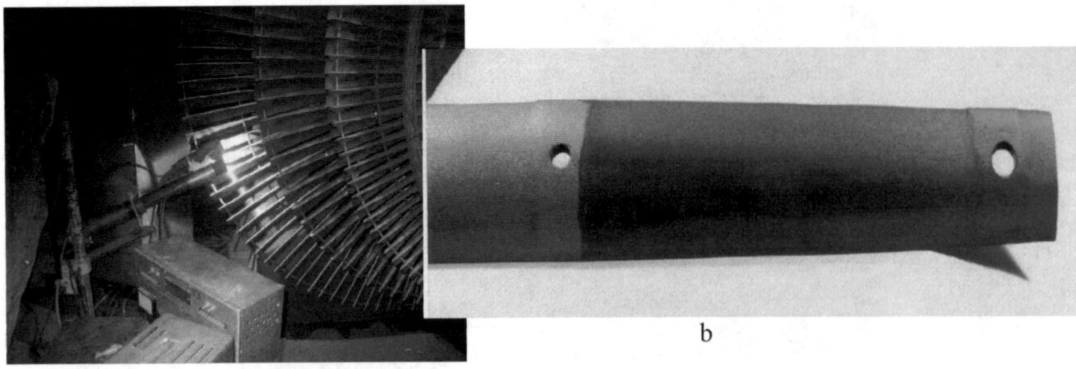

b

a

Figure 3. Recovering of operating blades of a steam turbine of a heat-power station using method of plasma deposition by means of a compact unit: a – recovery process: b – rotor blade recovered with plasma coating.

The first work dealing with recovering of leading edges of steam turbine blades was carried out more than 20 years ago on the turbine PT60 – 130/13. The last, 30th blade stage was removed because of its accident wear. Between overhauls of the steam turbine PT60 – 130/13 its observed life was 186,000 hours. When examining, it was discovered that 90% of all the blades of the 29th turbine rotor stage had total lack of stellite plates, and the remaining 10% of blades had only separate pieces of plates. In all cases, it was mentioned occurrence of partial erosive damages of leading edges at the places where stellite plates lacked. To prevent further erosion wear of leading edges, blade surfaces were cleaned from the remaining plate fragments. Then applying plasma gas-thermal method, we deposited two-layer protecting coating and recovered initial blade geometry by means of mobile equipment (Figure 3a), moreover, turbine rotor was not dismantled (Figure 3b).

In 218,544 hours of observed life, during the next overhaul visual inspection of operating blades of the 29th stage of the turbine rotor was done. It was noted that over 32544 hours plasma coating operated successfully protecting structural material. Erosion was not discovered on the material surface of blades, wear took place along the coating. Outer layer of the plasma coating was worn on all the areas excepting 'suction face' of blades, but the first layer was not worn completely. There were no deep cavities and sharp projections typical for erosion wear of the material on the surface of the remaining layer. As a result, operating blades of the 29th stage continued operating as turbine parts up to the next overhaul.

After examination of the turbine in 248,142 hours, it was discovered that surfaces of the blades were smooth and without erosion wear marks of structural material as before where the coating was deposited. Areas of outer layer of the coating at 'suction face' of blades preserved and had the same state as during the previous examination, in the last repair period.

4. Discussion of results

Finally, plasma coating deposited on the blades of the 29th stage of the steam turbine rotor provided efficient protection of structural material of the leading edge from erosion wear and destructive effect of fast steam-gas stream for 62,142 hours of operation (life time 10 years). Leading edges of 27th and 28th stages of turbine rotor that operated without coating within the same observed life had considerable erosion wear. It should be noted that in comparison with stellite plates the coating has much better protective properties. Obtained results and practical experience on providing efficient protection of structural material of steam turbine blades of a heat-power station for more than 62,000 hours do not have analogues in world practice due to using plasma coatings, so they are the pioneering ones.

Physical nature of such erosion resistance increase of plasma coatings in comparison with the material of blade base or stellite plates is connected in many respects with special meso-arranged structure of the coating that provides its damping on impact and another mechanism of micro-wear on erosion [1].

By now, works on deposition of erosion-resistant coating on two last stages (30 and 29 stages) have been completed on 9 turbines of different heat-power stations. Obtained beneficial results have predetermined for transition on more cost-beneficial operating procedure – on real situation, when during scheduled repair of the steam turbine without its dismantling, plasma coating is deposited to harden blade surfaces. Besides, cost of works does not exceed 10% of cost of a new blade set. During the next scheduled repair, blades with coating are examined, then the decision is made on coating recovery or extension of operation. The important condition on operation on real situation is absence of main material damages of a blade before coating deposition. Such damages being a concentrator of fatigue failure development will result in necessity of expansive replacement of an old blade with a new one. Obtained experimental results on operation mode on real situation prove possibility of service life increase of blades of 29th and 30th stages subjected to erosion wear in not less than 3 times.

Thus, developed technology of recovery and hardening of power unit parts applying plasma deposition allows to increase durability of power unit parts subjected to erosion wear in 3 times. It also decreases expenses on repair of separate parts and assemblies of power plants in dozens of times, reduces power equipment down time because of repair and failure in about 20 times, and reduces probability of man-caused impacts in case of equipment breakdown and failures.

Acknowledgement
The reported study was funded by the Ministry of Science and Higher Education of the Russian Federation.

References

[1] Barvinok V A 2005 Plasma in technology, reliability, resource *Moscow: Science and technology publisher* p 452
[2] Bobrov G V, Ilyin A A and Spektr V S 2014 Theory and technology of formation of inorganic coatings *Moscow: Alfa-M publisher* p 925
[3] Hawking M 2000 Wasantasri V., Sidqi P. Metal and ceramic coatings: Preparation, properties and application *Moscow: World publisher* p 518
[4] Ryzhenkov V A 2000 State of the problem and ways to increase the wear resistance of power equipment of TPP *Heat power engineering* (№6) pp 20-25
[5] Lisyansky A S, Lisyansky A S, Tikhomirov S A, Simin O N and Chizhik T A 2005 Providing resistance to drop-impact erosion and fretting corrosion of steam turbine blades due to vacuum ion-plasma surface hardening *Energomashinostroenie* (№2-3) pp 43-53
[6] Smyslova M K 2004 Research and development of combined ion-plasma technologies that improve the performance properties of gas and steam turbine blades *Vestnik UGATU.* (Vol.5 № 3(11)) pp 67-83
[7] Mann B S and Vivek Arya 2003 HVOF coating and surface treatment for enhancing droplet erosion resistance of steam turbine blades *Wear* (Vol. 254, Issues 7-8 DOI: 10.1016/S0043-1648(03)00253-9) pp 652 – 667
[8] Perelman R G and Pryakhin V V 1986 Erosion of steam turbine elements *Moscow: Ehnergoatomizdat publisher* p 184

[9] Barvinok V A and Bogdanovich V I 2012 Physical and Mathematical Simulation of the Formation of Mesostructure-Ordered Plasma Coatings *Technical physics* (Volume 57, Issue 2 DOI: 10.1134/S1063784212020053) pp 262 – 269

[10] Mann B S 2014 Water Droplet Erosion Behavior of High-Power Diode Laser Treated 17Cr4Ni PH Stainless Steel *Journal of Materials Engineering and Performance* (Volume 23, Issue 5 DOI: 10.1007/s11665-014-0927-6) pp 1861 – 1869.

[11] Barvinok V A, Bogdanovich V I and Dokukina I A 2014 Patent № 31897 RF MPK7 N05NS 1/26. Plasma torch for spraying metals and oxides *applicant and patent holder: Federal state Autonomous educational institution of higher education "Samara state aerospace University named after academician S. P. Korolev (national research University)»* (declared 21.11.2013, published 10.07.2014)

[12] Liu-xi Cai, Jing-ru Mao, Shun-sen Wang, Juan Di and Zhen-ping Feng Experimental investigation on erosion resistance of iron boride coatings for steam turbines at high temperatures *Proceedings of the Institution of Mechanical Engineers, Part J: Journal of Engineering Tribology* (Volume 229, Issue 5 DOI: 10.1177/1350650114557105) pp 636 – 645

[13] Pérez F J, Castañeda S I, Hierro M, Escobar R, Galindo J C and Sánchez-López S Mato 2014 Comparative Study of Micro- and Nano-structured Coatings for High-Temperature Oxidation in Steam Atmospheres *Oxidation of Metals* (Volume 81, Issue 1-2 DOI: 10.1007/s11085-013-9447-2) pp 227 – 236

[14] Shipway P H and Gupta K 2011 The potential of WC-Co hardmetals and HVOF sprayed coatings to combat water-droplet erosion *Wear* (Volume 271, Issue 9-10 DOI: 10.1016/j.wear.2010.12.058) pp 1418 – 1425

AUTHOR INDEX